Environmental Plant Physiology

Environmental Plant Physiology

Neil Willey

LONDON AND NEW YORK

First published 2016 by Garland Science

2 Park Square, Milton Park, Abingdon, Oxfordshire OX14 4RN
52 Vanderbilt Avenue, New York, NY 10017

Routledge is an imprint of the Taylor & Francis Group, an informa business

First issued in paperback 2019

ISBN 978-0-8153-4469-8 (pbk)

Neil Willey is Reader in Environmental Plant Physiology at the University of the West of England (UWE), Bristol, UK. He holds a BSc (Hons) in Biology & Geography and a PhD in Botany, both from the University of Bristol, UK. He's an active teacher of plant biology to undergraduate and postgraduate students from a variety of disciplines. His research focuses on the behavior and effects of pollutants, especially radioisotopes, in the soil-plant system. He's the Director of the UWE Graduate School and the Chair of the UK Coordinating Group for Environmental Radioactivity.

Front Cover.
Close-up of Rhododendron flower.

Library of Congress Cataloging-in-Publication Data
Names: Willey, Neil, author..
Title: Environmental plant physiology / Neil Willey.
Description: New York, NY : Garland Science, 2016.
Identifiers: LCCN 2015039662 | ISBN 9780815344698 (alk. paper)
Subjects: LCSH: Plant ecophysiology.
Classification: LCC QK717 .W55 2016 | DDC 571.2--dc23
LC record available at http://lccn.loc.gov/2015039662

Preface

Environmental plant physiology focuses on the foundations of life on land, with direct implications for the possibility of humans inhabiting Earth sustainably. I hope that readers of this book will gain an understanding of the importance of plant-environment interactions and be inspired to help humanity face some of its major challenges—in particular, global food security and the conservation of the natural world. This book is intended for upper-level undergraduate and graduate students, but will also be useful to some researchers. Several excellent textbooks inspired my interest in plant-environment interactions, but so significant are recent scientific advances that I felt a new textbook was needed. Its approach is intended to be useful across the biological, environmental, and agricultural sciences. In an era in which science is progressively focusing on the major challenges facing humanity, the ability to engage with a topic using both disciplinary and trans-disciplinary perspectives is increasingly important. In this book my aim is to provide a synthesis that is useful to individual biological, environmental, and agricultural disciplines, whilst also providing a framework for understanding across them all.

As is evident from the structure of the book, I think that a useful contemporary understanding of environmental plant physiology can be built by focusing on environmental variables, each of which is important to the biological, environmental, and agricultural sciences. These environmental variables are grouped into 'resources', 'stressors', and 'xenobiotics' according to whether they help synthesize biomass, limit biomass production, or poison plants. Each chapter is structured to focus on plant responses to progressively more profound variation in the environmental variable. This is used to suggest a hierarchy of responses from molecular to ecological scales in each chapter, and then as an overall framework in the final chapter. Throughout, examples are drawn from both unmanaged and managed ecosystems, and the importance of evolutionary history is acknowledged whenever possible to help to explain the occurrence of adaptations.

In each chapter the text is enhanced by large boxes that include additional stories of interest, small boxes in the margins that elucidate some key points, and by tables and figures. In particular, I have made an effort to provide a new set of figures for understanding environmental plant physiology. Further reading and references for each section are listed at the end of each chapter, and all bold terms are defined in the glossary. The book is complemented by resources available to students online: the glossary and flashcards; image gallery of the key species discussed in the text as indicated in bold green font; audio files that provide a commentary for each topic; and a set of multiple-choice questions. For instructors, all of the figures from the book are available to download in both PowerPoint® and JPEG format.

Overall, this textbook provides a heuristic framework that enables students, in the midst of the torrent of information published in scientific journals, not only to think about guiding principles and concepts in environmental plant physiology, but also to form their own perspectives. I would be interested in any feedback from anyone (contact me at Neil.Willey@uwe.ac.uk).

Acknowledgments

I am very grateful to the students who have taken my courses; the discussion that their interest stimulates has been vital to developing this book. I would also like to express my appreciation for the very helpful comments of colleagues and reviewers. Many thanks to staff from Garland Science: Dave Borrowdale for steering the project from beginning to end, Gina Almond for her role in getting it off the ground, and Georgina Lucas for all her efforts during the publication process. And finally, particular thanks must go to Lorna and the rest of my family without whose support this book, and much else, would never have happened.

Neil Willey

Reviewers

Habib-ur-Rehman Athar (Bahauddin Zakariya University); Juergen Burkhardt (University of Bonn); Ivan Couée (University of Rennes); Peter S. Curtis (The Ohio State University); Stephen Ebbs (Southern Illinois University); David E. Evans (Oxford Brookes University); Ros Gleadow (Monash University); Kevin Griffin (Columbia University); Stuart Lane (Plymouth University); Richard Leegood (University of Sheffield); Denis Murphy (University of South Wales); Bob Nowak (University of Nevada, Reno); Léon-Etienne Parent (Université Laval); David Tissue (University of Western Sydney); Alyson Tobin (University of St Andrews); Marcos Yanniccari (CONICET, Universidad Nacional de La Plata); Jianhua Zhu (University of Maryland).

Contents

Chapter 1

Contexts, Perspectives, and Principles

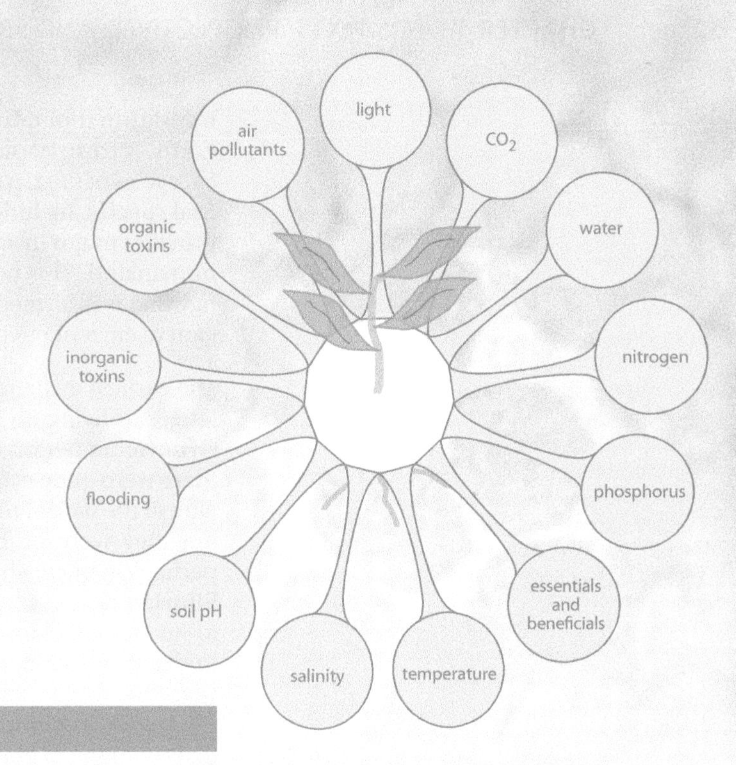

Plant growth is affected by variation in independent, interacting environmental variables.

Key concepts

- Plant–environment interactions are the foundation of terrestrial ecosystems.
- Environmental change and food security are significant challenges for humankind.
- Physiology is the study of how and why organisms function as they do.
- Biomass production and quality are dependent on resources, stressors, and xenobiotics.
- The environmental factors that affect plants are independent variables with overlapping effects.
- Major soil types embody the effects of many interacting variables that influence plant growth.
- Spatial and temporal variation provides key perspectives on plant–environment interactions.
- Plants detect environmental variation via signal transduction networks.
- There is a hierarchy of adaptations that underpin plant responses to environmental variation.
- Environmental plant physiology can elucidate key ecological processes.
- Agricultural systems can benefit from insights into plant–environment interactions.
- Models can be used to predict plant behavior in a changing environment.

Plant interactions with the atmosphere, hydrosphere, and geosphere underpin terrestrial ecosystems

The colonization of the land surface by multicellular plants was a momentous phase in the history of life on Earth, in significant part because multicellular plants in effect colonized the atmosphere, providing an unprecedented link between the subsurface, surface, and above-surface environments. It initiated perhaps the most significant ever increase in the number of niches and ecosystems. The evolution of organisms into these new niches transformed not only the domains of life but also the biogeochemistry of Earth. Terrestrial plants are therefore at a crucial nexus of the biogeochemical cycles of the Earth, and help to provide the life-support system for terrestrial species, including humans. Understanding plant function at this nexus provides major insights into many of the environmental challenges that face humankind. This book aims to provide an understanding of plant physiology that is informed by the development of terrestrial ecosystems and relevant to current environmental challenges.

The earliest evidence of multicellular plants that were adapted to the challenges of living on land is provided by spore **tetrad** microfossils from the **Ordovician** period. These suggest that, in some terrestrial locations at least, there were quite extensive stands of plants on land by 450 million years ago. Due to the lack of macrofossils, it is uncertain what these plants looked like, but they were probably **liverwort**-like and inhabited wet environments, perhaps living in shallow standing water. Macrofossils from the subsequent **Silurian** period suggest that, by 425 million years ago, plants on land were up to 10 cm tall and had **rhizoids**—they were beginning to function partly in the atmosphere and partly in the **regolith**. Between 425 and 300 million years ago there was a profound increase in the diversity of terrestrial plants (Figure 1.1). Complex terrestrial ecosystems began to develop in which a diverse range of plants, many of which have descendants in current ecosystems, adapted to the challenges of life on land. Fossils from the **Devonian** and **Carboniferous** periods show that some of these plants were many tens of meters tall, and although many of them were clearly swamp dwellers, some probably inhabited drier habitats.

Detailed understanding of the environmental physiology of early plants is difficult, but numerous features of extant terrestrial plants that are interpreted as adaptations to the challenges of life on land were evident in some of the early terrestrial plants. These include adaptations for gas exchange with the atmosphere, for water transport, and for nutrient uptake. It was the profound

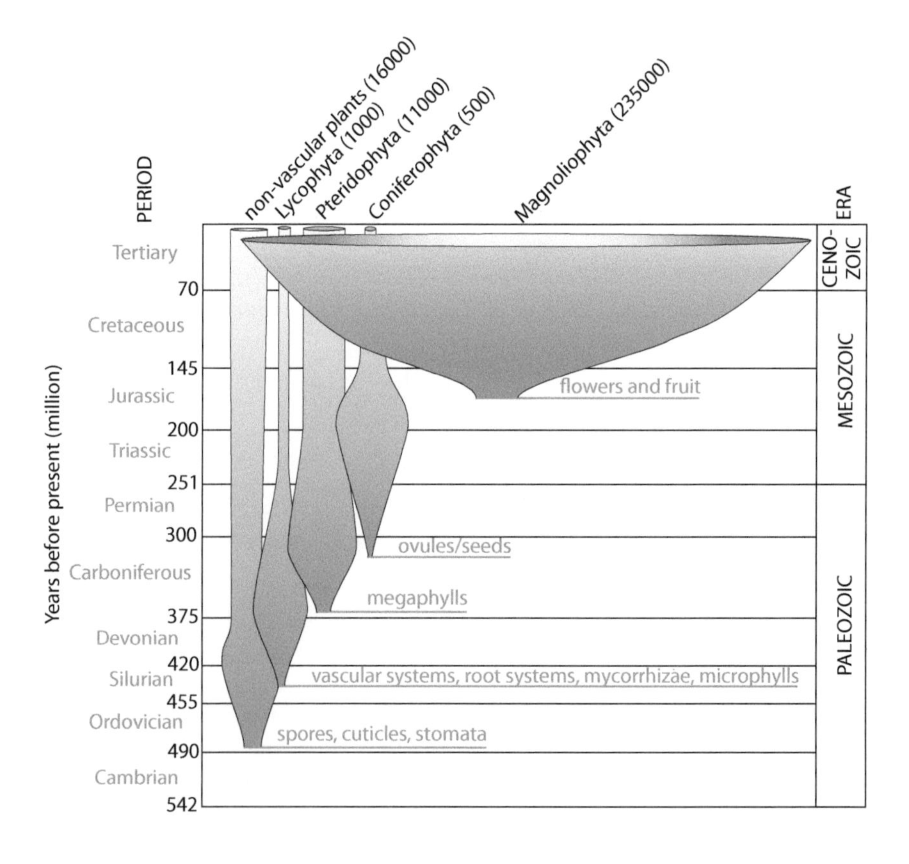

Figure 1.1. The evolution of major terrestrial plant phyla. Extant major groups are shown, but not the first 3.7 billion years of Earth's history (pre-Cambrian), nor many fossil and minor extant groups. Characteristics in green letters are major adaptations that affected plant–environment relationships. Mycorrhizae = root symbioses with fungi; microphylls = small simple leaves without extensive vascular systems; megaphylls = large true leaves with extensive vascular systems; ovules = structures that contain the female gametophyte. (Redrawn from Ridge I [2002] Plants. With permission from Oxford University Press.)

environmental changes which these adaptations eventually wrought that affected the biogeochemistry of Earth—for example, the composition of the atmosphere from this time on was affected by the activity of terrestrial plants and the decomposition of their dead biomass. The cycling of fresh water was transformed so that a significant proportion of all the water moving from the land to the atmosphere did so through plants. Following the injection of organic matter into the regolith, true soils became extensive for the first time, transforming the geochemistry of the land surface. Terrestrial plants are therefore integral to terrestrial ecosystems at perhaps the most important interface between the biosphere and the atmosphere, hydrosphere, and geosphere (Figure 1.2).

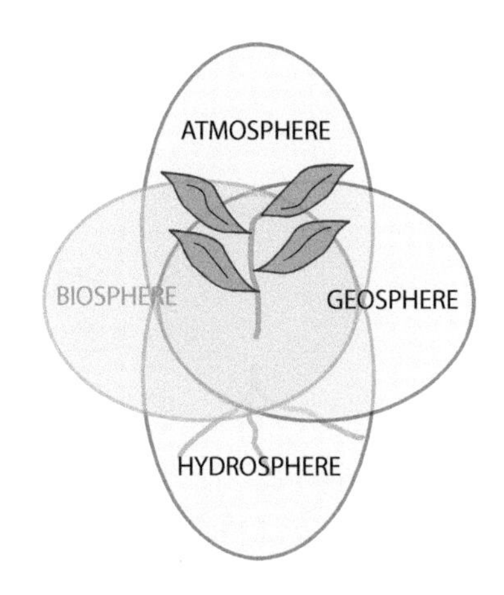

Figure 1.2. **Plants at a primary interface.** The soil is a primary interface between the atmosphere, hydrosphere, and geosphere. Plants function at this interface, affecting many of the most important biogeochemical cycles on Earth, and providing the foundation of terrestrial ecosystems.

An ecosystem can be defined as a community of organisms and the physical environment with which it interacts, via flows of energy and nutrients, to develop trophic relationships. This book focuses on the interaction of multicellular terrestrial plants with their abiotic environment—that is, one facet of terrestrial ecosystems. It does so in a way that is useful not only to those whose particular interest in terrestrial ecosystems relates to the plant and biological sciences, but also to those whose primary interests are in the environmental and agricultural sciences. To achieve this, a range of ecosystems are discussed in terms of "unmanaged" and "managed" ecosystems, to reflect some of the systematic differences between "natural" and "agricultural" terrestrial ecosystems. It is indisputable that, in addition to the topics covered here, understanding of both plant interactions with the biotic components of the environment and of ecosystem functioning will be vital to meeting the environmental challenges that face humankind.

Minimizing human impact on ecosystems and achieving global food security are significant challenges

If, as many geologists posit, humans are now a primary agent of environmental change, the current geological **epoch** can be defined as the **Anthropocene**. There is much evidence that hunting by early humans had significant adverse impacts on the **megafauna** of numerous regions on Earth—for example, large mammals were often hunted to extinction after the arrival of humans. However, it was with the initiation of agriculture that humans began to have effects on the environment that were detectable on a wide scale. The invention of agriculture between 10,000 and 8000 years ago is perhaps the most significant event in the human story, and it has been suggested that it defines the start of the Anthropocene. However, despite the environmental impact of early humans, it is generally accepted that in the twentieth century the human impact on terrestrial ecosystems was unprecedented and initiated "something new under the sun"—that is, a truly global scale of environmental impact on Earth by a single species. The Millennium Ecosystem Assessments, the United Nations Environment Program Global Environmental Outlook ("GEO") reports, and numerous other assessments have described the global scale of human impact on terrestrial ecosystems, which now constitutes one of the greatest challenges facing humankind. The social and natural sciences will both clearly have vital roles in meeting this challenge. A striking number of aspects of the latter and many of the potential solutions do, however, occur at the nexus of plant-environment interactions. Some of the most important impacts of changing CO_2 concentrations are, directly or indirectly, on plant growth, but plants also help to control atmospheric CO_2 concentrations. Threats to unmanaged ecosystems include uncontrolled exploitation by humans for food, fuel, and plant products, and contamination with nutrients and xenobiotics, but manipulation of plant growth might also help to solve some of these problems. Such current challenges and potential solutions mean that in order to minimize human impacts on the environment, an understanding of the interactions of plants with their abiotic environment is probably more important now than it has ever been.

In the 1970s it was difficult to predict when global population growth might stop, but now most credible predictions suggest that the global population will peak at 10–12 billion in the second half of the twenty-first century. This is a very significant increase in population when the strain on terrestrial ecosystems from the current population of about 7 billion is already so significant. The intensification of agriculture during the latter part of the twentieth century was one of the most profound and successful of the many applications of science that were developed in that century. In 1900 it was inconceivable that it would be possible to come anywhere close to meeting current demand for agricultural products used for food, forage, fuel, clothing, drinks, or raw materials. A modest proportion of the increase in agricultural productivity in the twentieth century was due to an increase in the total area of agricultural production, but the greatest proportion was due to an increase in the amount of resource per unit area per unit time being converted into yield. This was often achieved by adding resources and by using varieties that matured more quickly to enable production of more than one crop per year. Thomas Malthus famously suggested in 1798 that population growth would inevitably outstrip agricultural production, but the history of agriculture suggests that humans are adept at avoiding Malthus's prediction. They have done so in significant part by increasing the efficiency of plant–environment interactions in agricultural ecosystems.

The intensification of agriculture in the twentieth century, and in particular the **Green Revolution** in developing agriculture, probably saved hundreds of millions of lives. The food, forage, and plant products that its techniques produce now underpin, to a remarkable extent, twenty-first-century societies. In significant part because of its great success, intensive agriculture is now intimately connected to many of the world's environmental challenges. These include demand for water, nutrients, and agrochemicals, together with the problems that their production and use entail. It has been recognized since the later years of the twentieth century that in order to achieve global food security in the twenty-first century, while at the same time reducing the environmental impact of agriculture, there is a necessity for a "doubly green" revolution—that is, an increase in production similar to that of the Green Revolution, together with the minimization of agriculture's environmental impact. If this occurs, it is likely to involve a profound change in the way that plants are used to convert inorganic resources from the environment into food, forage, and plant products, and it is likely to be dependent on an appreciation of plant–environment interactions in both unmanaged and managed ecosystems. For example, a very significant pressure on agriculture is the scarcity of land on which to grow crops. This arises in part because of the restricted range of plants that are used to provide the majority of food. It has been long been recognized that many species of plants other than those currently used might be useful for producing food, and that natural ecosystems often have highly efficient resource-use patterns. The great variety of species that are used to produce forage and plant products is a reminder of the diversity that is potentially available for harvest. The history of **alternative crops**, **agroforestry**, **intercropping**, and numerous other agricultural systems runs deeper than is often supposed, and in the twenty-first century such systems might play a significant role in helping humans to harvest food and other plant products sustainably. If this happens, an understanding of plant–environment interactions is likely to aid their development. In this book, the minimizing of human impact on the environment, along with the sustainable harvesting of food, forage, and other plant products, are used as key contexts for understanding plant–environment interactions.

Proximate and ultimate questions elucidate how and why plants interact with the environment

Physiology is the study of function in living systems. Function is most fully understood by asking both proximate and ultimate questions. Molecular biology transformed the plant sciences, and it is now possible to describe in unprecedented detail the changes that occur in plants in response to a change in the environment. We are now more able than ever before to answer questions such as "How do plants respond to a decrease in soil pH?" This book provides an overview of how plants respond to variation in key environmental variables. The ability to describe the mechanistic details of how something happens—that is, its proximate cause—is vital to understanding plant-environment interactions. Understanding plant function necessitates investigations of proximate causes on a scale ranging from that of biomolecule to that of whole organism.

However, it is also possible to ask "Why do plants respond as they do to decreasing temperature?" Such questions are aimed at the ultimate cause of a phenomenon (Box 1.1), and in this book are regarded as integral to physiology as the study of function. Multicellular terrestrial plants have an evolutionary history. DNA sequences have an imprint of past events, via both genetic and **epigenetic** mechanisms, which legitimizes the asking of ultimate questions. Thus, for example, plants respond as they do to decreases in temperature in significant part because one of the most significant cold periods in the last 250 million years on Earth occurred very recently, ending only 10,000 years ago. In this book, plant physiology is defined as the study of "how and why a plant functions as it does."

Phylogenetics has transformed our understanding of the evolutionary history of many groups of organisms, including plants (Box 1.2), providing unprecedented answers to ultimate questions. Evolutionary relationships between many terrestrial plants are now known, providing a phylogenetic perspective on function. For example, despite the diversity and advantages of nitrogen-fixing symbioses, phylogenetics shows—perhaps surprisingly—that the propensity to form these symbioses in flowering plants evolved in only one group. The likely **phylogeny** of the angiosperms that constitute the majority of plant biomass on Earth and many of the groups mentioned in this book is shown in Figure 1.3. The perspectives derived from proximate and ultimate questions, informed by molecular biology and phylogenetics,

BOX 1.1. PROXIMATE AND ULTIMATE QUESTIONS ABOUT PLANT BEHAVIOR

The study of animal behavior is regarded as a discipline in its own right, namely ethology. Plants also react, respond, and adapt to their environment. They do this to enhance their chances of survival by, for example, foraging for resources, avoiding stress, and communicating between individuals. This is, in a very real sense, "plant behavior." Ethologists often distinguish between proximate and ultimate questions about behavior. These are "Tinbergen's questions," and they have echoes of some of the "causes" of animal characteristics distinguished by Aristotle. The biologist Ernst Mayr was the most prominent proponent of the utility of proximate and ultimate causes in explaining biological phenomena. Proximate questions focus on the mechanisms that explain *how* a behavior or phenomenon is produced within the lifespan of an individual. Ultimate questions focus on explaining why the particular behavior or phenomenon occurs in the species.

BOX 1.2. MOLECULAR PHYLOGENETICS

Centuries of study have identified many of the major groups of plants, so if the evolution of a particular character is being investigated, an **outgroup**, which is highly likely to be distantly related, can be chosen to root an analysis. Using an outgroup enables a rooted evolutionary tree to be constructed, rather than a simple cluster of similarity. A major complication in phylogenetic reconstruction is long-branch attraction, in which a small change in a character can produce a radical reorganization of the long branches of a phylogenetic tree. The use of algorithms to control for long-branch attraction and of multiple characters increases the likelihood of identifying the evolutionary pathways that actually occurred. When variance in a molecular character such as a gene sequence is known, **Monte Carlo methods** can be used to produce what are in effect infinitely large data sets from which numerous phylogenetic trees can be reconstructed. Consensus trees can then be used to estimate the probability of each node having occurred. Measures such as bootstrap and jackknife values quantify how often particular nodes occur in reconstructed consensus trees, and thus estimate their likelihood of having occurred (Figure 1).

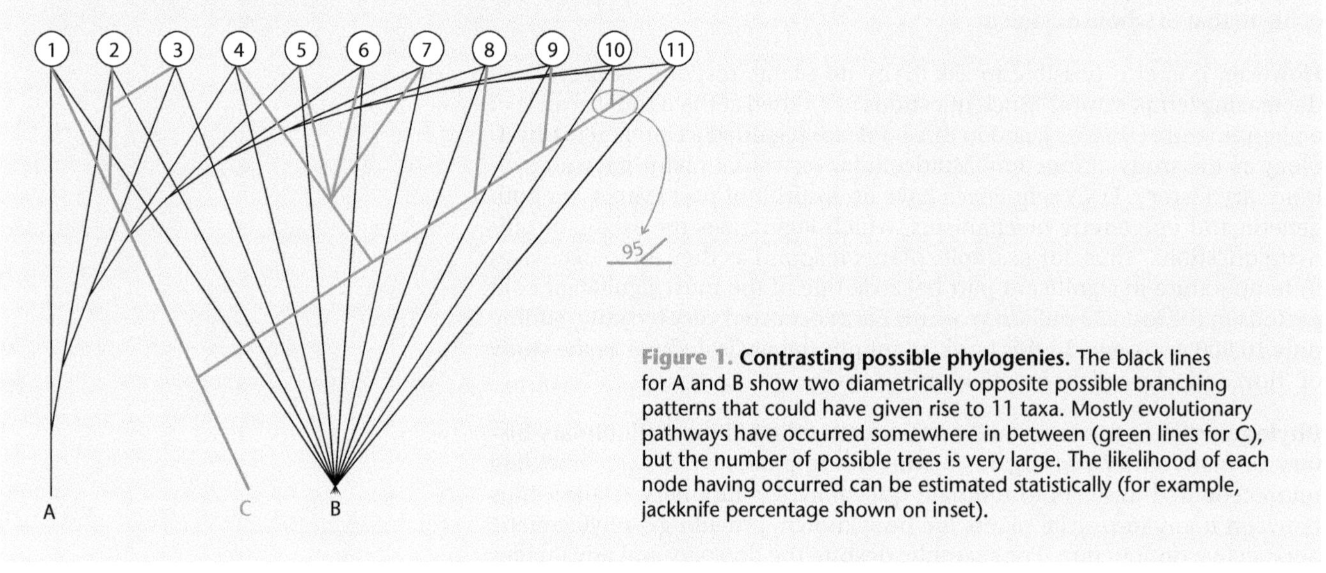

Figure 1. Contrasting possible phylogenies. The black lines for A and B show two diametrically opposite possible branching patterns that could have given rise to 11 taxa. Mostly evolutionary pathways have occurred somewhere in between (green lines for C), but the number of possible trees is very large. The likelihood of each node having occurred can be estimated statistically (for example, jackknife percentage shown on inset).

respectively, are used in each chapter of this book to elucidate plant function. The discussion of function in each chapter starts with molecular biology and proceeds towards function of whole plants in a phylogenetic context.

Resources, stressors, and toxins affect plant biomass production and quality

Plants, as **autotrophs**, utilize inorganic resources from the environment to synthesize the organic molecules necessary for life. The fundamental challenge facing terrestrial plants involves gathering the resources necessary to drive autotrophy. The plants of terrestrial ecosystems are almost all **photoautotrophs,** so light is their fundamental resource. Light energy is used to synthesize organic compounds based primarily on carbon (C), hydrogen (H), oxygen (O), nitrogen (N), sulfur (S), and phosphorus (P) atoms. These compounds include the carbohydrates, amino acids, and lipids that are essential requirements for almost all other organisms in terrestrial ecosystems. This book focuses first on plant capture of resources (Figure 1.4), because of their importance for all inhabitants of terrestrial ecosystems, including humans. Resources are defined by their necessity for the production of the biomolecules that are essential for life, and by the adverse effects that a sub-optimal supply can have. The use in ecology of the concepts underpinning "the capture of resources," with their implications of competition for and exploitation of resources, developed synergistically with their use in economics. However, they have provided a powerful perspective for understanding how

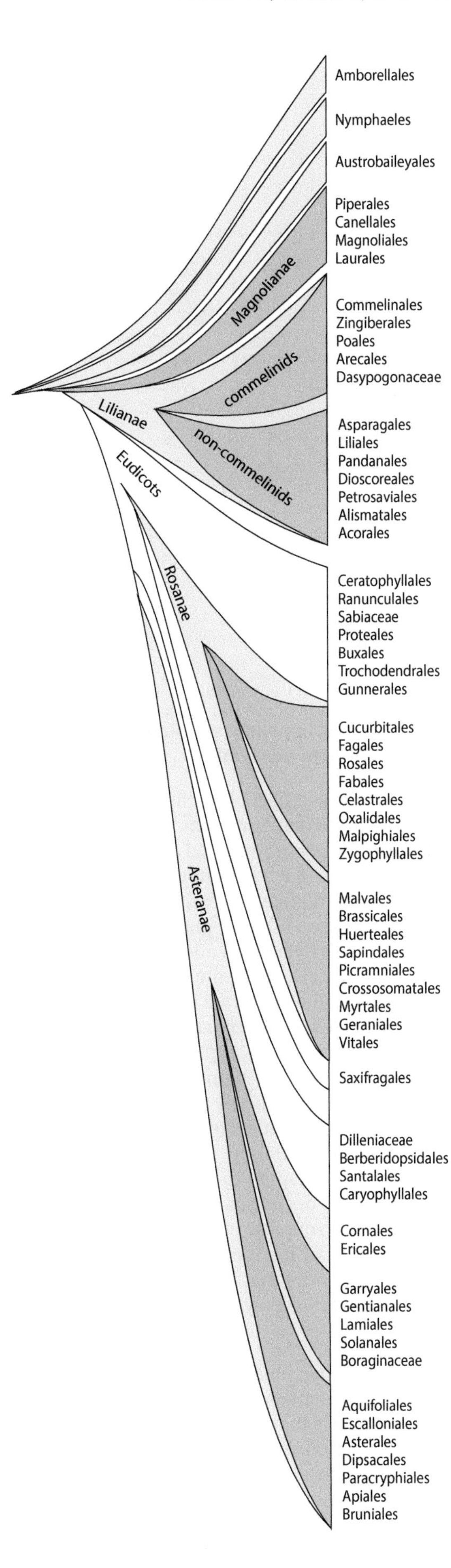

Figure 1.3. A phylogeny of the angiosperms. A phylogeny of angiosperm orders based on Angiosperm Phylogeny Group III phylogeny is shown. Primitive orders of dicotyledonous plants are at the top; the Lilianae are monocotyledonous plants and the Eudicots contain the remaining dicotyledonous plants. (Some orders are omitted and families, when named, are unplaced in an order). The Eudicots contain most dicotyledonous plants.

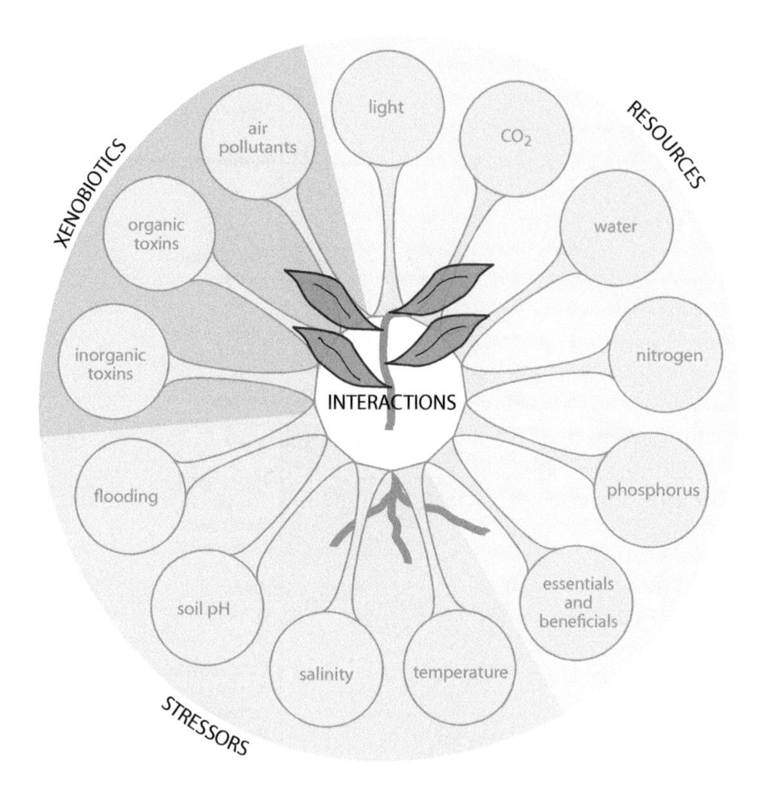

Figure 1.4. Resources, stressors, and xenobiotics. The environmental variables that affect plant growth can be classified as resources, stressors, or xenobiotics. Resources are needed both to build the biomolecules of life, and for them to function. Stressors produce sub-optimal conditions for growth. Xenobiotics are substances that either do not naturally occur in plants, or that can occur at unnaturally high concentrations.

plants interact with the environment to maintain the production of biomass, and are therefore often used in this book.

Environmental conditions can affect plant capture of resources. Most groups of plants evolved to suit particular environmental conditions, and significant departure from these produces sub-optimal conditions for resource capture. Thus environmental variables such as temperature, pH, salinity, and oxygen status, for which almost all multicellular organisms have a relatively narrow range of tolerance, affect the ability of plants to capture resources. Many plant physiologists regard the variables addressed in Chapters 8, 9, 10, and 11 as environmental stressors. Such terminology borrows directly from studies of human stress, which in turn derive from mechanics. In this book the stress perspective is regarded as useful, and is therefore used when appropriate. Stress is used *sensu* Levitt (1972) to denote an environmental condition that can inhibit growth and/or reproduction, and hence survival. The use of this analogy is not meant to imply that under sub-optimal conditions plants are necessarily "stressed" in a similar way to animals or mechanical systems. It is often difficult to quantify how "stressed" plants are, because they are in many instances adapted to potentially stressful environments, and the extent to which this decreases the production of biomass—that is, the extent of the stress—is frequently unclear. Nevertheless, in general, sub-optimal growth conditions, which include a lack of resources or the presence of environmental stressors that inhibit their capture, are regarded in this book as stressful in so far as they can potentially limit biomass production by plants.

In addition to the resources that plants capture and the compounds that they synthesize from them, plants are exposed to **xenobiotics**. Here a xenobiotic is defined as a toxic substance that does not naturally occur in plants or does not ordinarily occur at the concentrations observed. Many such substances have adverse effects on plants or are even inimical to plant growth. Plant uptake of xenobiotics can also lead to ecosystem contamination. In the final chapters of this book, some of the xenobiotics that currently have important impacts on plants in terrestrial ecosystems are discussed. Xenobiotics are included here because in the second half of the twentieth century it became clear that these substances could have impacts on a global scale. The control of ecosystem contaminants is vital both to the health of ecosystems and to food quality. In some parts of the world the challenges of controlling ecosystem contamination have begun to be met, but this is not the case in many other regions. Furthermore, a major challenge facing the management of water and soil resources is the rehabilitation of degraded and contaminated land, which frequently necessitates an understanding of the interaction of plants with xenobiotics. Indeed, some assessments have concluded that the rehabilitation of degraded land might be the key to saving natural ecosystems while at the same time increasing agricultural productivity. Toxicological studies in plants contribute fundamentally to ecotoxicology—an important discipline on a contaminated planet—and dose-response curves for xenobiotics are useful for assessing the effects not only of xenobiotics but also of some stressors on plants (Box 1.3). Human activities are

BOX 1.3. DOSE–RESPONSE CURVES AND PLANT–ENVIRONMENT INTERACTIONS

A central concept in toxicology is the dose-response curve, which describes the magnitude of the effect produced in an organism by an increasing dose of an effector. These dose-response curves are directly relevant to the study of the effects of xenobiotics on plants, but many of the concepts that underpin them are also useful for understanding a large number of the effects that environmental variables have on plants. It is frequently assumed that there is a linear no-threshold (LNT) relationship between the intensity of an environmental variable and plant response as measured by key endpoints, such as growth rate, morbidity, or mortality (Figure 1)—hence the prediction of low but significant numbers of "extra" cancers in humans at low radiation doses. However, there are many studies which show that low-intensity stress is not sufficient to have an impact on significant endpoints. In such instances there is a threshold below which there is no biologically significant effect, and above which there is an effect—a linear threshold (LT) response. It is likely that for many of the variables discussed in this book there is a threshold for impact on key endpoints—although this might occur at relatively low intensity—because of the homeostatic capacity of cells. There is also evidence that the dose responses for the effects of some metals, herbicides, and ionizing radiation on organisms, including plants, can be hormetic—that is, stimulatory at low intensity but inhibitory at high intensity. It is possible that for some plants there are **hormetic** effects of stressors. Hormetic effects probably occur because low-intensity stress helps to stimulate defense, repair, or adaptation. Overall, the general concepts of dose-response can be useful in discussions of plant-environment interactions. For each of the variables dealt with in this book there is probably variation that is below the threshold that evokes an effect on a key endpoint, and it is likely that for some variables the effects at high intensity might not be good predictors of the effects at low intensity.

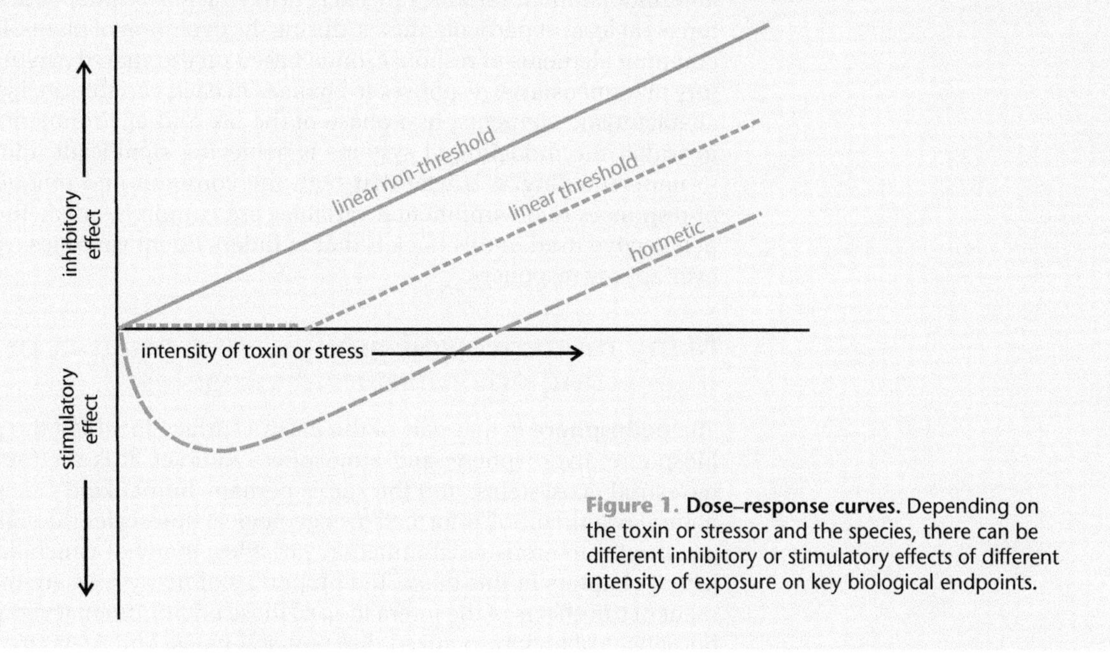

Figure 1. Dose–response curves. Depending on the toxin or stressor and the species, there can be different inhibitory or stimulatory effects of different intensity of exposure on key biological endpoints.

having profound effects on plant resources, stressors, and xenobiotics in the environment. This book focuses on how plants deal with variation in environmental resources, stressors, and xenobiotics, including much variation that is caused by humans.

Environmental factors that affect plant growth are interacting but independent variables

Plants must maintain internal conditions suitable for metabolic functioning despite having a variable external environment. Even under "optimal" conditions, minor variation in environmental conditions must often be counteracted. Plants therefore possess mechanisms that contribute to **homeostasis** by helping to maintain internal conditions within narrow limits and thus control resource supply or limit stress. Such homeostatic mechanisms are a defining feature of living systems, and across the kingdoms of life many of them have similarities. The responses to minor variation in many environmental variables overlap at the molecular level via homeostatic mechanisms. In addition, many physiological adaptations to stress have evolved from homeostatic mechanisms, so responses to different stressors can have common elements. In each chapter of this book the common features of responses to different environmental variables are discussed. This is visually represented at the beginning of each chapter using the template shown in Figure 1.4. Some of the many unexpected insights from molecular biology are the common features of plant physiological responses to stressors, and these are useful for understanding the effects of the multiple stresses to which plants can be subjected in the field.

A full understanding of the functioning of biological systems necessitates an understanding not only of the common elements of responses but also of the differences. This is particularly so for understanding responses to stress, because plant resources and stressors are independent environmental variables. Theoretically, temperature can change while all other variables remain unaltered. However, in the field this does not always happen—for example, when temperature changes, water availability often changes with it, but this is not necessarily the case. In this book it is regarded as of fundamental importance that the variables that affect plant growth are independent. It is for this reason that each chapter focuses on a single environmental variable. Independent variables are likely to have acted as independent selective forces, at least at particular times, during the evolution of plants. Despite the common elements of response, often based on the shared evolutionary history of homeostasis, responses to changes in each variable are likely to have characteristic elements. In a phase of the life and environmental sciences in which the modeling of systems is achieving significant improvements in understanding, it is vital that both the common and unique elements of responses to environmental variables are comprehended. Therefore the perspective used in this book is that of independent variables which evoke overlapping responses.

Many reference soil groups are a product of interacting environmental variables

The **pedosphere** is that part of the Earth's surface in which the geosphere, biosphere, hydrosphere, and atmosphere interact. It is the foundation of terrestrial ecosystems, and the soil is perhaps humankind's most valuable natural resource. Soil is formed over geological timescales through the interaction of numerous environmental variables, many of which are the subject of chapters in this book. The properties of many soils are the embodiment of the effects of the interaction of these environmental variables. Early humans probably recognized different soil types, and there are numerous taxonomies that are now used to classify soils. In some of them, tens of thousands of different soil types are recognized, but in almost all national and international soil classification systems there are major soil types in which properties are dominated by variables dealt with in this book.

The FAO's World Reference Base for Soil Resources is probably the most significant attempt to provide an overarching classification system for soils. It is based on measurable properties of soils, a large number of which directly affect plant growth. Many of its 32 reference soil groups (RSGs) are useful for understanding the effects of environmental variables on plant growth (Table 1.1). For example, organic soils (histosols) have characteristic properties, and frequently have characteristic floras. Other soils have properties dominated by water regime, iron/aluminum (Fe/Al) chemistry, or accumulation of salts, and particular plants grow or can be grown on them. Soils such as cambisols and chernozems are much more extensively used for crop production than other groups. Some of the major RSGs listed in Table 1.1 are used during the discussion of topics in this book.

Spatial and temporal analyses provide insights into plant responses to environmental variation

Field measurements of most of the environmental variables discussed in this book produce spatially dependent values. For example, the pH of a soil at one location is much more likely to be similar to that at a location nearby than to one at a significant distance—that is, it is very unlikely that soil pH at two adjacent locations will cover the whole possible range. Thus, although many statistical techniques assume that individual samples of a variable are independent, this is not true of the environmental variables that are the subject of each chapter in this book. **Geostatistical** methods can be used to analyze spatial dependence of a variable. In several chapters of this book the **semivariogram** (Box 1.4) is used in discussions because, for example, some molecular responses make most sense when viewed in this context.

Variation in some environmental variables, especially soil variables, has approximately **fractal geometry**. In fact, analyses of spatial variation in the earth sciences played a significant role in developing the mathematical concepts of **self-similarity** into the fractal geometry that is useful for analyzing the natural world. For example, soil-water release curves describe the dynamics of water release from soils, and depend on the geometry of soil pores. Soil pores therefore control the supply to plants of water, nutrients, and xenobiotics. However, they are very difficult to describe using **Euclidean geometry**. Many aspects of the environment, such as soil pores, have long been noted to have significant self-similarity—that is, when described at a variety of scales there is a similarity in the geometry observed. In many cases in the soil-plant system, self-similarity at different scales is not exact because it breaks down at very low and/or very high resolution. Nevertheless, over the range of scales relevant to plant-environment interactions, self-similarity is significant, and fractal geometry provides useful physiological insights. For example, the fractal nature of soil pores means that different root hairs, and probably different parts of the same root hair, can be expected to experience significant differences in water supply—just as can different secondary roots or parts of primary roots. At different scales in the soil-plant system the absolute magnitude of differences will not necessarily be the same, but differences can exist at different scales and, if they provide a significant challenge, have to be coped with physiologically. Thus when studying plant responses to variation in the environment, spatial perspectives are vital, and familiarity with, for example, semivariograms and fractal geometry provides useful insights. Understanding plant-environment interactions depends on understanding not only how a plant responds, but also where the responses occur across a range of scales.

Table 1.1. Reference soil groups (RSGs) with key properties that affect plant growth. The World Reference Base for Soil Resources defines 32 RSGs in total that cover about 13,000 million hectares of the Earth's surface.

Morphology	Reference soil group	Occurrence	Properties relevant to plant growth
Soils with thick organic layers	Histosols	325–375 mi ha. 90% boreal, sub-arctic, and arctic. Swamp peats in tropics, especially in SE Asia	Very slow mineralization. Wide variety of organic compounds. Mostly reducing. Acid or alkaline
Soils with high levels of organic matter and bases	Chernozems	230 mi ha. Mid-latitude tall-grass steppes of Eurasia and N America	Fertile, water-retaining, pH buffered at neutral or above, deep, dark upper horizon. Highly productive
	Kastanozems	465 mi ha. Short-grass steppes	Less organic, higher in carbonates, and drier than chernozems
	Phaenozems	190 mi ha. Wet grasslands, forests, in warm temperate Americas and China	Often on glacial deposits, high in organic matter but more leached of bases than chernozems and kastanozems. Low Ca
Soils predominantly influenced by water	Gleysols	700 mi ha. Azonal. All areas with excess groundwater	Waterlogged, reducing conditions that promote characteristic "gley" color development
	Fluvisols	350 mi ha. Azonal. At least 50% in tropics	Consist of alluvial material. Most waterlogged, many strongly sulfidic and acidic
	Solonetz	135 mi ha. Areas of steppe climate, dry summer up to 500 mm rainfall	Little or no organic surface horizons, high Na and Mg, often strongly alkaline (pH 8.5) because of $NaCO_3$
	Solonchak	260 mi ha. Arid and semi-arid areas of high evaporation. Coastal areas	High concentration of salts in surface horizons due to evaporation for a significant part of the year
	Vertisols	335 mi ha. Subtropical. On highly weathered or basaltic substrates	Churning, sticky when wet, hard when dry. Fertile, but physical properties due to swelling clays affect plant growth
Soils predominantly influenced by Fe/Al chemistry	Podzols	485 mi ha. Mostly boreal and temperate N hemisphere. Some humid tropics	Highly organic surface horizon, upper horizon bleached by loss of Fe/Al–organic complexes to a lower horizon, often an iron pan. Acidic
	Nitisols	200 mi ha. Subtropics, especially in Africa, often in highlands	Deep, red soils. Productive, but high Fe promotes low P availability
	Ferralosols	700 mi ha. Humid tropics on the continental shields	The classic, deep, red, Fe-dominated soils of the tropics. Low nutrients, low water-holding capacity, and high P-fixing capacity
Soils with accumulation of less soluble salts	Gypsisols	100 mi ha. Deserts	A common desert soil with accumulation of $CaSO_4$ (gypsum)
	Calcisols	Up to 1000 mi ha	High concentrations of lime dominate properties. Alkaline pH
Soils with clay-enriched subsoils	Acrisols	1000 mi ha. Regions of intense weathering. Many "red and yellow earths"	Highly acid soils. Low concentration of bases because of leaching with clays to subsoil
	Albeluvisols	350 mi ha. Mostly Eurasia over glacial till and Aeolian deposits	White, washed out soils. Acid, low in nutrients. Poor drainage
	Alisols	100 mi ha. Humid tropics to warm temperate zone	High Al clays in subsoil, strongly acidic
	Luvisols	500–600 mi ha. Temperate to cool temperate zone	Clay higher in subsurface, but only some leaching. High base saturation in upper horizons. Fertile. Widely used in agriculture
	Lixisols	435 mi ha. Dry tropics to warm temperate zone, especially savannahs and open forest	High in bases but low in clay and organic matter. Often fine textured and easily eroded
Soils with little or no profile development	Umbrisols	100 mi ha. Humid climates. Mostly mountainous	Dark, organic. Low base saturation. Often forest covered
	Arenosols	1300 mi ha. Azonal. Sandy substrates worldwide	Poor water-holding capacity, poor fertility. Often develop hydrophobicity
	Cambisols	1500 mi ha. Regions affected by Pleistocene glaciation. Also on alluvial deposits, such as Ganges basin	Leaching has not been extensive. High base saturation. Fertile. Widely used for agriculture

BOX 1.4. THE SEMIVARIOGRAM

Within, for example, in a 100 m × 100 m plot many environmental variables can differ significantly, and this can have effects on, for example, niche segregation or productivity. However, for most environmental variables their value at location x will be more likely to be similar to that at x plus a distance d if $x + d$ is nearer to rather than further away from x. Analysis of the spatial variation and comparison with that which provokes stress responses in plants is fundamental to understanding plant-environment interactions. The spatial dependence of variance can be described using the semivariance, calculated as:

$$\gamma(\mathrm{d}) = \frac{1}{2(n-d)}\sum_{i-1}^{n-d}(Z_{xi+h} - Z_{xi})^2$$

where $\gamma(\mathrm{d})$ = semivariance at lag d, d = lag (that is, distance between points), Z_i = value of the variable, and x = location. This equation can be used to construct a semivariogram between semivariance and lag (Figure 1). Measurements of variables are used to calculate semivariance and plotted against the distance between points. This describes the spatial correlation of values between points at given distances (lags) apart.

Different models can be used to describe the relationship between semivariance and lag, some being more appropriate for certain variables than others. In Figure 1 the nugget has a value of about 2. This means that there is no spatial correlation until the difference between values is 2 or more. This is a realistic situation with regard to pH in many soils—even at two very close locations in the soil, and thus the root surface, pH can vary by 2 units. It can be predicted that all plants can probably deal with variations in pH of up to 2 units because these could occur between, for example, adjacent root cells. Variation by 3 or 4 units is possible over fairly small distances, and differences in pH tolerance of this much might be expected between species from similar habitats—the pH of calcareous grassland does vary over lags of tens of meters. In general, over tens of kilometers pH is not spatially correlated, and traveling this distance could radically alter pH. Plants that live kilometers apart could have entirely different pH tolerances—that is, they are in the sill. The nugget, the range, and the sill vary with soil type and variable. Such geostatistical analyses are useful for understanding plant responses to environmental variation because they describe a fundamental aspect of the challenge that has driven the evolution of these responses.

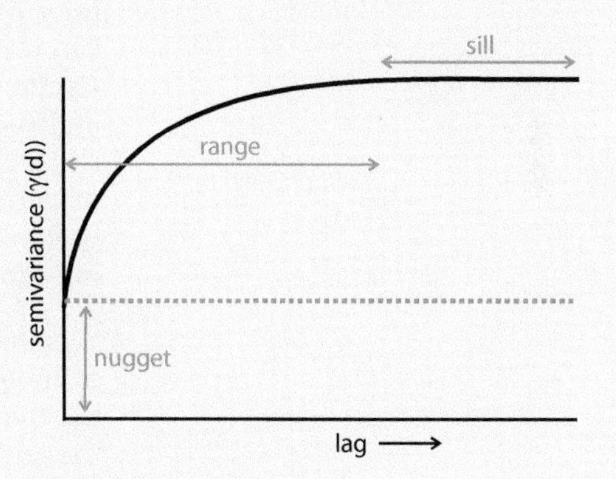

Figure 1. A semivariogram.

The environment varies not only in space but also in time. The **circadian clock** is used to match numerous aspects of a plant's physiology to daily variation in the environment. In the temperate zone in particular there are also seasonal events to which almost all plants are attuned. These regular changes that occur over daily and seasonal timescales have tended to deflect attention from important variation that occurs over a timescale of hours, or over timescales that are longer than a day but shorter than a season. For example, daily variation in light intensity compounded by underlying seasonal patterns provides an overall description of variation in light intensity in the temperate zone. However, for an individual plant, **phenological** patterns in, for example, leaf development of neighboring plants can significantly complicate this pattern over a period of weeks. Furthermore, in the summer in the shade of a forest floor, there is generally not much change in light intensity, but sunflecks can mean that some plants experience a change in light intensity over a few minutes that is almost as great as that ordinarily experienced daily. For many environmental variables, similarly irregular temporal changes can be superimposed on more regular changes. For example, rainfall can cause not only quite abrupt changes in temperature as water percolates down the profile, but also rapid—and sometimes very significant—changes in many soil properties, such as soil oxygen status, nutrient availability, and soil pH. Environmental variation over the full range of spatial and temporal scales is used as a perspective in this book for understanding plant-environment interactions.

Plants process information about environmental variation using signaling networks

In signaling networks, signals reflecting external conditions are detected by receptors that transduce them. In cells, many—though not all—receptors are membrane-bound proteins to which molecules from the environment can bind. All of the variables discussed in this book affect the physicochemical environment and can act as signals for plant cells—that is, plant cells can use changes in the concentration or state of ions and/or molecules to detect changing environmental conditions. Receptors transduce signal information about the environment into the cell, often by changing **conformation** and then changing, for example, the activity of ion transporters, **G-proteins, kinases**, or enzymes (Figure 1.5). Ionic balance in plant cells is kept within narrow limits, but conformational changes in receptors can provoke very rapid flows of ions down electrochemical gradients to transduce signal information. G-protein-coupled receptors are bound to **heterotrimeric** G-proteins which are uncoupled following ligand binding and can transduce information to specific signaling networks. Other receptors are linked to kinases through which responses are initiated. After signal information has been transduced it must be relayed to parts of the cell that can respond. This frequently involves amplification cascades (Figure 1.5). **Cyclic nucleotides**, **inositol lipids**, Ca^{2+}, **reactive oxygen species** (ROS), and **reactive nitrogen species** (RNS) can all be involved in complex cascades that transmit and amplify information from receptors. There is much evidence that the amplification of signals via many of these networks can be confined to certain parts of a cell, sometimes by the presence of scaffolding—for example, that provided by **14-3-3 proteins**.

Signaling cascades have to be controlled and attenuated. Phosphorylation of molecules by kinases and dephosphorylation by phosphatases is one key mechanism of control. Phosphorylation of proteins occurs primarily at the amino acids serine, threonine, or tyrosine, with the γ-phosphate of ATP being transferred to the OH group of the amino acid. Serine and threonine are generally phosphorylated by one class of kinase, and tyrosine by another. Phosphorylation can dramatically alter the free energy of subsequent reactions involving the protein (that is, "activate" it), or change its conformation to allow or prevent its action. Numerous elements of signal perception, transduction, and transmission in plants are controlled by kinases and phosphatases, sometimes via well-known cascades. There are kinases that interact specifically with G-proteins, cyclic nucleotides, **calmodulin**, and **mitogen**. In addition to the action of phosphatases, responses to signals can be attenuated by the action of the **ubiquitin–proteasome system**. There is a constant turnover of proteins in a plant cell that fine-tunes the proteome to the prevailing conditions. This requires not only synthesis of new proteins but also destruction of unwanted proteins, mostly achieved through the ubiquitin–proteasome system (Figure 1.6). The principles of cell signaling are therefore vital to an understanding of plant detection of environmental variation. In some instances there are relatively clear sequences of events between a particular signal and a particular response, but numerous responses have many features in common and can be used to understand how signaling pathways convey both specific and general information about the environment.

Differences in gene expression and in the genes expressed underpin a hierarchy of plant adaptations

A plant's metabolism, growth, and development are controlled by the activity of proteins. These are encoded in the **genome**, but the relationship between a cell's genome and **proteome** is complex—some genes do not encode

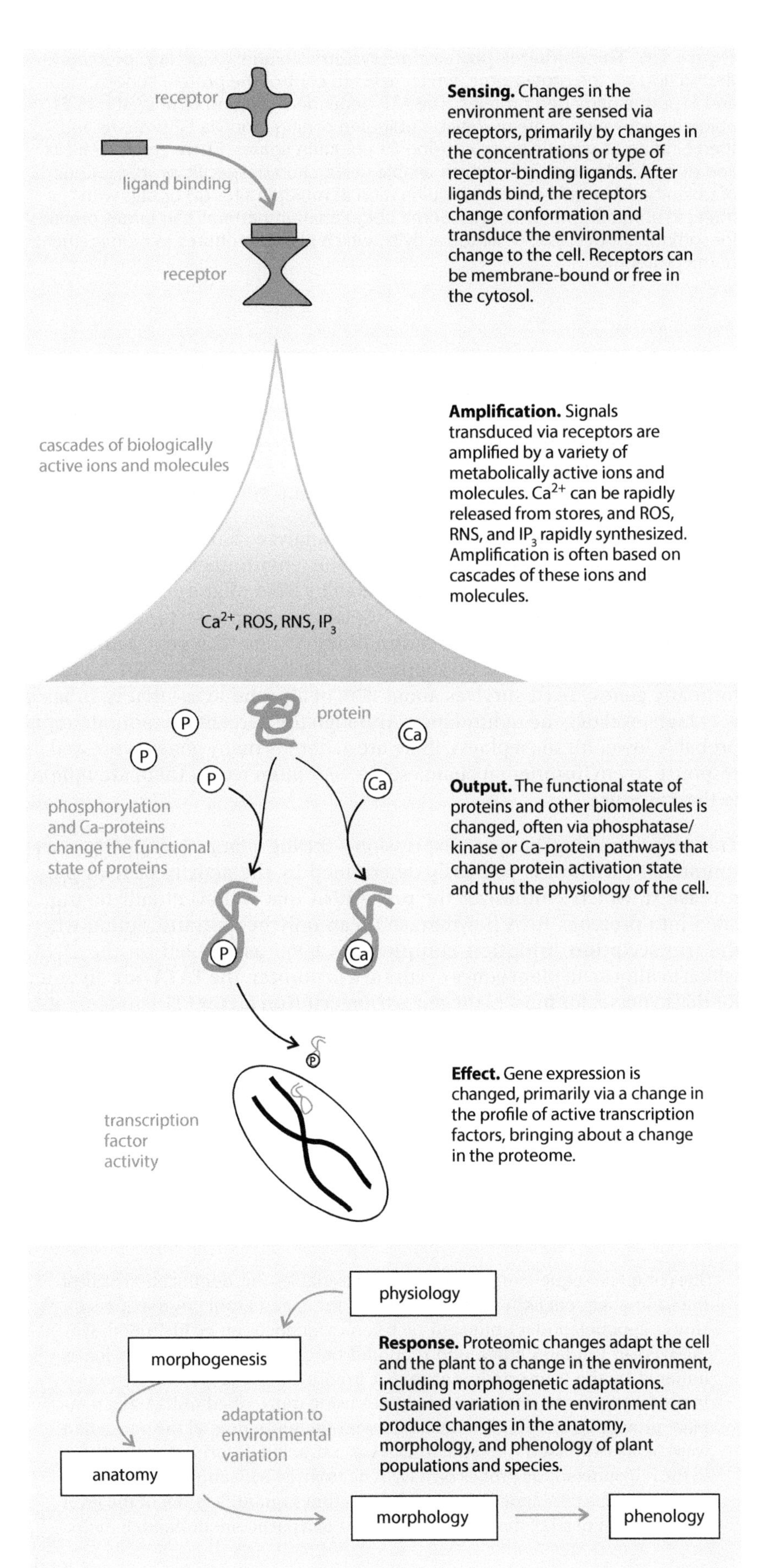

Figure 1.5. Plant cell signaling. A general outline of processes from detection of an environmental signal to the plant response. ROS = reactive oxygen species; RNS = reactive nitrogen species; IP_3 = inositol phosphate.

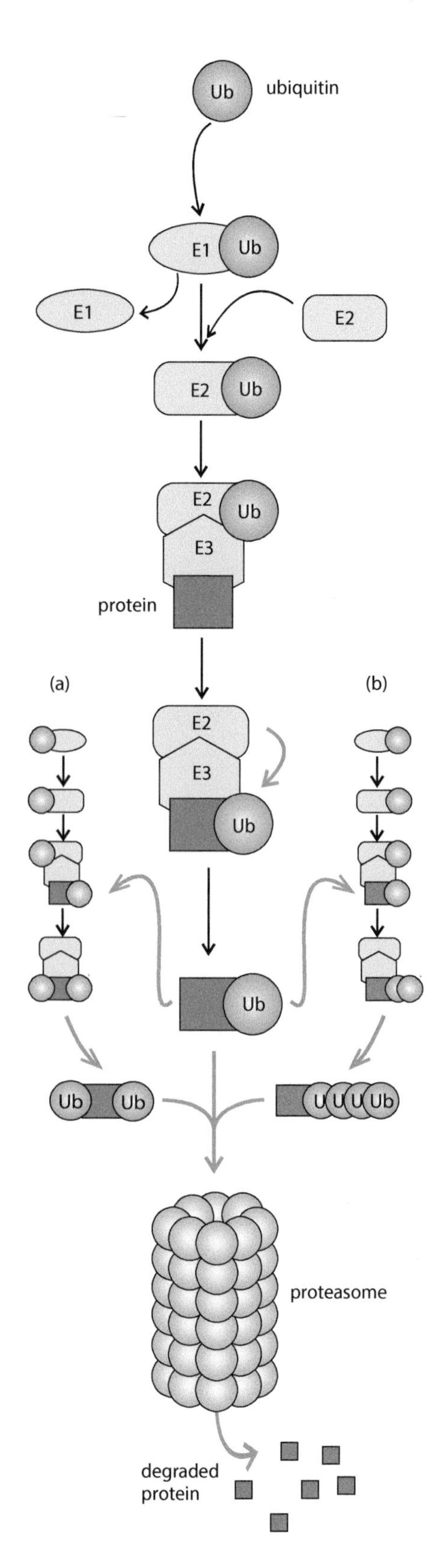

Figure 1.6. The ubiquitin–proteasome system. Ubiquitination tags proteins for destruction via the proteasome, which helps to control the protein concentration and expunge damaged proteins. The ATP-dependent binding of ubiquitin to E1 ubiquitin-activation proteins initiates ubiquitin conjugation via E2 proteins and then binding to specific proteins using E3 ubiquitin ligases. There are several E2s and many E3s in plant cells, which enable destruction of specific proteins. Rounds of ubiquitination can result in ubiquitination at multiple sites (a) or tails with multiple ubiquitin molecules (b). Some ubiquitination patterns can target proteins for sorting, endocytosis, or kinase activity, which also contributes to management of the proteome.

proteins, different genes can encode similar proteins, and proteins can be modified after translation. In addition, there is complex control of the activity of proteins in a plant cell. In general, however, in plants it is at the level of gene **transcription** that much control over the proteome is exerted, and the myriad mechanisms that regulate gene expression have almost all been shown to have significant roles in particular plant–environment interactions.

Microarrays have been used extensively to analyze changes in gene expression initiated in plants by changes in their environment (Box 1.5). The majority of these data have been generated by DNA **oligonucleotide** arrays. In general, the expression of genes in plants in response to environmental variables has a high level of **redundancy**. Plants with gene **knock-outs** (KOs)—that is, eliminated or suppressed gene expression—can be made for many genes. Yeast survives about 80% of all gene KOs—that is, it has a very high level of gene redundancy. Although the percentage redundancy is probably lower for most plants, there are generally many genes expressed in response to environmental changes, but very often few of them are unique to that response.

Transcription—that is, gene expression—during plant response to environmental variation is primarily determined by the activity of **RNA polymerase II**, which synthesizes the **pre-mRNA** that will eventually be translated into proteins. RNA polymerase II can only begin transcription when the **transcription initiation complex** has been assembled on the DNA, which in almost all plant genes occurs at a **promoter**, the **TATA box**. In order for this to occur, for most plant genes **transcription factor** (TF) proteins also have to be bound to the GC and CAAT boxes slightly further upstream. For almost all plant genes, even further upstream there are distal regulatory sequences to which further TFs can bind to regulate gene expression. TFs

BOX 1.5. DNA MICROARRAYS

The complete sequencing of a plant genome enables the identification of all of the coding sequences in a genome—that is, the genes. From these sequences, short oligonucleotides unique to each gene sequence can be identified, synthesized in the laboratory, and mounted on an array. Gene expression is initiated by the transcription of mRNA from a coding sequence. Synthesis in the laboratory of single-stranded DNA using transcribed mRNA as a template provides a set of DNA molecules with the sequences of the genes that were being expressed when the mRNA was extracted. Hybridization of these to the oligonucleotide probes on a DNA microarray with appropriate **fluorophors** enables the identification, and sometimes quantification, of the level of expression of all of the genes in a genome at a particular moment in time.

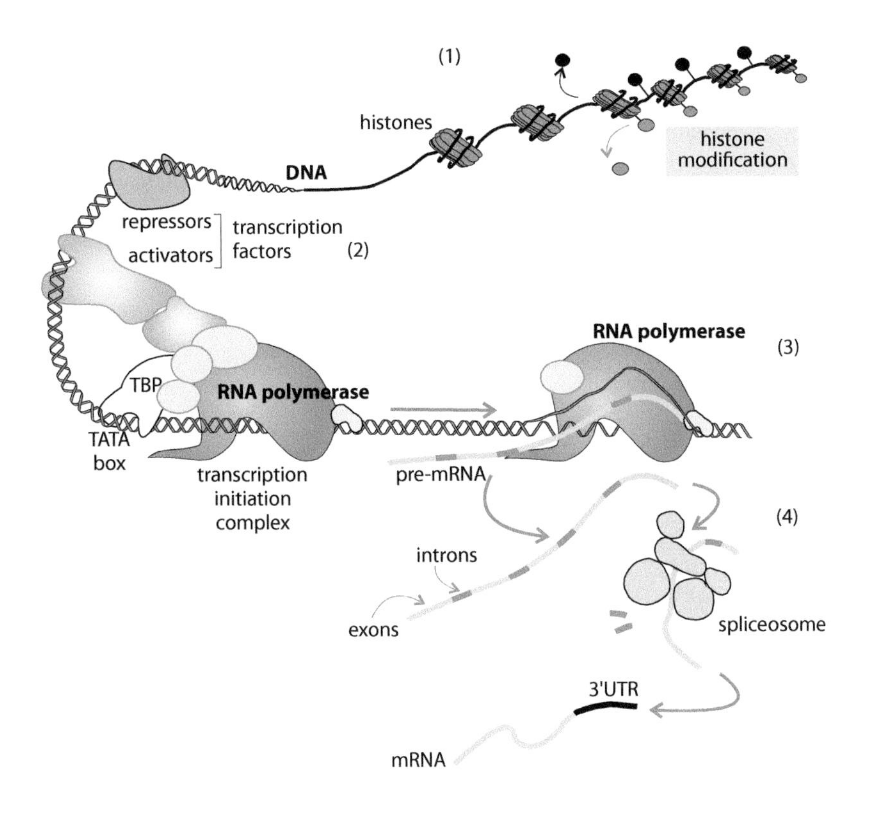

Figure 1.7. The control of gene expression in plants. The control of gene expression exerts significant control over the proteome in plant cells. (1) Epigenetic factors—that is, those that do not depend on gene sequence—affect expression via histone modification (acetylation inhibits unwinding of DNA) and DNA methylation (which prevents transcription). (2) Transcription factors have an important role in controlling gene expression in plants via their effects on the construction of the transcription initiation complex (TIC). TF activators of gene expression, which can enable or speed up transcription, bind between upstream regions of DNA and the TIC. Repressors can also bind to these regions of DNA, preventing the assembly of the TIC. (3) After transcription has been initiated, most accessory proteins dissociate and the RNA polymerase synthesizes the pre-mRNA. (4) Pre-mRNA is fed to the spliceosome, which excises introns (non-expressed sequence), leaving expressed sequences (exons) and a 3′ untranslated region (UTR). Alternative splicing can produce mRNA variants. mRNAs are translated into proteins in ribosomes. Alternative splicing to produce variant proteins is less important in plants than in other eukaryotes. Post-translational modifications of proteins can further regulate proteome activity, but again this is not as important in plants as in many other eukaryotes.

that initiate gene expression are known as "activators," whereas those that prevent it are termed "repressors" (Figure 1.7). The transcription initiation complex and the binding of TFs control transcription mostly via the physical orientation of all the molecules involved in transcription. Some important types of TFs that control gene expression in plants are listed in Table 1.2. TFs can be encoded at a variety of locations across the genome, and can control the expression of all genes that have a particular distal regulatory sequence. The eukaryotic genes of higher plants have many **introns** that need to be spliced out of the pre-mRNA, which also has to be modified for transport out of the nucleus, and to be resistant to **RNases** once in the cytoplasm. In plants there are some significant instances in which alternative splicing of RNA controls protein profiles.

Small RNAs are important for the control of some plant responses to environmental stressors. They include **microRNAs** (miRNAs) that are about 21 nucleotides long and prevent the translation of specific mRNAs.

Table 1.2. Key types of transcription factors in plants

Family	Key characteristic	Distribution and use
bZIP	Basic-region leucine-zipper DNA-binding domain	Some present in all plants
MYB	MYB-like DNA-binding domain	Many present in plants with many functions
bHLH	Basic-region helix-loop-helix DNA-binding domain	Present in plants, often with developmental function
HD proteins	Homeodomain for DNA binding	Many present in plants, often with developmental roles
MADS box	MADS motif for DNA binding	Several groups in plants
ZINC proteins	Various Zn-containing DNA-binding domains	
AP2/EREBP	AP2/EREBP DNA-binding domain	A few TFs with vital roles in development and ethylene responses

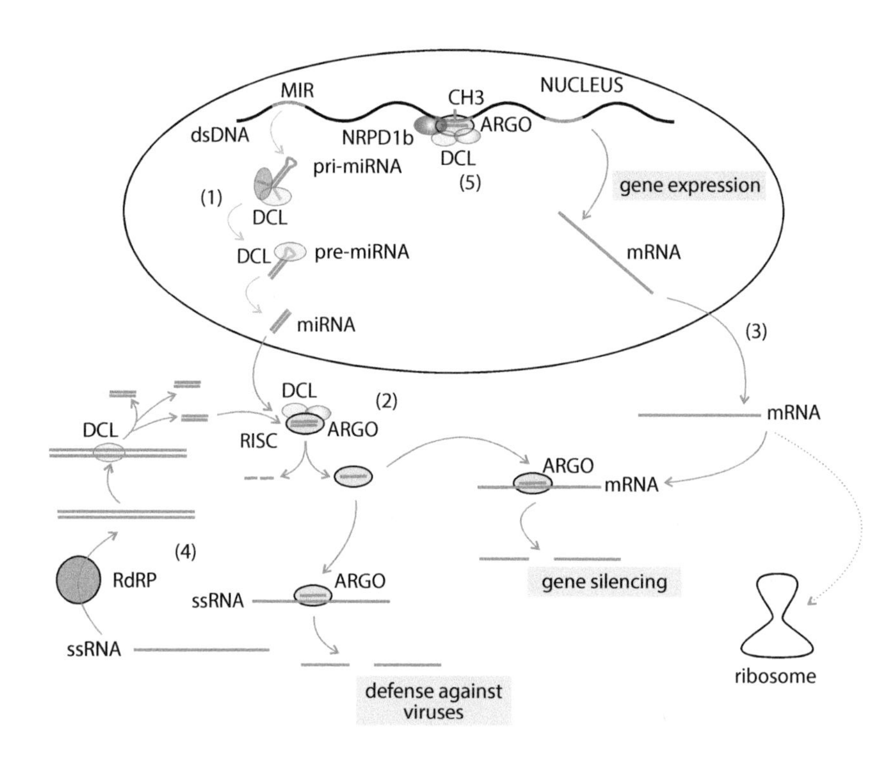

Figure 1.8. **The inhibition of gene expression by small RNAs.** (1) In plant genomes, in addition to protein-coding genes there are numerous DNA sequences that code for double-stranded microRNAs (miRNAs). These miRNA sequences are palindromic, producing hairpin primary miRNA (pri-miRNA) structures about 70 nucleotides long. These are processed, in part using Dicer-like proteins (DCLs) that can slice RNA, to produce double-stranded miRNAs about 19–23 nucleotides long that are exported to the cytoplasm. (2) In the cytoplasm these bind to the RNA-induced silencing complex (RISC), which includes ARGONAUT proteins (ARGO) with RNA-binding and slicing capacity. (3) ARGO-bound RNA binds to mRNA from gene expression, slicing it to prevent its translation in ribosomes. In plants, miRNA sequences can be very specific to 3′ UTRs, inhibiting the translation of specific genes, although some miRNAs can inhibit translation of several genes. (4) It is likely that miRNA inhibition of mRNA translation evolved from plant defense against the many single-stranded RNA (ssRNA) viruses that attack plants. ARGO can slice ssRNAs, or RNA-dependent RNA polymerases (RdRPs) convert ssRNA virus sequences into double-stranded RNA (dsRNA) that can be attacked by DCLs and attached to the RISC complex for slicing. (5) The RISC complex also has a role in methylating genes to exert epigenetic control over their expression.

The microRNAs are made from RNAs about 70 nucleotides long that are encoded in the genome. These have **palindromic sequences** towards their ends, form hairpin loops, and are then cleaved by Dicer-like enzymes to form miRNAs. A ribonuclease complex, the **RNA-induced silencing complex**, uses the miRNAs to attach to specific mRNAs and cleave them, initiating a cascade of RNA degradation that prevents translation (Figure 1.8). This mechanism of preventing mRNA translation probably evolved from defense against the single-stranded RNA viruses common in plants, and is also the basis for the **short-interfering RNAs** (siRNAs) that can be used to silence the expression of specific genes.

If translation is initiated, polypeptides begin to emerge from ribosomes. The conformation of most proteins is essential to their function, but cannot be finalized until translation is complete. There are therefore numerous mechanisms in plants that control protein folding and assembly. These can be important in the regulation of translation into a proteome response to particular stressors. In plants, low-intensity or short-term responses can induce significant changes in gene expression. The physiology of a plant at a given moment in time is therefore in part a product of the expression of its genes and the condition of its environment. However, there are also mechanisms through which the conditions that a plant or a population of plants has experienced in the past can affect gene expression and adaptation to the environment. Epigenetic differences in gene expression are heritable but do not arise from differences in gene sequence. The **methylation** of the cytosine and adenine bases in DNA, and the **acetylation** of the histones around which DNA is wrapped, and which can be altered by environmental conditions, can result in heritable differences in gene expression. To combat intense or sustained environmental variation, a combination of constitutive differences in gene expression and differences in the genes expressed also occurs, and can give rise to the different anatomies, morphologies, and phenologies that adapt plants to their environment. In general, anatomical and morphological plant adaptations to intense or chronic differences in environmental variables are complementary to biochemical and physiological adaptations to low-intensity or transient environmental variation. Thus a hierarchy of plant adaptation can be conceptualized based on progressively more profound differences both in gene expression and in the genes that are being expressed. This hierarchy is used to structure each chapter in this book (Figure 1.9).

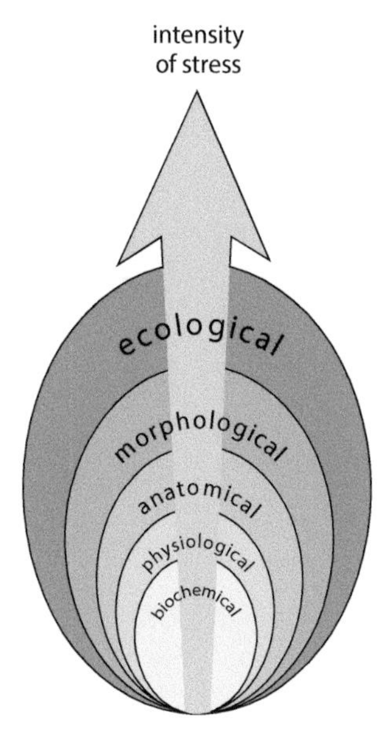

Figure 1.9. A stress response hierarchy for plants. Transient changes in gene expression, often in individual plants, can tune biochemistry and physiology to transient environmental variation. Sustained differences in gene expression and differences in the genes expressed, often between populations or species, can adapt plant anatomy, morphology, and phenology to sustained differences in intensity of environmental variables.

Environmental plant physiology is ecologically useful in defining plant traits and niches

A plant "trait" is a measurable characteristic. Suites of plant traits that are found in particular environments can be deemed "strategies." Some traits and strategies have been used to predict plant responses to the environment, and thus ecosystem processes. For trait-based ecological analyses, functional traits are important for understanding the ecology of terrestrial ecosystems. Many functional traits are biochemical, physiological, anatomical, and morphological features of plants dealt with in this book, and many of the chapters help to explain the quantitative values of some ecologically significant traits. For example, the worldwide leaf economic spectrum (WLES) arises from a significant correlation that occurs across widely divergent biomes and plant species (perhaps all plant species) between photosynthetic carbon fixation per unit leaf mass (A_{mass}), leaf nitrogen concentration per unit leaf mass (N_{mass}), leaf mass per unit area (LMA), and leaf longevity (LL) (Figure 1.10). The WLES has been used to predict numerous ecological processes in terrestrial ecosystems across a range of scales. The numerical values of A_{mass} and N_{mass} in an individual plant are determined by processes discussed in this book in Chapter 3 on CO_2 and Chapter 5 on nitrogen. There are numerous other examples in which traits are useful for ecological predictions, ranging from the invasiveness of alien plants to nutrient-use efficiencies. There are also traits that are not affected by environmental plant physiology, but for those that are it is hoped that this book will help to explain their quantitative values.

Niche concepts underpin many ecological analyses, but are more problematic when discussing plants than when considering animals. Plants require fewer, simpler resources from the environment than do animals, and in general have greater morphological uniformity. It is therefore easy to see how similar plant species compete with each other, but less easy to define different niches that can explain their diversity and elucidate ecosystem processes. There are numerous different niche concepts, but many of them, explicitly or implicitly, depend on features of the environment and how organisms interact with these. A particular challenge has been to explain the difference in diversity of plants in different ecosystems—models of non-diverse ecosystems can be based on simple niche concepts, but they work less well in biodiverse ecosystems.

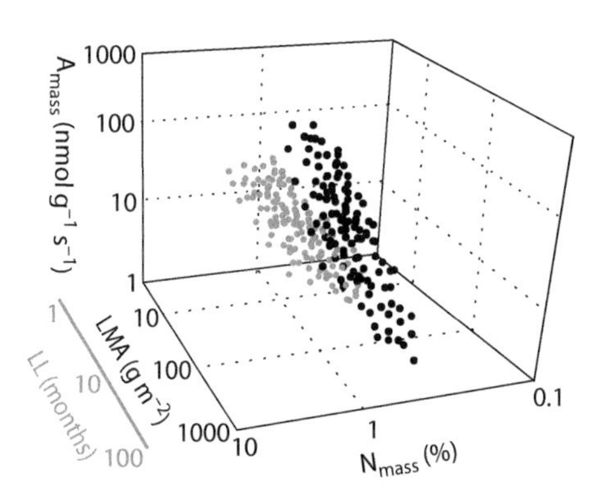

Figure 1.10. The worldwide leaf economics spectrum (WLES). Across very diverse species from different biomes there are significant correlations between photosynthetic carbon fixation per unit leaf mass (A_{mass}), leaf nitrogen concentration per unit mass (N_{mass}), leaf mass area (LMA), and leaf longevity (LL). This suggests that there are some fundamental influences driving plant adaptation across diverse environments.

There is an increasing number of well-analyzed instances in which the environmental variables covered in this book operate as **niche segregators** for plants—that is, there are fine-scale differences in environmental variables which provide distinct habitable environments for different species. This is, for example, particularly the case for water regimes in soils across divergent ecosystems. In contrast, the **neutral theory** of biodiversity posits that the diversity of plant species in a particular area is dependent on **stochastic** events, especially those that affect germination, migration, and death. In some ecosystems such events do explain plant biodiversity but, overall, biodiversity in ecosystems is most commonly held to occur along a niche-neutral continuum. Thus the responses of plants to many of the environmental variables dealt with in this book are contributors to niche segregation, and underpin processes that are important for explaining biodiversity and ecosystem functioning.

Studying plant–environment interactions can help to increase agricultural efficiency and sustainability

The increases in production of plant products and decreases in environmental impact that are necessary if sustainable production is to be achieved strongly suggest that current production systems will have to develop significantly in the coming decades. At the time of writing these developments look likely to be primarily in the **germplasm** used and in the ecological interactions that are encouraged. The majority of crop production is based on very few crop species (Figure 1.11). There are numerous minor food crops throughout the world, but the majority of production comes from 15 crops, and non-food products are dominated by just a few, such as timber, cotton, and coffee. In total, three crops provide more than 50% of total global crop yield, and six crops provide close to 75% of it. There are tens of thousands of **cultivars**, particularly of the cereals, but in general a restricted germplasm is used to produce the majority of crop yield. In the early 1990s, genetically modified (GM) crop plants began to transform the germplasm used for crop production, and by 2010 more than 10% of agricultural yield came from GM crops. In some agri-systems, GM crops dominate—for example, the majority of the cotton crop in Australia is GM, as is the majority of the soybean and maize crop in the USA. Although some long-term studies suggest that the benefits of GM crops have sometimes been overstated, it seems very likely that in the coming decades GM crops will increasingly contribute to a changed germplasm being used for production. Much of this change will be focused on crop relationships to the resource variables dealt with in this book. It is likely that an understanding of the responses outlined in the chapters on resources will inform the development of new germplasm for crop production.

Principles derived from stress physiology are also likely to contribute to the development of agriculture. For example, globally there are very significant problems with salinity in crop production systems. Biotechnology and marker-assisted breeding programs have increased the salinity tolerance of some crop plants, but this is only beginning to have an impact on commercial production. However, there are salt-tolerant angiosperms, and it has long been thought possible that these might be used for the production of food or at least forage. Investigations from the 1950s to the 1970s initiated large-scale attempts to develop saline agriculture using alternative species. None of these entered commercial production, but the principle of using "alternative crops" for production is experiencing a renaissance. Knowledge of how wild plants accumulate resources and tolerate stresses might aid not only understanding of but also the selection of species as alternative

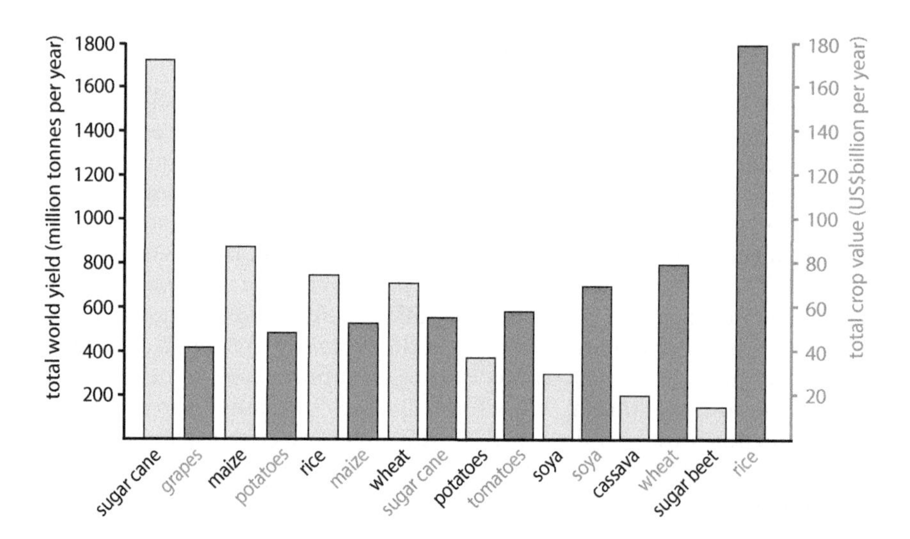

Figure 1.11. World crop production in 2010. Sugar cane is the crop with the highest harvested yield each year, but rice is the crop with the highest value each year. Harvesting techniques for sugar cane contribute to its high yield by weight, but so does its widespread use as a biofuel. The greater proportion of rice used for human rather than animal food than is the case with maize or wheat contributes to its high value. (Data from Food and Agriculture Organization of the United Nations, Statistics Division [FAOSTAT]; http://faostat3.fao.org/)

crops. In the early years of the twenty-first century the production of biofuels from crops increased enormously, and it is a now a major use of crop plants. The development of new uses for crops is likely to increase, and is frequently informed by an understanding of how they interact with the environment. Over many years agroecologists have developed a large number of systems in which ecological principles are incorporated into crop production—among these, intercropping and agroforestry systems provide established examples. Some intercropping systems in saline conditions mix wild halophytes with domesticated crops. It seems likely that in agriculture in the twenty-first century, increasing attention will be paid to the principles of plant-environment interactions and agroecology.

Modeling is improving our understanding of plant–environment interactions

Systems approaches to modeling in the earth and environmental sciences have inspired some of the most powerful models of ecological and molecular processes. Ultimately, modeling of plant-environment interactions will probably utilize those from both the environmental and biological sciences. In the chapters of this book, discussions are cognizant of the models developed in numerous relevant topics, and specific models are sometimes mentioned, but the mathematical details of these models are not discussed. Some of the most highly developed are **mechanistic models**. These are based on fundamental equations that mathematically describe processes. In recent years there have been dramatic improvements in many such models, primarily to make them dynamic in their description of events. There are many mechanistic models of the behavior of, for example, nutrients in the soil-plant system. However, mechanistic models are not available for many of the plant-environment interactions discussed. More common are **compartmental models**, in which **constants of proportionality** are used to describe the flow of resources, energy, or information between compartments, at scales ranging from the subcellular to whole ecosystems. **Systems models** and **network models** link many compartments, and sometimes sub-models, together. Most mechanistic models are sensitive to initial conditions—that is, values of the variables necessary for equations—and many compartmental models are very sensitive to exact measurement of constants of proportionality. Both can be restricted in the range of conditions under which they can accurately simulate plant-environment interactions. Each chapter in this book is a synthesis of knowledge on a topic using the contexts, perspectives, and principles outlined in this introductory chapter such that they might aid an understanding of models of plant-environment interactions.

Summary

Understanding the functioning of plants in an environment that varies in space and time is very helpful to efforts aimed at solving the challenges of environmental change and global food security. It necessitates understanding the variation in the environment together with the responses that it provokes in plants. Recent advances in the life sciences are contributing to an increased understanding of how and why plants respond to environmental variation. It is likely that knowledge of plant-environment interactions from both unmanaged and managed ecosystems will more often be seen as complementary in the future. Advances in systems analysis and modeling capability will increase the accuracy of predictions of the behavior of the plant-environment system. This book synthesizes recent knowledge from both the environmental and life sciences, from a wide range of spatial and temporal scales, and from unmanaged and managed ecosystems, to provide an understanding of environmental plant physiology that is useful in a twenty-first-century context.

Further reading

Atmosphere, hydrosphere, and geosphere

Jones VAS & Dolan L (2012) The evolution of root hairs and rhizoids. *Ann Bot* 110:205–212.

Le Hir G, Donnadieu Y, Godderis Y et al. (2011) The climate change caused by the land plant invasion in the Devonian. *Earth Planet Sci Lett* 310:203–212.

Pires ND & Dolan L (2012) Morphological evolution in land plants: new designs with old genes. *Philos Trans R Soc Lond B Biol Sci* 367:508–518.

Tomescu AMF, Wyatt SE, Hasebe M et al. (2014) Early evolution of the vascular plant body plan - the missing mechanisms. *Curr Opin Plant Biol* 17:126–136.

Anthropogenic environmental change

Lenton TM & Williams HTP (2013) On the origin of planetary-scale tipping points. *Trends Ecol Evol* 28:380–382.

McNeill JR (2001) Something New Under the Sun: An Environmental History of the 20th Century. Penguin.

Steffen W, Persson A, Deutsch L et al. (2011) The Anthropocene: from global change to planetary Stewardship. *Ambio* 40:739–761.

United Nations Environment Programme (2012) Global Environmental Outlook-5. Progress Press Ltd.

Global food security

Foley JA, Ramankutty N, Brauman KA et al. (2011) Solutions for a cultivated planet. *Nature* 478:337–342.

Godfray HCJ & Garnett T (2014) Food security and sustainable intensification. *Philos Trans R Soc Lond B Biol Sci* 369:20120273.

Huber V, Neher I, Bodirsky BL et al. (2014) Will the world run out of land? A Kaya-type decomposition to study past trends of cropland expansion. *Environ Res Lett* 9:024011.

Tilman D, Balzer C, Hill J et al. (2011) Global food demand and the sustainable intensification of agriculture. *Proc Natl Acad Sci USA* 108:20260–20264.

Proximate and ultimate causes

Laland KN, Odling-Smee J, Hoppitt W et al. (2013) More on how and why: cause and effect in biology revisited. *Biol Philos* 28:719–745.

MacDougall-Shackleton SA (2011) The levels of analysis revisited. *Philos Trans R Soc Lond B Biol Sci* 366:2076–2085.

McNickle GG, St Clair CC & Cahill JF Jr (2009) Focusing the metaphor: plant root foraging behaviour. *Trends Ecol Evol* 24:419–426.

Taborsky M (2014) Tribute to Tinbergen: the four problems of biology. A critical appraisal. *Ethology* 120:224–227.

Resources, stressors, and xenobiotics

Ghosh D & Xu J (2014) Abiotic stress responses in plant roots: a proteomics perspective. *Front Plant Sci* 5:6.

Levitt J (1972) Responses of Plants to Environmental Stresses. Academic Press.

Smirnoff N (2014) Plant stress physiology. In Encyclopedia of Life Sciences (Hetherington A ed.). John Wiley & Sons Ltd.

Romero LM (2004) Physiological stress in ecology: lessons from biomedical research. *Trends Ecol Evol* 19:249–255.

Suzuki N, Rivero RM, Shulaev V et al. (2014) Abiotic and biotic stress combinations. *New Phytol* 203:32–43.

Geostatistics and soils

Garcia-Palacios P, Maestre FT, Bardgett RD et al. (2012) Plant responses to soil heterogeneity and global environmental change. *J Ecol* 100:1303–1314.

International Union of Soil Sciences Working Group WRB (2006) World Reference Base for Soil Resources 2006. World Soil Resources Reports No. 103. Food and Agriculture Organization of the United Nations.

Oliver MA & Webster R (2014) A tutorial guide to geostatistics: computing and modelling variograms and kriging. *Catena* 113:56–69.

Xi N, Carrere P & Bloor JMG (2014) Nitrogen form and spatial pattern promote asynchrony in plant and soil responses to nitrogen inputs in a temperate grassland. *Soil Biol Biochem* 71:40–47.

Cell signaling

Conde A, Manuela Chaves M & Geros H (2011) Membrane transport, sensing and signaling in plant adaptation to environmental stress. *Plant Cell Physiol* 52:1583–1602.

Hancock JT (2010) Cell Signalling, 3rd ed. Oxford University Press.

Stone SL (2014) The role of ubiquitin and the 26S proteasome in plant abiotic stress signaling. *Front Plant Sci* 5:135.

Wrzaczek M, Brosche M & Kangasjarvi J (2013) ROS signaling loops - production, perception, regulation. *Curr Opin Plant Biol* 16:575–582.

Gene expression

Kaufmann K, Pajoro A & Angenent GC (2010) Regulation of transcription in plants: mechanisms controlling developmental switches. *Nat Rev Genet* 11:830–842.

Lyzenga WJ & Stone SL (2012) Abiotic stress tolerance mediated by protein ubiquitination. *J Exp Bot* 63:599–616.

To TK & Kim JM (2014) Epigenetic regulation of gene responsiveness in *Arabidopsis*. *Front Plant Sci* 4:548.

Vaahtera L & Brosche M (2011) More than the sum of its parts - how to achieve a specific transcriptional response to abiotic stress. *Plant Sci* 180:421–430.

Physiology and ecology

Araya YN, Silvertown J, Gowing DJ et al. (2011) A fundamental, eco-hydrological basis for niche segregation in plant communities. *New Phytol* 189:253–258.

Donovan LA, Maherali H, Caruso CM et al. (2011) The evolution of the worldwide leaf economics spectrum. *Trends Ecol Evol* 26:88–95.

Messier J, McGill BJ & Lechowicz MJ (2010) How do traits vary across ecological scales? A case for trait-based ecology. *Ecol Lett* 13:838–848.

Sack L, Scoffoni C, John GP et al. (2013) How do leaf veins influence the worldwide leaf economic spectrum? Review and synthesis. *J Exp Bot* 64:4053–4080.

Physiology, ecology, and agriculture

Balmford A, Green R & Phalan B (2012) What conservationists need to know about farming. *Philos Trans R Soc Lond B Biol Sci* 279:2714–2724.

Burger JR, Allen CD, Brown JH et al. (2012) The macroecology of sustainability. *PloS Biol* 10:e1001345.

Mayes S, Massawe FJ, Alderson PG et al. (2012) The potential for underutilized crops to improve security of food production. *J Exp Bot* 63:1075–1079.

Phalan B, Green R & Balmford A (2014) Closing yield gaps: perils and possibilities for biodiversity conservation. *Philos Trans R Soc Lond B Biol Sci* 369:0285.

Sayer J & Cassman KG (2013) Agricultural innovation to protect the environment. *Proc Natl Acad Sci USA* 110:8345–8348.

Systems and modeling

Chandran AKN & Jung K-H (2014) Resources for systems biology in rice. *J Plant Biol* 57:80–92.

Cramer GR, Urano K, Delrot S et al. (2011) Effects of abiotic stress on plants: a systems biology perspective. *BMC Plant Biol* 11:163.

Kluge M (2008) Ecophysiology: migrations between different levels of scaling. *Prog Bot* 69:5–34.

Tardieu F & Tuberosa R (2010) Dissection and modelling of abiotic stress tolerance in plants. *Curr Opin Plant Biol* 13:206–212.

Chapter 2
Light

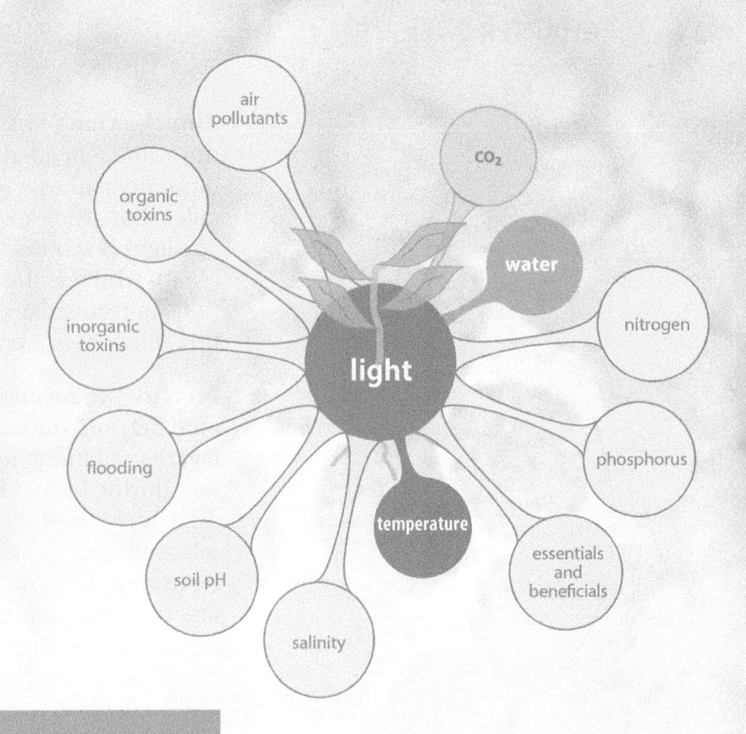

Key concepts

- Light energy transduced into chemical energy powers terrestrial ecosystems.
- The light-dependent reactions of photosynthesis catalyze energy transduction.
- Variation in irradiance in terrestrial ecosystems necessitates control over its acquisition.
- Non-photochemical quenching of light energy gives rapid control over its use in photosystems.
- Numerous electron sinks provide photochemical quenching capacity.
- Repair of photosystem II is vital to plant responses to high irradiance.
- Chloroplast movement exerts significant control over light capture.
- Thylakoid constituents and structure adapt to sustained differences in irradiance.
- Leaf optical properties optimize light capture from contrasting light regimes.
- Leaf movement and plant architecture help to control light capture in canopies.
- Tropical alpine plants overcome the most severe photoinhibitory conditions.

The challenge of managing light acquisition is particularly exacerbated by variation in temperature, water, and CO_2 levels.

In plants, ancient photosynthetic systems provide the chemical energy for terrestrial ecosystems

Terrestrial ecosystems developed when the Earth's surface was predominantly oxidizing. In oxidized environments, the essential elements of life occur mostly in forms that are metabolically unreactive. Thus the hydrogen (H), carbon (C), nitrogen (N), and sulfur (S) that are used for making biomolecules occur primarily as water (H_2O), carbon dioxide (CO_2), nitrate (NO_3^-), and sulfate (SO_4^{2-}), and must be reduced in order to be metabolically useful. The reducing agent **ferredoxin** (Fd) from the light-dependent reactions of photosynthesis is the dominant ultimate **reductant** in terrestrial ecosystems. It is used to reduce both the **nicotinamide adenine**

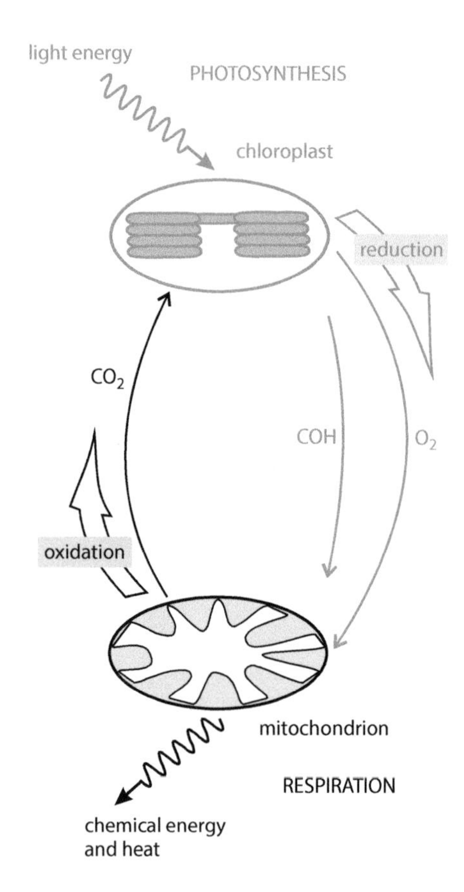

Figure 2.1. The relationship between photosynthesis and respiration. Photosynthesis reduces CO_2 to produce carbohydrates (COH) and O_2, whereas respiration oxidizes COH to produce CO_2.

dinucleotide phosphate (NADP) needed for carbohydrate synthesis, and the N and S needed for amino acid synthesis. The light-dependent reactions of photosynthesis therefore underpin the production of reduced biomolecules that are re-oxidized during respiration to release energy (Figure 2.1). The light reactions of photosynthesis also produce **adenosine triphosphate** (ATP), which is the primary source of chemical energy for metabolic reactions in terrestrial ecosystems. Light energy trapped during photosynthesis therefore powers both unmanaged and managed terrestrial ecosystems.

Primary producers of biomass are those organisms that can reduce inorganic CO_2 and nutrients to organic compounds. Across the kingdoms of life, a variety of inorganic substrates are used as sources of electrons for reduction during biosynthesis. For example, reduced forms of N, S, and iron (Fe) are widely used by microorganisms as electron sources. Often the energy released during oxidation of such substrates, although relatively low, can be used in biosynthesis. Chemosynthetic microorganisms, which were the first primary producers on Earth, oxidize reduced N, S, and Fe compounds to release the energy for primary production. Their most commonly used electron acceptor is now oxygen (O_2)—that is, many are **aerobes**—but some are **anaerobes**, which were probably among the Earth's original inhabitants. In most current ecosystems, chemosynthesis of biomass is generally a small component of primary production, an exception being **hydrothermal vent** ecosystems. Chemosynthetic organisms, which have no source of energy other than electron flows from their inorganic substrates, rely on the availability of high-energy electrons that are now relatively rare on the Earth's surface. In hydrothermal vents, hydrogen sulfide (H_2S), produced ultimately from unusually direct access to geothermal energy, provides such an electron source. On the Earth's surface, light energy is abundant and its use to increase the energy of electrons enabled alternative lower-energy sources of electrons to be used for reduction in biosynthesis. Photosynthetic organisms that use light energy and the widespread low-energy electron source H_2O dominate primary production in most ecosystems, but especially terrestrial ecosystems.

Purple sulfur bacteria were probably the first photosynthetic organisms. They use sunlight as an energy source and H_2S as an electron source. Green sulfur bacteria have been shown to also use hydrogen (H_2) and Fe^{2+} as sources of electrons. Sulfur bacteria use a single **photosystem** to trap light energy and catalyze the removal of an electron from H_2S, releasing protons and elemental sulfur. Electrons are cycled around the photosystem, via **quinone** and a cytochrome, to produce a transmembrane H^+ gradient. This proton gradient drives the formation of ATP via an **ATP synthase**. Frequently, an organic source of electrons is used to produce the reductant NADPH, a process that utilizes some of the ATP produced. The **bacteriochlorophylls**, the photosystem, and the ATP synthase of these microorganisms were the origin of oxygenic photosynthesis. **Cyanobacteria** have two photosystems in series and an ATPase. They use H_2O as a source of electrons, release O_2, and maintain a non-cyclic flow of electrons from water to NADPH to produce both the reducing power and the chemical energy necessary for primary production. The evolution of oxygenic photosynthesis was perhaps the most important event in the history of life on Earth, because the evolution of photosystems that could use light energy to remove electrons from water transformed the amount of reducing power available for primary production. It was also mostly responsible for oxygenating the Earth's environment, and thus transformed the evolution of life. Chloroplasts of terrestrial plants are essentially cyanobacteria within plant cells, so the oxygenic photosynthesis that they evolved enabled life not only to proliferate on the land surface but also, for the first time, to penetrate the atmosphere.

Photosystems, cytochromes, and ATP synthases transduce light energy into chemical energy

The light-dependent reactions of photosynthesis which produce the NADPH and ATP that power terrestrial ecosystems are mediated by two photosystems, a cytochrome and an ATPase (Figure 2.2). Photosystem I (PSI), which is descended from the photosystem of sulfur bacteria, was probably the progenitor of photosystem II (PSII). In the reaction centers of photosystems, light energy drives charge separation in chlorophyll (Chl) molecules—that is, Chl + light = Chl^+ + e^-. In PSII, Chl^+ drives the splitting of $2H_2O$ into $4H^+$, O_2, and $4e^-$. The electrons released in PSII are carried via a cytochrome to PSI, where they gain further energy from light and are then used

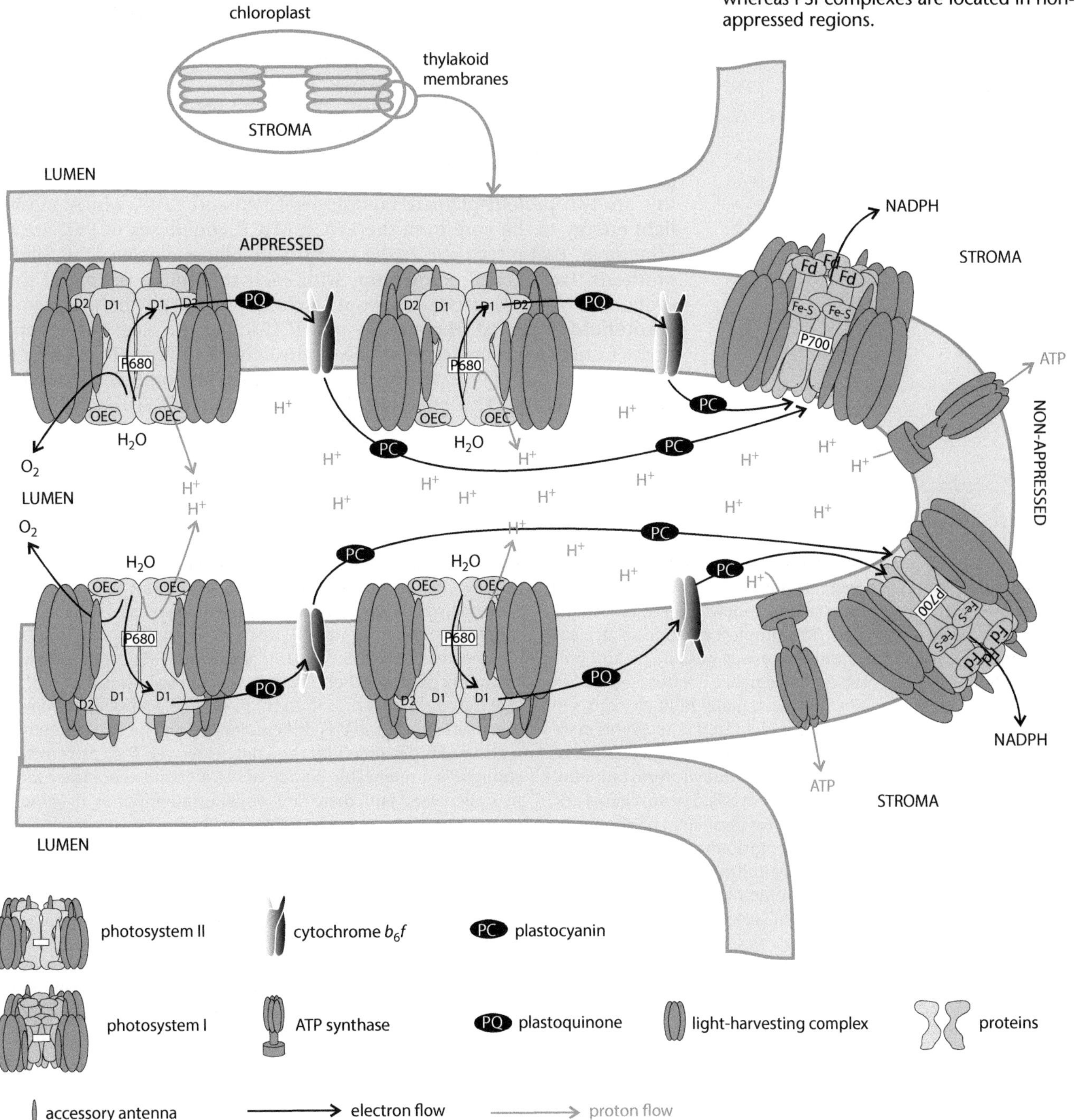

Figure 2.2. The light-dependent reactions of photosynthesis. Light-harvesting complexes focus energy on primary pigment molecules (P680) in photosystem II (PSII) complexes, producing chlorophyll$^+$ (Chl^+). Chl^+ drives the extraction of an electron from water, producing O_2 and H^+. Electron flows driven by light energy, and further oxidizing potential from primary pigment molecules (P700) in photosystem I (PSI) complexes, are used to synthesize the reductant NADPH. The proton gradient between the inside and outside of the stroma drives the formation of ATP. PSII complexes are located in appressed regions of thylakoids, whereas PSI complexes are located in non-appressed regions.

Table 2.1. The proportions of light reaction constituents relative to chlorophyll (mmol/mol chlorophyll)

PSII	3.0
LHCII	34
PSI	2.3
Cytochrome b_6f	1.3
ATPase	1.0
NDH	0.1

From Pribil M, Labs M & Leister D (2014) *J Exp Bot* 65:1955–1972. With permission from Oxford University Press.

by ferredoxin to produce NADPH. Thus, when these constituents occur in appropriate proportions (Table 2.1), the light reactions catalyze a light-driven linear flow of electrons from H_2O to NADPH. This is solar-powered electricity generation and photosystems are the focus of significant renewable energy research. Furthermore, the splitting of water in PSII is light-driven H^+ production, and because of the great potential of H_2-fuel cells, the coupling of photosystems to **hydrogenases** is also a potential renewable energy source (Box 2.1). The protons released at PSII acidify the thylakoid lumen because, unlike the O_2 produced, they cannot diffuse through the membrane due to their charge. The proton gradient between the lumen and the stroma drives the production of ATP via the ATP synthase. The PSII complexes are located in appressed regions of the thylakoids, and PSI complexes are located in the non-appressed regions because ferredoxin needs access to the stroma for NADPH synthesis (Figure 2.2). In a process that echoes the origins of photosynthesis, a cyclic flow of electrons focused on PSI and cytochrome aids ATP production.

Photosystems consist of a reaction center core surrounded by **light-harvesting complexes** (LHCs). The core of PSII has about 20 protein subunits and relatively few pigment molecules. At its heart is a pair of primary chlorophyll molecules that absorb light of wavelength 680 nm (P680). Two proteins, D1 and D2, provide the scaffolding of the PSII core. Mounted on this are two protein–pigment complexes, CP43 and CP47, which conduct light energy to the core from the LHCs. LHCII complexes of PSII are the "antennae" that often capture the majority of light energy entering photosynthesis. Each LHCII is a **trimer**, with each subunit consisting of three proteins (*LHCb*1–3), each with about 14 chlorophyll *a* or *b* molecules and 4 **carotenoids** attached. The number of LHCII trimers on each PSII complex varies from two upwards, depending on the light conditions. The core of PSI is different to that of PSII, and its primary pigment chlorophylls absorb light at 700 nm, but the LHCs of PSI are the same as those of PSII. Each PSII core

BOX 2.1. PHOTOSYNTHESIS AND RENEWABLE ENERGY

Annual global energy consumption is 130–140 TWh. The amount of energy trapped in photosynthesis is about 750 TWh per year, so its potential as a renewable energy source has long been recognized. The simplest method of accessing energy trapped by photosynthesis is in biofuels and biomass. Biofuels from sugar cane and maize are now significant renewable energy sources. However, many biofuels compete with crop plants for productive agricultural land, so biomass production on marginal land for burning or digestion is of increasing interest. Microalgae that can directly form biodiesel or that synthesize combustible oils have also been extensively researched. Knowledge gained about these uses of photosynthesis as an energy source is also contributing to research into industrial-scale "artificial photosynthesis" in which the reactions of photosynthesis are combined with engineering systems to produce, on an industrial scale, energy-rich organic molecules from sunlight.

The photosystems of photosynthesis produce electricity, at relatively low efficiency. For example, isolated photosystems have been attached directly to platinum electrodes and mounted in gels of high electrical conductivity. The incorporation of photosystems, parts of photosystems, or structures inspired by photosystems into solar electricity generation is a real possibility. There is also much ongoing research into engineering photosystems to provide more efficient converters of light energy into chemical energy that can be mass produced. In addition, photosynthesis produces H^+, and H_2 might be a significant energy source in the near future. Many organisms, including photosynthetic ones, have hydrogenases that catalyze the conversion of H^+ into H_2. In general, hydrogenases are inactivated by O_2, so it is difficult for them to be used directly with the H^+ produced by the OEC. However, PSII driven by sunlight is a renewable source of the electrons necessary for hydrogenases, and there are ongoing attempts to develop systems based on extracted PSIIs and hydrogenases in which electrons from PSII can be used by hydrogenase without the O_2 causing inactivation. Microbial fuel cells that produce electricity from H_2 using electrons from PSII of algae have been developed. Given the numerous possibilities for using photosynthesis to provide human societies with energy, ranging from the use of biofuels to advanced artificial photosynthesis, the study of the light reactions seems very likely to contribute to the production of renewable energy during the twenty-first century.

Further reading: Sherman BD, Vaughn MD, Bergkamp JJ et al. (2014) *Photosynth Res* 120:59–70.

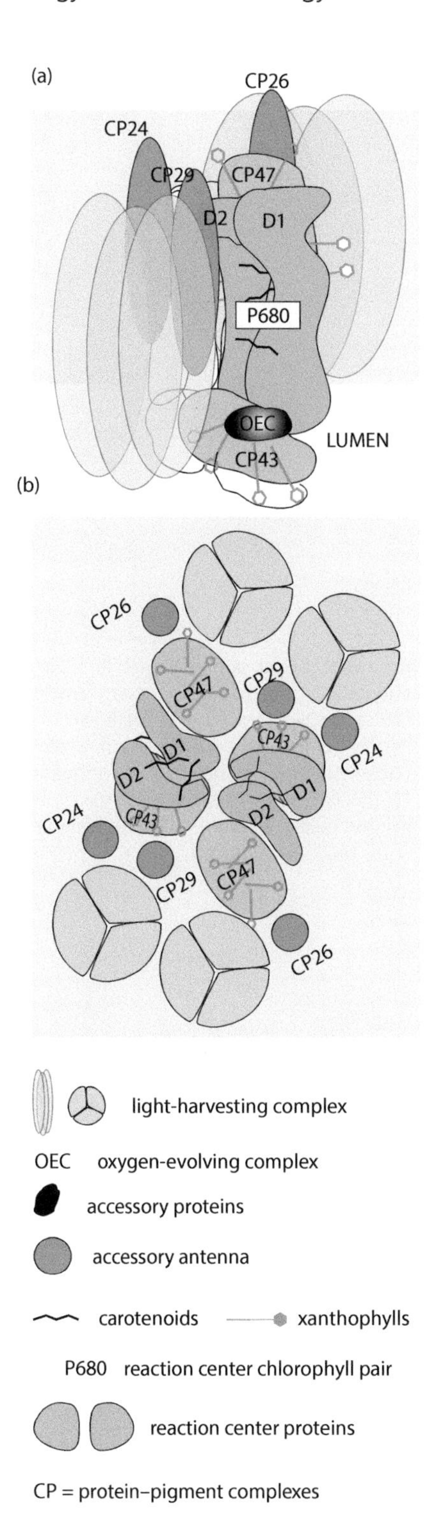

Figure 2.3. Photosystem II of spinach. (a) Monomer longitudinal section. A pair of pigment molecules (P680) forms the reaction center with two primary proteins (D1 and D2). Light-harvesting complexes focus energy through protein–pigment complexes (CP43 and CP47) to the reaction center. (b) Dimer transverse section. PSII mostly occurs as a dimer, with each subunit having not only two or more LHCs but also three accessory antenna protein–pigment complexes (CP24, CP26, and CP29).

also has three small antenna proteins, namely CP24, CP26, and CP29. PSII is generally visualized as a **dimer** of complexes (Figure 2.3), but depending on the light conditions a variety of megacomplexes of PSII units can be formed. LHCIIs and small antennae harvest and then, by transmission of excitation energy, focus light energy through CP43 and CP47 to cause charge separation alternately in the two primary pigment chlorophylls. The strongly oxidizing Chl^+ drives the splitting of water in the oxygen-evolving complex (OEC). At the center of the OEC are four manganese (Mn) atoms and one calcium (Ca) atom that holds a water molecule from which an electron is extracted, releasing $2H^+$ and O. The electrons are transferred to a **tyrosine** residue on D1 and from there to Chl^+, forming Chl, and then, if light energy is being focused on Chl, charge separation occurs again, forming Chl^+ and e^-. The electron is transferred to a **phaeophytin** molecule. The mobile electron carrier **plastoquinone** takes electrons from phaeophytin and transfers them to **cytochrome b_6f** and the electron carrier **plastocyanin,** and the electron carrier plastocyanin from the cytochrome to PSI complexes. The ATP synthase (Figure 2.4), driven by the proton gradient across the thylakoid membrane, occurs mainly in the non-appressed regions of the thylakoids.

On the stromal side of PSI, ferredoxin is used not only to produce NADPH for carbohydrate synthesis, but also in the assimilation of N and S. NO_3^- cannot be utilized metabolically because N must be assimilated into amino acids in its reduced form, ammonia (NH_3). In the cytoplasm, NADPH provides the reducing power for **nitrate reductase**, which converts NO_3^- into NO_2^- (nitrite), but NO_2^- is reduced to NH_3 in the chloroplast using electrons from ferredoxin. SO_4^{2-} is metabolically inactive and has to be reduced for the synthesis of sulfolipids and S-containing amino acids. The reducing power to convert SO_4^{2-} to SO_3^{2-} (sulfite), which is used to synthesize sulfolipids, comes

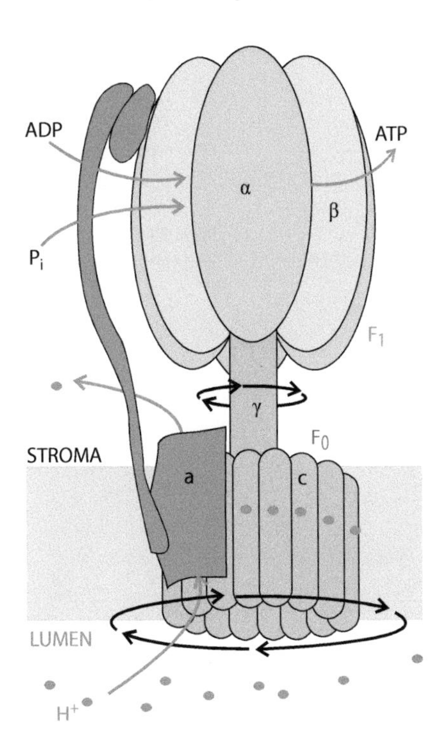

Figure 2.4. ATP synthase. The synthesis of ATP is driven by the proton gradient down which protons flow. Their flow through the F_0 subunit spins it round. The γ subunit of F_0 is asymmetrical, causing the α and β subunits of F_1 to change conformation in sequence, combining ADP with P_i to form ATP.

from glutathione, which is reduced by NADPH. Sulfite reductase, which is used to further reduce S for the production of S-containing amino acids, is a ferredoxin-dependent chloroplast enzyme. Thus in terrestrial ecosystems the light reactions of photosynthesis are the ultimate source of both the reducing power and the energy necessary for C, N, and S assimilation and hence the rest of metabolism.

Terrestrial plants have to adapt to a generally high and very variable light regime

A relatively constant flux of light energy from the sun, known as the **solar constant**, currently reaches Earth at about 1370 J s^{-1} m^{-2}. On its passage through the atmosphere, about 13% of incoming light is absorbed by gases (especially O_2, H_2O, and O_3) and 13% is scattered by particulates, so that at a surface perpendicular to the flux about 1000 J s^{-1} m^{-2} reach the Earth's surface. The wavelengths of light arriving at the Earth's surface peak at about 500 nm, and the absorption and scattering of light filter out particular wavelengths, producing a characteristic spectrum at the surface (Figure 2.5a).

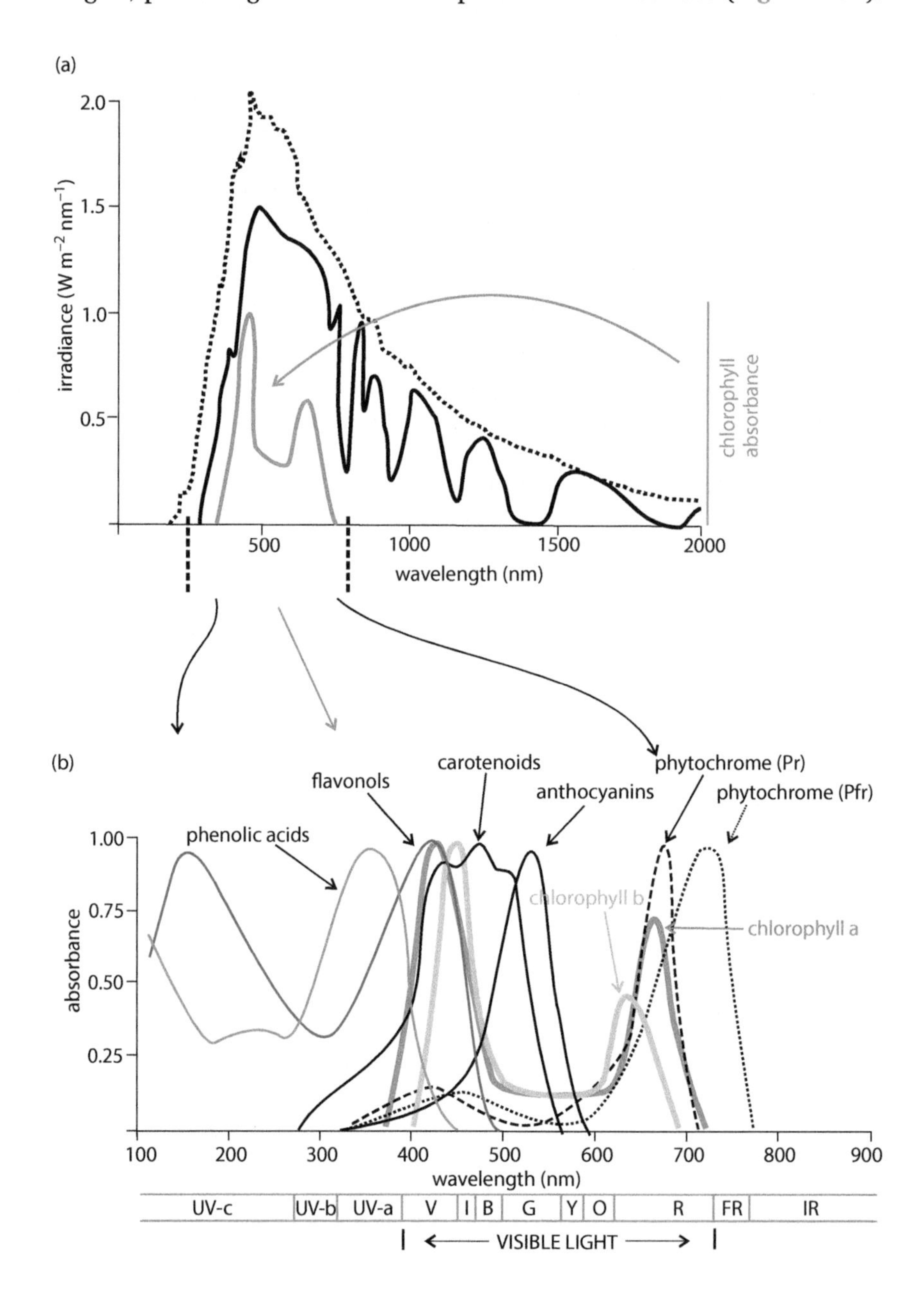

Figure 2.5. Light spectra and absorption spectra. (a) The spectrum of the solar constant (......), the light arriving at the Earth's surface (—), and the absorption spectrum of chlorophylls (—). (b) The absorption spectra of pigments found in plants. UV = ultraviolet; FR = far red; IR = infrared.

Photosynthetically active radiation (PAR) is determined by the absorption spectra of photosynthetic pigments that evolved to capture incoming light (Figure 2.5b). Maximum **irradiance** varies with latitude and with daily and annual cycles, determining the maximum amount of solar energy available in different regions of the Earth. On a day of average length, the PAR on a plant varies from zero to a maximum within about 6 hours. The differences in PAR available at different latitudes on Earth, and the very rapid change in PAR that can occur at any given spot, constitute profound variation in an essential resource, and often produce sub-optimal or excess irradiance for plants.

The **fluorescence** properties of the light reactions provide an insight into their functioning and into the stress caused by variation in light levels (Box 2.2). Exposure to excess light decreases the proportion of functional PSII complexes and hence the amount of light energy entering photosynthesis. Measurements of this "**photoinhibition**" of photosynthesis show that it can occur in plants well below maximum irradiances—that is, irradiances

BOX 2.2. FLUORESCENCE PARAMETERS IN STRESS PHYSIOLOGY

If a leaf that has been in the dark is transferred directly into the light, there is a transient increase in chlorophyll fluorescence over approximately 1 s, and then a gradual quenching of fluorescence, known as the **Kautsky effect** (Figure 1). Photosystems are said to be open if electrons are flowing from them into photochemical reactions, but closed if they are not. In the dark, the capacity of the photochemical reactions to process electrons is negligible because their components are only activated in the light and NPQ does not operate. A pulse of light in a dark period (pulse 1 in the figure) occurs to closed photosystems and produces a burst of maximum fluorescence (F_m). Some fluorescence is produced by non-photosynthetic light, so minimum fluorescence (F_o) is measured in its presence but without PAR. The difference between F_m and F_o is the maximum variation in fluorescence (F_v). F_v/F_m directly reflects the maximum possible quantum yield of PSII, and in unstressed plants is almost always about 0.83. However, if PSII is damaged, as is the case with many stresses, F_v/F_m is lowered, providing an easy way to estimate the stress that a plant is experiencing.

If a leaf is then transferred into PAR (Figure 1), a burst of fluorescence is produced and then quenched over time as the photochemical reactions and NPQ provide sinks for electrons and energy, respectively. It takes around 20 min to reach steady-state fluorescence (F′). At this point a pulse of light (pulse 2 in the figure) induces maximum steady-state fluorescence (F_m'), but this is less than F_m because of NPQ. $(F_m-F_m')/F_m'$ therefore provides an estimate of NPQ capacity. The difference between F_m' and F′ is F_q'. If the leaf is then transferred into the dark, fluorescence decreases further to F_o'. The difference between F_m' and F_o' is F_v'. F_q'/F_v' provides an estimate of photochemical quenching, and F_q'/F_m' gives an estimate of the quantum efficiency of PSII. The NPQ/PQ ratio provides an estimate of NPQ activity, with most plants having values in the range 0.1–1.0, but those adapted to high light environments having values much greater than 1. These relatively easily measured parameters are very important in relation to measuring the performance of the light reactions in plants and the effects of stress on them.

Further reading: Murchie EH & Lawson T (2013) Chlorophyll fluorescence analysis: a guide to good practice and understanding some new applications. *J Exp Bot* 64:3983–3998.

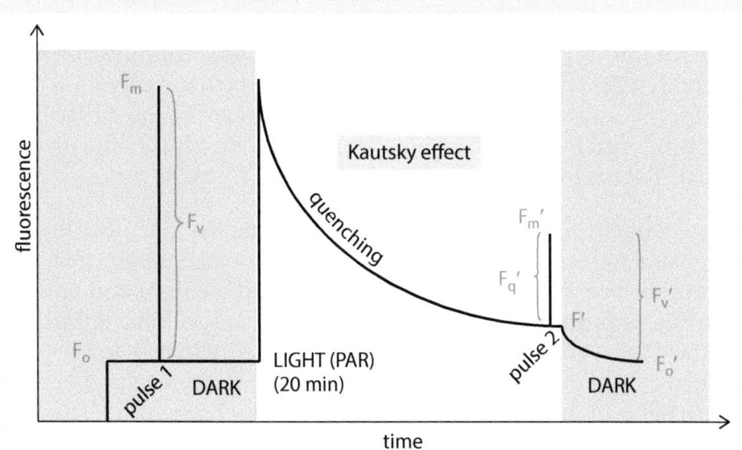

Figure 1. Leaf fluorescence responses and their quantifiable parameters. PAR = photosynthetically active radiation.

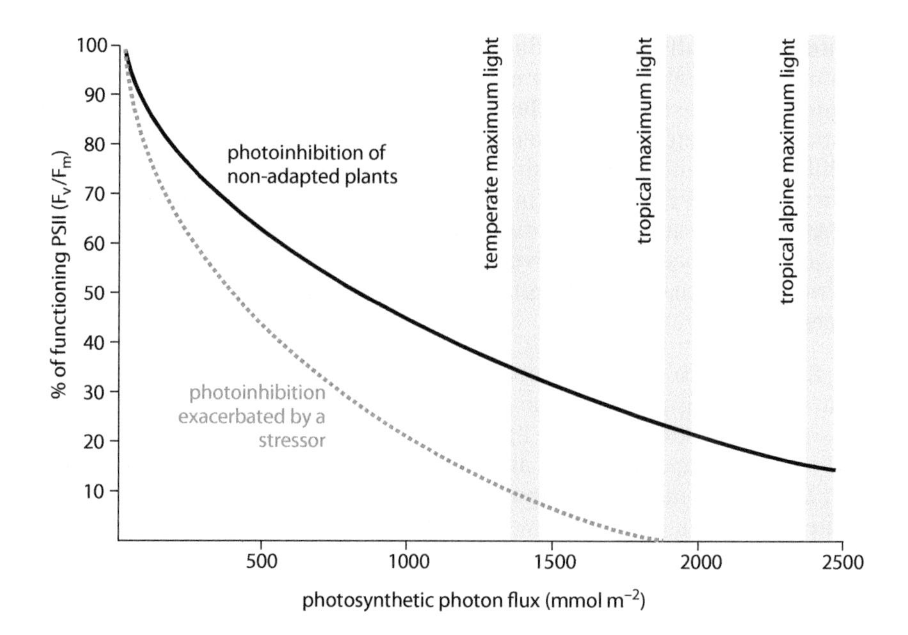

Figure 2.6. Photoinhibition of photosynthesis. High light conditions decrease the proportion of functioning PSII. Under cold or other stress conditions the inhibition of photosynthesis increases in severity. Plants suited to different light regimes have numerous adaptations that minimize photoinhibition of photosynthesis. Almost all plants must adapt to light intensities that are sometimes excessive, and some plants have to adapt to very challenging light regimes.

on the Earth's surface are often much higher than is optimal for plants (Figure 2.6). A proportion of PSII complexes are in fact damaged at almost any irradiance, but at high light levels most plants do not have the capacity to repair the PSII complexes at the rate that they are being damaged, increasing photoinhibition. PSI can also be damaged by light, but at a much lower frequency than PSII. Fluorescence measurements also show that variation in numerous plant stressors, but especially in temperature, drought, and CO_2, exacerbates photoinhibition. In all oxygenic photosynthetic organisms, PSII complexes have to be repaired, but high, variable irradiance compounded by other stressors presents a particular challenge to plants on the terrestrial surface, and photoinhibition can be a significant cost.

Photosynthetic organisms are subjected daily to a dark–light transition, so upon first light there are significant changes in the photosynthetic machinery that prepare photosystems for increased light. The full expression of this capability was probably vital to the evolution of plants on the Earth's terrestrial surface. The development of terrestrial plant canopies, in part enabled by the high light energy per unit area on the Earth's surface, produced further variation in light regimes, and in particular a "shade with sunfleck" regime that can change the conditions of a plant or leaf from deep shade to full sunlight within minutes, often many times a day. Sunflecks are found not only on forest floors, but can also occur throughout the vegetative canopy. In some ecosystems the majority of energy trapped is from sunflecks. The evolution of woody plants dramatically changed plant canopies, probably to the extent that the shade they produced became the main driver for the evolution of ferns, and many plants have evolved to exploit the particular light regimes produced by plant canopies, from shade dwellers of the forest floor to epiphytes of the high canopy.

Light incident on plants varies not only in quantity but also in quality. Cloud cover significantly affects not only the amount of light that reaches the Earth's surface, but also the ratio between **diffuse light** and **collimated light.** In the uniform canopies of agricultural crops, diffuse light has long been known to be an important contributor to productivity—many places in the canopy cannot be penetrated by collimated light but can be penetrated by diffuse light. In montane "cloud forests" in which plants are surrounded by cloud, for prolonged periods only diffuse light can be captured. The leaves of plants that ordinarily inhabit particular light environments are adapted to particular ratios of diffuse to collimated light, so changes in the ratio can affect productivity. For example, modeling of global productivity after the massive volcanic eruption at Mount Pinatubo in 1992 revealed a particulate-induced increase in the proportion of diffuse light reaching the Earth's surface, and an increase in productivity in many forest ecosystems. For plants, the proportion of red to far-red light (R:FR ratio) in incident radiation is also important. In full sunlight the R:FR ratio is 1.2, whereas on forest floors it can drop to 0.2 because significant amounts of light in the red part of the spectrum (Figure 2.5b) are absorbed during photosynthesis, whereas far-red light is not. Within canopies the R:FR ratio therefore decreases with the depth of shade, and in broadleaved canopies the decrease in R:FR tends to be greater than in coniferous canopies (Figure 2.7). **Phytochromes** are used by many plants to detect R:FR ratios, and **cryptochromes** are used to detect blue light, and hence the blue:red (B:R) ratio. The latter tends to increase down the canopy, but on days with a high proportion of diffuse light the opposite trend can occur (Figure 2.7). Many plants respond to both the R:FR and B:R ratios in order to optimize light capture.

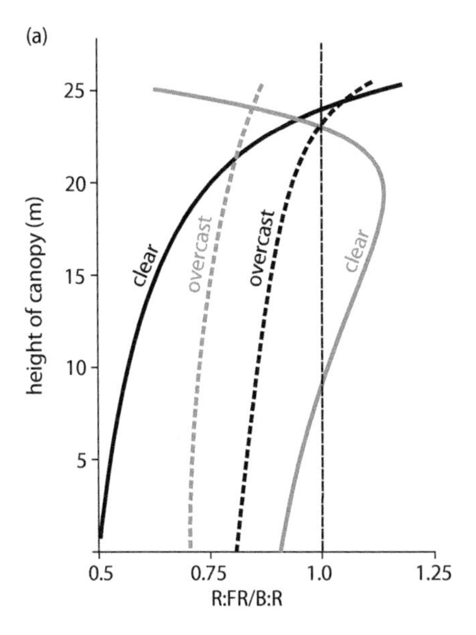

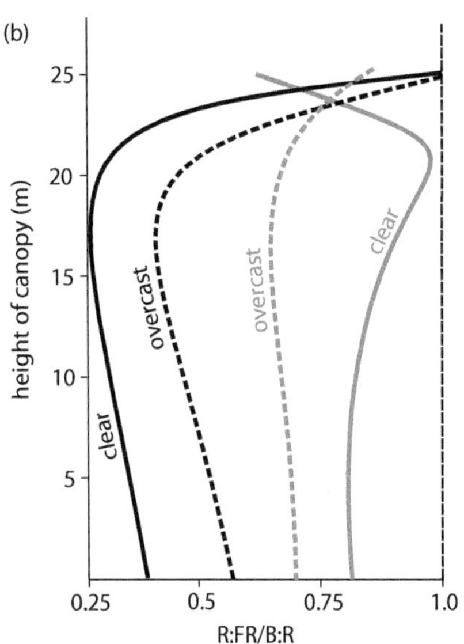

Figure 2.7. Yearly averages for the R:FR and B:R ratios in canopies. (a) *Picea abies* (Norway spruce), (b) *Fagus sylvatica* (beech). The proportion of red to far-red light decreases down the canopy because photosynthetic pigments absorb red light. On overcast days with a greater proportion of diffuse light more red light penetrates down the canopy than on clear days with more collimated light. The B:R ratio can peak in the upper canopy due to reflection, although this does not always occur in overcast conditions. The changes in R:FR and B:R ratios vary between gymnosperms (needle leaf, spruce) and angiosperms (broad leaf, beech), and at different times of the year. Black lines, R:FR; green lines, B:R. (From Hertel C, Leuchner M & Menzel A [2011] *Agric For Meteorol* 151:1096–1105. With permission from Elsevier.)

Human activities are changing the light environment at the Earth's surface. Global climate change models predict an increase in temperature during the twenty-first century that will change atmospheric water vapor concentrations and thus alter cloud cover. This in turn will change the proportion of diffuse light that reaches the Earth's surface, with potential global consequences for the productivity of ecosystems. Human activities have also considerably increased the particulate load in the Earth's atmosphere, in some places very significantly, and this has decreased irradiance and changed the light spectra in the lower atmosphere and at the surface ("global dimming"). In addition, light pollution can be a significant problem in many parts of the world. Human-induced changes in light regime add variation to what is, for most plants, already a highly variable supply of a key resource.

Plants can adjust quickly to variation in PAR using non-photochemical quenching

The absorption of energy by atoms can produce "excited" states in which electrons occupy orbitals with a higher energy than those that the atoms occupy when they are in the "ground" state. Excitation energy can be emitted in the form of light—chlorophyll molecules excited by absorption of light can fluoresce (that is, emit light energy) as they revert to their ground state. Photosystems are structured so that fluorescence out of them is usually minor. At high light levels, excess excitation energy can be quenched through the emission of heat and the transfer of electrons, via chemical reactions, to other molecules. Excited molecules can be reactive (for example, the damaging **singlet oxygen** ($^1O_2$) that can be formed in chloroplasts during photosynthesis under excess light), so their production must be minimized via the quenching of excitation energy.

The photochemical reactions of photosynthesis are driven by charge separation (that is, by an electron leaving a chlorophyll molecule). This requires more energy than excitation, and more energy than can be absorbed from natural light by a single chlorophyll molecule. In LHCs, numerous chlorophyll molecules focus excitation energy on the primary chlorophyll molecules in the reaction center. The final transfer to the primary chlorophyll molecules occurs from excited **triplet-state** chlorophylls (^{3}Chl), and the accumulated, focused energy is sufficient to form Chl^+ and e^- (Figure 2.8). The transfer of energy between adjacent chlorophylls in LHCs occurs via **resonant transfer**. Resonant transfer of energy occurs because pigment molecules are situated closer together than the wavelengths of fluoresced light, so although the transfer of energy is in effect fluorescence and absorption, no light is actually emitted. However, triplet chlorophyll can

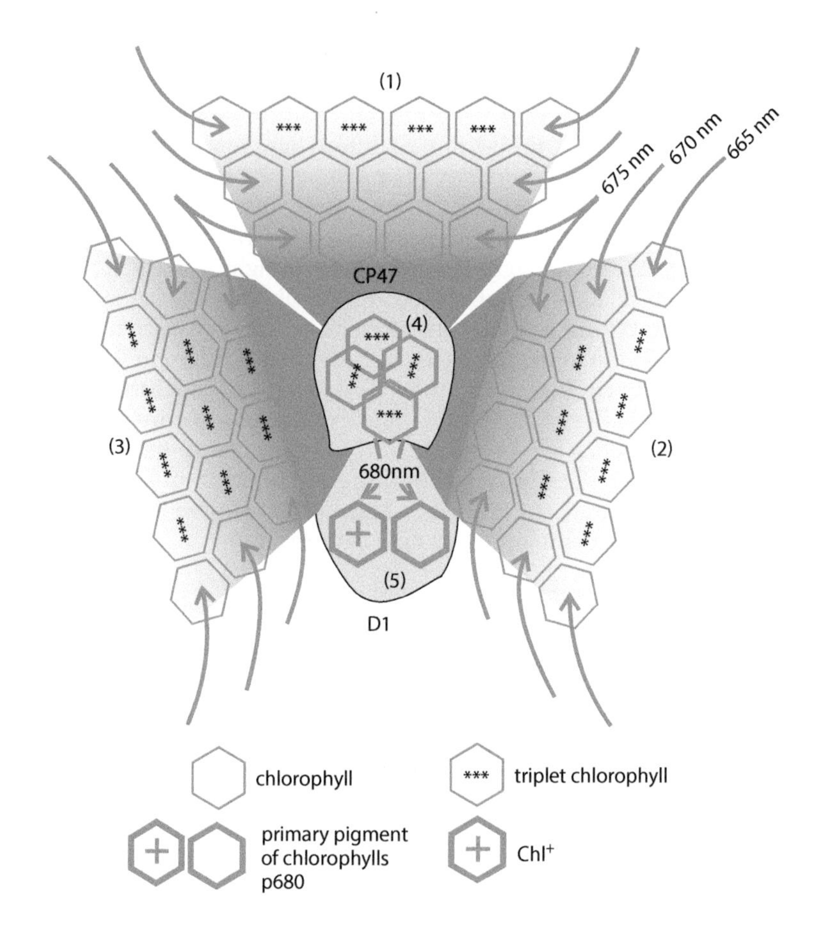

Figure 2.8. The energy cascade to P680. Light of particular wavelengths is captured by chlorophyll molecules, exciting them to the triplet state (1). This causes them to fluoresce, but other chlorophyll molecules are closer than the wavelength of emitted light, so energy is transferred by resonant energy transfer (RET) (2, 3, and 4). RET to molecules harvesting light at progressively longer wavelengths focuses energy on a pair of primary chlorophyll molecules (P680) via the CP47 complex (4 and 5). Together with the light that the primary chlorophyll molecules absorb at 680 nm, the energy transferred is sufficient to cause charge separation to form Chl^+. The primary chlorophyll molecules in P680 are ionized in turn. (The RET sequence is shown sequentially for clarity, but occurs simultaneously in functioning PSII. The phytol tails of chlorophyll molecules are not shown.)

also excite O_2, producing $^1O_2$. It is management of the transfer of excitation energy to molecules which can dissipate it as heat that provides the quickest response to changing light intensity. This non-photochemical quenching (NPQ) of excess energy provides a release valve that works on a timescale of seconds to minutes.

Hydrolysis at PSII produces electrons and protons. If the production of electrons at PSII is greater than acceptor capacity then there are not only excess electrons in the photosystems but also excess protons in the thylakoid lumen. Changes in pH in the lumen trigger changes in PSII within seconds. The thylakoid membrane protein PsbS (*Arabidopsis* mutants of which have no NPQ capacity) probably acts as a sensor of lumen pH, but direct protonation of PSII proteins may also be important. PsbS has similarity to proteins of the LHC, and probably evolved from them, but has no pigment-binding capacity. With the exception of the freshwater **charophytes**, from which all terrestrial plants evolved, algal photosynthesis uses the LHCR protein for pH detection—changes in the lumen pH sensor were probably very important for plants evolving in the high light environment of the Earth's surface. When a drop in lumen pH is detected, peripheral LHCs detach from reaction centers, decreasing the energy flowing to them. **Xanthophyll** pigments are integral to the LHC complexes of PSII. Resonance energy transfer can occur from chlorophylls to xanthophylls such as zeaxanthin and lutein that do not fluoresce, but dissipate the energy as heat. Chlorophyll fluorescence studies describing the kinetics of NPQ reveal that much NPQ capacity disappears quickly in the dark. It seems likely that this quenching, which is primarily associated with the first events precipitated by decreasing lumen pH and that can be a significant fraction of NPQ, might have evolved to aid the dark–light transition, because its fast dynamics will track changes in light intensity over a period of seconds to minutes, providing some homeostatic control over light absorption.

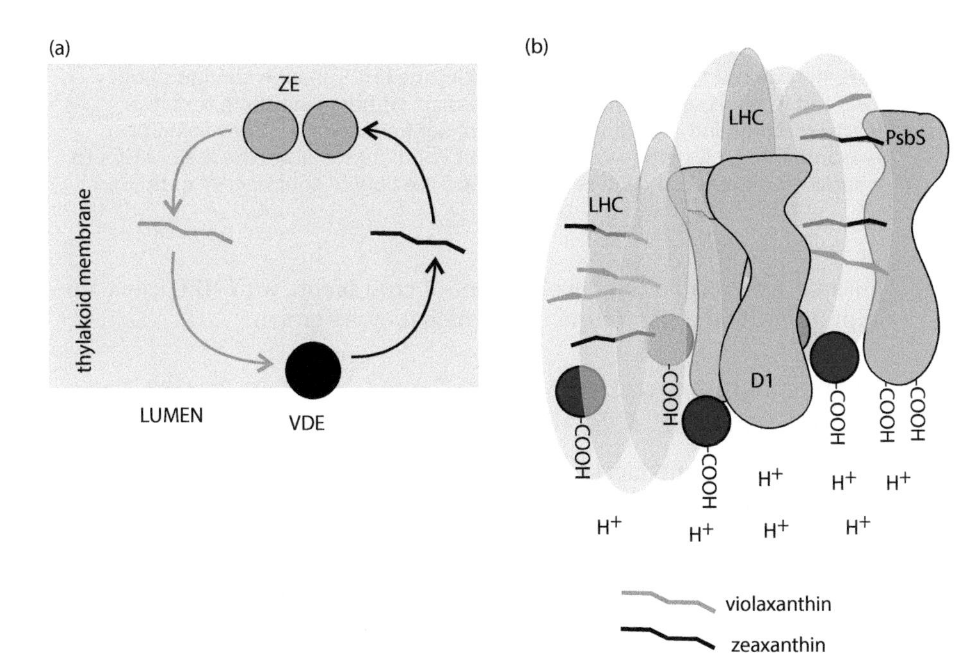

Figure 2.9. The xanthophyll cycle. (a) Violaxanthin de-epoxidase (VDE) and zeaxanthin epoxidase (ZE) interconvert violaxanthin and zeaxanthin. Zeaxanthin dissipates light energy as heat energy after resonance transfer from chlorophyll molecules. Violaxanthin fluoresces most of the light that it absorbs. Zeaxanthin thus decreases non-photochemical quenching, whereas violaxanthin tends to increase energy input into PSII. (b) The protein PsbS is probably a pH sensor, and VDE is pH sensitive. Increasing light increases the proton concentration in the lumen. This protonates PsbS and VDE, initiating the conversion of violaxanthin to zeaxanthin. Increasing amounts of zeaxanthin increase the capacity for NPQ at high light intensities. LHC = light-harvesting complex.

Some plants also have NPQ capacity with dynamics of the order of minutes to hours. In such quenching, zeaxanthin is synthesized from violaxanthin by the low-pH-activated violaxanthin de-epoxidase, using reducing power from ascorbate in the lumen. The pH-dependent interconversion of violaxanthin and zeaxanthin at different light levels is known as the xanthophyll cycle (Figure 2.9). Violaxanthin emits absorbed energy as light, whereas zeaxanthin dissipates it as heat. There is some evidence that zeaxanthin can also absorb free electrons and quench ^{3}Chl. Changes in lumen pH therefore trigger changes in xanthophyll pigments that change NPQ capacity. There are also xanthophyll-induced changes in the structure of LHCs that change the surface area of the light-harvesting antennae (Figure 2.10). At low light intensity, the integral xanthophyll in LHCs is violaxanthin, which produces a distributed light-scavenging LHC structure, whereas at high light intensity the conversion of integral xanthophylls to zeaxanthin produces a detached, clustered LHC structure that decreases the flow of light to the reaction center.

During NPQ, xanthophylls are not only interconverted but can also be synthesized *de novo*—the concentration of xanthophylls often increases, contributing to a sustained NPQ. Plants such as alpines, from environments with the particularly damaging combination of low temperatures and high light intensity, are often rich in xanthophylls. Long-term studies of quenching in winter evergreen species such as *Pinus sylvestris* have shown that NPQ increases significantly during the winter, and thus contributes to long-term adaptation to excess light. The **anthocyanins** that are common in alpine plants and very young plants probably also play a role in sustained quenching. In plants with a sub-optimal supply of resources other than light, the intensity at which light becomes excessive is lowered. This is the case for many—perhaps most—plants in unmanaged systems, which are therefore dependent on mechanisms of NPQ for survival. In environments that promote chronic susceptibility to photoinhibition, mechanisms of NPQ incur a significant cost to growth. Many crop plants have significant NPQ capacity, and the vigor of some varieties of, for example, *Triticum aestivum* (wheat) is thought to be due to significant NPQ capacity. However, NPQ does decrease CO_2 assimilation per unit of absorbed light, so it can be argued that it evolved primarily to protect photosynthesis in resource-poor environments, and might therefore be an unnecessary cost to some crop plants. Understanding NPQ in both unmanaged and managed ecosystems is ecologically important,

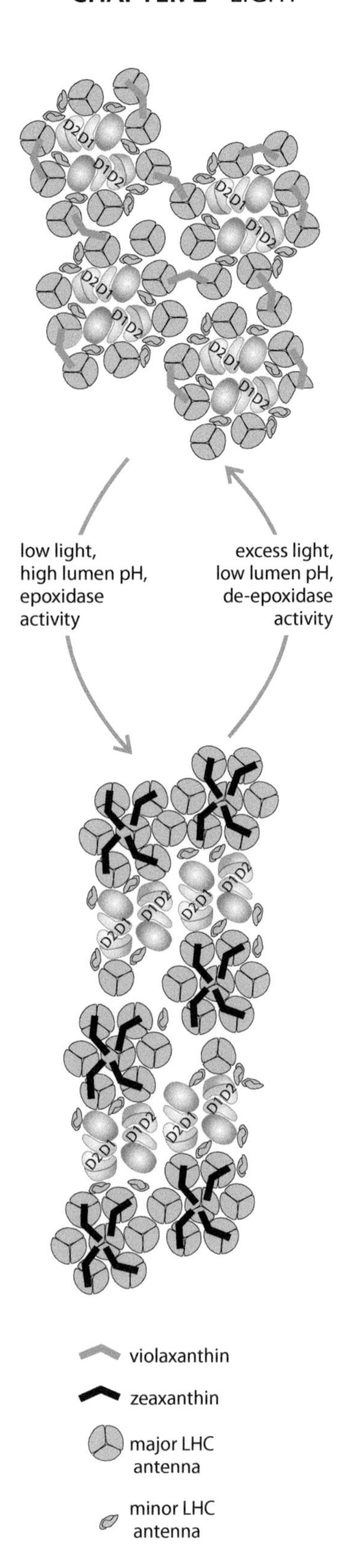

Figure 2.10. Reconfiguration of the light-harvesting complexes (LHCs) of PSII. In response to different light intensities the LHCs change configuration. Zeaxanthin binds several major LHC antennae, withdrawing them from the reaction center and decreasing the flow of light energy to it. Conversion of zeaxanthin to violaxanthin under low light conditions enables the major LHCs to associate with reaction centers and increase the flow of light energy to them.

but may also enable the development of crop plants with NPQ capacities appropriate to the environments in which they are grown.

Plants can adjust electron flows to help them to withstand variable light intensities

To achieve photochemical quenching (PQ) of light energy, plants adjust the size of the sinks that receive electrons from the light reactions. This helps to control the utilization of electrons following the key photochemical process, namely charge separation in PSII. For electrons to be able to flow to sinks, the primary electron acceptor, namely the plastoquinone pool in the thylakoids, must be kept oxidized (that is, ready to receive electrons), and for this to be achieved the electrons have ultimately to be moved into their final sinks. If electron sinks are insufficient and plastoquinone is in a reduced state then not only is it easy to detect unquenched light energy (that is, fluorescence), but also damaging reactive species can be produced at PSII—for example, superoxide ($O_2^{\bullet-}$) formed from O_2 and e^-.

Cyclic electron flow (CEF) around PSI occurs when electrons from ferredoxin are inserted back into the linear electron flow (LEF) via plastoquinone, rather than being used to reduce NADP (Figure 2.11). Studies of mutants have shown that PGR5 proteins and/or the NADH oxidase-like (NDH) complex are involved in CEF, and that plants which lack these molecules have a significantly reduced capacity for PQ. The cyclic flow of electrons also contributes significantly to a decrease in pH in the lumen, helping to activate NPQ. CEF increases with temperature and drought stress, and is thus probably an important mechanism for occupying a pool of excess electrons in a way that activates NPQ. The change in lumen pH produced by CEF helps

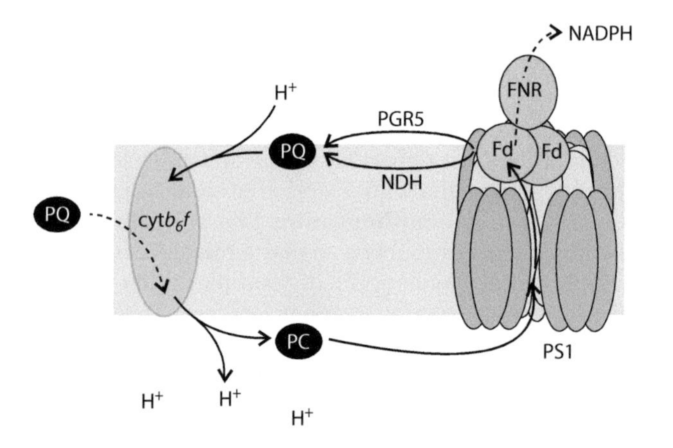

Figure 2.11. Cyclic electron flow. During cyclic electron flow (black arrows), ferredoxin (Fd) passes electrons to a plastoquinone (PQ) pool in the non-appressed thylakoid membranes. These electrons are passed to cytochrome b_6f (cytb_6f) and then passed back to PSI via plastocyanin (PC). In linear electron flow (dashed arrows) the PQ pool of the appressed thylakoids feeds electrons to cytochrome b_6f, and they are passed from Fd to NADPH via ferredoxin NADP oxidoreductase (FNR). PGR5 = PGR5 protein; NDH = NADH oxidase-like complex.

to continue the production of ATP without the production of NADPH. The Calvin–Benson cycle requires, on average, an ATP:NADPH ratio of 3:2. The ratio in which these molecules are produced by LEF alone is about 2.6:2, so CEF can help to increase the ratio to 3:2. However, there are other significant metabolic demands on reducing power (for example, N assimilation), and on ATP demand (for example, the CO_2 pumps of C_4 plants), which change the ratio in which ATP and NADPH must be produced. CEF therefore has a role in adjusting ATP/NADPH ratios to suit demand while keeping the plastoquinone pool oxidized to maximize PQ. Some studies have detected supercomplexes of PSI and its LHCs that might help to maximize CEF when necessary. In chloroplasts, **plastid terminal oxidase** (PTOX) helps to control the redox potential of the stroma and hence the balance between CEF and LEF. PTOX can oxidize plastoquinone directly, and probably has a role in keeping it oxidized in developing seedlings and plants such as alpines, which inhabit conditions that promote photoinhibition.

Electrons can flow from PSI to O_2 because ferredoxin and reduced Fe-S centers of PSI can add an electron directly to O_2 to produce $O_2^{\bullet -}$ (the **Mehler reaction**), contributing to $O_2^{\bullet -}$ produced at PSII. Chloroplasts have significant antioxidant capacity that can reduce $O_2^{\bullet -}$ to H_2O. Thus direct reduction of O_2 in PSI provides a route for the linear flow of electrons from H_2O, through PSII and PSI, to H_2O. In this "water-water" cycle, one electron from $2O_2^{\bullet -}$ undergoes **dismutation** to H_2O_2 by superoxide dismutases, and another is transferred to ascorbate (Figure 2.12). This water-water cycle provides an alternative sink for electrons, aiding PQ, but also contributes to a lowering of pH in the thylakoid lumen, enhancing NPQ. The proportion of electrons that enter the water-water cycle is usually low, but many studies suggest that it increases at high light intensities and in stressed plants, and that there is a significantly higher electron flux to the water-water cycle in C_4 plants.

In many organisms, malate valves provide sinks for excess electrons from a variety of metabolic processes. In chloroplasts, the malate valve has a light-activated NADP-dependent malate dehydrogenase that converts oxaloacetate into malate using, ultimately, electrons from PSI. The export of malate from chloroplasts therefore exports reducing equivalents (that is, electrons), providing a sink for electrons. The chloroplast malate valve has been shown in many studies to contribute to PQ. An additional and often significant sink for electrons is photorespiration. The majority of **rubisco** in the stroma drives the Calvin–Benson cycle, which starts with the fixation of CO_2 to ribulose bisphosphate, and uses ATP and NADPH to produce carbohydrate.

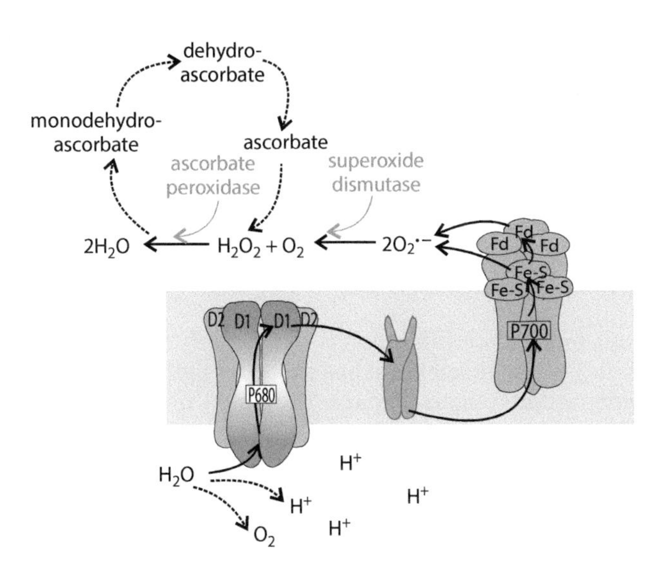

Figure 2.12. The water–water cycle. The sink for electrons from H_2O via PSI is usually NADPH, but in some plants electrons can also flow from H_2O and, after combining with O_2, and H_2O_2 form H_2O (black arrows), providing an additional sink to help to minimize photoinhibition. Fd = ferredoxin; Fe–S = iron–sulfur center.

However, rubisco can also fix O_2, which results in the net production not of carbohydrate but of CO_2—hence the term "photorespiration." Photorespiration uses ATP and NADPH, but also produces NH_3, which is toxic and must be assimilated using electrons from ferredoxin. Photorespiration decreases the efficiency of photosynthesis because more ATP and NADPH is used to produce each unit of carbohydrate, but at high light levels this is an electron sink that helps to keep plastoquinone oxidized and maintain PQ. Plants that have been engineered to have reduced photorespiration are very susceptible to photoinhibition at high light levels. Finally, in the long term an increase in the production and storage of carbohydrate via the Calvin–Benson cycle increases the demand for ATP and NADPH, keeping the light reactions quenched at high light levels. There is therefore a range of mechanisms that plants can use to balance the sinks for electrons exiting the light reactions with their production in PSII, minimizing the production of reactive species that can cause photoinhibition. This PQ capacity is often matched to the plant's natural habitat and is, for example, enhanced in winter evergreens during the cold winter months.

PSII repair is important in plants that tolerate high light intensities

If radiolabeled amino acids are fed to plants under high light intensities, a significant proportion appear quite rapidly in D1 polypeptides of PSII complexes, indicating a high turnover due to rapid damage and replacement. The precursor of the D1 polypeptide occurs abundantly in the stroma, and is encoded by the *psbA* gene on the chloroplast genome. PSII complexes are located in the appressed regions of the thylakoids and have no direct contact with the stroma (Figure 2.2). PSII repair replaces damaged D1 polypeptides with new ones from the stroma. The PSII complex is a dimer. When a D1 polypeptide is damaged, PSII is cleaved into monomers and the oxygen-evolving complex is removed, together with CP43. The damaged reaction center (including D1, D2, and CP47-associated components) is moved to non-appressed regions of the thylakoids that are in contact with the stroma (Figure 2.13). Chlorophyll molecules joined to reaction centers via their phytol tails are disconnected, and LHCs are removed and stored. There is some evidence that, in cyanobacteria, small CAB-like proteins are used to hold chlorophyll molecules while reactor center repair occurs.

FtsH2/3 proteases are used to degrade damaged D1 polypeptides. *FstH* KO mutants of various plants can still have some D1 degradation capacity, so other enzymes such as the **Deg proteases** probably contribute to degradation. FstH2/3 is a hexamer through which D1 is degraded starting from the N-terminus. FtsH proteases are ATP-driven and have a **zinc-binding domain** that is vital to protein hydrolysis. They are also active in heat-damaged PSII complexes, and probably have a general role in dismantling dysfunctional PSII subunits under a variety of stressors. Carboxy-terminal-processing peptidase (CtpA) removes a C-terminal portion of D1 precursor to provide a new D1 from the stroma. After the new D1 is associated with the remaining reactor center components (D2 and CP47), CP43 and the OEC are joined back on to complete the reassembly. This reassembly is essentially the same process that is used when PSII complexes are first assembled.

The action spectrum for photoinhibition is significantly different to that for photosynthesis, indicating that it is not just excessive light absorbed by photosynthetic pigments that triggers damage. The action spectrum for damage suggests that absorption of light by Mn compounds, such as those in the OEC, is involved. The extraction of electrons from H_2O in the OEC is driven by the high oxidation potential of $P680^+$ (*c.* 1.2 V). $P680^+$ has to be very rapidly quenched with an electron, otherwise it is likely to oxidize other nearby

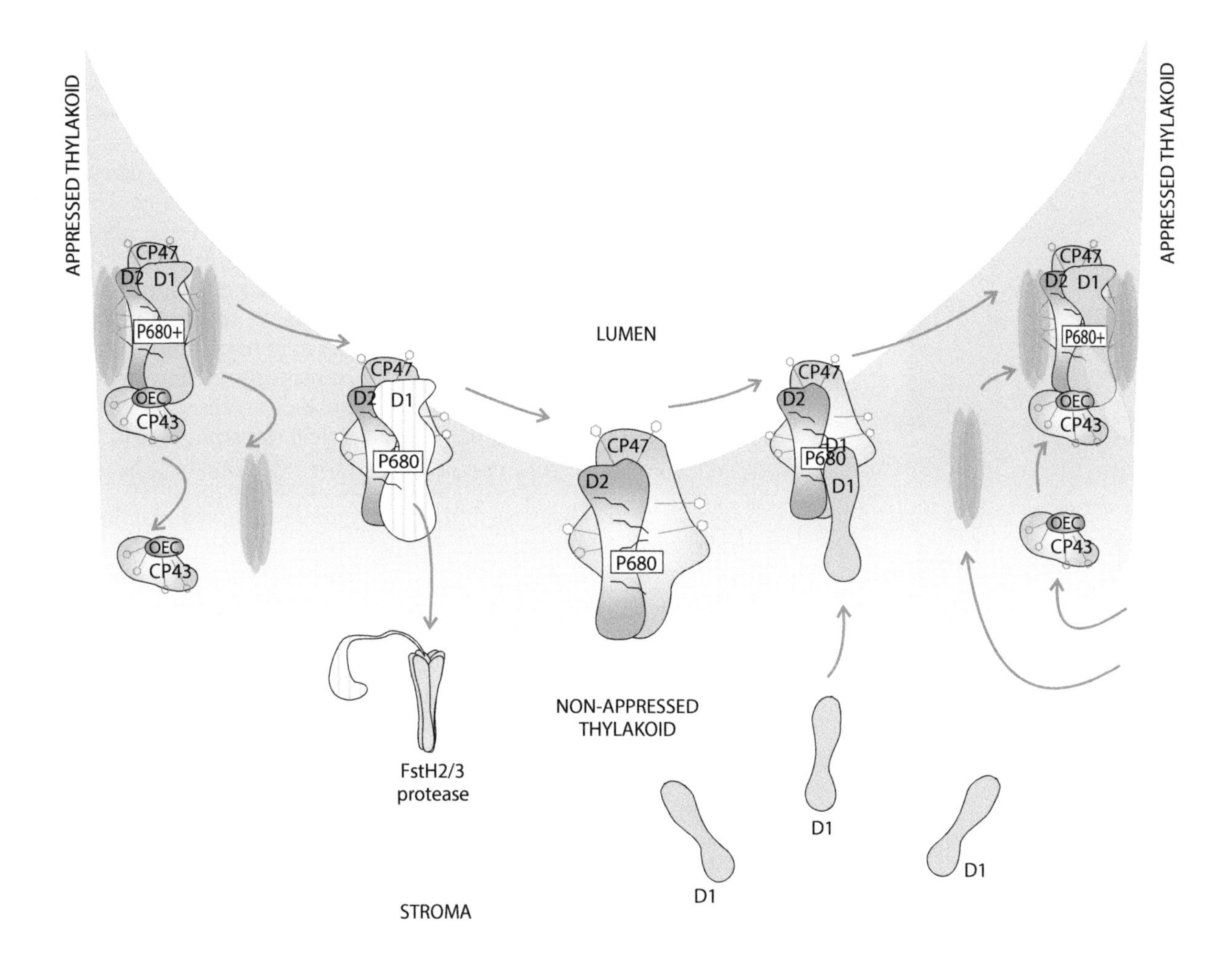

Figure 2.13. **Repair of the D1 polypeptide.** PSII is disassembled, the damaged D1 polypeptide is digested by the FstH2/3 protease, and the reaction center is moved to the non-appressed region of the thylakoids, where a new D1 polypeptide encoded in the chloroplast genome is inserted. The reaction center is returned to the appressed region, and PSII is reassembled.

molecules, and free electrons produced by its formation must be removed rapidly because they can combine with O_2 and produce $O_2^{\bullet-}$. Damage to the D1 polypeptide is caused by oxidation, primarily by P680$^+$ and/or $O_2^{\bullet-}$, produced by an imbalance in electron flow through P680$^+$. The flow of electrons through PSII can become uncoupled in numerous ways, depending on the molecular environment of the chloroplast, the light conditions, and other environmental stresses. It seems likely that at high light intensity a combination of enhanced production of P680$^+$ and damage to the OEC, which supplies electrons to quench P680$^+$, enhances the oxidation of D1. High temperatures can be associated with enhanced photoinhibition, but it is particularly pronounced at low temperatures. At cold temperatures, protein movement through membranes is slowed, so the removal of electrons via PQ is decreased, promoting the formation of $O_2^{\bullet-}$, and the repair of D1 is slowed dramatically because of decreased rates of movement of reaction centers during repair. Many plants which are adapted to conditions that promote photoinhibition (for example, winter evergreens and alpines) can have enhanced PSII repair capacity that extends their tolerance of light stress.

Chloroplast movements can be used to adjust fairly rapidly the amount of light absorbed

In eukaryotic cells, the spatial organization of organelles is often altered in response to cellular or environmental conditions. One of the most dramatic examples is the dynamic alteration of chloroplast position in the cells of static plants. In some species this can produce significant changes in leaf color within less than an hour. Studies comparing the responses of mutants

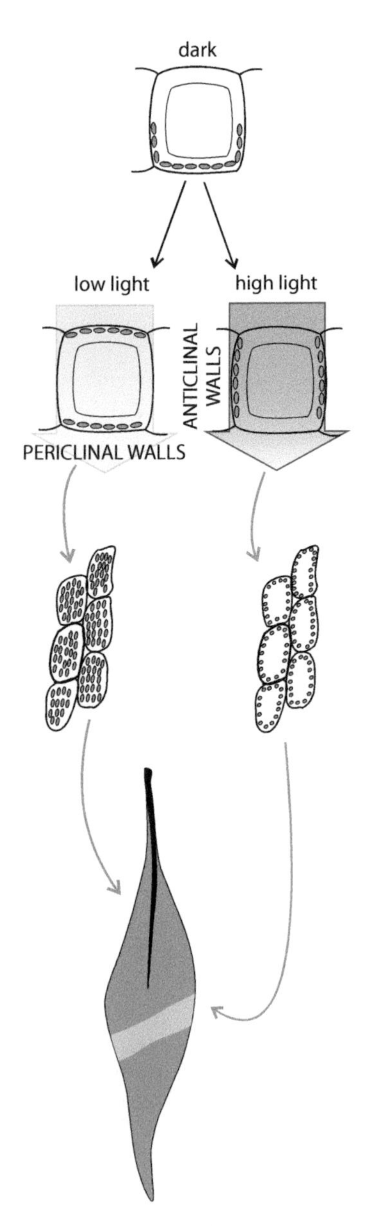

Figure 2.14. The movement of chloroplasts. Chloroplasts in many species are in a resting position in the dark, but move quickly and significantly in response to changes in light intensity. At low light intensity, chloroplasts in many species can move to the periclinal walls, thus maximizing light capture. At high light intensity, in many species they can move to the anticlinal walls, thus minimizing light capture and photoinhibition.

with regard to NPQ and chloroplast positioning have shown that the latter adapts plants to more profound variation in light intensity. With sudden exposure to high light intensity, the chloroplasts of many species have been shown to quickly gather on the **anticlinal** walls of cells, whereas at low light intensity they gather on the **periclinal** walls (Figure 2.14). These changes of position produce a change in the light transmittance of whole leaves that is measurable in the field during, for example, the day-night cycle. At dawn, leaf transmittance is at its minimum, whereas at peak PAR it is often at its maximum. Periclinal positioning maximizes light capture at low, sub-saturation irradiance (an accumulation response), whereas anticlinal positioning allows excess, potentially damaging light to pass through the leaf at above-saturating irradiance (an avoidance response). There are differences in light intensity at which the chloroplast positioning response occurs in species adapted to shade and sun, but it alters leaf transmittance in both cases.

Chloroplast movements are an ancient response in photosynthesizing eukaryotes. Many algae use diurnal movements up and down the water column to control light exposure, but others maintain their position in the water column and move their chloroplasts to adjust light capture. Mosses and ferns also display pronounced chloroplast movements. In many shade-adapted angiosperms (for example, *Tradescantia albiflora*) the PSII complexes are not much damaged by sudden high irradiance, because chloroplast movements avoid photoinhibition. A number of mutants with non-existent or altered chloroplast movement are known, and they are very susceptible to photoinhibition. In the field, light is incident not just perpendicular to the leaf and its periclinal cell walls, but is diffuse and often reflected or channeled within the leaves. Chloroplast position in many plants is continually adjusted to variation in irradiance and to the direction of arrival of light in individual cells within a leaf. Narrow microbeams of light can induce movement of chloroplasts within individual cells.

Intracellular movement of organelles in eukaryotes generally occurs via the action of **myosin** motors on the **actin** filaments (AFs) of the cytoskeleton. AFs form "baskets" around the chloroplasts, and there are significant rearrangements of the AFs in a cell when chloroplasts move in response to light. In general, the net length of AFs is determined by the difference between the joining of ATP-actin trimers at the "barbed" end and the leaving of ADP-actin trimers at the "pointed" end. When chloroplast movement is activated there are, uniquely in the chloroplast, numerous short AFs formed ("cp-actin"), often between the chloroplast and the plasma membrane, in the direction in which the chloroplast will move (Figure 2.15). The concentrations of actin, associated proteins, and metabolites can all affect net AF length. The chloroplast unusual positioning (*chup1*) mutant of *Arabidopsis*, in which the chloroplasts do not respond to changing irradiance, lacks the CHUP1 protein that extends AFs by polymerization (Figure 2.16). Thus, in contrast to the myosin-driven movement of other eukaryotic organelles, chloroplast movement is driven by CHUP-mediated changes in AFs and binding to them. Analyses of mutants have shown that KAC1, KAC2, and JAC1 are also involved. KAC1 and KAC2 are **kinesins** that usually act as molecular motors between myosin microtubules of the cytoskeleton, but those of plant cells have no myosin-binding activity. There is evidence that in plant cells they bind instead to actin, perhaps helping to propel the chloroplast. JAC1 is an **auxilin**-like protein usually associated with binding to the chaperone HSP70 and to **clathrin,** which is normally associated with the regulation of vesicle dynamics. Other mutations have shown that a WEB1-PMI2 complex regulates the involvement of JAC1. Myosin inhibitors can clearly affect chloroplast responses, and thus a complex interaction of many parts of the cytoskeleton is involved in producing chloroplast movements in plant cells.

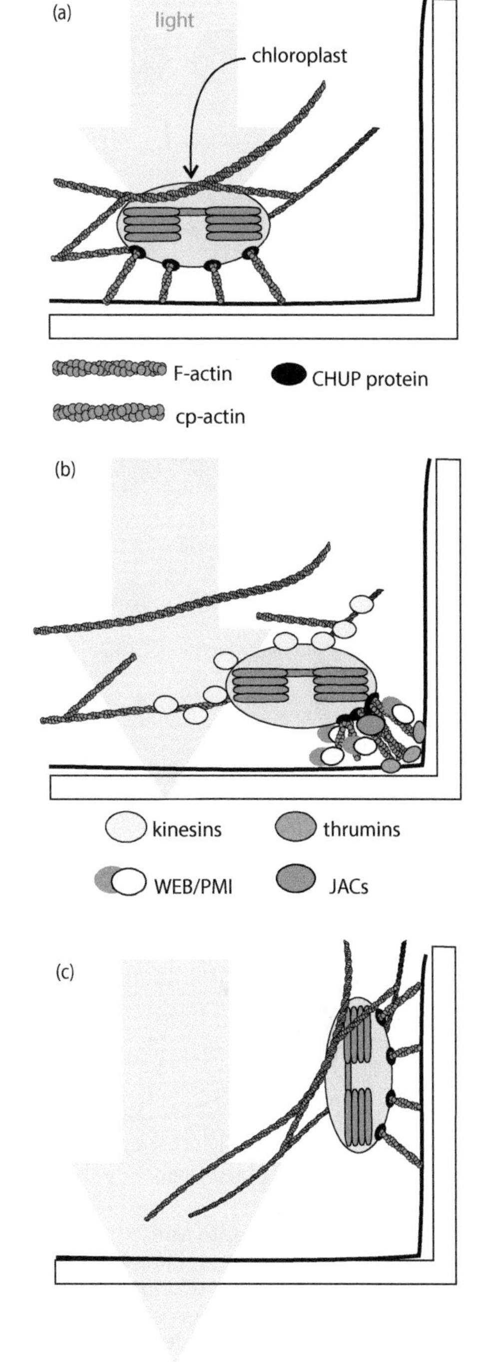

Figure 2.15. Mechanism of chloroplast movement. (a) The chloroplast is cradled in actin filaments (F-actin) and anchored by cp-actin via CHUP proteins. (b) In excess light, F-actin can be loosened and cp-actin directionally reconnected. Kinesins probably catalyze chloroplast movement along F-actin. A WEB/PMI complex and JAC proteins are necessary for cp-actin-dependent movement, and thrumins anchor the cp-actin. (c) The chloroplast is recradled and anchored in its new position.

In most plant species it is blue light detected by **phototropin** that triggers chloroplast movement—*phot* mutants of phototropin do not show chloroplast movements in response to variation in irradiance. In some mosses and ferns, red light detected by **phytochrome** can induce chloroplast movement, and although red light does not induce movements in higher plants, *phyt* mutants without active phytochrome do have an altered capacity for chloroplast movement. Phototropin is an auto-phosphorylating protein activated by blue light through **flavin mononucleotide** (FMN). It has a **serine/threonine kinase** towards its C-terminus, and two "light, oxygen, and voltage" (LOV) domains towards its N-terminus, to which FMN binds. Blue light absorbed by FMN changes the structure of LOV2, releasing the kinase activity and causing phosphorylation of numerous sites on the phototropin (Figure 2.17). Phototropin is primarily membrane-bound in the dark, but a proportion of it is mobilized into the cytoplasm in the light. The events initiated by phototropins have well-established roles in a variety of biological signaling cascades, including the Ca^{2+} fluxes and H_2O_2 increases that have been detected during chloroplast movement. There are two phototropins in plant cells, with PHOT2 being primarily responsible for the chloroplast avoidance response. Both PHOT1 and PHOT2 are involved in the accumulation response, which is generally much slower. At low light intensity, such as in shade, the proportion of diffuse radiation is high, whereas in a sunfleck, for example, radiation is initially incident on the leaf primarily from one direction. Rapid photoinhibition in mutants and responses at high irradiance in non-mutants indicate that chloroplast movement for avoidance at high irradiance is probably more important to plants than movement for accumulation at low light levels, perhaps because of the proportion of diffuse radiation, but chloroplast movement is likely to be a ubiquitous response in plants to variation in irradiance.

Photosystems, grana, and thylakoids adapt to differences in light regime

The chloroplasts of many **sun leaves** have less extensively developed thylakoid grana and a higher proportion of PSI to PSII complexes than those of **shade leaves**. They also have a higher concentration of chlorophylls,

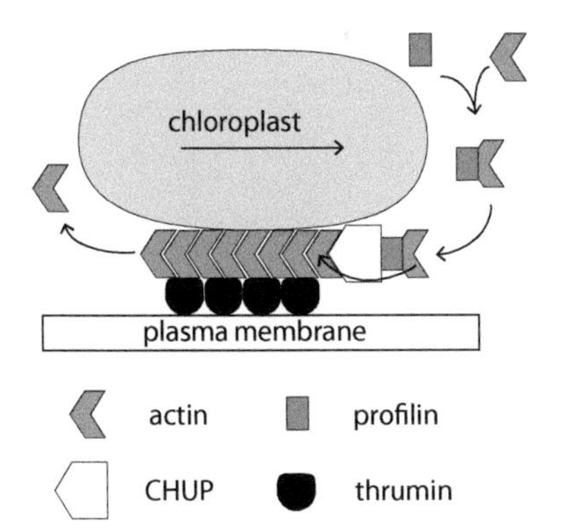

Figure 2.16. The role of CHUP proteins in chloroplast movement. CHUP proteins recruit profilin–actin complexes and add them to the barbed end of an actin filament that is anchored to the membrane by thrumin. Individual actin molecules progress towards the pointed end, and are then released, moving the chloroplast along the membrane. (Adapted from Wada M [2013] *Plant Sci* 210:177-182. With permission from Elsevier.)

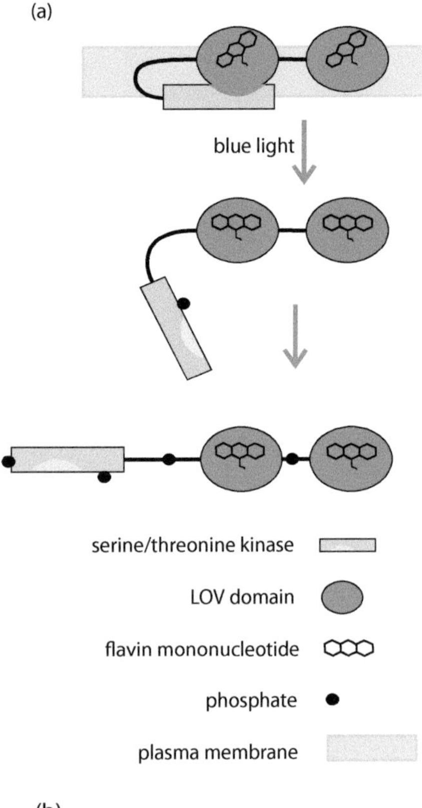

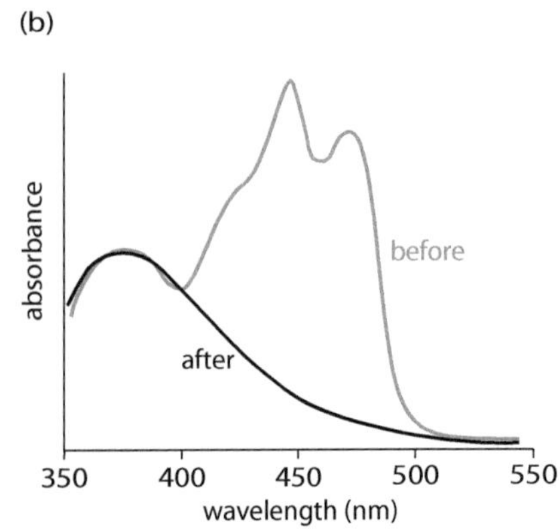

Figure 2.17. Phototropin and chloroplast movement. (a) Phototropin is plasma membrane bound at low light intensity. The absorption of blue light by flavin mononucleotide alters its attachment to the LOV2 domain. This releases the serine/threonine kinase domain that phosphorylates numerous serine residues on the phototropin molecule. The phosphorylated molecule is released to the cytoplasm, initiating cascades of signaling molecules, such as Ca^{2+} and 14-3-3 proteins, and the events that move chloroplasts. (b) These changes are reflected in the absorption spectra for phototropin.

a higher ratio of chlorophyll *a* to chlorophyll *b*, and a higher ratio of chlorophylls to most light reaction components (Table 2.2). These differences reflect not only light intensity but also light quality—light that reaches the understory is depleted in short-wavelength PAR because of the absorption spectrum of PSII in particular. The higher ratio of PSII to PSI complexes, and concomitant changes in chlorophyll *a* and *b*, help shade leaves to maintain linear electron flow. These differences help to explain the light-use efficiency of different leaves. However, photosystem, granum, and thylakoid characteristics are also dynamic, and can be adjusted within a day or so in individual mature leaves.

In chloroplasts, protein phosphorylation is used to adapt numerous aspects of thylakoid structure to photosynthetic conditions. In *Arabidopsis*, the redox state of both cytochrome b_6f and ferredoxin controls, via plastoquinone and thioredoxin concentrations, the activity of the STN7 transmembrane protein kinase. STN7 phosphorylates the *lhcb*1 and *lhcb*2 proteins of the LHC complex and some of its accessory proteins. With these proteins in the phosphorylated state, many LHCII complexes associate with PSI complexes rather than with PSII complexes. In shade, a phosphatase, TAP38, dephosphorylates LHCII proteins, reversing the process. Phosphorylation of numerous other proteins in the thylakoids also changes diurnally and with changes in light regime. For example, in *Arabidopsis,* another protein kinase, STN8, phosphorylates the scaffold proteins of PSII and PSI, initiating changes in photosystem structure. Such changes in LHCs and their distribution help to control electron flow during medium-term changes in light intensity and quality. Changes initiated by STN7, if sustained, also change granal structure. **Tomography** suggests that the most likely three-dimensional arrangement of grana is a right-handed helix of lamellae with prominent junctions between the granal and stromal thylakoids. The LHC proteins control granal stacking, with the movement of LHCIIs to PSI in non-appressed regions of thylakoids decreasing granal stacking (State 2), and their movement back to appressed regions increasing it (State 1) (Figure 2.18). During these state transitions, up to 80% of LHCIIs can transfer between photosystems. In State 1, unphosphorylated PSII complexes help to hold appressed thylakoids together, forming granal stacks, whereas in State 2 they do not, and the grana become less pronounced. The CURT1 protein is located in non-appressed

Table 2.2. Differences in the ratio of chlorophylls to other light reaction constituents in plants grown at high light (HL) and low light (LL) intensities

Property	HL	LL
Chlorophyll/leaf area	High	Low
Chlorophyll *a*/*b*	High	Low
Assimilation/leaf area	High	Low
Chloroplast number/cell	High	Low
LET/chlorophyll	High	Low
PSII/chlorophyll	High	Low
LHCII/chlorophyll	Low	High
Cytochrome b_6f/chlorophyll	High	Low
Plastocyanin/chlorophyll	High	Low
PSI/chlorophyll	Same	
ATP synthase/chlorophyll	High	Low

LET, linear electron transfer.
From Schöttler MA & Tóth SZ (2014) *Front Plant Sci* 5:188. With permission from Frontiers Publishing.

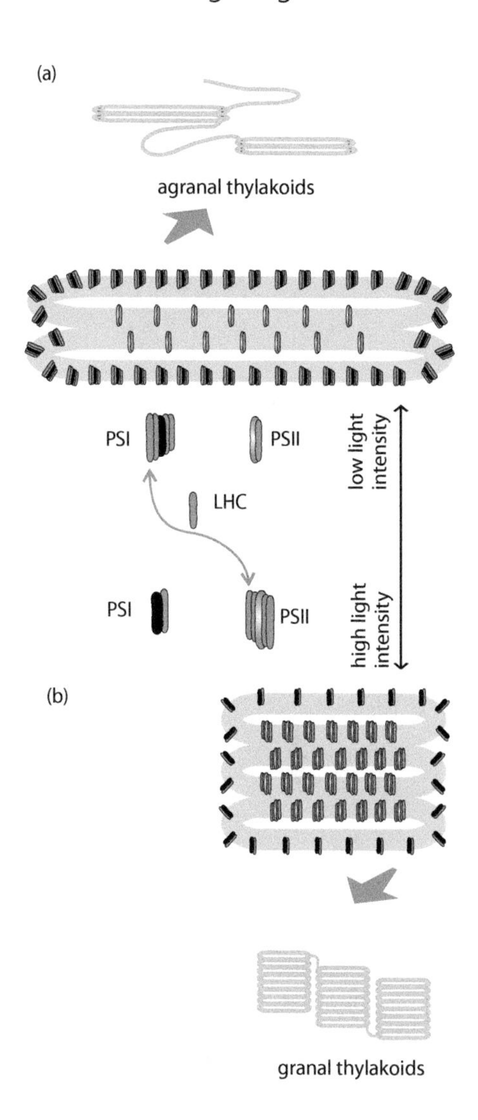

Figure 2.18. Light intensity and thylakoid grana. (a) Thylakoids in chloroplasts at low light intensity have a high PSI:PSII ratio and PSI complexes with many light-harvesting complexes (LHCs). (b) Thylakoids in chloroplasts at high light intensity have a much lower PSI:PSII ratio, and PSII complexes with more LHCs are characteristic. PSI and PSII complexes structure the thylakoids, with low PSI:PSII ratios producing strongly granal thylakoids. Some plants can rapidly change the relative proportions of PSI and PSII proteins in the thylakoids, and can migrate LHCs between photosystems. The relative proportions of particular chlorophylls and of photosystems help plants to maintain electron flow under conditions of changing light intensity and quality.

regions and probably has a role in remodeling grana. In State 2, thylakoids can also swell, changing their surface area. Cyanobacteria do not form grana with their thylakoids whereas terrestrial plants do, so grana and the flexibility in light capture that they allow were probably very important in the development of terrestrial ecosystems.

Many proteins of the LHCs are chlorophyll *a*/*b* binding (CAB) proteins. Numerous CAB-like proteins are expressed in plants, particularly in the thylakoids in response to stress. They form a family of proteins that can probably be traced back to the earliest light-harvesting systems in plants. One set, the early light-induced proteins (ELIPs), was discovered in **de-etiolating** seedlings, but is now widely credited with a role in photoprotection. In *Vitis vinifera* (Figure 2.19), a crop that generally grows in high light intensity, the expression of ELIPs is developmentally controlled, occurring in developing leaves when photoinhibition is at a maximum. In evergreens, maximum photoinhibition occurs in the winter and coincides with an increase in ELIPs in numerous tree and shrub species. Short-term experiments with *Arabidopsis* mutants in which ELIP expression is affected have not always found a role for ELIPs in photoprotection, indicating that their role might be more significant in longer-term developmentally regulated photoprotection. ELIPs in the thylakoids bind chlorophyll molecules and can dissipate energy trapped by them. This might have been their original function in **phycobilisomes** of cyanobacteria before some of them evolved into the closely related proteins of LHCs that bind chlorophyll but can also capture energy. High light-induced proteins (HLIPs) that can bind chlorophylls also occur in thylakoids. They have a single transmembrane helix and, after two gene duplications, probably evolved into the ELIPs that have four transmembrane helices. There are also a large number of proteins found in the thylakoid lumen, many with unknown roles, but some of which probably help to adapt the thylakoid constituents and their properties. Overall, changes to photosystems, grana, and thylakoids help plants to significantly adapt their photosynthetic capacity to medium-term variation in the light regime.

Leaf optical properties are adapted to long-term variation in light regimes

The anatomy and morphology of leaves are adapted to capture light and deliver it to chloroplasts. Light that is incident on a leaf can be reflected, absorbed, or transmitted. In many leaves, about 15% of collimated light that is incident on the **adaxial** surface from above is reflected. However, a

Figure 2.19. *Vitis vinifera* (grape). The wild species from which cultivated *V. vinifera* cultivars (shown here) have been domesticated grow in full sunlight as a liana, and produce new leaves regularly. In cultivated varieties, early light-induced proteins (ELIPs) still have an important role in protecting new leaves from photoinhibition in full sunlight.

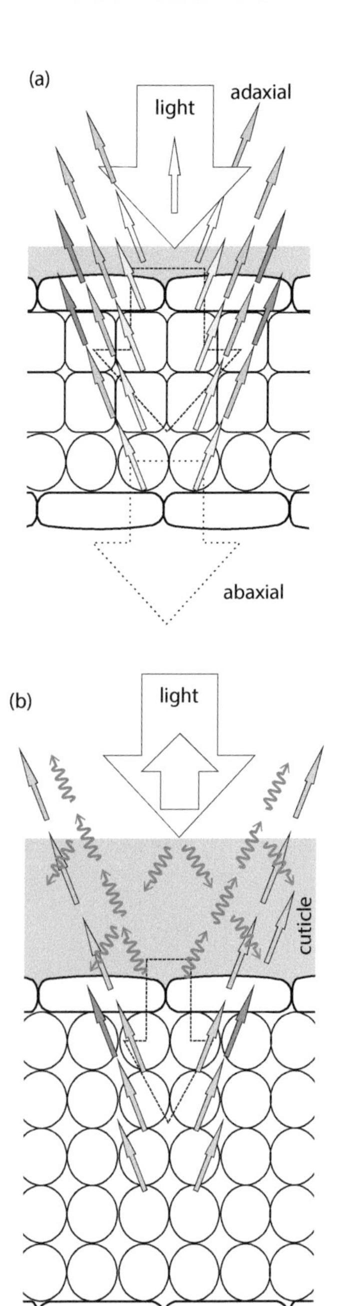

Figure 2.20. Leaves and the reflection of light. (a) Of the collimated light that is incident perpendicular to a thin leaf with a thin cuticle, 80–90% penetrates into the mesophyll. As light penetrates the leaf it becomes depleted in the non-green visible wavelengths. There is also reflection off the interfaces between cell wall and water, including the abaxial epidermis, although some light is transmitted. Some of the reflected light escapes, giving the leaf its green color. (b) In the leaves of plants adapted to excess light the cuticle is often very thick, increasing reflection and reducing penetration of light. In thick cuticles, Rayleigh scattering at shorter, blue wavelengths occurs, and the emitted blue-green light gives the leaves a glaucous appearance.

significant proportion of the total reflection of such light from a leaf often occurs from the **abaxial** surface—that is, much reflected light can penetrate through to the lower surface of the leaf and then be reflected back up through the leaf. As light penetrates to the abaxial surface, pigments remove much of the non-green light, so light reflected from the abaxial surface is greener than that reflected from the adaxial surface. This helps to explain the green color of leaves and why the adaxial surfaces are sometimes greener than the abaxial ones (Figure 2.20a). The cuticle is the primary reflector at the adaxial surface, reflecting light that is spectrally unchanged but more polarized than incident light. Light reflected from plant cuticles can be used by remote sensing techniques to determine numerous characteristics of vegetation—for example, tree species—because it is affected by cuticular properties. Plants from environments in which light can be in excess, such as alpine, boreal, saline, and dry ecosystems, often have deep cuticles and can reflect up to 50% of the radiation that is incident on their leaves—a significant factor in limiting damage to their photosynthetic apparatus. The microarchitecture of cuticular waxes, and epidermal structures such as trichomes and salt bladders, provide further control over the penetration of light into leaves. Light that penetrates the cuticle can be subject to **Rayleigh scattering** within it, which is more pronounced at shorter, blue wavelengths, giving many plants with thick cuticles a blue-green **glaucous** color (Figure 2.20b).

In some plant species, leaves that develop under chronic exposure to high light intensity have deep **palisade mesophyll** layers, whereas those that develop in low light have hardly any palisade mesophyll and consist almost entirely of **spongy mesophyll**. In plants that are adapted to chronically high or low light regimes these differences can be constitutive and pronounced. At high light intensity with good water availability, high photosynthetic rates per unit area of ground are possible, but not within a single thin leaf. High light intensities can only be converted into high photosynthetic rates if the energy dissipated by NPQ and PQ is minimal. A thin leaf that captures all light at high irradiance will often trap too much energy for photosystems to cope with (Figure 2.21), so the optimizing of photosynthetic rate depends on the dispersal of light throughout a leaf and on allowing some transmission of light to lower leaves. Ultimately, high photosynthetic rates per unit area require high rates of CO_2 fixation per unit area and, given that CO_2 has to diffuse to rubisco, this means a significant depth of rubisco-containing cells per unit area, and hence a significant depth of cells to which light has to be delivered. Elongated palisade cells not only produce a greater number of chloroplasts per unit area of leaf, but also—because air-water, water-cell wall, and water-membrane boundaries reflect light—channel a proportion of light down into the spongy mesophyll (Figure 2.22). The delivery of light in sun-adapted leaves is an adaptation that allows the almost collimated radiation of full sunlight to penetrate deep into the leaves.

In spongy mesophyll the cells are loosely packed, so the variety of angles of the water-air, water-cell wall, and water-membrane interfaces promotes reflection of light within the leaf. This increases the path length of light in this tissue, helping to scavenge light that penetrates to the spongy mesophyll. Shade leaves are thin and consist almost entirely of spongy mesophyll, often with cells that are smaller than those of sun-leaf mesophyll, so are adapted to light scavenging. Leaf epidermal cells, in particular those of some shade species, are lens-like and, together with trichomes and other epidermal structures, can have a role in concentrating light in particular locations in the leaf. The radiation that is incident on shade leaves can be predominantly diffuse, and the anatomy of the epidermis and mesophyll of these leaves increases the efficiency of the use of such light. Leaves that are adapted to promote the internal reflection of light also make more use of green light.

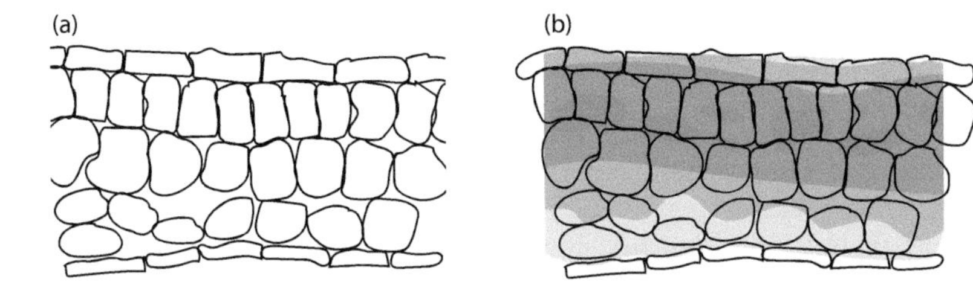

Figure 2.21. Fluorescence in a leaf at high light intensity. (a) A cross section of a young *Arabidopsis* leaf with relatively little differentiation into palisade and mesophyll tissue. (b) Fluorescence at high light intensity in the same leaf. The dark patches are areas of significant fluorescence (that is, excess light), even though the bottom of the leaf exhibits very little fluorescence. Strongly developed palisade tissue helps to channel the light to the lower part of the leaf.

To make efficient use of light, photosynthetic pigments that did not absorb in all wavelengths of light, but that allowed some penetration of light into light-harvesting structures, were probably necessary. Although chlorophylls absorb less green than blue or red light, many leaves absorb 80% of the green light that is incident upon them. The proportion of green light harvested depends on the path length of light in the leaf, with those that promote internal reflection harvesting all wavelengths more fully. These effects of shade are important not only ecologically but also agriculturally. Coffee (*Coffea arabica, C. canephora;* **Figure 2.23**), cocoa (*Theobroma cacao*), and tea (*Camellia sinensis*) are used to produce some of the most valuable of all plant products, and have wild ancestors that are shade-tolerant understory trees. Their cultivation often includes maintaining appropriate light conditions—for example, via the use of taller shade trees. The production of leaves of *Piper betle* that are used in the Indian subcontinent for rolling "paan" is induced by depriving them of light by wrapping the vine in **raffia**, and the production of large thin leaves of *Nicotianum tabaccum* for rolling into cigars is also induced by shading.

Photoprotective pigments in particular cells are also used by many plants to make long-term adaptations to light regimes. Juvenile and mature plants have different susceptibilities to excess light stress, and often have different pigments. Perennial plants often have different pigment distributions in their leaves and stems at different times of year. Plants grown in greenhouses, in which light is significantly reduced in shorter wavelengths, have a different pigment composition to those grown in open sunlight. Plant cuticles can contain **phenolic** acids that absorb ultraviolet-A (UV-A) (Figure 2.5), and such pigments are probably widely used to provide protection from UV radiation. Other phenolic compounds that are water soluble, including flavonols and anthocyanins, frequently occur in epidermal cell vacuoles, in some species only at particular times, and give leaves and/or stems a red-purple color. Carotenoids, including xanthophylls, are lipid soluble and can occur in extra-thylakoidal lipids. In algae, lipid bodies with carotenoids have an established photoprotective role, and it is likely that they also have such a role in land plants. It is notable that, in concert, these photoprotective pigments absorb wavelengths of light in the range 400–575 nm, in which the majority of the energy that drives photosynthesis is harvested, and therefore where protection is most needed if light is in excess. Numerous shade plants in the tropics have abaxial epidermal cells with a high concentration of anthocyanins, giving the underside of the leaf a red color. In shade

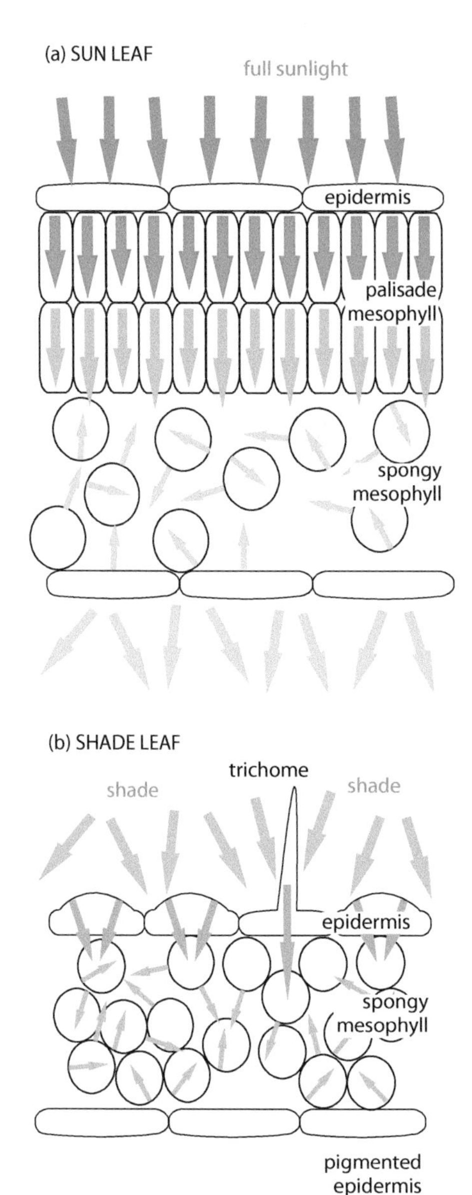

Figure 2.22. Sun leaves and shade leaves. (a) Full sunlight, which tends to be collimated, can overwhelm photosystems, so one function of the palisade mesophyll is to channel some of the incident light into the spongy mesophyll. The cells of the spongy mesophyll provide a variety of reflectance angles, scavenging light by increasing its path length in this tissue. (b) Light incident on shade leaves tends to be diffuse in quality and insufficient in quantity, so these leaves have only light-scavenging spongy mesophyll. They can also focus light with lens-like epidermal cells and with trichomes. Many shade-tolerant plants have pigmented lower epidermal cells, which probably decrease the transmittance of light that reaches them.

Figure 2.23. The fruits of *Coffea arabica* (coffee). *C. arabica* is an understory shrub native to the Ethiopian highlands. It produces "arabica" coffee and in many places grows best when shaded. *C. canephora* produces "robusta" coffee and is less shade demanding. The figure shows green, unripe coffee fruits. These turn red when ripe, and each fruit contains two coffee beans, which are harvested and then dried and roasted.

leaves in low light conditions, internal reflection in extensive spongy mesophyll means that it is unlikely that much light reaches the abaxial surface. However, in a sunfleck a large amount of light reaches the abaxial surface of the leaf and, if reflected, would further increase the already excessive leaf light intensity. Anthocyanins in the abaxial epidermis absorb excess light, perhaps preventing excessive internal reflection in leaves that are adapted to normally low light intensity. The anatomy and morphology of leaves are thus adapted to deliver an appropriate amount of light to the photosynthetic apparatus despite chronic differences in light regime. A disproportionate number of shade-tolerant species have iridescent leaves. This is probably to disguise the leaf and thus guard the investment in resources that a leaf adapted to a shady environment represents.

Adjustments in leaf position and plant architecture adapt plants to different light regimes

Many plant species alter the position of their leaves to adapt to the radiation regime via **nastic** (non-directional) and/or **tropic** (directional) responses. **Nyctinastic** movements ("sleep movements") of organs are quite common in plants—species of many plant families close their petals at night, and sleep movements of the leaves occur in many species of some families—for example, the Fabaceae. Nyctinastic leaf movement in the Fabaceae is regulated by the circadian clock and driven by **pulvini**, which in several species are also used during the day to orientate the leaf to optimize light capture. In the Fabaceae, pulvini are distinct bands of tissue at the base of the leaf petiole and/or leaflet **petiolule**. Rapid changes in the concentration of K^+, Cl^-, Ca^{2+}, and other osmolytes in the adaxial and abaxial layers cause influx or efflux of water through **aquaporins**, and the swelling of particular cells to adjust leaf orientation. In legumes, leaf movements to increase light capture at low irradiance and decrease it at high irradiance are also quite common. Pigeon pea (*Cajanus cajan*) is an important pulse crop in the Indian subcontinent in particular, and its well-known leaf movements significantly adjust light capture. Many species of grass (in the family Poaceae) have long leaves, and leaf rolling varies with the environmental conditions. In rice, for example, there are well-characterized **bulliform** cells along the primary vasculature on both the adaxial and abaxial surfaces. The expansion and contraction of these cells determines the extent of leaf rolling. Many of the highest yielding rice varieties have a V-shaped leaf lamina that is in effect partly rolled, and which provides the highest light capture per unit area in a monoculture. The full leaf rolling that is characteristic of drought conditions in many grasses not only helps to reduce water loss, but also decreases light interception and photoinhibition. Adjustable orientation and rolling of leaves is important in many natural canopies, but is also an aspect of crop manipulation that is often considered to have the potential to increase light capture.

In order to capture the light that they need, many plants can also adapt their growth and development to the light regime. Large seeds that can draw upon significant reserves of food often germinate and begin development in the absence of light (**skotomorphogenesis**). The seedlings that are produced in the dark have elongated, hooked **hypocotyls** and small cotyledons surrounding an **epicotyl** suspended in the early stages of development—these are adaptations for pushing up through the soil. When the seedling reaches the light, if the R:FR ratio is high (that is, near that of full sunlight) it radically alters its development and enters **photomorphogenesis**, in which epicotyl development is dominant. In small seeds with few reserves, germination and development often occur only after exposure to light with a high R:FR ratio. Thus detection of the R:FR ratio of light provides plants with important clues about their light environment, and in some plants can initiate, for example, a shade avoidance response (SAR). In shade, red light is depleted compared with far-red light. Some species are unable to respond to shade, but in others it can promote not only the formation of shade leaves but also the SAR. In plants that exhibit the SAR, a low R:FR ratio induces **etiolation** in stems and petioles, a growth form underpinned by similar pathways to skotomorphogenesis. Using auxin-mediated pathways, etiolated seedlings undergo expansive growth of epidermal and cortical cells concurrent with a decrease in cambial growth and cell differentiation. They also show dramatically decreased expression of everything necessary for photosynthesis, including a decrease in the photoprotective adaptations that are expressed in full sunlight. Root growth also generally ceases, so etiolation represents a near total commitment to vegetative growth in order to avoid shade and reach the light. The SAR interacts significantly with phototropic responses that orientate plants to light direction using blue light detection.

Plants detect the R:FR ratio by using phytochromes, which are proteins with attached **tetrapyrrole** chromophores. These chromophores change reversibly between R and FR forms depending on the R:FR ratio of incident radiation. Three phytochromes (PHYA, PHYB, and PHYC) are common to all angiosperms, but several additional forms are known. Particular forms and combinations of phytochrome control different aspects of the R:FR response, with many variations occurring in ecotypes adapted to particular light regimes. In full sunlight, the majority of phytochrome is converted to the FR form (P_{fr}), which accumulates in the nucleus and helps to control photomorphogenesis. A combination of kinase and phosphatase activity and ubiquitin-mediated protein degradation helps to control the effects of P_{fr}. In shade, phytochrome occurs primarily as the R form (P_r), which promotes the production of phytochrome interacting factors (PIFs) that in species with the SAR induce etiolation. PIFs are a major target of degradation during the P_{fr} response. Many species with the SAR detect changes in the R:FR ratio that occur well before the **leaf area index** (LAI) of the canopy reaches 1 (that is, before the leaves start to overlap), because the P_r:P_{fr} system is sensitive enough to detect changes in R:FR ratio in reflected and diffuse light from neighboring plants before overtopping occurs. In natural ecosystems, the SAR is vital to canopy development because it determines the growth of plants that compete for light, separating them spatially from shade-tolerant plants. In some forests it has been linked to niche separation, especially via the ability of plants to adjust to disturbances in the canopy. In crop plants the SAR can be linked to significant decreases in yield, and for many crops it determines the maximum planting density because it promotes vegetative growth, which decreases **harvest index**. Negative regulators prevent an exaggerated SAR, and include both a suite of COP/DET/FUS proteins that degrade proteins of the SAR, and transcription factors such as the bHLH HFR1/SICS1 that oppose the SAR. In selecting for high-yielding varieties that can be planted at high densities (maize being a good example), crop breeders have significantly decreased the SAR response in many crops. Various further manipulations of the SAR have been achieved, and such manipulation, including that achieved via the use of RNA interference (RNAi) and microRNAs (miRNAs), is considered to have significant potential for increasing light capture and yield per unit area in crops.

Changes in plant architecture were vital to the Green Revolution, in significant part because of their effects on light capture. In natural ecosystems, ecotypes from different light regimes often have different architectures—for example, *Rhamnus catharticus* grown in shade has a taller tree-like architecture that is distinct from the shrub-like architecture of plants grown in full sunlight. The architecture of plants is a product of a branching morphology, from crown to shoot to branch, with light being one of the important environmental factors affecting it. **Strigalactones** are hormones that tend to be transported upward through plants, and that suppress branching and tillering in eudicots and monocots, respectively. They adjust plant architecture not only to light regime but also to the availability of nutrients such as P and to temperature regime. In grasses, including cereal crops, pulvini at the origin of tillers, at nodes, at lamina origins, and at the bases of inflorescence branches also help to determine plant architecture. The balance between adaxial and abaxial growth in the pulvinus can control the openness of the culm, leaves, and inflorescence. In grasses, pulvini are also the site of gravitropic responses that help stems that have been beaten down to reorientate to the light. Plants thus possess numerous mechanisms for adjusting leaf morphology, leaf position, and plant morphology to optimize light capture. For many plants in natural ecosystems these adaptations are vital for surviving systematic differences in light regimes, whereas in crop plants there is still potential for manipulating them to increase light capture.

Photoinhibition is most severe in alpine environments

The most challenging night–day transition for plants occurs in alpine ecosystems, which are located above the **treeline**. At high altitude the nights are cold, and during the day solar radiation is intense—at high altitudes it is significantly nearer the solar constant than the light incident on lowland plants. At a solar angle of 90°, radiation is almost twice as intense at 3000 m as it is at sea level because the air mass through which light has passed is less. There is also a shift towards shorter, high-energy wavelengths as altitude increases. Plants in tropical alpine environments experience more intense light than any other plants on Earth. As the leaves start each day frozen or cold, and midday radiation is most intense, for much of the day alpine conditions favor photoinhibition rather than photosynthesis. Tropical alpine plants experience diurnal variation in their physical and light environment that is so significant they effectively experience winter and summer conditions every 24 hours. Temperate alpine plants experience not only significant diurnal variation but also seasonal variation in their light and temperature environment. Alpine plants have characteristic physiological, anatomical, and morphological adaptations to maintain photosynthesis. For photosynthesis in alpine plants a significant additional challenge is a drop of about one-third in the partial pressure of CO_2 at 3000 m.

Alpine ecosystems not only have unique floras, but also form the vegetation of the upper watersheds of many of the most important rivers in the world. The most extensive alpine ecosystems occur in and around the Tibetan Plateau, which covers an area of about 2.5 million km^2 at or above 4500 m, and is probably the most extensive high-altitude environment that has ever existed on Earth. Alpine ecosystems in the temperate zone have plants long known to be tolerant of photosynthetic photon flux densities (PPFDs) of up to 2000 μmol m^{-2} s^{-1}. In general, temperate alpine plants have maximal rates of CO_2 fixation that are less than those of lowland plants (averaging around 15 μmol CO_2 m^{-2} s^{-1} rather than 30 μmol CO_2 m^{-2} s^{-1}), and that occur at relatively high PPFDs of up to more than 1000 μmol m^{-2} s^{-1}. Their capacity to quench excess excitation energy is very significant, with NPQ ratios in the range of 3–4 often being measured. Enhanced xanthophyll cycling, antioxidant activity, and cyclic electron transport have all been reported in temperate alpine plants, producing inefficient photosynthesis for much of the day. NPQ is especially significant in the morning, but photoinhibition is rapidly reversible, enabling maximum rates of photosynthesis to be reached quickly as conditions warm. In many alpine plants there is a decrease in photosynthesis at midday because, despite the warm temperatures, the solar intensity is too great for the photochemical reactions of photosynthesis. Many temperate alpines have an **erinaceous** (hedgehog-like or cushion-like) morphology, while others have primarily erect leaves, and some have rosette leaves. In plants with a cushion-like habit, temperature variation is minimized, expanding the photosynthetic window. Erect-leaved and rosette-leaved plants avoid direct sunlight at midday and in the morning, respectively, but photoinhibition is favored in the morning and at midday, respectively. These adaptations enable alpine plants to take advantage of brief periods of the day when the combination of leaf temperature and light intensity is conducive to photosynthesis.

The equator crosses the Andes and is very close to mountains in Africa and Hawaii that are high enough to be above the treeline, and therefore to have tropical alpine ecosystems. Regular diurnal temperature changes of 40°C and light intensities of up to 2500 μmol m^{-2} s^{-1} occur. Seasonality in temperature is minimal, but each night exposes plants to a freezing "winter" regime and each day subjects them to a hot "summer" regime, resulting in an extremely photosynthetically challenging night–day regime. Many grasses in tropical alpine ecosystems occur as tussocks, whereas shrubs are large and have small, fine leaves. Most notably, however, in the Andes, in Africa, and in Hawaii large **caulescent** rosette plants are an important part of the flora at the very highest zones of vegetation. These rosette plants occur above the treeline, but have tall stems, which make them the highest living caulescent plants. Acaulescent plants with a similar morphology include the yuccas, agaves, and aloes of low-altitude deserts (Figure 2.24). The highest point of the Rwenzori Mountains of Uganda (Mount Stanley, at 5109 m) is higher than that of the European Alps (Mont Blanc, at 4810 m), but the Rwenzori Mountains are only 30 km north of the equator. Their glaciated, snow-capped peaks are very probably Ptolemy's "Mountains of the Moon," believed in ancient times to be the source of the Nile. Together with the other high mountains of eastern Africa (Figure 2.25) they are home to

Figure 2.24. ***Agave parryi.*** Agaves, such as this species, grow as giant acaulescent (stemless) rosettes, as do many other plants that inhabit dry environments, such as species of *Yucca* and *Aloe*. They have a morphology similar to that of caulescent tropical alpine rosette plants.

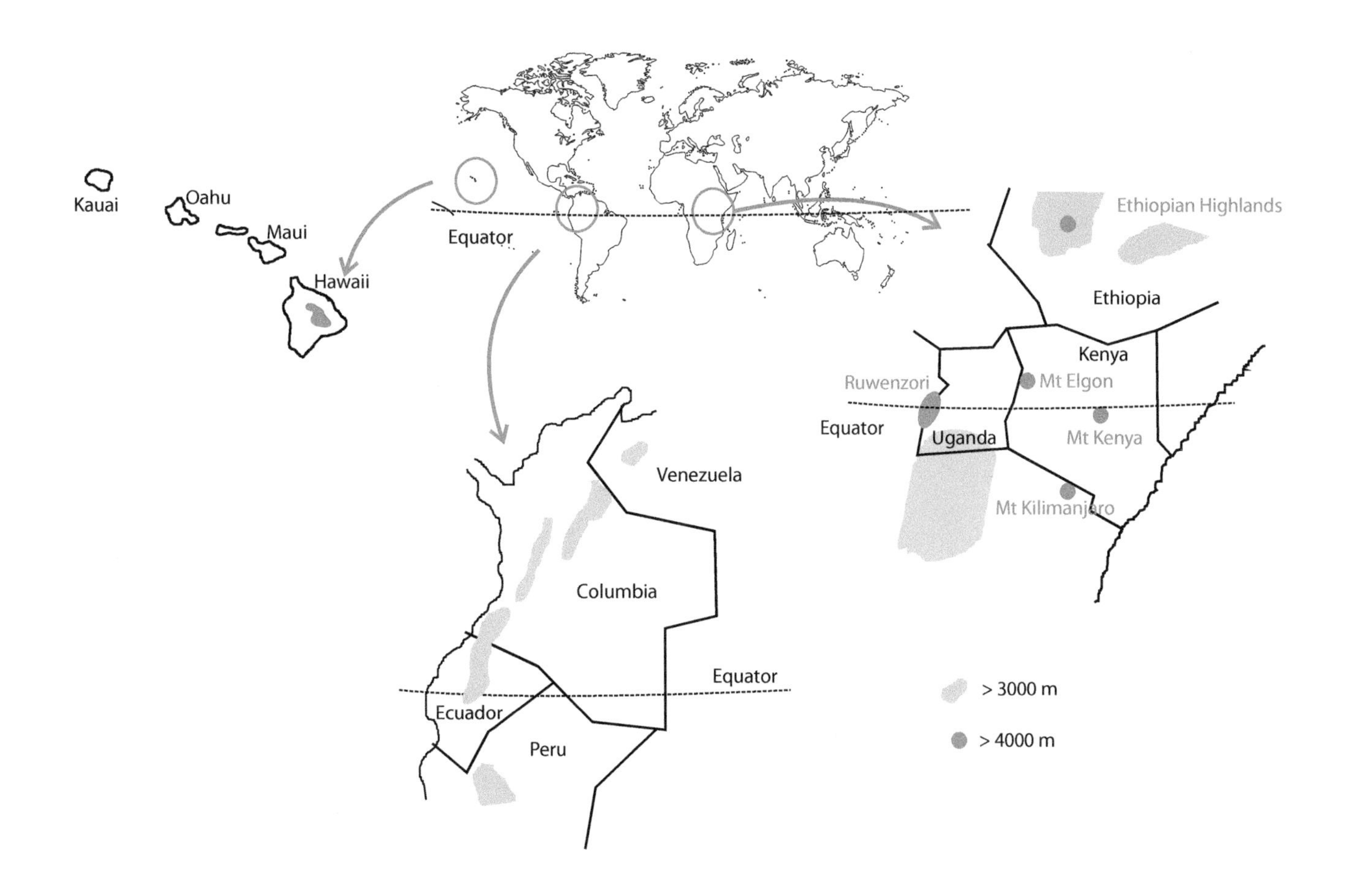

Figure 2.25. **Tropical alpine environments.** These environments occur in only a few localities on Earth. Their vegetation often consists of giant caulescent rosette plants together with shrubs and grasses.

several giant caulescent rosette species of *Lobelia* (Campanulaceae) and *Dendrosenecio* (Asteraceae). These plants characterize the Afro-alpine ecosystems that occur above about 3500 m. In high-altitude **paramo** systems of the North-East Andes, caulescent species of *Espeletia* (Asteraceae) occur. On Hawaii, the rosette plants known as "silversword," which are species of the genus *Argyroxiphium* (Asteraceae), inhabit the highest vegetated zones. The tropical alpine species in the Asteraceae and Campanulaceae provide a remarkable example of convergent evolution driven by light and temperature regimes (Figure 2.26).

The bunched leaves of caulescent rosette plants minimize the freezing of centrally located young leaves and meristems. Tropical alpine environments, which are located in stable tropical weather systems, have much lower wind speeds than temperate alpine environments, which are located in frequently turbulent temperate weather systems. At high altitude, air at a height of 1–3 m above the ground experiences less profound diurnal changes in temperature than does the air directly above the soil surface. The elevated rosettes that can evolve in low-wind-speed tropical alpine environments help to minimize temperature variation in the leaves. The leaves often fold up at night, trapping large quantities of water in which freezing occurs first in the upper surfaces, decreasing the likelihood of the meristem, in deeper water, freezing. Copious quantities of mucus can also be exuded into trapped water stores to prevent them from freezing. In *Espeletia*, the size of the water-filled pith in the stem increases with altitude and cold temperatures, probably helping to buffer these plants against changes in temperature. In the older leaves of *Lobelia rhynchopetalum*, exceptionally high NPQ ratios of 10 have been measured, with a relaxing of NPQ only either side of midday. In the younger, more vertical, inner leaves, NPQ is less significant because these leaves are shaded in the morning and vertical to the sunlight at midday. Giant caulescent plants have very low CO_2 uptake rates of 5–6 $\mu mol\ m^{-2}\ s^{-1}$, and photosynthetic maxima that occur only at very high PPFDs of around 1500 $\mu mol\ m^{-2}\ s^{-1}$ and at temperatures of about 15°C. Low-altitude vegetation that surrounds the "islands in the sky" of Afro-alpine vegetation has a high proportion of C_4 plants, but as altitude increases up Mount Kenya, the proportion of C_3 plants increases until they dominate the Afro-alpine zone just below the snowline. This is likely to be because photorespiration in C_3 plants can help to decrease photoinhibition. The leaves of many giant caulescent rosettes are **pubescent**. In general this decreases incident radiation, reduces temperature fluctuations, and minimizes water loss from the leaves. Thus the adapted physiology of photosynthesis within the characteristic morphology of tropical alpine plants has enabled them to be net fixers of CO_2 in an environment which has a night-day change that is extremely photosynthetically challenging. These plants are the most extreme example of the important interaction between the effects of light intensity and temperature. Their morphological similarity to yuccas, agaves, and aloes is striking, although these plants are generally not caulescent. In desert environments a lack of water decreases CO_2 uptake, and hence photochemical quenching, promoting photoinhibition, so giant rosette morphology is probably a response to chronic excess light.

Dendrosenecio keniensis
(East Africa)

Argyroxiphium sandwicense
(Hawaii)

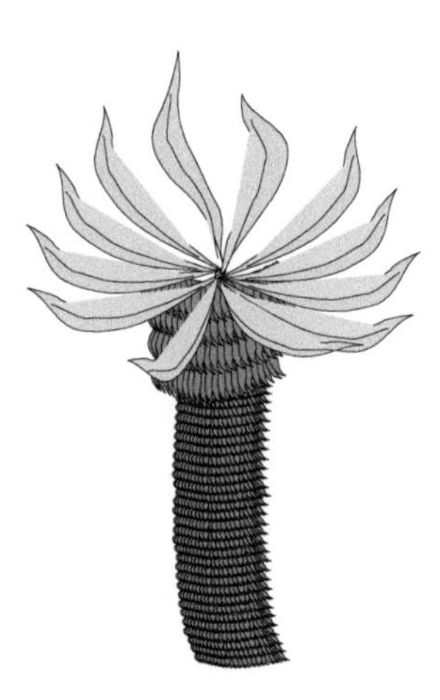

Espeletia pycnophylla
(Northern Andes)

Figure 2.26. Giant caulescent rosette plants. The challenging conditions of tropical alpine environments have led to the evolution of a characteristic plant morphology—giant caulescent rosettes. The three species shown in the figure are all members of the Asteraceae that have independently evolved this morphology in different parts of the world. Species in other families (for example, *Lobelia* species in the Campanulaceae) have also evolved this morphology.

Summary

Land plants have adapted to optimize the transduction of light energy into chemical energy. The night-day transition that is experienced by all plants necessitates a flexible light-capturing apparatus. This has probably evolved into the biochemical and physiological adaptations shown by most plants to the short-term variation in light intensity that occurs on land. In canopies or in understories, plants that experience sustained differences in light regime have anatomical and morphological adaptations to regulate light

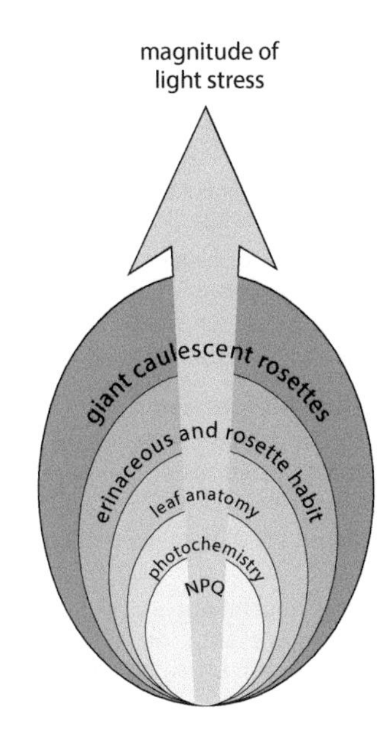

Figure 2.27. Light and the stress-response hierarchy. Plants show a range of adaptations to progressively greater variation in light intensity. Although adaptations at higher levels can sometimes supersede those at lower levels, they are often additive. NPQ = non-photochemical quenching.

capture, and these complement the physiological flexibility of photosynthesis (Figure 2.27). In most plants, such adaptations generally match the delivery of light to the photosynthetic apparatus, so stresses such as temperature, water deficit, and lack of CO_2 that inhibit plant responses to excess light can promote photoinhibition. The most lethal combination of stressors to the photosystems consists of the high light intensity and extremely cold temperatures of tropical alpine ecosystems, but any of the environmental factors that are dealt with in this book can restrict adaptation to the light regime and thus disrupt photosynthesis.

Further reading

Evolution of photosynthesis

Gupta RS (2012) Origin and spread of photosynthesis based upon conserved sequence features in key bacteriochlorophyll biosynthesis proteins. *Mol Biol Evol* 29:3397–3412.

Hohmann-Marriott MF & Blankenship RE (2011) Evolution of photosynthesis. *Annu Rev Plant Biol* 62:515–548.

Niedzwiedzki DM, Fuciman M, Frank HA et al. (2011) Energy transfer in an LH4-like light harvesting complex from the aerobic purple photosynthetic bacterium *Roseobacter denitrificans*. *Biochim Biophys Acta* 1807:518–528.

Light reactions

Busch A & Hippler M (2011) The structure and function of eukaryotic photosystem I. *Biochim Biophys Acta* 1807:864–877.

Nelson N & Yocum CF (2006) Structure and function of photosystems I and II. *Annu Rev Plant Biol* 57:521–565.

Pan X, Liu Z, Li M et al. (2013) Architecture and function of plant light-harvesting complexes II. *Curr Opin Struct Biol* 23:515–525.

Renger T, Madjet ME, Knorr A et al. (2011) How the molecular structure determines the flow of excitation energy in plant light-harvesting complex II. *J Plant Physiol* 168:1497–1509.

Variation in incident light

Raven JA (2011) The cost of photoinhibition. *Physiol Plant* 142: 87–104.

Takahashi S & Murata N (2008) How do environmental stresses accelerate photoinhibition? *Trends Plant Sci* 13:178–182.

Way DA & Pearcy RW (2012) Sunflecks in trees and forests: from photosynthetic physiology to global change biology. *Tree Physiol* 32:1066–1081.

Zhang M, Yu G, Zhuang J et al. (2011) Effects of cloudiness change on net ecosystem exchange, light use efficiency, and water use efficiency in typical ecosystems of China. *Agricult Forest Meteorol* 151:803–816.

Non-photochemical quenching

Arnoux P, Morosinotto T, Saga G et al. (2009) A structural basis for the pH-dependent xanthophyll cycle in *Arabidopsis thaliana*. *Plant Cell* 21:2036–2044.

Johnson MP, Goral TK, Duffy CDP et al. (2011) Photoprotective energy dissipation involves the reorganization of photosystem II light-harvesting complexes in the grana membranes of spinach chloroplasts. *Plant Cell* 23:1468–1479.

Lambrev PH, Miloslavina Y, Jahns P et al. (2012) On the relationship between non-photochemical quenching and photoprotection of Photosystem II. *Biochim Biophys Acta* 1817:760–769.

Verhoeven A (2014) Sustained energy dissipation in winter evergreens. *New Phytol* 201:57–65.

Adjustment of electron flows

Johnson GN (2011) Physiology of PSI cyclic electron transport in higher plants. *Biochim Biophys Acta* 1807:384–389.

Miyake C (2010) Alternative electron flows (water-water cycle and cyclic electron flow around PSI) in photosynthesis: molecular mechanisms and physiological functions. *Plant Cell Physiol* 51:1951–1963.

Schoettler MA & Toth SZ (2014) Photosynthetic complex stoichiometry dynamics in higher plants: environmental acclimation and photosynthetic flux control. *Front Plant Sci* 5:188.

Voss I, Sunil B, Scheibe R et al. (2013) Emerging concept for the role of photorespiration as an important part of abiotic stress response. *Plant Biol (Stuttg)* 15:713–722.

PSII repair

Nath K, Jajoo A, Poudyal RS et al. (2013) Towards a critical understanding of the photosystem II repair mechanism and its regulation during stress conditions. *FEBS Lett* 587:3372–3381.

Nath K, Poudyal RS, Eom J-S et al. (2013) Loss-of-function of OsSTN8 suppresses the photosystem II core protein phosphorylation and interferes with the photosystem II repair mechanism in rice (*Oryza sativa*). *Plant J* 76:675–686.

Nickelsen J & Rengstl B (2013) Photosystem II assembly: from cyanobacteria to plants. *Annu Rev Plant Biol* 64:609–635.

Takahashi S & Badger MR (2011) Photoprotection in plants: a new light on photosystem II damage. *Trends Plant Sci* 16:53–60.

Chloroplast movements alter light capture

Kong S-G & Wada M (2014) Recent advances in understanding the molecular mechanism of chloroplast photorelocation movement. *Biochim Biophys Acta* 1837:522–530.

Morita MT & Nakamura M (2012) Dynamic behavior of plastids related to environmental response. *Curr Opin Plant Biol* 15:722–728.

Shimazaki K-I & Tokutomi S (2013) Diverse responses to blue light via LOV photoreceptors. *Plant Cell Physiol* 54:1–4.

Wada M (2013) Chloroplast movement. *Plant Sci* 210:177–182.

Adaptation of thylakoids and photosystems

Alvarez-Canterbury AMR, Flores DJ, Keymanesh K et al. (2014) A double SORLIP1 element is required for high light induction of ELIP genes in Arabidopsis thaliana. *Plant Mol Biol* 84:259–267.

Iwai M, Yokono M & Nakano A (2014) Visualizing structural dynamics of thylakoid membranes. *Sci Rep* 4:3768.

Pribil M, Labs M & Leister D (2014). Structure and dynamics of thylakoids in land plants. *J Exp Bot* 65:1955–1972.

Tikkanen M & Aro E-M (2014) Integrative regulatory network of plant thylakoid energy transduction. *Trends Plant Sci* 19:10–17.

Leaf optical properties

Gorton HL, Brodersen CR, Williams WE et al. (2010) Measurement of the optical properties of leaves under diffuse light. *Photochem Photobiol* 86:1076–1083.

Hughes NM (2011) Winter leaf reddening in 'evergreen' species. *New Phytol* 190:573–581.

Hughes NM, Vogelmann TC & Smith WK (2008) Optical effects of abaxial anthocyanin on absorption of red wavelengths by understorey species: revisiting the back-scatter hypothesis. *J Exp Bot* 59:3435–3442.

Zakharov VP, Bratchenko IA, Sindyaeva AR et al. (2009) Modeling of optical radiation energy distribution in plant tissue. *Opt Spectrosc* 107:903–908.

Adjustments in leaf position and morphology

Chen J, Moreau C, Liu Y et al. (2012) Conserved genetic determinant of motor organ identity in *Medicago truncatula* and related legumes. *Proc Natl Acad Sci USA* 109:11723–11728.

Hersch M, Lorrain S, de Wit M et al. (2014) Light intensity modulates the regulatory network of the shade avoidance response in *Arabidopsis. Proc Natl Acad Sci USA* 111:6515–6520.

Liscum E, Askinosie SK, Leuchtman DL et al. (2014). Phototropism: growing towards an understanding of plant movement. *Plant Cell* 26:38–55.

Ruberti I, Sessa G, Ciolfi A et al. (2012) Plant adaptation to dynamically changing environment: the shade avoidance response. *Biotechnol Adv* 30:1047–1058.

Alpine plants and photoinhibition

Fetene M, Nauke P, Luttge U et al. (1997) Photosynthesis and photoinhibition in a tropical alpine giant rosette plant, *Lobelia rhynchopetalum. New Phytol* 137:453–461.

Givnish TJ (2010) Giant lobelias exemplify convergent evolution. *BMC Biol* 8:3.

Montgomery RA, Goldstein G & Givnish TJ (2008) Photoprotection of PSII in Hawaiian lobeliads from diverse light environments. *Funct Plant Biol* 35:595–605.

Sanchez A, Posada JM & Smith WK (2014) Dynamic cloud regimes, incident sunlight, and leaf temperatures in *Espeletia grandiflora* and *Chusquea tessellata*, two representative species of the Andean paramo, Colombia. *Arct Antarct Alp Res* 46:371–378.

Chapter 3
Carbon Dioxide

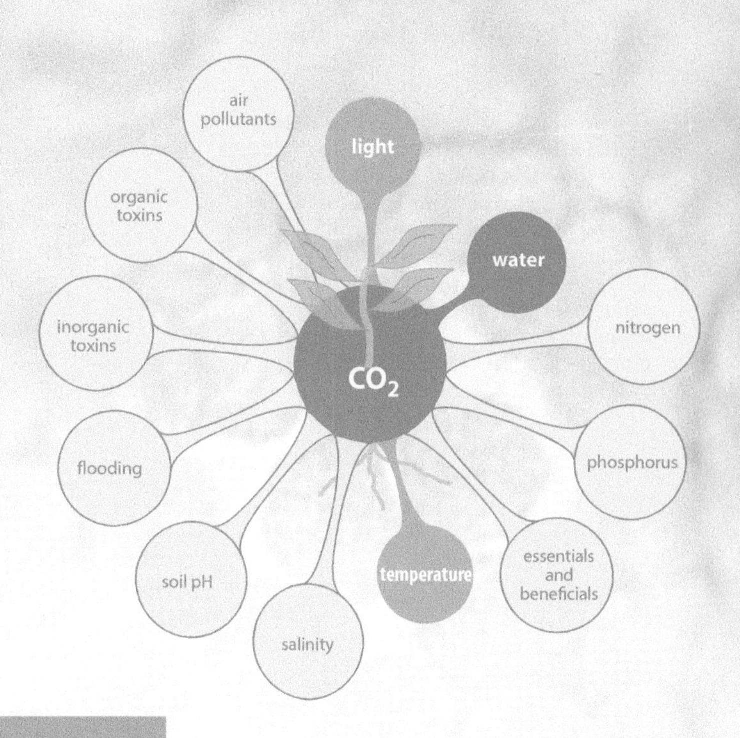

The challenge of variation in CO_2 levels interacts, in particular, with variation in water, temperature, and light levels.

Key concepts

- Carbon dioxide (CO_2) fixation underpins the process of biomass production on Earth.
- Human activities are causing an increase in atmospheric CO_2 concentrations.
- Plants control CO_2-fixation kinetics to adapt to short-term variation in CO_2 supply.
- Photorespiration can decrease net carbon (C) fixation but increase photosynthetic flexibility.
- C_4 photosynthesis is an adaptation to limited CO_2 supply.
- C_4 and C_3 plants respond differently to variations in CO_2 supply.
- CAM photosynthesis adapts plants for CO_2 uptake in environments that are periodically dry.
- Sustained increases in CO_2 levels affect plant productivity and ecosystems processes.
- Increased CO_2 fixation in plants has long-term effects on climate.
- Sustainable human habitation of Earth depends on the management of CO_2 fixation.

CO_2 fixation underpins the primary production of biomass

CO_2 is the primary carbon source for the organic compounds of terrestrial ecosystems. It was discovered in the eighteenth century as "fixed air" released when plant material, for example, was combusted. The process of carbon assimilation, often referred to as "CO_2 fixation," underpins the primary production of biomass. In unmanaged ecosystems, the dynamics of primary production in plants help to determine ecosystem **carrying capacity**, biodiversity, resilience, biogeochemical cycling, and the provision of ecosystem services. In managed agricultural and forestry ecosystems, yield is dependent on primary productivity. Terrestrial CO_2 fixation by plants also plays an important role in determining atmospheric CO_2 concentrations—the almost real-time effects of CO_2 fixation can be seen in atmospheric monitoring data (Figure 3.1). The greenhouse effect in the Earth's atmosphere, to which CO_2 contributes very significantly, helps to

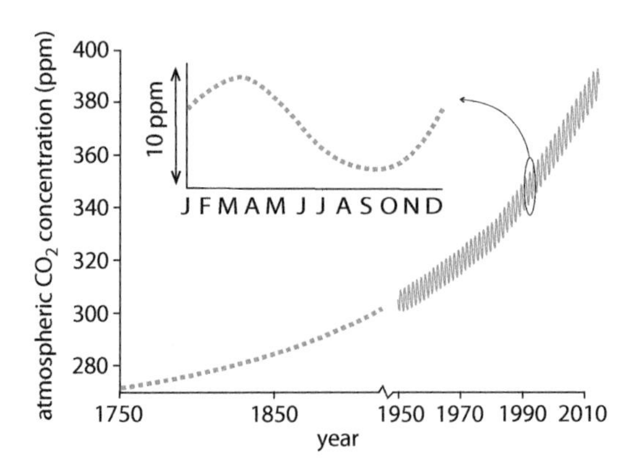

Figure 3.1. Post-Industrial Revolution global atmospheric CO_2 concentrations. Since the early 1950s, monitoring at locations such as Mauna Loa in Hawaii has shown not only a long-term increase but also a significant annual variation in CO_2 concentration (inset). The annual variation reflects northern hemisphere summer photosynthesis, and is detectable globally because of the rapid mixing of the troposphere.

determine the suitability of Earth for life. If there was no greenhouse effect, Earth's mean temperature would be about 30°C lower than it is at present, but high CO_2 concentrations can induce temperatures that are too high for most life forms. Uncertainties about future CO_2-fixation rates are a significant challenge to prediction of the climate for the next few centuries. Thus CO_2 fixation is among the most fundamental processes on Earth.

Global C fluxes can be estimated from C inventories, C isotope ratios, and satellite remote sensing. C inventories for leaves, plants, or sometimes ecosystem types measure C pools to estimate **net primary productivity** (NPP). C isotope ratios in the **troposphere**, which is well mixed, can be used to make estimates of global C uptake and emission. The **Dole effect** describes the enrichment of atmospheric oxygen with ^{18}O compared with that of seawater, and reflects global primary production (Box 3.1). The **Suess effect** describes the reduction in the proportion of ^{13}C and ^{14}C relative to ^{12}C in fossil-fuel-derived CO_2, which enables estimates of anthropogenic CO_2 emissions. Remote sensing can be used to estimate parameters that relate directly to net primary production (NPP), such as light use efficiency (LUE) of vegetation and absorbed photosynthetically active radiation (PAR). Together these techniques are generally used to suggest that, from the 750 Gt of C in the atmosphere, there is global C fixation of about 120 Gt $year^{-1}$, with about half of this (56 Gt $year^{-1}$) achieved by land plants. On a dry weight basis, about 50% of terrestrial biomass is C, forming a labile C reserve of 600 Gt (that is, approaching the total amount of atmospheric C). Anthropogenic C emissions were just over 10 Gt $year^{-1}$ by 2014, the majority being derived from fossil fuels, but this was not inducing quite as large an increase in C fluxes as expected, suggesting an underestimation of C dynamics or sinks. Some ^{18}O studies do indicate that global C fixation might be as high as 150 Gt $year^{-1}$. Including forestry, it is estimated that humans now appropriate about 25% of total terrestrial NPP, mostly from the 38% of the ice-free surface that is managed land, of which about one-third is arable and the rest is pasture or plantation.

BOX 3.1. THE DOLE EFFECT

About 0.2% of oxygen atoms are the ^{18}O isotope, which is slightly heavier than the ^{16}O isotope that makes up about 99.75%. Water in the atmosphere is significantly depleted in ^{18}O compared with that in the sea (the Dole effect). This is mainly because the enzymes of respiration discriminate slightly between ^{16}O and ^{18}O due to their different weights, and water from terrestrial ecosystems, for example transpiration, is therefore enriched in ^{18}O. Rain falling from the atmosphere to the land surface therefore provides water for terrestrial photosynthesis with a reduced ^{18}O content compared with water used in marine photosynthesis. Measurements of the $^{16}O/^{18}O$ ratio can therefore be used to estimate marine and terrestrial contributions to respiration and primary productivity.

Anthropogenic increases in CO_2 levels generally have a CO_2-fertilization effect on plants, making them C sinks, although past natural changes show that, under particular CO_2 regimes, land plants can overall be sinks or sources. Change of land use by humans has consequences for global C sinks and sources. Deforestation decreases both the storage of C per unit area and NPP on regional scales, which can have global consequences. Reforestation can have the opposite effect in the short term, but its effects, and those of increasing CO_2 levels, are dependent on the availability of other resources and on environmental stressors. Biomass produced under higher than current ambient CO_2 concentrations has higher C:N, C:P, and C:S ratios—that is, CO_2 fixation affects **stoichiometry** and hence the biogeochemistry of

N, P, and S on regional and global scales. A significant driver of the hydrological cycle is transpiration by plants, which accounts for about 35% of all **evapotranspiration** from the land surface. Transpiration is affected by atmospheric CO_2 concentrations because H_2O is transpired when stomata are open to collect CO_2—a process that is reduced at higher CO_2 concentrations, due to stomatal closure. An understanding of the global ramifications of CO_2 fixation in land plants and its interaction with the availability of other resources, on all spatial and temporal scales in both unmanaged and managed ecosystems, will be of increasing importance if humans are to cope with the global scale of the impact of their activity on Earth.

Management of NPP underpins a variety of initiatives to ameliorate the effects of human activity on the environment. A major challenge for sustainable global food security is to produce more food without increasing the land area under production, in particular closing the gap between potential and actual yield on much arable land. This means increasing total agricultural production—a challenge that has up to now been met primarily by relieving limiting factors on production, increasing croppings per year, and increasing the area under cultivation. However, in the future the challenge might also have to be met by increasing, sustainably, the capacity for crop CO_2 fixation. In addition, the use of biofuels is increasing rapidly, its major limitation being that it competes with food production. The development of plants for biofuels that do not compete with food production will help to determine their future use. It is also possible that artificial C fixation will be used in energy production. Non-fossil-fuel energy can be used to produce methanol from CO_2 and H^+, and from methanol it is possible to produce fuels, synthetic hydrocarbons, and, based on microbial growth, proteins that are C neutral. Artificial enzymatic CO_2 fixation is also possible. This chapter deals with the plant CO_2-fixation processes that, in a variable environment, underpin primary production, interact with numerous environmental cycles, and need to be managed and/or supplemented to enable sustainable human habitation of Earth.

Variation in the supply of CO_2 to plants is significant and affected by human activity

Plants, particularly in canopies, are challenged by short-term and small-scale variations in CO_2 concentration. Such variation is a product of photosynthetic uptake and respiratory output of CO_2 from both plants and soil inhabitants, modified by airflow and leaf conductivity to CO_2. Monoculture often exaggerates the pronounced diurnal changes of ambient CO_2 concentrations in canopies (Figure 3.2). Above a crop canopy at night, a CO_2 level of 500 ppm is possible, whereas in the day it can drop to 200 ppm. CO_2 fixation depends on CO_2 reaching the enzyme ribulose-1,5-bisphosphate carboxylase/oxygenase (rubisco) in chloroplasts, and there are numerous potential resistances to flux from outside the leaf to the chloroplast. This is particularly so given that leaf anatomy and morphology are primarily driven by water and light regimes such that leaves can be, for example, **hypostomatous** or **amphistomatous**, and can have very different surface area to volume ratios. Leaf intercellular CO_2 to ambient CO_2 (C_i/C_a) ratios are mostly significantly less than 1 but, in a given species, C_i is often fairly constant, indicating that plants exert significant control over it, with plants of different growth types and from different habitats having significantly different C_i/C_a ratios. Short-lived plants, including many arable crops, have higher than average C_i/C_a ratios. Plants from drier ecosystems have lower than average C_i/C_a ratios, mostly reflecting the influence of water availability on the conductivity of leaves. The distribution of rubisco in leaves is frequently uneven, probably reflecting spatial variation in the supply of CO_2.

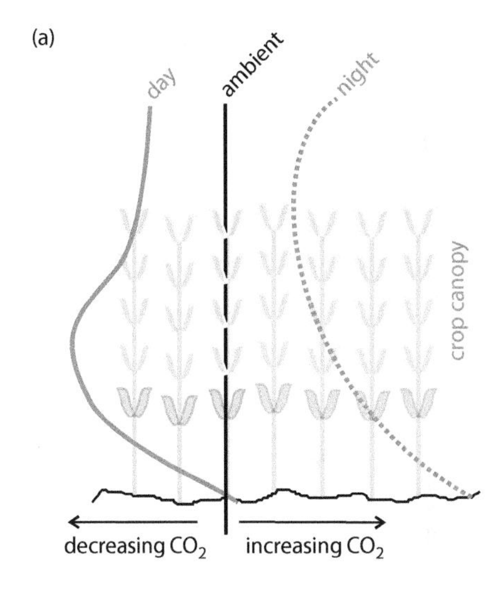

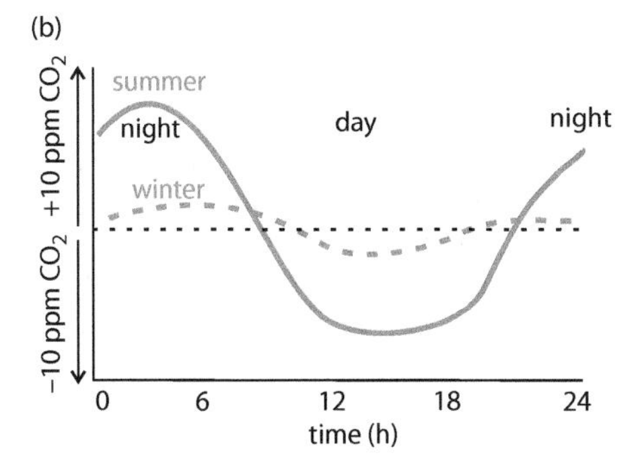

Figure 3.2. CO_2 concentration and crop canopies. (a) Generalized diurnal variation in CO_2 concentration in a vertical section of crop canopy between day and night. (b) Diurnal variation in CO_2 concentration directly above a canopy in winter and summer. Net uptake of CO_2 during the day (due to photosynthesis) and net efflux of CO_2 at night (due to respiration) produce significant spatial and temporal variation in CO_2 concentration around crop canopies.

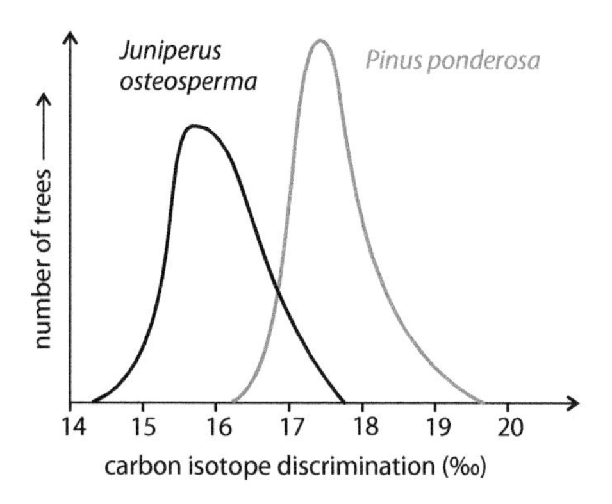

Figure 3.3. Carbon isotope discrimination and water availability. During the production of biomass, trees that inhabit arid environments, such as *Juniperus osteosperma*, discriminate less between ^{12}C and ^{13}C than do those from semi-arid environments, such as *Pinus ponderosa*, because their photosynthesis is more CO_2 limited. This is because their morphology is adapted primarily for conserving water rather than for supplying CO_2 to chloroplasts, which are therefore often exposed to quite low CO_2 concentrations. (Data from Ehleringer JR & Cerling TE [1995] *Tree Physiol* 15:105–111.)

BOX 3.2. CARBON ISOTOPES AND CO_2 FLUXES

About 1% of carbon atoms are the ^{13}C isotope. The fixation of CO_2 is catalyzed by the enzyme rubisco. Fixation of $^{12}CO_2$ is slightly more rapid than that of $^{13}CO_2$ because it is lighter and diffuses to, and is processed at, the fixation site at a slightly higher rate. This also produces glucose molecules with different $^{12}C/^{13}C$ ratios on individual carbons, namely the fourth and sixth carbon atoms, both of which tend to be enriched in ^{13}C. These carbon atoms are the origin of the carbon in CO_2 from respiration, which is therefore enriched in ^{13}C. Other biotic and abiotic processes can affect the $^{12}C/^{13}C$ ratio, but it is still a very useful tracer of carbon fluxes in cells, plants, and canopies.

C fixation by rubisco depletes CO_2 of ^{12}C, enriching the remaining CO_2 with ^{13}C, an effect that is enhanced by respiration preferentially producing $^{13}CO_2$ (Box 3.2). The $^{13}C/^{12}C$ ratio of CO_2 can therefore be used to gain a better understanding of fluxes of CO_2 in canopies. In whole leaves it is a summation of the CO_2 and H_2O regimes experienced, and a 1‰ change approximates to a 15 ppm shift in intercellular CO_2 concentration.$^{13}C/^{12}C$ ratios confirm that plants in drier locations can have significantly lower intercellular CO_2 concentrations (Figure 3.3). $^{13}C/^{12}C$ ratios can also differ significantly in deep canopies, tending to become lower as height up the canopy increases. This indicates that at the top of the canopy C fixation can be CO_2 limited, reflecting the more pronounced hydraulic challenges of the upper canopy. Plants are therefore adapted to variation in CO_2 concentrations, but water availability in particular has a significant effect on the ability of plants to maintain CO_2 supply to the chloroplasts. Agricultural canopies and anthropogenic changes in environmental variables, such as water, that affect CO_2 uptake exacerbate the variation in CO_2 concentration to which chloroplasts are exposed.

In many plant species, during the development of a leaf the stomatal density is affected by ambient CO_2 concentrations, with higher CO_2 levels inducing fewer stomata to develop per unit area. This effect underpins an important reconstruction of long-term atmospheric CO_2 concentrations (Box 3.3). Atmospheric CO_2 concentrations are currently low compared with most that have previously existed, including those that prevailed during the development of terrestrial ecosystems (Figure 3.4). On geological timescales, low CO_2 concentrations often coincide with episodes of glaciation. During the Pleistocene glaciation (2.6 million to 11,500 years ago), at glacial maxima CO_2 concentrations were as low as 190 ppm, and during interglacial periods were up to 270 ppm. For most of the **Holocene** (from 11,500 years ago until today) the increase in CO_2 concentrations was characteristic of an interglacial, and atmospheric CO_2 concentrations were fairly stable at about 270 ppm for several centuries up until the Industrial Revolution in the eighteenth century. Since then they have been increasing progressively more rapidly, primarily because of additions of ancient CO_2 produced from burning fossil fuels. Contemporary CO_2 concentrations of about 400 ppm are higher than

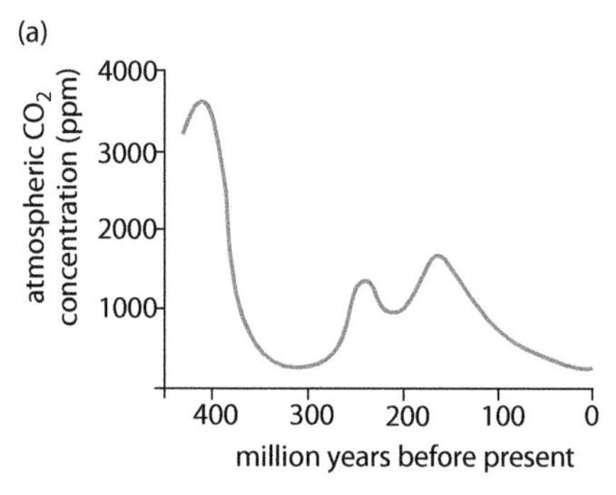

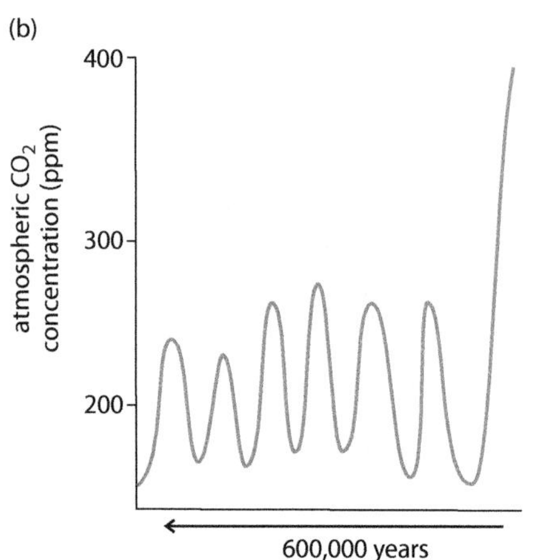

Figure 3.4. Changes in atmospheric CO_2 concentration during the existence of land plants. Atmospheric CO_2 concentrations (a) since multicellular plants colonized the land and (b) during the recent Quaternary ice ages. Stomatal indices on fossil leaves are a primary source for (a) and Antarctic ice cores are a primary source for (b). In general, the periods of low CO_2 concentration occur during significant global glaciations, and the periods of high CO_2 concentration during hot periods. (From Hetherington AM & Raven JA [2005] *Curr Biol* 15:R406–R410. With permission from Elsevier.)

BOX 3.3. STOMATA AS CO_2 MONITORS

For many plant species, herbarium specimens from before the recent rise in atmospheric CO_2 concentration have significantly more stomata on their leaves than specimens collected at the present time from the same locations. This effect can be reproduced in the greenhouse by growing plants at different CO_2 concentrations. It is now widely recognized that, for a significant number of species at least, the CO_2 concentration under which leaf development occurs affects the number of stomata that are formed. Stomatal density (SD)—that is, the number of stomata per unit leaf area—is related to CO_2 concentration, but epidermal cell density (ED), which is not affected by CO_2 concentration, differs between species and constrains the variation in SD. Stomatal index (SI), measured as $SI = SD \times 100/(SD + ED)$, is generally found to correlate more closely with CO_2 concentration than does SD. Many studies have reported decreasing SI over the last few centuries and decades. Some studies have shown decreasing SI on an individual tree over recent decades. Decreasing SI has led to a reduction in leaf stomatal conductance in various ecosystems, providing a link between CO_2 concentrations and changes in ecosystem processes.

In peat bogs at high latitudes, macroscopic remains of plants enable SIs to be used to chart CO_2 concentrations over many thousands of years. Using extant trees that have long fossil records it is possible to infer CO_2 concentrations over millions of years. Species of *Ginkgo* have very distinctive leaves (Figure 1 a, b) that are recognizable in the fossil record going back almost 300 million years. *G. biloba,* the one extant species, is native to China and now only grows in cultivation. The relationship between CO_2 concentration and SI in *Ginkgo* has been widely established, and fossil leaves have been used to reconstruct the CO_2 record with the longest time span (Figure 1c). Another living fossil tree, *Metasequoia glyptostroboides* (dawn redwood) (Figure 1 d, e), is also widely used to reconstruct ancient CO_2 concentrations. *Metasequoia* fossils were known and the species was assumed to be extinct before, in one of the major botanical finds of the twentieth century, "living fossil" *Metasequoia* were discovered in China in the late 1940s.

Figure 1. Plant fossils and reconstruction of past CO_2 concentrations. (a, b) *Ginkgo biloba* has numerous fossil relatives, all of which have very distinctly characteristic leaves. (c) Stomatal index data from *Ginkgo* have been used to reconstruct the CO_2 record. (d, e) Other trees with long fossil records, such as *Metasequoia glyptostroboides*, can also be used for this purpose. (From Retallack GJ [2001] *Nature* 411:287–290. With permission from Macmillan Publishers Ltd.)

have existed for many millions of years on Earth. Humans evolved, and the terrestrial ecosystems on which they are dependent developed, under CO_2 concentrations much lower than those that are probably now inevitable by the end of the twenty-first century and beyond. On geological timescales, primary production of biomass has occurred during great variation in CO_2 levels, with terrestrial plants probably adapting to the variation in CO_2 concentration. However, primary production in ecosystems of the past was very different to that of today, in significant part because of different atmospheric CO_2 concentrations. Geological changes in CO_2 concentrations also occurred much more gradually than the current anthropogenic increase.

Human activity is therefore exacerbating significantly the Holocene increase in CO_2 levels, with potentially global consequences for the primary production of biomass. The existence in plants of some capacity to adapt to changing CO_2 concentrations increases the complexity of predicting the consequences of increased atmospheric CO_2 concentrations, and the legacy of the prolonged period of low CO_2 levels for the Earth's terrestrial ecosystems is a vital context for predicting their response to variable CO_2 supply under rapidly increasing CO_2 concentrations.

The regulation of rubisco activity controls CO_2 entry into the Calvin–Benson cycle

The oxidation state of C in CO_2 is +4, whereas in most organic molecules it is 0. The fixation of C therefore requires four reducing equivalents. There are six known pathways by which autotrophs can use reducing equivalents to produce organic molecules from CO_2. An example is the reductive tricarboxylic acid (TCA) cycle—that is, the formation of acetyl CoA from $2CO_2$ by reversal of the oxidative TCA cycle. This occurs in green sulfur bacteria and other anaerobes. However, it is the Calvin–Benson cycle that dominates C fixation, especially on land. Fixation via the Calvin–Benson cycle occurs when the **pentose** ribulose-1,5-bisphosphate (RuBP) undergoes **carboxylation** to form **phosphoglycerate** in a reaction catalyzed by rubisco. The **phosphorylation** of three ribulose-5-phosphate molecules and the production of six glyceraldehyde-3-bisphosphate molecules, five of which are used to regenerate RuBP and one to synthesize carbohydrates, requires nine ATP molecules and six NADPH molecules (Figure 3.5). These are supplied by the light-dependent reactions of photosynthesis. The enzymes of the Calvin–Benson cycle are subject to circadian regulation by cellular redox via the effects of light on thioredoxin. Rubisco is the most abundant protein on Earth, constituting up to 30% of the protein in

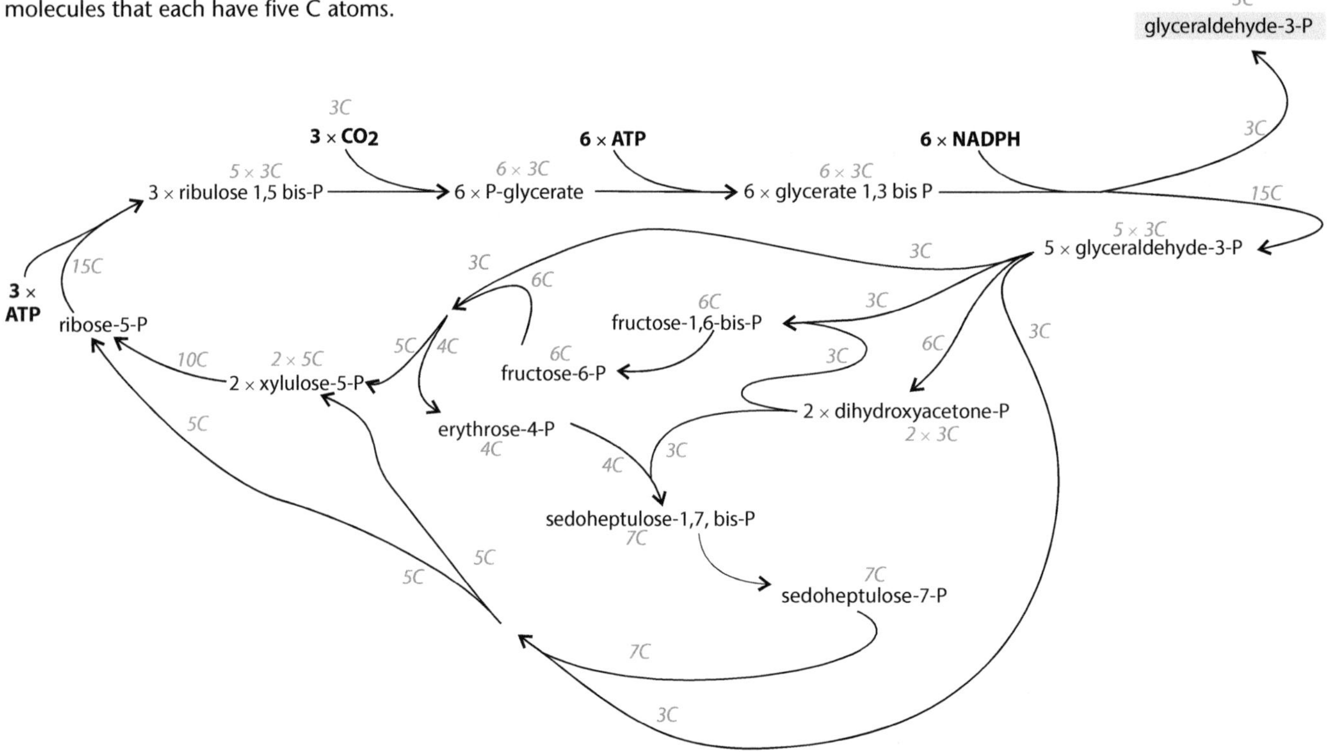

Figure 3.5. The Calvin–Benson cycle. The net fixation of three CO_2 molecules to three ribulose-1,5-bisphosphate molecules and reduction of the products using six ATP molecules and six NADPH molecules produces six molecules of glyceraldehyde-3-phosphate. One in every six molecules of glyceraldehyde-3-phosphate enters the glucose synthesis pathways, which require two molecules of glyceraldehyde-3-phosphate to synthesize each molecule of glucose. The five glyceraldehyde-3-phosphate molecules have 15 C atoms in five 3-C molecules, so a shuffling of carbon atoms and an active phosphorylation that uses three ATP molecules is necessary to regenerate three ribulose-1,5-bisphosphate molecules that each have five C atoms.

leaves, so its activity, together with that of **phosphoribulokinase,** which regenerates ribulose-1,5- bisphosphate, helps to control not only the global C cycle but also the global cycling of N and H_2O. The total activity of rubisco is adapted to short-term fluctuations in CO_2 concentration, and controls C fixation. Increased yields in crop plants have not generally been achieved by increased CO_2-fixation efficiency, so there is significant interest in manipulating not only the activity of rubisco but also its efficiency, in order to increase the productivity of food, forage, and biofuel crops.

In land plants, rubisco is assembled and located in the chloroplast stroma. It is a **hexadecamer** with eight large subunits (LSUs) coded by *rbc*L in the chloroplast genome and eight small subunits (SSUs) encoded by *rbc*S in the nucleus. The LSUs occur as four dimers, each with oppositely orientated partner molecules, with SSUs on the top and bottom (Figure 3.6). Rubisco is primarily active in CO_2 fixation only when sufficient PAR is available, and in general the amount of rubisco in leaves correlates with photosynthetic rate—for example, in trees the expression of rubisco has been shown to be greater at the top of the canopy than at the bottom. The relative rates of biosynthesis and degradation of rubisco are affected by numerous translational and post-translational processes, but there are also many effector molecules, including **chaperones, activators**, and **inhibitors**, that determine its activity. The chaperonins GroEL/GroES and RbcX2 control the assembly of the large subunits of rubisco. Each large subunit has two structural domains, with the large C-terminal domain having an α/β barrel in which the active site is located.

The active site of rubisco (Figure 3.6d) includes loops from the C-terminal domain and from the N-terminal domain of the partner molecule. Rubisco has to form a **carbamate** at a lysine residue in the active site and have Mg^{2+} present at this site to be able to fix CO_2 (Box 3.4). RuBP first interacts with rubisco to produce an **enediol** intermediate form, and if CO_2 is present a loop from the C-terminal domain moves to join with one from the N-terminal domain of the partner molecule, closing the site. Completion of the reaction between rubisco, RuBP, and CO_2 opens the site, releasing two phosphoglycerate molecules. The SSUs probably help to hold CO_2 before it diffuses to the active sites on the LSUs. There are a number of sugar phosphates that act as inhibitors of rubisco by binding to the active site, either before or after carbamylation, closing it and inactivating the enzyme. Inactivation by inhibitors is regulated by, for example, light intensity. The best-known inhibitor is carboxyarabinitol 1-phosphate (CA1P). The *rca* mutant of *Arabidopsis* cannot fix CO_2 at atmospheric concentrations because it lacks the enzyme rubisco activase that promotes the removal of sugar-phosphate inhibitors from the active site of rubisco. Rubisco activase is one of the AAA^+-ATPases—a family of proteins that actively regulate the conformation, assembly, and disassembly of a variety of proteins. This enzyme provides an important regulatory link between the light reactions and CO_2 fixation because its activity correlates with illumination, the ATP/ADP ratio and redox in the stroma, and electron transport through PSI. Rubisco activase is also notably sensitive to high temperatures, leading to a down-regulation of C fixation.

Models of the Calvin–Benson cycle and the delivery of substrates for carbohydrate metabolism are an important focus of systems biology. Many of these models are directly constructed using the kinetics of rubisco-mediated

(a)

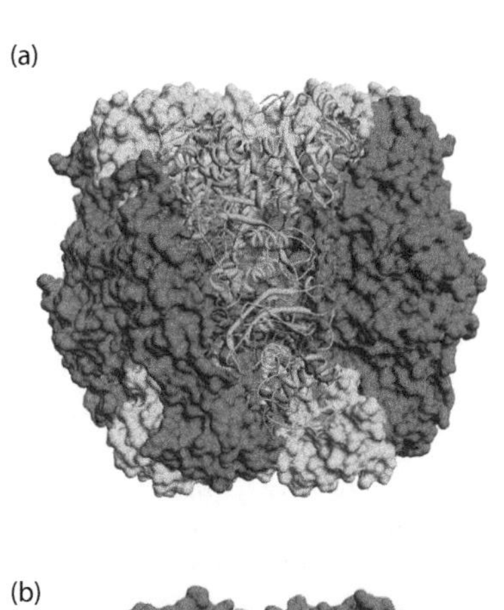

(b)

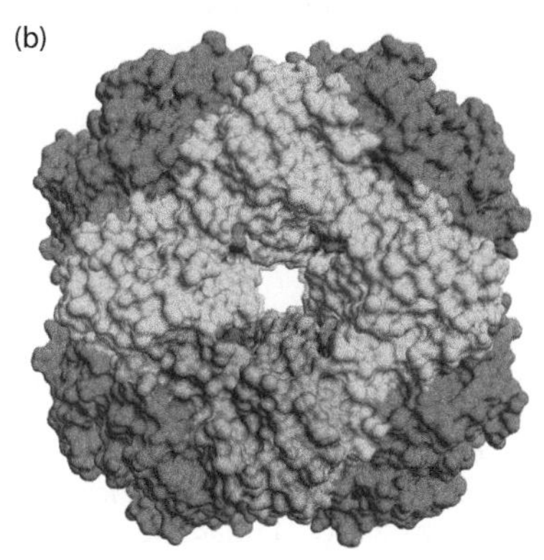

(c)

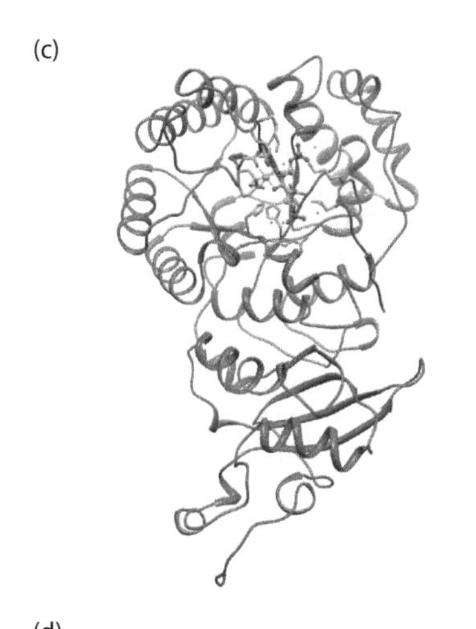

(d)

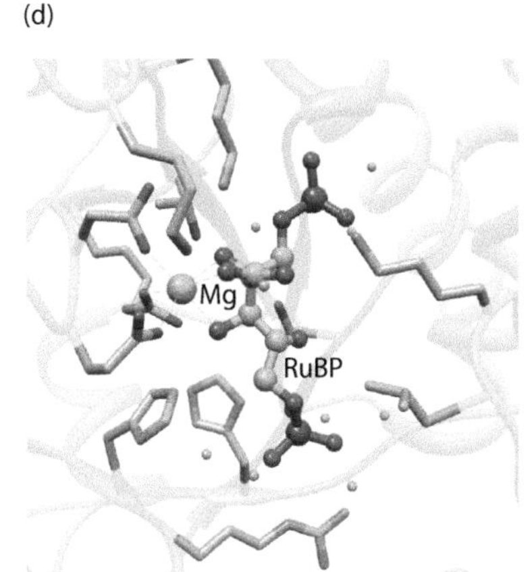

Figure 3.6. The structure of rubisco. Rubisco has eight large subunits (LSUs) and eight small subunits (SSUs), which each occur as four dimers. (a) Side view with one LSU visualized as a ribbon structure to show α-helices and β-sheets. (b) Top view with one LSU highlighted. (c) Ribbon structure of LSU. (d) Close-up of active site with RuBP. (Data for spinach rubisco visualized using Chimera software.)

BOX 3.4. CARBAMATE FORMATION AND RUBISCO

Carbamates are formed when CO_2 reacts with an amine. Rubisco has a lysine residue that must form a carbamate before it can become involved in CO_2 fixation. The carbamated rubisco is usually inactive, but only carbamated rubisco molecules can be activated by rubisco activase. The CO_2 molecule that forms the carbamate is not the same CO_2 molecule that is fixed.

CO_2 fixation, and many others quantify this process indirectly. Such models can be used to investigate the behavior of the Calvin–Benson cycle under varying conditions, and are an important aid to understanding the response of plants to variation in CO_2 supply and to the manipulation of CO_2 fixation. Together with previous research, these models indicate that for an enzyme of such importance, with at least 2 billion years of evolutionary history, rubisco does not seem to be very efficient. Meta-analyses of enzyme activities indicate that rubisco is not as unusual in this respect as is often assumed, and that together with the majority of enzymes it is in fact moderately efficient (Figure 3.7). This level of efficiency of most enzymes seems to arise from either moderate selection pressure for high efficiency (so enzymes of secondary metabolism are about 30 times slower than those of primary metabolism), or limits to efficiency because of substrate properties, especially low molecular mass. The variants of rubisco that occur in different photosynthetic organisms indicate that there has been selection for efficiency over the long term. However, CO_2 has low molecular mass and a relatively uniform topology, so rubisco in land plants probably has moderate efficiency in part because of its challenging substrate. Overall, adaptation of CO_2 fixation to short-term environmental variation in CO_2 concentrations therefore occurs primarily via spatial and temporal variation in expression and activation of rubisco.

Oxygenation of RuBP decreases growth but provides rapid metabolic flexibility

The photosynthetic organisms from which all land plants are descended evolved to fix CO_2 in the Earth's early atmosphere, which had high CO_2 levels and was almost devoid of O_2. The global scale of both CO_2 fixation and O_2 production by oxygenic photoautotrophs contributed significantly to changing the Earth's atmosphere to its present composition, low in CO_2 and high in O_2. Not only is CO_2 a challenging substrate for an enzyme, but also O_2 can enter rubisco's active site and be added to RuBP. This **oxygenase** activity produces **phosphoglycolate**, a 2-carbon molecule that is not only incapable of entering the Calvin–Benson cycle, but also inhibits many of its enzymes. The carboxylase activity of rubisco is greater than its oxygenase activity, but overall fixation depends on the CO_2:O_2 ratio at the active site. CO_2 is more soluble in water than O_2 at moderate Earth surface temperatures, so the approximately 500-fold higher O_2 than CO_2 concentrations in the current atmosphere only result in carboxylase:oxygenase activity of about 4:1 in most temperate plants. As the temperature increases, the solubility of CO_2 decreases more rapidly than that of O_2, so in warmer regions of Earth (above 25°C) the carboxylase:oxygenase activity can decrease to 2:1. During

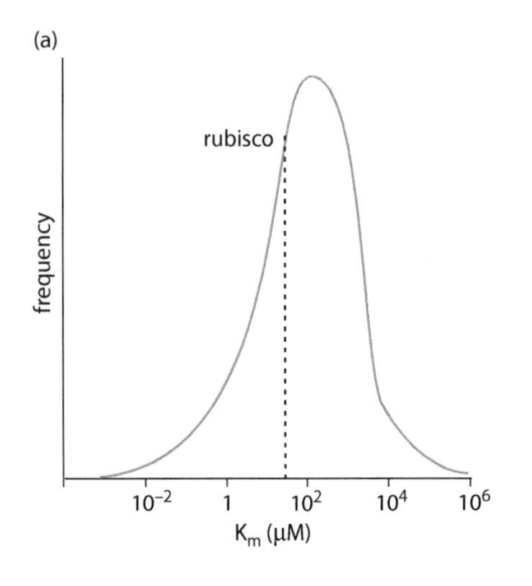

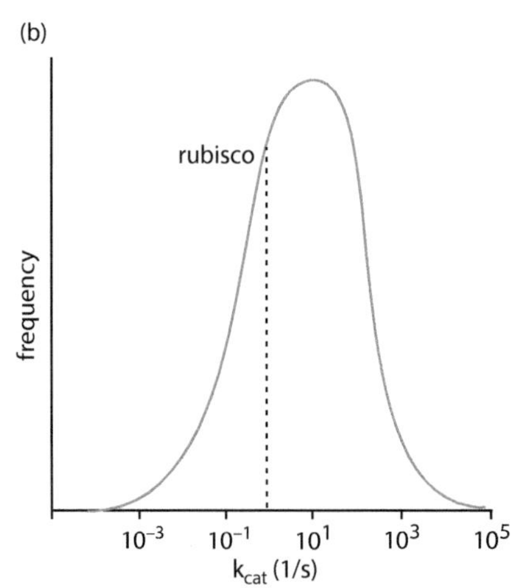

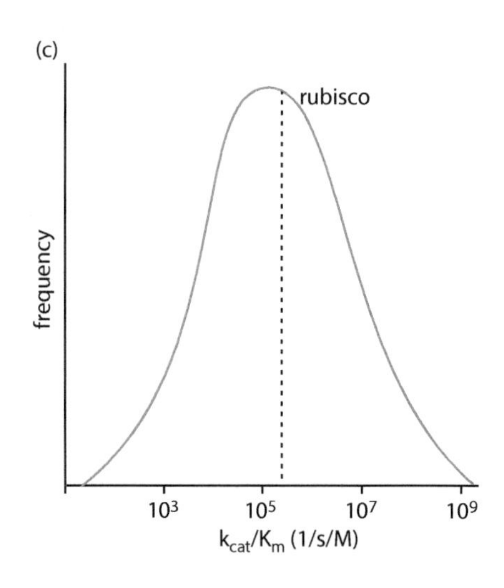

Figure 3.7. **The kinetic properties of rubisco.** Histograms for K_m, k_{cat}, and k_{cat}/K_m are shown for numerous enzymes in biological systems. (a) The Michaelis–Menten constant (K_m) quantifies the affinity of an enzyme (n = 5194 different enzymes). (b) The k_{cat} is the number of reactions that each molecule of enzyme catalyzes per second—that is, the turnover rate (n = 1942). (c) The k_{cat}/K_m ratio is a measure of enzyme efficiency (n = 1882). Enzymes of secondary metabolism have significantly lower efficiencies than those of primary metabolism. Rubisco has moderate efficiency, which is lower than that of many other key enzymes of primary metabolism. (From Bar-Even A, Noor E, Savir Y et al [2011] *Biochemistry* 50:4402–4410. With permission from ACS Publications.)

dry periods in many environments, the closure of stomata limits water loss, resulting in decreased CO_2 influx and decreased O_2 efflux, favoring oxygenase activity. In the current atmosphere, the oxygenase activity of rubisco can have significant adverse effects on primary productivity.

The inhibition of Calvin–Benson cycle enzymes by phosphoglycolate probably helped to drive the evolution of pathways to utilize it, but they also serve to recycle 75% of the carbon atoms of phosphoglycolate, enable some flexibility in C and N reduction, and participate in redox homeostasis. They consume O_2 and produce CO_2—oxygen fixation thus drives the production of CO_2, prompting the use of the general term "photorespiration" for the process. The generation of 3C phosphoglycerate for insertion into the Calvin–Benson cycle from 2C phosphoglycolate also releases NH_4^+ (Figure 3.8). Natural variation in photorespiration and the phenotypes of numerous photorespiration mutants suggest that, despite its adverse effects, it is useful to plants in the current atmosphere. This is in part because, under high light and numerous other stressful conditions, photorespiration provides an overspill for excess reducing equivalents from the light reactions. Photorespiration has probably contributed significantly to driving the recent evolution of C_3 plants. Stomatal and mesophyll conductance to CO_2, adaptations to minimize leaf temperature, and the evolution of rubisco in many species have probably helped to decrease photorespiration. A number of C_3 plants have also been shown to release CO_2 from glycine in particular compartments in which it can be scavenged, and/or to have chloroplasts in cell peripheries surrounding the mitochondria from which photorespiratory CO_2 is released.

The variation in CO_2 concentrations to which plants are exposed, together with stressful conditions that decrease plant capacity to take in CO_2, mean that many plants are intermittently exposed to conditions that produce a low CO_2:O_2 ratio in intercellular spaces and chloroplasts. At low CO_2:O_2 ratios, the high reductant capacity necessary to achieve high CO_2-fixation rates becomes a liability. Under conditions of varying CO_2 concentration, the ability to metabolize phosphoglycolate provides, when the CO_2:O_2 ratio is

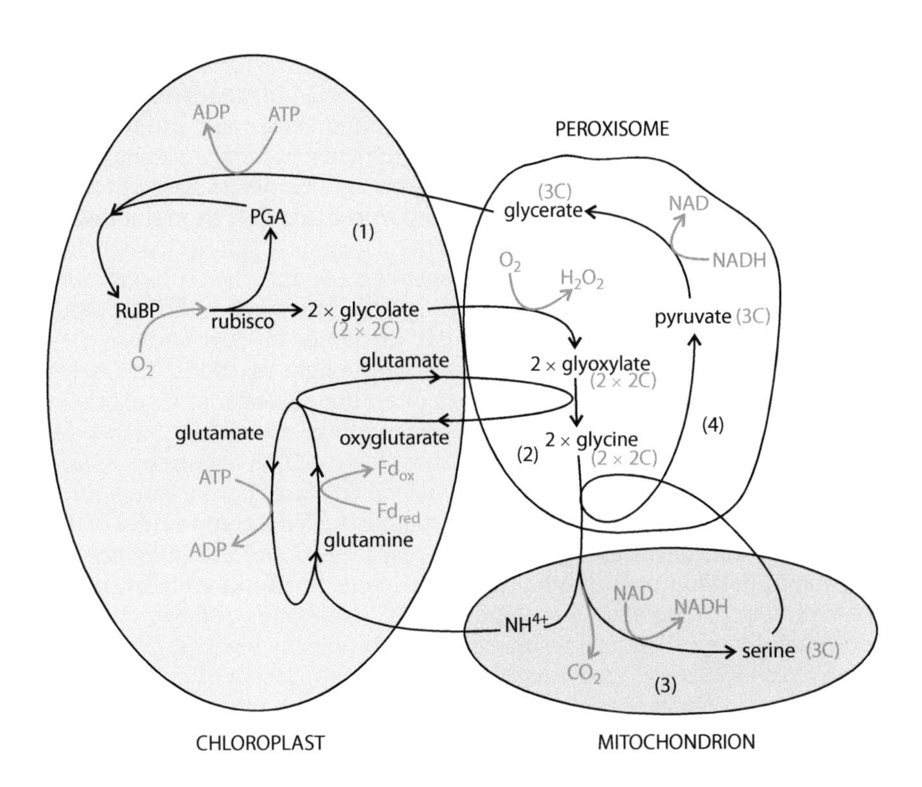

Figure 3.8. The photorespiratory pathway. The fixation of O_2 produces phosphoglycolate, which is dephosphorylated (not shown) to glycolate. (1) Glycolate is transported into the peroxisome, where glycolate oxidase catalyzes the production of glyoxylate and H_2O_2. Peroxisomal catalase then breaks down the H_2O_2. (2) Glyoxylate is converted into the amino acid glycine via aminotransferases linked to the N-assimilation pathways of the chloroplast. (3) The glycine decarboxylase complex decarboxylates and deaminates glycine, resulting in the production of CO_2, NH_4^+, NADH, and methyltetrahydrofolate (not shown), which is used in the synthesis of serine. (4) Serine is transported into the peroxisome, where aminotransferases and hydroxypyruvate reductase catalyze the synthesis of glycerate. This is transferred back into the chloroplast, phosphorylated, and then enters the Calvin–Benson cycle. Overall, two molecules of glycolate (2 × 2C) and two ATP molecules are used to synthesize one 3C glycerate molecule, with the loss of one C atom.

lowest, an outlet for the high reducing capacity of the light reactions necessary to take advantage of the highest $CO_2:O_2$ ratio. The chronic relatively low atmospheric $CO_2:O_2$ ratio of the last 50 million years or so has favored photorespiration, which occurs in all groups of oxygenic autotrophs, suggesting that they are using it to help them to adapt to such conditions.

When there is a sustained low CO_2 supply, C_4 plants maintain a high $CO_2:O_2$ ratio in the vicinity of rubisco

Under chronically hot conditions with a low $CO_2:O_2$ ratio, the advantageous short-term flexibility that the photorespiratory pathway provides can be outweighed by the losses that it causes, so many plants which inhabit such conditions have a carbon-concentrating mechanism that increases the $CO_2:O_2$ ratio in the vicinity of rubisco. Such plants have in effect simulated the atmosphere of the early Earth for their rubisco. ^{14}C-labeling experiments with sugar cane in the 1950s and 1960s showed that in some plants the first product of C fixation was 4C malic acid, rather than 3C phosphoglycerate. It is now estimated that almost 8000 species of angiosperm are "C_4" plants. The majority of these are monocots in the Poaceae (4600 species, representing almost 50% of grass species) and Cyperaceae (1600 species, representing about 25% of sedge species), with about 10% of monocots having the pathway. About 1600 species in 16 families of eudicots (that is, 0.5% of eudicots) are C_4. The majority of these are in the Chenopodiaceae, Amaranthaceae, Euphorbiaceae, and Asteraceae. C_4 plants are generally characteristic of environments that are hot, dry, and light, the exemplar being the tropical and subtropical "C_4 grasslands" in which more than 90% of the grasses can be C_4 (and in which hominids largely evolved). There is evidence that the prevalence of C_4 plants is increasing in some parts of the world, probably because of increasingly hot dry climates. Sugar cane and maize are both C_4 plants and often provide, by weight, more than 50% of the yield of the world's major crops. In drier areas of the world, *Sorghum bicolor* (sorghum) and *Pennisetum glaucum* (pearl millet) are the most important cereals, and are also C_4 plants. Although only about 4% of angiosperms and no forest trees are C_4 plants, and the pathway does not seem to exist in other land plants, C_4 plants probably produce about 30% of terrestrial biomass.

C_4 photosynthesis is **polyphyletic**, having evolved independently over 60 times. It first appeared in monocots over 50 million years ago, and in dicots about 30 million years ago, but came to prominence in the grasslands of the late **Miocene** between 7 and 4 million years ago. C_4 plants have very low photorespiration rates because they pump to rubisco CO_2 from the decarboxylation of C_4 acids, maintaining high $CO_2:O_2$ ratios. C_4 plants have higher N and water use efficiency than C_3 plants because they use rubisco more efficiently and lose less water in obtaining CO_2. In cold conditions, photoinhibition rates in C_4 plants can be high, in significant part because they have no overspill for excess reductants from the light reactions. This helps to explain the reduced prevalence, and often the absence, of C_4 plants in temperate and cold ecosystems. The origin and expansion of C_4 plants are generally explained by decreases in atmospheric CO_2 concentrations, and a threshold of CO_2 concentration below which C_4 plants have a competitive advantage over C_3 plants is commonly hypothesized. The complexities of the advantages and disadvantages of C_4 photosynthesis, and the influence on ecosystem development of other factors, such as fire and herbivory, make a full understanding of the development of C_4 grasslands difficult, but they are generally regarded as a response to the chronically low $CO_2:O_2$ ratio in Earth's recent history. Predicting the fundamental process of primary productivity in unmanaged ecosystems, and perhaps manipulating aspects of C_4 into C_3 crops to increase food production in managed ecosystems, depends

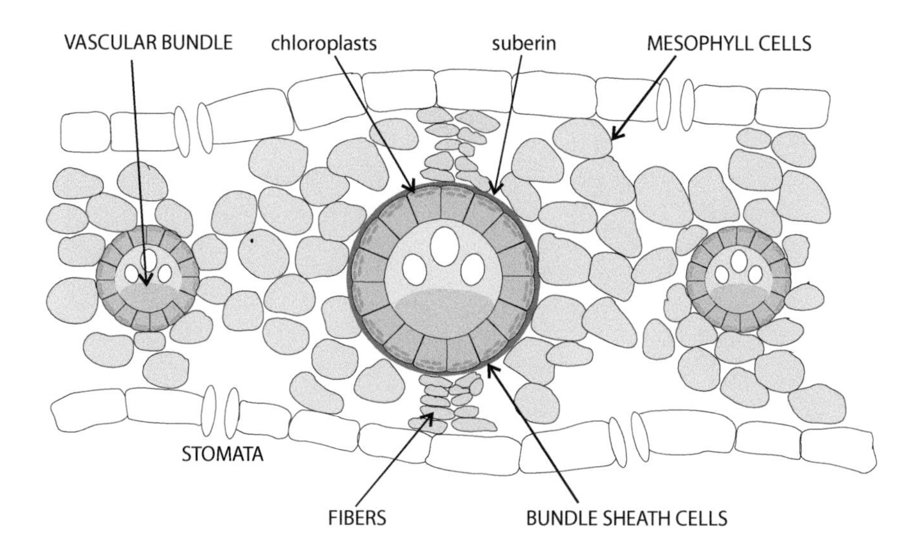

Figure 3.9. Generalized Kranz anatomy of a C_4 leaf. The vascular bundles are surrounded by a wreath ("Kranz") of bundle sheath (BS) cells. The mesophyll cells are loosely packed and, in contrast to the anatomy of most C_3 leaves, are not divided into adaxial palisade and abaxial spongy mesophyll. In some C_4 species there is a suberin layer around the BS and the chloroplasts are located centrifugally (as shown), whereas in others there is no suberin layer and the chloroplasts are located centripetally (that is, closely packed around the wall of the BS cells closest to the vascular tissue). The chloroplasts of BS cells in some species can be distinct and have very few grana compared with those of mesophyll cells. In C_4 species such as maize, fibrous tissue suspending the major vascular bundles divides the leaf into compartments that can have different CO_2 regimes. Stomata in C_4 leaves are often located on both the adaxial and abaxial surfaces.

on predicting the response of C_4 photosynthesis to current simultaneous changes in CO_2, temperature, water, and N regimes in the Earth's ecosystems.

Most C_4 plants have a characteristic leaf anatomy, generally with no distinction between palisade and spongy mesophyll. The internal path lengths of light are short, which probably in part explains the prominence of C_4 plants in high light conditions. Vascular bundles are numerous and prominent, located midway between the adaxial and abaxial surfaces, and with major veins surrounded by a tightly packed "wreath" of cells (Figure 3.9). These C_4 "bundle-sheath" (BS) cells are anatomically distinct from the mesophyll cells, and have chloroplasts that tend to be large, agranal, and located centripetally. Through their outer walls, BS cells often have numerous plasmodesmatal connections to the nearest mesophyll cells. Their outer walls are also sometimes suberized. However, C_4 anatomy is variable, and plants are known that use C_4 photosynthesis but do not have the characteristic "Kranz" (meaning "wreath") anatomy. In a few plant species, C_4 photosynthesis occurs not in two cell types, but in two chloroplast types within a single cell. In hot conditions, which mean that dissolved gases diffusing to the stroma have a particularly low CO_2:O_2 ratio, the coexistence of photosystem II (PSII) complexes (which produce O_2) and rubisco (which has oxygenase activity) in the same chloroplasts presents a significant challenge to C fixation. C_4 plants have adapted to this by first fixing C in mesophyll cells using **phosphoenolpyruvate carboxylase** (PEPc), which has no oxygenase activity, and has a higher affinity for its substrate, bicarbonate (HCO_3^-), than rubisco has for its substrate (CO_2). The 4C acids produced are then actively transported into the BS cells, where they are decarboxylated to release CO_2 for rubisco. In all C_4 plants, **carbonic anhydrase** is used in the mesophyll cells to enhance the dissolution of CO_2 to HCO_3^-, increasing the rate of C fixation by PEPc. High CO_2 concentrations are thus maintained in the BS cells, which are the primary site of C entry into the Calvin–Benson cycle via rubisco (Figure 3.10), or in the CO_2-fixing chloroplasts of single-celled C_4 plants (Figure 3.11). The agranal chloroplasts have few PSII complexes, significantly decreasing the O_2 concentrations in BS cells. Variants of the C_4 pathway exist, and there are C_3–C_4 intermediates, some of which are constitutively C_3–C_4 and some of which are flexible in their use of C_3 or C_4 pathways. In all of these variants, PEPc is used for the initial fixation of C, but the 4C compounds that are pumped into the BS cells, the pathway of decarboxylation, and the cell types in which it occurs vary (Figure 3.12). These variants and intermediates provide unique insights into the role of C_4 photosynthesis in response to environmental variation, and into potential routes for manipulating aspects of C_4 into C_3 plants.

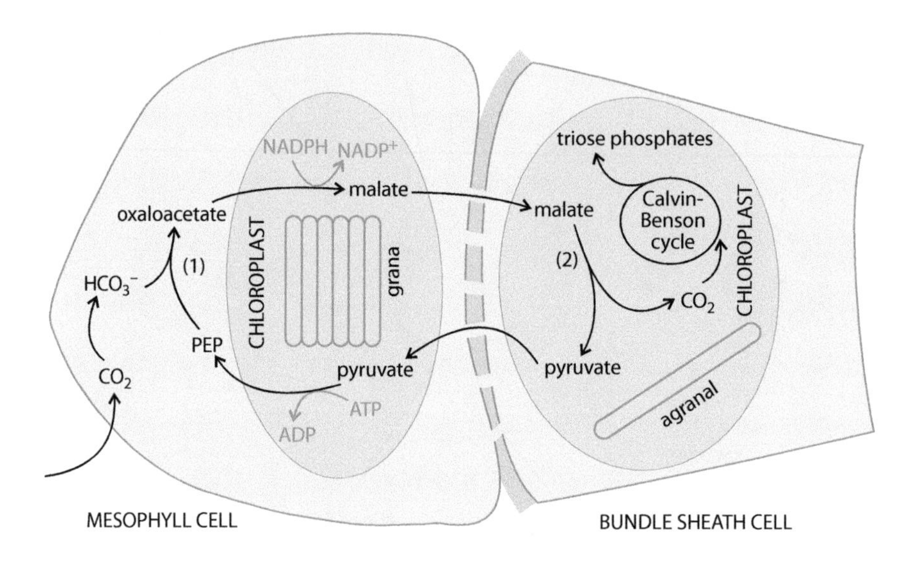

Figure 3.10. The physiology of CO_2 fixation in C_4 plants. CO_2 dissolves in the cytoplasm of the mesophyll cell to form HCO_3^-, a process that is enhanced by the action of carbonic anhydrase, which is an important enzyme for the initiation of C accumulation in many C_4 plants. (1) Phosphoenolpyruvate carboxylase (PEPc) catalyzes the fixation of HCO_3^- to PEP, producing a 4C oxaloacetate molecule. PEPc fixes HCO_3^- efficiently and has no oxygenase activity. In most C_4 plants, oxaloacetate is reduced to malate in the chloroplast and then transported into the bundle sheath cells. (2) Most commonly malate is then decarboxylated by $NADP^+$-dependent malic enzyme (NADP-ME) to release CO_2, which enters the Calvin–Benson cycle. In most C_4 plants, pyruvate is returned to the chloroplasts of the mesophyll cells, where it is phosphorylated to PEP and released into the cytoplasm. The pumping of malate into the BS cells and the release of CO_2 can produce very high concentrations of CO_2 in the BS cells, suppressing photorespiration. Many C_4 plants have suberin around the outside of the bundle sheath, which helps to minimize the escape of CO_2. Chloroplasts in BS cells often have few grana and few PSII complexes, thereby minimizing non-cyclic photophosphorylation and the generation of O_2 by photolysis.

In plants, PEPc evolved a role in the TCA cycle long before its role in C_4. The TCA cycle provides not only the electrons with which mitochondria generate ATP, but also some compounds from which essential metabolites are synthesized. The withdrawal of compounds for metabolism is potentially disrupting to the supply of electrons to mitochondria, so **anaplerotic** reactions are used to synthesize TCA cycle intermediates from alternative substrates and feed them back into the TCA cycle to maintain electron flow to mitochondria. In animal cells, pyruvate carboxylase is used to synthesize oxaloacetate directly from pyruvate to replenish oxaloacetate when it is limiting. In bacteria, yeast, and plant cells, oxaloacetate is replenished by PEPc. In response to chronic low CO_2:O_2 ratios it has proved easier to recreate the Earth's early atmosphere in BS cells by using an existing enzyme than to adapt rubisco to the current atmosphere. In significant part this is likely to have been not only

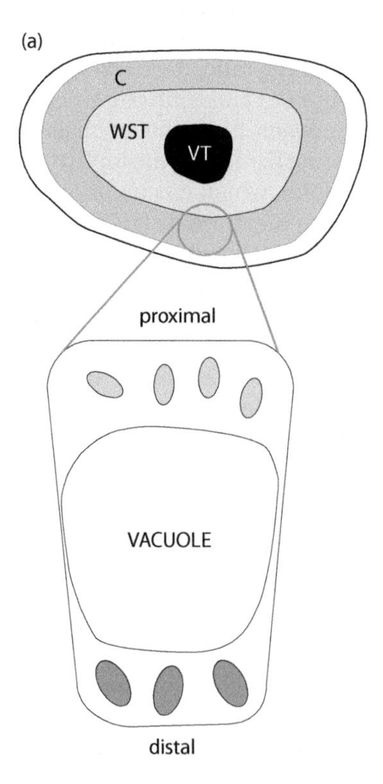

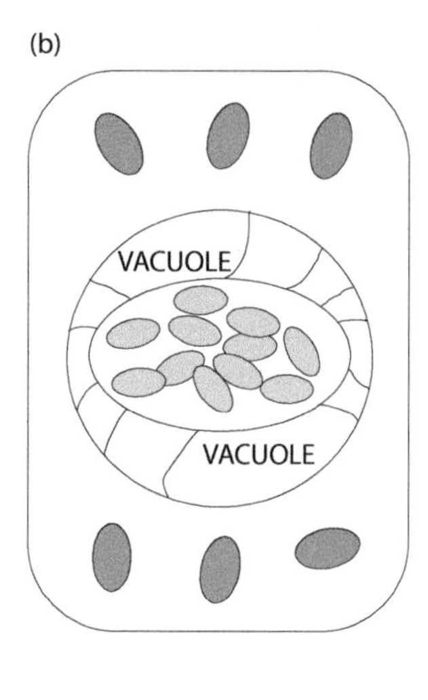

Figure 3.11. Single-celled C_4 plants. In a few plants, C_4 functions are compartmented within different chloroplasts of a single cell. In one chloroplast type, rubisco activity is suppressed and there is PEPc activity, while in the other the opposite occurs. (a) In *Suaeda aralocaspica,* chloroplasts in chlorenchyma cells (C) distal to the water storage tissue (WST) and vascular tissue (VT) fix CO_2 using PEPc (dark green chloroplasts in the figure), and it is transported to proximal chloroplasts for decarboxylation and fixation (light green chloroplasts). (b) In *Bienertia* species, decarboxylation and CO_2 fixation occur in chloroplasts in a separate but connected cytoplasmic compartment in the middle of the vacuole. (Data from Sharpe RM & Offerman S [2014] *Photosynth Res* 119:169–180.)

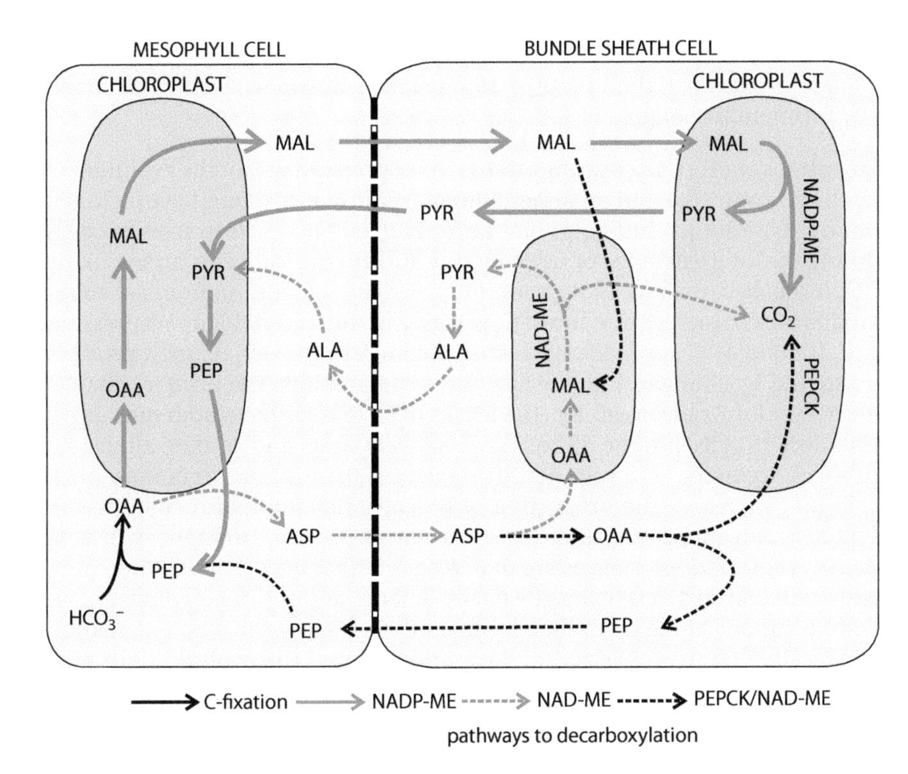

Figure 3.12. The variants of C_4 photosynthesis. All C_4 plants use PEP carboxylase for the initial fixation of C, but a variety of enzymes are used in decarboxylation to release CO_2 in the bundle sheath chloroplasts. C_4 plants are often divided into three types according to the enzyme used for decarboxylation, but the C_4 pathways probably do not fall as neatly into these three categories as this suggests. These different pathways occur in a wide variety of leaf anatomies, so across the many C_4 plants the C_4 syndrome probably defies very clear subdivision. The NADP-ME subtype is most common. It is generally associated with centrifugally located, agranal, low PSII chloroplasts, and often with suberized bundle sheath cells. The NAD-ME subtype is also common, but tends to have centripetally located granal chloroplasts. The PEPCK/NAD-ME subtype is least common, but is an interesting derivative of parts of the other two variants. OAA = oxaloacetate; MAL = malate; PYR = pyruvate; ALA = alanine; ASP = aspartate; PEP = phosphoenolpyruvate; NADP-ME = NADP-malic enzyme; NAD-ME = NAD-malic enzyme; PEPCK = phosphoenolpyruvate carboxykinase. Other enzymes and all transporters are not labeled.

because of rubisco's limits as an enzyme, but also because control of CO_2:O_2 ratios in the vicinity of rubisco is advantageous to meeting the challenges of variable CO_2 concentrations on all spatial and temporal scales.

C_3–C_4 intermediates and C_4 plants show distinct responses to chronic differences in the environment

The Poaceae is perhaps the most important plant family on Earth. It directly supplies at least 50% of calories consumed by humans, it provides grazing for a significant proportion of animals that have been domesticated by humans, and it dominates significant unmanaged ecosystems—and about half of its species are C_4 plants. Variation in atmospheric CO_2 concentrations, water availability, temperature, and light intensity contributed to natural selection for C_4 photosynthesis. An understanding of the role of these environmental variables in the evolution and ecology of C_4 plants is necessary to predict the effects of anthropogenic environmental change on some of Earth's most important ecosystems, because they are among the environmental variables that are being most significantly affected by human activities. During C fixation, rubisco discriminates significantly between $^{13}CO_2$ and $^{12}CO_2$, producing biomass that is depleted in ^{13}C ($\delta^{13}C$) by -28‰. PEPc also discriminates against ^{13}C, but to a lesser extent than rubisco does, and the full expression of C_4 photosynthesis under optimal conditions produces biomass with a $\delta^{13}C$ of -14‰. A $\delta^{13}C$ that is intermediate between those of rubisco and PEPc reflects intermediate use of C_3-C_4 photosynthesis, or sub-optimal conditions for C_4 photosynthesis. In order to operate efficiently, C_4 pathways have to balance with C_3 fixation of C. At low light intensities that cannot generate the extra ATP necessary for pumping 4C acids into the BS cells, there is significant leakage of CO_2 back into the mesophyll cells, which increases photorespiration and decreases $\delta^{13}C$ in C_4 plants. During the development of carbonaceous structures that form animal fossils, although there is discrimination between ^{13}C and ^{12}C during C assimilation, the $\delta^{13}C$ of dietary plants endures. The $\delta^{13}C$ of plant and animal fossils can therefore be used to infer the proportion of C_4 plants in animal diets and thus in vegetation.

The current environmental correlates of C_4 distribution together with other climate proxies can then be used to recreate the responses of C_4 plants to environmental changes.

Low atmospheric CO_2 concentrations were a prerequisite for the evolution of C_4 photosynthesis. Land ecosystems for most of their existence have included hot dry conditions, but C_4 photosynthesis does not seem to have evolved during the long existence of angiosperms until, in the Miocene, atmospheric CO_2 levels became low enough for it to be advantageous. Anatomical adaptations are characteristic of most C_4 plants, with full physiological expression of C_4 (that is, $\delta^{13}C$ approaching –14‰). An increase in vein density, possibly prompted by changed water availability, was probably the primary change necessary for Kranz anatomy. The leaves of C_3 grasses have both mesophyll cells and BS cells, but the BS cells are relatively small and contain almost no rubisco. An increase in vein density decreases the proportion of mesophyll cells and increases the proportion of BS cells, making their use in photosynthesis advantageous, and perhaps even necessary. Differences in vein density of *Arabidopsis* leaves are due to early differences in auxin transport, with both leaf surfaces being abaxial by default, and the development of an adaxial surface being controlled by auxin flow. Changes in auxin regulation can feasibly be hypothesized to have underpinned the evolution of Kranz anatomy. Leaf vein density is primarily controlled by the demands of leaf hydraulics—in a single tree, leaf vein density varies significantly with height and the associated challenges of water transport. It seems likely that during the evolution of C_4 pathways, low water availability played a significant role in selecting plants with increased leaf vein density triggered by changes in leaf auxin concentration.

In C_3 plants, about 2.5% of genes are expressed differently in mesophyll and BS cells, but in a C_4 plant such as maize the difference can be as much as 18%. The physiological changes necessary for the expression of C_4 physiology in Kranz anatomy depend on changes in enzyme and transporter expression in both mesophyll and BS cells. C_3–C_4 intermediates show a range of altered levels of these proteins, and the evidence indicates that they can occur quite frequently in some clades, particularly in those in which genome duplications have occurred. When such changes in expression occur in C_3–C_4 intermediates with weak or no expression of Kranz anatomy there is competition between C fixation via PEPc and rubisco, which would favor their anatomical separation. $\delta^{13}C$ values indicate that there are relatively few C_3–C_4 intermediate plants compared with C_4 plants, which suggests that the evolution from C_3–C_4 to C_4 is frequent and advantageous. In many C_3–C_4 intermediates the glycine produced by photorespiration is transported into BS cells, where it is decarboxylated in the mitochondria, and CO_2 is refixed by rubisco. Given that decreasing photorespiratory loss of CO_2 is the ultimate advantage of C_4 photosynthesis, changes in transport and organelle activity derived from its pathways are likely to have initiated changes in protein expression necessary for C_4 photosynthesis. The increased expression of PEPc in the surrounding mesophyll cells to scavenge CO_2 leaking out of the BS cells probably helped to make the use of PEPc advantageous. In C_3 plants, in vascular tissue the expression of NADP-malic enzyme (NADP-ME) and NAD-malic enzyme (NAD-ME) is higher than in other tissue, indicating that all plants can exert differential control over the enzymes used in C_4 photosynthesis. Genome expression studies suggest that up to 3% of all genes may show differential expression between C_3 and C_4 plants. This includes differences in protein and starch synthetic pathways, and in nitrogen metabolism. In most C_4 plants, rubisco is slightly different to that of C_3 plants, and has a higher k_{cat} but lower stability. The PEPc of C_4 plants is also slightly different, with a greater affinity for HCO_3^- compared with that in C_3 plants, and shows less allosteric inhibition by malate and aspartate.

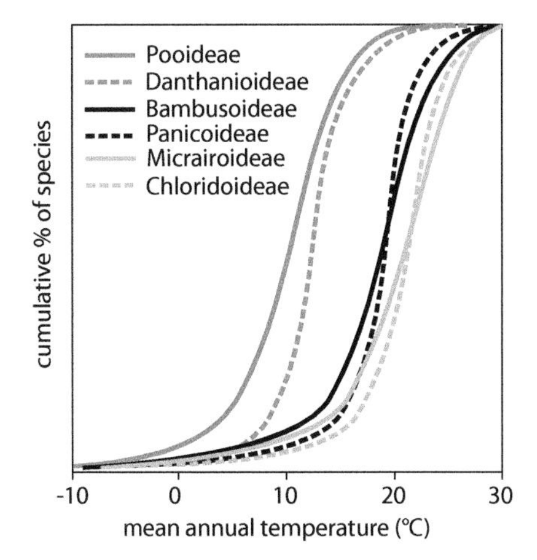

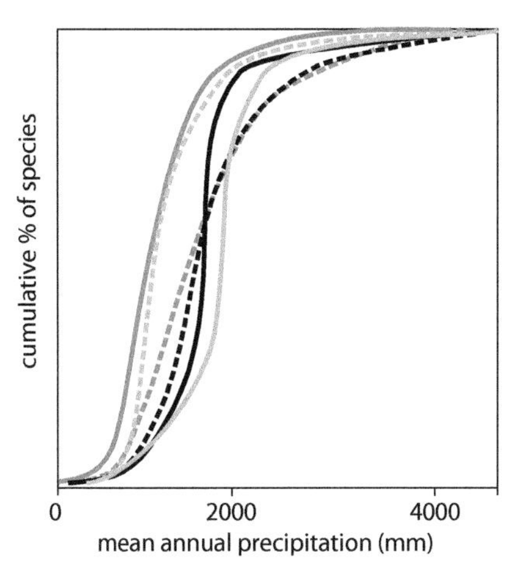

Figure 3.13. Grass tribes and environmental conditions. Compilations of the conditions under which grass species live suggest that most tribes, not just those with many C_4 species, are adapted to fairly hot temperatures and relatively low precipitation. The Pooideae inhabit predominantly cool temperate regions, but most other tribes, some but not all of which have C_4 members, inhabit predominantly warmer and drier conditions. Thus grasses generally, not just C_4 species, are adapted to warm dry environments. C_4 photosynthesis is certainly a major adaptation to warm dry conditions, but data such as those shown in the figure suggest that other environmental factors (for example, the lack of light in forest environments) might also have contributed to the evolution of C_4 photosynthesis in many grasses. (From Edwards EJ & Smith SA [2010] *Proc Natl Acad Sci USA* 107:2532–2537. With permission from the National Academy of Sciences. More than 1.5 million values from over 10,000 taxa were used in the data compilation.)

The increased efficiency of CO_2 fixation in C_4 plants increases their water use efficiency because stomata can be kept more closed, so less water is lost in gaining the necessary CO_2. The reservoir of 4C acids produced by PEPc also means that C_4 plants can photosynthesize for much longer after the stomata have closed than can C_3 plants. Overall, C_4 grasses inhabit environments with average precipitation of about 1200 mm, whereas for C_3 grasses the corresponding figure is 1800 mm. In many instances, C_4 photosynthesis is associated with particularly seasonal precipitation. Water stress is thus a significant selective force for C_4 photosynthesis, but temperature is also important, with enhanced photoinhibition at low temperature being disadvantageous. Light use efficiency is greater in C_4 plants, but the generation of the extra ATP necessary to drive the extra transport steps in NADP-ME (which is achieved mostly via increased cyclic photophosphorylation) is not possible at low light levels. The evolution of C_4 photosynthesis, especially in grasses, was therefore probably triggered, under warm conditions with low CO_2 levels, by the development of open, light habitats with precipitation of less than 1500 mm, the minimum value normally necessary for the development of a closed forest canopy under subtropical and tropical conditions. However, there are C_4 plants that have been a significant element of the flora of the Tibetan Plateau, for millennia at least. Here the high light levels and low partial pressures of CO_2 can still perhaps be advantageous for C_4 photosynthesis, despite the cold temperatures. All grass clades except the Pooideae occur mostly under hot conditions (Figure 3.13), so the transition to open habitats was perhaps more important than temperature per se in the evolution of C_4 grasses. C_4 plants occur with higher frequency in environments with many abiotic stressors (for example, salinity), and that are highly disturbed. At a given N availability, C_4 plants fix more CO_2 than C_3 plants because they devote only 10–15% of total leaf N to PEPc plus rubisco, whereas C_3 plants devote up to 30% of N to rubisco. Under N stress the pumping of CO_2 into BS cells and the N advantage of C_4 decline, but C_4 plants convert increasing N availability into increasing productivity more effectively than do C_3 plants. The numerous independently evolved C_3-C_4 intermediates and C_4 plants probably show many differences in the way that they react to chronic variation in numerous key environmental variables, but in general C_4 photosynthesis is a pathway for overcoming the effects of photorespiration under warm, low-CO_2 conditions that is enabled by high light conditions and that reduces the effects of abiotic stressors on productivity (Figure 3.14). The effects of anthropogenic environmental change on C_4 plants are difficult to predict. Increasing temperature is affecting the distribution of some C_4 plants, as might changes in water or N availability, and increasing CO_2 concentrations might decrease the advantage of C_4 photosynthesis. As the photosynthetic pathway that is most efficient in its use of resources, C_4 is very significant in unmanaged and managed ecosystems, but anthropogenic environmental change might be reducing this advantage, with global consequences for primary productivity.

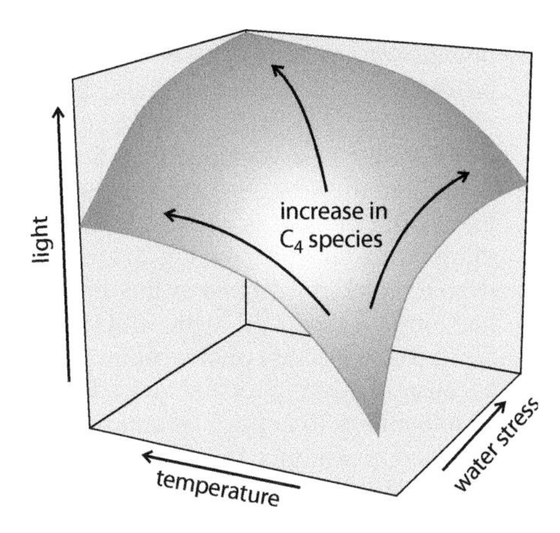

Figure 3.14. The conditions that favor C_4 photosynthesis. C_4 plants, in particular those that inhabit the C_4 grasslands, characteristically grow in hot, dry, and light conditions.

Crassulacean acid metabolism adapts plants to chronically difficult CO_2-fixation conditions

In C_3 plants the stomata are closed at night and, if water is available, open during the day. Prolonged periods of low water availability necessitate stomatal closure during the day. Such conditions present a long-term challenge for CO_2 uptake because, in C_3 photosynthesis, CO_2 uptake mostly occurs during the day in order to utilize the reductants that are being produced by the light reactions. In semi-arid and arid environments, several thousand **xerophytic** species, primarily in the Cactaceae, Agavaceae, Crassulaceae, Aizoaceae, Euphorbiaceae, and Didiereaceae, use Crassulacean acid metabolism (CAM) to take up CO_2 at night. The stomata of CAM plants have a daily pattern of opening that is the exact opposite to that of C_3 plants—that is, they open at night and are closed for at least a significant part of the day. More than 50% of epiphytes, including almost all species in the Bromeliaceae and many in the Orchidaceae, also use CAM. Epiphytes, which colonized forest trees to exploit the higher light levels above the forest floor, face chronic water-availability challenges, but in forest canopies the CO_2 levels are also notably low during the day and high at night. CAM also occurs in freshwater plants in which CO_2 supply is limited by low CO_2 concentrations and slow diffusion to plants. In total, probably at least 30,000 species of angiosperm, some lycopsids and ferns, and probably a cycad and *Welwitschia* use CAM. *Ananas comosus* (pineapple) is the most significant CAM crop, but in dry areas species such as *Agave sisalana* (sisal) and *A. tequilana* (tequila agave) can produce significant plant products. CAM plants produce a significant proportion of the biomass in some semi-arid and forest systems, and some can be extremely problematic weeds (Box 3.5). For most plants the advantage of CAM is that it enables them to take up CO_2 under water regimes that are very challenging for C_3 photosynthesis.

BOX 3.5. *OPUNTIA*: A CAM ALIEN IN AUSTRALIA

The Cactaceae and many other families with CAM evolved after Australia separated from South America and South Africa during the break-up of Gondwana, leaving Australia with few native CAM species. The invasion of up to 30 million hectares of eastern Australia within less than a century by the introduced CAM cactus *Opuntia* ("prickly pear") (Figure 1) provides one of the starkest examples of the problems that can be caused by alien weeds. It is also a reminder of the biomass potential of CAM plants under suitable conditions. *Opuntia* was introduced to Australia from South America in the early 1840s, and the area it invaded increased logarithmically until the 1930s, when it is estimated that there were 1.5 billion tonnes of biomass occupying an area greater than that of cultivated crops in Australia at that time. In Australia, *Opuntia* had found a semi-arid niche with intermittent summer rainfall in which its CAM photosynthesis was a distinct advantage because of the high water use efficiency that it conferred. Heroic efforts to control *Opuntia* up until the 1920s, which included techniques inspired by experiences of World War One, such as gassing, flamethrowers, and even the suggestion that tanks might be used, indicate the scale of threat that it posed. After the testing of numerous cactus predators, the moth *Cactoblastis cactorum* from South America was found to be extremely effective, and by the mid-1930s its introduction had brought the alien under control. Together with the detection of certain volatile compounds, night influxes of CO_2 (which contrast with the effluxes in non-CAM species) are used by *C. cactorum* females to identify *Opuntia* for egg laying. The use of *C. cactorum* to control *Opuntia* provides one of the best examples of biological control.

Further reading: Osmond B, Neales T, Stange G (2008) Curiosity and context revisited: crassulacean acid metabolism in the Anthropocene. *J Exp Botany* 59:1489–1502.

Figure 1. ***Opuntia ficus-indica.*** Opuntias are cacti native to the Americas. They have green photosynthetic cladodes (enlarged, flattened stems) that bear spines (modified leaves). The fruits, which are produced on the cladodes as shown in the figure, have long been used as a human food source and for preparing beverages.

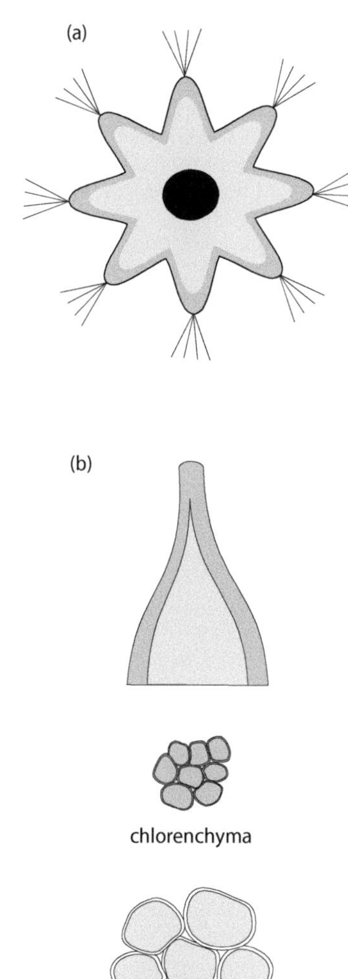

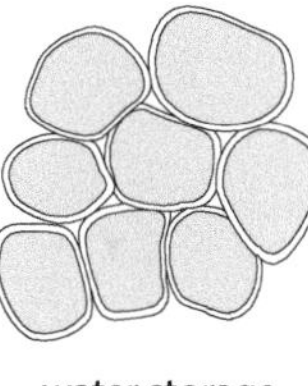

Figure 3.15. The anatomy of succulent plants. Succulent plants are generally either (a) stem or (b) leaf succulents. Almost all stem succulents have much reduced leaves and near-surface chlorenchyma (green) surrounding extensive water-storage parenchyma (gray). Plants that tend towards leaf succulence sometimes just have chlorenchyma cells with particularly large vacuoles, but full leaf succulents tend to show a division into near-surface chlorenchyma and internal achlorophyllous, water-storage parenchyma.

CAM plants are almost all perennial succulents. Some of them are tree-like but are monocots without true wood, the only true CAM trees being in the eudicot genus *Clusia*. Succulent plants have low surface area to volume ratios, and have an internal anatomy dominated by large vacuolated cells adapted for water storage. Some are leaf succulents and some are stem succulents, some have **chlorophyllous parenchmya** that also stores water, and some have separate chlorophyllous and **water storage parenchyma** (Figure 3.15). Large water-storing cells often have few spaces between them and a small surface area of contact with the spaces, which can slow down the internal diffusion of CO_2 and necessitate the development of a characteristic parenchyma. The daily cycle of CO_2 fixation in CAM plants can be divided into four phases (Figure 3.16). In phase I, during the night, the stomata are open and CO_2 uptake occurs via the action of PEPc on HCO_3^- and PEP. The malate produced is transported into vacuoles where it is stored as malic acid. In this phase rubisco is deactivated whereas PEPc is activated. In phase II, CO_2 fixation occurs when the stomata remain open into the light period, enabling both C_3 and CAM pathways, because rubisco is activated before PEPc is deactivated. In phase III the stomata are closed and malic acid is decarboxylated, increasing internal CO_2 concentrations very significantly. During this phase, rubisco fixes CO_2 under conditions that minimize both water loss and photorespiration. Phase IV is characterized by the stomata opening before nightfall, as malate reserves are exhausted, to allow direct uptake of CO_2 for rubisco. The length of these phases in CAM plants is variable, with phases II and IV becoming less pronounced as water becomes more scarce. In some plants, "CAM cycling" occurs, in which the stomata are closed at night, and CO_2 from respiration is fixed by PEPc and malate is decarboxylated to supply CO_2 to C_3 photosynthesis with open stomata

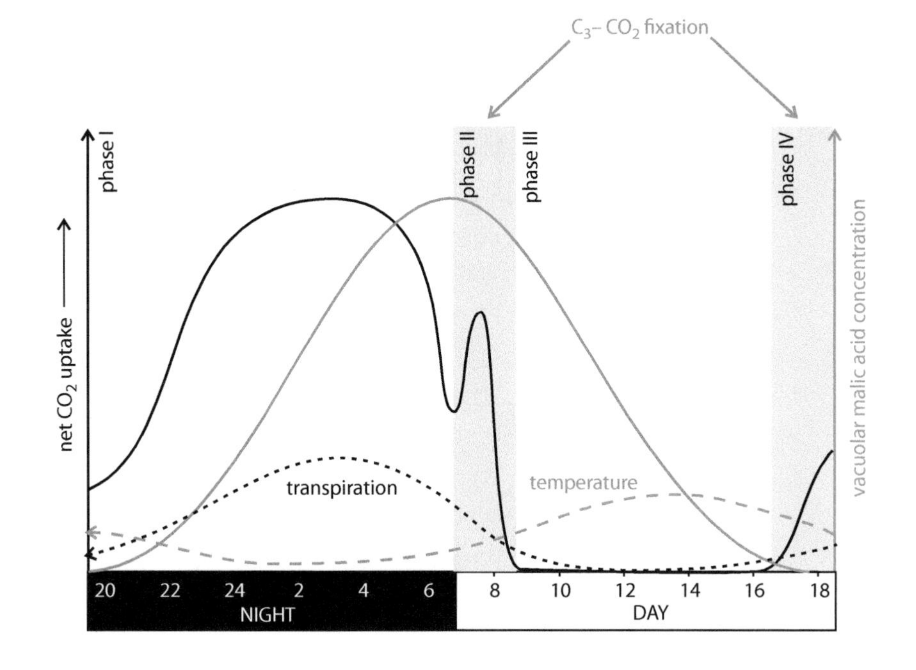

Figure 3.16. The phases of photosynthesis in CAM plants. There can be four phases of photosynthesis in CAM plants. In phase I, which occurs at night, CO_2 fixation occurs via phosphoenolpyruvate carboxylase (PEPc), and malic acid concentrations increase (CAM photosynthesis). Phase II, which varies in length, and does not occur under drought conditions , takes place during the first part of the day before the stomata close, and CO_2 fixation occurs directly, catalyzed by rubisco (C_3 photosynthesis). CO_2 uptake does not occur in phase III, and phase IV can occur if there is sufficient water for the stomata to be open before the end of the day, to enable direct CO_2 fixation by rubisco.

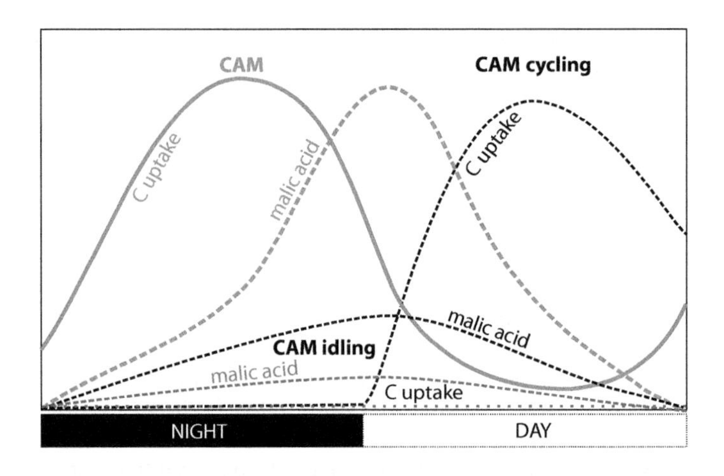

Figure 3.17. **CAM under different conditions.** Some plants do not have full expression of CAM, but use some "CAM cycling" to supplement C capture by converting CO_2 produced during respiration at night into malic acid that is decarboxylated during the day. In plants that inhabit hot, semi-arid environments, and in epiphytes, full expression of CAM to fix CO_2 from the atmosphere into malic acid at night is quite common. In extreme conditions, some plants can close their stomata completely and use "CAM idling" to capture CO_2 produced by respiration at night to form malic acid, which is the only source of C for photosynthesis during the day.

during the day. Under very severe conditions, CAM idling occurs, in which the stomata are permanently shut and CO_2 from respiration is released and fixed by PEPc at night, and the malate produced is the only source of C for rubisco (Figure 3.17). In some taxa (for example, *Clusia*), CAM is inducible in plants that otherwise use C_3 photosynthesis. Despite this variation, $\delta^{13}C$ studies reveal that there are few true C_3-CAM intermediates, so it is likely that C_3 and CAM pathways tend to be mutually exclusive. The duration of CAM and its phases varies according to the environment, but it is always based on the same physiological pathways (Figure 3.18). The PEPc on which it is based is characteristic of CAM, and sometimes of particular lineages of CAM, and it originated, along with other aspects of the syndrome, during the Miocene decrease in CO_2 levels.

CAM plants have a high water use efficiency, with **transpiration ratios** in the range 20–600, compared with 600–1400 in C_3 plants. In many CAM plants the ratio of carboxylation to oxygenation by rubisco is 10–30, whereas

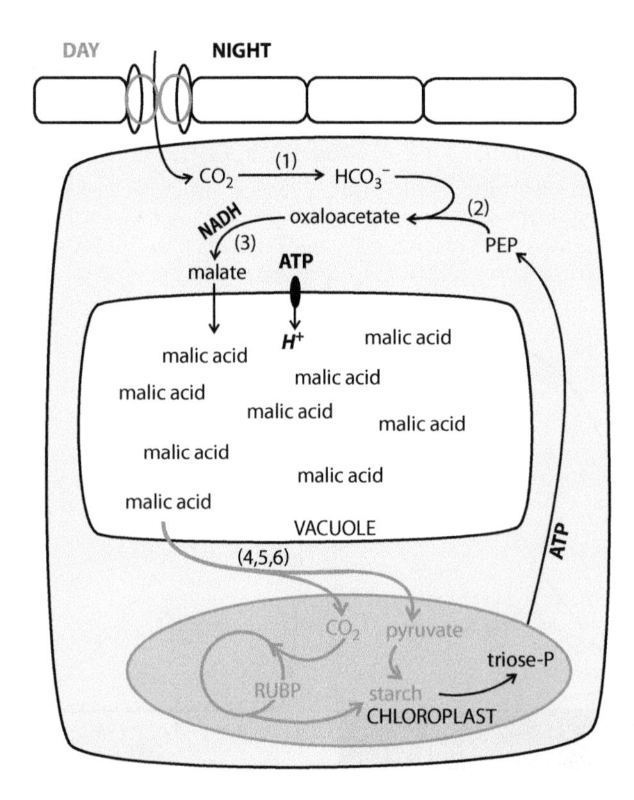

Figure 3.18. **The physiology of CAM.** During the day the stomata are mostly closed (shown in green), and the primary source of CO_2 for the Calvin–Benson cycle is the decarboxylation of malic acid stored in the vacuole, which is catalyzed by NADP-malic enzyme (4), NAD-malic enzyme (5), and/or PEP carboxykinase (6). At night the stomata tend to be open, allowing the entry of CO_2. Carbonic anhydrase (1) helps to form HCO_3^-, which is fixed to phosphoenolpyruvate (PEP) by the action of PEP carboxylase (2). Malate dehydrogenase catalyzes the reduction of oxaloacetate to malate (3).

in C_3 plants it is about 3. In C_3 plants, as the temperature increases, CO_2 fixation by rubisco becomes diffusion limited, an effect that is exaggerated by a succulent anatomy. CAM allows CO_2 fixation to occur under water and temperature regimes that necessitate morphologies and anatomies which are chronically challenging for CO_2 fixation. The evolution of CAM in plants was driven by water stress, but CAM is notably heat tolerant, and CAM plants can have quite high N use efficiency. In their natural environments, CAM plants tend to have low net C fixation rates, but the supply of some water can increase growth rates significantly. Human use of fresh water is one of the key environmental challenges for the twenty-first century, so CAM plants, which grow on land that is currently agriculturally marginal, have long been recognized as offering significant potential for biomass and food production under the environmental constraints that are likely to occur during the twenty-first century and beyond. They are also finding significant roles in water-recycling schemes in urban environments, including green roofs and green walls. CAM plants are notably absent from some of the truly arid areas of Earth. Despite the significant build-up of osmotica such as malate, they tend not to generate ψ in their roots as low as the values in true desert plants. Instead they have root morphologies that take advantage of periodically available water, in epiphytes often including dew or, especially in cloud forests, mist. CAM plants are also not generally salt tolerant. However, the semi-arid conditions that they inhabit are very likely to become more widespread if anthropogenic environmental changes occur as currently predicted.

Long-term increased CO_2 levels can increase plant growth, but limiting factors can moderate this effect

The effects of sustained increases in CO_2 concentration on plant growth exemplify why an understanding of the effects of spatial and temporal variation in interacting abiotic factors is vital to understanding plant-environment interactions. As CO_2 concentrations rise in the coming decades, this perspective will be vital to predicting food production and food quality, the harvest of non-food plant products, the state of the Earth's vegetation, and the success of environmental technologies based on biomass production. Overall, since CO_2 and its assimilation by plants were discovered over 200 years ago it has been evident that, under certain conditions, increasing CO_2 concentrations can result in increased plant growth. This "CO_2-fertilization" phenomenon has for decades been used to boost production in greenhouses. Thus, in the short term, growing plants in CO_2 concentrations that they might experience during the rest of the twenty-first century as compared with those experienced in the Earth's recent history has a significant effect on their growth (Figure 3.19).

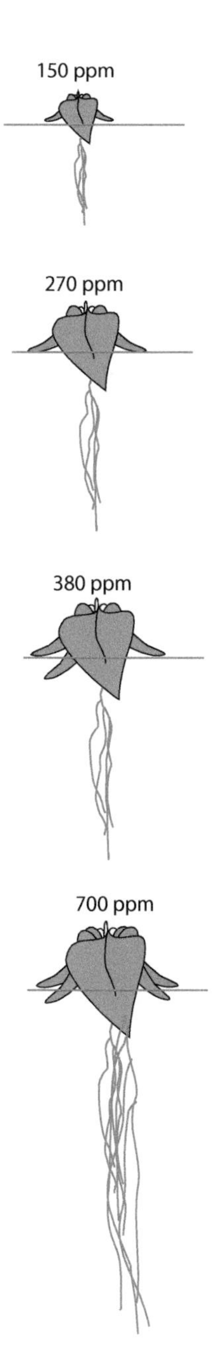

Figure 3.19. Plants grown in low and high CO_2 concentrations. Plants grown under optimum conditions but with different CO_2 concentrations produce, in the short term, significantly different amounts of biomass. In recent Pleistocene glacial periods (1.5 million to 10,000 years ago), CO_2 concentrations reached 150 ppm, whereas during interglacial periods, such as that from 10,000 years ago until the Industrial Revolution, CO_2 concentrations of 270 ppm occurred. Post-Industrial Revolution CO_2 concentrations are now almost 400 ppm and rising rapidly. (Data from experiments with *Abutilon theophrasti* in Gerhard & Ward [2010] *New Phytol* 188:674–695.)

Meta-analyses of data from annual plants grown under optimal conditions—that is, with no resource, pest, or pathogen limitations—in general indicate that an approximate doubling of CO_2 levels from current concentrations can increase the photosynthetic rate by up to 50% in some species. In general, the use of more realistic sub-optimal conditions—for example, a realistic daily light fluctuation—reduces the increase in photosynthetic rate to 30%. The relationship between photosynthetic rate and growth rate is complex, with the increase in growth rate in high CO_2 experiments being significantly less (up to 10%) than the increase in photosynthetic rate. High CO_2-grown annuals produced under optimal conditions have a higher concentration of many carbohydrates. They also have a lower concentration of rubisco and often N, but a higher proportion of N in structural proteins. This can significantly affect biomass quality, including the concentration of Fe and Zn, with potentially global implications for human nutrition. Effects on many other fundamental aspects of plant metabolism have also been reported,

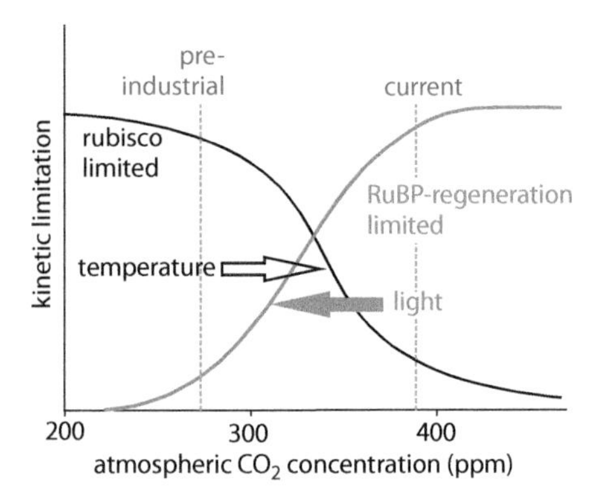

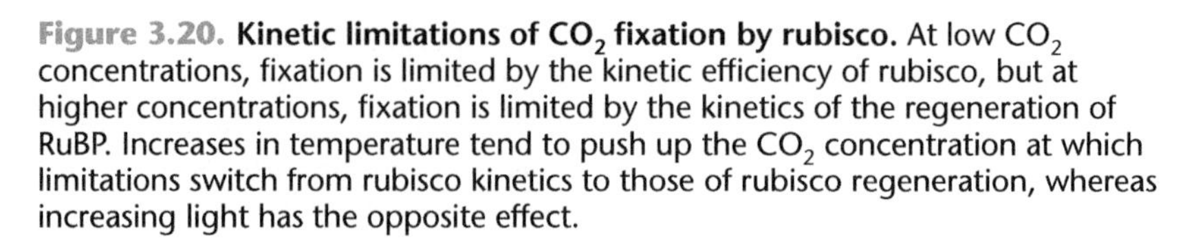
Figure 3.20. Kinetic limitations of CO_2 fixation by rubisco. At low CO_2 concentrations, fixation is limited by the kinetic efficiency of rubisco, but at higher concentrations, fixation is limited by the kinetics of the regeneration of RuBP. Increases in temperature tend to push up the CO_2 concentration at which limitations switch from rubisco kinetics to those of rubisco regeneration, whereas increasing light has the opposite effect.

including hormone signaling pathways, germination, flowering time, and harvestable yield. Overall, high CO_2-grown plants generally have higher water use efficiency, which can be related directly to the decreases in stomatal conductance induced by high CO_2. Within chloroplasts, the rate of CO_2 fixation is limited either by rubisco kinetics or by the regeneration of RuBP, depending on conditions (Figure 3.20). Models of CO_2 fixation by rubisco predict changes of net photosynthesis under high CO_2 concentrations that are within the range of increases found for annual plants if there are no other limiting factors. Experiments with enhanced CO_2 have in general confirmed the expectation that photosynthesis in C_4 plants is unaffected. Thus, under optimal growth conditions, which occur infrequently in the field, in annuals there is a CO_2-fertilization effect underpinned by the kinetics of rubisco-mediated CO_2 fixation, but which only applies to C_3 plants, and that alters plant metabolism and biomass quality.

In contrast to those grown under optimal conditions, annual plants grown in high CO_2 concentrations but under field conditions, and even those grown in small pots, do not generally exhibit photosynthetic rates that can be directly related to CO_2-fixation kinetics. This has revealed much about the subtle but significant differences between the physiology of annuals grown under optimal and field conditions in both unmanaged and managed ecosystems. For example, the differences in gene expression due to high CO_2 levels are a small proportion of those induced by field as compared with controlled conditions. Open-topped chambers that enclose crops or natural vegetation under field conditions, soil-plant-atmosphere research (SPAR) units, and free-air CO_2 enrichment (FACE) facilities have all been used to investigate the effects of sustained high CO_2 levels on plant growth under simulated field conditions (Figure 3.21). For annual monocultures, many years of experiments with these field simulation systems have emphasized the limits to CO_2 fertilization that are induced by interacting abiotic variables, especially water, and for unmanaged ecosystems have highlighted the importance not only of limiting environmental factors, but also of complex long-term ecosystem effects. In general, experiments with soya and other crops indicate that legumes show the greatest increases in growth in the field at high CO_2 levels, because they are less likely to be N limited, although the increases in growth are seldom as great as those in highly controlled conditions—water, sunlight, pests, and other factors are usually limiting for at least part of the growth period. In annual crops, some of the most important effects relate to competition with weeds, which frequently changes (either increasing or decreasing) much more than crop growth per se. C_4 weeds, many of which are currently problematic (Table 3.1), might decrease under high CO_2 conditions, but many C_3 weeds seem likely to increase. Changes in weed management induced by sustained high CO_2 levels and related to

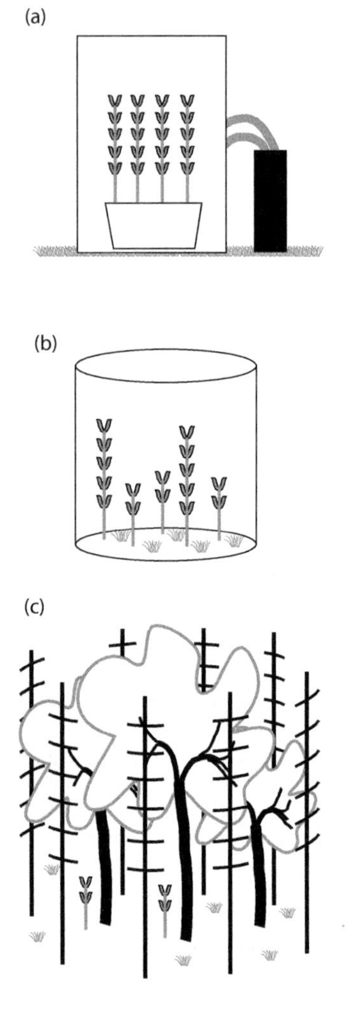

Figure 3.21. Experimental techniques for high CO_2 studies. (a) Soil–plant–atmosphere research (SPAR) protocols enable control of variables such as CO_2, but under the light and temperature conditions of the field. (b) Open-topped chambers can be used to enhance CO_2 concentrations while exposing plants to field conditions. (c) In free-air CO_2 enrichment (FACE) facilities, CO_2 can be sprayed into sections of natural ecosystems, including significant areas of forest. CO_2 sensors that control CO_2 input can produce tight control over CO_2 concentrations in FACE facilities.

plant CO_2-fixation pathways are likely to be an increasing challenge to agriculture in the twenty-first century.

Many, although not all, experiments with unmanaged vegetation ranging from grasslands to forests have found changes induced by sustained high CO_2 levels that are indirectly related to changes in CO_2 fixation. In general, in the short term there can be increases in growth, but the effects of these depend on the availability of other resources and on species–species interactions. In the longer term, many plants acclimate to increased CO_2 levels, and growth rate decreases towards pre-enhancement rates (Figure 3.22). Woody species, especially trees, have generally acclimated less to sustained high CO_2 levels than have other species over the timescales of years for which the experiments have been run, but known effects of high CO_2 levels on stomatal density in trees suggest that, over long time periods, trees too may acclimate to high CO_2 levels. Plot experiments in temperate and tropical forests, including large-scale studies in the Amazon and across the tropical forests of Africa, show that in recent decades forests have been C sinks. During this time, terrestrial vegetation has removed almost 50% of anthropogenic CO_2 emissions from the atmosphere and been a net C sink. In most years, CO_2 fixation in forests (ranging from boreal to tropical) is almost entirely responsible for this net C sink, but in El Nino years with changed global rainfall patterns other ecosystems also contribute.

Table 3.1. Some problematic C_4 weeds

Scientific name	Family
Amaranthus hybridus (smooth pigweed)	Amaranthaceae
Cyperus rotundus (purple nut sedge)	Poaceae
Cynodon dactylon (Bermuda grass)	Poaceae
Echinochloa crus-galli (cockspur)	Poaceae
Eleusine indica (goosegrass)	Poaceae
Euphorbia esula (green spurge)	Euphorbiaceae
Portulaca oleracea (purslane)	Portulacaceae

Most of the carbon in the C sink in forests resides in woody growth, some of which contributes to increases in soil organic C. However, there is evidence that the forest sink is weakening, perhaps as tree populations consisting of long-lived individuals acclimate to high CO_2 levels. In addition, although it does not contribute significantly to the C sink, the cycling of labile C (that is, leaves, roots, and seedlings) increases, with consequences for forest dynamics. The effects of limiting factors such as low N, extreme soil pH, water deficit, or salinity all have a significant impact on unmanaged ecosystem outcomes under high CO_2 levels. There are species-specific responses to such environmental variables that over the course of years significantly limit increases in growth, but that can induce significant change in species composition and hence species–species interactions (Figure 3.23). In ecosystems that exist under conditions of chronic resource limitation (for example, grass swards), and ecosystems that are highly disturbed by herbivores or natural events such as fire, these limiting factors are generally more powerful than any CO_2-fertilization effect, and high CO_2 levels induce few changes in growth. Thus, in some unmanaged ecosystems, sustained high CO_2 levels can at first induce an increase in CO_2 fixation, but there is significant acclimation over time. In almost all unmanaged

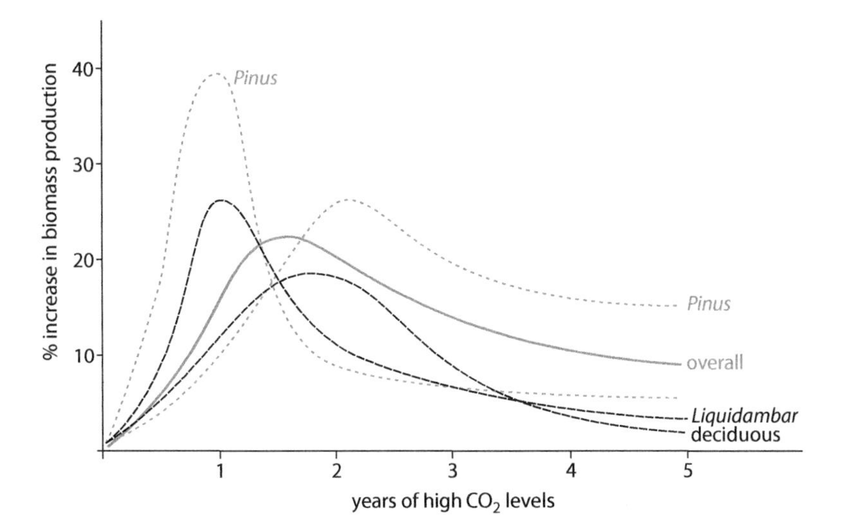

Figure 3.22. The acclimation of tree biomass production to high CO_2 levels. In the first year of a variety of high-CO_2 FACE experiments, many trees show significantly enhanced biomass production. The increases differ between species, and between different conditions for the same species, but after a few years there is significant acclimation to enhanced CO_2 levels. Short-lived herbaceous plants acclimate more quickly, and in the long term many long-lived woody species acclimate completely to enhanced CO_2 levels. (From Körner C [2006] *New Phytol* 172:393–411. With permission from John Wiley and Sons.)

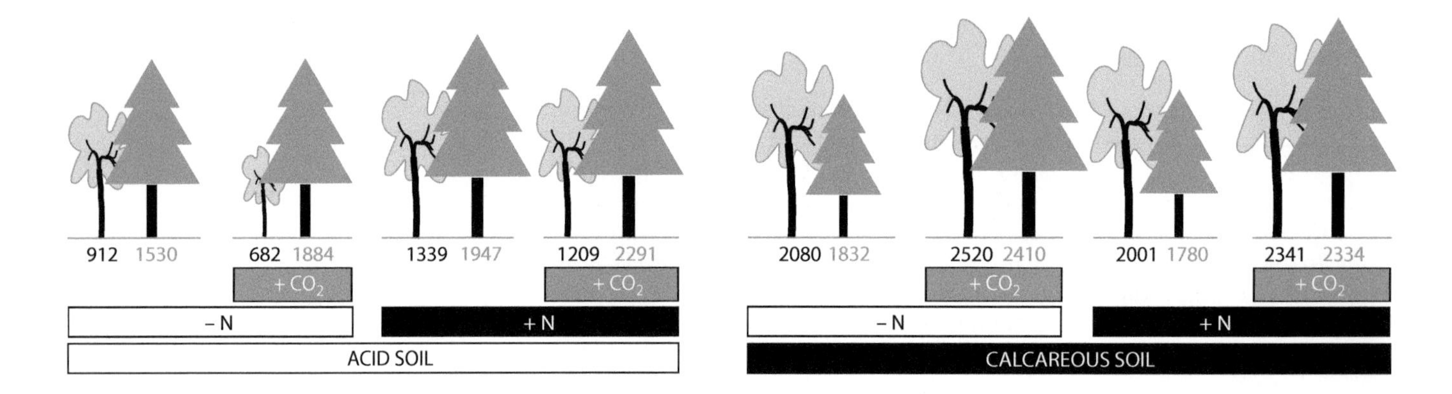

Figure 3.23. The effects of high CO_2 concentrations depend on other environmental variables. FACE experiments show that different soil pH and N fertilizer regimes have a significant effect on the biomass produced under high CO_2 conditions, and that the effects vary, for example, between deciduous and evergreen trees. The blocks show experimental treatments and biomass (in kg) produced per hectare. (Data from Körner C [2006] *New Phytol* 172:393–411.)

ecosystems, conditions are seldom optimal and species-specific responses to high CO_2 levels in sub-optimal conditions can change the composition and interactions of ecosystems, even if there are no changes in ecosystem productivity. Ecosystems depend on species–species interactions, and a change in these, based on species-specific responses to CO_2 fixation under particular sets of conditions, is likely to be a significant effect of increased CO_2 levels during the twenty-first century.

Plant responses to increasing CO_2 levels will affect the hydrological cycle and Earth's climate

In many FACE experiments, the acclimation of CO_2 fixation is underpinned by reduced stomatal aperture and, in the longer term, leaves that develop under high CO_2 levels having reduced stomatal density. At high CO_2 levels, leaves with a lowered stomatal density can supply sufficient CO_2 to meet demand, especially if other factors are limiting. The evaporation of water from leaves that drives transpiration is the mechanism that many plants use to moderate their temperature, so reduced stomatal conductivity reduces transpiration and increases temperature sensitivity. Many plant species, including crops, herbs, and trees in unmanaged ecosystems, have been shown to be more sensitive to increases in temperature under high CO_2 regimes, probably because they are less able to control heat accumulation. For many crop species there is evidence that increased temperatures at high CO_2 levels reduce yield. Such experiments have highlighted the fact that transpiration is often not only a necessary consequence of having stomata open for CO_2 accumulation, but also part of a vital nexus of CO_2, water, and temperature interactions in plants. In fact there is increasing evidence that terrestrial plant life is so important for the global cycling of CO_2 and water that this nexus in plants can have a significant "physiological forcing" effect on global climate.

Overall, more water is transpired from the land surface than evaporates. In heavily vegetated ecosystems, more than 75% of the water that is delivered to the atmosphere can be from transpiration. Data on precipitation and river flow in many parts of the globe are extensive for the last century or more. Compilation of such data suggests that in the years since 1960 there has been an overall decrease in precipitation but an increase in river flow across the Earth, although the effects vary from one location to another. This is probably because with increased CO_2 levels there is less transpiration and therefore less precipitation, and consequently a larger amount of water left in soils to flow through into rivers. Models that incorporate physiological interactions with the atmosphere predict that increasing CO_2 concentrations will decrease the relative humidity over land via decreases in transpiration. Thus there might be not only an increase in run-off into rivers but also, because of a decrease in heat capacity with decreased humidity, an increase in land-surface temperature over and above increases resulting from direct radiative forcing by CO_2. This might be compounded by the effect of decreased humidity on cloud cover. Such additional "physiological forcing" of climate change with increased CO_2 levels is of some global significance, but in some regions is predicted to be a major factor in temperature increases. Given the food production, nature conservation, and water resource challenges that humans will face in the twenty-first century, long-term physiological forcing of climate change via managed and unmanaged ecosystems may be an important impact of the plant CO_2-fixation response under high CO_2 concentrations.

An understanding of CO_2 fixation by plants is important for sustainable food production and ecosystem conservation

The demand for food and fuel is predicted to continue to increase very significantly, throughout the twenty-first century. Given that human impacts on Earth, which arise in significant part because of attempts to meet current demand for food, fuel, and other plant products, are currently unsustainable, the devising of ways to meet future demand sustainably while conserving unmanaged ecosystems is a truly "grand challenge." Furthermore, the major agronomic changes and huge increases in yield resulting from the twentieth-century intensification of agriculture were based primarily on overcoming limitations other than increasing C-fixation capacity, and the potential for achieving increases in yield with this strategy has plateaued. Global estimates of yield gaps—that is, the gap between potential yield using current technologies and actual yield based on lack of their implementation—suggest that closing these gaps could produce great increases in yield globally, but this would not necessarily address the issue of sustainability. Given that even if global atmospheric CO_2 concentrations reach, for example, 600 ppm, which is relatively low compared with the levels under which the photosynthetic apparatus evolved, there is generally acknowledged to be significant potential to manipulate photosynthesis in order to increase the production of food, fuel, and other plant products. This is particularly so given the dramatically increased capabilities for the identification and manipulation of genotypes and phenotypes in the post-genomic era. Experiments with enhanced CO_2 concentrations reveal starkly the **law of limiting factors**, and suggest that only under optimal conditions will enhanced photosynthetic rates translate into enhanced growth, so numerous agroecological approaches to attaining a sustainable supply of food and fuel are also being explored.

A common measure of photosynthetic capability is photosynthetic rate per unit leaf area—a characteristic in crops that has not changed anywhere near as significantly as many other phenotypes, and in some instances has actually decreased. Attempts to increase photosynthetic rate, ranging from alterations of rubisco to engineering of C_4/CAM or CO_2-concentrating mechanisms from cyanobacteria into C_3 plants, are currently under way. There is a greater diversity of rubisco, even within its three main types, than was previously thought, with each form probably adapted to the conditions under which it functions. It is likely that research into the targeted manipulation of rubisco will soon provide mechanisms for enhancing the kinetics and selectivity of rubisco in plants. Much of the activity and synthesis of rubsico are controlled by rubisco activase, and the manipulation of this enzyme is also advancing rapidly. As CO_2 concentrations increase, CO_2 fixation becomes limited by the regeneration of RuBP, and the potential for manipulating this is also the subject of ongoing research. In fact, systems analyses of metabolic fluxes associated with all aspects of CO_2 fixation, including the central role of sinks in driving it, are identifying key metabolic bottlenecks and suggesting ways in which these might be overcome. Under many conditions, photorespiration is the key limiting factor for net CO_2 fixation, so it is the focus of intensive efforts to reduce it. The natural genetic variation in photosynthetic rates, produced by myriad variations in all of the previously mentioned limits to photosynthesis and more, and the ability of cultivars to grow at elevated CO_2 concentrations, are being explored and are likely to have a significant role in efforts to manipulate CO_2 fixation for the production of food and fuel.

Predicted trends in conditions on Earth make the capacity of C_4 plants to maintain high CO_2 fixation with higher water use efficiency and N use efficiency at higher temperature than C_3 plants a particularly attractive option for meeting yield demands. Although in most instances C_4 photosynthesis requires not only physiological adaptations but also anatomical adaptations, the transfer of C_4 photosynthesis into crops that are currently C_3 is deemed realistic. For example, concerted efforts to develop C_4 rice have been under way for some time (Box 3.6). C_4 transfer into C_3 crops seems to

BOX 3.6. ENVIRONMENTAL SOLUTION: CAN C_4 RICE HELP TO FEED THE WORLD?

For over 3.5 billion people on Earth, rice (Figure 1) is a staple food—that is, it provides at least 20% of their calorie intake. Since the dawn of agriculture, humans have derived more calories from rice than from any other crop, and it now provides about 25% of all calories consumed by humans. It is also one of the most important sources of protein for human consumption, providing about 15% of total intake. Rice was domesticated in Asia, which currently dominates production, particularly from wet regions, with more than 75% of world production being grown in flooded paddy soils. Paddy rice can produce multiple crops per year, and it can be grown without rotation. Some paddies in Asia have produced rice continuously for centuries, and probably longer. Demand for rice is increasing rapidly with the growth of the world population.

The increases in rice yields per hectare achieved during the twentieth century were remarkable—50 years ago, 4 tn ha^{-1} was regarded as a high yield, whereas now the corresponding figure is 10 tn ha^{-1}. If the increasing demand for rice is to be met, further increases in yield per hectare will probably be necessary. Since 1960 the International Rice Research Institute (IRRI) in the Philippines has been a global focus of rice research, and has enabled many of the recent changes in global rice production systems. Many IRRI projects have focused on new varieties of rice that can help to increase yield, but one important project has been to carry out the fundamental science that might enable rice, a C_3 plant, to be transformed into a C_4 plant. Such an achievement has the potential to increase rice yields significantly, if the appropriate growth conditions can be provided.

When C_4 photosynthesis was first discovered it seemed that its anatomical peculiarities would preclude the development of C_4 in C_3 crops such as rice. The discovery of C_3–C_4 intermediates and then the operation of C_4 photosynthesis in single cells (that is, without Kranz anatomy) began to suggest that the transfer of C_4 photosynthesis into C_3 crops might not be impossible. The insight that much Kranz anatomy was a consequence of a relatively simple increase in density of vascular bundles provided further impetus. In the genomic era it was then possible to determine exactly which genes were expressed differently in C_3 and C_4 plants. The importance of rice and the very large number of varieties of this species, including those with different anatomies and gene expression levels, led to it being the first C_3 crop to be the focus of C_4 development. In 2008 a major international initiative coordinated by IRRI began to identify all of the gene expression patterns that would need to be changed in rice for it to utilize C_4 photosynthesis. The first phase suggested that there were perhaps 12 key genes whose expression would have to be changed—a major challenge, but one which suggested that there was enough potential for a second phase of the project to be initiated with the aim of ultimately developing C_4 rice.

Figure 1. ***Oryza sativa*** **(rice).** Rice is a member of the Poaceae (grass family). After fertilization, grasses produce a caryopsis ("grain")—that is, a fruit which consists almost entirely of a single large seed, with the fruit reduced to just a few layers of cells fused around the seed. The many cultivars of rice can be divided into two main ecogeographic subtypes, *O. sativa* Japonica Group and *O. sativa* Indica Group. Although the varieties of each vary greatly in both their form and their preferred growing conditions, the most important long-grain rices are from the Indica varieties, whereas the most important short-grain sticky rices are from the Japonica varieties.

be possible because of its multiple origins, the existence of C_3-C_4 intermediates, the existence of intracellular compartmentation of C_4 photosynthesis in some species, and rapid advances in the understanding of the development of Kranz anatomy. It seems very likely that increases in photosynthetic rate per unit area will be achieved in crop plants in the first half of the twenty-first century, and that if complemented by optimal growing conditions these will translate into increased growth rate. However, it is also the case that a further intensification of monocultures, even if the resources supplied to them are derived more sustainably than at present, might meet the challenge of increasing yield but not necessarily translate into a contribution to sustainable production methods. In some unmanaged ecosystems, sustainable CO_2-fixation rates per unit area are higher than those of even the most intensive unsustainable monocultures. The limits on agriculturally useful land are in significant part defined by the limited range of crop plants that are cultivated. The potential usefulness of manipulating increases in photosynthesis per unit area in crop plants highlights the fact that other strategies for managing CO_2 fixation for food and plant products, probably based on agroecological methods such as intercropping and agroforestry, should be envisaged as playing a significant role in developing sustainable production systems. The methods of synthetic biology are already being brought to bear on the CO_2-fixation challenge, and could dramatically alter the current dependence of humans on terrestrial plants for the production of food and other plant products.

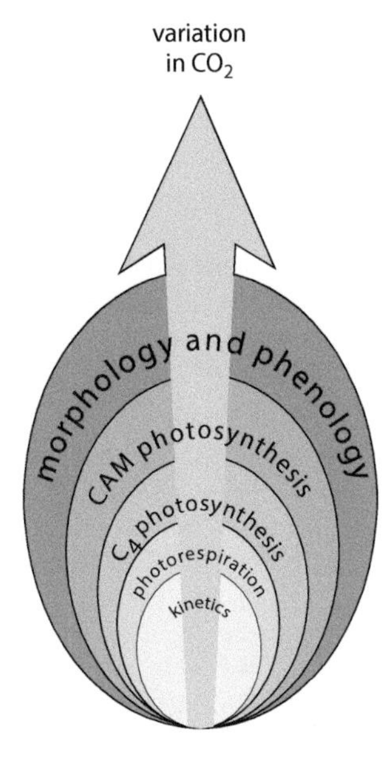

Figure 3.24. Variation in CO_2 availability and the stress-response hierarchy. There are many differences in the kinetics and distribution of rubisco in leaves that adapt plants to minor variations in CO_2 availability. Anatomical and morphological differences that control leaf conductance can help plants to adjust to variation in CO_2 concentration in canopies. Changes in water availability and temperature have a significant impact on the ability of plants to accumulate CO_2, so adaptations driven by these variables, ranging from biochemical to morphological levels, are necessary to maintain CO_2 accumulation, and are more important than variation in CO_2 concentration per se. There are few phenological characteristics that are driven by variation in CO_2 levels.

Summary

CO_2 fixation in multicellular plants is the primary provider of biomass in terrestrial ecosystems. Plants maintain CO_2 fixation during short-term variations in external CO_2 concentration by changing and controlling the expression of enzymes in the Calvin–Benson cycle. Long-term variation in CO_2 levels, and in particular the low atmospheric CO_2 concentration of the last 20–30 million years combined with the inhibition of CO_2 uptake induced by water shortage, have led to the evolution of new physiological pathways which, when allied to anatomical and morphological adaptations, enable CO_2 fixation to be maintained (Figure 3.24). Increases in CO_2 concentration can induce a CO_2-fertilization effect on productivity, but plants often acclimate to this or show a limited response because of constraints imposed by other environmental variables. Even when productivity is not enhanced by increased CO_2 levels, there can be numerous effects on ecosystem functioning, ranging from interspecies competition to temperature tolerance, as well as hydrological and climatic effects. The management of CO_2 fixation, from C sinks to biofuel production and green roofs, and the manipulation of CO_2 fixation in agricultural crops and in synthetic systems, are likely to develop significantly as humans focus on sustainable habitation of Earth in the twenty-first century.

Further reading

CO_2 fixation and primary production

Ito A (2011) A historical meta-analysis of global terrestrial net primary productivity: are estimates converging? *Global Change Biol* 17:3161–3175.

Slade R, Bauen A & Gross R (2014) Global bioenergy resources. *Nature Clim Change* 4:99–105.

van der Giesen C, Kleijn R & Kramer GJ (2014) Energy and climate impacts of producing synthetic hydrocarbon fuels from CO_2. *Environ Sci Technol* 48:7111–7121.

Welp LR, Keeling RF, Meijer HAJ et al. (2011) Interannual variability in the oxygen isotopes of atmospheric CO_2 driven by El Nino. *Nature* 477:579–582.

Human activities and variation in CO_2 levels

Desai AR (2014) Influence and predictive capacity of climate anomalies on daily to decadal extremes in canopy photosynthesis. *Photosynth Res* 119:31–47.

Leakey ADB & Lau JA (2012) Evolutionary context for understanding and manipulating plant responses to past, present and future atmospheric [CO_2]. *Philos Trans R Soc Lond B Biol Sci* 367:613–629.

Prentice IC, Harrison SP & Bartlein PJ (2011) Global vegetation and terrestrial carbon cycle changes after the last ice age. *New Phytol* 189:988–998.

Zscheischler J, Mahecha MD, von Buttlar J et al. (2014) A few extreme events dominate global interannual variability in gross primary production. *Environ Res Lett* 9:035001.

Rubisco activity and the Calvin–Benson cycle

Bar-Even A, Noor E, Savir Y et al. (2011) The moderately efficient enzyme: evolutionary and physicochemical trends shaping enzyme parameters. *Biochemistry* 50:4402–4410.

Michelet L, Zaffagnini M, Morisse S et al. (2013) Redox regulation of the Calvin–Benson cycle: something old, something new. *Front Plant Sci* 4:470.

Tabita FR, Hanson TE, Satagopan S et al. (2008) Phylogenetic and evolutionary relationships of RubisCO and the RubisCO-like proteins and the functional lessons provided by diverse molecular forms. *Philos Trans R Soc Lond B Biol Sci* 363:2629–2640.

van Lun M, Hub JS, van der Spoel D et al. (2014) CO_2 and O_2 distribution in Rubisco suggests the small subunit functions as a CO_2 reservoir. *J Am Chem Soc* 136:3165–3171.

Oxygenation of RuBP

Bauwe H, Hagemann M & Fernie AR (2010) Photorespiration: players, partners and origin. *Trends Plant Sci* 15:330–336.

Hofmann NR (2011) The evolution of photorespiratory glycolate oxidase activity. *Plant Cell* 23:2805–2805.

Moroney JV, Jungnick N, DiMario RJ et al. (2013) Photorespiration and carbon concentrating mechanisms: two adaptations to high O_2, low CO_2 conditions. *Photosynth Res* 117:121–131.

Sage RF (2013) Photorespiratory compensation: a driver for biological diversity. *Plant Biol (Stuttg)* 15:624–638.

C_4 plants maintain high CO_2:O_2 ratios in the vicinity of rubisco

Covshoff S, Burgess SJ, Knerova J et al. (2014) Getting the most out of natural variation in C_4 photosynthesis. *Photosynth Res* 119:157–167.

Edwards EJ & Smith SA (2010) Phylogenetic analyses reveal the shady history of C_4 grasses. *Proc Natl Acad Sci USA* 107:2532–2537.

Taylor SH, Ripley BS, Martin T et al. (2014) Physiological advantages of C_4 grasses in the field: a comparative experiment demonstrating the importance of drought. *Global Change Biol* 20:1992–2003.

Wang L, Peterson RB & Brutnell TP (2011) Regulatory mechanisms underlying C_4 photosynthesis. *New Phytol* 190:9–20.

C_4 and C_3–C_4 responses to the environment

Bellasio C & Griffiths H (2014) Acclimation to low light by C4 maize: implications for bundle sheath leakiness. *Plant Cell Environ* 37:1046–1058.

Osborne CP & Beerling DJ (2006) Nature's green revolution: the remarkable evolutionary rise of C_4 plants. *Philos Trans R Soc Lond B Biol Sci* 361:173–194.

Still CJ, Pau S & Edwards EJ (2014) Land surface skin temperature captures thermal environments of C_3 and C_4 grasses. *Glob Ecol Biogeogr* 23:286–296.

Thomas EK, Huang Y, Morrill C et al. (2014) Abundant C_4 plants on the Tibetan Plateau during the late glacial and early Holocene. *Quaternary Sci Rev* 87:24–33.

CAM and CO_2 accumulation

Cheung CYM, Poolman MG, Fell DA et al. (2014) A diel flux balance model captures interactions between light and dark metabolism during day-night cycles in C_3 and crassulacean acid metabolism leaves. *Plant Physiol* 165:917–929.

Hernandez-Hernandez T, Brown JW, Schlumpberger BO et al. (2014) Beyond aridification: multiple explanations for the elevated diversification of cacti in the New World Succulent Biome. *New Phytol* 202:1382–1397.

Ogburn RM & Edwards EJ (2010) The ecological water-use strategies of succulent plants. *Adv Bot Res* 55:179–225.

Quezada IM, Zotz G & Gianoli E (2014) Latitudinal variation in the degree of crassulacean acid metabolism in *Puya chilensis*. *Plant Biol (Stuttg)* 16:848–852.

Increasing CO_2 levels and primary production

Kirschbaum MUF (2011) Does enhanced photosynthesis enhance growth? Lessons learned from CO_2 enrichment studies. *Plant Physiol* 155:117–124.

Myers SS, Zanobetti A, Kloog I et al. (2014) Increasing CO_2 threatens human nutrition. *Nature* 510:139–142.

Phillips OL & Lewis SL (2014) Evaluating the tropical forest carbon sink. *Global Change Biol* 20:2039–2041.

Poulter B, Frank D, Ciais P et al. (2014) Contribution of semi-arid ecosystems to interannual variability of the global carbon cycle. *Nature* 509:600–603.

Physiological forcing of climate change

Cao L, Bala G, Caldeira K et al. (2010) Importance of carbon dioxide physiological forcing to future climate change. *Proc Natl Acad Sci USA* 107:9513–9518.

Gedney N, Cox PM, Betts RA et al. (2006) Detection of a direct carbon dioxide effect in continental river runoff records. *Nature* 439:835–838.

Pu B & Dickinson RE (2014) Hydrological changes in the climate system from leaf responses to increasing CO_2. *Clim Dynam* 42:1905–1923.

Willeit M, Ganopolski A & Feulner G (2014) Asymmetry and uncertainties in biogeophysical climate-vegetation feedback over a range of CO_2 forcings. *Biogeosciences* 11:17–32.

Improving photosynthesis for agriculture

Borland AM, Hartwell J, Weston DJ et al. (2014) Engineering crassulacean acid metabolism to improve water-use efficiency. *Trends Plant Sci* 19:327–338.

Chen CP, Sakai H, Tokida T et al. (2014) Do the rich always become richer? Characterizing the leaf physiological response of the high-yielding rice cultivar Takanari to free-air CO_2 enrichment. *Plant Cell Physiol* 55:381–391.

McGrath JM & Long SP (2014) Can the cyanobacterial carbon-concentrating mechanism increase photosynthesis in crop species? A theoretical analysis. *Plant Physiol* 164:2247–2261.

Peterhansel C (2011) Best practice procedures for the establishment of a C_4 cycle in transgenic C_3 plants. *J Exp Bot* 62:3011–3019.

Chapter 4
Water

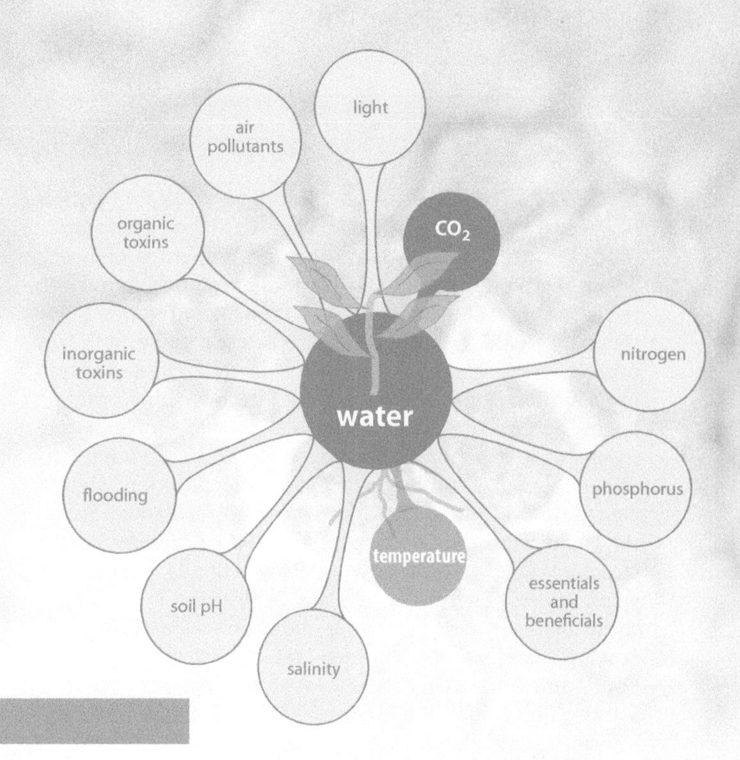

The challenges of variable water supply overlap, in particular, with the challenges of CO_2 acquisition and temperature variation.

Key concepts

- Plant–water relations are important from a cellular to a global scale.
- Human use of fresh water is dominated by irrigation of crops.
- Water potential explains the movement of water in the soil–plant system.
- Plants control resistance to flow to manage minor fluctuations in available water.
- Physiological adjustments help plants to adapt to transient water deficit.
- Roots are used to forage for water when it is in short supply.
- Leaf adaptations help many plants to adapt to sustained dry conditions.
- Xerophytes that inhabit chronically dry ecosystems adapt their stems, leaves, and life cycles.
- Resurrection plants are the largest organisms that can survive complete desiccation.
- A "blue revolution" in agriculture is probably necessary for global food security.

Plant–water relations affect physiological processes from a cellular to a global scale

The ubiquity of water as the primary biochemical solvent reflects life's origin in aqueous solutions, and enables the enhanced reaction rates that underpin life on Earth. Animals first evolved in the marine environment, and their sodium-requiring physiology means that, even in terrestrial ecosystems, they maintain a saline solution that bathes their cells. All terrestrial plants evolved from freshwater algae and therefore have to maintain simulated fresh water in and around their cells. However, fresh water is relatively scarce on the land surface. Terrestrial plant-water relations therefore exert significant control over the dynamics of unmanaged terrestrial ecosystems and the productivity of agricultural ecosystems.

In addition to being a solvent, water acts as a matrix—that is, it provides a semi-structured environment—for life. The structures and properties of water under the conditions on Earth are unusual (Box 4.1). The hydrogen

BOX 4.1. THE PROPERTIES OF WATER

Metabolism occurs in aqueous solutions, so the properties of water are fundamental to life. Many earth and atmosphere processes are also dependent on the properties of water, which therefore probably have a more profound impact on the biotic and abiotic environment than the properties of any other molecule. Oxygen is more **electronegative** than any other element except fluorine, and in water molecules an oxygen atom bonds to two hydrogen atoms with a bonding angle of about 104°. The extreme electronegativity of oxygen means that, in water molecules, electrons are preferentially pulled towards it, producing a polar molecule with a significant **dipole moment**. Angular water molecules with a significant dipole moment can form tetrahedral supramolecular structures (Figure 1). Although hydrogen bonds are relatively weak, those between oxygen and hydrogen—for example, those between adjacent water molecules—are among the strongest, and every water molecule can form four of these bonds simultaneously. These hydrogen bonds are transient, forming for a few femtoseconds, and there has been significant debate about a variety of supramolecular structures that might form in water. Ionic compounds dissolve rapidly in water because water molecules are attracted to their ions more strongly than to other water molecules. Non-polar compounds do not attract water molecules sufficiently strongly to dissolve in them, whereas polar compounds do. C-H and C-S bonds are essentially non-polar, so, for example, sulfolipids containing only these bonds do not dissolve in water. C-O and C-N bonds are polar because of the electronegativity of O and N, so any molecules containing these bonds, such as carbohydrates and amino acids, dissolve in water. The functions of the molecules of life, including the readily dissolvable ions that are used to carry charge, the non-polar molecules that form membranes, and the dissolvable polar molecules of metabolism, all arise from their chemical interactions with water.

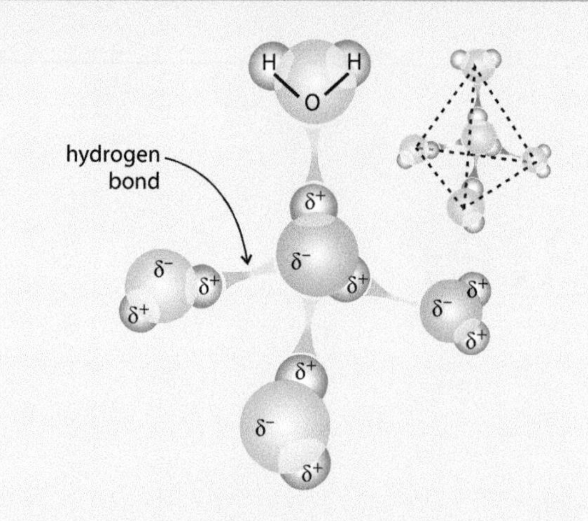

Figure 1. **The tetrahedral structure of water molecules.**

Intermolecular hydrogen bonding in water is sufficient for it to be a liquid under many conditions on Earth (in contrast, H_2S, which is chemically similar and almost twice as heavy, is usually a gas). The very large amount of hydrogen bonding between molecules gives water a high tensile strength and a high bulk modulus—that is, a large amount of energy is required to pull water molecules apart or to compress them. It also gives it a high specific heat capacity. The high tensile strength and specific heat capacity of water are vital for transpiration and its cooling effect. Its high bulk modulus underpins the turgidity of plant cells.

Further reading: Ball P (2008) Water: water—an enduring mystery. *Nature* 452:291–292.

bonds that water molecules can form, and the supramolecular structures that arise from them, are vital to many reactions in plant physiology—for example, the structure and function of many enzymes are dependent on direct interactions with semi-structured water solutions. Water is also vital as a reagent in many hydrolysis/condensation and redox reactions in plants, and its high specific heat capacity means that evaporation of water requires much heat—hence transpiration can contribute significantly to heat loss. Plant cells have walls that generally inhibit their expansion, but water is drawn into their cells down solute concentration gradients, so plant cells mostly have positive hydrostatic pressures, as do many fungal cells—this is in contrast to most animal cells, which are under atmospheric pressure (Figure 4.1). The general terrestrial plant form, consisting of growing roots and shoots that bifurcate to provide a large surface area of interaction with the environment, is dependent on **turgid** cells because root and shoot growth is driven by hydrostatic pressure. Support for shoots in particular is also dependent on turgid cells. In addition, hydrostatic pressure controls stomatal aperture (Figure 4.2). On land, plants face a significant challenge in obtaining sufficient fresh water to meet these demands.

The significant relationship between ecosystem productivity and water availability highlights the fact that water is frequently a limiting resource for

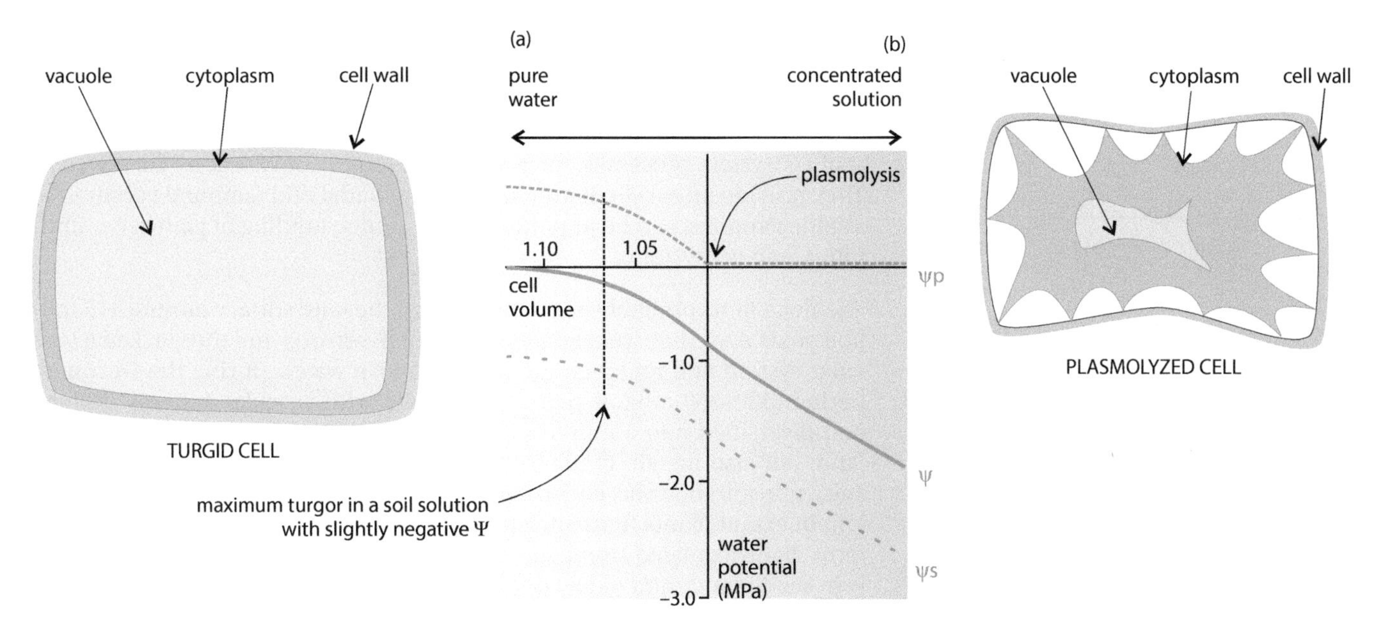

Figure 4.1. Water relations of plant cells. A plant cell immersed in pure water (a) takes up water and increases in volume only a small amount, causing an increase in ψ_p, ψ_s, and ψ. Water uptake stops when ψ_{cell} is equal to $\psi_{external}$. With pure water this is at $\psi = 0$, but in dilute solutions such as soil solution (black broken line) this equilibrium is reached at a slightly negative ψ. Thus, in contrast to most animal cells, because plant cells have a cell wall they do not swell up and burst if immersed in dilute solutions, but become hydrostatically pressurized (that is, turgid). If plant cells are immersed in a concentrated solution (b), such as that of saline soils, they lose water, ψ and ψ_s decrease, and they become flaccid. If water loss is severe, the cells plasmolyze. During plasmolysis the vacuole and cytoplasm dehydrate and the cell decreases in volume by a relatively small amount because of the rigid cell wall. Plasmolysis occurs under atmospheric pressure ($\psi_p = 0$) because the bathing solution penetrates the cell wall.

plants, suggesting that competition for it might be intense. Competition-driven niche separation has underpinned much ecological theory, particularly explanations for the coexistence of species. The concept of the niche was developed primarily from the study of animals, in which competition frequently occurs through differences in food sources. Plants do not compete for food but for a relatively limited range of inorganic resources, which has meant that investigations of niche separation in plants have been problematic. In fact, **neutral theory**, which posits that all species have equal fitness in competing for the same resources, can in some environments predict the coexistence of plant species from **stochastic** processes rather than niche separation. However, there is also evidence of true niche separation in coexisting plant species, in particular for ecohydrological niches—in some

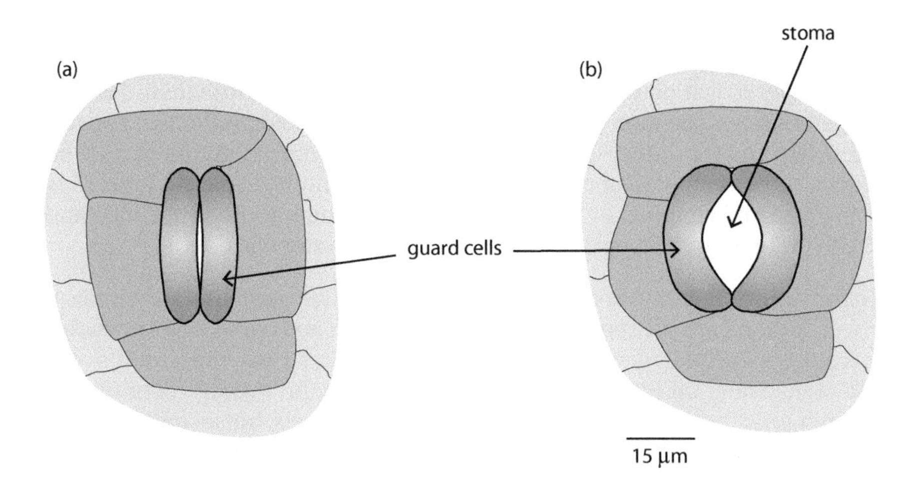

Figure 4.2. Generalized stomatal opening. (a) When the stomata are closed, the guard cells are flaccid. (b) If the guard cells become turgid, the stomatal aperture increases, allowing the flux of gases into and out of the leaf.

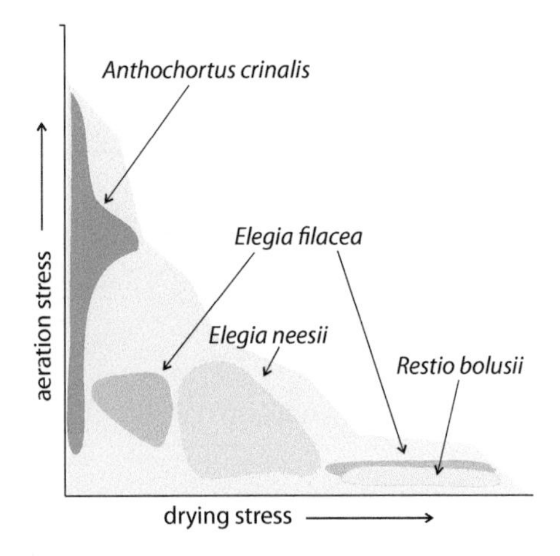

Figure 4.3. Ecohydrological niche separation in the Restionaceae. In the fynbos of Southern Africa, species in the Restionaceae are common. Despite their relatedeness and morphological similarity, they coexist in significant part because they inhabit distinct ecohydrological niches. Too little or too much water can produce drying or aeration stresses, respectively. Common restios of the fynbos have a significant tendency to inhabit niches with distinct combinations of water stresses. (Adapted from Araya YN, Silvertown J, David J. Gowing DJ et al [2011] *New Phytol* 189:253–258. With permission from John Wiley & Sons.)

contrasting ecosystems, coexisting plants have been shown to be exploiting spatially or temporally separated differences in soil hydrological regime (Figure 4.3). It is likely that ecohydrological niches are an important driver of niche separation in plant communities and thus underpin many terrestrial ecosystem processes, including, for example, invasion by alien weeds. They provide an excellent example of why spatial and temporal variation in abiotic variables can be important for an understanding of plants, communities, and ecosystems.

The first plants probably began to colonize the land surface around 475 million years ago. They were related to extant liverworts and thus lacked a vascular system and relied on diffusion to move water. During the Devonian period (420–360 million years ago), plants with vascular systems evolved and diversified into a number of plant groups that colonized not only the land, but also the air. By the Carboniferous period (360–295 million years ago), transpiration was occurring up through vascular plants tens of meters tall. In extant plants from ancient groups, such as lycopods, horsetails, and ferns, light-regulated stomatal control of transpiration is not as precise as it is in seed plants, and in low CO_2 conditions the stomata of conifers are not as responsive as those of flowering plants. The evolution of plant groups and hence the development of terrestrial ecosystems has been accompanied by increased control over stomatal conductance. Transpiration by terrestrial plants now plays a major role in the hydrological cycle. There are, on average, about 15,000 km^3 of water vapor in the Earth's atmosphere. Each year about 32,000 km^3 of water are transpired into the atmosphere from terrestrial plants, and this therefore has a major impact on hydrological processes of the atmosphere. Of the 110,000 km^3 of water that are precipitated on to the land surface each year, about 40,000 km^3 become run-off or through-flow, so transpiration makes up about 50% of the remaining 70,000 km^3. The movement of water in the soil–plant–atmosphere system is important not only for plant cells, plant communities, and ecosystems, but also for the climate of Earth.

Water management is vital for ensuring global food security and minimizing the impact of human activity on the environment

Many of the earliest civilizations developed water management techniques to boost food production. By 1900, large-scale irrigation schemes had been widely initiated (Box 4.2), and irrigated agriculture contributed a significant proportion of global food production. During the twentieth century, regional-scale water management schemes became common and the land area under irrigation increased sevenfold—it now produces about 35% of all food. Between 1960 and 2010 the global area under rain-fed cropping remained constant at about 1200 million hectares, and the increase in cropland that helped to feed a burgeoning population was due to an increase in area under irrigation to more than 300 million hectares. Humankind is now dependent on irrigation for food. It underpins the most productive agricultural systems, it produces many of our most high-value plant products (and hence more than 50% of production by value), and its expansion has long been a key mechanism for meeting increased demands. Between 2000 and 2050 it is likely that global demand for irrigation water will double.

In the twenty-first century, efforts to expand irrigated agricultural systems to meet demand are substantial, but their expansion is slowing, most probably because of the limited availability of fresh water. Fresh water represents only 1% of the water on Earth. Globally, its availability and management underpin not only the production of food and plant products, but also key aspects of development. Currently about 1000 litres of water are needed to produce

BOX 4.2. CONTROLLING THE WORLD'S RIVERS

The development of agriculture, which occurred independently in many different places, often took place in river valleys on the flood plain. Small dam schemes have been used to control water flow for agriculture almost from its very beginning. Ancient civilizations across the world had extensive schemes to control, capture, and distribute water, in significant part for agriculture. In more recent times, but before the development of hydroelectricity, there were extensive regional schemes to provide irrigation water. For example, in the late nineteenth and early twentieth century in the Punjab, a province which then covered an area in the northwest of the subcontinent that is now split between North-East Pakistan and North-West India, a regional-scale water management system was constructed based on canals carrying water to almost 2.5 million hectares of land (Figure 1). Water management in the Punjab had for centuries made it one of the most agriculturally productive regions in the subcontinent, but the scheme transformed agriculture in the Punjab and included the development of entire towns (the "canal colonies"). The social, economic, and environmental ramifications of regional water management schemes such as that in the Punjab are the subject of ongoing debate. The numerous huge dam schemes of the twentieth century were often aimed not only at hydroelectricity generation and flood prevention, but also at providing reservoirs for irrigation of crops. The Three Gorges Dam, which spans 2 km of the Yangtze river in China, is part of the largest civil engineering project in history.

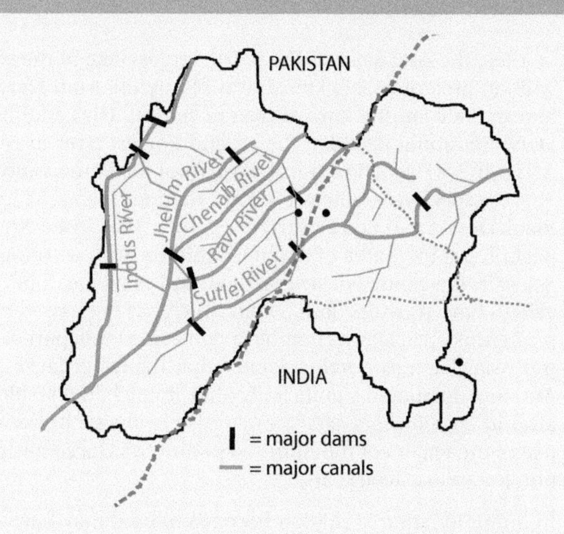

Figure 1. The waterways of the Punjab.

The immediate benefits of hydro schemes seem clear in most instances, but there is much evidence that the long-term environmental impacts are significant. Taken together, the benefits and costs of all the hydro schemes on Earth represent one of the most significant human impacts on the natural environment. Water management, driven in significant part by demand for irrigation water, will be a key element of human environmental impact in the twenty-first century. The Three Gorges Dam, massive regional schemes such as that for the Narmada River in India, and current discussions about linking waterways across some whole countries suggest that water management on regional scales and beyond will continue to be attempted.

each kilogram of wheat. In many instances, and for many other cereals, even more water is necessary. About 10,000 litres of water are used to produce each kilogram of beef. Of the fresh water that is used by humans, around 80% is for agriculture, most of this for irrigation. About 3000 km^2 of water are extracted each year from rivers, lakes, and **aquifers** (a value that is predicted to rise beyond 4000 km^2 by 2050), and around 70% of this is used for irrigation, although in some parts of the world this figure is more than 90%. Many water extraction regimes are unsustainable, some spectacularly so (Box 4.3). Globally, because of excessive extraction, up to 10% of rivers sometimes never reach the sea, and during the twentieth century many aquifers were significantly depleted by extraction. As indicated by many measures, by 2010 the world had entered a global freshwater crisis because supply could not meet demand. Hydrologists define the fresh water in rivers, lakes, and aquifers as **blue water**, which represents 20–40% of fresh water. **Green water** is water in the unsaturated zone of soils. Almost all water management focuses on blue water, whereas most crop production is dependent on green water, and in the 3000 million hectares of pasture that feed animal livestock, essentially all water is green water. Advancing the management of blue and green water in the soil–plant system is vital for the future of agriculture.

About 40% of Earth is covered in "drylands"—that is, areas with arid, semi-arid, or sub-humid climates. About 2 billion people depend on drylands, often on the plants that grow in them. **Desertification** is a major challenge,

BOX 4.3. IRRIGATION WATER AND THE ARAL SEA

In 1960, the Aral Sea was the fourth largest lake in the world, with an area of 65,000 km^2. It was fed by the Amu Darya (at almost 2500 km the longest river in central Asia) and the Syr Darya (at about 2200 km the second longest river in central Asia). These rivers flow north-west, from the Pamirs and Tian Shan, respectively, across the Kara Kum and Kyzyl Kum (the Black Desert and Red Desert) (Figure 1). The closed Aral Sea basin has a total area of about 6.5 million km^2, with the Aral Sea at its low point. For several thousand years the upper river valleys have provided fertile land which has been used by the numerous civilizations that have dominated this part of central Asia. There is much evidence that the water level of the Aral Sea fluctuated significantly after it reached a stable size around 20,000 years ago. There is also evidence of extensive use of the waters of the Amu Darya and Syr Darya for irrigation for almost 3000 years.

In antiquity, up to 3 million hectares of land may have been irrigated using the waters of these rivers. However, by the 1950s, Soviet irrigation schemes were diverting unprecedented flows of water to almost 8 million hectares of crops. Between 1960 and 2000 the inflow to the Aral Sea dropped from 55 km^2 $year^{-1}$ to 2.3 km^2 $year^{-1}$, and its water level dropped by more than 25 m, compared with natural fluctuations of 4–5 m. In the years between 1960 and 2010 the Aral Sea essentially disappeared, primarily because of the unsustainable extraction of water for irrigation (Figure 2). Since 1960 the Aral Sea has lost over 90% of its volume and almost 90% of its area, and its salinity has increased from about 10 g l^{-1} to at least 80–90 g l^{-1}, and in some instances 200 g l^{-1}. The remaining waters are not only saline but also have high concentrations of agrochemicals from upstream. Modeling of the hydrological changes has been difficult because the many monitoring points for inflows, water height, and other parameters have in many cases been above the water line for several decades. The once economically and ecologically important aquatic ecosystems of the Aral Sea have collapsed, and only a remnant of the once extensive delta of the Amu Darya now exists. Problems of desertification are severe over tens of thousands of km^2, and the unprecedented dust storms of the central Aral basin spread sand, which is often contaminated, for thousands of miles. The head waters of the Amu Darya and Syr Darya continue to provide vital irrigation water for the nations of central Asia, but for the Aral Sea and the many humans and other species that are dependent on it, water diversion for agriculture caused one of the worst ecological disasters of the twentieth century. It provides a type example of the ecological impact of over-extraction of water for irrigation. The agricultural benefits for central Asia of water extraction for irrigation have been clear for millennia. However, in the late twentieth century, extraction exceeded the level that was sustainable for many of the most important ecosystems of the Aral basin.

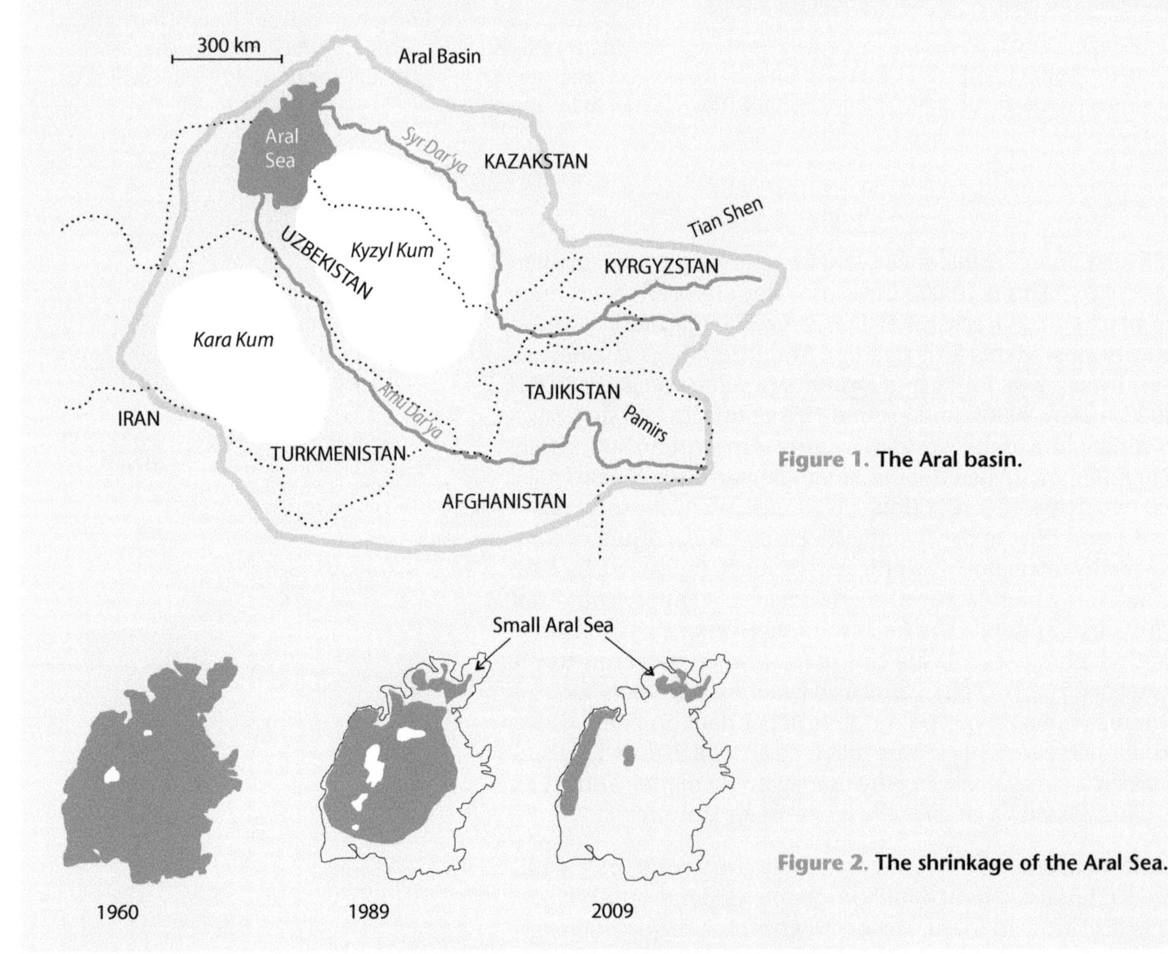

Figure 1. The Aral basin.

Figure 2. The shrinkage of the Aral Sea.

and the United Nations Convention to Combat Desertification (UNCCD) is a significant response that aims to combat it. Definitions of desertification all include a decline in usefulness of land to humans, almost always accompanied by a change in vegetation cover. Desertification is most common in, but not restricted to, drylands. At least 250 million people are affected by desertification, and about 1 billion people are probably at risk. Changes in climate are implicated, but also changes in land use and water management. An understanding of the use of water by crop and pasture plants, together with an understanding of the ecophysiology of water relations in wild plants, will be vital for meeting some of humankind's greatest challenges in the twenty-first century. Changing climate and increasing competition for water are likely to exacerbate these challenges, increasing the importance of water in the soil-plant system.

Water potential gradients drive water movement, including transpiration in trees over 100 m tall

The potential energy of water is defined as "water potential" (ψ), and its gradient is used to predict water movement. Due to their polarity, water molecules are attracted to surfaces that have an unequal distribution of charge, such as those which form soil pores. In fine soil pores, surface attraction is sufficient to drag water significant distances. This "capillary action" is a matric potential (ψ_m), and dominates ψ_{soil}. Water above **field capacity** moves out of soils under gravity, but water below field capacity has to be extracted down gradients of water potential, so soils of different textures have different water-release curves (Figure 4.4). Variation in ψ_m in the surface soil is dominated by relatively short-term fluctuations, whereas in deeper horizons variations occur either over seasons or longer, or may not occur at all (Figure 4.5). In general, ψ_m becomes less negative (an effect that is driven by rainfall or irrigation events) faster than it becomes more negative (an effect that is driven by drying). In plant roots, the effects of solutes (ψ_s) and hydrostatic pressure (ψ_p) dominate ψ. The total water potential is the sum of factors that affect it, and its gradient between root and soil drives water

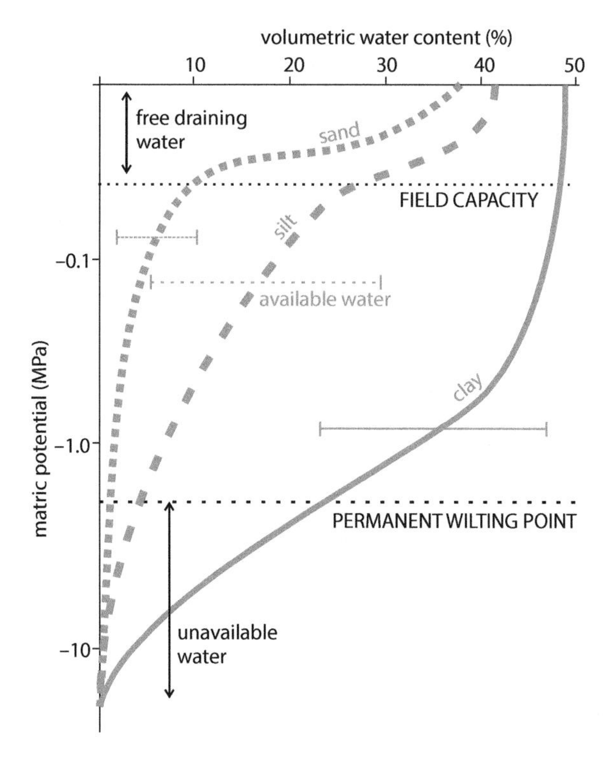

Figure 4.4. The dynamics of water release from different soil types. If almost any soil acquires, by volume, a high water content, a large amount of water runs out under the force of gravity. The volumetric water content of a soil after water has drained freely from it is the field capacity. In general, most water with a matric potential lower than about –0.03 MPa cannot run out under gravity. In a coarse-textured sand with a relatively low surface area to volume ratio, field capacity is less than 10% water by volume. In a fine-textured clay it can approach 50%. Depending on the species, plants cannot generate sufficiently negative ψ_{root} to extract water from less than about –2 to –5 MPa, and they wilt permanently if the soil water has ψ less than this. In general, sands hold a small amount of water but on average release it quickly and easily, whereas clays hold a large amount of water but its release is slower and more difficult. Silts, which are the most common agricultural soils, hold the most water, and it is moderately quickly and easily available.

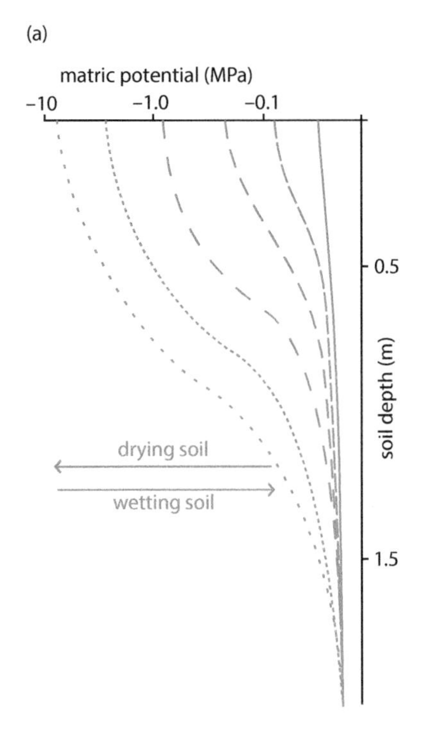

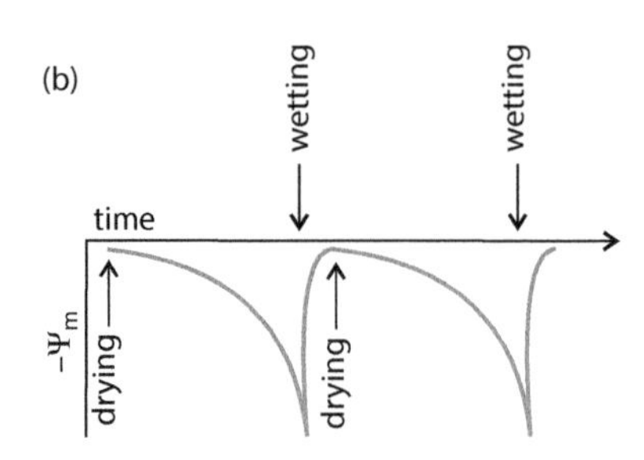

Figure 4.5. **Variation of soil ψ_m with changing water content.** (a) In most soils it is the surface layers that dry out first, producing a rapid decrease in ψ_m at the surface, while the deeper layers take much longer to dry out. Conversely, increased water contents from surface precipitation take much longer to change ψ_m in the deeper layers than at the surface. (b) In general, decreases in ψ due to drying take longer to occur than increases in ψ due to wetting. Soil properties, such as texture and pore structure, affect the rate and depth of change, but the spatial and temporal variation in ψ_m is almost always greatest in the surface layers.

extraction. Differences in ψ_m within and between soils and variation in ψ_m over time provide the water extraction challenge for crops and wild plants.

Water potential in the atmosphere drives evapotranspiration, but varies (Figure 4.6). During rainfall, the ψ_{atmos} can approach zero, but as the relative humidity declines it becomes very negative. This can be sufficient to drive the rapid extraction of essentially all water from soil surface layers and pull water to great heights in plants. Water flow across roots to the central vascular strand occurs when there is lower ψ in the roots than in the soil, and it can occur via the **apoplast**, the **symplast**, and transmembrane pathways (Figure 4.7). The existence of an apoplastic pathway is sometimes important for plants during periods of low relative humidity, and the entry of some tracers and ions to the xylem via the apoplast has been shown, but often their exclusion from the root is essentially complete. When there is no apoplastic flow to the vascular tissue, passage through the membrane via the symplast and transmembrane pathways is necessary.

In order to colonize the air, plants needed systems for delivering water to cells significantly above soil height. The xylem of plants is a conduit for water flow to the atmosphere down a gradient of ψ. In non-angiosperms, xylem is composed of **tracheids**, which often have the dual function of conducting water and providing mechanical support for the plant. In angiosperms, xylem cells are differentiated into **vessels** that conduct water, and fibers that enhance structural support. Although tracheids present much more resistance to flow than vessels, this is often not the most important source of resistance. The tension in xylem water caused by it being pulled up into the atmosphere induces **metastability** and promotes **embolism** by **cavitation**, which can significantly decrease water flow. **Bordered pits** of xylem elements play a significant role in controlling cavitation and in maintaining water flow. The bordered pits of gymnosperms have less resistance to water flow than those of ferns or the vessels of angiosperms, and the production of secondary xylem in wood can provide new unblocked conduits for water flow.

The geometry of the angiosperm vascular system is generally fractal, and the diameter of the xylem elements at each layer of the fractal hierarchy often tapers towards the top of tall plants. The low CO_2 concentrations of the

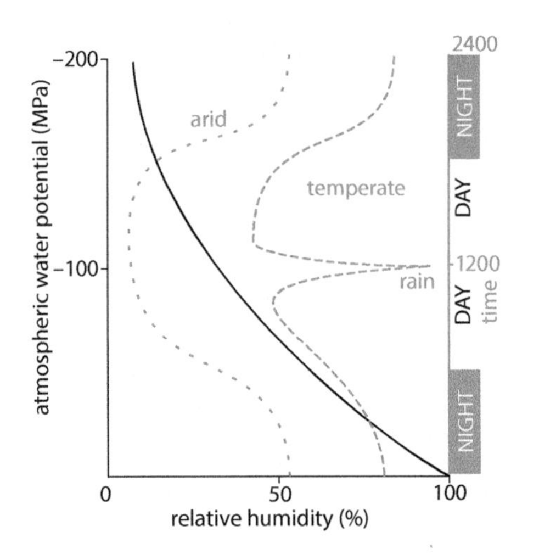

Figure 4.6. **Atmospheric water potential and relative humidity.** The very low water potentials that generally prevail in the atmosphere produce a massive driving force for transpiration, which is directly related to vapor pressure deficit (VPD). Relative humidity (RH), which takes into account temperature and pressure effects on VPD, is generally only 100% when it is raining. Even at 75%, a common RH in the temperate zone, the ψ of the atmosphere is very much lower than that of soil, producing a significant gradient in ψ. In arid zones the gradient in ψ between soil and atmosphere is very large indeed. In all but the very coolest and most humid conditions, atmospheric ψ is sufficient to drive transpiration.

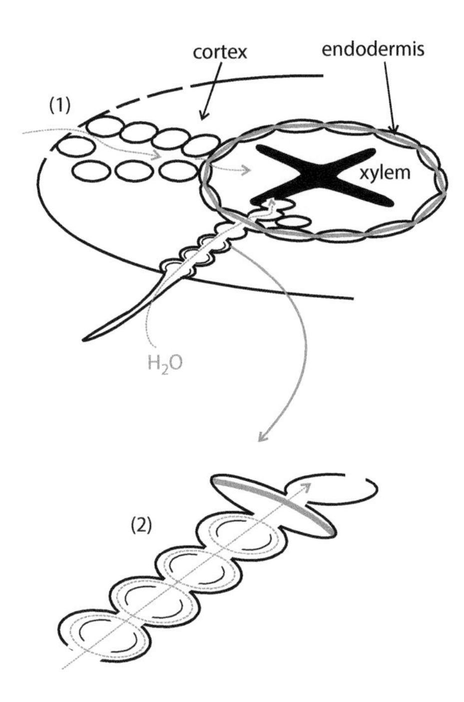

Figure 4.7. **Water flow into plant roots.** (1) Water moves into roots through the apoplast (that is, in the cell walls and intercellular spaces). The Casparian strip of the endodermis is sometimes a complete seal that prevents apoplastic radial flow of water, but in many instances there is water penetration of the endodermis that contributes to water flow. (2) Water can also move in the symplast around vacuoles through the cytosol, and from cell to cell, via a transmembrane route. These water flows avoid the Casparian strip and move uninterrupted into the vascular strand. They require transmembrane transport through aquaporins. Together these pathways constitute the "composite transport model" for water in roots.

last 80 million years have driven increases in stomatal conductance. Plants have had to evolve hydraulic systems that can deliver water at increasing leaf conductance. Tapering xylem elements, which can deliver water under the increasing tension at greater heights because they are more resistant to implosion, and adjustments in the fractal hierarchy towards higher-level (that is, smaller) elements, have helped to optimize xylem for water transport under low CO_2 conditions. The tallest gymnosperm is *Sequoia sempervirens*, native to the west coast of North America, with extant specimens over 110 m tall, and the tallest angiosperm is *Eucalyptus regnans*, native to Australia, with specimens over 100 m tall. Several other tree species have recorded specimens over 90 m tall. Measurements at the top of the world's tallest trees reveal that, at a height of 100 m, xylem water is close to the point of spontaneous embolism. Furthermore, at low water tensions the generation of the turgor necessary for cell growth is seldom possible, so cytoplasmic volume is minimal and the ratio of cell wall to cytoplasm, and biomass per unit leaf area, is high. This means that the top leaves on the highest trees are ineffective exchange surfaces, and it is unlikely that trees could ever grow much taller than 120 m, because the water and CO_2 fluxes and the cell turgor necessary to sustain growth would be unattainable. However, the world's tallest trees do all inhabit a distinct climate, indicating that not only hydrological but also temperature and biomechanical considerations may limit tree height. The decline in net primary production as forests age and the decrease in plant height with altitude are often related to changes in hydraulic capacity. Hydraulic properties have been vital to the evolution of plants, and provide the context for understanding how plants react to variations in water availability.

Short-term adjustments of resistance to water flux allow water homeostasis

The soil-plant-atmosphere flux of water is a product of the gradient of ψ and the resistances to flow. At a given gradient of ψ, a plant's water status is determined by the difference in resistances to loss (transpiration) and to influx (uptake). Resistances to flow can be rapidly adjusted by plants in both leaves and roots. The difference in vapor pressure deficit between the inside and outside of a leaf drives transpiration from the leaf. The resistance to transpiration is determined by leaf conductivity (G_s). There is some water flux through the cuticle covering epidermal cells, but in most instances the majority of water flux occurs through stomata. In the absence of a leaf **boundary layer** there is an approximately linear relationship between stomatal aperture and G_s. Plants can adjust stomatal aperture rapidly in order to change resistance to transpiration. Mechanistic models of generalized leaves show that the effect of altered G_s on transpiration depends on vapor pressure deficit (Figure 4.8). Stomatal responses vary between species and are often driven by demands for CO_2. However, in general, soil water deficit, salinity, and low temperature—all of which reduce water availability—rapidly initiate a decrease in stomatal aperture which increases resistance to

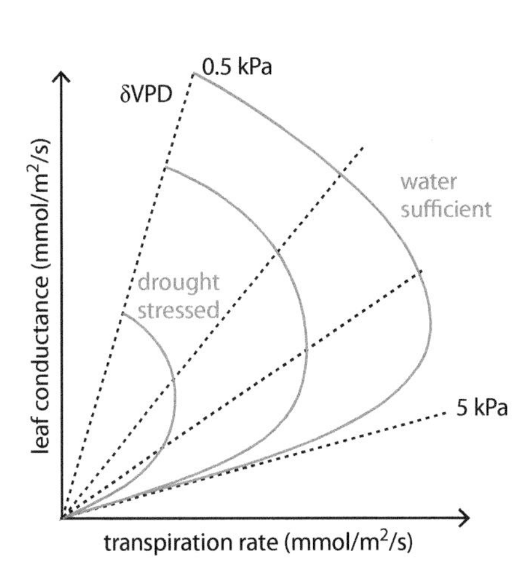

Figure 4.8. **Vapor pressure deficit and transpiration.** The difference in vapor pressure deficit (VPD) between the inside and the outside of stomata drives transpiration. At low δVPD (for example, 0.5 kPa), a large change in stomatal conductance (that is, a large increase in stomatal aperture) only increases transpiration by a small amount. At high δVPD (for example, 5 kPa), small changes in conductance dramatically increase transpiration. Drought-stressed plants maintain sufficiently low leaf conductance that even at high δVPD, transpiration rates are very low.

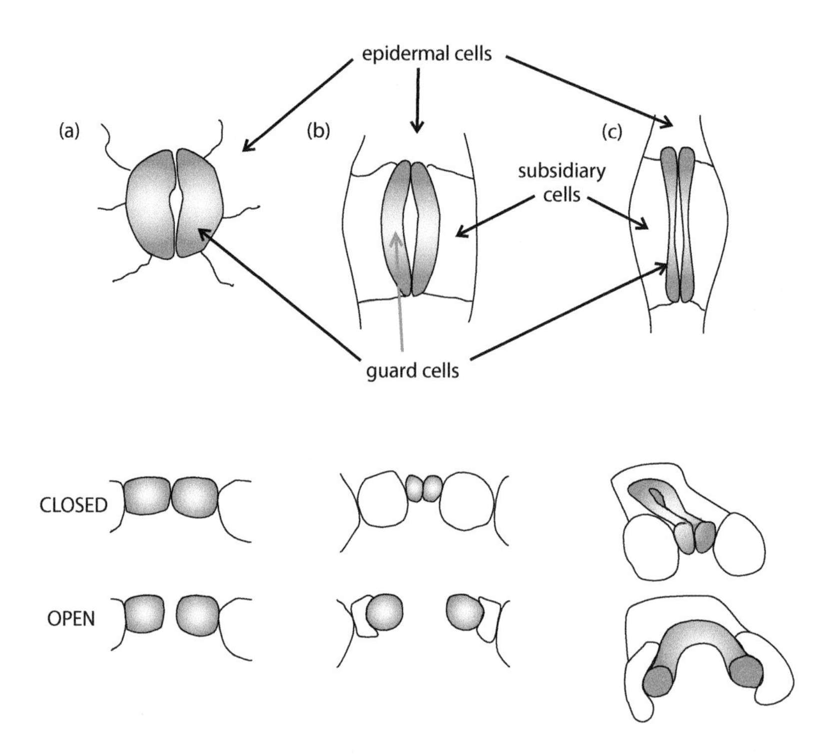

Figure 4.9. Stomatal guard cell types. Guard cells and their effect on stomatal aperture differ between the major groups of plants. (a) In primitive plants such as *Huperzia prolifera* (Lycopodiales), guard cells have no dedicated subsidiary cells, and turgid guard cells form a relatively small stoma. (b) In the dicotyledonous angiosperms such as *Tradescantia virginiana*, the guard cells and subsidiary cells work together to produce a large stoma. (c) In the grass *Triticum aestivum* (Poaceae), the dumbbell-shaped guard cells work with the subsidiary cells to open along their length, maximizing the area of the stoma. (Data from Franks PJ & Farquhar GD [2007] *Plant Physiol* 143:78–87.)

water loss. Stomatal anatomy varies (Figure 4.9), with more primitive plants having less refined stomata. In gymnosperms and angiosperms the guard cells are similar, but in angiosperms the subsidiary cells decrease in size when the guard cells become turgid, enhancing changes in stomatal aperture. Grasses have characteristic guard cells and rapidly responding accessory cells. The ecological success of angiosperms, and of grasses in particular, has been attributed to their enhanced stomatal responsiveness because it enables high **water use efficiency** (WUE) in environments with variable water regimes. Thus rapid changes in stomatal aperture play a major role in fine-tuning resistance to water loss and maintaining water status.

The **axial** transport of water through the xylem from root to shoot generally meets much less resistance than **radial** transport across the root. Water flux across roots has to meet transpiration demand, and control of it is a major determinant of plant susceptibility to dehydration. The transport of water through plant cell membranes is mediated by **aquaporins**. These are members of the **major intrinsic protein** (MIP) family that have six membrane-spanning α-helices which function in tetramers to transport an array of neutral molecules across membranes (Figure 4.10). Aquaporins catalyze water transport across membranes in all the kingdoms of life. Plants are particularly rich in aquaporins, with some species having been shown to have up to 70 different types. These include plasma-membrane intrinsic proteins (PIPs), tonoplast intrinsic proteins (TIPs), and several other types. Orchestration of aquaporin activity exerts significant control over water flow into and out of plant cells, and between their compartments, including radial transport of water in roots. Aquaporin expression in root cells influences root hydraulic conductivity (L_r). Changes in aquaporin expression have been related to diurnal changes in L_r, to drought, salt, and temperature-induced changes in L_r, to CAM-related changes in L_r, and to changes in radial water transport across the branches and stems of trees. Experiments with anti-sense RNA, KOs, and over-expression of aquaporins have often produced changes in L_r. In some plant species, and under particular conditions, apoplastic radial water flows are important, but in the many species and many instances in

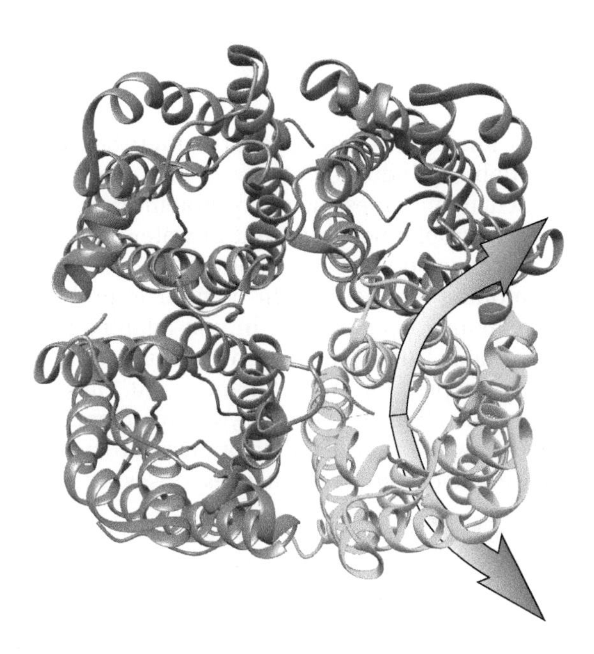

Figure 4.10. The structure of a plant aquaporin. The aquaporin shown in the figure functions as a tetramer. The figure shows the ribbon structure of SoPIP2;1 from spinach viewed as looking at the membrane surface, with one molecule highlighted in light green. Water can move in either direction through each individual molecule. One of the loops (loop D) on the cytoplasmic side moves to either block or open the pore, depending on physiological conditions. (Drawn in Chimera from PDB 2B5F: Tornroth-Horsefield S, Wang Y, Hedfalk K et al [2006] *Nature* 439:688–674.)

which symplastic and transmembrane water flows dominate radial water transport, expression of aquaporins is probably used to adjust L_r and minimize dehydration. Aquaporins are also important in the plant response to spatial variation in water availability across the root system—compensatory adjustments in L_r in parts of root systems exposed to different water regimes have been related to aquaporin expression.

The transpiration stream is clearly a vital conduit for nutrients and some other molecules from root to shoot. Given that the distances from root tips to shoot tips in tall trees are greater than the distances within any other extant organism, the rapid transport system that the transpiration stream provides can be advantageous. However, under certain conditions, including those in experiments that use anti-transpirants, herbaceous plants can grow well in the absence of transpiration. Despite this, under many field conditions transpiration is very useful not only to trees but also to most other plants. In plants that inhabit the wet tropics, up to 80% of heat loss occurs via evaporative cooling during transpiration, and on a global scale stomatal densities correlate not only with water availability but also with air temperature. Biophysical models have been used to show that many plant leaves under a wide range of conditions would overheat if they transpired less, and the venation of angiosperms is designed to deliver much more water to leaves than is needed for growth and photosynthesis. Plants with reduced abscisic acid (ABA) synthesis, and hence less control over transpiration, can overheat. However, transpiration does sometimes occur at times when cooling is not necessary—for example, at night. Mechanistic models of nutrient movement in the soil–plant system have long suggested that a significant proportion of many nutrients must be delivered to the roots via mass flow of soil solution driven by transpiration—movement by diffusion is not sufficient. In some instances, the nutrient status of a plant has been shown to affect the L_r and hence the arrival of nutrients at the root by mass flow. Nutrient-dependent effects on aquaporins—for example, via changes in cellular pH—have been demonstrated. The evolution of stomata was probably driven by low CO_2 concentrations, but plants also evolved to deliver water to leaves to take advantage of transpiration for both heat loss and nutrient delivery. Loss of more water than is necessary to fulfill the above-mentioned functions can occur, but the regulation of hydraulic conductivity in both leaf and root enables significant homeostatic control of water status.

Many plants adapt physiologically to short-term water deficit

Efficient metabolic functioning is dependent on there being enough water to minimize hydrophobicity-induced changes in biomolecules. Under conditions of water deficit, membranes and many proteins reorganize or denature due to increasing hydrophobicity per unit of water. Dehydration disrupts photosynthesis and redox poise, closes stomata, and alters protein–protein and protein–membrane interactions. In crops it is probably responsible for more losses than any other stressor. In non-adapted plants, prolonged dehydration can be fatal. During adaptation to dehydration there are significant changes in gene expression and cellular content. These tend to be most profound in those plants that are most dehydration tolerant. Dehydration is caused not only by low water availability, but also by low temperature and high salinity. Many dehydration avoidance responses are mediated by ABA.

In guard cells an early response to water deficit is an increase in ABA concentration, the release of Ca^{2+} and K^+ from guard cell vacuoles, depolarization of the guard cell membranes, the efflux of K^+, NO_3^-, and Cl^-, and the conversion of malate to starch (Figure 4.11). This results in a less negative ψ, a loss of water, and thus a decrease in stomatal aperture and transpiration. There are a number of environmental cues other than dehydration that can trigger stomatal closure against the usual circadian rhythms of opening, so the signaling pathways that underpin closure are complex. Nitric oxide (NO) is produced in guard cells in response to ABA, and improves dehydration avoidance. Its production, mostly from nitrate reductase, is dependent on H_2O_2. NO and H_2O_2 decrease influx through K^+ channels and increase efflux. Their action is dependent on classical signaling molecules, including protein phosphatases, mitogen-activated protein (MAP) kinases, and cyclic guanosine monophosphate (cGMP). Carbon dioxide (CO_2) and hydrogen sulfide (H_2S) also interact with these pathways and affect stomatal closure.

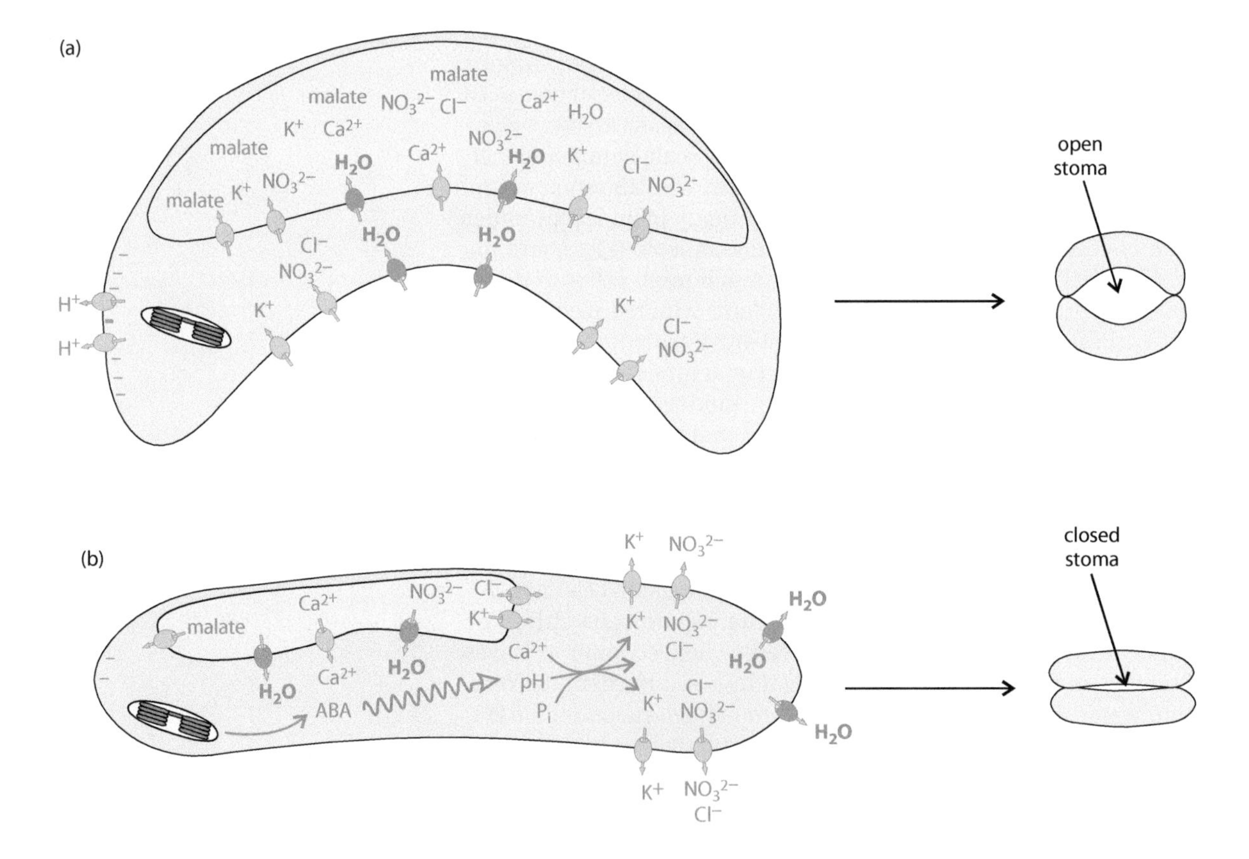

Figure 4.11. The physiology of guard cell opening. (a) A turgid guard cell. The membrane is polarized by H^+ efflux. K^+ channels and NO_3^-/Cl^--H^+ co-transporters drive the influx of ions. A battery of transport proteins drives the accumulation of K^+ and NO_3^-/Cl^- in the vacuole. Water is drawn into the vacuole down ψ gradients opening the stoma. (b) A flaccid guard cell. From left to right: Drought or other stimuli, such as CO_2, cause depolarization of the membrane and increase ABA concentrations. ABA can originate from the stroma of chloroplasts as photosynthesis slows down and it becomes more acid, but much ABA is synthesized *de novo*, including that in the surrounding cells. ABA triggers Ca^{2+} release and a signal transduction via numerous molecules to increase pH and P_i. Malate ions are transported out of the vacuole, eventually being used to synthesize starch. Ca^{2+}, pH, and P_i all help to activate the efflux of K^+ and NO_3^-/Cl^-. The decrease in ψ drives water efflux through aquaporins and stomatal closure.

Stomata evolved to respond to their gaseous environment, and although CO_2 and water are probably the most important constituents, the numerous gases that affect stomatal opening might together provide clues about the gaseous environment. In many plants, stomatal responses are vital for the avoidance of dehydration.

The concentration of active ABA in the cytosol of dehydrating tissues increases rapidly. In fully hydrated tissues, the highest concentrations of ABA are mostly found in chloroplasts. This is because ABA is a weak acid with a carboxylic end group, and the chloroplast stroma is more alkaline than the cytosol, so the protonated form (ABAH) from the cytosol permeates into the chloroplast, deprotonates to ABA^-, and cannot escape. During dehydration-induced slowing of photosynthesis the stroma becomes less alkaline, and ABA equilibria favor its location in the cytosol. This release of ABA to the cytosol and to the transpiration stream causes a rapid increase in ABA concentrations in cells, especially in guard cells. Under conditions of dehydration, ABA is also synthesized *de novo*. It is a terpenoid synthesized from β-carotene in the cytosol. The regulatory step is catalyzed by 9-*cis*-epoxycarotenoid dioxygenase (NCED), many manipulations of which have been shown to alter ABA concentration and dehydration tolerance in plants. Exogenously applied ABA acts as an anti-transpirant, and mutants of ABA synthesis transpire excessively and wilt. ABA is produced in many parts of plants, and also has important roles in the dehydration of tissues necessary for seed and bud dormancy. **Pyrabactin** resistance (PYR) and PYR-like (PYL/RCAR) proteins act as receptors for ABA. The binding of ABA to PYR/PYL/RCAR initiates events that lead to an increase in the transcription factors (TFs) that change gene expression in response to dehydration (Figure 4.12). There are binding elements for TFs that are up-regulated in an ABA-dependent manner. In addition, ABA-independent pathways that are often also initiated by cold or low temperature induce the expression of other TFs, some of which can bind to the same genes as those of ABA-dependent pathways. At the molecular level the environmental variations that cause osmotic stress to plant cells therefore have significant overlap (Figure 4.13), with modeling revealing several interacting and **emergent** characteristics.

Altered gene expression produces numerous physiological changes in dehydrating plants. Dehydration-tolerant plants synthesize compatible solutes in response to osmotic stress. This is a common response across the kingdoms of life. Many plants have been shown to increase the synthesis of proline, mannitol, and a variety of polyols. These compatible solutes lower

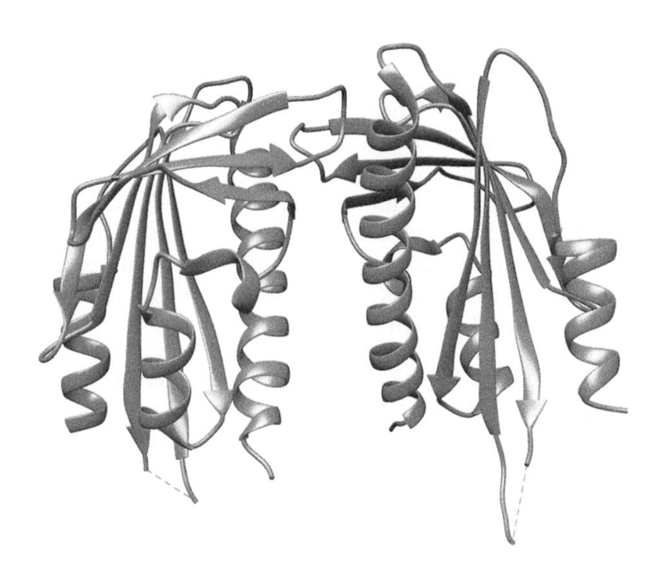

Figure 4.12. **ABA receptors.** Soluble pyrabactin-resistance/pyrabactin-like (PYR/PYL) proteins act as soluble receptors for ABA. In the PYL1 dimer, shown as a ribbon structure in the figure, when ABA is present it binds deep inside the pocket of one molecule, changing the conformation of the dimer. This can then bind to and inhibit the protein phosphatase type 2Cs (PP2Cs) that suppress ABA-dependent pathways, releasing their expression. PYL1, PYL2, and PYR2 are very similar molecules used in ABA perception. (Drawn in Chimera from PDB 3KAY: Santiago J, Dupeux F, Betz K et al [2012] *Plant Sci* 182:3–11.)

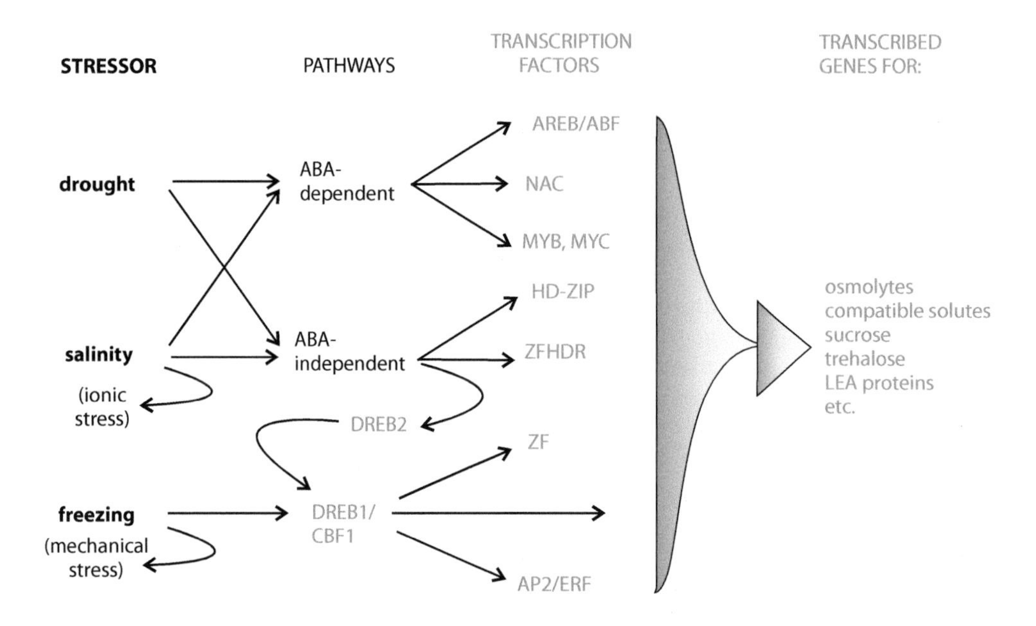

Figure 4.13. Interaction of osmotic stress pathways in plants. Although salinity and freezing cause ionic and mechanical stresses, respectively, they also both cause major osmotic stress to plants. There is significant overlap in the pathways that plants use to respond to osmotic stress caused by different environmental stressors. AREB/ABF, ABA responsive element (ABRE) binding/ABRE binding factor; NAC, NAM, ATAF and CUC; MYB/MYC, myeloblastosis related protein/myelocytomatosis related protein; HD-ZIP, homeodomain-leucine zipper; ZFHDR, zinc finger homeodomain recognition; ZF, zinc finger; DREB, dehydration responsive element binding; CBF, core binding factor; AP2/ERF, APETELLA2/ethylene responsive factor.

the ψ of the plant cells, helping them to manage the soil-plant gradient for water influx. In addition, many of them have been shown to reduce the ROS produced during dehydration, and minimize protein-protein interactions. Many plants with enhanced dehydration tolerance synthesize them readily or even constitutively. In many plants there are changes (sometimes profound) in carbohydrate content. Changes in the concentrations of many sugars have been reported, including raffinoses, fructans, and trehalose. **Gluconeogenesis** can be activated, increasing the capacity to alter carbohydrate pools. Changes in carbohydrate pools also affect ψ, but in most instances they minimize the adverse effect of decreasing free water on protein and membrane structure. **Late embryogenesis abundant (LEA) proteins** and trehalose, which are found in desiccated seeds, are also detected in some plants during dehydration.

Extended water deficit induces changes in root growth

The morphology of a root system is a product of soil properties and root capacities. Not only ψ_{soil}, but also soil compaction, aeration, and nutrient concentrations affect the final form of a root system. Experiments that eliminate the effects of gravity show that roots are not only **gravitropic** but also **hydrotropic**, responding to gradients of ψ. The root tip is the primary sensor for hydrotropism, with the MIZ proteins playing a key role. *miz* mutants, and manipulation of MIZ proteins, have shown that during hydrotropism there are MIZ-dependent changes in auxin, cytokinin, Ca^{2+}, and ROS that induce root curvature towards areas of favorable ψ by altering growth in the elongation zone. These changes are similar to many induced during gravitropism, but during hydrotropic responses the gravitropic response is specifically suppressed. Phospholipase D (PLD), a common signaling molecule, has been shown to have a role in this suppression, allowing roots even to grow upwards towards areas of wet soil.

In **mesophytic** plants exposed to extended water deficit, shoot growth stops but root growth continues. Depending on the species, responses to water deficit can include differences in the proportion of lateral roots, in length per unit biomass, and in root biomass production. These developmental changes have often been linked to changes in auxin, cytokinin, and ABA levels. Meta-analyses of the allocation of biomass between plant parts have

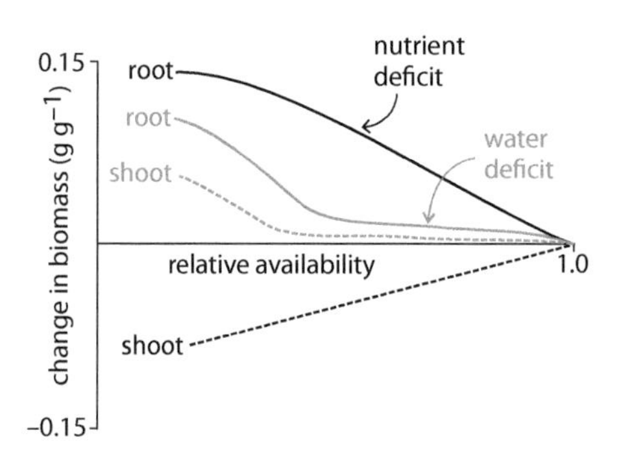

Figure 4.14. Changes in biomass allocation with water and nutrient deficit. Meta-analyses of responses in numerous species show that, overall, when nutrient or water availability decreases, plant biomass allocation changes. Although plant biomass allocation is particularly responsive to nutrient and water availability, it is less sensitive to decreases in water availability. (From Poorter H, Niklas KJ, Reich PB et al [2012] *New Phytol* 193:30–50. With permission from John Wiley & Sons.)

shown that this change in allocation takes longer to occur than changes induced by nutrient deficit, and that the water deficit has to be proportionally greater than that of nutrients to induce the change (Figure 4.14). This is probably because the variation in soil water potential can ordinarily be quite profound—especially at the surface, soils can become dry and then be recharged by rainfall within a short period of time. Although adjustments of hydraulic conductivity are probably sufficient for plants to adjust to variation in ψ, in order to survive a period of extended water deficit an allocation of resources that favors the roots is necessary.

The **Casparian strip** of the endodermis provides a sealed inner layer around the vascular strand of the root, but under conditions of water sufficiency the seal is sometimes imperfect, and bypass of the endodermis by water or nutrients is commonly reported. The Casparian strip is formed by **lignin** that impregnates the radial walls of endodermal cells. The CASP proteins, which are unique to plants, play a key role in this process by directing the band of deposition. The performance of **polarized epithelium** is crucial to multicellular organisms and, because of its role in regulating water relations of primary producers, the endodermis is one of the most important polarized epithelia in terrestrial ecosystems. Under conditions of water deficit, Casparian strips in a number of species have been reported to become more extensive, and the endodermis less leaky to apoplastic flow. In addition, under water deficit many plants have an **exodermis** with a Casparian strip that supplements the function of the endodermis (Figure 4.15). Pronounced endo- and exodermal layers, which can include extensive secondary **suberin** deposition not only in the Casparian strip but also in lamellae on the inner surface of all the walls of the endodermis, enhance the proportion of water entering the root symplast through aquaporins, increasing plant control over root hydraulic conductivity. Perhaps more importantly, however, they enable control of the efflux of water from the root. During prolonged dry periods the ψ in the soil can be much less than that of the root, and therefore, unless it is prevented from doing so, water will flow out of the root. The presence of a pronounced Casparian strip does make high hydraulic conductivities difficult to achieve, but root growth is quite rapid, and it has been suggested that it, and other adjustments, prolong root life to ensure that the plant can take advantage of rewetting when it occurs. Many species also develop **rhizosheaths** in response to water deficit. These are formed by root hairs trapping small particles around the root, often using exuded mucilage. Rhizosheaths, and exuded mucilage in general, can significantly affect water flux dynamics in the rhizosphere, aiding survival in times of water deficit. Changes in root anatomy and morphology are therefore important to plants that can survive prolonged periods of water deficit.

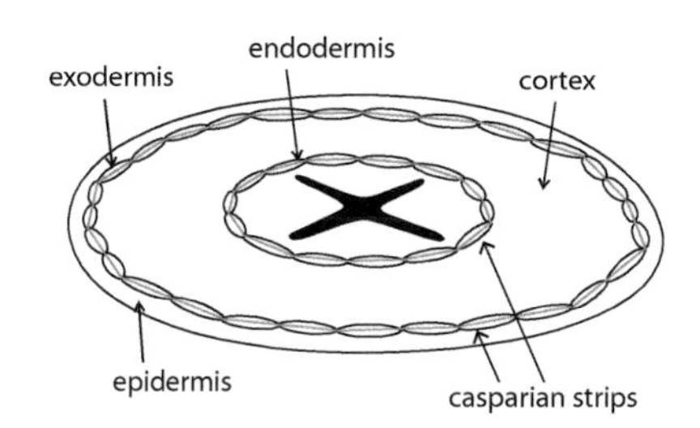

Figure 4.15. Root exodermis. Some plants can produce an exodermis in response to drought, whereas in others this tissue is always present. The exodermis often has a fully developed Casparian strip. In some plants there can be a layer of cortex between the exodermis and the epidermis.

In dry conditions, many plants develop deep roots, sometimes spectacularly so. Where deep layers of soil occur they have less pronounced fluctuations in ψ than do shallow ones, and act as reservoirs of soil moisture. When surface soils are dry and deep layers are moist, roots are presented with some severe hydrological challenges—at the surface roots dehydrate, but the low ψ in the atmosphere, especially during the day, drives water up from deep roots through the plant and past shallow roots because the ψ is lowest in the atmosphere. At night, or under other conditions that decrease the plant–atmosphere gradient of ψ, significant amounts of water have been shown

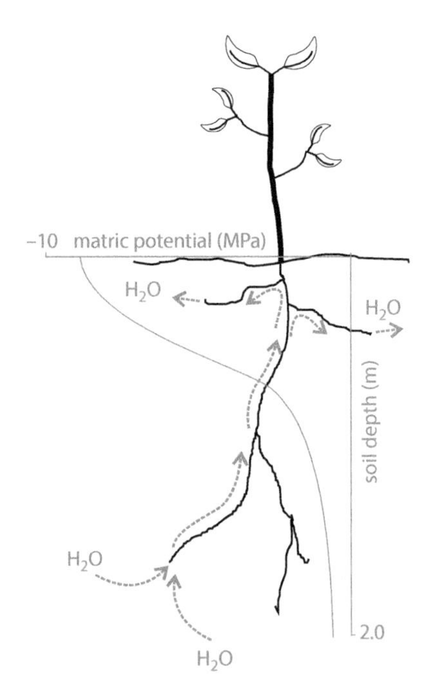

Figure 4.16. Hydraulic lift by plants. Deep-rooted plants often have their root extremities in soil with a small matric potential, so water enters these roots. If the soil surface layers are dry their matric potential will be very low, so water can move out of roots near the surface, and in effect the water is lifted by the plant from the deep soil layers to the surface.

to move from deep roots up into shallow ones. In many instances there can then be a significant leakage of water out into the soil (Figure 4.16). This hydraulic redistribution of water, although it is often only a small proportion of total water transpired, can be very significant to the water and nutrient balance of surface soil layers in a wide variety of ecosystems. In fact, terrestrial biosphere models have been shown to make more accurate predictions of water and nutrient flows when they take it into account.

Leaf adaptations aid drought survival and provide alternative ways of capturing water

Many plants that inhabit drought-prone ecosystems have leaves with specialized anatomies and morphologies. These are often interpreted as adaptations to maximize water use efficiency, but in recent years it has become apparent that they often function in other ways, too, in particular to increase nutrient use efficiency. Drought occurs not only in Mediterranean and semi-arid ecosystems but also in sub-arctic ecosystems, where water is often frozen. In all of these ecosystems the stomata in many species occur in crypts (Figure 4.17) or grooves. Advanced three-dimensional models of transpirational flow suggest that crypts only decrease water flow through stomata when the stomata are fully open. Under many conditions, therefore, crypts might not increase resistance to transpiration, so other functions have been suggested, including the delivery of CO_2 into leaves that tend to have low mesophyll conductance. In many dry areas, vegetation is **sclerophyllous** and characterized by long-lived, leathery leaves with a high mass per unit area, a low nutrient content, and often a marked bilateral symmetry characterized by abaxial pubescence. Relatively low stomatal densities and their abaxial location among hairs increase resistance to water loss. The removal of hairs from pubescent plant surfaces significantly increases their rate of water loss (Figure 4.18). Many sclerophyllous leaves roll downwards significantly from the adaxial surface, especially when conditions are dry. Such rolling also occurs in many sub-arctic plants and in many grasses of semi-arid zones. It serves to provide a humid environment surrounding the stomata, reducing the driving forces for water loss through the stomata when they are open. Such adaptations increase WUE but create problems for both CO_2 acquisition and temperature control. Sclerophyllous leaves are long-lived and of low palatability—a strategy that helps to conserve captured resources, including not only water but also C, P, and N. The **albedo** of sclerophyllous vegetation is often high, or the leaves are highly reflective, which helps to reflect incident radiation and control heat gain. Drought-tolerant plants have different hydraulic properties, including a decreased tendency to xylem cavitation, but leaf adaptations provide the most important control over water loss.

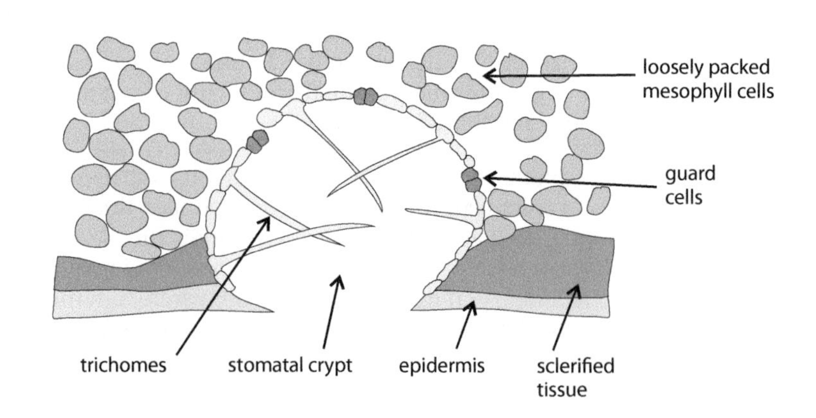

Figure 4.17. Stomatal crypts and grooves. A common xerophytic adaptation in plants is to have stomata located in crypts. Trichomes are common in stomatal crypts. The crypts and trichomes trap air, decreasing water loss when the stomata are open for CO_2 uptake. In many plants with sclerified leaves the chlorophyllous mesophyll is very loosely packed, to allow CO_2 penetration. Trapped air in crypts probably slows down water loss in many plants, but might be at least as important in helping CO_2 to penetrate into the mesophyll of sclerophyllous leaves.

In some ecosystems, or at some times of the year, more atmospheric water occurs as fog (drops of *c.* 20 μm in diameter) than as rain (drops greater than 100 μm in diameter). Many plants that inhabit these ecosystems are adapted to harvest water from fog. In the tropics and subtropics, "cloud forests" occur at particular altitudes on many mountain ranges, and the west coasts of the continents are characterized by a particularly dry but foggy climate associated with upwelling cold-water currents. Plants of cloud forests, which have prolonged cloud presence, have long been known to intercept significant quantities of water from fog. Cloud forests often occur in areas with seasonal rainfall, and during the dry season most of the water that drips through the canopy to the soil can be derived from fog harvesting. In general, the water in rain is enriched in the stable isotopes ^2H and ^{18}O, whereas the water in fog is depleted of them. If the $^1H/^2H$ and $^{16}O/^{18}O$ ratios in water are measured at a particular location, the proportion of water in a plant derived from rain and fog can then be calculated. The coast redwood (*Sequoia sempervirens*) is restricted to the coast of western North America. Stable isotope studies show that, during the dry summer, coast redwoods derive a significant proportion of their water from fog, and this increases up the canopy, relieving some of the largest plants on Earth from their dependence on soil water.

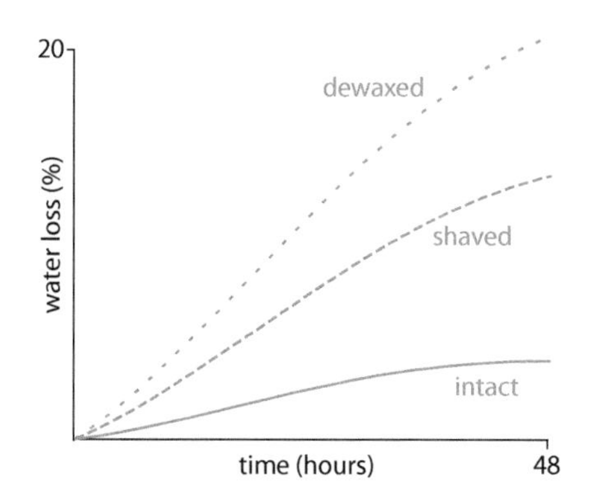

Figure 4.18. Pubescence, waxes, and water loss from plants. At 24°C and 40% relative humidity, peaches lose water relatively slowly. Shaving or dewaxing them increases water loss significantly. Peaches have a low surface area to volume ratio, and are only moderately pubescent and waxy, so in other plants and plant organs pubescence and waxes could have even more significant effects. (Data from Fernández V, Khayet M, Montero-Prado P et al [2011] *Plant Physiol* 156:2098–2108.)

Many studies have shown that, during periods of water deficit, fog droplets trapped by plants can coalesce and run or drip down into the soil, from where this water can be taken up. Many species have been shown to have extensive surface root systems that can take advantage of this water intercepted from fog. However, it is also clear that in some instances plant leaves directly absorb water from fog. Many of the plant species that inhabit the "redwood forests" have been shown to do this. The leaf cuticle is generally most impermeable to water when leaves are dry and slightly shrunken. Upon leaf wetting and swelling in fog the permeability of the cuticle increases, which increases the capacity for water absorbance. **Hydathodes** are often associated with water guttation, but they can also have a role in absorbing water from fog (Figure 4.19).

Many species of *Tillandsia* (Bromeliaceae), the "air plants," have roots only for attachment to the substratum, and obtain their water entirely from the atmosphere. Water uptake occurs through complex multicellular trichomes that channel water from the atmosphere directly into the mesophyll cells (Figure 4.20). Rosette plants with long thin leaves are common in many areas where fog harvesting occurs. Calculations using a range of artificial plant morphologies that vary in leaf size, number, and shape have shown that rosette plants with numerous long thin leaves provide the best fog-harvesting morphology. Fog droplets often do not penetrate boundary layers very fast. Long thin leaves minimize boundary layers, maximize the surface area of contact, and can "whisk" easily in the wind, breaking down boundary layers and collecting fog. Morphological indices of length/width in the field have shown that plants with a whisk-like morphology are more prominent where fog is more common in both South America and South Africa. The Namib Desert of South-West Africa and the Atacama Desert along the west coast of South America have two of the driest climates on Earth. *Stipagrostis sabulicola* (Poaceae) grows in a hyper-arid environment of the Namib Desert that can receive less than 20 mm of rainfall per year. However, it can harvest fog and this enables it to deliver water to the soil around its roots equivalent to rainfall of up to 1000 mm a year. Grooves on the underside of the erect leaves provide a site for water drop formation, and channel droplets down to the leaf bases, thus avoiding water dropping off on to dry soil away from the plant. In the Atacama Desert, significant patches of vegetation occur only where there are regular fog episodes. Harvesting of fog in such environments can produce enough water to be useful for plants, and inspired the

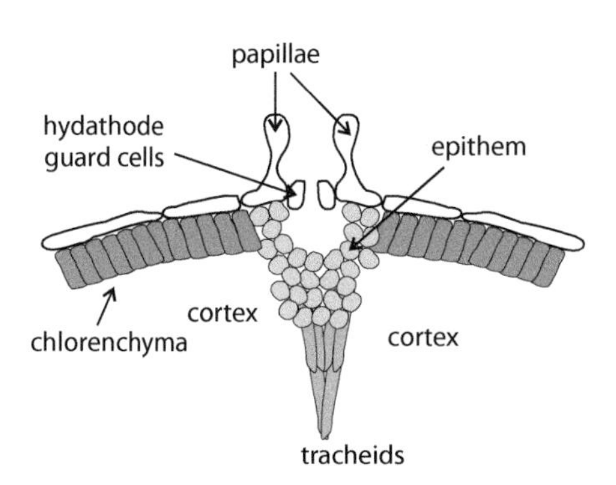

Figure 4.19. Laminar hydathodes of the xerophyte *Crassula*. Hydathodes are usually associated with guttation, and many mesophytes have them at the end of vascular strands at the leaf edges. Xerophytes such as the *Crassula* species of the Namib Desert have them on the leaf surfaces, where they may aid the uptake of dew. The epithem cells help to transfer water to and from the vascular tissue. The papillae and guard cells help to control water flux. (From Martin CE & DJ von Willert [2000] *Plant Biol* 2:229–242. With permission from John Wiley & Sons.)

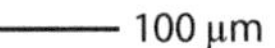

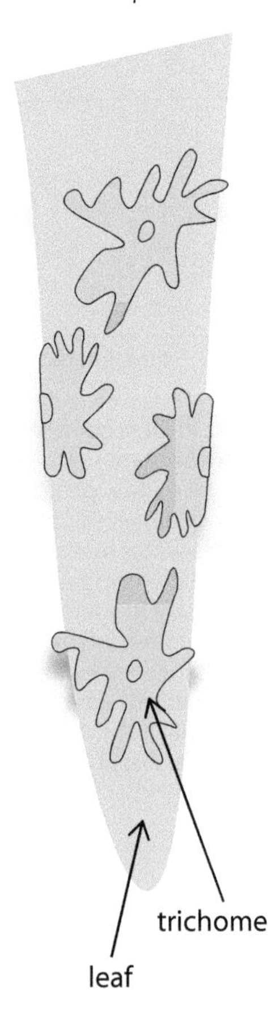

Figure 4.20. The trichomes of *Tillandsia* species (Bromeliaceae). Many *Tillandsia* species are "air plants" that take up their water directly from the atmosphere. The trichomes on their narrow needle-like leaves provide the surface area to capture water and conduct it down into the leaf cortex. *Tillandsia usneoides* forms the "Spanish moss" that cascades down from trees in humid areas of the Americas. (Data from Papini A, Tani G, Di Falco P & Brighigna L [2010] *Flora* 205:94–100.)

fog-harvesting technologies that are used to collect water for human use. Overall, these examples demonstrate that anatomical and morphological adaptations can significantly decouple plants from soil-water regimes, and that water can sometimes move from atmosphere to plant to soil.

Succulent xerophytes are physiologically decoupled from their chronically arid environments

The plants that are most decoupled from their soil-water regime are xerophytic succulents. In the very driest ecosystems, succulents represent a small proportion of most floras, and, for example, they are relatively rare in the native flora of Australia. Rather than exhibiting succulence, the few plants that can withstand the driest conditions on Earth characteristically show extreme resistance to transpirational water loss, extreme resistance to low tissue ψ, and deep rooting systems. For example, the creosote bush (*Larrea tridentata*), which is typically found in the driest areas of the deserts of the South-West USA and Northern Mexico, has these characteristics. Succulence is most common in arid and semi-arid ecosystems with episodic rain. It is a radical anatomical and morphological adaptation that allows plants to collect water and use it to enable CO_2 fixation during prolonged drought—that is, the succulent tissues act as water capacitors. Most stem and leaf succulents have an extensively developed cortex consisting of achlorophyllous water-storing parenchyma cells. In *Echinocactus platyacanthus* (Figure 4.21) the cortex can be up to 300 mm across, whereas in *Arabidopsis* it is generally around 0.05 mm across. Chlorenchyma forms the outer layers, often below an extensive **hypodermis** and epidermis (Figure 4.22). In some stem succulents, water-storing parenchyma also occurs in the pith, which is sometimes the dominant water storage tissue. In such anatomies the two-dimensional vascular system that is found in flat leaves would be unsuitable, and mesophyll conductance can be decreased. Increasing succulence correlates with a switch to a three-dimensional vascular system, especially in stem succulents. For example, in the Cactaceae there is a unique three-dimensional network of cortical bundles that has enabled the development of a gigantic cortex while providing transport routes through it. Subtle anatomical variety has been shown to help to maintain mesophyll conductance in several succulents.

Figure 4.21. The giant barrel cactus *Echinocactus platyacanthus*. The barrel cacti are almost spherical when full of water—that is, they are the shape of the solid with the minimum surface area to volume ratio.

In contrast to the epidermis of animals, that of plants does not contain continually dividing cells. In most perennials, bark forms in its place, with the exception of most succulents, which can have the same epidermis for many decades. The stem wood of the baobab is used for water storage, a strategy of stem succulence that might also be used in other species, but most stem succulents do not form bark. Because the succulent organ also has the function of taking in CO_2 and photosynthesizing, an impermeable opaque layer such as bark would not be feasible. The outer layers of succulents, which are almost all perennials, must be permeable to CO_2 and light but also able to undergo massive cortical development while reducing water loss. The extensive cuticular waxes of many desert succulents help to prevent water loss, and also prevent chronic UV radiation from causing cumulative damage to long-lived epidermal cells. The many halophytes that are succulent are characterized by a consistently very negative ψ based on a relatively unchanging

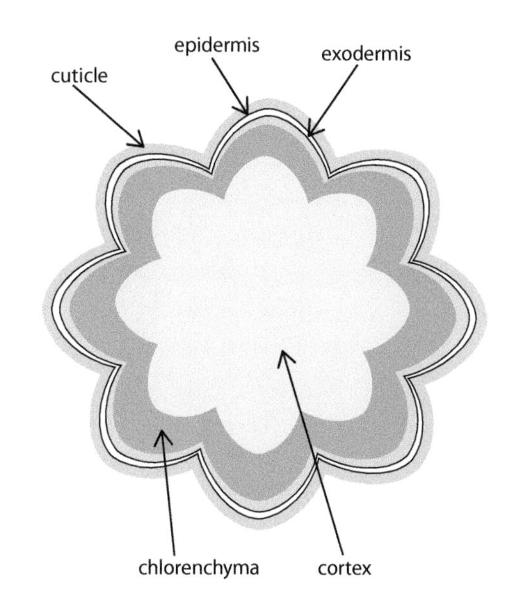

Figure 4.22. Xerophyte stem anatomy in cross section. Many xerophytes have thick cuticles and an exodermis, which help to reduce water loss. Photosynthesis occurs primarily in a thin chlorenchymatous layer that surrounds a massive cortex composed of loosely packed water-storage parenchyma. These layers occur in a variety of forms in the leaves and stems of many xerophytes.

water and ionic content. In contrast, many xerophytic succulents have a high ψ and water content that changes significantly. The changes in water content demand changes in volume that must occur without damaging the long-lived epidermis and hypodermis. Hence many succulents have ribs or tubercles that allow significant changes in volume without damage to the outer layers of the plant (Figure 4.23). In some cacti there are collapsible cortical layers that aid significant volume reductions. During dry periods some cacti have been shown to lose 75% or more of their volume and survive. In essence, the use of particular cells to store and supply water allows ψ in the chlorenchyma to be maintained within physiological limits. In contrast to halophytic succulents, most xerophytic succulents do not have high concentrations of compatible solutes, but are rich in sugars and polysaccharide mucilages. This unusual range of sugars and mucilages has been linked to resisting the effects of low ψ in cells and to the storage and supply of water in the cortical apoplast, respectively.

The development of these morphologies in cacti is initiated by the largest apical meristems in any angiosperm—the large-volume morphology is more securely achieved by a few divisions of a large number of meristematic cells than by many divisions of a few cells. When water is available, xerophytic succulents take it up quickly in large quantities into water storage cells. In the often sandy desert soils this is possible because all water is available at relatively low ψ. It is enabled by the often strikingly extensive but shallow root systems of desert succulents, and often enhanced by the development of "rain roots" within hours of rainfall. The stores of water are then supplied to chlorenchyma over long dry periods, enabling the plant to photosynthesize essentially independently of water supplies from the environment. Isolation from the water regime of the environment is exemplified by the roots of many desert succulents acting as rectifiers—that is, they only allow conduction of water in one direction. Dry soils have a much lower ψ than that of the water storage tissue, and water should be drawn out of the plant into the soil. In dry soils, if roots dehydrate, their hydraulic conductivity decreases massively, preventing water from moving back out into the soil and helping to decouple the plant from the soil–water regime. On soil rewetting, the roots are rewetted and their hydraulic conductivity is increased. These rectifier properties are produced by dehydration of the root cortex while the stele maintains hydration. The concentration of stomata in furrows and other areas of high humidity also plays an important role in minimizing water loss during photosynthesis.

Resurrection plants cope with complete desiccation

Desiccation removes all physiologically useful water from an organism, and in almost all species results in death. It is generally defined as a water content of less than 10% by weight. In most plants, the loss of water even to values well in excess of 10% is fatal because of irreversible mechanical effects,

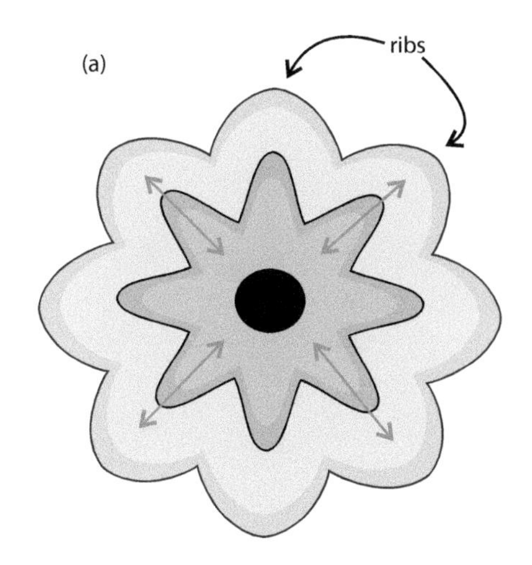

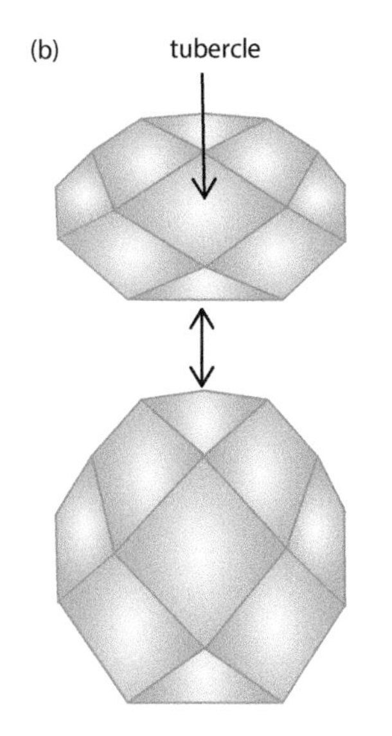

Figure 4.23. Expansion and contraction of xerophytes. Large changes in the volume of stored water in xerophytes necessitate morphological adaptations. Cacti and other xerophytes have (a) ribs or (b) tubercles that enable them to expand and contract according to their water status.

problems of oxidative stress, loss of membrane structure, and increased protein aggregation. However, several non-vascular plants and about 350 vascular plants are desiccation tolerant, and after months, years, or even decades in a quiescent state, rehydrate and within hours, or at most a few days, can be fully functional again. Plants with this **anhydrobiosis** capability are often referred to as "resurrection plants." The early plant colonizers of the land were probably desiccation tolerant, because quite a number of extant species in early plant groups are so today, and this capability was likely to have been very advantageous on the early land surface.

The vegetative tissues of many mosses can survive rapid desiccation. Evidence from flash-dried mosses shows that most desiccation damage is caused on rehydration. It is the repair of damage, rather than adaptations that avoid damage, which is the key to the ability of mosses to survive desiccation. In contrast, vascular plants that have desiccation-tolerant vegetative tissues have adaptations that enable them to avoid the damaging effects of desiccation, and cannot survive flash-drying but need a drying period to prepare for desiccation. It is likely that during the evolution of vascular plants the primitive form of desiccation tolerance was lost and another form evolved. Although there are desiccation-tolerant ferns, there are no desiccation-tolerant gymnosperms. The spores, seeds, and pollen of vascular plants all have the capacity to survive desiccation. It seems likely that in ferns, and then again in angiosperms, vegetative desiccation tolerance evolved from mechanisms used by spores and seeds. There are about 60 desiccation-tolerant angiosperms that occur in genera which are widely spaced taxonomically (Figure 4.24). In some of them chlorophyll persists during desiccation, but in others chlorophyll is removed before desiccation occurs. On the relatively dry land surface, desiccation tolerance has clear advantages, reproductive tissues in all vascular plants possess it, and it has evolved several times independently—yet it is rare, with less than 0.02% of vascular plants having desiccation-tolerant vegetative tissue.

Several genera of bacteria and yeast are desiccation tolerant, but multicellular organisms face particular mechanical stresses caused by anhydrobiosis. Desiccation-tolerant animals are all less than 5 mm in length and have no skeleton. It is thought that the mechanical stresses induced by desiccation in larger animals, especially those with skeletons, are too great for important structures to survive intact. The largest desiccation-tolerant plant is the shrub *Myrothamnus flabellifolius*, found in Southern Africa, which can grow to a height of 3 m (Figure 4.25). Plant cell walls help to maintain morphological integrity at low water contents, but at less than 10% water content, major mechanical stresses are produced. Perhaps the best-investigated resurrection plant, *Craterostigma plantagineum*, undergoes dramatic leaf rolling and folding during desiccation. Many other desiccation-tolerant plants make similar morphological adjustments, and if prevented from doing so are unable to resurrect. There are anatomical adaptations in, for example, the mesophyll, that enable these changes, and there is high **expansin** activity during desiccation in *C. plantagineum*, which suggests that there are significant changes in cell wall flexibility. This includes, perhaps uniquely, changes in cell wall properties in xylem vessels. Changes in tensile strength and other properties have also been noted in plant leaves during desiccation. These morphological changes not only prevent mechanical stresses from damaging important structural components of the leaves, but also decrease photoinhibition, which is an important cause of damage to resurrection plants if leaf rolling is prevented. In *M. flabellifolius* and other resurrection plants the grana of thylakoids change significantly during drying. In some resurrection plants the machinery of the chloroplasts is dismantled to produce "desiccoplasts" that lack chlorophyll, but from which chloroplasts can later be redeveloped.

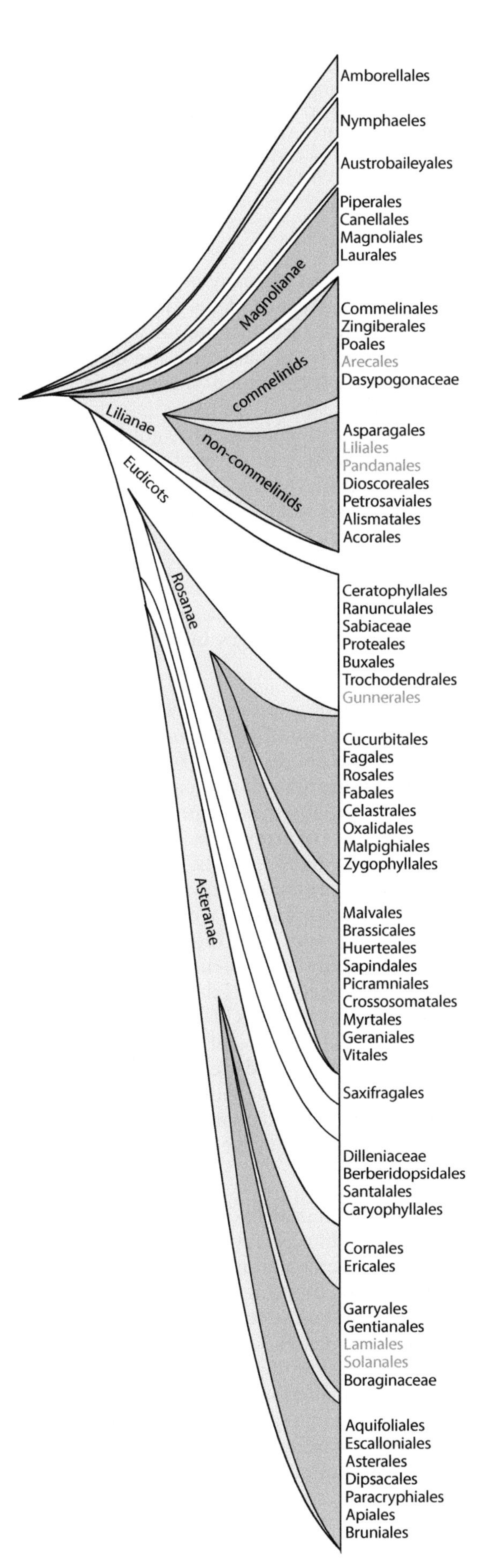

Figure 4.24. The taxonomic distribution of desiccation-tolerant angiosperms. Desiccation tolerance has evolved independently in eight different families of angiosperms. Orders with desiccation-tolerant species are shown in green in the figure.

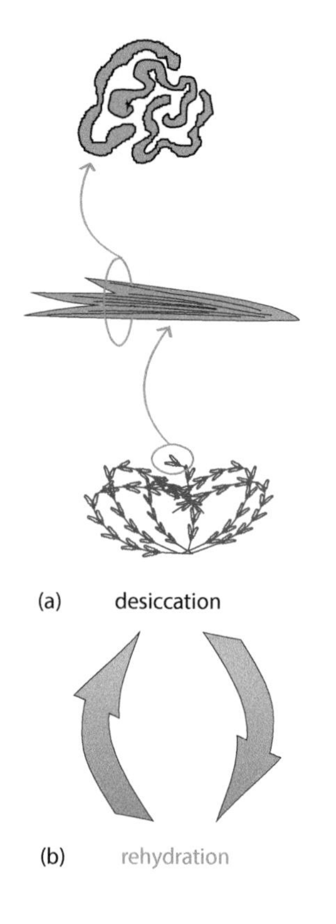

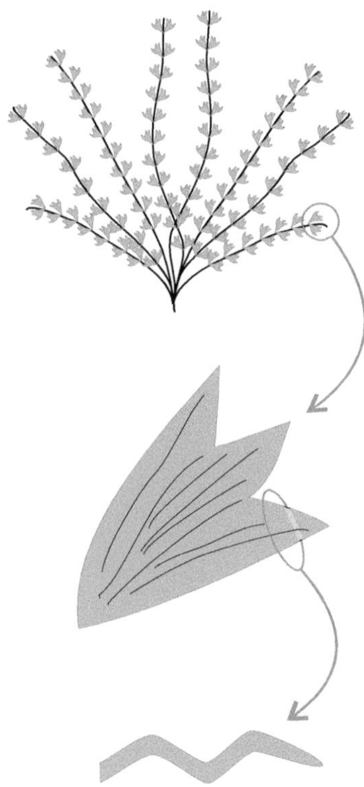

Figure 4.25. The resurrection plant *Myrothamnus flabellifolia*. (a) Dramatic morphological changes occur during desiccation. These are enabled by anatomical adaptations and complemented by physiological adaptations. The plant can survive for many years in a desiccated state. (b) On rehydration the stems unfurl, the leaves open out, and chlorophyll concentrations increase. (From Moore JP, Lindsey GG, Farrant JM & Brandt WF [2007] *Ann Bot* 99:211–217.)

Compatible solutes are produced by many organisms in response to low external ψ, and are found in resurrection plants. However, in addition, during preparation for desiccation the carbohydrate metabolism of resurrection plants changes very significantly. In particular, the disaccharides sucrose and **trehalose** can increase in concentration—for example, *C. plantagineum* can reach up to 40% dry weight sucrose. Trehalose is widely accumulated by other organisms during desiccation stress. At relatively low levels of water loss these non-reducing sugars in part physically replace lost water, but at very low water concentrations they vitrify—that is, they form an amorphous solid (bioglass). Such bioglasses hold cell contents in a state of suspended animation, and are also important in seed tolerance of desiccation. Trehalose can also bind to many biomolecules, which probably helps to stabilize them. **Raffinose** and **stachyose** have similar roles in some plants, and *C. plantagineum* synthesizes the unique sugar, **octulose**. However, for survival of complete desiccation, changes in osmolyte and carbohydrate metabolism are insufficient.

Desiccation causes reorganization of the proteome. This underpins morphological changes, reorganization of photosynthesis, and changes in carbohydrate metabolism. It also contributes to other metabolic changes, in particular a dramatic change in antioxidant capacity, including many effects mediated by glutathione. Perhaps most importantly it involves a major increase in the concentration of **hydrophilins**. These extremely hydrophilic proteins are intrinsically disordered—that is, because of their sequences they have no stable secondary structure in aqueous solution. They occur primarily as **random coils**, and associate with a higher number of water molecules than almost any other proteins. By far the most important type in plants are the late embryogenesis abundant proteins (LEAs). Based on sequence differences, seven different categories of LEAs can be described. They are vital for desiccation tolerance of embryos in seeds, and also widely occur in overwintering buds, resurrection plants, and many desiccation-tolerant animals. LEAs contain over 6% glycine, and are rich in small amino acids such as alanine and serine. In desiccation the most important LEAs are probably group 2 LEAs, the **dehydrins**. These contain a lysine-rich K-segment that is repeated 1–12 times, a Y-segment at the N-terminus, and sometimes an S-segment rich in serine.

The hydrophilicity of LEAs such as dehydrins suggests that they might act as water buffers, but there is not much evidence that they affect the overall kinetics of water loss during dehydration. Many of them have a high binding capacity for metal ions, and they might contribute to redox homeostasis by binding some reactive divalent metals. They significantly enhance the formation and stability of bioglasses with sugars. Most significantly, however, they prevent the aggregation of proteins that occurs during desiccation. Computer models and many measurements *in vitro* show that LEAs gain structure during desiccation, especially at less than 20% relative water content. These structures, including α-helices, some β-sheets, and often some polyproline-II helices, probably help them to prevent protein aggregation. Dehydrins do not bind to proteins in the way that stress-induced molecular chaperones such as heat shock proteins do. Molecular chaperones, in contrast to dehydrins, have a definite structure and often actively reverse aggregation and denaturation. During desiccation, LEAs associate loosely with proteins, providing a hydrophilic environment and preventing aggregation because they reduce the rates of collision between molecules. In dehydrins the Y-segment is very flexible, which probably allows K-segments to maintain their associations with proteins during the changes induced by desiccation. When LEAs associate with proteins within a bioglass, proteins can be kept unchangingly intact in a state of suspended animation until rehydration. Many dehydrins associate with membranes, some having been shown to form linear structures that can be inserted into membranes to help to stabilize them, while others affect the formation of membrane vesicles as water concentrations drop to 10%. Some LEAs specifically stabilize nucleic acids, and they also help to reconfigure the cytoskeleton. In addition to hydrophilins, there are numerous intrinsically disordered proteins (IDPs), some of which are important in cytoskeletal construction, so LEAs probably play the key role in managing the cytoskeleton during desiccation.

The survival of desiccation therefore requires changes in the morphological, anatomical, and cellular structure of plants together with a change in the cellular solvating environment. The structural changes avoid damage that would be caused by massive shrinkage, and the solvation changes vitrify cellular contents. Numerous plant groups have evolved desiccation tolerance independently, so in view of the fact that it solves such a fundamental problem it is perhaps surprisingly uncommon. Desiccation-tolerant plants all grow very slowly. Many of them are constitutively prepared for desiccation, which probably inhibits normal growth, and the cost of tolerance seems to be very high, because such plants generally outcompete others only in the most marginal environments. Desiccation tolerance also does not seem to be possible in plants that grow taller than about 3 m. Desiccation empties xylem vessels that can only be refilled by capillary action, which cannot reach a height of more than 2–3 m.

Interactions between water and other stressors provide important environmental insights

Plant–water relations are at a nexus of interactions that underpin terrestrial ecosystems. The capture of water and CO_2 by plants is central to their physiology and hence to terrestrial ecosystems. Rainfall patterns on Earth are likely to change significantly in the next few decades, and CO_2 concentrations will continue to rise. The effects of this are difficult to predict with certainty. Decreases in rainfall are expected to increase soil water deficit in many regions, but in the short term an increase in CO_2 levels in the atmosphere of plants increases their WUE, which might offset the adverse effects of increased water deficit on plants. However, increased CO_2 concentrations tend to increase biomass, increasing total transpirational flux, and all of these changes can affect temperature balance and hence growth. By area, about one-third of all terrestrial ecosystems are grasslands, and almost all grazing livestock depend on grass. Debates about the effects of changing water, temperature, and CO_2 regimes on grasslands highlight the importance of interactions in predicting the effects of plant–water relations on a large scale and over the long term. Large-scale experiments on the effects of simultaneous increases in temperature and CO_2—in large part mediated by the effects of available water—on some grasslands suggest that soil moisture might not be greatly affected and that productivity might increase above and below ground. This effect is produced primarily because the environmental changes favor C_4 grasses. Predicting such effects in the longer term, for all grasslands and for all ecosystems, will be vital for understanding the ecosystems of Earth in the coming decades. Studies of sclerophyllous Mediterranean oaks show that their leaf characteristics, rather than their physiology per se, change with drought. This is a reminder that in some species it is not physiological but anatomical and morphological changes that occur during chronic water stress. Knowledge of a wide range of adaptations is likely to play an important role in making, and making sense of, long-term predictions of the effects of drought on plants.

Soil salinity is an increasing challenge to terrestrial plants, and a global problem in agriculture. Salinity causes osmotic and ionic stress. Freezing water causes osmotic and mechanical stress. It is very likely that water deficit drove the evolution of osmotic adaptations in the early land plants, and that these adaptations were derived from those of early life forms. In terrestrial plants, adaptations to ionic and mechanical stresses have supplemented those to osmotic stress, and enabled the evolution of salinity and freezing tolerance, respectively. Thus adaptations to manage water are central to many of the stress combinations that plants encounter in terrestrial ecosystems. In agricultural plants, **heterosis** (that is, hybrid vigor) has been perhaps the most important contributor to increased yields during the twentieth century, and is in part often explained by improved WUE. The latter is associated with increases in photosynthetic and nutrient use efficiency. Changes in WUE and the consequent subtle but significant effects on other aspects of growth that contribute to heterosis exemplify the importance of water for numerous aspects of plant growth.

The WUE and drought tolerance of many major crop plants have been improved markedly since about 1960. Improvements in water management have probably had effects at least as significant. In many areas it is estimated that twice as much "crop per drop" is now possible as could be achieved a few decades ago. However, there is general consensus that a "blue revolution" is probably necessary in order to sustain the production of food and other plant products during the twenty-first century. Fluorescence imaging of photosynthesis, measurements of isotope discrimination for H, C, and O, and thermal imaging of crop temperatures have all been used to select crops with higher WUE. They are all likely to find increasing use in improving water management in agriculture as water resources become increasingly scarce. The potential of biotechnology to provide crops with enhanced drought tolerance is starting to be realized, and is likely to be significant in the coming decades. Many irrigation techniques exist that increase water use efficiency, and the pressure to develop many more is intense. If there is to be a blue revolution in agriculture it will probably depend on integrating insights from plant-water relations across all temporal and spatial scales from a variety of ecosystems.

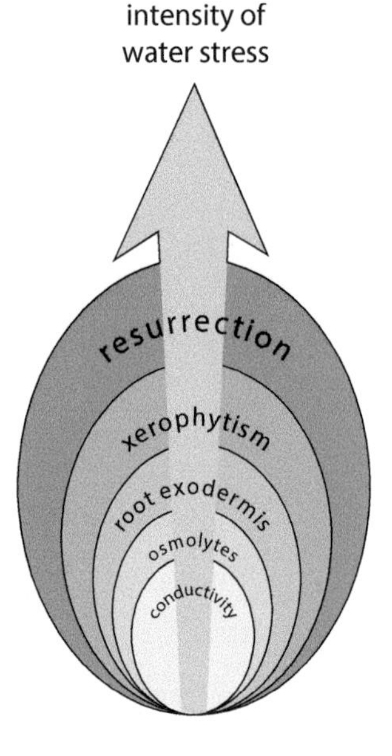

Figure 4.26. Water and the stress-response hierarchy. Short-term fluctuations in water availability are counteracted primarily by physiologically induced changes in root and shoot conductivity. The concentration of solutes that generate a water potential (osmolytes) can also help plant cells to maintain ψ. Adaptation to progressively more profound variation in water availability involves a hierarchy of anatomical and morphological responses. A few plants can survive the complete absence of water by attaining a state of suspended animation and resurrecting when water is available again.

Summary

Water use by plants is vital for sustainable human habitation of Earth, because of its central role in unmanaged terrestrial ecosystems, its role in climate, and its role in the production of food and other plant products. The adaptations that plants have evolved in order to maintain a freshwater environment in their cells constrain many aspects of life in terrestrial ecosystems. On the cellular scale, mild water deficit can be withstood by physiological adjustment. In plants that are tolerant of intense or prolonged water deficit, physiological adjustments are complemented by anatomical and morphological adjustments. Anhydrobiosis is possible in some multicellular land plants, but is rare. In most species, anhydrobiosis in spores or seeds rather than in vegetative tissues enables plants to survive in the absence of water (Figure 4.26). Recognition of the global importance of plant-water relations is leading to new ways of thinking about water on land. An integrated understanding of the molecular biology, physiology, agronomy, and ecohydrology of plant-water relations is likely to be required if the insights necessary for sustainable human habitation of Earth are to be gained.

Further reading

The importance of water to plants

Chater C, Gray JE & Beerling DJ (2013) Early evolutionary acquisition of stomatal control and development gene signalling networks. *Curr Opin Plant Biol* 16:638–646.

Schlesinger WH & Jasechko S (2014) Transpiration in the global water cycle. *Agricult Forest Meteorol* 189:115–117.

Sheil D (2014) How plants water our planet: advances and imperatives. *Trends Plant Sci* 19:209–211.

Silvertown J, Araya YN, Linder HP et al. (2012) Experimental investigation of the origin of fynbos plant community structure after fire. *Ann Bot* 110:1377–1383.

Irrigation and water resources

Haddeland I, Heinke J, Biemans H et al. (2014) Global water resources affected by human interventions and climate change. *Proc Natl Acad Sci USA* 111:3251–3256.

Pfister S, Bayer P, Koehler A et al. (2011) Projected water consumption in future global agriculture: scenarios and related impacts. *Sci Total Environ* 409:4206–4216.

Schewe J, Heinke J, Gerten D et al. (2014) Multimodel assessment of water scarcity under climate change. *Proc Natl Acad Sci USA* 111:3245–3250.

Wada Y, van Beek LPH & Bierkens MFP (2012) Non-sustainable groundwater sustaining irrigation: a global assessment. *Water Resour Res* 48:W00L06.

Water potential gradients and transpiration

Boyce CK, Brodribb TJ, Feild TS et al. (2009) Angiosperm leaf vein evolution was physiologically and environmentally transformative. *Proc R Soc Lond B Biol Sci* 276:1771–1776.

Knipfer T & Fricke W (2010) Root pressure and a solute reflection coefficient close to unity exclude a purely apoplastic pathway of radial water transport in barley (*Hordeum vulgare*). *New Phytol* 187:159–170.

Larjavaara M (2014) The world's tallest trees grow in thermally similar climates. *New Phytol* 202:344–349.

Pittermann J, Limm E, Rico C et al. (2011) Structure-function constraints of tracheid-based xylem: a comparison of conifers and ferns. *New Phytol* 192:449–461.

Adjustment of resistance to water flux

Aroca R, Porcel R & Manuel Ruiz-Lozano J (2012) Regulation of root water uptake under abiotic stress conditions. *J Exp Bot* 63:43–57.

Chaumont F & Tyerman SD (2014) Aquaporins: highly regulated channels controlling plant water relations. *Plant Physiol* 164:1600–1618.

Cramer MD, Hawkins H & Verboom GA (2009) The importance of nutritional regulation of plant water flux. *Oecologia* 161:15–24.

McLean EH, Ludwig M & Grierson PF (2011) Root hydraulic conductance and aquaporin abundance respond rapidly to partial root-zone drying events in a riparian *Melaleuca* species. *New Phytol* 192:664–675.

Physiological response to water deficit

Antoni R, Gonzalez-Guzman M, Rodriguez L et al. (2012) Selective inhibition of clade A phosphatases type 2C by PYR/PYL/RCAR abscisic acid receptors. *Plant Physiol* 158:970–980.

Bertolli SC, Mazzafera P & Souza GM (2014) Why is it so difficult to identify a single indicator of water stress in plants? A proposal for a multivariate analysis to assess emergent properties. *Plant Biol (Stuttg)* 16:578–585.

Hancock JT, Neill SJ & Wilson ID (2011) Nitric oxide and ABA in the control of plant function. *Plant Sci* 181:555–559.

Lee SC & Luan S (2012) ABA signal transduction at the crossroad of biotic and abiotic stress responses. *Plant Cell Environ* 35:53–60.

Noctor G, Mhamdi A & Foyer CH (2014) The roles of reactive oxygen metabolism in drought: not so cut and dried. *Plant Physiol* 164:1636–1648.

Root and water deficit

Cassab GI, Eapen D & Eugenia-Campos M (2013) Root hydrotropism: an update. *Am J Bot* 100:14–24.

Geldner N (2013) The endodermis. *Annu Rev Plant Biol* 64:531–558.

Roppolo D, De Rybel B, Tendon VD et al. (2011) A novel protein family mediates Casparian strip formation in the endodermis. *Nature* 473:380–383.

Sardans J & Penuelas J (2014) Hydraulic redistribution by plants and nutrient stoichiometry: shifts under global change. *Ecohydrology* 7:1–20.

Leaf adaptations and drought

Eller CB, Lima AL & Oliveira RS (2013) Foliar uptake of fog water and transport belowground alleviates drought effects in the cloud forest tree species, *Drimys brasiliensis* (Winteraceae). *New Phytol* 199:151–162.

Hassiotou F, Evans JR, Ludwig M et al. (2009) Stomatal crypts may facilitate diffusion of CO_2 to adaxial mesophyll cells in thick sclerophylls. *Plant Cell Environ* 32:1596–1611.

Johnstone JA & Dawson TE (2010) Climatic context and ecological implications of summer fog decline in the coast redwood region. *Proc Natl Acad Sci USA* 107:4533–4538.

Nardini A, Lo Gullo MA, Trifilo P et al. (2014) The challenge of the Mediterranean climate to plant hydraulics: responses and adaptations. *Environ Exp Bot* 103:68–79.

Roth-Nebelsick A, Hassiotou F & Veneklaas EJ (2009) Stomatal crypts have small effects on transpiration: a numerical model analysis. *Plant Physiol* 151:2018–2027.

Water relations of xerophytes

Arakaki M, Christin P, Nyffeler R et al. (2011) Contemporaneous and recent radiations of the world's major succulent plant lineages. *Proc Natl Acad Sci USA* 108:8379–8384.

Kumar GN & Srikumar K (2014) Molecular and computational approaches to characterize thermostable laccase gene from two xerophytic plant species. *Appl Biochem Biotechnol* 172:1445–1459.

Ogburn RM & Edwards EJ (2013) Repeated origin of three-dimensional leaf venation releases constraints on the evolution of succulence in plants. *Curr Biol* 23:722–726.

Ripley BS, Abraham T, Klak C et al. (2013) How succulent leaves of Aizoaceae avoid mesophyll conductance limitations of photosynthesis and survive drought. *J Exp Bot* 64:5485–5496.

Resurrection plants

Aidar ST, Meirelles ST, Oliveira RF et al. (2014) Photosynthetic response of poikilochlorophyllous desiccation-tolerant *Pleurostima purpurea* (Velloziaceae) to dehydration and rehydration. *Photosynthetica* 52:124–133.

Lyall R, Ingle RA & Illing N (2014) The window of desiccation tolerance shown by early-stage germinating seedlings remains open in the resurrection plant, *Xerophyta viscosa*. *PLoS One* 9:e93093.

Rakic T, Lazarevic M, Jovanovic ZS et al. (2014) Resurrection plants of the genus *Ramonda*: prospective survival strategies – unlock further capacity of adaptation, or embark on the path of evolution? *Front Plant Sci* 4:550.

Rascio N & La Rocca N (2005) Resurrection plants: the puzzle of surviving extreme vegetative desiccation. *Crit Rev Plant Sci* 24:209–225.

Drought and the future of agriculture

Hu H & Xiong L (2014) Genetic engineering and breeding of drought-resistant crops. *Annu Rev Plant Biol* 65:715–741.

Limousin J, Misson L, Lavoir A et al. (2010) Do photosynthetic limitations of evergreen *Quercus ilex* leaves change with long-term increased drought severity? *Plant Cell Environ* 33:863–875.

Luis Araus J, Sanchez C & Cabrera-Bosquet L (2010) Is heterosis in maize mediated through better water use? *New Phytol* 187:392–406.

Morgan JA, LeCain DR, Pendall E et al. (2011) C_4 grasses prosper as carbon dioxide eliminates desiccation in warmed semi-arid grassland. *Nature* 476:202–205.

Chapter 5
Nitrogen

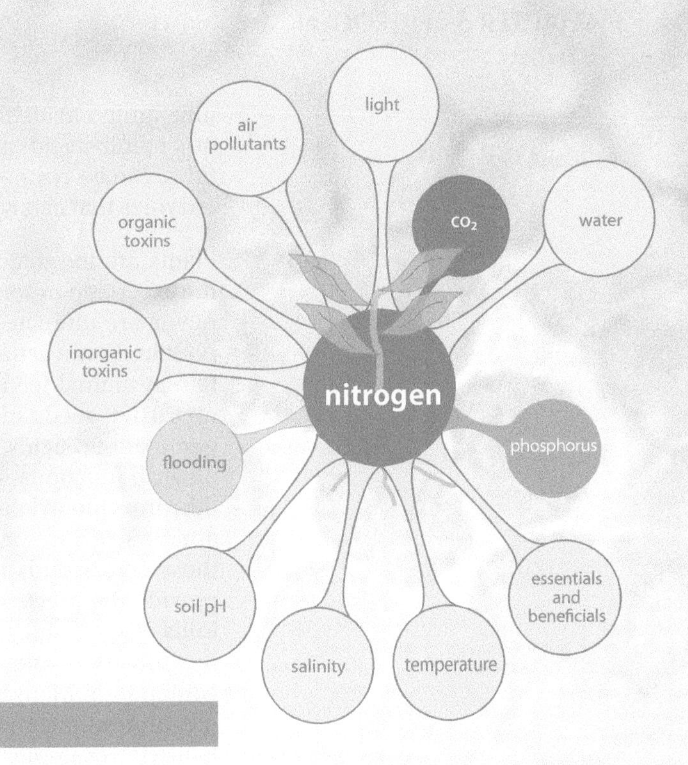

The challenge of variable N supply interacts, in particular, with variation in CO_2 levels, P availability, and waterlogging.

Key concepts

- Plant assimilation of N is its primary conduit into terrestrial biomass.
- The release of artificially fixed N into the environment is having global consequences.
- Microbially mediated transformations produce different forms of N in soils.
- A battery of transporters mediate N uptake and aid N homeostasis in plants.
- N assimilation pathways help to maintain plant N status across a variety of sources.
- Physiological adjustment helps plants to combat significant differences in N supply.
- During N shortages, some plants adjust their root morphology significantly.
- Symbioses with N-fixing microbes aid N capture in N-deficient environments.
- Carnivorous plants opportunistically capture N from an unreliable erratic supply.

Nitrogen assimilated in plants is vital for the production of biomolecules in terrestrial organisms

Proteins and nucleic acids, both of which contain N, are fundamental to all life on Earth. The amino acids and **nucleotides** that, respectively, polymerize to form them were probably among the first biomolecules. Amines are derivatives of NH_3 in which an organic group, or groups, replaces hydrogen. In amino acids these groups are carboxylic acids, and the amino groups can form **peptide** bonds. The purine and pyrimidine bases of nucleotides have **heterocyclic rings** (with C and N atoms) and complementary structures, so double strands with complementary sequences can form. Only autotrophs can convert inorganic NH_3 into the amine-containing compounds necessary to synthesize amino acids and then nucleotides. Essentially all proteins contain just 22 amino acids, and autotrophs can, if they have sufficient NH_3, synthesize all of them entirely from inorganic substrates. Heterotrophs and saprotrophs cannot do this, and have an essential requirement for at least nine amino acids that they must obtain from autotrophs, and from which other amino acids and N-containing molecules, such as the bases of nucleotides, can be synthesized. Many proteins are structural, but many others are enzymes that catalyze almost all of the reactions of life.

Plants are the major source of essential amino acids in ecosystems. With the exception of some microbial sources, almost all human protein requirements are ultimately met by plants, supplied either directly or via animal protein. Some estimates suggest that, with the increased use of seed protein in animal feed, up to 70% of global human protein requirements are met using seeds of terrestrial plants. There are numerous non-**proteinogenic** amino acids that serve as **co-factors** or that have particular physiological functions—for example, γ-aminobutyric acid (GABA) in plants, or dihydroxyphenylalanine (L-DOPA) in mammals—but non-proteinogenic amino acids are particularly numerous in plants, from which over 200 of these compounds have been isolated. The diverse amino acids in plants provide the precursors for the synthesis of over 12,000 N-containing **alkaloids** of **secondary metabolism.** All of these are probably unique to plants, in which they often act as anti-herbivore, antimicrobial, and **allelopathic** compounds. Many have effects on the nervous system of animals (for example, morphine), and many are poisons (for example, strychnine) or stimulants (for example, caffeine). Aromatic amino acids are the precursors for over 10,000 **phenolic** compounds. The phenolics are frequently volatile and vital for flavor chemistry (for example, coumarin), pigments (for example, anthocyanins), and the lignin used to make wood, as well as functioning as herbivore-deterrent and antimicrobial compounds. The chlorophyll of photosynthetic organisms contains four N atoms in each heme group. Plant N metabolism, much of which is unique, therefore not only synthesizes the essential amino acids used by most organisms in terrestrial ecosystems, but also initiates the synthesis of some of the most ecologically and economically important compounds on Earth.

The Earth's crust is on average composed of less than 0.1% N, and provides little N for living systems. N fixation is the production of reactive N (NH_3 and molecules made using it) from N_2. In the O_2-poor, CO_2-rich, and N_2-rich atmosphere of the **Hadean** and **Archean**, the action of lightning on the atmosphere probably produced sufficient fixed N for life. By the late Archean, around 2.2 giga-years (Gy) ago, the decrease in CO_2 and increase in O_2 concentrations in the atmosphere had probably decreased the production of fixed N from lightning by two orders of magnitude. It is likely that this provoked a crisis of N supply for early life, and favored the evolution of biological N fixation. Almost all of the ecosystems on Earth for at least the last 2 billion years developed because microbial N fixation supplied NH_3, which autotrophs could convert into amino acids. The natural cycling of N on Earth is therefore particularly dependent on biological activity, with plants being the conduit for N into most unmanaged and managed terrestrial ecosystems (Figure 5.1). In general, the cycling of organic N is more important than was once thought, with plants now known to compete with microorganisms for soil organic N. In addition, marine anaerobic ammonium oxidation (**"anammox"**) is now an established part of the N cycle, and might be important in some terrestrial ecosystems.

Artificially fixed nitrogen significantly affects the biosphere and atmosphere

Nitrogen makes up 2.5% of the average human, and over 50% of this is from inorganic N fertilizers (Box 5.1). Total global N fixation is about 410 Tg $year^{-1}$, including anthropogenic input of about 210 Tg $year^{-1}$, which could double by 2050. There are significant anthropogenic inputs from the burning of fossil

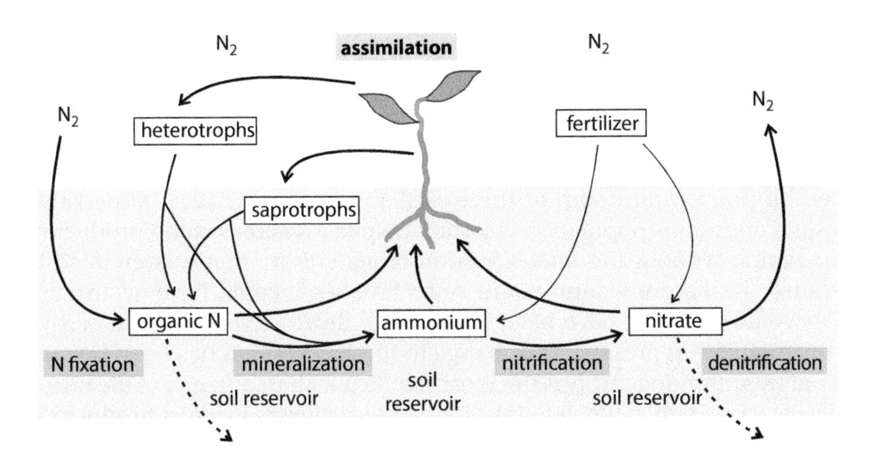

Figure 5.1. Plants and the N cycle. Plants can take up N from all three of the significant soil-N reservoirs. They then play a key role in assimilating this N into the biological molecules of terrestrial ecosystems. Together with biological N fixation, a significant proportion of which occurs in symbiosis with plants, the mineralization of plant material returns N to inorganic forms. Plant uptake of fertilizer and organic N helps to determine the size of the soil N reservoirs, and thus the potential both for N leaching out of the soil into watercourses, and for denitrification.

BOX 5.1. THE HABER–BOSCH PROCESS AND BIOINSPIRED N_2 FIXATION

The Haber-Bosch process for atmospheric nitrogen (N_2) fixation has had consequences for humans and the Earth almost as profound as those of any other scientific innovation of the twentieth century. It underpins the production of nitrogenous fertilizer that is used to grow at least 50% of the food that humans eat. The consequence, unforeseen by Haber and Bosch, has been disruption of the N cycle on a global scale. The development of N_2 fixation was driven by constraints on the supply of saltpetre to Germany in the early twentieth century. Saltpetre was extensively used for making not only explosives but also fertilizer.

N_2 is unreactive because its triple bond is so strong. N≡N bonds can be activated by coordination with transition metal complexes, and direct protonation of some N_2-metal complexes (including those of Fe, Mn, and W) is then possible, but almost exclusively under fairly extreme conditions of temperature and pressure. The generation of high temperature and pressure makes N_2 fixation relatively expensive, so there is intensive ongoing research to identify catalysts that can work under less extreme conditions. Among the most successful are those based on Zr, but Fe-Co alloys and other bimetallic nitrides and RuO can also act as catalysts. In Fe alloys, three Fe atoms form a nitride with N_2, activating it to allow direct protonation, which is probably how all the metal catalysts work. The equation for the Haber-Bosch process is as follows:

$$N_{2(g)} + 3H_{2\,(g)} \leftrightarrow 2NH_{3(g)}\ \Delta H\ -92\ kJ\ mol^{-1}$$

Generally a porous Fe catalyst with Ca, Al, and K impurities is used. The presence of Ca and Al helps to maintain porosity, with K increasing electron density and N_2 activation. The reaction is extremely slow, so high temperatures are used to speed it up. However, the reaction is exothermic, so high temperatures drive the equilibrium to the left, decreasing yield. The four gas molecules on the left of the equilibrium occupy twice the volume of the two molecules on the right, so increasing the pressure can drive the equilibrium to the right. Temperatures of around 400°C and pressure of 200 atmospheres are a compromise that gives a product yield of approaching 20%. If the product is cooled, it liquefies and can be drawn off, so successive rounds of heating and recycling of reactants are used to eventually yield over 95% product from reactants. The H_2 is generally derived from natural gas, and the energy for maintaining high temperatures and pressures is mostly, although not necessarily, derived from fossil fuels—leading to the suggestion that in effect much of humankind eats fossil fuel.

Nitrogenase complexes catalyze a stepwise addition of H_2 to N_2 to yield NH_3. The Lowe-Thorneley scheme has been used to describe these steps for many years. Only very recently have some of the details of N_2 fixation by nitrogenase started to become clear, and they might be very useful for developing bioinspired N_2 fixation. The nitrogenase complex contains Fe and MoFe proteins. Fe proteins contain an Fe_4S_4 that donates electrons to the MoFe protein, whose P cluster (Fe_8S_7) mediates the transfer. The active site is the FeMo co-factor ($FeMo_{co}$: $MoFe_7S_9X$·homocitrate). $FeMo_{co}$ can be isolated at various stages of fixation, and much evidence suggests that X is a C atom. A central core of cysteine and histidine residues, Fe, Mo, and C atoms, and a citrate molecule are now known to provide the reaction center. These structures are consistent with biological N fixation occurring not by direct protonation of N_2, but via coupled electron and proton transfers to $FeMo_{co}$. Protons are probably bound to S sites in $FeMo_{co}$ and then reduced with electrons to H, which then reacts with active N_2. This must occur in the absence of water, and the central core of $FeMo_{co}$ has been shown to be hydrophobic due to the physical exclusion of water. The elucidation of the details of N fixation by nitrogenase could inspire processes that significantly increase humankind's capacity to carry out the process of N fixation. These processes might be not only chemical but also biotechnological, or even based on synthetic biology. Nitrogenase activity also produces H_2, and the elucidation of the operation of this complex might also have an impact on biohydrogen production. As human populations increase during the twenty-first century, the process of N fixation is likely to continue to be central to the interface between food production and the environment.

Further reading: Pool JA, Lobkovsky E & Chirik PJ (2004) Hydrogenation and cleavage of dinitrogen to ammonia with a zirconium complex. *Nature* 427: 527–530.

Harris TV, Szilagyi RK (2011) Comparative assessment of the composition and charge state of nitrogenase FeMo-cofactor. *Inorg Chem* 50: 4811–4824.

Maurakami J & Yamaguchi W (2012) Reduction of N_2 by supported tungsten clusters gives a model of the process by nitrogenase. *Sci Rep* 2: 407.

fuels and the growing of legume crops, but the main anthropogenic N inputs are fertilizers. The transformation of global agriculture in the twentieth century occurred in significant part because of changes in crop varieties, in the capacity to multiple crop, and in pest management, but a key factor was the delivery of increased N to crops and pastures. Artificially fixed N helped to fuel the population boom of the twentieth century because, remarkably, despite enormous population growth it helped to increase food production per capita. Without the intensification of agriculture that was enabled by fertilizer N, the population boom since 1900—if it could have occurred at all—would probably have been much more destructive to natural ecosystems than it is at present. The change in the N cycle that has been brought about by anthropogenic N is the most significant change in any of the terrestrial cycles on which life depends. Inorganic fertilizers in effect produce the food of about 3.5 billion people, so human populations are now beholden to a dramatically altered terrestrial N cycle, the management of which will be a key determinant of the habitability of Earth in the twenty-first century.

In 2010 it was estimated that only 30–50% of inorganic fertilizer used in agriculture since 1900 had reached crops. The majority of N applied as fertilizer is now in soil, watercourses, and the atmosphere. A significant proportion of the most important freshwater courses on Earth are suffering, or have suffered, adverse effects of increased N. The rate of biological N fixation in the oceans is about 140 Tg $year^{-1}$, but up to 100 Tg $year^{-1}$ of artificially fixed N reaches the world's oceans from rivers and the atmosphere. From space, blooms of phytoplankton—often fueled by artificially fixed N—in water bodies are one of the most visible effects of humans on Earth. The burning of fossil fuels in particular produces gaseous reactive nitrogen oxides ("NO_x"—that is, NO, NO_2, and N_2O), and animal production can produce large amounts of NH_3. In anaerobic conditions, which occur in almost all soils at some point, NO_x can also be produced by the denitrification of added NO_3^-. These emissions enhance the deposition of N in many natural ecosystems, and can change them significantly because they are often naturally N-limited. Reported changes across the globe arising from N deposition include changes in biomass production, biodiversity, ecosystem dynamics, and acidity. By 2030, a significant proportion of the land surface will probably have deposition rates in excess of those that initiate changes, and in many areas a great deal more (Figure 5.2). The contamination of land, water,

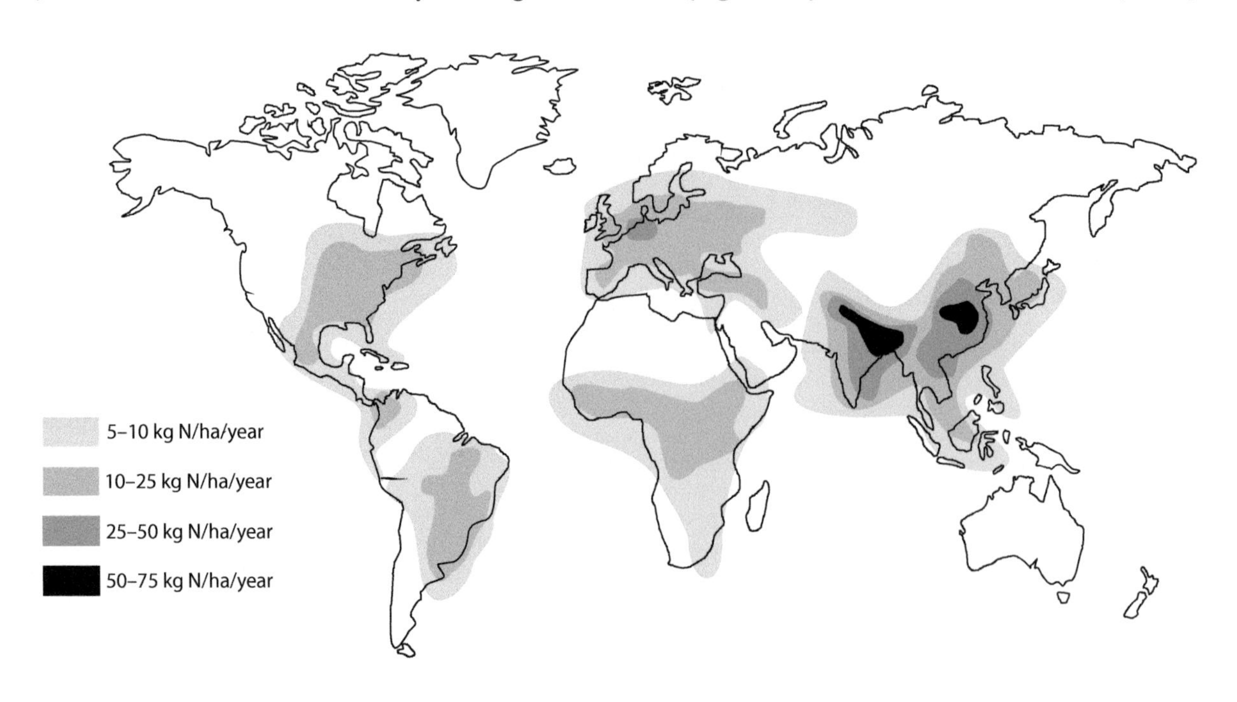

Figure 5.2. Predicted global N deposition by 2030. Current trends suggest that N deposition rates will stabilize or even decrease across Europe by 2030. In the Americas and Africa they are likely to increase slightly from current levels, and across India and China they will probably increase significantly to the rates shown in the figure. At chronic deposition rates of 5–10 kg N ha^{-1} $year^{-1}$ effects on species can occur, at rates of 10–25 kg N ha^{-1} $year^{-1}$ ecosystem effects are frequently reported, and at rates of over 25 kg N ha^{-1} $year^{-1}$ such effects are almost certain to occur. At higher resolution, many localities have much higher N deposition rates than those shown. (Based on Bleeker A, Hicks WK, Dentener E et al. [2011] *Environ Pollut* 159:2280–2288).

and air with artificially fixed N in the coming decades may become more severe because N use has always increased more rapidly than the population, primarily because of increased per capita consumption of meat and fossil fuel as the population increases (Figure 5.3).

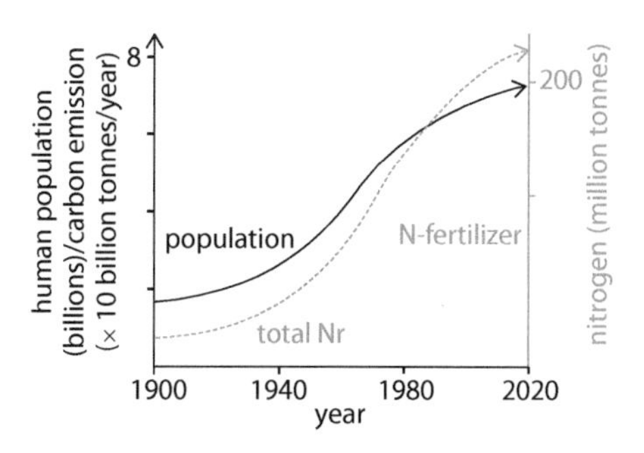

Figure 5.3. Population growth and increases in reactive N production. Since 1900, massive population growth has been accompanied by, and perhaps enabled by, huge increases in the use of nitrogenous fertilizers. The total reactive N (total Nr) from fertilizer and other sources released into the environment has increased more rapidly than the population, and will probably continue to do so for the foreseeable future.

In many terrestrial ecosystems, N availability is second only to water availability as the limiting factor for biomass production. In general, in almost all N-limited ecosystems, N-fueled increases in biomass production decrease biodiversity. Rare plants are frequently disproportionately affected (Figure 5.4), decreasing the stabilizing effect of diversity, as are ecosystems in early stages of development. Alien invasive species are often favored by enhanced N inputs. The Convention on Biological Diversity identifies areas of high conservation value and emphasizes that enhanced N deposition is one of the most significant threats to them. The effects of increased CO_2 levels are also clearly dependent on N availability—when N is limiting, increased CO_2 concentrations have few effects, but if it is not limiting there can be significant effects on biomass production. Furthermore, the current global enrichment in both C and N is unprecedented, and although it might increase the global C sink, it can cause plants to put more resources into P acquisition in an effort to maintain the C:N:P ratio, which affects the cycling of P. If water is limiting, enhanced N has fewer effects, and excess water favors denitrification rather than biomass production. N enrichment also significantly changes the flow of many contaminants in ecosystems. Furthermore, the discharge of N_2O and NH_3 to the atmosphere is implicated in climate change. N_2O is 300 times more potent as a greenhouse gas than CO_2, and when denitrification is significant, as it is in many temperate ecosystems (both unmanaged and managed), it can be a major contributor to greenhouse gas emissions. NH_3 in the atmosphere often reacts to form ammonium sulphate $(NH_4)_2SO_4$, which is a major contributor to particulates in the atmosphere. Some models suggest that the climate-warming effects of N_2O are less than the **global-dimming** effects of $(NH_4)_2SO_4$. NO_x can increase the ground-level production of O_3, where it adversely affects perhaps 30% of cereal production. The management of N use in agriculture and the prediction of the effects of altered N supply on ecosystems constitute two of the key environmental challenges facing humankind.

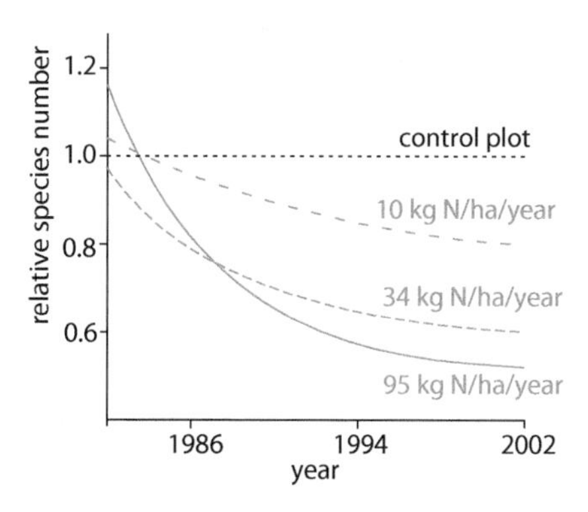

Figure 5.4. The effect of addition of N on the biodiversity of grasslands. Over a period of 20 years, the addition to grasslands in Minnesota, USA, of N at 10 kg N ha^{-1} $year^{-1}$ and above (in addition to regional wet deposition) significantly decreased biodiversity, and indicated a critical load of about 5 kg N ha^{-1} $year^{-1}$. The loss of rare species was almost entirely responsible for the decrease in biodiversity, tending to leave plots with fewer, dominant species (species number is relative to that of control plots at 1.0). (Adapted from Clark CM & Tilman D [2008] *Nature* 451:712–715. With permission from Macmillan Publishers Ltd.)

The concentration of different forms of soil nitrogen varies significantly

Much of the N on Earth occurs in the atmosphere as N_2, which is unreactive and biologically inert. However, N does have valence states ranging from −3 (NH_3) to +5 (NO_3^-), so if it can be "fixed" from N_2 to NH_3 it can be reactive. Biological assimilation of N uses NH_3 as a substrate, so the processes that fix N_2 into NH_3 underpin life on Earth and are vital for terrestrial ecosystems. There are some minerals that contain N, but in general only a very small proportion of soil N (and often none) is derived from mineral weathering. In some localities, KNO_3 and $NaNO_3$ (**saltpetre**) occur in sufficient concentrations to have been mined for fertilizer or gunpowder production, but the primary natural source of N in soils is biological N fixation. This is carried out by bacteria and archaea with the **nitrogenase complex**, and defines them as **diazotrophs** (Figure 5.5). Almost all diazotrophs switch off N fixation both if N is already available and if O_2 is present. Some, such as *Azotobacter*, can fix N in aerobic conditions, but only if they respire at extremely high rates. Bacteria in anaerobic microsites carry out much N fixation. Some microbes have nitrogenases with different transition metals (Mo, Fe, and V) that function at different temperatures. In all diazotrophs, N fixation is coordinated by the *nif* group of genes in response to low N levels and under low O_2 concentrations. There are 20 or so *nif* genes that synthesize the components of

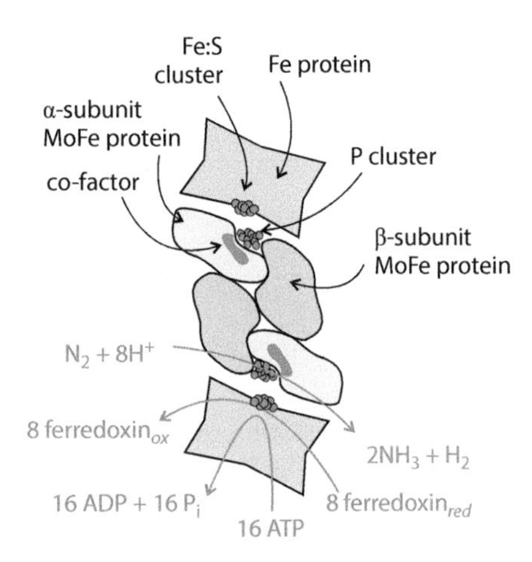

Figure 5.5. The nitrogenase complex. Nitrogenase occurs only in diazotrophic microorganisms. An Fe protein holds a 4Fe:4S cluster. Electrons, often from ferredoxin but sometimes from other electron carriers, are passed to the Fe:S cluster. MgATP is then used to move the Fe:S cluster close to the P cluster. Electrons are passed to the P cluster (8Fe:7S) on an MoFe protein, whose action to transfer electrons to N_2 necessitates a co-factor. This co-factor often contains FeMoCo, but can contain FeVCo or FeFeCo. The Fe protein is coded by *nifH*, and the MoFe protein is coded by *nifD* (α) and *nifK* (β).

the nitrogenase complex, assemble it, and regulate its expression. They are often located on the chromosome, but in some bacteria are on **plasmids**—for example, in *Rhizobium*, which has a Sym plasmid. Generally, *nif*A regulates the expression of genes that encode the nitrogenase complex, and is controlled by proteins that are sensitive to N and O_2 concentrations. The NH_3 produced by nitrogenase is assimilated into glutamate, and from this all of the other N-containing biomolecules are made. Ultimately, death and decomposition of diazotrophs result in the release of fixed N into the soil and, especially in ecosystems without anthropogenic inputs, represent the primary N input. Nitrogenase is able to reduce not only N_2 but also many other molecules so, for example, the reduction of C_2H_2 to C_2H_4 by nitrogenase provides a way of measuring its activity.

The rate of **mineralization** of organic N to NH_3 in soils is a key determinant of the total supply of usable N. In general, almost irrespective of the conditions that exist in a soil, there are groups of heterotrophic bacteria or archaea that can mineralize organic compounds containing N to release NH_3. In many ecosystems, complete mineralization to NH_3 is probably less important than was once thought, because many plants have been shown to take up amino acids and small peptides in significant quantities. The N in NH_3 is highly reduced, and its oxidation can release a large amount of energy. A few genera of chemoautotrophic nitrifying bacteria use the energy released from the oxidation of NH_3 to NO_3^- to drive carbon fixation (Figure 5.6). They are responsible for almost all of the naturally occurring NO_3^- in soils. **Nitrification** involves three steps and different genera of bacteria. In soils the first two steps, which produce NO_2^-, are mostly catalyzed by species of *Nitrosomonas*, and the second step, which converts NO_2^- to NO_3^-, by species of *Nitrobacter*. There are other genera that can contribute to nitrification in soils and that dominate nitrification in the sea. However, NO_3^- is highly oxidizing, so if O_2 is in short supply and a food source is available, some heterotrophic bacteria use it as an electron acceptor to release energy in the process of **denitrification**. There are four reductase enzymes that make up the denitrification proteome (Figure 5.7), with many bacteria having nitrate reductase and nitrite reductase, but far fewer having NO reductase and N_2O reductase. Denitrification can result in N losses from soils, but N_2O production is particularly important because it is a significant contributor to the greenhouse effect, and about 50% of anthropogenic emissions of this gas are from agricultural soils.

The energy that can be extracted by bacteria from redox reactions of N, and their sensitivity to environmental conditions, means that N form and concentration vary significantly in soil solutions. In warm aerobic soils of

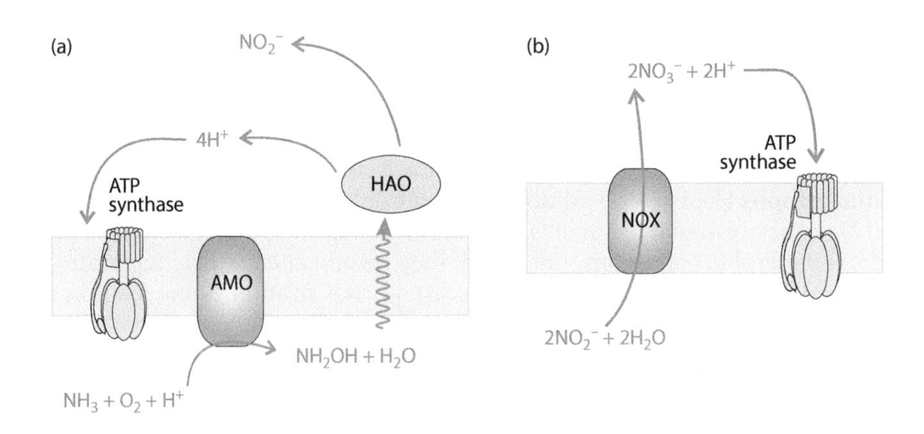

Figure 5.6. Nitrogen transformations of nitrification. (a) A few genera of β-proteobacteria (in soils mostly *Nitrosomonas*) have the enzymes ammonium monooxidase (AMO) and hydroxylamine oxidoreductase (HAO). These enzymes couple reactions that oxidize NH_3 to NO_2^- to the generation of proton gradients that energize the synthesis of ATP. (b) A few genera of α-proteobacteria (in soils mostly *Nitrobacter* species) have the enzyme nitrite oxidoreductase (NOX), which oxidizes NO_2^- to NO_3^- and generates a proton gradient. In both groups of bacteria these reactions underpin the fixation of CO_2 (not shown).

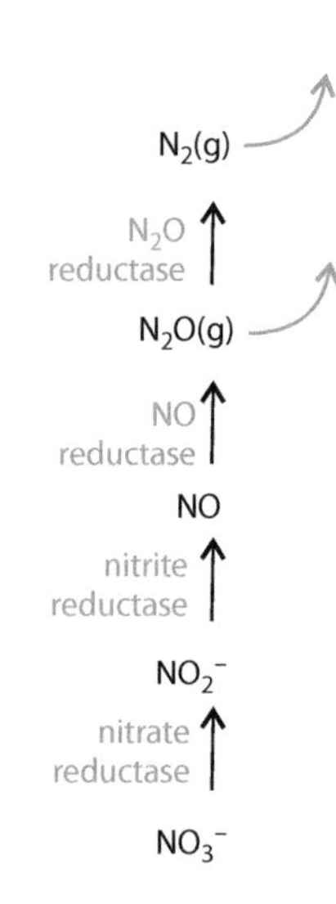

Figure 5.7. Denitrification of soil nitrate. Under anaerobic conditions, many microorganisms use terminal electron acceptors other than O_2 during respiration. The oxidized forms of soil N are among the most energetically favorable electron acceptors in anaerobic soils. Many microorganisms possess some of the enzymes of denitrification, but relatively few possess the entire denitrification proteome necessary to produce N_2. *Paracoccus denitrificans* is a β-proteobacterium common in soils, and most strains have the entire denitrification proteome. The final gaseous products of denitrification diffuse out of soils and can produce significant losses of NO_3-N. N_2O is an active and significant greenhouse gas. The mixture of microorganisms, and the variants of enzymes that they possess, determine the ratio of gaseous N produced from anaerobic soils. Most soils have anaerobic microsites in which some denitrification occasionally occurs, but many soils are anaerobic for long periods during which denitrification can be very significant. (g) = gas.

around neutral pH, mineralization occurs most rapidly and most completely. In cold anerobic acid soils, mineralization is slowest—often slower than organic matter input, which builds up as organic horizons or even peat. In boreal ecosystems, N in the soil solution is predominantly organic N, which is also present in temperate and tropical soils. Nitrification is limited by the rate of NH_3 oxidation, but this is generally much faster than the rate of mineralization, so if the warm oxidizing conditions that are generally necessary for nitrification pertain, most N occurs as NO_3^-. Most soils have a cation-exchange rather than anion-exchange capacity, so NO_3^- occurs almost entirely in solution, which makes it available for uptake by plants but also for leaching out of the soil. The susceptibility of NO_3^- to denitrification depends on redox and pH conditions in the soil—it only occurs under reducing conditions, and tends to be faster under more acidic conditions. These conditions occur in soil microsites or with waterlogging, so denitrification can occur, at least at times, from almost any soil. In some forest ecosystems and in temperate grasslands, nitrification rates have long been known to be lower than predicted from prevailing conditions. **Polyphenols** exuded from plant roots can act as nitrification inhibitors, and in species of the grass *Brachiaria*, which naturally dominate African savannah grasslands but have been very widely planted elsewhere, cyclic **diterpenes** have been shown to have similar effects. Sorghum, the world's fifth most widely planted cereal (Figure 5.8), exudes sorgoleone, which also has similar effects. These biological nitrification inhibitors promote a mix of N sources in the soil by decreasing nitrification and hence denitrification. Nitrification inhibitors are used in traditional and intensive agricultural systems to limit N losses from leaching and denitrification (Box 5.2). In the absence of nitrification, mineralization means that most N in the soil solution occurs as NH_4^+. In aqueous solution, NH_3 occurs in equilibrium with NH_4^+ at a **pKa** of 9.24, so at neutral pH and below only a very small fraction occurs as NH_3. The cation-exchange capacity of most soils, and occasionally specific interactions with particular clays, mean that soils can bind NH_4^+, sometimes very strongly. The soil solution concentration of NH_4^+ therefore tends to be much more highly buffered than that of NO_3^-. Semivariograms for soil N tend to have quite large nuggets in managed systems—that is, soil N varies widely at a particular place, but in both unmanaged and managed ecosystems the lag is significant because of large-scale landscape effects on the variables that affect soil N processes (Figure 5.9).

Figure 5.8. *Sorghum bicolor.* Sorghum is native to Africa but is now grown around the world. It is one of the most water-use-efficient cereals, and has the ability to inhibit nitrification in its rhizosphere.

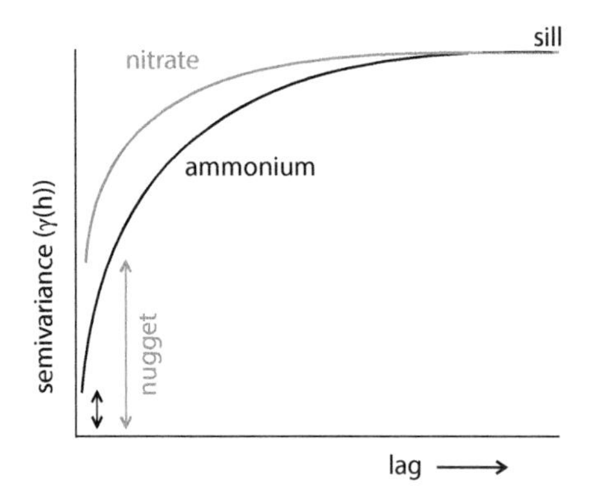

Figure 5.9. **Semivariograms for available soil nitrogen.** The semivariogram for soil solution NO_3^- tends to have a greater nugget than that for NH_4^+—that is, NO_3^- concentrations can vary significantly on a very small scale. The sill occurs at a significant lag, often many km, reflecting the effect of landscape-scale processes, especially hydrological ones, on NO_3^- concentrations. NO_3^- concentrations also vary very significantly over time, generally making this the nutrient ion with the most variable availability. NH_4^+ concentrations are more highly buffered, but also vary on landscape scales.

BOX 5.2. NITRIFICATION INHIBITORS AND NEEM TREES

Synthetic nitrification inhibitors (NIs) have been used for several decades in commercial agriculture. They work by inhibiting the bacteria, especially *Nitrosomonas* species, that nitrify NH_3 to NO_3^-. Dicyandiamide (DCD) has found quite widespread use in Europe and Australasia, whereas N-Serve (nitrapyrin, 2-chloro-6-trichloromethylpyridine) has been used in North America. DMMP (dimethylpyrazolphosphate) has for many years also been added to fertilizer preparations as an NI. These chemicals can decrease NO_3^- leaching, but there are many conditions under which they do not do so, and they are relatively expensive.

Figure 1. ***Azadirachta indica*** **(Meliaceae).** The neem tree grows widely on the Indian subcontinent, and has long had numerous uses based on the biologically active molecules that it produces. The figure shows the pinnate leaves and small white flowers.

NIs provide an interesting example of how the processes of natural ecosystems and insights of agroecology might prove useful in large-scale agricultural production. The neem tree (*Azadirachta indica*) (Figure 1) has been used in many ways for millennia in the Indian subcontinent. It produces numerous limonoid compounds, a large number of which are very biologically active. Many of them have long-established insecticide, fungicide, and pesticide properties. Preparations from neem trees and extracts of neem oil are widely used to control a variety of pests, and have found significant use in organic production systems. Between 1995 and 2005 there was a high-profile case in which the European Patent Office had granted a patent for a fungicide based on neem developed between Indian (PJ Margo) and US (WR Grace) companies. The European patent was ultimately not granted, and the furore aided the passing of the Indian Biological Diversity Act (2002) and the development of the Traditional Knowledge Digital Library, which now help to protect the biological resources and establish precedence of use of indigenous knowledge in India.

In India, as in many other parts of the world, urea is the most widely used nitrogenous fertilizer. There are probably approaching 15 million tonnes of nitrogenous fertilizers applied to crops each year in India, the majority as urea. Much of the urea added to soils is converted to NH_3 by microorganisms with urease, and thence to NO_3^- by nitrifying bacteria. The use of neem cake, neem leaves, and ground neem kernels has often, although not always, been found to increase N-use efficiency because neem limonoids decrease both urease and nitrification activity. Neem leaves have traditionally been added to soils in a variety of agricultural systems in India, perhaps in part because of benefits produced by NI activity. Organic NIs based on neem extracts are now widely available. Many naturally occurring compounds have been shown to have NI properties, and in many natural ecosystems NI capacity in the soil is significant. It seems likely, given the soil-N challenges that humankind will face in the twenty-first century, that experience with NIs such as neem is likely to inspire advances in soil-N management.

Further reading: Mohanty S, Patra AK & Chhonkar PK (2008) Neem (*Azadirachta indica*) seed kernel powder retards urease and nitrification activities in different soils at contrasting moisture and temperature regimes. *Bioresource Technol* 99: 894–899.

Sanz-Cobena A, Sanchez-Martin L, Garcia-Torres L et al. (2012) Gaseous emissions of N_2O and NO and NO_3- leaching from urea applied with urease and nitrification inhibitors to a maize (*Zea mays*) crop. *Agric Ecosys Environ* 149: 64–73.

Plant nitrogen-transporter uptake capacity is tuned to variation in soil nitrogen supply

In many warm aerobic soils, and hence in many agriculturally important soils worldwide, NO_3^- is an important form of soil N. The plant species that grow, or can be grown, on such soils can meet much of their N demand by taking up NO_3^-. In agriculture and horticulture, plants that can do this are sometimes referred to as "nitrate feeders." In soils in which it predominates, NO_3^- concentrations can vary by at least three orders of magnitude. Short-term variations in soil availability are countered by significant changes in root transporter capacity. Significant vacuolar stores of NO_3^- are used by many plants as a reservoir of N that is drawn upon to help to regulate N supply to metabolism.

NO_3^- transport proteins in *Arabidopsis* and other model plants show similarity to those in a variety of other species and organisms. In most plant species that take up NO_3^-, uptake capacity is adjusted to short-term variation in NO_3^- supply. Analyses of NO_3^- uptake kinetics in plants reveal that the latter constitutively express both high-affinity (K_m of up to 20 μM) and low-affinity uptake systems. The high-affinity uptake system is rapidly up-regulated in response to NO_3^-, and probably dominates NO_3^- uptake when plants are N limited. NRT2 proteins mediate induced high-affinity NO_3^- uptake in plants. On average, plants have four *NRT2* genes encoding these $2H^+/NO_3^-$ co-transporters. NRT2 transporters occur primarily in the epidermis of roots, and, depending on species or nutritional status, are expressed in both primary and lateral roots, often localizing preferentially towards the root tips with increasing N deficiency of the plant. In *Arabidopsis* and some other species, NRT2.1–7 have been shown to mediate NO_3^- transport at specific locations in the plant (Figure 5.10). Low-affinity uptake is mediated by NRT1 transporters, which probably dominate uptake under N-sufficient conditions but also, unusually, change affinity with different concentrations of NO_3^-. There are more than 50 NRT1 proteins expressed in different plant parts, and they have evolved independently several times in land plants. NRT1 transporters are $2H^+/NO_3^-$ transporters that catalyze transport against electrochemical gradients. They evolved from peptide transporters, and hence transport a variety of organic molecules, leading to the nitrate/protein family (NPF) designation. Many plant cells also have NO_3^- efflux capacity, and NAXT transporters have been shown to enable NO_3^- efflux down electrochemical gradients. SLAC and other closely related transporters mediate efflux of NO_3^- from guard cells in particular. NO_3^- transport into vacuoles occurs down electrochemical gradients. Chloride channels (CLCs) generally catalyze Cl^- transport, but CLC transporters in the tonoplast membrane of *Arabidopsis* can mediate at least 50% of NO_3^- transport into the vacuole. Across the angiosperms there are numerous variants of NRT1, NRT2, NAXT, and CLC that take up NO_3^- and distribute it within plants.

Significant components of the NO_3^- transport systems of plants are tuned to short-term variations in external supply. Transcriptional control of NRT transporters can be exerted by ANR1, NPL7, and LBD transcription factors. Concentrations of glutamine, sucrose, and light are used as cues to tune NO_3^- uptake not only to external supply, but also to N and C status and to circadian rhythms of metabolism. NRT1.1 (Figure 5.11) probably acts as a **transceptor** that not only transports NO_3^-, but is also the receptor used to detect external NO_3^- concentrations. NRT1.1 changes from low-affinity to high-affinity uptake by phosphorylation of a threonine (Thr101). This is catalyzed by CIPK23 kinase, which is an important component of plant responses to changes in N status. Many NRT2 transporters require association with NAR2 proteins, providing an additional mechanism of control. NO_3^- is clearly used as a signal in plants for numerous aspects of physiology and development.

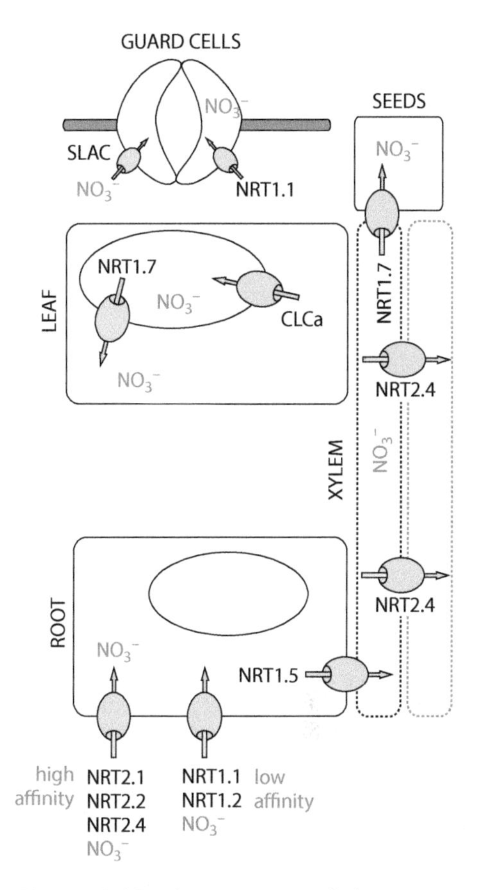

Figure 5.10. The transport of nitrate in *Arabidopsis*. The kinetics of NO_3^- uptake into roots indicate that there are both high- and low-affinity mechanisms. NRT1.1 is probably involved in both, but in general NRT2 transporters mediate high-affinity uptake. NRT2.4 is expressed particularly during N starvation, and has an expression pattern in the root system that is generally complementary to that of NRT2.1. NRT1.5 mediates xylem loading and NRT2.4 is involved in phloem loading throughout the plant. CLC channels mediate NO_3^- transport into the vacuoles of shoot cells, which are often a significant store of NO_3^-. NRT1.7 aids the release of NO_3^- from vacuolar stores. NO_3^- fluxes into seeds can be significant, and use NRT1.7. NO_3^- fluxes help to generate changes in guard cell turgor, and are mediated by NRT1.1 and SLAC channels. NRT transporters are well characterized in other species, such as rice, where they also have specific expression patterns. Other proteins, such as NAR2 in rice, can also be involved in NO_3^- transport. SLAC = slowly activating anion channel.

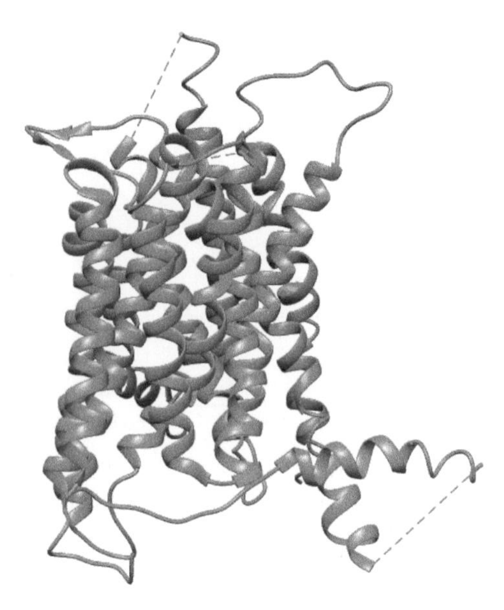

Figure 5.11. **The nitrate transporter NRT1.1.** *Arabidopsis thaliana* NRT1.1, shown as a ribbon structure in the figure, changes affinity with NO_3^- concentration, a process that is mediated by a threonine residue located on the intracellular side (bottom) of the central pore. There are 12 membrane-spanning domains, and NRT1.1 probably occurs in the membrane in dimers. (Drawn in Chimera from PDB 4CL4: Parker & Newstead [2014] *Nature* 507:67–71.)

Point mutations of Thr101 do not affect transport ability but do affect signaling capacity. Conversely, other point mutations eliminate transport capacity but not signaling capacity. There is some evidence that NRT2.1 may also have transceptor capacity. Transceptors are well documented in other organisms, but NRT proteins are among the first to have been proposed in plants. There is also significant post-translational control of NO_3^- uptake. For example, significant cleavage activity on the internal C-terminus of NRT2.1 has been reported to affect transport activity. At sustained high NO_3^- concentrations, NRT2.1 activity is repressed, but there are mutants without this capacity. The activity of RNA polymerase II is often controlled by **histone** modification, and NRT2.1 repression mutants have altered histone modifications at the *NRT2.1* locus. Overall, numerous physiological mechanisms work in concert to achieve N homeostasis in plants that are challenged by short-term variation in NO_3^- supply.

Across the domains of life, NH_4^+ is transported across membranes by the AMT/MEP/Rh transporters. In almost all plant species that have been investigated, genes in the *AMT1* and *AMT2* family have been found. AMT1 transporters are similar to those of prokaryotes, whereas AMT2 transporters originated in fungi. Particularly in plants characteristic of soils in which NH_4^+ predominates, AMT transporters are used to take up a significant proportion of the plant's N. In many plants that are defined as NO_3^- feeders there is significant AMT1 activity in the roots. The converse is also true—in many "ammonium feeders," such as rice, there is significant NRT activity. The ubiquity and activity of NH_4^+ transporters in plants, together with much evidence from stable isotope studies that many plants grow optimally while taking up a mixture of NO_3^- and NH_4^+, suggest that the contribution of NH_4^+ to plant N uptake has often been underestimated. AMT functions as a homotrimer. Each protein has a central pore with a highly NH_4^+-selective mid-section based on the position of two histidines. The cytosolic carboxy-terminal domain is involved in allosteric control of activity between subunits aided by phosphorylation events. In *Arabidopsis* there are five AMT1 transporters. AMT1.1-1.5 have been shown to be localized, with AMT1.1, AMT1.3, and AMT1.5 in the rhizodermis and root hairs, and AMT1.2 in the cortex and endodermis. In some plants, coordinated expression of AMT transporters in the roots probably dominates N uptake, and in many others contributes significantly to N uptake. AMT transporters operate up to the millimolar (mM) range, and their expression is also affected by N and C status, so regulation of their activity makes a significant contribution to N homeostasis during variation in NH_4^+ supply from the soil (Figure 5.12).

It has been known for more than a century that some plants can absorb amino acids through their roots. Amino acid transporters are found in all organisms, and in sequenced plant genomes there are many genes (up to 100) that encode for amino acid transporters. The transport of amino acids is vital for regulating protein synthesis in all organisms, and their rich variety in plants probably reflects the diversity of plant amino acid metabolism.

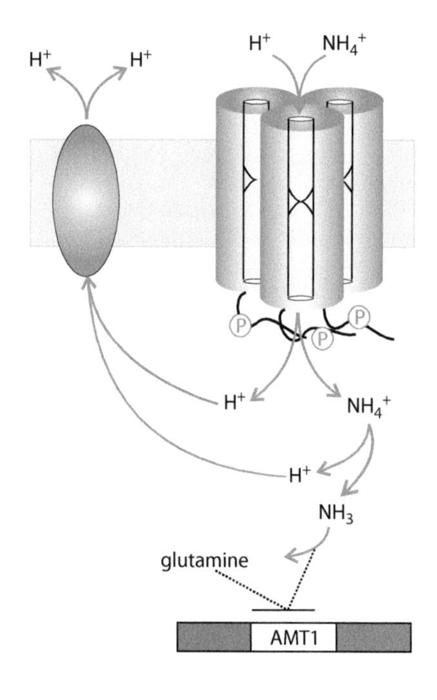

Figure 5.12. **Ammonium transport in plants.** All plant species that have been investigated have AMT genes. The genes that encode AMT transporters, or very similar proteins, occur throughout the kingdoms of life. AMT transporters are H^+/NH_4^+ co-transporters that function as homotrimers, with opening and closing coordinated via phosphorylation of threonine residues in cytoplasmic C-terminals. The external and internal concentrations of NH_4^+ help to control the expression of AMT1 in roots, as does the concentration of glutamine, the assimilation product of NH_3. NH_4^+ uptake results in net efflux of H^+. Many species that are generally regarded as NO_3^- feeders have significant AMT activity in their roots, and may use NH_4^+ to a greater extent than is often assumed. Many species find it difficult to use NO_3^-, and depend extensively on AMT1 transporters for NH_4^+ uptake.

Plants from a wide variety of habitats, including many arable crop plants, grow well in hydroponic systems in which the only N source is an amino acid such as glycine. In roots, LHT and AAP transporters mediate amino acid uptake (Figure 5.13). Lysine and arginine are basic, and under soil conditions are often cationic, so are mostly transported by AAP transporters. The neutral and acidic amino acids are transported by LHT. Most amino acids have L- and D-**enantiomers** and, as is the case in most biological processes, plant uptake and utilization of amino acids are dominated by L-enantiomers. Glycine is neutral, non-enantiomeric, has the highest diffusion coefficient in soil of any amino acid, and has been most successfully used in hydroponics. Many plants also have a significant capacity to take up peptides and quaternary ammonium compounds that commonly occur in soils.

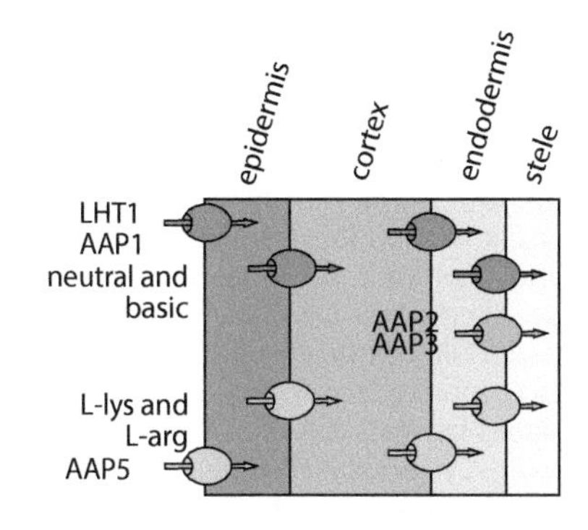

Figure 5.13. **Uptake of amino acids in roots.** All plants have amino acid transporters, which are essential for N metabolism. Most plants that have been investigated also express amino acid transporters in root plasma membranes, and can use amino acids as a sole N source. There are many amino acid transporters in plant roots, but the two amino acids that are cationic under most soil conditions, namely lysine and arginine, are probably transported via a separate set of transporters of AAP type. Transporters similar to LHT and AAP types are widely used throughout the kingdoms of life. Many plants in the field, especially where mineralization and nitrification are slow, probably use amino acid uptake extensively to meet their N requirements.

It has been suggested that plants have amino acid transporters in their roots solely to recapture amino acids that undergo efflux from the roots. However, many studies with ^{15}N have indicated that amino acids can contribute significantly to N uptake in the field. Furthermore, there is a wider variety of microbes that can break down proteins into peptides, and then into amino acids, than there is of microbes that can mineralize amino acids into NH_4^+. Particularly in colder soils, the production of amino acids is much more substantial than their mineralization to NH_4^+, so the peptide/amino acid pool of soil N, and flux through it, can be substantial. In the rhizosphere it is difficult to disentangle the microbial fluxes of amino acids from direct plant uptake, but in arctic, boreal, and alpine systems it is likely that amino-acid N makes a very substantial contribution to N nutrition. There are numerous examples from temperate and Mediterranean systems which suggest that, at least in certain instances, amino acids also make a contribution to plant N nutrition in these ecosystems. Mycorrhizal fungi can break down proteins and take up amino acids, which can also provide a significant route for the entry of amino N into plants. Overall, plant roots have a battery of transport proteins that can take up both inorganic and organic forms of N. Their expression is spatially and temporally variable, both within roots and between species, which helps plants to regulate N supply from a variable soil solution. In general, on any particular soil the capacity of plants to take up different forms of N corresponds to the forms that are produced in the soil.

Plants integrate nitrogen from different sources by converting it to NH_3 for assimilation

Assimilation of inorganic N into organic compounds occurs via the glutamate synthetase/glutamine oxoglutarate aminotransferase (GS/GOGAT) pathway. N assimilation via this pathway in plants dominates the entry of N into the organic molecules of terrestrial ecosystems, and is central to the uptake, translocation, and utilization processes that determine the N-use efficiency of plants. N assimilation is initiated by the action of GS, which joins NH_3 to the γ-carboxyl group of glutamate (Figure 5.14). NH_3 is provided to GS/GOGAT by conversion from several sources, including NH_4^+ uptake, nitrate reduction, deamination of phenylalanine, photorespiration, and mobilization of N from protein stores. GS is cytosolic, with GS1 activity dominating N assimilation from NH_4^+ uptake by roots and from the action of **phenylalanine ammonia lyase** (PAL). GS2 activity predominates in chloroplasts, especially in developing leaves, where NH_4^+ is primarily derived from NO_3^- reduction and, in C_3 plants, photorespiration. The GS **holoenzyme** has a **decameric** structure, with two **pentamers** face to face, in which N fixation occurs between molecules. The activity of GS depends on plant physiological state, and genetic manipulations of it produce significant changes in plant growth. One of the two glutamate molecules produced by the action of GOGAT is the starting point for the synthesis of all the molecules in terrestrial ecosystems that contain N.

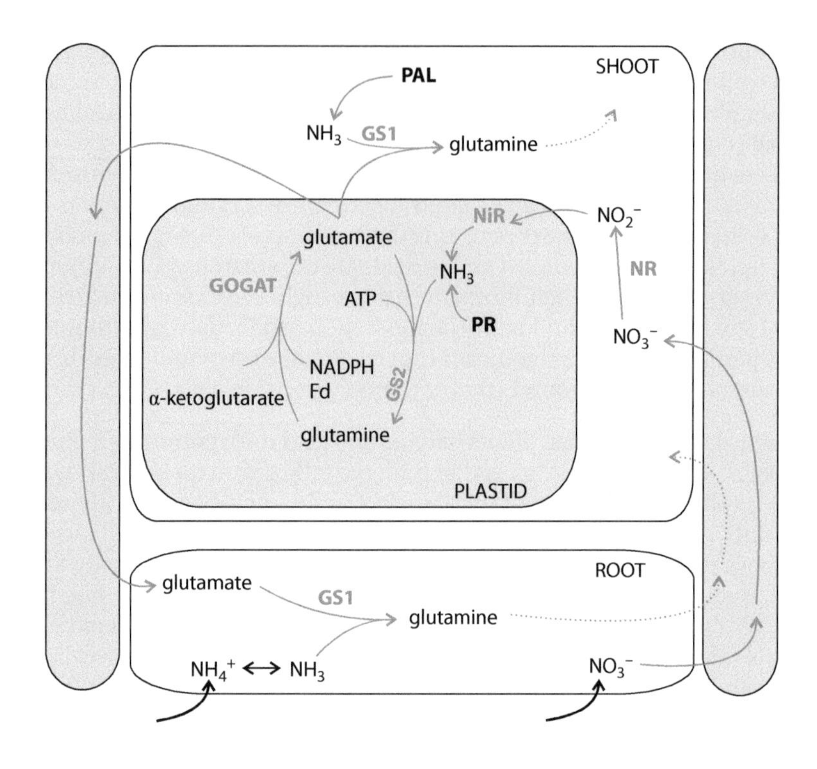

Figure 5.14. Pathways of nitrogen assimilation. If NH_3 is taken up into plant roots it is assimilated by the action of cytosolic glutamine synthetase (GS1, encoded by *GLN1*), mostly in the roots. If NO_3^- is taken up it is transported to the shoots, where it is reduced to NH_3 by nitrate reductase (NR) and nitrite reductase (NiR). Photorespiration (PR) can also produce NH_3 in plastids, which is assimilated by plastidic GS (GS2, encoded by *GLN2*). Cytosolic NH_3 in leaves, and sometimes in roots, can be produced by the action of phenylalanine ammonia lyase (PAL) during secondary metabolism. Glutamine is converted into nitrogenous compounds for transport and entry into N metabolism (broken lines). The ATP, NADPH, and ferredoxin (Fd) used for most NH_3 assimilation are derived directly from the light reactions of photosynthesis. GOGAT = glutamine oxoglutarate aminotransferase.

The highly oxidized N in NO_3^-, although useful in the vacuole as an osmoticum and as an N store, must be reduced to NH_3 for N assimilation. Nitrate reductase (NR) catalyzes the reduction of NO_3^- to NO_2^- (nitrite) driven by NADPH. NO_2^- is highly oxidizing, and is toxic to animals because it binds to hemoglobin. Animals that consume NO_3^--rich plants can suffer NO_2^- toxicity, but in plants NO_2^- is rapidly reduced to NH_3 by nitrite reductase (NiR) to prevent toxicity. NR in most plants occurs as a cytosolic homodimer with three prosthetic groups. The activity of NR in plants is often dependent on NO_3^- uptake, but there are many species known with either constitutive expression or no expression at all, even in the presence of NO_3^-. In general, NR activity is low in those plants that grow on soils in which NH_4^+ and organic forms of N predominate. In those plants that use it, NR activity is tuned to the circadian clock and switched on during daylight and when sucrose concentrations are high, but switched off by high concentrations of glutamine. NR is encoded by *NIA* genes, which have NO_3^--dependent expression. NR activity is significantly modulated by phosphorylation. In *Arabidopsis*, the phosphatase PP2A removes P from a serine between the heme and the Mo-co-factor binding domains, activating NR. NR kinase uses ATP to phosphorylate NR, enabling the binding of 14-3-3 protein and thence inactivation. The activity of NR is also increased by SUMO ligation using SIZ1. Plant NiR is a plastidic enzyme with three co-factors, encoded by *Nii* genes, of which there are only one or two copies in most plants. In chloroplasts it binds reduced ferredoxin, which provides electrons for the reduction of NO_2^-. The electrons, of which six are necessary for the reduction of each NO_2^- to NH_3, are passed from each of six ferredoxin molecules to an Fe–S cluster and thence to an NO_2^- molecule which is held in close proximity to the Fe–S cluster by a **siroheme**. NO_2^- is toxic, so the activity of NiR is generally much greater than that of NR. Most NO_2^- is reduced in leaves in the light, but reduced ferredoxin can be generated using NADPH in non-photosynthetic plastids. These regulatory mechanisms control the activity of NR and hence the assimilation of much of the NO_3^- fertilizer that reaches arable crops, making them important for fertilizer-use efficiency. The reduction and assimilation of both CO_2 and

NO_2^- therefore occur in plastids. The concentrations of sugar phosphates and reduced N compounds reflect the rates of C and N assimilation, respectively, and they and numerous other signals help to coordinate the assimilation of C and N from a variable supply.

Whole-plant physiological adjustments help to use different patterns of nitrogen supply

Even quite small changes in N availability can produce significant changes in the physiology of a plant. Studies on different accessions of *Arabidopsis* have emphasized the variety of responses to N availability in a single well-characterized species, and many lines of many crop plants are well established as being more or less N-use-efficient. Plants also respond differently to different forms of N. In general, all plants can take up NO_3^-, but some species only assimilate it very slowly. This is primarily because of constitutively low expression of NR and NiR, even when they are exposed to NO_3^-. These plants tend to inhabit ecosystems in which soil NO_3^- levels are very low. If supplied as the sole N source, NH_4^+ tends not to produce optimal growth in any species, and in many species it significantly inhibits growth, especially primary root growth. This is often referred to as NH_4^+ toxicity. In general, there are fundamental differences in cellular redox and energetics between NO_3^--grown and NH_4^+-grown plants. NH_3 assimilation from NO_3^- does not acidify the cytoplasm, whereas NH_3 assimilation from NH_4^+ does. NH_4^+-derived NH_3 tends to be mostly assimilated in the roots, producing excess H^+. Acidification of the rhizosphere is therefore often characteristic of NH_4^+-grown plants because of enhanced efflux of H^+. Many studies have shown alteration of root growth due to rhizosphere acidification, but this does not wholly explain NH_4^+ toxicity. In many plants, NO_3^- supply moderates NH_4^+ toxicity. Numerous explanations for this effect have been suggested, including less energy expenditure on cellular redox control and the increased sink for electrons from the light reactions that NO_3^- reduction can provide. In many plants, up to 25% of the reductant from photosynthesis can be used for NO_3^- reduction, which provides a very useful sink in photoinhibitory conditions. In addition to challenges related to cellular redox, cellular energetics fundamentally change with N supply because of the necessity to reduce NO_3^- (Figure 5.15).

In crop plants such as maize, shoot demand for N is a fundamental driver of growth—if N can be supplied, the capacity of the shoot to utilize it is an important determinant of eventual biomass. In many NO_3^- feeders, the detection of NO_3^- invokes the primary NO_3^- response in which there are fundamental changes in physiology, which drive the uptake, assimilation, and utilization of NO_3^-. A NIN-like protein (NLP) family of transcription factors is the master regulator of this response, which can involve more than 1000 genes. There is also a significant role for miRNAs, especially miR169, miR171, and miR398, in regulating whole-plant responses to different patterns of N supply. In perennial plants, N concentration is fundamental to the leaf economics spectrum. At lower N concentrations, leaf life span tends to be longer, increasing the residence time of N. Longer leaf life spans are associated with a sclerophyllous leaf morphology, which has implications for the dynamics of whole ecosystems. The sclerophylly of many vegetation types is attributed in significant part to N availability. The recycling of N from old to new leaves by some plants decreases the effects of low N on leaf life span, but it is still the whole-plant N physiology that drives important aspects of ecosystem dynamics (Figure 5.16). Physiological adjustments that plants make to differences in N supply are thus fundamental to plant growth and hence ecosystem dynamics in both managed and unmanaged ecosystems.

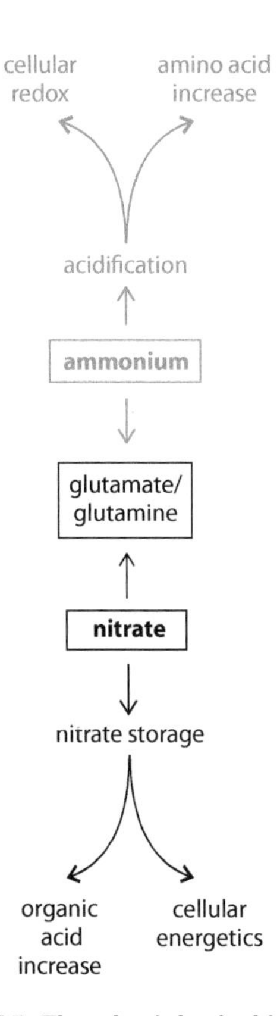

Figure 5.15. The physiological impact of nitrate and ammonium supply to plants. Some plants can grow well on pure NO_3^- or pure NH_4^+, but they must cope with the physiological effects of pure N sources. Particular problems stemming from acidification are caused by a pure NH_4^+ source. Many plants grow best on a mixed N source (although the optimum ratio varies according to the species), probably because NO_3^- helps to decrease acidification, provides a convenient N store, and provides a sink for reducing power during adjustments of cellular energetics, whereas NH_3 provides a more immediately assimilated form of N.

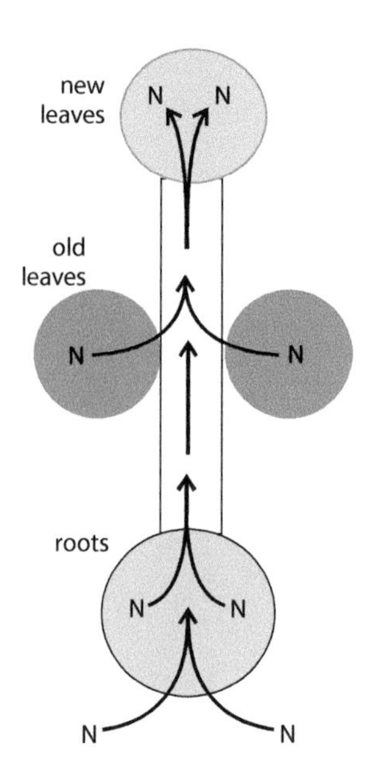

Figure 5.16. Retranslocation of nitrogen in plants. Demand for N is high, and growing shoots are the primary sink for N in plants. Interactions between the availability and retranslocation of N determine fundamental plant properties and have ecosystem-wide implications. N is highly mobile in plants, as NO_3^- or in organic acids, and can be retranslocated from senescing leaves into new growth. In the short term, low N levels in the soil promote retranslocation from old leaves to new ones. In general, chronic low N levels are associated with an increased leaf life span (LLS), which decreases demand for N, producing the relationship between leaf N and LLS in the leaf economics spectrum. In some plants, however, chronic low N levels are associated with rapid leaf turnover, which maximizes retranslocation. Extremes of LLS affect biomass quality, C fixation rates, and transpiration rates.

Plants adjust their root morphology in response to shortages of nitrogen

The tuning of transport and assimilation capacity to soil solution concentration is apparently insufficient to optimize the utilization of N in the long term, because there are morphological responses of root systems to N deficiency. These responses not only position transport systems for optimum function in space and time, but are also mediated by them. In general, many herbaceous plants decrease the root proportion of biomass when N availability is high and increase it when N availability is low. Many models that successfully predict root:shoot allocation are equilibrium models that maximize relative growth rate (RGR) and assume a positive relationship between RGR and N content of biomass. Empirical studies of responses to nutrient availability highlight differences between species and deviate from models at particularly low or high N, but N status is clearly linked to root:shoot biomass ratio in herbaceous plants. The dynamics of many processes in terrestrial ecosystems are driven by the root:shoot biomass allocation in plants, and hence by soil N availability.

Under conditions of N deficiency, investment in roots aids foraging behavior. When nutrients are in short supply, stores are often expended in an effort to restore their capture. Monocotyledonous root systems develop from adventitious roots, and dicotyledonous root systems develop from a primary root. On adventitious roots, first-order lateral roots (LRs) develop, whereas on a primary root, first-, second-, and third-order LRs can be common. Split-root and nutrient partitioning experiments with NO_3^-, NH_4^+, and amino acids have shown that their deficiency induces specific local changes in root system architecture consistent with an **optimal foraging strategy**. In plants that are deficient in P, root growth quickly stops, but in N-deficient roots the growth capacity is more sustained. LRs also respond dramatically if the root grows through a patch of high N concentration. Especially in N-deficient plants, an NO_3^- patch stimulates the growth of pre-existing first-order LRs, and in some cases the initiation of new ones. Some evidence suggests that exposure to NH_4^+ tends to increase LR initiation and, in dicots, to stimulate the growth of initials on second- and third-order LRs, resulting in a short-root phenotype. The uptake of amino acids by roots has numerous indirect effects on root growth because of altered N metabolism, but glutamate can have direct effects. It stops primary root growth and promotes the early stages of LR development. In those plants that can use a combination of N sources there may be complementary root system architecture responses arising from different N forms. This is likely to be necessary given the different dynamics of supply of different N sources. Plant root systems undergoing LR responses to an N patch can increase root hair length, in particular in response to NH_4^+. Changes in LRs and root hairs are accompanied by changes in expression of N transporters, presumably integrating responses to variation in N supply at both fine and medium scales. Overall, many N-deficient plants are able to restore N supply through root foraging.

The localized foraging for N by roots indicates that there are separate "modules" of the root system that provide a flexible response at an important scale of variation in soil N supply. However, the response of individual modules is coordinated. The response that a particular soil N concentration elicits depends on the N status of the whole plant and on the NO_3^- concentrations to which other modules of the root system are being exposed. Experiments that remove shoot systems suggest that systemic root-to-shoot-to-root signals, and in particular cytokinin, coordinate the response of root modules. Responding to high N concentrations in one part of the root system suppresses the growth of roots in positions where there is low N. The sensitivity of the response also declines progressively as plants get older. Root system architecture is also affected by factors other than N concentration, including mechanical resistance, water availability, and temperature. The sensors for N responses are NRT and AMT acting as transceptors—analyses with numerous mutants of both N and auxin metabolism indicate that NO_3^- and NH_4^+ are directly used to signal root responses. The responses converge on the antagonistic effects of auxin and cytokinin on the development of LRs. The MADS-box TF ANR1, and other TFs that mediate responses to auxin, help to control the expression of genes that induce LRs. Proteins involved in auxin influx (that is, AUX) and efflux (that is, PIN) help to establish the gradients of auxin that regulate LR development. In *Arabidopsis*, LR initiation occurs in a zone of cells behind the root tip, and is preceded by a dramatic oscillation in the expression of several thousand genes, which is linked to the oscillation of auxin response factors and the circadian clock. miR167 targets an auxin response factor (ARF8), and miR393 targets an auxin receptor (AFB3), helping to control LR initiation. Several kinases characteristic of N and auxin response pathways have also been shown to respond during changes in root morphology. ABA produced during drought conditions can suppress the initiation of LRs, but ethylene tends to promote their development.

Symbioses contribute significantly to plant nitrogen uptake in nitrogen-deficient environments

The only organisms on Earth that are able to fix both C and N are cyanobacteria. In one group of these organisms, N fixation occurs in specialized **heterocysts**, which provide an anaerobic site for this process, whereas in non-heterocystous cyanobacteria N fixation can only occur in anaerobic environments. Heterocysts probably first evolved about 2–2.5 billion years ago, when O_2 partial pressures in the atmosphere rose significantly and inhibited N fixation. The monophyletic heterocystous clade of cyanobacteria is generally regarded as the origin of the chloroplast. Heterocysts are degenerate photosynthetic cells with PSI complexes that provide ATP for N fixation but, in contrast to the other cells, they have no PSII and thus no O_2 production. N fixation in the plastid of a plant cell has never been reported, despite the origin of the chloroplast in a group of N-fixing organisms. It seems that the ability to form N-fixing heterocysts was lost during chloroplast evolution. Instead, a number of plant groups have, in the generally chronically low-N environments of the Earth's surface, gained an advantage by evolving separate N-fixing symbioses with diazotrophs. In most instances these symbioses are suppressed when inorganic N is available.

Cycads first evolved in the Permian, but "the age of the cycads" was during the Jurassic (Figure 5.17a). From about 250 million until 65 million years ago, in many ecosystems cycads probably contributed very significantly to biomass production, but from the Cretaceous onward conifers and then angiosperms dominated, so that there are only about 250 extant cycad species in three families. Extant cycad families evolved relatively recently, but all of them form N-fixing symbioses with heterocystous cyanobacteria in the genus *Nostoc*, or close relatives. It seems likely that all extinct cycads

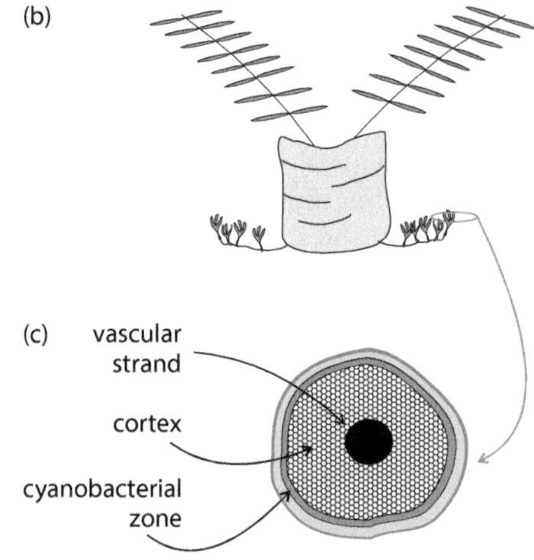

Figure 5.17. Nitrogen fixation in coralloid roots of cycads. (a) Cycads have whorls of tough leaves that are often edged with spines—a morphology that perhaps evolved to deter herbivorous dinosaurs in the Mesozoic. They produce cones (not shown) from the center of the whorl of leaves. (b) Cycads can produce negatively geotropic coralloid roots that penetrate up through the soil around the stem. They also produce a positively geotropic root system that anchors the plant and takes up water and nutrients. (c) In cross section there is a distinct zone of cells with symbiotic N-fixing cyanobacteria (*Nostoc* species) in the outer layers of cortex.

also had this symbiosis, and that N-fixing symbioses have contributed to biomass production in terrestrial ecosystems for much of their history. The symbiosis occurs in coralloid roots that often emerge from the soil surface (Figure 5.17b). Some water ferns in the Salviniales have an N-fixing symbiosis with cyanobacteria in the genus *Anabaena* (Box 5.3), but although ferns

BOX 5.3. WATER FERNS AND N FERTILIZERS

Water ferns in the genus *Azolla* (Figure 1) can be serious aquatic weeds, but have also been widely used to provide N for crops, especially rice. In South-East Asia, from China through to Vietnam, water ferns have been used in rice production for many centuries. Water ferns have an N-fixing symbiont, *Anabaena* (a cyanobacterium). Water ferns are unusual ferns not only because of their aquatic habit, but also because they are heterosporous—that is, they produce spores of different sizes. They produce a megaspore carrying a female gametophyte that has floats and a colony of *Anabaena*. The cyanobacteria are therefore passed down the maternal line.

Azolla pinnata was the primary water fern in rice paddies of South-East Asia for centuries before the introduction of *A. filiculoides*. *Azolla* can be grown on the surface of paddy water when rice is flooded, where it not only aids N input but also suppresses the growth of aquatic weeds. It is also commonly grown in separate ponds to produce high-N biomass for use as fertilizer in non-flooded soils. When flooded, rice paddies can have significant populations of free-living cyanobacteria. General estimates are that free-living cyanobacteria can fix 20–30 kg ha^{-1} per season, whereas those with symbiotic *Azolla* can potentially fix 50–90 kg ha^{-1}, although rates of 25 kg ha^{-1} per season are common. *Azolla-Anabaena* is present in about 3 million hectares of rice paddy, where it contributes the majority of the several Tg of N fixed each year by cyanobacteria in rice paddies.

Azolla filiculoides, native to North America, has now been introduced to most parts of the world. It grows very fast in a wide variety of aquatic systems, and is a serious weed. However, its rapid growth does mean that it can accumulate xenobiotics from water. There has been a great deal of research into the use of *Azolla* in phytoremediation of a very wide range of inorganic and organic contaminants. Its rapid C-fixing, N-fixing symbiosis, and water-cleansing capacity have led to serious investigation of its use in Controlled Ecological Life Support Systems for space flight.

Figure 1. ***Azolla filiculoides.*** An aquatic N-fixing fern (top view of fern floating on water is shown).

Further reading:
Herridge DF, Peoples MB & Boddey RM (2008) Global inputs of biological nitrogen fixation in agricultural systems. *Plant and Soil* 311:1–18.

Rahman MA & Hasegawa H (2011) Aquatic arsenic: Phytoremediation using floating macrophytes. *Chemosphere* 83:633–646.

Chen M, Deng S, Youquan Y et al. (2012) Efficacy of oxygen-supplying capacity of *Azolla* in a controlled life support system. *Adv. Space Res* 49:487–492.

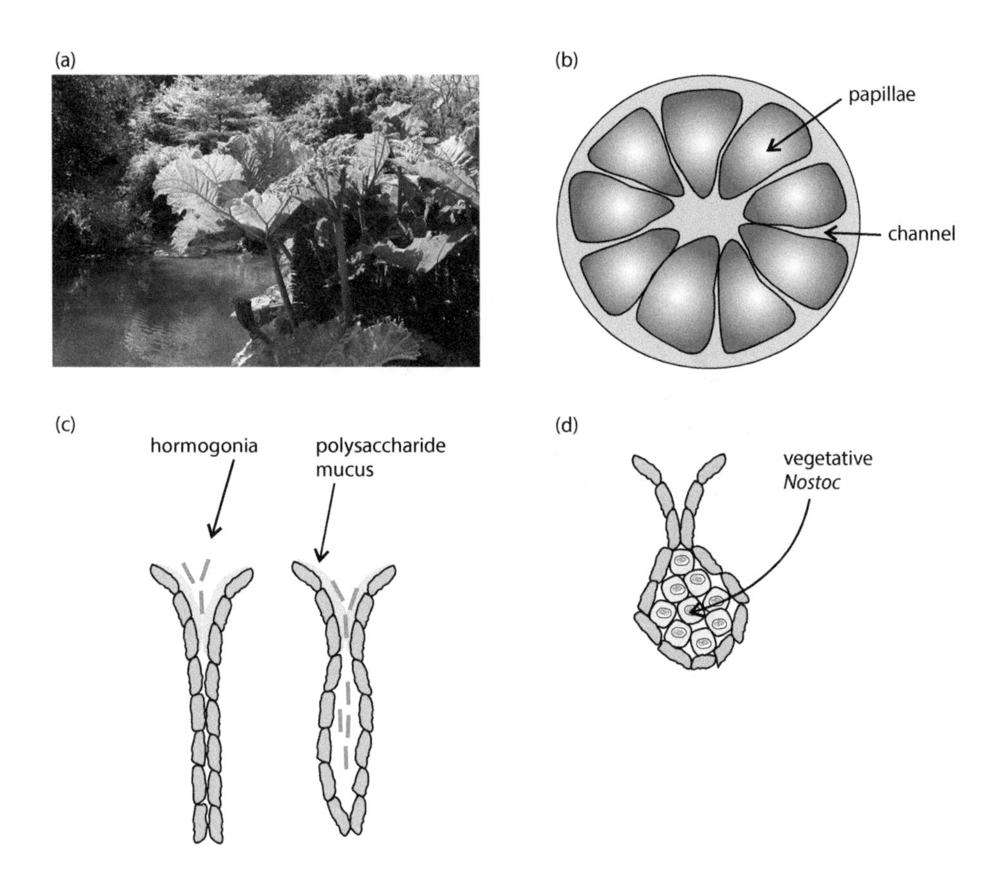

Figure 5.18. Nitrogen fixation in *Gunnera* species. (a) *Gunnera* is a herbaceous, primarily neotropic genus, with some species having leaves up to 2 m wide. (b) After germination it forms glands on the hypocotyl that persist at the base of each petiole. The papillae are red due to anthocyanin. (c) *Nostoc* gains entry through channels between the papillae. It enters the channels as hormogonia, which are short gliding filaments that move through a polysaccharide mucus. (d) Once the *Nostoc* has penetrated the gland and is mature, it produces vegetative filaments that are engulfed by plasma membrane. Breakdown of starch and other carbohydrates in the surrounding cells provides glucose and fructose to the cyanobacteria, feeding them but also maintaining them in the vegetative N-fixing state.

evolved before cycads, their N-fixing symbiosis is more recent. No N-fixing symbioses have been found in conifers, but there are three in angiosperms. Uniquely for an angiosperm, *Gunnera* species form a symbiosis with a cyanobacterium (*Nostoc punctiforme*) in glands that develop on the stem at low N concentrations (Figure 5.18). About 200 species from eight angiosperm families form, in root nodules, N-fixing symbioses with *Frankia*, a filamentous actinobacterium (Figure 5.19). Many species with this symbiosis flourish in low-N soils—for example, alders (*Alnus* species) and she-oaks (Casuarinaceae) in temperate and subtropical climates, respectively. In all the genera of the Fabaceae there are species that form N-fixing symbioses in root nodules with proteobacteria. Referred to collectively as "rhizobia," members of both α- and β-proteobacteria form the symbiosis. The Fabaceae, which are the third most speciose plant family (with nearly 20,000 species), are especially important in many low-N ecosystems, and second in agricultural importance only to the Poaceae. Ecologically and agriculturally their N-fixing symbiosis is the most important. All angiosperms with N-fixing symbioses in root nodules are closely related on one "N-fixing" rosid clade (Figure 5.20), so the propensity to develop N-fixing root nodules seems only to have evolved in this group once.

The root symbiosis with Glomeromycota that form arbuscular mycorrhizas (AM) is 400 million years old, and occurs in perhaps 90% of extant angiosperms. It can significantly aid plant uptake of inorganic and organic N. Strigolactones from plant roots stimulate fungal growth and the production of lipo-chitooligosaccharides ("myc factors") from fungal cell walls, which are detected by receptor kinases in plant roots. These receptor kinases initiate reactive oxygen species (ROS) and mitogen-activated protein kinase (MAPK) cascades, leading to ion fluxes across the plasma membrane and into the nucleus, where they activate specific transcription factors. This results in significant intracellular rearrangement to make a pre-penetration apparatus through which root infection occurs. These pathways overlap with

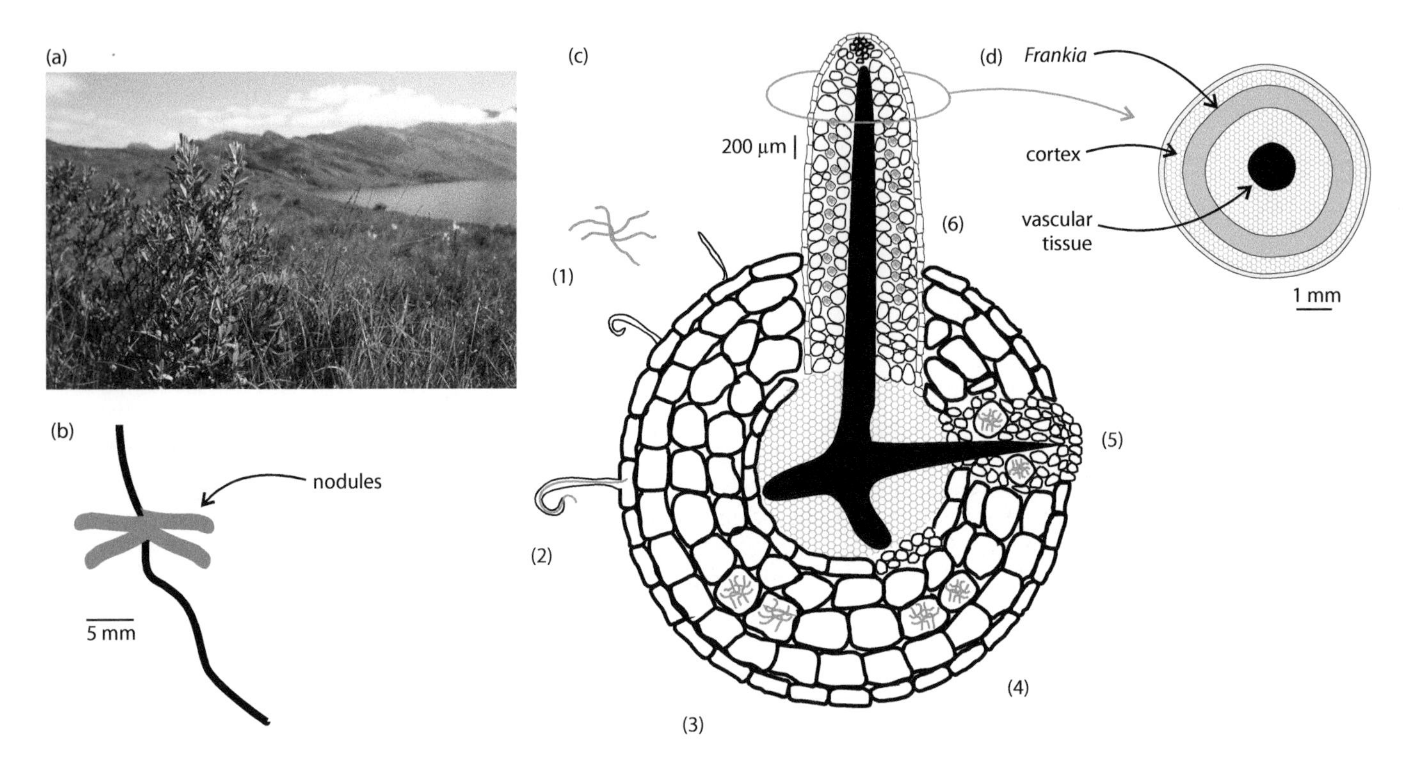

Figure 5.19. The N-fixing nodules of the symbiosis with *Frankia*. (a) *Myrica gale* (bog myrtle), a shrub with N-fixing *Frankia* in root nodules, growing on low-N blanket peat in Scotland. Most species that have a symbiosis with *Frankia* are woody. (b) Nodules with *Frankia* often occur in clusters on roots. (c) The development of a *Frankia* nodule: (1) Root hairs deform in response to chemicals exuded by *Frankia*. (2) *Frankia* enters the root through deformed root hairs. (3) *Frankia* proliferates in cortical cells. (4) Cells in the pericycle opposite a xylem pole begin to differentiate in a similar manner to the formation of a lateral root. (5) The developing nodule grows outward to engulf the cells infected with *Frankia* which begin to show hypertrophy. (6) A root nodule, which is a modified lateral root with many hypertrophic infected cells between uninfected cortical cells. The nodule has a meristem and continues to grow. (d) In cross section there is a distinct zone of cells infected with *Frankia*. The infected cells contain a high concentration of auxin, which the surrounding cortical cells help to maintain.

those initiated during infection by pathogenic fungi, which might have first driven their evolution. Seed germination in many parasitic plants is triggered by strigolactones, and the physiological changes that occur in host cells during the initial phases of parasitic infection have much in common with those of AM. It is likely that parasitic plants have co-opted elements of the AM signaling pathway to gain entry to plant roots, but perhaps the most important adaptation of this pathway is the initiation of nodules for rhizobia by species in the Fabaceae. Flavonoids, and sometimes other compounds, exuded from plant roots trigger the production by rhizobia of "Nod factors."

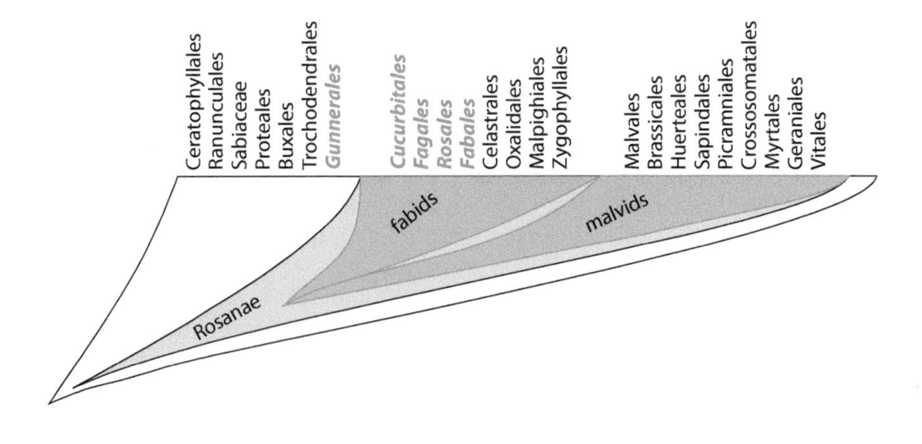

Figure 5.20. The N-fixing clade. All of the angiosperm orders which contain plants that form N-fixing symbioses with *Frankia* or rhizobia are located in the fabid clade in the Rosanae (orders shown in green text). Rhizobia are known to infect one genus outside the Fabales, *Parasponia*, which is in the Ulmaceae in the Rosales. The Gunneraceae that form a symbiosis with *Nostoc* are classified as a basal rosid on a sister clade to the fabids.

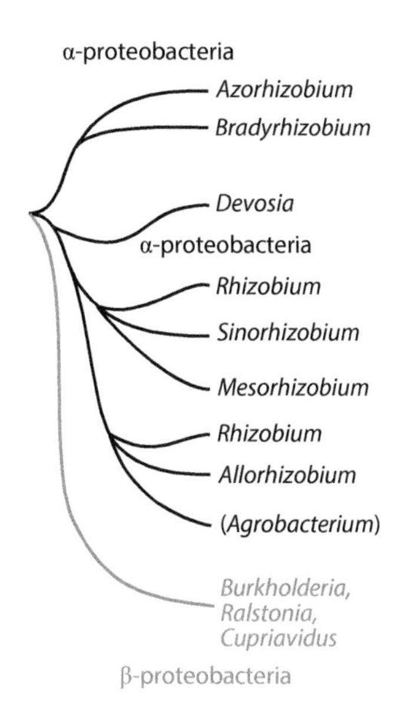

Figure 5.21. **The phylogeny of N-fixing symbionts in the Fabaceae.** "Rhizobia" originally referred to organisms in the genus *Rhizobium*, but the term is now used to refer to related bacteria that also form symbioses with members of the Fabaceae. However, molecular studies have shown that N-fixing symbionts in the Fabaceae are paraphyletic and arise from distinct groups within the α-proteobacteria. Some species in the genus *Rhizobium* are closely related to *Agrobacterium*. Isolation of bacteria from nodules of diverse members of the Fabaceae has shown that there are also quite distantly related β-proteobacteria that can be N-fixing symbionts (Data from Willems A [2006] *Plant Soil* 287:3–14.)

Rhizobia come from several genera (Figure 5.21) that do not correlate with host specificity, although the structure of the Nod factors that they produce generally does. Up to 34 *nod*, *nol*, and *noe* genes, variously organized on plasmids in different rhizobia and under the control of NodD, control the synthesis and export of Nod factors. These are lipo-chitooligosaccharides that initiate the same biochemical pathways as AM, which are thus called the "common symbiosis signaling (Sym) pathway" (Figure 5.22). Peas (*Pisum sativum*) form AM and N-fixing nodules, and are infected by the root parasitic plant *Orobanche crenata* and by the root-gall-forming nematode *Meloidogyne*. In all cases, interaction is initiated in significant part through the Sym pathway.

In the rhizobia–legume symbiosis, the Sym pathway activates TFs that control the expression of early nodulation (*enod*) genes. In a short zone behind the root tip these trigger the formation of hooked root hairs, through which an infection thread forms by the loosening of cell walls and invagination of the plasma membrane. Pre-infection changes in the cytoplasm and cell

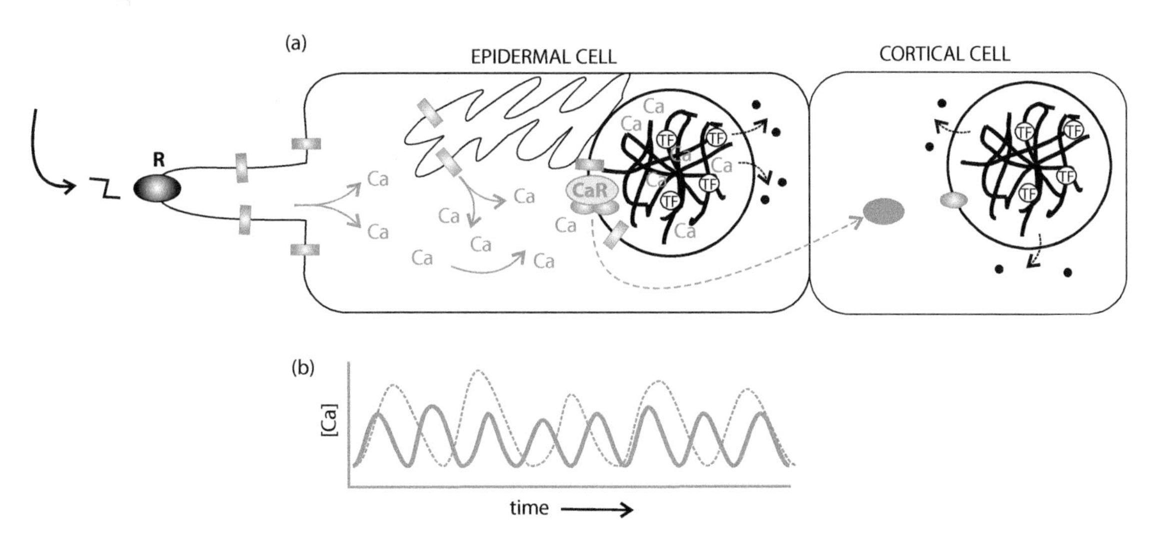

Figure 5.22. **The Sym pathway.** Many legumes can form both nodules and arbuscular mycorrhizas (AM). The detection of symbiotic partners and the transcription factors (TFs) that control root anatomical changes to accommodate them are specific to each association (shown in black letters), but the signaling events that connect detection and development share the Sym pathway (shown in green). (a) Nodulation factors (NFs) from rhizobia, which are lipo-chitooligosaccharides (LCOs), are detected by Lys-M-like receptor-like kinases (R) in the root hairs of epidermal cells. Mycorrhization factors are also LCOs, but often have different receptors. The common Sym pathway involves the activation of channels that produce waves of Ca in the cytoplasm. These waves are detected by a Ca-receptor kinase (CaR) which, together with nuclear K influx channels and elements of the circadian clock, controls waves of Ca in the nucleus. If NF-induced patterns of Ca are detected in the cytoplasm (denoted by solid line in (b)), TFs are expressed which induce the production of proteins (•) that control infection by rhizobium. If Ca patterns characteristic of mycorrhizas are detected (denoted by broken line in (b)), TFs that control mycorrhizal development are expressed. In both associations, cytokinin receptors in cortical cells detect events downstream from CaR (green filled ellipses), initiating the expression of TFs that control anatomical changes in the cortex. The AM association is ancient (over 400 million years old) and the rhizobial association is more recent (around 60 million years ago), so early events in the rhizobium–legume association probably evolved directly from those controlling AM.

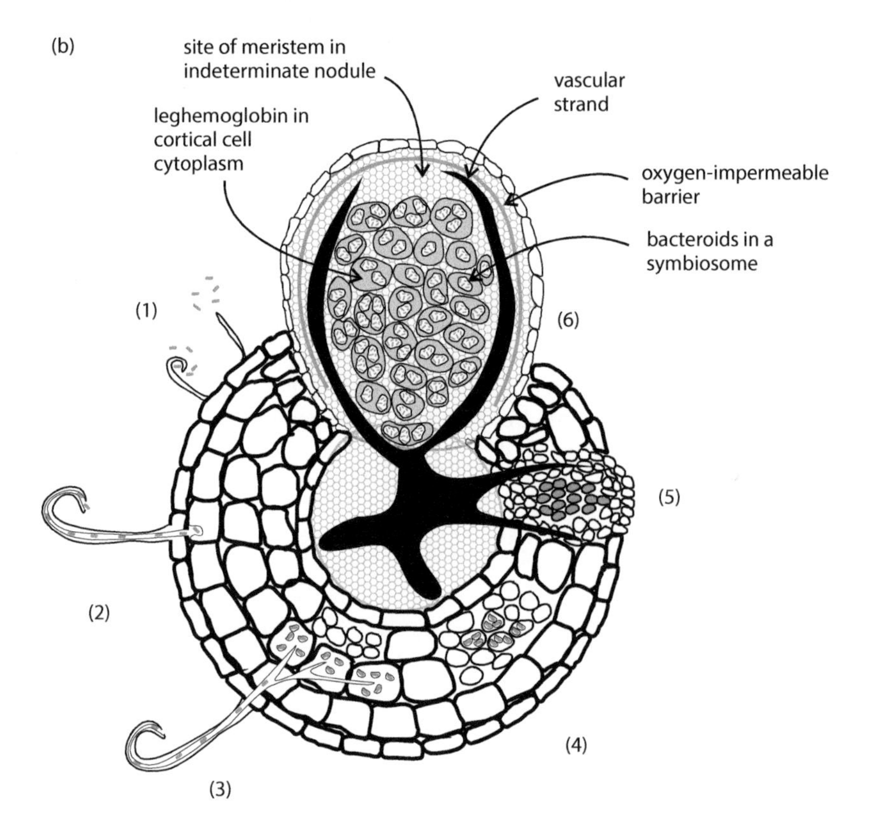

Figure 5.23. The nodules of the Fabaceae. (a) *Ulex europaeus* (European gorse) with the characteristic zygomorphic flower of the Papilionoideae subfamily, and *Acacia* species with the characteristic actinomorphic flowers of the Mimosoideae. (b) The development of a nodule in the Fabaceae: (1) Rhizobia attracted to the root by flavonoids induce root-hair curling with Nod factors. (2) Rhizobia enter the root through an infection thread. (3) The infection thread distributes membrane-bound rhizobia into the cortical cells, and a nodule primordium begins to divide. (4) The nodule primordium begins to surround the rhizobia-containing cells. (5) A vascular strand penetrates the periphery of the developing nodule. (6) A functioning nodule with a peripheral barrier layer and leghemoglobin to aid control of O_2 concentrations so that rhizobia in the symbiosomes can fix N.

walls prepare cells for penetration by the infection thread, which forms an entry route into the cortex for rhizobia. In the cortex opposite the protoxylem poles, ENODs initiate the formation of a nodule primordium (Figure 5.23). Nodules that arise in the outer cortex, as is characteristic of many legume species, are determinate and spherical in structure. In other species, nodules arise in the inner cortex and are indeterminate and cylindrical. In some species, polar auxin transport is inhibited below the nodule primordium, and high auxin concentrations are characteristic of most mature nodules. Thus gradients of cytokinin and auxin probably determine which cortical cells become nodule primordia. During entry into nodule primordial cells from the infection thread, rhizobia are surrounded by plant-derived peribacteroid membranes, which form symbiosomes. It is in these that bacteria will divide to produce the N-fixing **bacteroids** that fill the bulk of an active nodule. Legume root nodules are anatomically unique and adapted to provide the conditions necessary for N fixation. An O_2 permeability barrier in the outer layer of the nodule helps to decrease the entry of O_2 into the nodule, and leghemoglobin in the cytoplasm of infected cells helps to minimize the entry of O_2 into bacteroids. Leghemoglobin is synthesized by the plant and binds single molecules of O_2, so its mode of action is like that of myoglobin rather than hemoglobin. Its protein **moiety** is similar to those of other O_2-binding proteins in plants, and there is some evidence that it can also bind N_2, so it might play a role in regulating N_2 flow to bacteroids. Penetration of the plant vascular system into the nodule provides a route for carbohydrates to enter the nodule, mostly from the phloem, and for nitrogenous compounds to exit the nodule via the xylem. Sucrose is transported from the phloem into the cytoplasm of infected cells, and broken down into fructose and UDP-glucose, from which PEP carboxylase and malate dehydrogenase synthesize malate to be transported into the bacteroids. The rhizobial plasmids include genes for carboxylic acid transporters that mediate malate influx. In many legumes the amides glutamine and asparagine are exported from the nodule via the xylem, having been synthesized by glutamine synthetase in the cytosol and GOGAT/aspartate aminotransferase in the plastid, respectively. In some legumes, ureides derived from glutamate and aspartate via purines are exported to the plant.

As in other diazotrophs, in rhizobia N fixation is controlled by *nif* genes and, in response to O_2, the fixL–fixJ protein complex regulates their expression. Ethylene inhibits the Sym pathway, and legume mutants deficient in ethylene signaling pathways hypernodulate by producing nodules all around the cortex, not just opposite the protoxylem poles. Prolonged exposure to Nod factors increases ethylene concentrations in roots switching off nodulation, and during normal development ethylene probably inhibits nodule primordia between the xylem poles. Increased ABA concentrations also inhibit nodulation via the Sym pathway, so particularly wet or dry conditions can inhibit nodulation. There is significant autoregulation of nodulation across root systems, because nodule formation on one part of the root system can inhibit nodulation on another part. Nitrate has also long been known to inhibit nodulation, and might do so through the autoregulation pathway. Enhanced soil P concentrations increase nodulation at low N concentrations but not at high concentrations. These reactions are probably integrated with the suppression of parasitic plants which, given the common Sym pathway, might invade instead of rhizobia.

In many unmanaged ecosystems, symbiotically fixed N is the most important N input. This is especially so in chronically low-N ecosystems, such as savannas and other grasslands, chaparral and other Mediterranean vegetation types, and tropical forests where legumes and other species with N-fixing symbioses are relatively common. It is also generally common in many early successional stages when N concentrations are low but light, space, and other nutrients are non-limiting. Established temperate forests, which are low-N ecosystems, are strikingly depauperate in plants with N-fixing symbioses. This is most probably because of light limitations in established temperate forests that inhibit the growth of plants with N-fixing symbioses. Around 50–70 million tonnes of N are biologically fixed in agricultural systems each year which, despite the declining proportion of legume-derived N in agriculture since 1960, is a significant amount of N when compared with the 120 million tonnes from inorganic fertilizers.

Carnivorous plants are mixotrophs that can obtain nitrogen opportunistically from an erratic supply

There are more than 600 species of plants that can obtain nutrients by trapping and digesting animals. Feeding experiments with many carnivorous plant species have shown that they produce more biomass and reproduce more successfully when they catch prey. N is the most important nutrient obtained by carnivorous plants. Carnivory has evolved independently at least six times across five different orders of flowering plants (Figure 5.24). Fossilized remains of plants suggest that it first evolved at least 85 million years ago, and that some carnivorous genera may have been more common in the past than they are now. There are four different types of prey trap, all made from adapted leaves, namely pitfall traps (for example, *Sarracenia*, *Nepenthes*, *Heliamphora*, *Darlingtonia*), suction traps (for example, *Utricularia*, *Genlisea*), adhesive traps (for example, *Drosera*, *Drosophyllum*, *Pinguicula*), and snap traps (for example, *Dionaea*, *Aldrovanda*). In general, insects are the predominant prey items, but carnivorous plants that have traps in aquatic or semi-aquatic environments (for example, *Utricularia*, *Aldrovanda*, *Genlisea*) have been shown to trap crustaceans and protozoa. The insect taxa caught by carnivorous plants vary significantly. Perhaps the most remarkable are the snap traps, which exhibit the fastest movements in the plant kingdom. They evolved from the adhesive traps, primarily driven by prey size. The snap traps, which catch a few large prey items and after successful capture prompt significant changes in an individual, epitomize a strategy geared towards catching a few erratic bursts of N. Adhesive traps are less effective for trapping large prey species, and more easily damaged by

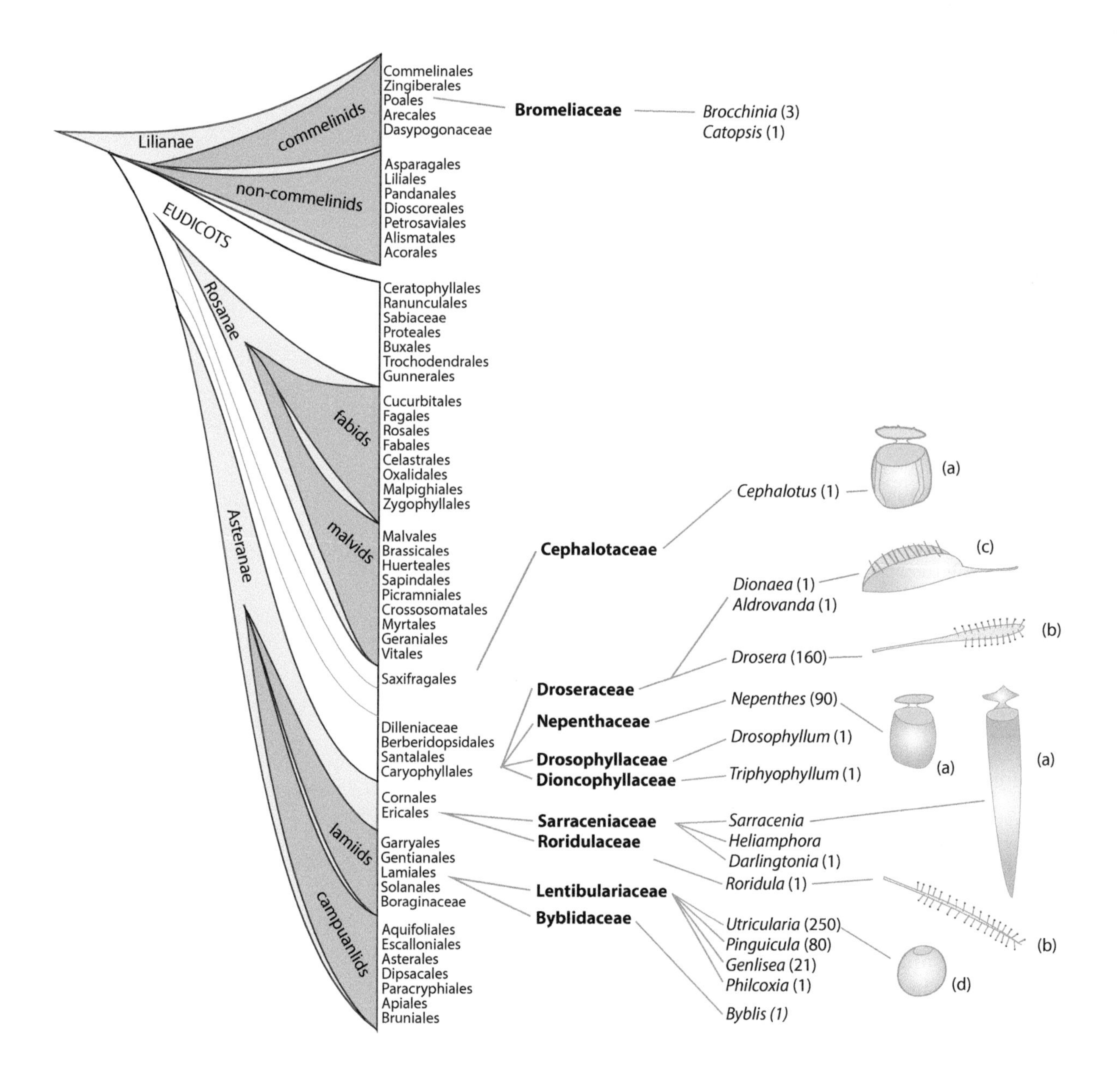

Figure 5.24. The phylogenetic distribution of carnivorous plants. Carnivorous plants belong to five different orders. There is only one family of monocotyledonous carnivorous plants (Bromeliaceae). There are 480 carnivorous species, although the majority (300 species) are in the Lentibulariaceae in three genera—*Pinguicula*, *Utricularia*, and *Genlisea*. A variety of traps are used by carnivorous plants. Examples are shown of (a) pitfall traps, (b) adhesion traps, (c) snap traps, and (d) bladder traps.

them, than are the snap traps. A number of plant species are known to trap insects but have not been shown to benefit from this in the way that truly carnivorous species do. The recent discovery of subterranean carnivory via leaves adapted to catch nematodes in *Philcoxia* is a reminder of a common suggestion that carnivory may be more common in plants than is generally assumed. There is much evidence that pitfall traps, adhesive traps, and snap traps actively lure their prey. Traps are adapted leaves, as are petals, and in many species they have been shown to utilize many of the same strategies for luring insects as are used by insect-pollinated flowers—both extra-floral **nectaries** and significant production of volatile attractants are common in pitfall and adhesive traps. In some species but not others, visual cues play a role in directing prey into the trap, as they do in petals.

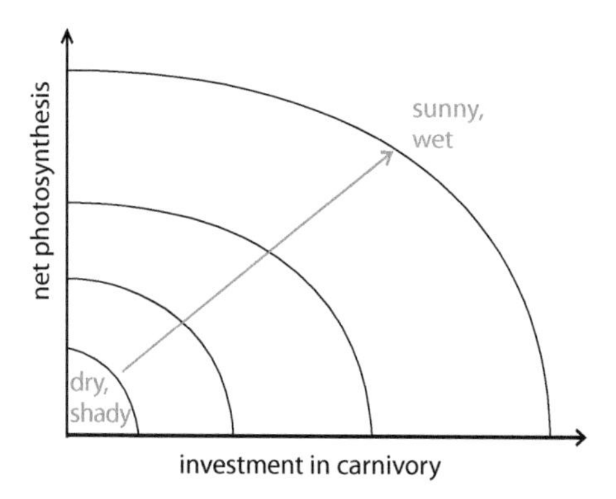

Figure 5.25. Advantageous conditions for photosynthesis in carnivorous plants. The development and maintenance of traps decreases net photosynthesis. Traps are modified leaves which photosynthesize less than unmodified leaves, and in most species waxes, nectar, and enzymes have to be synthesized for traps to function. In shady and/or dry conditions, investment in carnivory can decrease net photosynthesis to zero. Only in sunny and wet conditions is net photosynthesis sufficient to allow investment in carnivorous traps to supplement nutrient supplies.

Most carnivorous plants inhabit **oligotrophic** substrates, but only where water and light are readily available. Leaf adaptation for carnivory decreases photosynthetic capacity, and photosynthate is expended in trap construction and maintenance. Many traps maintain some photosynthetic capacity, but traps are more complex than simple leaves and have significantly higher construction costs. Pitfall traps have zones lined with particularly low-friction epicuticular waxes that have significant synthesis costs, and extra-floral nectaries have to be supplied with sucrose, in the case of the sundews (*Drosera*) in significant amounts. General cost-benefit analyses (Figure 5.25), suggest that the sacrifice of photosynthetic leaf area can only be of benefit if water and light are not limiting. In some carnivorous species, shading experiments that decrease growth have supported this assertion. In many carnivorous species, including *Utricularia*, *Drosera*, and *Sarracenia*, nutrient enrichment has been shown to decrease plant investment in carnivory. Before traps are formed, carnivorous plants derive all their nutrients from the soil, and it is clear that many of these plants continue to utilize soil N when they are mature. Experiments with ^{15}N-labeled nitrogen sources in the field have shown that in some instances at least 40–60%, and sometimes up to 80% or more, of plant N is derived from captured prey. It is not generally essential for carnivorous plants to obtain N from prey, even in the field, and in many instances prey capture does not seem to completely overcome N limitation. In oligotrophic conditions where interspecies competition is weak, the investment in traps seems generally to provide a valuable supplementary nutrient supply to aid **mixotrophic** growth.

Analysis of elements in the leaves of carnivorous plants indicates that they are particularly low in N and P (Figure 5.26), and that supplying these nutrients to them relieves this limitation. The analysis of enzymes secreted by carnivorous plants and of fluids in pitfall traps confirms that carnivorous plants have the capacity to digest not only N-containing but also P-containing molecules. *Heliamphora* and *Darlingtonia* rely on microbes in their pitchers to digest caught prey items, but many carnivorous plants exude enzymes, often copiously, into their traps. These exuded enzymes have long been known

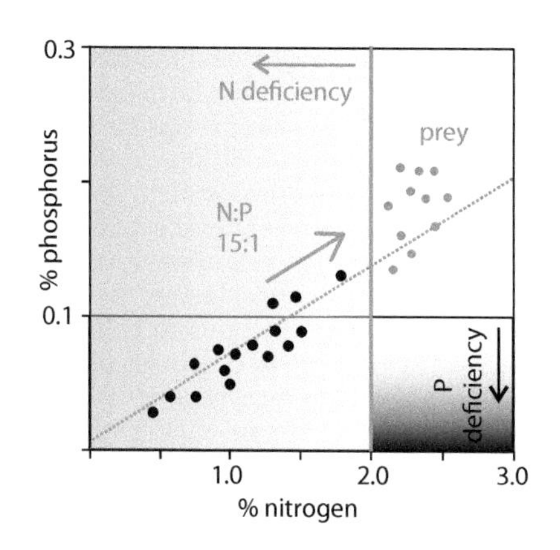

Figure 5.26. N:P stoichiometry in carnivorous plants. The N:P ratio in plants is generally stable at 15–16 when both nutrients are available. In carnivorous plants this N:P ratio is generally found, but below the deficiency thresholds of N and P (2% and 0.1% of biomass, respectively). In carnivorous plants this N:P ratio is maintained, mostly below the deficiency thresholds of N and P (2% and 0.1% of biomass, respectively) (black dots). Feeding inorganic N or P to carnivorous plants can relieve either N or P deficiency, but feeding them their usual prey species tends to relieve deficiencies of both nutrients (green spots). In some species, P deficiency is relieved at least as much as N deficiency by feeding on prey. (From Ellison AM [2006] *Plant Biol* 8:740–747. With permission from John Wiley and Sons.)

BOX 5.4. THE PROTEASES OF *NEPENTHES*

The proteases of *Nepenthes* have similar activity to those of other carnivorous species. Nepenthesins are a unique type of aspartate protease produced by *Nepenthes*. Pepsin, a proteolytic enzyme used extensively in vertebrate digestive systems, is an aspartate protease. Two opposing aspartate groups are vital to the proteolytic action of these enzymes. Nepenthesin sequences are quite different from those of other aspartate proteases, and include regions of high cysteine concentration. It has been suggested that these provide the structural stability over a wide range of conditions that is necessary for a protease to function in the variable environment of a pitcher rather than a digestive system.

to include **protease** activity (Box 5.4). Chitinase activity is often significant, and not only is the ability to digest the **chitin** of insect exoskeletons a key first step in the breakdown of prey, but also chitin contains a significant proportion of the N content of an insect. A number of acid phosphatases that release PO_4^{2-} have been found in carnivorous plants, as has cellulase activity, which might help to digest plant litter that is blown into traps. The excretions and fluids of carnivorous plants are often acidic, helping to provide optimal conditions for the action of proteases and phosphatases, and thus they really do simulate the digestive tract of many animals. The proteins excreted by all plants into their extracellular spaces are being shown to have an increasingly important biological role, and they might have a role in plant N nutrition that goes far beyond their use by carnivorous plants.

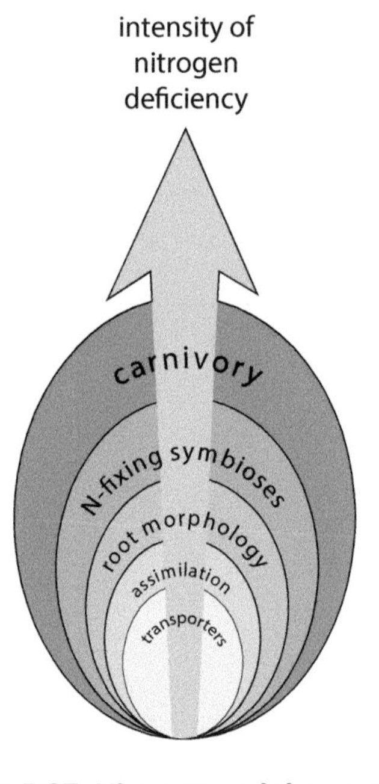

Figure 5.27. Nitrogen and the stress-response hierarchy. Plants have numerous N transporters that help to regulate N uptake during fluctuations in availability and form. Prolonged exposure to differing availability and forms of N necessitates differences in reduction and assimilation capacity. Some plants can adapt to chronic N limitation by housing N-fixing prokaryotes in specialized nodules or, in extreme cases, by radical changes in leaf morphology to form traps that catch and digest animals.

Summary

Some of the fundamental molecules of life contain N, so all organisms require it in significant amounts. N_2 is by far the most common form of N on Earth, but is unreactive. Only microorganisms with nitrogenase can fix N_2 into NH_3, which both plants and microbes can assimilate. Archaea and bacteria mediate the interconversion of numerous N species in the environment. N availability is often the limiting factor in unmanaged systems, so higher plants have adaptations to increase its supply. Regulation of uptake during short-term fluctuations in N availability tends to be via physiological adjustments, whereas adaptation to increasingly chronic low N tends to involve anatomical and morphological adjustment (Figure 5.27), including strategies for capture of N from erratic supplies. The widespread occurrence of the uptake of organic molecules, of symbioses, and of carnivory, all aimed at N capture, shows that many plants in terrestrial ecosystems are not successful purely as autotrophs. During the twentieth century, inorganic N fertilizers were instrumental in the greatest ever transformation in the production of food and plant products, and perhaps therefore of human societies. The development of a sustainable N supply for food and other crops is vital to sustainable human habitation of Earth, because of the global environmental consequences of unsustainable use of inorganic fertilizer. Advanced agronomic regimes have been developed for increasing the efficiency of N supply to crops, but N use in agriculture is still generally inefficient compared with that in unmanaged ecosystems, and utilizes autotrophic biomass production. It is likely that features of N utilization in unmanaged ecosystems will have to be incorporated into food production systems if global food security is to be achieved.

Further reading

The importance of N in the soil–plant system

Fernandez-Martinez M, Vicca S, Janssens IA et al. (2014) Nutrient availability as the key regulator of global forest carbon balance. *Nature Clim Change* 4:471–476.

Fowler D, Coyle M, Skiba U et al. (2013) The global nitrogen cycle in the twenty-first century. *Philos Trans R Soc Lond B Biol Sci* 368:20130164.

Moore BD, Andrew RL, Kuelheim C et al. (2014) Explaining intraspecific diversity in plant secondary metabolites in an ecological context. *New Phytol* 201:733–750.

Tilman D, Balzer C, Hill J et al. (2011) Global food demand and the sustainable intensification of agriculture. *Proc Natl Acad Sci USA* 108:20260–20264.

The impact of enhanced N in the environment

Bodirsky BL, Popp A, Lotze-Campen H et al. (2014) Reactive nitrogen requirements to feed the world in 2050 and potential to mitigate nitrogen pollution. *Nat Commun* 5:3858.

Hautier Y, Seabloom EW, Borer ET et al. (2014) Eutrophication weakens stabilizing effects of diversity in natural grasslands. *Nature* 508:521–525.

McLauchlan KK, Williams JJ, Craine JM et al. (2013) Changes in global nitrogen cycling during the Holocene epoch. *Nature* 495:352–355.

Penuelas J, Poulter B, Sardans J et al. (2013) Human-induced nitrogen–phosphorus imbalances alter natural and managed ecosystems across the globe. *Nat Commun* 4:2934.

Variation in forms of soil N

Cameron KC, Di HJ & Moir JL (2013) Nitrogen losses from the soil/plant system: a review. *Ann Appl Biol* 162:145–173.

Saggar S, Jha N, Deslippe J et al. (2013) Denitrification and N_2O:N_2 production in temperate grasslands: processes, measurements, modelling and mitigating negative impacts. *Sci Total Environ* 465:173–195.

Tesfamariam T, Yoshinaga H, Deshpande SP et al. (2014) Biological nitrification inhibition in sorghum: the role of sorgoleone production. *Plant Soil* 379:325–335.

Warren CR (2014) Organic N molecules in the soil solution: what is known, what is unknown and the path forwards. *Plant Soil* 375:1–19.

Plant N transporters

Leran S, Varala K, Boyer J-C et al. (2014) A unified nomenclature of nitrate transporter 1/peptide transporter family members in plants. *Trends Plant Sci* 19:5–9.

Parker JL & Newstead S (2014) Molecular basis of nitrate uptake by the plant nitrate transporter NRT1.1. *Nature* 507:68–72.

von Wittgenstein NJJB, Le CH, Hawkins BJ et al. (2014) Evolutionary classification of ammonium, nitrate, and peptide transporters in land plants. *BMC Evol Biol* 14:11.

Warren CR (2013) Quaternary ammonium compounds can be abundant in some soils and are taken up as intact molecules by plants. *New Phytol* 198:476–485.

The assimilation of N

Heidari B, Matre P, Nemie-Feyissa D et al. (2011) Protein phosphatase 2A B55 and A regulatory subunits interact with nitrate reductase and are essential for nitrate reductase activation. *Plant Physiol* 156:165–172.

Nemie-Feyissa D, Krolicka A, Forland N et al. (2013) Post-translational control of nitrate reductase activity responding to light and photosynthesis evolved already in the early vascular plants. *J Plant Physiol* 170:662–667.

Park BS, Song JT & Seo HS (2011) *Arabidopsis* nitrate reductase activity is stimulated by the E3 SUMO ligase AtSIZ1. *Nat Commun* 2:400.

Xu G, Fan X & Miller AJ (2012) Plant nitrogen assimilation and use efficiency. *Annu Rev Plant Biol* 63:153–182.

Physiological responses to N supply

Krapp A, David LC, Chardin C et al. (2014) Nitrate transport and signalling in *Arabidopsis*. *J Exp Bot* 65:789–798.

Liu Y, Lai N, Gao K et al. (2013) Ammonium inhibits primary root growth by reducing the length of meristem and elongation zone and decreasing elemental expansion rate in the root apex in *Arabidopsis thaliana*. *PLoS One* 8:e61031.

Marty C, Lamaze T & Pornon A (2009) Endogenous sink–source interactions and soil nitrogen regulate leaf life-span in an evergreen shrub. *New Phytol* 183:1114–1123.

Patterson K, Cakmak T, Cooper A et al. (2010) Distinct signalling pathways and transcriptome response signatures differentiate ammonium- and nitrate-supplied plants. *Plant Cell Environ* 33:1486–1501.

Root morphology and N availability

Jones B & Ljung K (2012) Subterranean space exploration: the development of root system architecture. *Curr Opin Plant Biol* 15:97–102.

Li B, Li Q, Su Y et al. (2011) Shoot-supplied ammonium targets the root auxin influx carrier AUX1 and inhibits lateral root emergence in Arabidopsis. *Plant Cell Environ* 34:933–946.

Mounier E, Pervent M, Ljung K et al. (2014) Auxin-mediated nitrate signalling by NRT1.1 participates in the adaptive response of *Arabidopsis* root architecture to the spatial heterogeneity of nitrate availability. *Plant Cell Environ* 37:162–174.

Ruffel S, Krouk G, Ristova D et al. (2011) Nitrogen economics of root foraging: transitive closure of the nitrate–cytokinin relay and distinct systemic signaling for N supply vs. demand. *Proc Natl Acad Sci USA* 108:18524–18529.

N-fixing symbioses

Downie JA (2014) Legume nodulation. *Curr Biol* 24:R184–R190.

Gherbi H, Markmann K, Svistoonoff S et al. (2008) SymRK defines a common genetic basis for plant root endosymbioses with arbuscular mycorrhiza fungi, rhizobia, and *Frankia* bacteria. *Proc Natl Acad Sci USA* 105:4928–4932.

Menge DNL, DeNoyer JL & Lichstein JW (2010) Phylogenetic constraints do not explain the rarity of nitrogen-fixing trees in late-successional temperate forests. *PLoS One* 5:e12056.

Mortier V, Holsters M & Goormachtig S (2012) Never too many? How legumes control nodule numbers. *Plant Cell Environ* 35:245–258.

Schmitz AM & Harrison MJ (2014) Signaling events during initiation of arbuscular mycorrhizal symbiosis. *J Integr Plant Biol* 56:250–261.

Venkateshwaran M, Volkening JD, Sussman MR et al. (2013) Symbiosis and the social network of higher plants. *Curr Opin Plant Biol* 16:118–127.

N acquisition by carnivorous plants

Ellison AM (2006) Nutrient limitation and stoichiometry of carnivorous plants. *Plant Biol (Stuttg)* 8:740–747.

Kruse J, Gao P, Honsel A et al. (2014) Strategy of nitrogen acquisition and utilization by carnivorous *Dionaea muscipula*. *Oecologia* 174:839–851.

Millett J, Svensson BM, Newton J et al. (2012) Reliance on prey-derived nitrogen by the carnivorous plant *Drosera rotundifolia* decreases with increasing nitrogen deposition. *New Phytol* 195:182–188.

Schmidt S, Raven JA & Paungfoo-Lonhienne C (2013) The mixotrophic nature of photosynthetic plants. *Funct Plant Biol* 40:425–438.

Takeuchi Y, Salcher MM, Ushio M et al. (2011) *In situ* enzyme activity in the dissolved and particulate fraction of the fluid from four pitcher plant species of the genus *Nepenthes*. *PLoS One* 6:e25144.

Chapter 6
Phosphorus

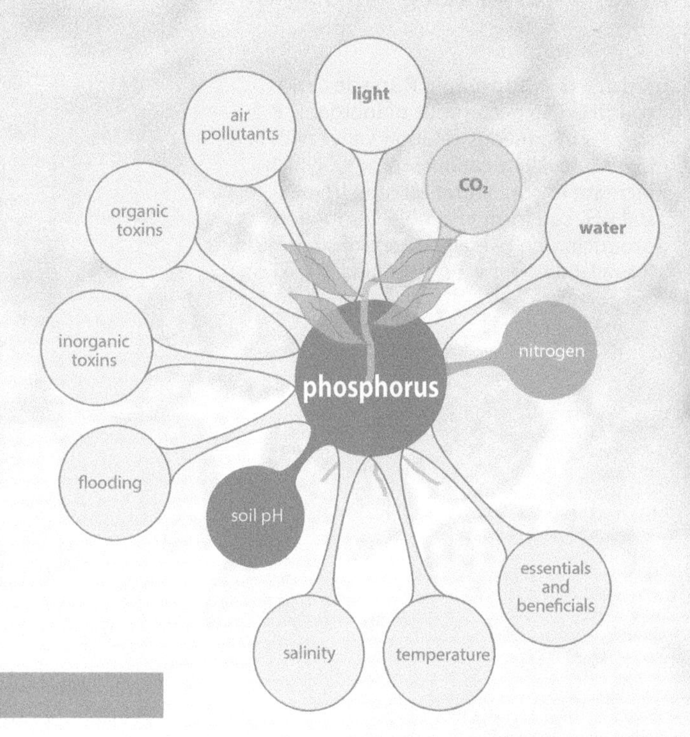

The availability of soil P is influenced by soil pH, and its uptake is influenced by C and N availability.

Key concepts

- P availability controls key processes in terrestrial ecosystems.
- Human activity is increasing P concentrations in the environment, with global consequences.
- P homeostasis in plants is vital for biomass production.
- A battery of transport proteins mediates P uptake and transport in plants.
- Plants can actively change rhizosphere P availability using root exudates.
- In P deficiency, root morphology can be adapted to optimize P capture.
- Mycorrhizal symbioses are widely used by plants to aid P capture from low-P soils.
- Plants use cluster root systems to intensively mine chronically low-P soils.
- Sustainable P use in agriculture is vital for food security and ecosystem conservation.

Phosphorus availability often controls terrestrial biomass production and ecosystem processes

In most terrestrial ecosystems, P is the macronutrient that has the lowest concentration in the soil solution, and for most soil types this is highly buffered (Figure 6.1). In aqueous solution, the formation of inorganic P species is pH-dependent, with maximum solubility at pH 3–7, when the $H_2PO_4^-$ ion predominates (Figure 6.2, Table 6.1). Phosphates of monovalent cations tend to be soluble, but multivalent cations mostly form insoluble precipitates with the $H_2PO_4^-$, HPO_4^{2-}, and PO_4^{3-} ions found in soil solution. Thus in acid soils the presence of Al^{3+} and/or Fe^{2+} can limit P availability, whereas in calcareous soils Ca^{2+} can do so. In highly weathered acidic tropical soils such as oxisols, ultisols, and andisols, P availability is particularly low. The chronic low availability of P in soils was a challenge to the colonization of the land, and root biology is in part a response to the challenge of P availability in soil. Plant growth in unmanaged ecosystems is limited by P in perhaps 70% of soils, competition with microbes for P is intense, and P is the second most

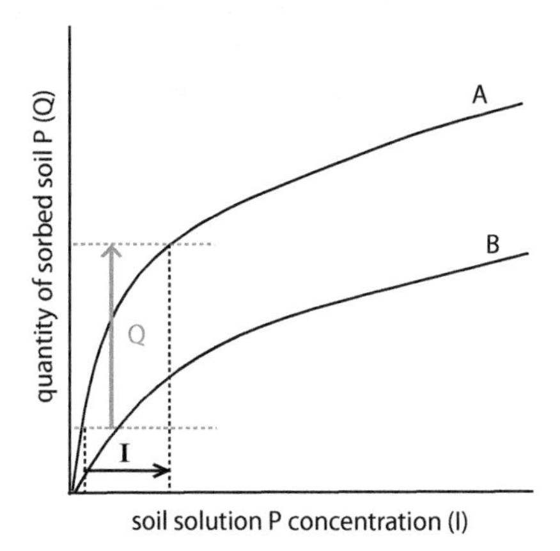

Figure 6.1. Generalized quantity/intensity (Q/I) plots for soil P. A large proportion of the world's soils (A) have low P intensity, and significant increases in the quantity of sorbed P increase intensity by a small amount (that is, low P is highly buffered). Some soils (B) that are sandy or low in multivalent cations are less highly buffered, but significant quantities of P are still necessary before additions are reflected by increases in intensity.

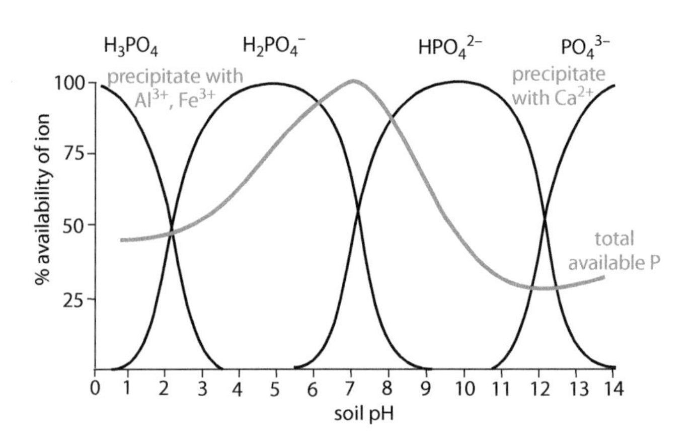

Figure 6.2. Inorganic P species and soil pH. The occurrence of inorganic P species in aqueous solution varies with soil pH. Multivalent cations, some of which increase in concentration at extreme soil pH, can form insoluble compounds, producing an overall pattern of maximum P availability between pH 6 and 7.

widely limiting nutrient to crop growth. Around 30–40% of the world's arable land is P deficient, and for some crops, especially many N-fixing legumes, P is generally the limiting nutrient.

Low P availability often controls not only primary productivity but also biodiversity, fecundity, and resource allocation. For example, across the temperate Eurasian land mass, more rare plant species probably exist under P limitation than under N limitation. In addition, because of low P availability, plant biomass generally has a high C:P ratio, making it a challenging food substrate for most eukaryotes. **Phytate** is widely used by plants to store P, especially in seeds, but can only be broken down by phytases, which are most commonly found in prokaryotes and plants. The evolution of digestive tracts in terrestrial animals was in part driven by the high C:P of plant biomass, and in the ruminant stomachs of the **Artiodactyla** not only cellulase but also, especially in grain-fed livestock, the **phytase** activity of microorganisms is vital to digestion. **Monogastric** animals, especially birds, produce P-rich excreta because of the challenge of accessing P from substrates with a high C:P ratio.

The N:P stoichiometry in marine environments is used as a biogeochemical baseline. N:P ratios both in deep waters and in organisms (that live primarily in surface water) tend to approach 16:1 (the **Redfield ratio**), which suggests that there is homeostatic control of N:P, with deviations reflecting N or P deprivation. The control of N:P ratios in terrestrial plants, although dependent on plant type, is probably more significant than was once thought, with links, for example, to water-use efficiency, life strategy, latitude, and altitude. It has been suggested that the tendency to approach a particular N:P ratio across the kingdoms of life reflects N:P demand during protein synthesis. Synthesis of ribosomal RNA (rRNA) for **ribosomes**, for which all organisms share genes, demands more P than does amino acid synthesis, which is mostly limited by N, which perhaps explains why organisms tend to approach a particular N:P ratio. Stoichiometric analyses of P in plants are useful for modeling global P, C, N, and S cycles. Such analyses, together with widespread agreement that many measures of available soil P do not very closely reflect the actual uptake of P by plants, suggest that plants mobilize P from soils on a global scale. Changes in P biogeochemistry in ocean sediments that coincide with plant colonization of the land surface indicate that this has been an important process for 400 million years. The mechanisms that plants have evolved to extract P from soils influence not only the availability of P to managed and unmanaged terrestrial ecosystems, but also the biogeochemistry of Earth. In fact, some global biogeochemical analyses suggest that P might be the most limiting nutrient for life on Earth, and that plants play a vital role in cycling it.

Table 6.1. Ionic species of "phosphate" found in aqueous solution under physiological conditons

Molecule	Name
H_3PO_4	Phosphoric acid
$H_2PO_4^-$	Dihydrogen phosphate
HPO_4^{2-}	Hydrogen phosphate
PO_4^{3-}	Orthophosphate
$[PO_4]_n$	Polyphosphate

Current phosphorus fertilizer regimes are unsustainable, inefficient, and often polluting

Before the nineteenth century all P fertilizers were organic—mostly manures, **guano**, and human excreta. For many of the world's farmers this is still the case. Justus von Liebig's demonstration that P was an essential macronutrient initiated the production of inorganic P fertilizers, the use of which helped to fuel global agricultural production after 1900 (Figure 6.3). Essentially all inorganic P fertilizers are made from **rock phosphate**. In the nineteenth century it was discovered that the reaction of $Ca_3(PO_4)_2$ with H_2SO_4 produces a "superphosphate" fertilizer that includes $Ca(H_2PO_4)_2$ and P_2O_5, which have relatively high availability in soils. Triple superphosphate produced by the reaction of H_3PO_4 and $Ca_3(PO_4)_2$ (both derived from rock phosphate) has even higher levels of available P (Figure 6.4), was the most important P fertilizer in many agricultural systems until the late 1960s, and is still desired by, but beyond the reach of, many of the world's farmers. In many intensive agricultural systems, $(NH_4)_3PO_4$ fertilizers now supply much P, but they too are made using H_3PO_4 produced from rock phosphate.

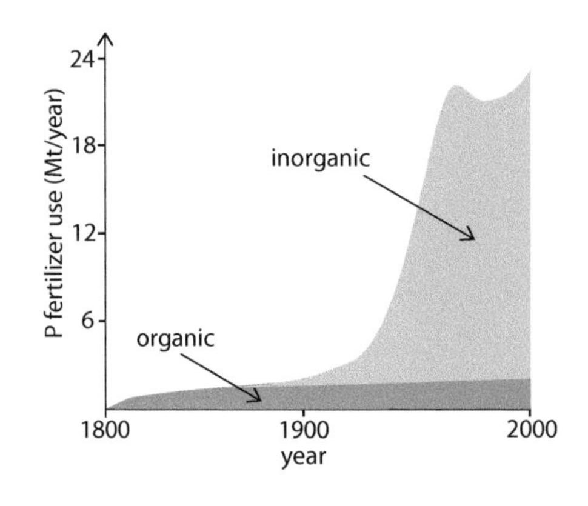

Figure 6.3. The use of P fertilizers. Organic forms include guano, manures, and excreta. Inorganic forms are almost all made from rock phosphate.

Guano is a very effective fertilizer because it contains H_3PO_4, N, and organic matter. It was mined in large amounts before the twentieth century in Peru, Nuaru, and elsewhere, and provides a type example of the unsustainable use of a finite natural resource culminating in near exhaustion (Box 6.1).

BOX 6.1. GUANO EXTRACTION FOR FERTILIZER

The Incas highly valued the guano of the Chincha Islands off the coast of what is now Peru, and there were restrictions on its use and on visiting the islands. Alexander von Humboldt, following his famous trip to South America, was the first to suggest that guano should be used as a fertilizer in Europe, presumably after he had seen its use in Peru. In the nineteenth century, guano became one of the world's most valuable commodities, with expansionist nations purloining much of the world's supply through, for example, the Guano Islands Acts under which the US Congress undertook to support any US citizen who took possession of any guano island for the USA. Many fortunes, such as those of the Gibbs family of Tyntesfield estate, Somerset, in the UK, were thus made from bird droppings.

Guano includes N and P in a readily available, easily soluble form, so is an excellent plant fertilizer, but it can only accumulate in a climate with very little rain. The cold Humboldt Current (Figure 1) that runs up the coast of South America brings huge amounts of nutrients to the surface, supporting one of the most productive marine ecosystems on Earth, which includes the vast numbers of birds that produce the guano on islands off the Peruvian coast in particular. The Humboldt Current also helps to produce a dry climate that allows the accumulation of guano.

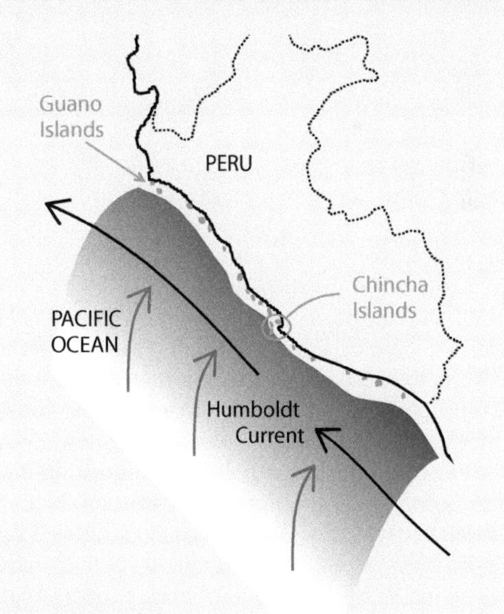

Figure 1. Upwelling caused by the Humboldt Current.

The rapid exploitation and virtual exhaustion of guano from, for example, the Chincha Islands in the nineteenth century, and Nauru in the Pacific in the twentieth century, together with the environmental, economic, and social impacts of this, now serve as a type example of unsustainable exploitation of a natural resource—and one that is technically renewable. It is also a reminder of the lengths to which humans have gone in the past to supply P to their crop plants. The use of guano as a fertilizer has been superseded by the use of inorganic P and urea, but because it is organic there are ongoing efforts to develop more sustainable ways to exploit what remains of the world's guano.

superphosphate:

$$CaF_2 . 3Ca_3(PO_4)_2 + 7H_2SO_4 + 14H_2O \rightarrow 3Ca(H_2PO_4)_2 + 7CaSO_4 . 2H_2O + 2HF$$

triple superphosphate:

$$CaF_2 . 3Ca_3(PO_4)_2 + 14H_3PO_4 \rightarrow 10Ca(H_2PO_4)_2 + 2HF$$

Figure 6.4. **The production of superphosphate and triple superphosphate.** The most common apatite in rock phosphate is fluorapatite, so the primary reaction is listed. Hydrogen fluoride (HF) reacts with silicates and is removed to leave soluble $Ca(H_2PO_4)_2$. Much rock phosphate also includes P_2O_5, which increases the proportion of available P in the end product. Phosphoric acid for the production of both triple superphosphate and ammonium phosphate is derived from the dissolution of $Ca(H_2PO_4)_2$ made via the superphosphate reaction.

Analyses of rock phosphate production and consumption increasingly have parallels with those for guano production. Rock phosphate reserves are finite, annual use is significant, the costs of extraction and processing are increasing as lower-quality reserves are accessed, there is a global production-consumption system in which alternatives are hard to develop, it is a market mostly driven by a single use (fertilizer), and reserves are primarily located in a few localities (the Western Sahara, South Africa, the USA, and China). The confluence of these factors has been used to predict, depending mostly on the definition of reserves, that there are only between 30 and 300 years of P reserves left for food production. Predictions of imminent P exhaustion are a reminder that global food production is dependent on the unsustainable use of a finite natural resource that occurs in relatively small amounts, and that the supply of P to crops is a key challenge in the medium term, and perhaps even in the short term. In addition, one of the world's most significant anthropogenic environmental radioactivity challenges also arises from the production of inorganic P fertilizer—the billions of tonnes of **phosphogypsum** waste that it has produced, and is still producing, are difficult to use because phosphogypsum can be significantly toxic and mildly radioactive (Box 6.2).

Large-scale analyses of farming systems as diverse as those in the North-East USA, Finland, the Netherlands, and China have shown that a significant proportion, probably about two-thirds, of the P added in fertilizers to agricultural soils in the twentieth century did not enter crop plants. Fertilization has increased the quantity of P in many agricultural soils more than it has increased P availability. In contrast to the high P-use efficiency of many natural ecosystems, there is therefore low P-use efficiency in many agricultural

BOX 6.2. PHOSPHOGYPSUM FROM P FERTILIZER PRODUCTION

The primary waste product of superphosphate production is gypsum ($CaSO_4$), which contains significant $H_2PO_4^-$ impurities—hence the term "phosphogypsum." About 300 million tonnes of phosphogypsum are produced each year, almost all from fertilizer production, and most of this is stored in giant "stacks." Phosphogypsum can be used instead of gypsum in a variety of construction materials. However, its use is often restricted because it is mildly radioactive, containing trace amounts of ^{238}U, particularly in rock phosphate deposited under marine conditions. ^{238}U is the dominant U isotope (over 99%), is not fissile, has a half-life of about 4.5 billion years, and thus has a very low activity. However, it is the start of the "radium-decay" series (Figure 1) that includes ^{226}Ra and ^{222}Rn, both of which have significantly higher activity than ^{238}U, with the gaseous ^{222}Rn being of particular concern with regard to human health. Much phosphogypsum waste has no current use, primarily because of its radionuclide content. The level of activity depends on the rock source, and it is heterogeneous within rocks and stacks, but most analyses record activities of several 100 Bq kg^{-1} ^{226}Ra. Such activities will produce ^{222}Rn of at least 10 times above background levels. Restrictions on the use of phosphogypsum, and on where it is used, are generally applied at around 40 Bg kg^{-1} ^{226}Ra, which means that the majority of phosphogypsum is essentially useless. The US Environmental Protection Agency (EPA) estimates that there are currently more than 1 billion tonnes of unusable phosphogypsum in the USA, stored in massive stacks. Restrictions on the use of phosphogypsum and its management in stacks mean that there is little evidence of its adverse radiological effects on humans or other living organisms. However, phosphogypsum stacks are a substantial reminder that current P-fertilizer use depends on a non-renewable resource that results in a significant waste problem—3 billion tonnes of potentially radioactive phosphogypsum waste are produced globally each decade, which is a significant challenge to economies and industries that aspire to "green" credentials.

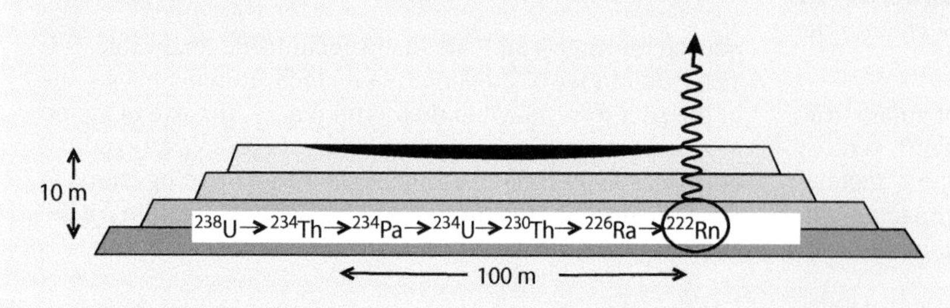

Figure 1. **Phosphogypsum from P fertilizer production.**

systems. A consequence of this has been eutrophication, especially of fresh water. The low solubility of P ensures that excess P rarely contaminates ground water, but through-flow and the erosion of soil colloids mean that P is a significant non-point source pollutant from many agricultural soils. In many parts of the world, costly schemes to control eutrophication are now in place—for example, in the Norfolk Broads in the UK (Box 6.3)—and significant legislation in North America and the European Union is now aimed at controlling it. The Millennium Ecosystem Assessment confirmed the global extent of eutrophication of freshwater bodies, in which it is often regarded as the primary factor affecting water quality. Many studies of the marine environment have also described significant dead zones caused by excess P—for example, in Chesapeake Bay, the Gulf of Mexico, the Baltic Sea, and the Great Barrier Reef. The P cycle is one of the nine "planetary boundaries" that might limit sustainable habitation of Earth, and one that humankind might soon exceed.

Phosphorus homeostasis is a key challenge for plants in terrestrial ecosystems

Plants need PO_4^{3-} to synthesize **phospholipids** for membranes and to make nucleotides for DNA and ATP. PO_4^{3-} is also necessary for energizing and coordinating metabolism, including, respectively, the sugar-phosphate intermediates necessary for photosynthesis and respiration, and signaling cascades based on **kinases**. These demands require the overall P concentration in plants to be tightly controlled in the millimolar (mM) range. P deficiency stunts the growth first of primary roots and then of shoots, and in *Ricinus communis* and many other species it induces the production of

BOX 6.3. EUTROPHICATION AND THE NORFOLK BROADS

The Norfolk Broads (Figure 1) are one of the most extensive freshwater wetlands in the UK, and have National Park status. In total they cover a relatively small area across the valleys of the rivers Thurne, Ant, Bure, Yare, and Waveney, but they provide a good example of recognition of the important impact of nutrient enrichment on the natural environment. Between the 1940s and the 1960s it became clear that significant changes in the aquatic systems of the Norfolk Broads were taking place. The Broads had a long history of drainage, water abstraction, hunting, and boating, the impacts of which were relatively well known, but many of the changes in aquatic systems that were observable by the 1960s were by then being widely recognized as due to nutrient enrichment. In the 1970s, numerous studies showed the "nutrient enrichment hypothesis" to be true for the changes in the Broads. A significant part of the management of the Broads is now focused on controlling and reversing nutrient enrichment. The reversal of nutrient enrichment is a longer-term challenge than many others, because it requires a decrease in inputs and also purging systems of nutrients. Both nitrate and phosphate from a variety of sources were important in nutrient enrichment of the Broads, with phosphate often being a particular challenge. Nutrient enrichment, or "eutrophication," arising from human activity not only in wetlands but also in many ecosystems around the world is now widely recognized as having severe adverse impacts. In many natural ecosystems, P is the limiting nutrient, and enrichment with P is particularly problematic. P inputs from sewage treatment, detergents, but frequently primarily from fertilizers are responsible for most of the adverse effects.

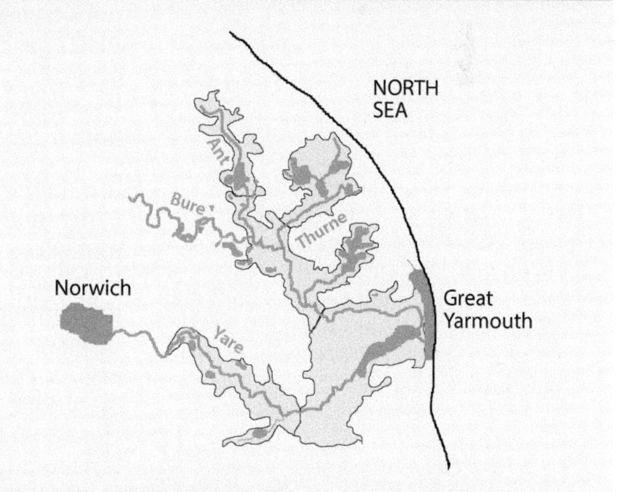

Figure 1. The Norfolk Broads. The county of Norfolk on the east coast of England was often under glaciers during the Quaternary glaciations. The rivers of the Broads are in low-lying land (shown in gray), much of it fen peat formed after the retreat of the ice when the sea level, relative to the isostatically rebounding land, was higher than at present. The Broads (lakes—shown in green) were, in significant part, formed by excavation of peat in medieval times.

anthocyanins, so plants tend to become purple-green in color. In order to control internal P concentrations during minor fluctuations in P supply, all plants have a homeostatic network.

The transcription factor PHR1 is vital to P homeostasis in plants ranging from algae to angiosperms. It is an MYB-related TF that binds to DNA as a dimer. MYB TFs are not common in animals, but are the most common family of TFs in plants. PHR1 is part of a small group of them with a GARP-binding domain in which the helices of each of two GARP domains enter the major groove of DNA in opposite directions to bind to the target sequence. Microarray experiments have described changes in thousands of genes under a variety of P regimes, but a significant proportion of them have in their promoter region the P1BS domain to which PHR1 binds. Genes with P1BS domains have many roles related to P nutrition, including carbohydrate metabolism, anthocyanin production, and P transporter activity. The expression of PHR1 is insensitive to P regime, and it is located in the nucleus at about the same concentration irrespective of the P status of the plant, but overexpression of PHR1 produces plants with elevated P concentrations. However, the activity of PHR1 is controlled—by SIZ1, a small ubiquitin-like modifier (SUMO) E3 ligase. In the *siz1* mutant of *Arabidopsis*, P-deficiency-induced expression of PHR1-regulated genes is minimal. The **sumoylation** of PHR1 thus helps to control PHR1 activity and therefore the expression of PHR1-regulated genes. The expression of SIZ1 is not directly affected by P deficiency, but is indirectly affected by aspects of sugar metabolism. Other MYB TFs, such as AtMYB62 and PHL1, help to control P status, but their overexpression induces deficiency symptoms. PAP1 and PAP2 play an important role in anthocyanin production via the **phenylpropanoid pathway**, and other similar MYB TFs have an important role in the quality of wine, because the color of red wine is determined primarily by anthocyanin concentration. The *pho1* mutant of *Arabidopsis* has shoot P deficiency because it cannot transfer P from root cells into the xylem. PHO1 has an SPX binding domain, is induced by P deficiency and, although necessary for xylem loading, is not a transporter so is probably a signal-transducing protein. The activity of PHO1 probably regulates the flow of P from roots to shoots. The *pho2* mutant of *Arabidopsis* over-accumulates P. The discovery of the role of MYB TFs in P nutrition and of the effects of *pho* mutations presaged the elucidation of P homeostasis pathways in plants.

A number of miRNAs are involved in plant responses to P deficiency, but PHR1 indirectly controls the activity of a subset of P-responsive proteins via miR399 in particular. miR399 is phloem mobile and is an internal signal of P status. It is induced by P starvation, which results in increased levels of P transporters in roots and shoots via inhibition of the **ubiquitin**-conjugating E2 enzyme UBC24—an enzyme that targets proteins, probably P transporters, for degradation and that is deficient in *pho2* mutants. Overexpression of miR399 in plants increases the shoot P concentration. Although it is not clear how P is sensed in plants, a network that increases the expression of P-related genes, increases the activity of transporters, and thus increases flow to the shoot in response to P deficiency is present (Figure 6.5). However, some proteins that show increased expression in response to P deficiency in fact decrease P uptake by plants, and some of the many TFs that are expressed in response to P deficiency (for example, WRKY75, bHLH32, and MYB62) repress the P-starvation response. SPX3 proteins directly inhibit the P response via their effects on SPX1, which affects regulation of PHR1-dependent gene expression. Similarly, the non-coding transcripts AT4 and IPS1 down-regulate the miR399 response by mimicking its intended target. There is thus not only a network that increases P uptake in response to deficiency, but also one that attenuates the P-deficiency response. This is characteristic of a homeostatic fine-tuning network that has evolved to reach a

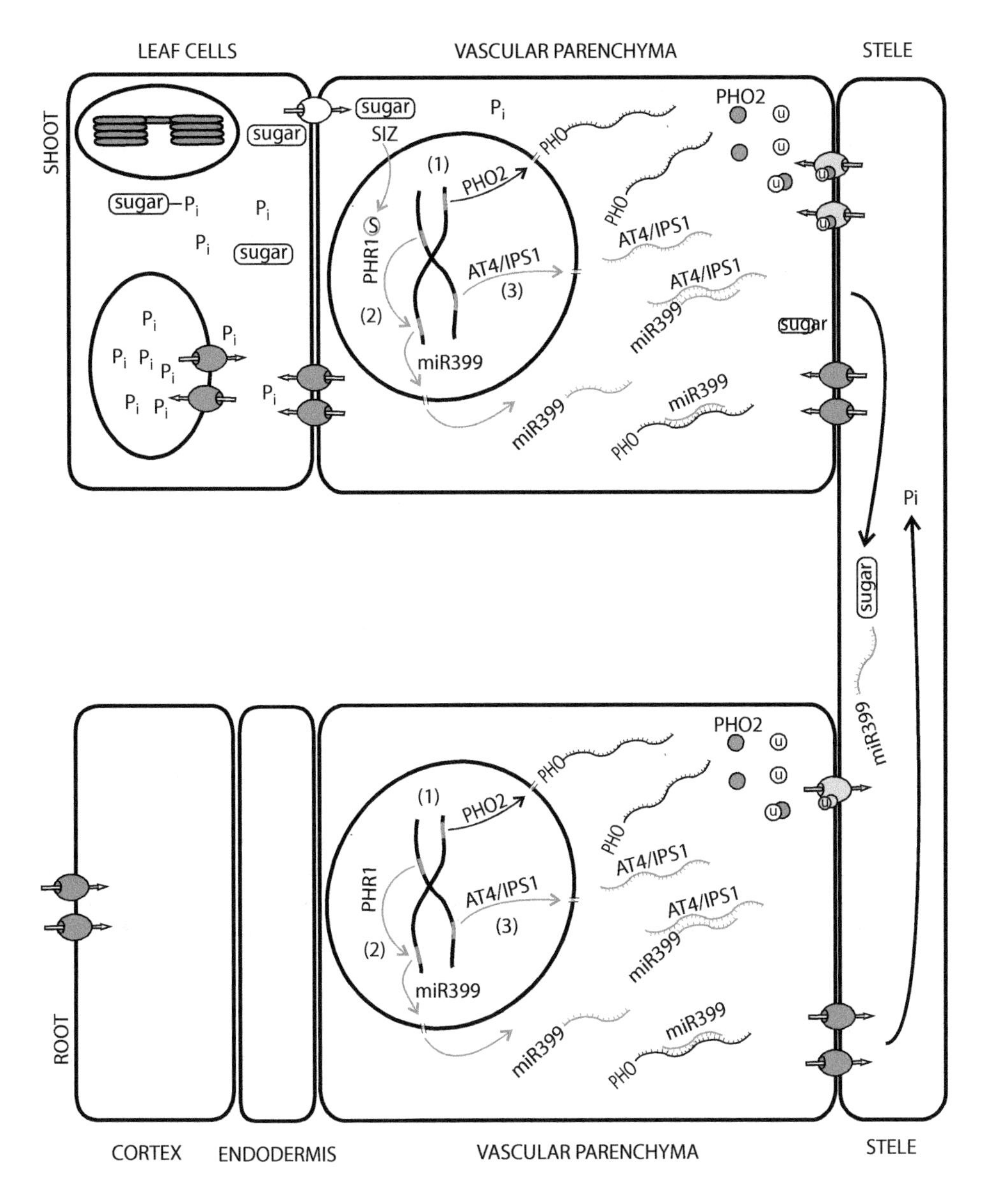

Figure 6.5. Homeostatic control of P concentrations in plants. (1) If excess P arrives from the roots, PHO2 is expressed, enhancing the ubiquitination and degradation of transporters for P influx. (2) In P deficiency, sugar-induced sumoylation of PHR1 induces the expression of miR399, which blocks the translation of PHO2 and the degradation of P transporters, increasing P influx to root and shoot cells. (3) AT4/IPS1 transcripts mimic PHO2, interfering with the action of miR399 and increasing the degradation of P transporters.

set point without overshooting it. Although the role of many P-deficiency-induced genes and the primary P-sensing mechanisms have yet to be clarified, it seems clear that plants have a homeostatic network that quickly adjusts their physiology to meet P demand during fluctuations in external supply. In cells of P-sufficient plants, 85–95% of total P is located in vacuoles, where its concentration can vary, whereas cytoplasmic P concentrations are much lower and are tightly controlled. P homeostasis must therefore interact with transporters that both take up and compartmentalize P.

Plants have numerous transporters that regulate uptake and translocation

P uptake from the soil solution must occur against its electrochemical gradient, because soil solution concentrations are very low (0.1–10 μM, about three orders of magnitude lower than those required in the plant), the predominant P species are anionic, and the root membrane potential inside is negative. P starvation increases the P-uptake capacity of many plants by increasing the numbers of transporters with a high affinity for P, whereas

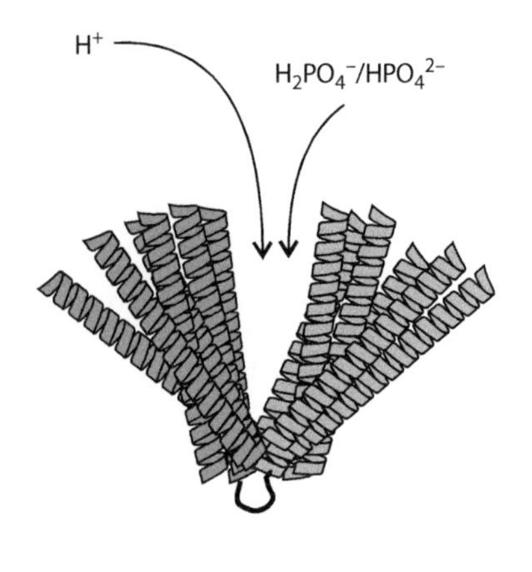

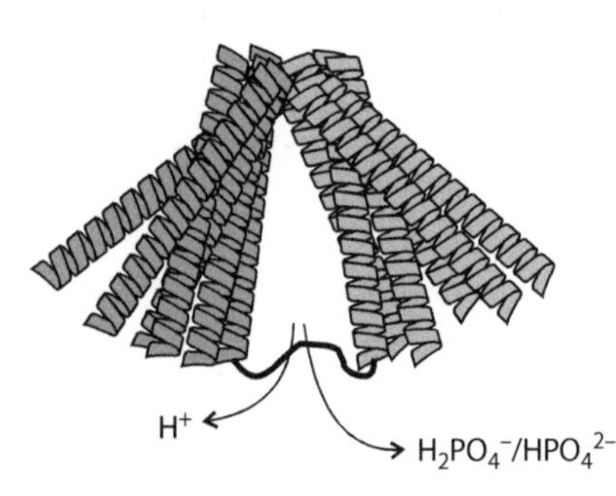

Figure 6.6. Phosphate transporters in plants. The PHT proteins that facilitate transmembrane transport of P in plants are major facilitator superfamily (MFS) transporters that operate via a "rocker-switch" mechanism energized by H^+ gradients generated by H^+-ATPases.

low-affinity transporters are generally constitutively expressed. Studies of many species indicate that the transporters which mediate P uptake by plants are H^+/P symporters driven by the inward electrochemical gradient for H^+ created by H^+-ATPases that enable efflux of H^+. H^+/P symporters belong to the major facilitator superfamily transporters (**MFS transporters**), which consist of single polypeptides that mediate the transport of small molecules in response to electrochemical gradients and, in contrast to the **ABC transporter** superfamily, have no ATP binding sites. They have 12 membrane-spanning domains with a long hydrophilic loop in the middle, and probably evolved by duplication of a 6-membrane-spanning domain transporter (Figure 6.6).

In several species, investigations of high-affinity uptake by the PHT1 family of proteins have indicated that they are symporters responsible for significant P influx in roots. PHT1 transporters are induced by P deficiency, have a K_m of 2–4 μM, and several are targeted to the plasma membrane of epidermal cells, especially root hairs, and cortical cells. **Heterologous expression** of barley HvPHT1 in *Xenopus* oocytes has been used to characterize in detail its high-affinity uptake capacity, which is similar to that found at whole-root level. Most PHT transporters selectively transport $H_2PO_4^-$ and HPO_4^{2-}, which are the phosphate ions that plants take up and that often predominate in soil solution. Four families of PHT transporters and two families of PT homologs have been found which regulate P uptake and translocation in different plant organs. In *Arabidopsis,* phosphate transporter traffic facilitator (PHF1) plays an important role in controlling plasma-membrane activity of PHT1 in response to soil P fluctuations. Low P levels induce transcription of PHT1, but it can only exit the endoplasmic reticulum and travel to the plasma membrane while it is bound to PHF1 and is phosphorylated at the C-terminal Ser 514. When P levels are sufficient, neither the action of PHF1 nor the phosphorylation at Ser 514 occurs, and PHT1 is found in **endosomes**, which are probably involved in moving it from the plasma membrane to the vacuole, where it is degraded. In *Arabidopsis*, some members of the PHT1 family are expressed in a variety of shoot tissues. It is likely that they play a role in the remobilization of P during deficiency. Low-affinity P transporters in other PHT families have a role in the internal transport of P. Split-root experiments and P applications to leaves have shown that if part of the root system or shoot system is P sufficient, P transporters in a P-deficient root zone are not affected. Links to homeostatic networks ensure that transporter activity is only increased when deficiency occurs at a whole-plant level. The numerous P transporters that have been isolated from a range of plants are closely related to the P transporters that are found across the living world.

Plants can increase the availability of inorganic phosphorus and the breakdown of organic phosphorus

The availability of P can affect ecosystem processes on a large scale (Box 6.4), and can be actively increased by plants. In natural ecosystems, such as the **fynbos** of the Western Cape in South Africa, mobilized P can represent the majority of P taken up. P availability is highest within a relatively narrow pH range (Figure 6.2), so plants control rhizosphere pH. H^+ efflux in response to P deficiency has been demonstrated in many plant species that inhabit neutral and alkaline soils, and can explain differences in low P tolerance. In contrast, *Aspalathus linearis* (rooibos), a bush characteristic of the acid soils of the fynbos, has a capacity for OH^- and HCO_3^- exudation. In plants, H^+ and HCO_3^- efflux is usually determined by N form. For plants to mobilize P by efflux of H^+ under alkaline conditions, which favor NO_3^- and therefore HCO_3^- efflux, or by efflux of OH^-/HCO_3^- under acid conditions, which favor NH_4^+ and H^+ efflux, significant expenditure of energy is necessary.

BOX 6.4. DOES P HELP TO DRIVE MIGRATIONS IN THE SERENGETI?

Ecosystems that have migrations of large herds of herbivores are spectacular and rare. One such ecosystem is the Serengeti in Tanzania (Figure 1). In 2011, a scheme to upgrade a gravel road to a paved dual carriageway across the northern Serengeti was scrapped because of the harm that such developments had caused to other herbivore migrations. The Serengeti's annual migration of up to 2 million herbivores, including zebra, Thomson's gazelles, and over 1 million white-bearded wildebeest, remains one of the wonders of the natural world. The migration is ultimately driven by annual rainfall patterns, but it is the interaction between water availability, grass growth, and nutrients that helps to explain the great migration.

The Serengeti Plain is populated by short to medium-height grasses and, compared with many other savannah grasslands, has few trees. The plain is surrounded by higher ground with much open woodland (shown in dark green on the map). Ash from volcanoes on the edge of the rift valley, blown on prevailing easterly winds, forms the substrate of most of the plain. Much of the ash is from Mount Kerimasi, although Mount Oldonyo Lengai has erupted more often in the last 150 years. The largest volcano in the chain, Ngorongoro, is now dormant. Olduvai Gorge, located between Ngorongoro and the Serengeti, cuts down through volcanic ash in which have been found some of the most important hominid fossils ever discovered. It is possible that some of the first bipedal hominids walked in these landscapes.

The volcanoes in this part of the Rift Valley produce unique carbonaceous ash that is rich in Ca, Na, and P. The lack of trees on the Serengeti is often attributed to the inability of tree roots to penetrate the hard pan formed by this ash. By about May the Serengeti starts to dry out, and over the next 7 or 8 months grass growth and nutrient content decline to well below that which can support large herbivore populations. The wildebeest and other herbivores migrate northward, following an increasing gradient of rainfall that supports grass growth. From August to October, at a time when the Serengeti grasslands are dry, many of the herbivores migrate to the Masai Mara in Kenya. The herds migrate south to the Serengeti (where they calve) for its wet season from December to April.

There is an optimum grass height (medium to short) for grazing, and the Serengeti herbivores in general follow the grasses as they reach this height. The N content of the grasses is also significant, and, especially in the northward migration via the western corridor, the herbivores migrate to grasses that are taking up N from water-sufficient soils. The surrounding areas of open woodland have taller grasses with a slightly lower nutrient value, but still provide good herbivore habitat, so there is something particularly attractive about the Serengeti in the wet season. It is probably significant that the open plains make it more difficult for predators to conceal themselves, which might be especially important during calving. However, it is almost certainly the nutrient content of the grasses that grow in the Serengeti wet season that enables such a spectacular number of herbivores to be supported. The grasses, unusually and as a result of their volcanic ash substrate, are relatively rich in Ca, P, and other nutrients, and low in Si. Some studies have suggested that the Ca and P are particularly important for lactation, helping to support the many hundreds of thousands of calves that are born each wet season in the Serengeti, and to explain the migration of such huge herds.

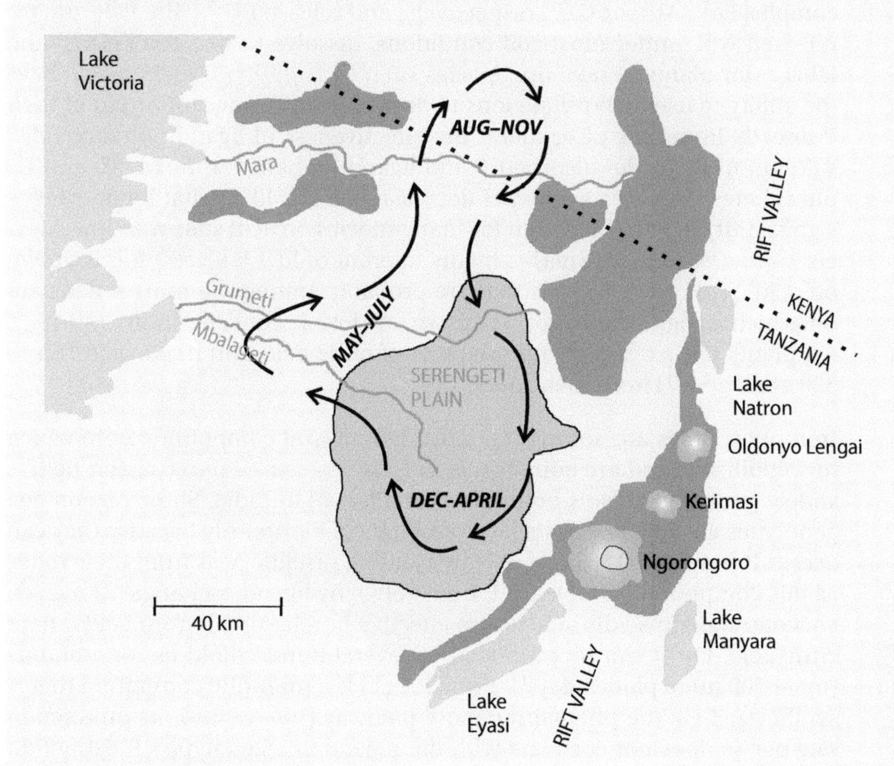

Figure 1. The migrations of the Serengeti. Herbivore movements are denoted by black arrows, higher ground is shown in dark green, and volcanoes are denoted by cones.

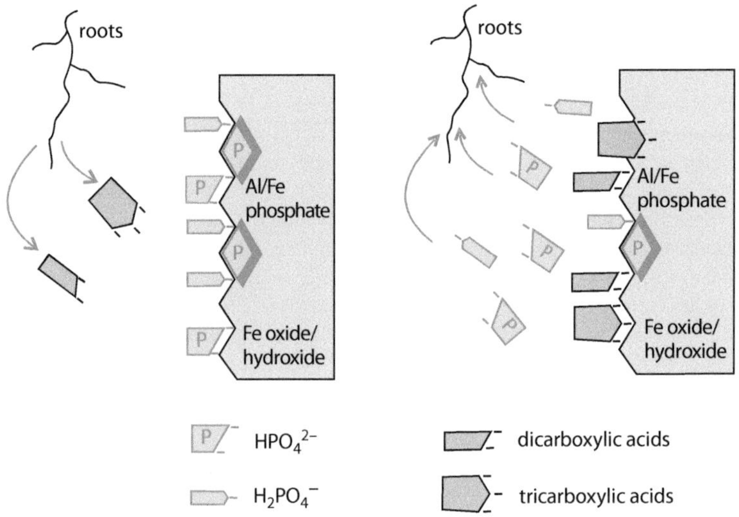

Figure 6.7. The ligand-induced release of inorganic phosphate from soils. Phosphates from the soil solution can bind to a variety of iron oxides and hydroxides in various crystalline states. Carboxylic acids, especially the tricarboxylic citrate but also, depending on the particular iron oxide or hydroxide, other carboxylic acids, bind as ligands, displacing phosphate into the solution. Depending on the pH, HPO_4^{2-} and $H_2PO_4^-$ form in solution and are taken up by plants. In the longer term, enhanced dissolution of iron oxides and hydroxides by these organic ligands decreases the soil adsorption capacity for phosphate. Fe and Al phosphates can also have phosphate displaced by some organic acids to release it into the soil solution.

In response to P deficiency, many low-P-tolerant plants exude carboxylic acids into the rhizosphere. In alkaline soils these acids may play a role in adjusting soil pH for maximal P availability, but their conjugate bases can also mobilize P. In acid to neutral soils, much inorganic P is either in ions adsorbed to oxides and hydroxides of Fe and Al, or in Fe and Al phosphates. In calcareous soils, much of it is in Ca phosphates. Exudation of carboxylic acids can induce significant ligand exchange of P adsorbed to oxides and hydroxides of Fe and Al (Figure 6.7). Furthermore, carboxylate-ligand-enhanced dissolution of a variety of P-adsorbing soil compounds, but especially of Fe/Al oxides and hydroxides, has been shown to slowly release P in significant quantities. For Fe oxides, citrate and malate are the most effective carboxylate anions for ligand-enhanced exchange and dissolution. For Fe, Al, and Ca phosphates, carboxylate anions can act directly as ligands to complex Fe^{3+}, Al^{3+}, or Ca^{2+}, respectively, and release PO_4^{3-}. The PO_4^{3-} that is released will, under most soil conditions, dissolve to become $H_2PO_4^-$ and HPO_4^{2-} for plants to take up. Species such as *Fagopyron esculentum* have the ability to use carboxylate ions to derive a significant proportion of their P directly from rock phosphate. The effectiveness of ligand-enhanced dissolution of Fe-oxide-adsorbed P and ligand exchange from Fe, Al, and Ca phosphates increases as the pH decreases, so it is likely that it contributes significantly to P mobilization for many species on acid soils with low P levels. *Camellia sinensis* (tea) is highly tolerant of low P levels, and is grown on acidic (pH 4.5–5.5) soils that are probably among the most P deficient on which a significant crop is grown. At low P concentrations (40 µM), tea plants release significant amounts of citric acid and malic acid (up to 500 nmol g^{-1} h^{-1}) from their roots.

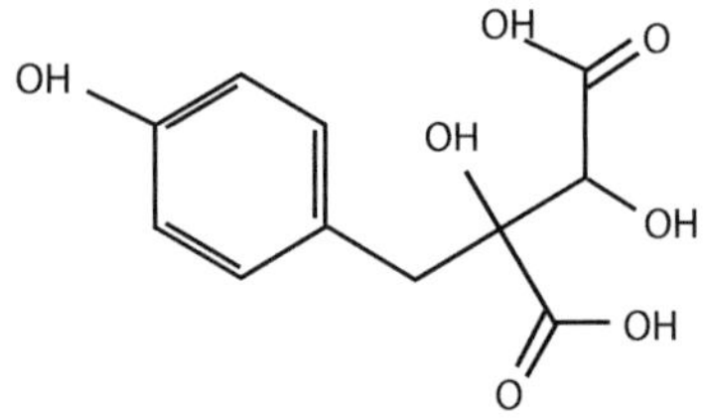

Figure 6.8. The structure of piscidic acid. The root exudates of some plant species, especially *Cajanus cajan* (pigeon pea), can include piscidic acid, which can solubilize iron phosphate, helping to increase plant available P.

Iron phosphates are among the most difficult soil compounds from which to solubilize P, and are common in alfisols. *Cajanus cajan* (pigeon pea) is widely grown on alfisols on the Deccan Plateau in India. Some pigeon pea genotypes are able to extract large amounts of P primarily because they can access P from $FePO_4$ (Table 6.2) by exuding piscidic acid from their roots, as this compound can release P from $FePO_4$ by ligand exchange. At a given concentration, piscidic acid is less effective in releasing P from $FePO_4$ than citric acid, but it can be produced in several hundredfold larger amounts (up to 700 nmol $plant^{-1}$ day^{-1}). Piscidic acid is a **phenolic** compound that is synthesized via the phenylpropanoid pathway (Figure 6.8). Its production rate per se does not correlate with the pigeon pea genotype's P-extraction capability, which also depends on other physiological and morphological

Table 6.2. Biomass production and P uptake by two crops on two different soils in India. (Data from Ae et al. [1990] *Science* 248: 477-481)

Crop	Soil	Dry matter (kg ha^{-1})	P uptake (kg ha^{-1})
Sorghum	Alfisol	1384	2.00
	Vertisol	3976	6.21
Pigeon Pea	Alfisol	2284	3.18
	Vertisol	2053	2.46

adaptations, but piscidic acid is nevertheless an important part of a strategy that maximizes the extraction and use of P.

It is unlikely that life could exist without the unique properties of **phosphoesters** (Box 6.5), as exemplified by ATP (Figure 6.9). Their ubiquity in organisms means that they are common in soils. Depending on the soil type, 30–80% of the P in soil can be organic, with much of it occurring as phosphoesters. Phosphatases that catalyze the breakdown of monoesters to release Pi are found in all the kingdoms of life, and are vital for controlling phosphate metabolism, and are notably diverse in plants. Plants exude from their roots not only phosphatases but also phosphodiesterases, which break down diesters, and phytases, which break down phytate. The importance of monophosphatases in root exudates to the mobilization of P is well established, but mobilization by phosphodiesterases and phytases may also be significant.

Different phosphatases have different pH optima. Many, though not all, of the phosphatases found in plants are acid phosphatases. Many of them in pure solution absorb light at 560 nm and are therefore purple in color. These purple acid phosphatases (PAPs) are especially important in plant nutrition. Root exudates in many plant species have phosphatase activity that increases in response to P deprivation, and that increases access to organic P. Molecular experimentation with certain species, especially *Arabidopsis*, has confirmed their importance. In *Arabidopsis,* about 30 of the 50 acid phosphatase genes are actively transcribed, with two proteins, AtPAP12 and AtPAP26, being particularly important in root exudates. Knockouts for *AtPAP26* grow normally in P-replete conditions, but die under P-deprived conditions to which WT (wild-type) plants respond by exuding AtPAP26. *Arabidopsis* is also able to grow normally when the only P source in the external medium is glycerol-3-phosphate, which is a phosphate ester and AtPAP26 substrate. The TFs that regulate P homeostasis affect phosphatase activity—often directly, because some have promoter regions for these TFs. Post-translational regulation of PAP gene expression is also achieved by miR399 and UBC24, integrating PAP enzymes into P homeostasis. PO_4^{3-} is an inhibitor of many PAPs, so that as P concentrations increase after resupply their activity decreases.

Orthologs of AtPAP12 and AtPAP26 have been described in many species, including monocots, eudicots, crop plants, wild plants, and some gymnosperms. The correlation between root phosphatase activity and P-use efficiency in crop genotypes, and between root phosphatase activity and

BOX 6.5. BIOLOGY AND PHOSPHOESTERS

Phosphoesters are very stable in aqueous solutions (the spontaneous hydrolysis half-life of the phosphodiester DNA is 31 million years), but **phosphatases,** which catalyze the breakdown of phosphoesters, have some of the highest rate accelerations of any enzyme. This combination, which underpins the stability of vital structural molecules in cells and the very fast reaction times needed for energy release from ATP or signaling reactions involving PO_4^{3-}, can probably not be replicated by organic compounds of any other element. **Esters** are formed when the OH group on an **oxyacid** is replaced with an *O*-**alkyl** or *O*-**aryl** compound. Esters can be formed from inorganic or organic oxyacids. Carboxylic acids form many biologically important esters, but phosphoesters, which are formed from phosphoric acids, predominate in living organisms.

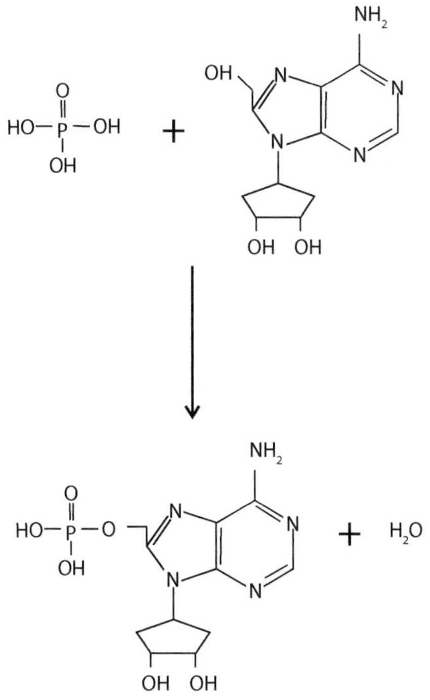

Figure 6.9. The formation of esters. Esters are products of condensation reactions between an acid and a hydroxyl group. Phosphoric acid reacts with the OH group on adenosine to form the ester adenosine monophosphate (AMP). Addition of two further phosphate groups gives adenosine triphosphate (ATP). The properties of phosphate esters underpin key life processes, such as the storage and release of energy from ATP.

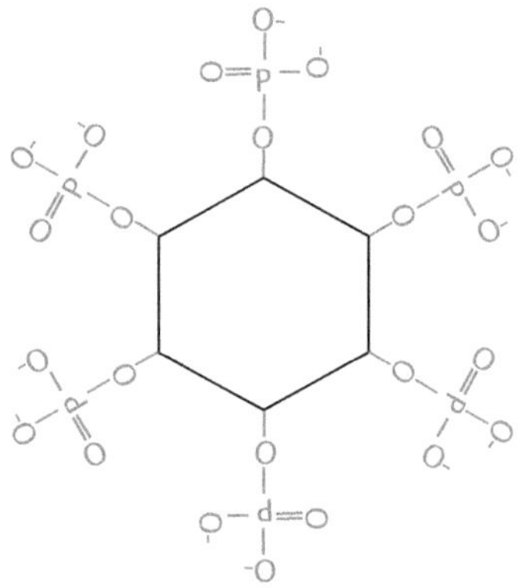

Figure 6.10. The structure of phytic acid. Inositol hexakisphosphate, or phytic acid, is an important P storage compound in plants. When phytic acid dissociates in aqueous solution, salts in which cations such as Ca, Mg, Fe, or Zn bind to the phytate anion (shown in the figure) can form. These salts are widely used by plants, especially in seeds, to store P.

contrasting P regimes in wild plants, suggest that it is an important determinant of tolerance of low P availability in terrestrial plants. The manipulation of PAPs in crop plants clearly has the potential to increase crop utilization of organic P, which has provoked detailed molecular investigations of these enzymes. They are **metalloenzymes** that are similar to phosphoprotein phosphatases and **exonucleases.** Plant PAPs contain active sites with Fe^{3+}/Zn^{2+} or Fe^{3+}/Mn^{2+}, in contrast to mammalian representatives, which contain Fe^{3+}/Fe^{2+}. In contrast to many of the PAPs found in other kingdoms, in *Arabidopsis* most of these enzymes are **oligomers** with a C-terminal catalytic domain homologous to the PAP monomers of other organisms, fused with an N-terminal non-catalytic domain. Most of the PAPs that are exuded from roots occur as homodimers, with some exuded as monomers, and have quite broad specificity, but the variety of PAPs, their oligomeric forms and their **glycosylation** have probably evolved to fine-tune the specificity of phosphatase activity.

In the organic P fraction in the soil solution, phosphomonoesters (likely to be derived from ATP and sugar phosphates) have mostly been found to be less common than phosphodiesters (likely to be derived primarily from phospholipids and DNA). The PAPs exuded from plant roots can clearly break down the phosphomonoesters that occur in soils, but the role of the phosphodiesterases that occur in root exudates is less well known. Inositol hexakisphosphate (IP6), also known as phytic acid, the commonest P storage compound in many plants (Figure 6.10), is also often reported in significant (and sometimes the highest) amounts in soil organic-P fractions. Some root exudates have phytase activity, partly from PAPs, and probably have a significant role in breaking down IP6. Overall, it is likely that plants adapted to particular soils have evolved the specific capacity to break down the phosphoesters that occur most commonly in their soil solution. There are numerous phosphatases and phosphodiesterases in plants, most of which are involved in internal mobilization of P during homeostasis or germination, and which might have evolved into root exudates in particular species. Furthermore, ribonucleases are a common constituent of root exudates, with gene expression and activity in exudates being highly responsive to P deprivation in some species. If nucleic acids are supplied to *Arabidopsis* plants as their sole P source they grow normally, so in many ecosystems nucleic acids could also be a significant source of P for plants.

Plants can adjust their root system morphology to optimize phosphorus uptake

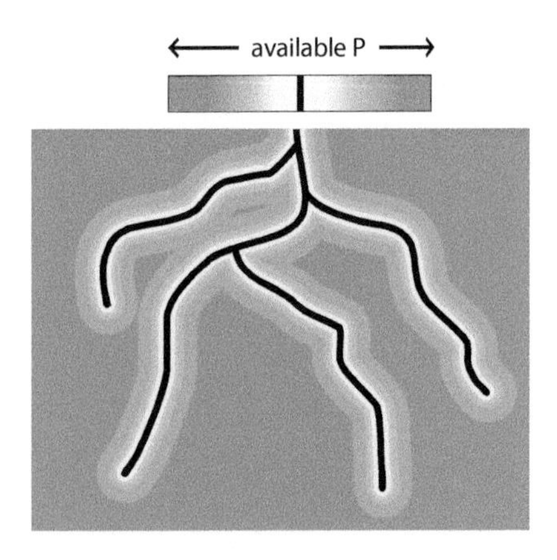

Figure 6.11. P depletion zones around plant roots. Frequently neither mass flow of P in the transpiration stream nor diffusion down concentration gradients is sufficient to meet plant demand, so distinct P depletion zones develop around roots.

Ions arrive at the root surface by a combination of **mass flow** (mostly driven by the transpiration stream) and diffusion down concentration gradients. At the low P concentrations that occur in almost all soils, including those in which significant mobilization of P occurs, the mass flow to plant roots of $H_2PO_4^-$ and HPO_4^{2-} is very low. Using **Fick's first law**, soil solution concentrations, and transpiration flows, it can be calculated that the majority of P in most plants cannot have reached the root surface by mass flow. The homeostatic control of P ensures that P deficiency increases P-uptake capacity, whilst root exudates increase availability. This rapidly depletes the rhizosphere of P, producing a significant concentration gradient between the bulk soil and the rhizosphere (Figure 6.11). The diffusion coefficient in soils of the ionic species of P that are taken up by plants is the lowest of any macronutrient (Table 6.3), and P movement to the root by diffusion down concentration gradients is severely limited. Thus, under many circumstances, P movement to plants roots by either mass flow or diffusion is insufficient to meet demand. In the long term, to overcome the limitations imposed by depletion zones, plants actively forage for P by sending out roots that are active in P uptake into new parts of the soil. Plant species and crop lines clearly differ in their ability to adjust root architecture, and this can be directly linked to their P-uptake efficiency in the field.

Roots, like other plant organs, undergo most of their development post-embryonically. The development of their final form is in large part a product of their environment acting on processes mediated by auxin and cytokinin. Many plants that are tolerant of low P concentrations, including well-known P-efficient lines of important crop plants, have root systems that not only mobilize P with exudates, but also respond morphologically to low P levels. Cessation of primary root growth, increased **lateral root** formation, increased **root hair** length and density, and increased **adventitious root** growth have all been reported in plants responding to P deficiency. Many of these responses are similar to those induced when plants are manipulated to overproduce auxins in the root. The effect of the responses is to increase both the overall surface area of the root and the active area of root hairs, thereby increasing the volume of soil being exploited. In many species, increases in root hair length correlate directly with increased soil volume exploited and increased P uptake. Up to 90% of P can be taken up via root hairs, which have the key role as the interface between the root and the volume of exploitable soil. *Arabidopsis lpr* mutants do not cease growth in the primary root meristem in response to low P levels. Quantitative trait loci (QTLs) and molecular analyses have shown that LPR1 and LPR2 are copper oxidases located in the root tip which are involved in early sensing of P deficiency. The *pdr2* mutant of *Arabidopsis* produces a root system with P-deficient morphology even when there is sufficient P. PDR2 is a P_5-type ATPase, probably a cation pump, located in the endoplasmic reticulum of root tip cells. Together with LPR it controls, via the endoplasmic reticulum, the activity of SCARECROW—one of the TFs through which auxin regulates meristem activity. PDR2 is thus involved in optimizing root architecture to meet P supply. LPR1 is necessary for auxin transport, so is likely to be involved in the redistribution of auxin (Figure 6.12).

Table 6.3. The diffusion coefficients for macronutrient ions in aqueous solution

Ion	Range of diffusion coefficients ($m^2\ s^{-1}$)
Cl^-	$2–9 \times 10^{-10}$
NO_3^{2-}	1×10^{-10}
SO_4^{2-}	$1–2 \times 10^{-10}$
K^+	$1–28 \times 10^{-12}$
PO_4^{3-}	$0.3–3.3 \times 10^{-13}$

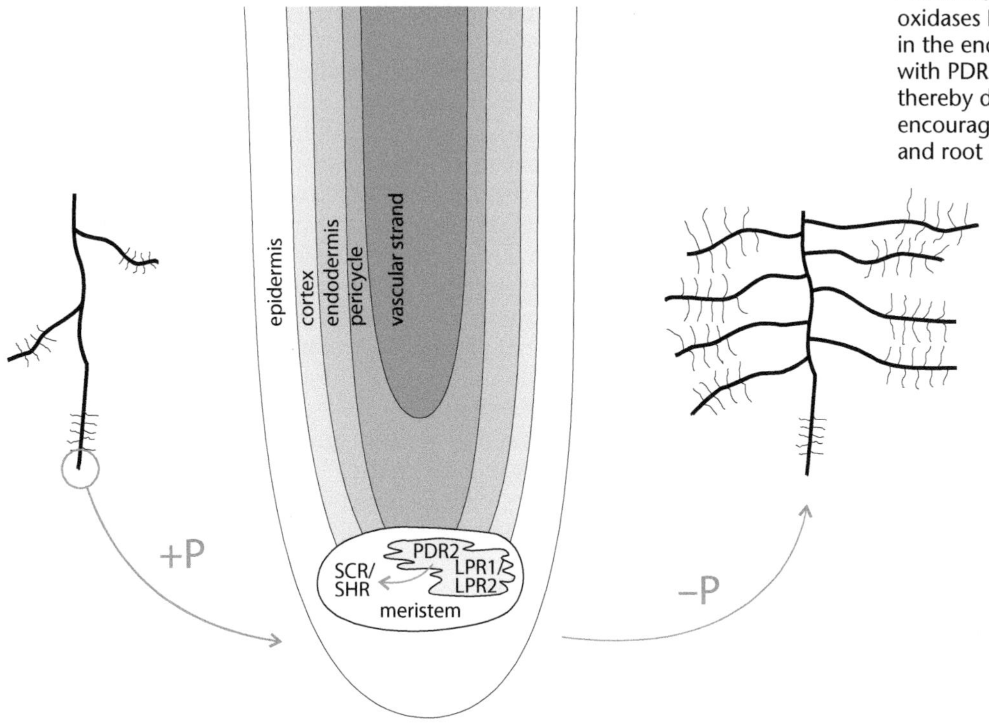

Figure 6.12. Plant P status and root morphology. The expression of genes that control meristem activity is regulated by the SCARECROW (SCR) and SHORT ROOT (SHR) transcription factors. Their activity is maintained via a P5-type ATPase, PDR2. During P deficiency the copper oxidases LPR1 and LPR2, which are located in the endoplasmic reticulum, together with PDR2, decrease SCR/SHR activity, thereby decreasing meristem activity and encouraging the growth of lateral roots and root hairs.

Elements of the P homeostasis network also directly affect root morphology. *siz1* mutants have an exaggerated response to low P. SIZ1 sumoylates PHR1 and also affects auxin distribution between the primary and lateral roots. It is a negative regulator of plant response to low P levels, and probably the trigger for the initiation of low-P root architecture. WRKY75 has also been shown to negatively regulate the low-P root response. ZAT6 overexpression leads to an exaggerated P response and is likely to be involved in repressing the expression of genes in the primary root. Early studies using radioactive ^{32}**P** established that P uptake occurred primarily through root hairs. The reconfiguring of root architecture, the expression patterns of key proteins, and the established links with well-known auxin pathways all suggest a coordinated strategy by P-efficient plants to increase the interception and uptake of P through root hairs. The cessation of primary root growth and increase in lateral and adventitious root growth also leads to more exploitation of the near-surface layers of soil. These layers can contain more P than deeper horizons, mainly because of biological activity, but they may have less advantageous water supply characteristics. The evolution of root systems with foraging capacity, which are vital for the supply of nutrients in terrestrial ecosystems, was probably driven in significant part by the low availability of $H_2PO_4^-$ and HPO_4^{2-} in soils.

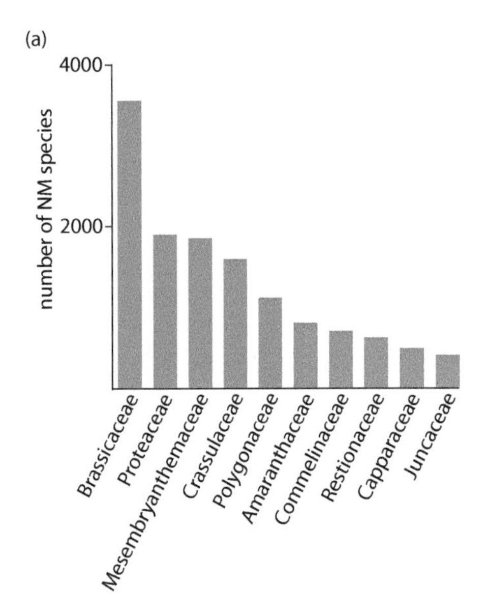

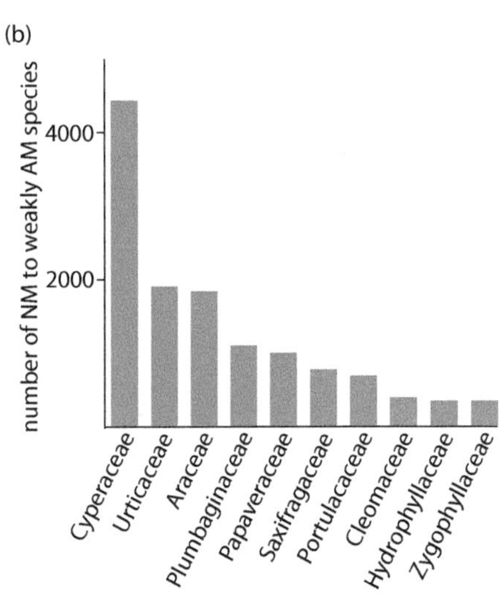

Figure 6.13. The occurrence of non-mycorrhizal and weakly mycorrhizal plants. The ten families with the highest number of species confirmed as being (a) non-mycorrhizal (NM) and (b) non-mycorrhizal to weakly arbuscular mycorrhizal (AM). Families with many parasitic, carnivorous, and epiphytic species that are rarely mycorrhizal are excluded.

Mycorrhizas are major adaptations for phosphorus acquisition in low-phosphorus environments

A wide range of fungi form **mycorrhizas** with plant roots, although fungal partners are less diverse than the plant partners, and the range of possible partners varies significantly in both fungi and plants. The most common type is the endomycorrhizal arbuscular mycorrhiza (AM) between fungi in the **Glomeromycota** and members of almost all groups of terrestrial plants. The next most important are ectomycorrhizas (EcM) between a range of **Basidiomycota**, **Ascomycota**, and **Zygomycota** and woody plants. Species in the Ericales form a number of distinct endomycorrhizas with Basidiomycota, and most species in the Orchidaceae also form distinct ectomycorrhizas. By 2010, the mycorrhizal status of about 10,000 terrestrial plant species had been described (about 3% of the total), which allowed the mycorrhizal status of over 65% of families, including essentially all those of ecological importance, to be established. From mycorrhizal occurrence at the family level it is estimated that 75% of all angiosperm species can form AM, 2% can form EcM, 1% can form ericoid mycorrhizas, and 9% can form orchidaceous mycorrhizas. About 8% of species are either non-mycorrhizal or weakly AM, and about 6% are entirely non-mycorrhizal. Of the non-mycorrhizal species, many have adaptations to aid nutrient capture, are aquatic, or are parasitic. The Amaranthaceae, Polygonaceae, Crassulaceae, Mesembryanthemaceae, Proteaceae, and Brassicaceae are the families of plants with by far the greatest number of truly non-mycorrhizal species (Figure 6.13). Many of the earliest known fossil terrestrial plants—for example, in the **Rhynie chert**—have AM. Extant members of some of the earliest plant phyla form AM even though they do not form true roots. The prevalence of mycorrhizas in extinct and extant plants suggests that they were integral to the evolution of terrestrial ecosystems. Mycorrhizas in extant plants have been shown to increase resistance to drought and root pathogen attack, increase uptake of numerous nutrients, and decrease uptake of toxic metals. In general, however, their most significant effect, especially for AM fungi, is to enhance P uptake. Mycorrhizas have been shown to determine nutrient flows, species diversity, and plant productivity in numerous terrestrial ecosystems.

Experiments with numerous plant species have shown that AM formation is inhibited by high P supply to roots, and that ^{33}**P** or ^{32}P fed to the fungal partner reaches the plant in significant amounts. In some experiments it was shown that almost all of the P that reaches the plant is derived from AM fungal uptake. In many AM crops, varieties that form more extensive mycorrhizal associations grow better in low-P conditions. The Glomeromycota that form AM are obligate **biotrophs**. The spores of AM fungi germinate in the presence of **strigolactones** secreted from plant roots, and AM fungi obtain carbohydrate from host plants. During the establishment of AM, fungi form an appressorium between epidermal cells, which anchors them, and they then push the epidermal cells apart and grow into the apoplast of the root cortex, forming an **intra-radical** network. AM fungi can penetrate the exodermis if present. In most species, hyphae spread through the cortical apoplast, producing a linear pattern through much of the cortex. As the intra-radical network spreads it promotes the growth of **extra-radical** hyphae in the soil. Intra-radical hyphae form **haustoria** on some root cortical cells, penetrate the cell wall, and force the **protoplast** of the cell to invaginate. Numerous studies at the genetic, biochemical, and physiological level have found plant-derived activity that aids the formation of a channel through the cell wall during haustorium formation. Once the fungus is inside the cell, in most plant species it forms arbuscules, which do not penetrate the plasma membrane (Figure 6.14). The junction between the cell wall and the haustorium is generally tight, and a variety of techniques have shown that the peri-arbuscular membranes are the primary site of nutrient and carbohydrate (mostly glucose) exchange during the symbiosis. Arbuscules vary somewhat in architecture, and in most plants are transient structures that last up to a few weeks—in profusely mycorrhizal roots there is a turnover of arbuscules. In general, AM form behind root tips and are most profuse in newly differentiated roots. In some species, fungal lipids can build up in the arbuscules to form vesicles through the conversion of glucose to triacylglycerides.

Due to the immobility and low concentration of P in many soils, and the rapid development of depletion zones around roots, the major determinant of plant access to P is the volume of soil exploited by the roots. The extra-radical hyphae of AM increase dramatically the volume of soil exploited, and thus increase the total amount of potentially accessible P. Individual root hairs are generally 5–15 μm in diameter and 0.3–0.5 mm in length, although in some species P deficiency causes an increase in their length. The extra-radical hyphae of AM spread great distances through the soil, often by the formation of runner hyphae, and in many species they also proliferate at particular points to produce thin absorptive hyphae. Importantly, the absorptive

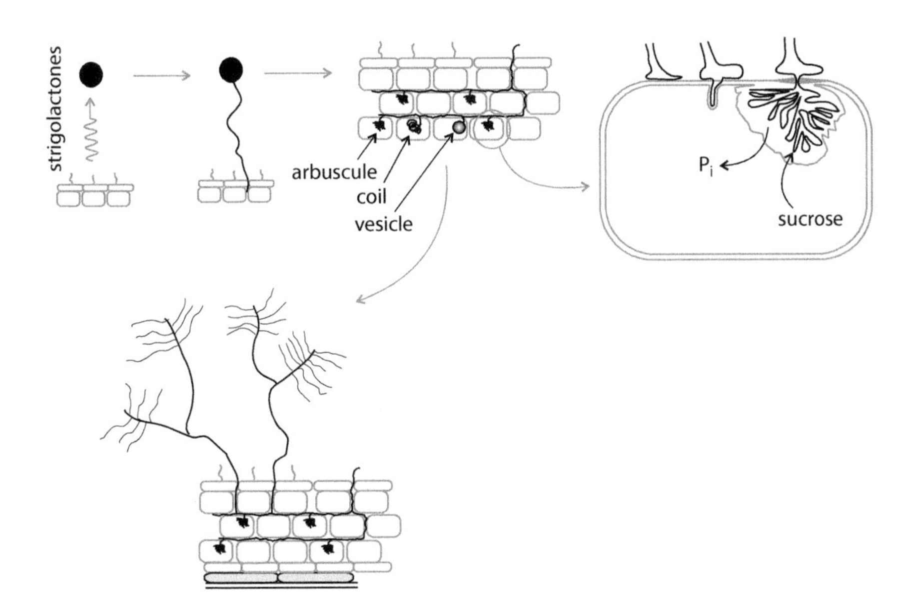

Figure 6.14. The development of arbuscular mycorrhizas (AMs). Strigolactone-induced germination produces hyphae that penetrate the root cortex, forming an intra-radical network. In most instances, arbuscules are formed which invaginate the cortical cell membranes and produce a peri-arbuscular space for exchange of nutrients. Sometimes intracellular hyphae coil or form vesicles that store triacylglycerides. The fungus grows an extra-radical hyphal network, often with very fine nutritional hyphae, which explore a much greater soil volume than could be explored by the root alone.

hyphae are significantly thinner than root hairs, and can penetrate smaller soil micropores. AM fungi have infrequently been shown to excrete phosphatases and other enzymes more characteristic of saprotrophic fungi, and thus do not access organic P to a significant extent. ^{32}P supplied to the soil has been shown to move very large distances through AM hyphae, primarily in cytoplasmic streaming, at speeds vastly in excess of those at which P can ordinarily move through soil. The extra-radical hyphae of AM therefore provide both a mechanism for exploiting a large volume of soil, and a conduit to rapidly move P to the plant root.

P transporters from a number of Glomeromycota that form AM have been described. Fungi including the Glomeromycota have high inside negative membrane potentials (–150 to –200 mV) generated by efflux of H^+ via an H^+-ATPase. P, probably predominantly in the form of HPO_4^-, is transported from the soil solution into the fungal hyphae via H^+/P symporters driven by inward H^+ gradients. These fungal P transporters show high similarity to the PHT/PT family of transporters found across the kingdoms of life. P is in most instances then used to synthesize polyphosphate (Figure 6.15), the form in which P is stored and transported. Polyphosphates can rapidly move significant distances to arrive in the arbuscules in root cortical cells. In the arbuscules the polyphosphates are broken down by alkaline phosphatases to release phosphates. These then undergo efflux across the arbuscular membrane. Less is known about the transporters responsible for efflux of P from the fungus than about those through which it is taken up in extraradical hyphae, but anion channels are the most likely type of transporter. Phosphate that effluxes from arbuscules is taken up through plant P transporters located on the peri-arbuscular membrane. These belong to a distinct group of PHT1 transporters. In a number of species, PT1 transporters of the peri-arbuscular membrane are only expressed during AM formation. The P concentration and solution environment at the peri-arbuscular membrane are significantly different from those of the soil solution, and it is likely that its PT1 transporters are optimized for them. In a number of AM associations, root infection with any fungus other than the AM partner will not induce changes in expression of PHT proteins. Association with the AM fungal partner can change the expression of PHT proteins not only in the peri-arbuscular membrane but also throughout the plant, presumably reflecting the particular P-uptake dynamics of the AM condition. The arbuscule is also the site of carbohydrate uptake by the fungal partner. Glucose is the predominant monosaccharide taken up, probably primarily through an H^+/monosaccharide co-transporter. There is much evidence that sucrose is the predominant saccharide in the apoplast of plants, and this includes root cortical cells with fungal arbuscules. Many AM fungi, and perhaps all of them, seem to have low uptake capability for sucrose, and **invertase** excreted by the root cortical cells is necessary to hydrolyze it and release glucose to the fungus. There is evidence that the excretion of invertase can provide a mechanism for the plant to control the extent of the AM association. The ericoid mycorrhizal association has clear similarities to the AM association. However, in general it seems to be less important for P nutrition and more important for tolerance of toxic metals, because the major activity is often the production and excretion of organic acids by the fungus to form stable complexes with the toxic metals in the acidic environments that many ericaceous plants inhabit.

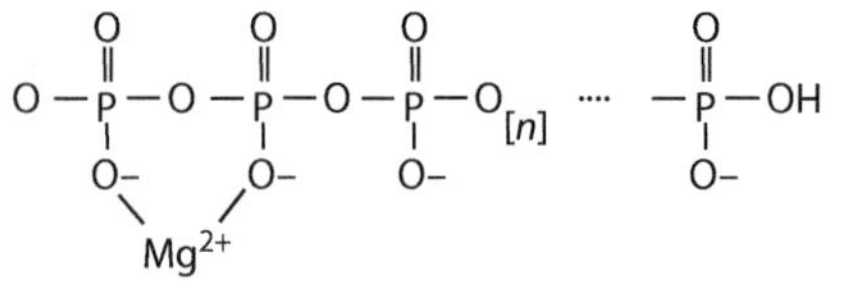

Figure 6.15. **The structure of polyphosphates.** Polyphosphates can be cyclical, and a great variety have been found in fungi, but those that are transported in mycorrhizal fungi are mostly linear, consisting of up to several hundred phosphates in a linear arrangement, most frequently as a magnesium salt.

EcMs are formed in the roots of woody plants, including both **gymnosperms** and **angiosperms**, most often with fungi in the **Basidiomycota**, but also with members of the **Ascomycota** and **Zygomycota**. EcMs are a significant feature of boreal and temperate forests, in which the majority of trees can be mycorrhizal and the fungi can account for a significant proportion of ecosystem biomass. EcMs probably evolved independently several times, but on each occasion much later than AMs evolved. Their first appearance seems to pre-date that of the mushroom-forming fungi, highlighting the fact that although they are not as ancient as AMs, they are a very long-established association. They certainly have a long-established role in P uptake. In contrast to AM, in which root morphology is not changed much by the association, the morphology of plant roots with EcM is distinct. In both gymnosperms and angiosperms, EcM association occurs primarily toward the root tips. Infected root tips have a short, fat appearance characteristic of EcM (Figure 6.16). In gymnosperms these short infected roots are **dichotomously branched**, whereas in angiosperms they are **sympodially branched**. A high proportion of the biomass in short infected roots is fungal. To form these short roots the fungus enters at or near the root cap, which ultimately results in cessation of root growth and proliferation of fungus around the entire root. In most mature infections the fungal biomass consists of a fungal mantle sheathing the outside of the plant root, together with a network of hyphae that enter the plant root. In general, in gymnosperms the mantle is thin, but the hyphae penetrate not only the root epidermis but also the root cortex. In angiosperms the mantle tends to be much thicker, and the network of hyphae is confined to the epidermis. The network of intra-radical hyphae that ramify through the apoplast of the epidermis and cortex, known as the **Hartig net**, provides a large area of contact between the fungus and the plant, thus forming the main site for exchange of nutrients and metabolites. Extra-radical hyphae extend from the mantle into the soil, often via mycelial strands or specialized **rhizomorphs**, and can have proliferating structures for nutrient absorption. In general, EcMs are found in ecosystems in which soils have well-developed organic horizons, and the extra-radical hyphae of EcMs extend primarily into these organic horizons rather than into the mineral horizons. Numerous mushroom-forming fungi form EcMs, sometimes with a restricted range of plant species, the association of *Amanita muscaria* (fly agaric) with *Betula* (birch) species being a well-known example in the northern temperate zone.

In many experiments the majority of P has been shown to enter trees through the fungal partner of EcMs. As in AM associations, the fungal hyphae of EcM significantly increase the volume of soil exploited and provide a conduit for the diffusion of P to the root. Plant-derived carbohydrate is transferred

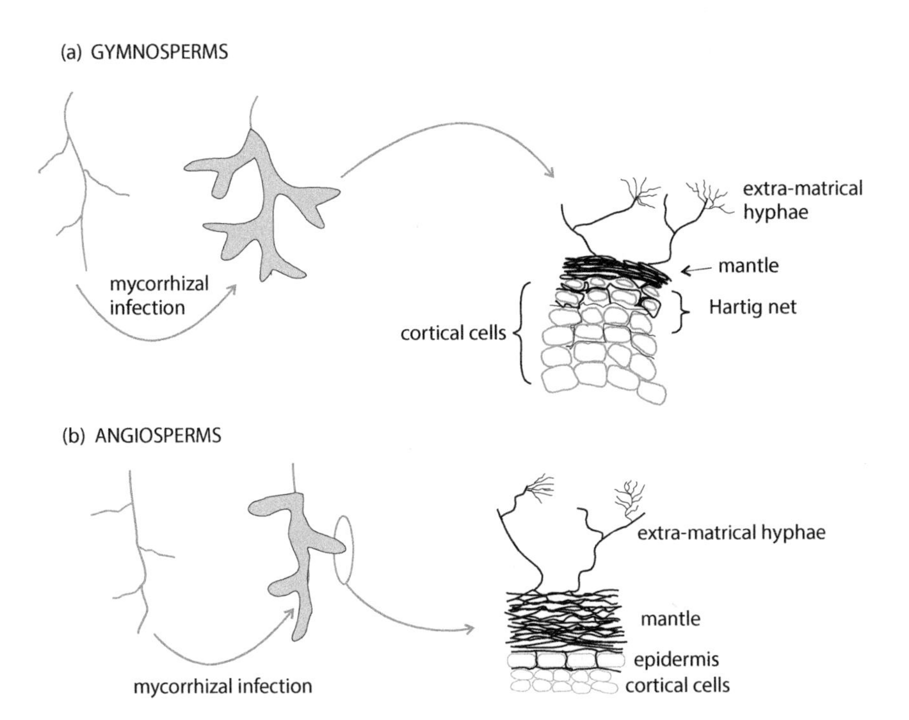

Figure 6.16. The formation of ectomycorrhizas. Ectomycorrhizas (EcMs) form thick hyphal mantles, particularly around lateral root tips. A network of hyphae forms in the root (the Hartig net), and a mantle of hyphae forms on the outside of the root. A network of extra-matrical hyphae ramifies through the soil, often with fine, P-absorbing proliferations of growth. (a) In gymnosperms, infected roots are stubby and dichotomously branched, and the mantle is relatively shallow but there is an extensive Hartig net, with the outer layers of root becoming indistinct. (b) In angiosperms with EcM the roots branch sympodially, the mantle is thick, and the Hartig net is mostly confined to the epidermis.

to EcM fungi in significant amounts, and has been shown to transfer from tree to tree through fungal hyphae. Fungal networks can connect numerous individual trees and transfer both nutrients and metabolites between them, which led to the concept of the "wood-wide web" based on EcMs as an important factor in forest processes. Phosphate transporters have been identified in a number of fungal species that form EcM. These are responsive to soil or fungal P status, contribute to plant P uptake, and have great similarity to P transporters isolated from AM fungi. They are proton-coupled transporters of a type found in most fungi. Although it has long been known that EcM fungi access organic P in significant amounts, phosphatases that break down phosphomonoesters in soil organic matter have only relatively recently been isolated from EcM fungi. A number of these enzymes are now known, and their role in solubilizing organic P into inorganic P for uptake by the fungus is well established. Tracer studies have shown that P from a wide range of compounds can be transferred to plants with EcM, which is probably explained by the wide range of substrates with phosphomonoesters that phosphatases can break down. However, the fungi that form EcM associations with plants are **saprotrophs**, and thus excrete a range of enzymes capable of attacking a very wide range of organic compounds. Although phosphatases are likely to have the primary role, it is quite possible that a range of fungal enzymes excreted from EcM associations have a role in mobilizing P. Inorganic P taken up by the fungus mostly travels to the plant root at much higher rates than are possible by diffusion or mass flow through the soil. In EcM fungi there is evidence to support active transport via tubular vacuoles that extend significant distances through extra-radical hyphae. The primary site of exchange with plant cells is the Hartig net. The transport processes that mediate efflux from the fungus and uptake by the plant are likely to be similar to those of the AM association, but the details are less well known. It is also clear that many EcM fungi excrete significant quantities of organic acids. Acids such as oxalic acid can solubilize P from apatite, and it is thought that the "rock-eating" capacity of some EcM fungi might be significant for plant P nutrition in some ecosystems.

Many benefits of AM associations have evolved during their 400-million-year history, but in herbaceous angiosperms they are probably the most important response to the very significant challenge of chronically low-P terrestrial environments, especially those in which P reserves are primarily inorganic. EcMs predominate in environments in which there are significant organic P reserves in the soil. The biotrophic and saprotrophic fungi involved in AM and EcM associations, respectively, in large part determine access to soil P reserves. However, mycorrhizas significantly alter the rhizospheres of plants, and in particular their microflora. The full complexity of interactions at the soil–mycorrhiza–microbe interface in the rhizosphere presents a significant systems analysis challenge, but one that it might be vital to address for a full understanding of the P nutrition of plants in terrestrial environments.

Some species use cluster root systems to intensively mine phosphorus from the soil

Mediterranean climatic regimes promote the development of vegetation dominated by sclerophyllous, often woody vegetation, well-known examples being the **maquis** of the Mediterranean, the **chaparral** of the western USA, and the **matorral** of Chile. Sclerophylly is in part an adaptation to low availability of nutrients, including P (Box 6.6). The fynbos of the Cape in South Africa and the **kwongan** of South-West Australia have similar climatic regimes and vegetation but, in addition, more severely P-impoverished soils. The concentration of detectable P in soils of the fynbos and kwongan can be below the nanomolar range. The Proteaceae are a **Gondwanan**

BOX 6.6. SCLEROPHYLLY AND PHOSPHORUS

Sclerophyllous leaves are hard and leathery. They contain a high ratio of carbohydrate to protein, and have a low P content. They are long-lived, resistant to herbivory, and tend to be small and frequent on branches. Sclerophylly helps to increase both water-use and nutrient-use efficiency. However, it can make it difficult for CO_2 to penetrate the leaf quickly, thus slowing down productivity. Sclerophylls are common among vegetation of the Mediterranean region, where their unique habit helps them to survive a challenging P regime.

family now distributed almost entirely in the southern hemisphere, with centers of diversity in South Africa and Australia. Species of the Proteaceae are prominent in fynbos and kwongan vegetation, and many have specialized proteoid roots adapted for intensive mining of P from soils. There are around 1600 species in the Proteaceae. About 50 other species, especially in the Fabaceae, Myricaceae, and Casuarinaceae, are also known to develop proteoid roots. Descriptions of occasional proteoid root formation in other distantly related angiosperms suggest that their formation might be possible in quite a variety of plants. Low Fe levels have sometimes been shown to induce cluster root formation, but it is P deficiency, often in the shoot, that most often initiates the development of proteoid roots. The Proteaceae and most other proteoid root-forming species are generally not mycorrhizal. Proteoid roots are an amalgam of many other low-P adaptations that enable non-mycorrhizal plants to inhabit soils with some of the most challenging P regimes in the world.

In the roots of Proteaceae grown in low-P conditions, the development of tertiary roots is initiated in the **pericycle** at each **protoxylem** initial. Tertiary lateral roots are unusual in plant root systems, which are usually based on secondary laterals that develop only on alternate or even more widely spaced protoxylem initials. To form proteoid roots, the multicellular tertiary laterals, or "rootlets," develop over a few days to 3–4 cm in length, and form a section of the root system in which they can be profuse enough to have a bottle-brush appearance (Figure 6.17). This section normally extends 1–3 cm along the root, but in some *Hakea* species there have been reports of proteoid roots extending for 20 cm. In some species of *Banksia*, complex branched proteoid roots are formed. The densely packed rootlets are very rich in root hairs and can provide up to 75% of the surface area of the root, and in young root systems they represent the majority of the root biomass. The rootlets of proteoid roots are **determinate,** and although they initially develop a meristem and root cap, when mature they have no meristem and root hairs around their tip. Compared with other roots, the rootlets have much more cortex and a smaller stele, but they do have a well-defined endodermis. The rootlets are transient structures that disappear about 10–14 days after they are formed. Under sustained low-P conditions, proteoid roots are formed sequentially, both along a root as the primary and secondary roots grow, such that apical clusters are developing while basal ones are dying off, and often in different parts of the root system. After the cluster roots have disappeared they leave the primary and secondary root systems intact. The distribution of proteoid roots is affected by a number of factors, including water availability, but these roots generally occur in the upper soil layers, sometimes including the litter layers. Plants in the monocotyledonous Cyperaceae form **dauciform** roots that are initiated by similar environmental cues, and that are functionally analogous to cluster roots (Figure 6.18). The Cyperaceae are a large cosmopolitan family of around 5000 species, and include *Cyperus papyrus* (papyrus). Members of this family are common in low-P environments, including the fynbos, and are especially prominent in wetlands, many of which have substrates with low P availability. Dauciform roots develop on distinct, carrot-shaped lateral roots about 1–2 cm long. The epidermis of these lateral roots has large elongated cells, and from almost every one of these a very long root hair develops, extending several millimeters and giving the root the appearance of a tapered cotton-wool ball. The cortex of the lateral at the heart of the dauciform root is loosely packed, with abundant air spaces, and only a few cells thick. The Restionaceae are a monocotyledonous Gondwanan family of about 500 species that are common in low-P ecosystems of the southern hemisphere, including the fynbos. Species in this family are known to form capillaroid roots in which the lateral roots have dense clusters of exceptionally long root hairs that can hold a significant amount of water by capillary action.

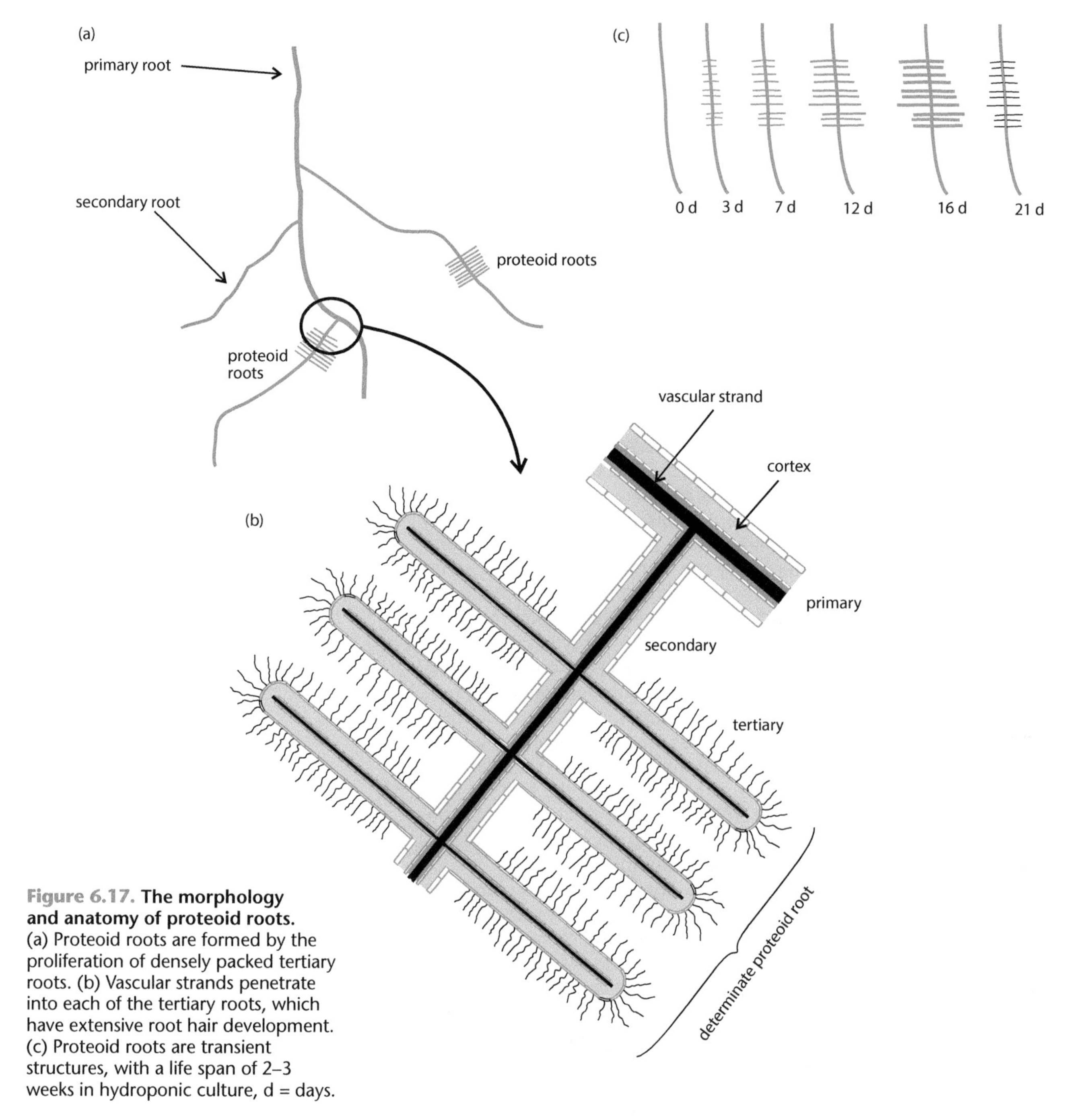

Figure 6.17. The morphology and anatomy of proteoid roots. (a) Proteoid roots are formed by the proliferation of densely packed tertiary roots. (b) Vascular strands penetrate into each of the tertiary roots, which have extensive root hair development. (c) Proteoid roots are transient structures, with a life span of 2–3 weeks in hydroponic culture, d = days.

Proteoid roots are the site of intense metabolic activity during their short life. This includes a massive burst of exudation of organic acids, which in the most intensively investigated proteoid roots, namely those of *Lupinus albus*, consist mainly of citrate. Complexation of Al, Fe, and Ca from their phosphates in soils releases P into solution. In general, the tricarboxylic citrate has a much higher stability constant with the multivalent ions that precipitate P in soils than do other carboxylic acids. Malate has been detected in exudates of proteoid roots, but generally at lower concentrations than citrate, and in most species its exudation is less responsive to P regime than is that of citrate. However, there is some evidence that citrate:malate ratios might be optimized to different soil pH values. In general, citrate exudation occurs in a concentrated short burst over about 2 days in mature cluster roots. In *Hakea prostrata* the exudation of carboxylic acids is coordinated by the activation of PEPc (phosphoenolpyruvate caboxylase) in the

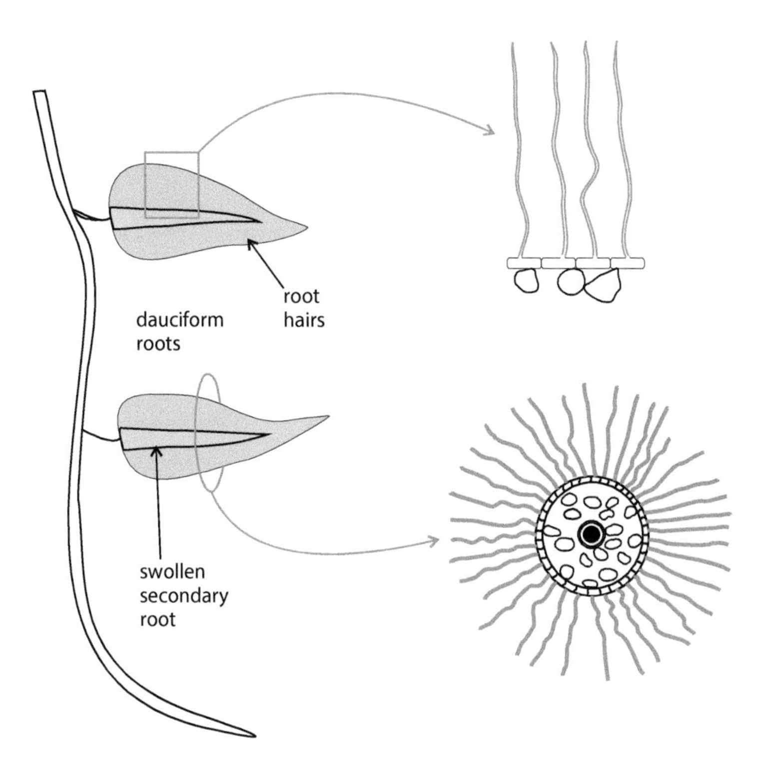

Figure 6.18. The structure of dauciform roots. These specialized secondary (lateral) roots are dauciform (carrot-shaped), with an expanded cortex and a very long root hair on almost every epidermal cell.

roots. An important function of PEPc is to produce oxaloacetate to balance the TCA cycle when carboxylic acids are removed from it. There is increasing evidence that PEPc has an important role in responses to P deficiency in other plant families, and that changes in its activity are responsible for the high exudation rates in proteoid roots. There is also a massive increase in H^+ efflux, which is sufficient to decrease the rhizosphere pH by 2 units or more. A decrease in pH tends to increase inorganic P availability, but probably in addition enhances the activity of acid phosphatases that are also exuded from the root, and that significantly increase the hydrolysis of organic P. A number of **flavonoids** have been reported in cluster root exudates, and these almost certainly inhibit microbial activity that might otherwise break down the exuded organic acids in the cluster root. Simple phenolic flavonoids have also been suggested to reduce mineral bound phosphates, thus mobilizing them. The plant P transporters that are utilized in P uptake have high activity in proteoid roots. The rapidly formed, large surface area of the proteoid roots promotes rapid efflux of exudates, and concentrates them in a small volume, but also provides a large surface area for rapid uptake of P mobilized from both mineral and organic sources. ^{32}P studies have confirmed that *Lupinus albus* with proteoid roots can access significant soil P reserves that are inaccessible to other species. Proteoid roots are therefore an adaptation that intensively and rapidly mines one particular volume of soil for P. Although the functioning of dauciform and capillaroid roots is much less well understood than that of proteoid roots, patterns of exudation in dauciform roots in particular seem to be similar to those in proteoid roots, and it seems likely that all of these root types are functionally analogous.

Manipulation of auxin concentrations in proteoid-root-forming species can induce proteoid roots in P-sufficient plants, or prevent proteoid root formation at low P levels. Nitric oxide (NO) is part of the signaling cascade that initiates proteoid root formation, because its production is related to proteoid root formation, inhibitors of NO prevent their formation, and NO producers increase their formation. Several of the TFs that mediate the response to P in normal roots have altered expression in proteoid roots, in particular the MYB-CC and R2R3-MYB TFs, some of which link P status to transport activity through Ca-dependent pathways. Although the SCR ("scarecrow") TFs that determine endodermal formation and quiescent center maintenance in normal roots do not change in expression in P-starved Proteaceae, they are necessary for cluster root formation. They are also clearly located in the primordia for tertiary root formation and in their endodermis, probably helping to determine their anatomy. A number of aspects of carboxylate metabolism and the TCA cycle are changed during the formation of proteoid roots. These changes presumably prepare the root for the exudative burst of carboxylic acid. Grafting experiments with *Hakea* have shown that, in this species at least, it is signaling of P deficiency in the shoot which is conveyed to the roots that initiates proteoid root formation. Proteoid roots and their analogs therefore help plants to mine P from severely P-deficient soils by a combination of physiological, anatomical, and morphological adaptation.

Carnivorous plants digest organic phosphorus using phosphatases

The evolution of leaves into carnivorous traps, which are dedicated in part to capturing P, decreases the surface area available for photosynthesis and transpiration, restricting carnivorous plants to environments that tend to be light and wet. A number of carnivorous species in the Lentibulariaceae and in other families have been shown to derive up to 50% of their P from the digestion of trapped prey. In carnivorous species with pitcher traps there is high phosphatase activity in the fluid. However, the constituents of the fluid are of complex origin because of its numerous microbial inhabitants and decomposing organisms. Primarily by using enzyme fluorescence techniques and growing plants in **axenic** conditions it has been established that, in some of the most ecologically important carnivorous plant groups, phosphatases derive primarily from excretion by the plants. In species with a variety of traps, including snap traps with no fluid, secretory glands have been shown to excrete nutritionally significant quantities of phosphatases (Figure 6.19). Some carnivorous species have also been shown to use hydathodes to excrete phosphatases onto the soil surface, where they may increase the availability of P. Carnivorous plants thus represent an interesting adaptation of the phosphatase-based P-mobilization system in plants. It should be pointed out that although the use of phosphatases to release P from organic compounds, even in carnivorous plants without fluid-filled traps, is readily termed "digestion," the use of the similar—perhaps identical—enzymes in root exudates to release organic P in soil does not attract the same terminology. Carnivorous plants provide an example of physiological components of P-acquisition systems evolving in conjunction with anatomical and morphological traits to adapt plants to chronically low-P environments.

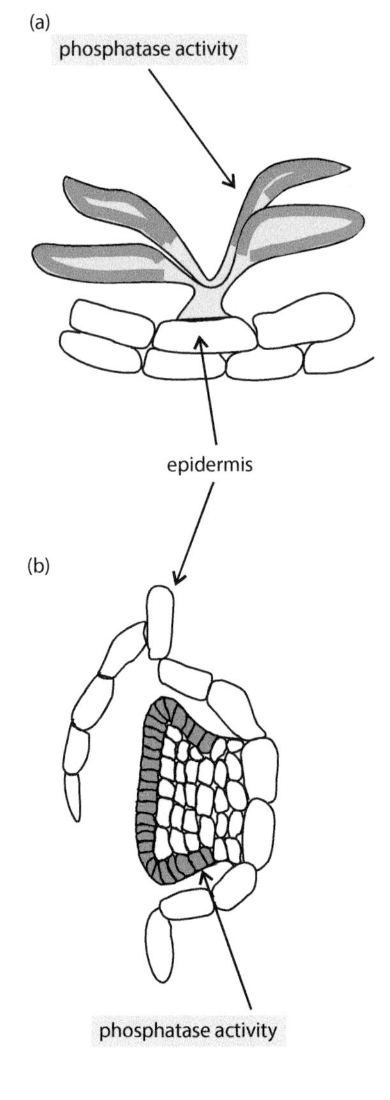

Figure 6.19. Phosphatase activity of the secretory glands of some carnivorous plants. (a) Quadrifid glands of the inner surface of bladder traps of *Utricularia* species. (b) Multicellular glands from the inner wall of the pitcher trap of *Nepenthes* species. The covering flap probably prevents insects from using the gland to climb up and out of the trap.

Summary

In order to inhabit the land surface, plants evolved mechanisms for accumulating P from the low concentrations found in soil solution. A network of transporters achieves homeostatic control of P concentrations during minor fluctuations in P supply. By changing the rhizosphere pH, mobilizing inorganic P and breaking down organic P in the rhizosphere, and

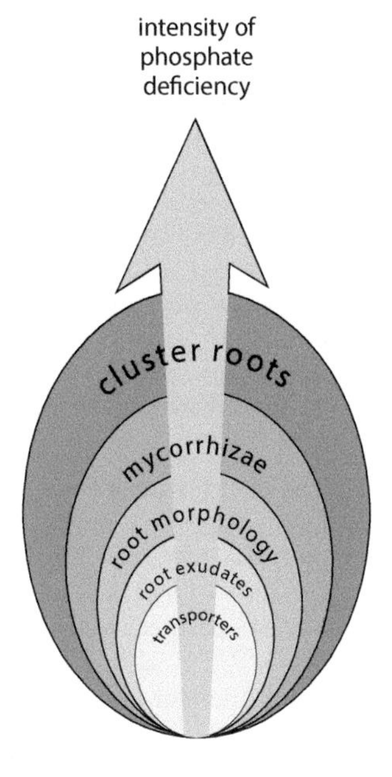

Figure 6.20. Phosphate deficiency and the stress-response hierarchy. Plants that are tolerant of increasingly severe P deficiency adapt biochemically initially, and then physiologically, anatomically, and morphologically. Adaptations to increasingly severe P deficiency can supersede those to less severe deficiency, but adaptations often occur in concert as the severity of deficiency increases. P deficiency tends to be chronic and without occasional bursts of available P, so there are no significant phenological adaptations to aid P capture.

adapting their root morphology, plants can increase P capture from soils. Mycorrhizas significantly alter root function and are vital for plants in many soils that are chronically low in P. Cluster roots and carnivorous plants are adaptations that necessitate dramatic physiological, anatomical, and morphological changes, but that enable plants to inhabit some of the lowest-P environments on Earth (Figure 6.20). The limited P supply of many unmanaged ecosystems means that they are susceptible to P pollution, which often occurs because P fertilization of crop plants is inefficient. Dwindling P supplies on Earth and problems of P pollution mean that an understanding of the environmental physiology of plant P nutrition will be important both for the preservation of natural ecosystems and for global food security in the twenty-first century.

Further reading

P in terrestrial ecosystems

Ceulemans T, Merckx R, Hens M et al. (2013) Plant species loss from European semi-natural grasslands following nutrient enrichment – is it nitrogen or is it phosphorus? *Global Ecol Biogeogr* 22:73–82.

Devau N, Le Cadre E, Hinsinger P et al. (2010) Soil pH controls the environmental availability of phosphorus: experimental and mechanistic modelling approaches. *Appl Geochem* 25:1094–1095.

Fujita Y, Venterink HO, van Bodegom PM et al. (2014) Low investment in sexual reproduction threatens plants adapted to phosphorus limitation. *Nature* 505:82–86.

Stock WD & Verboom GA (2012) Phylogenetic ecology of foliar N and P concentrations and N:P ratios across mediterranean-type ecosystems. *Global Ecol Biogeogr* 21:1147–1156.

P and the environment

Carpenter SR & Bennett EM (2011) Reconsideration of the planetary boundary for phosphorus. *Environ Res Lett* 6:014009.

Dodds WK, Bouska WW, Eitzmann JL et al. (2009) Eutrophication of U.S. freshwaters: analysis of potential economic damages. *Environ Sci Technol* 43:12–19.

Penuelas J, Poulter B, Sardans J et al. (2013) Human-induced nitrogen-phosphorus imbalances alter natural and managed ecosystems across the globe. *Nat Commun* 4:2934.

Withers PJA, Sylvester-Bradley R, Jones DL et al. (2014) Feed the crop not the soil: rethinking phosphorus management in the food chain. *Environ Sci Technol* 48:6523–6530.

P homeostasis

Cao Y, Yan Y, Zhang F et al. (2014) Fine characterization of *OsPHO2* knockout mutants reveals its key role in Pi utilization in rice. *J Plant Physiol* 171:340–348.

Lin W-Y, Huang T-K, Leong SJ et al. (2014) Long-distance call from phosphate: systemic regulation of phosphate starvation responses. *J Exp Bot* 65:1817–1827.

Shi J, Hu H, Zhang K et al. (2014) The paralogous SPX3 and SPX5 genes redundantly modulate Pi homeostasis in rice. *J Exp Bot* 65:859–870.

Zhang Z, Liao H & Lucas WJ (2014) Molecular mechanisms underlying phosphate sensing, signaling, and adaptation in plants. *J Integr Plant Biol* 56:192–220.

P transport

Chen A, Chen X, Wang H et al. (2014) Genome-wide investigation and expression analysis suggest diverse roles and genetic redundancy of Pht1 family genes in response to Pi deficiency in tomato. *BMC Plant Biol* 14:61.

Guo C, Guo L, Li X et al. (2014) *Ta*PT2, a high-affinity phosphate transporter gene in wheat (*Triticum aestivum* L.), is crucial in plant Pi uptake under phosphorus deprivation. *Acta Physiol Plant* 36:1373–1384.

Pedersen BP, Kumar H, Waight AB et al. (2013) Crystal structure of a eukaryotic phosphate transporter. *Nature* 496:533–536.

Wang X, Wang Y, Pineros MA et al. (2014) Phosphate transporters OsPHT1;9 and OsPHT1;10 are involved in phosphate uptake in rice. *Plant Cell Environ* 37:1159–1170.

Enhancement of soil P availability

Devau N, Le Cadre E, Hinsinger P et al. (2010) A mechanistic model for understanding root-induced chemical changes controlling phosphorus availability. *Ann Bot* 105:1183–1197.

Kong Y, Li X, Ma J et al. (2014) GmPAP4, a novel purple acid phosphatase gene isolated from soybean (*Glycine max*), enhanced extracellular phytate utilization in *Arabidopsis thaliana*. *Plant Cell Rep* 33:655–667.

Tran HT, Hurley BA & Plaxton WC (2010) Feeding hungry plants: the role of purple acid phosphatases in phosphate nutrition. *Plant Sci* 179:14–27.

Wang E, Ridoutt BG, Luo Z et al. (2013) Using systems modelling to explore the potential for root exudates to increase phosphorus use efficiency in cereal crops. *Environ Modell Softw* 46:50–60.

Wang L, Lu S, Zhang Y et al. (2014) Comparative genetic analysis of *Arabidopsis* purple acid phosphatases AtPAP10, AtPAP12, and AtPAP26 provides new insights into their roles in plant adaptation to phosphate deprivation. *J Integr Plant Biol* 56:299–314.

Adaptation of root morphology

Karthikeyan AS, Jain A, Nagarajan VK et al. (2014) *Arabidopsis thaliana* mutant *lpsi* reveals impairment in the root responses to local phosphate availability. *Plant Physiol Biochem* 77:60–72.

Miura K, Lee J, Gong Q et al. (2011) SIZ1 Regulation of phosphate starvation-induced root architecture remodeling involves the control of auxin accumulation. *Plant Physiol* 155:1000–1012.

Niu YF, Chai RS, Jin GL et al. (2013) Responses of root architecture development to low phosphorus availability: a review. *Ann Bot* 112:391–408.

Peret B, Clement M, Nussaume L et al. (2011) Root developmental adaptation to phosphate starvation: better safe than sorry. *Trends Plant Sci* 16:442–450.

Mycorrhizas and P uptake

Bucher M (2007) Functional biology of plant phosphate uptake at root and mycorrhiza interfaces. *New Phytol* 173:11–26.

Lekberg Y & Koide RT (2014) Integrating physiological, community, and evolutionary perspectives on the arbuscular mycorrhizal symbiosis. *Botany* 92:241–251.

Louche J, Ali MA, Cloutier-Hurteau B et al. (2010) Efficiency of acid phosphatases secreted from the ectomycorrhizal fungus *Hebeloma cylindrosporum* to hydrolyse organic phosphorus in podzols. *FEMS Microbiol Ecol* 73:323–335.

Nazeri NK, Lambers H, Tibbett M et al. (2014) Moderating mycorrhizas: arbuscular mycorrhizas modify rhizosphere chemistry and maintain plant phosphorus status within narrow boundaries. *Plant Cell Environ* 37:911–921.

Plassard C & Dell B (2010) Phosphorus nutrition of mycorrhizal trees. *Tree Physiol* 30:1129–1139.

Wang E, Yu N, Bano SA et al. (2014) A H^+-ATPase that energizes nutrient uptake during mycorrhizal symbioses in rice and *Medicago truncatula*. *Plant Cell* 26:1818–1830.

Cluster roots

Lambers H, Bishop JG, Hopper SD et al. (2012) Phosphorus-mobilization ecosystem engineering: the roles of cluster roots and carboxylate exudation in young P-limited ecosystems. *Ann Bot* 110:329–348.

Shane MW & Lambers H (2005) Cluster roots: a curiosity in context. *Plant Soil* 274:101–125.

Shane MW, Cawthray GR, Cramer MD et al. (2006) Specialized 'dauciform' roots of Cyperaceae are structurally distinct, but functionally analogous with 'cluster' roots. *Plant Cell Environ* 29:1989–1999.

Shane MW, Fedosejevs ET & Plaxton WC (2013) Reciprocal control of anaplerotic phosphoenolpyruvate carboxylase by *in vivo* monoubiquitination and phosphorylation in developing proteoid roots of phosphate-deficient harsh hakea. *Plant Physiol* 161:1634–1644.

Carnivorous plants

Adamec L (2013) Foliar mineral nutrient uptake in carnivorous plants: what do we know and what should we know? *Front Plant Sci* 4:10.

Ellison AM (2006) Nutrient limitation and stoichiometry of carnivorous plants. *Plant Biol (Stuttg)* 8:740–747.

Chapter 7

Essential and Beneficial Elements

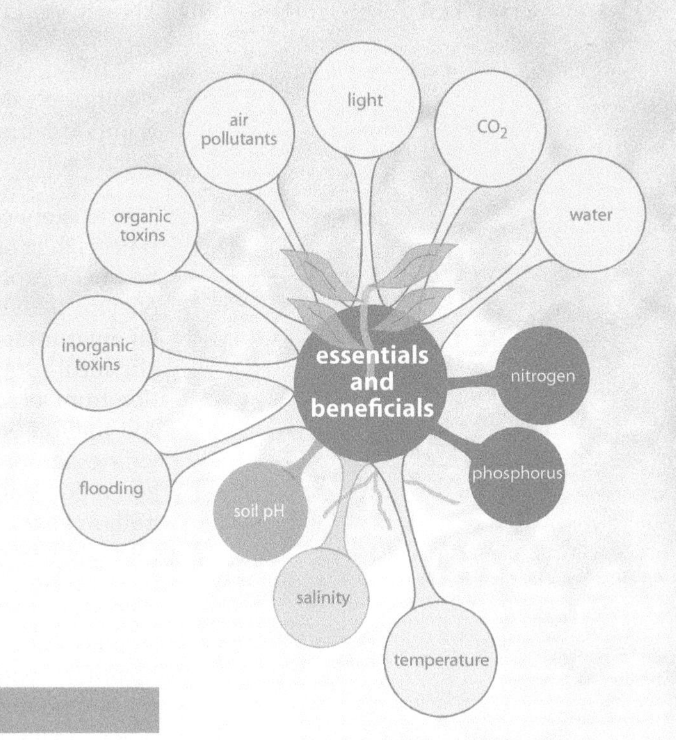

The uptake of mineral elements is influenced in particular by the availability of N and P, and by salinity and soil pH.

Key concepts

- Terrestrial life depends on plants mining a suite of essential nutrient elements from the soil.
- In terrestrial ecosystems, biomass production and quality are often nutrient limited.
- Under conditions of fluctuating nutrient availability, concentrations in plants are homeostatically controlled.
- Some non-essential elements can have beneficial effects on plants.
- S is involved in many plant–environment interactions, and is often a useful fertilizer.
- K fertilization enhances crop growth but has modest effects on global K cycles.
- Ca and Mg availability can affect crops and has widespread effects on ecosystems.
- Chronic lack of macronutrients can lead to anatomical and morphological changes in roots.
- Plant–microbe interactions aid nutrient capture in chronically infertile unmanaged ecosystems.
- The concentrations of many elements in plants are interrelated.

Terrestrial plants evolved to mine the soil for an ancient suite of available elements

The most important resources for plants are generally light, CO_2, H_2O, N, and P, so the availability of these has significantly influenced the evolution of terrestrial ecosystems. In addition, however, there are numerous other elements that are essential for plant growth, and several that are beneficial. The availability of these elements is currently frequently limiting to plant growth, and probably was so in the past. In unmanaged and managed terrestrial ecosystems the availability of these elements has consequences not only for plants but also for other organisms, because they too have an essential requirement for these elements, and for most organisms plant uptake from the soil is their ultimate source. Studies of **watersheds** generally indicate that the presence of plants increases the flux of nutrients by two- to five-fold, and significantly decreases their movement into watercourses, so plant nutrition is important not only for food chains but also for the biogeochemistry of the Earth's surface.

The existence of a set of mineral elements essential for all organisms reflects the composition of the environment in which life evolved. The very first cells probably lacked complex proteins with membrane transport or enzyme activity, so early metabolism probably occurred in an intracellular environment similar to the extracellular one, and made direct use of its chemical properties. After chemical buffers became saturated, the transition from a reducing to an oxidizing environment occurred on Earth about 750 million years ago and was a key geochemical transition for life, providing significant selection pressure to control the internal concentration and redox state of essential elements. The colonization of the land 470–420 million years ago presented life with new mineral supply challenges, and ultimately selected plants that could actively mine the soil for nutrients in order to provide a cytoplasmic composition redolent of the Earth's early environment. During evolution the elements fulfilling some functions have changed, and animals such as humans have evolved an essential requirement for a wider range of mineral elements than have plants, so the whole set of essential elements can differ between taxa (Figure 7.1), as can the ratios in which they are needed. For plants and other organisms there are several elements that are not essential but which are beneficial. Even for these elements, plant uptake from the soil is often their primary point of entry into terrestrial ecosystems.

Plants use roots to mine the soil for nutrients. Roots evolved at least twice in land plants, once in **lycophytes** and at least once in **euphyllophytes**. In land plants, the rhizoids of **gametophytes** and the roots of **sporophytes** are both controlled by a similar set of genes, whose evolution in a variety of plants probably underpinned the increase in plant diversity during the Devonian period. The soil is quite a homogeneous medium, so across the entire range of land plants the anatomical and morphological diversity of roots is generally low. The elemental composition of plants infrequently reflects that of the soil solution—plants are, overall, accumulators of essential elements and excluders of non-essential and toxic elements. The process of selection is in significant part a product of root anatomy. Most fine roots of plants are anatomically similar. They have an epidermis and a loosely packed cortex into which the soil solution can sometimes penetrate, although it does not do so unaltered because of the properties of the root cell walls, especially their cation-exchange capacity. Some epidermal cells, especially in the **maturation zone**, have long extensions known as root hairs. Together, the epidermal cells, especially their root hairs, and the cortical cells expose a significant surface area to the soil solution, as might be expected of a biological surface adapted for uptake of substances from a fluid. In general, however, the soil solution cannot penetrate to the vascular system at the center of the root because of the endodermis (Figure 7.2). This "inner skin" is a polarized epithelium—that is, it has distinct properties on each side, and the cells are

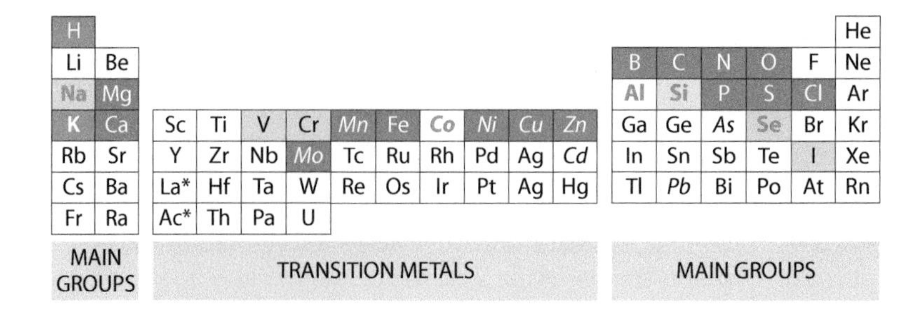

Figure 7.1. The essential and beneficial elements for plants. The elements with dark green cells shown in the periodic table are essential for plants and other forms of life. Essential elements tend to be lighter elements. With the exception of B, main group elements are required in large amounts and transition metals in small amounts. Na is only essential for a few plants, although it is a beneficial element for some plants (shown in bold green type), but it is essential for all other organisms. Si is beneficial for many plants, but is essential only for some. Se, Co, and Al are also beneficial for some plant species. There are hyperaccumulators of elements (shown in italics) that often grow best in the presence of these elements. V, Cr, Se, and I (gray cells) are essential for animals but not for plants. A very few organisms have been discovered with essential requirements for Br, Cd, Sn, Sr, Ba, or W. *Elements of the lanthanide (La) and actinide (Ac) series are not shown.

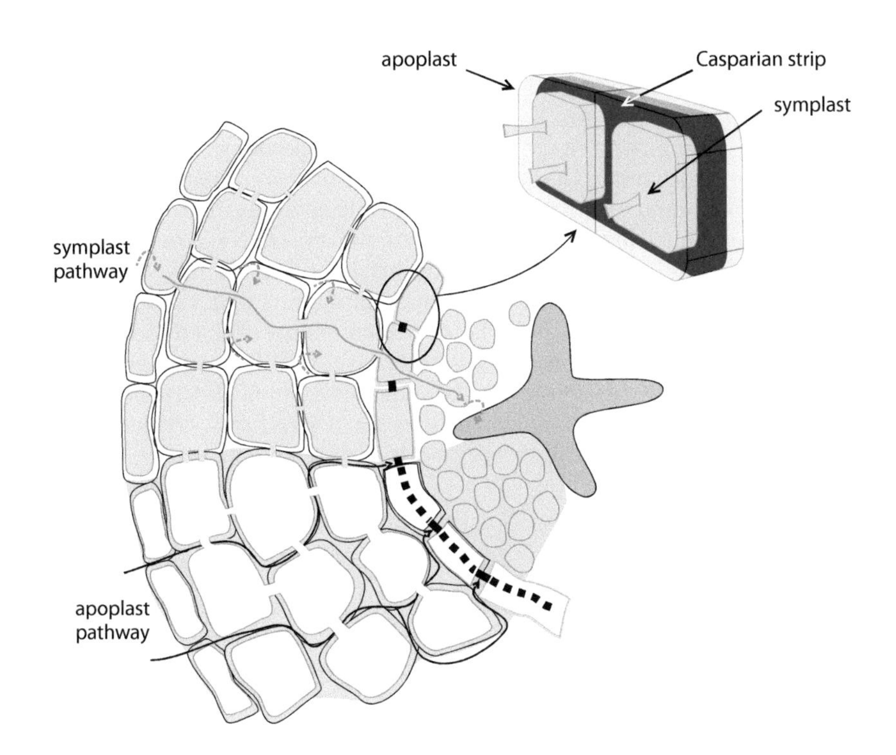

Figure 7.2. Root anatomy and the endodermis. Most plant roots have outer layers of loosely packed cortex and a central vascular strand (the stele) surrounded by an endodermis. Dissolved ions can move through non-living spaces in cell walls, between cortical cells, and in the xylem—that is, through the "apoplast" (shown in gray). The Casparian strip is a lignin layer that impregnates the walls of endodermal cells and which, when fully developed, blocks the apoplastic movement of ions between the cortex and the stele (black line). The symplast (shown in green) is the root-cell continuum of cytoplasm interconnected via plasmodesmata through which ions can move from the cortex, through the endodermis and into the xylem (green solid line). In order to move from the apoplast into the symplast in the cortex, or from the symplast into the apoplast at the xylem, ions have to be transported across the plasma membrane (green broken line).

sealed together. The Casparian strip of the endodermal cell walls begins a process of sealing cells together that forms a barrier to the flow of solution beyond the cortex. Therefore, in order to reach the vasculature and be available for transport to the shoot, ions generally have to cross the plasma membrane. In roots this can occur at the epidermal, cortical, or endodermal cells. In plant cells the presence of the plasmodesmata, which are common in root cortical cells, means that once the plasma membrane has been crossed the symplast can provide a route to the vascular system. In addition, plant uptake of nutrients through the leaves can be significant. For plants such as submerged aquatic macrophytes this can be particularly important, as can uptake of some elements in gaseous form, but many elements when applied to the leaves can slowly permeate to the leaf cells. As in roots, they then have to cross the plasma membrane. In some ecosystems, leaf uptake can result in significant cycling of nutrients in the canopy.

Uncharged substances can cross plasma membranes, but almost all essential elements form ions in the aqueous solution of the soil, and can only cross plasma membranes with the aid of specific transport proteins. For this reason the anatomy of the root promotes a selective flux of ions, mediated by specific transporters, across the plasma membrane. The selective flux occurs to a greater or lesser extent in different taxa and at different times. For example, Si and Al are the ions that occur at the highest concentrations in many soil solutions, but they are found in significant amounts in only a few plant taxa, and in many taxa they are virtually absent. However, flux selectivity can vary because the endodermal seal takes time to develop in roots and environmental conditions can affect its development, and in some species also promote the development of an exodermis beneath the epidermis that increases flux selectivity. The endodermis also helps to prevent root penetration by pathogens, and in some plants it minimizes radial O_2 loss, but it probably first evolved to promote the selective accumulation of elements. Roots have also evolved morphologically. For example, during the **Cretaceous** period there seems to have been a significant increase in the specific root length (SRL) of the fine (that is, non-woody) root systems of most plants, and probably an increase in root hair density. This would have significantly increased plant uptake of low-availability nutrients from soils. Overall, the ramifying, elongating root system of plants is adapted for selective nutrient foraging in soil. It has also been suggested that the perennial woody habit of many plants, which became particularly prominent in the Cretaceous period, might have been prompted by low availability of nutrients. Certainly many woody plants recycle nutrients internally to a significant extent and have lower nutrient construction costs, which in effect frees them from uncertain soil supplies of nutrients. In addition, it is woody plants that show some of the most dramatic root adaptations to aid nutrient capture. Overall, as life colonized the land surface, plants evolved to mine the nutrients that their evolutionary past made essential. The continuing challenge of mining these essential and beneficial elements affects many processes in unmanaged ecosystems, and presents significant challenges for the management of agricultural ecosystems.

The availability of essential nutrients limits biomass production and quality in many ecosystems

A key principle of ecology is that low availability of nutrients can limit primary productivity. N and/or P availability is frequently a limiting factor in ecosystems, but any of the other essential nutrients can also be a limiting factor, although generally their effects on ecosystems have been less well investigated. On a global scale, the primary soil factors that affect essential nutrient availability are rock type, pH, organic matter, and weathering intensity/soil exchange capacity. Geological sources of S and interactions with organic matter affect S availability, so in areas of, for example, sub-Saharan Africa, soil S can limit productivity. K^+ is susceptible to leaching, so in sandy soils and in soils of the subtropics and tropics that are subject to prolonged weathering it is often limiting to ecosystem productivity. Ca and to a similar extent Mg concentrations in soils are affected by rock type, with high concentrations in the calcareous soils that cover 30% of the terrestrial surface, and low concentrations in the acidic soils that cover a similar proportion of land area. Some soils, such as those over **serpentine**, have particularly low Ca:Mg ratios and exemplify the ecosystem-wide implications of soil-to-plant transfer of particular metals. Soil pH has significant effects on the availability of many micronutrients, with high-pH soils often having limiting concentrations of Fe, Zn, Cu, and Ni. Thus some general effects on essential nutrient availability in unmanaged ecosystems can be identified, but often as important are small-scale effects, with patches of limiting availability occurring on scales of as little as a few meters.

One of the greatest challenges facing humankind in the twenty-first century is ensuring the supply of nutritious food to an increasing population. This depends on uptake of essential elements by plants, not only because the elements themselves need to be present in food, but also because they are necessary for plants to synthesize the amino acids, carbohydrates, lipids, and vitamins that are essential in the human diet, as well as numerous other compounds that increasing evidence is showing are beneficial for human health. Some estimates suggest that two-thirds of the human population is at risk of developing mineral deficiencies. Actual deficiencies of Fe, I, Zn, and Se probably occur in 40%, 35%, 33%, and 15% , respectively, of the world's population. These deficiencies contribute directly to well over 1 million deaths each year, but also to general ill health for many millions more. For similar reasons, human deficiencies of Ca, Mg, and Cu are also widespread. These deficiencies occur because of low concentrations in food, which can often be traced directly back to low uptake by food and forage crops. Most crop plants are not grown on soils to which their ancestors were adapted, are increasingly being grown on marginal soils, and have not been bred with micronutrient concentrations in mind. For example, the Green

Revolution that occurred between approximately 1960 and 1980 increased yields of cereal crops in particular. Many high-yielding cereal varieties have lower concentrations of many micronutrients than did the old varieties, and their success has promoted their use in food in preference to dicotyledonous crops with a higher nutrient content. For example, in India and China, because of a combination of low soil availability and crop uptake characteristics, about 50% of soils produce Zn-deficient crops. There are numerous significant ongoing efforts, based on manipulation of soil–plant interactions, to fortify crop plants with essential nutrients, including agronomic regimes, traditional breeding, and biotechnology. In addition, the pathways of uptake of essential nutrients provide a conduit for the contamination of food chains. For example, Cd is a significant contaminant that enters food chains as an analog of Zn, and after nuclear contamination ^{137}Cs and ^{90}Sr enter food crops as analogs of K and Ca, respectively.

Elemental homeostasis is achieved using both ion-binding compounds and transport proteins

By fresh weight, more than 98% of most plants consists of C, H, and O, and even by dry weight other elements only account for 10%. In addition to N and P, the elements K, S, Ca, and Mg are required in relatively large amounts by all terrestrial plant taxa, and are often referred to as macronutrients, whereas Cl, Fe, B, Mn, Zn, Cu, Ni, and Mo are required in smaller amounts and are referred to as micronutrients. For most of these elements, although their concentrations in plants are low, these levels are higher than in the soil solution and occur in different ratios. Some terrestrial plants require, and others benefit from, a supply of Si, Na, and Se, and some plants have adapted to grow on soils with high concentrations of elements such as Al. Plants have no requirement for V and I, but can be vital conduits for the supply of these elements to other organisms, and in addition can contain small amounts of almost any other element that occurs in soil. However, overall, plants have significant control over their elemental composition.

The essential plant nutrients can be categorized by their function. The macronutrients are primarily used structurally, with S being utilized in amino acids, Ca in cell walls, and Mg in chlorophyll. Alternatively, K^+, Ca^{2+}, Mg^{2+}, Cl^-, and Mn^{2+} can be grouped together because they often occur in a single ionic form in plants, the Mn in the O_2-evolving complex being an exception. The micronutrients Fe, Zn, Cu, Ni, and Mo occur in multiple ionic forms (that is, they take part in redox reactions), and are also bound in the cell by **ligands**. The micronutrient metals in particular have a major role in protein function—they can bind and affect protein structure, often as cofactors that activate enzymes, or act directly in the **active sites** of enzymes. Although it has long been suggested that more than 45% of proteins have metal-binding sites that affect their structure, and over 40% have active sites containing metals, large-scale combinations of genomics and mass spectrometry suggest that these estimates, which are based on known metal-binding sequences, probably underestimate the proportion of **metalloproteins** in organisms. The chemistry of the elements has not changed since the origin of life, and these roles are ancient. Hadean environments were reducing, so metals in aqueous solutions occurred as sulfides, not sulfates. In particular Fe and Mg, but also Ni and Mn, sulfides are soluble, making these transition metals available to function, as they do today, in early redox reactions and in binding to organic molecules. The key role in all organisms of Fe-containing proteins such as cytochromes and of Mg bound to ADP/ATP dates from this time.

Cu and Mo, and to a significant extent Zn, are not soluble in reducing water, but they are in oxidizing conditions. These elements became more widely available, and more widely used by living organisms, when the Earth's environment became oxidizing. In fact, ancient organisms may have used elements such as W and V quite widely in enzymes that now bind Mo, which, in contrast to W in particular, is available under oxidizing conditions. Thus the utilization of Zn, Cu, and Mo increases with biological complexity, whereas that of W, V, Ni, and Co decreases. Plants contain a very wide array of organic ligands that bind many essential and beneficial metal and metalloid ions, which evolved under either reducing or oxidizing conditions. During both the oxidation of the Earth and the colonization of the land, plants were faced with biogeochemical challenges with regard to the availability of the elements that life evolved to use. Organisms, including terrestrial plants, adapted to these challenges at least in part by utilizing ligands to help to control the ionic environment of cells. Thus in extant terrestrial plants, when there are fluctuations in the availability of many elements, transporters work in conjunction with ligands that chaperone, translocate, or sequester elements in order to homeostatically control cellular ion activity.

Almost all of the elements that are essential or beneficial for plants occur primarily as ions in the soil solution. The exceptions are boric acid and silicic acid. However, mineral ions do not just have to be transported into the root cells. In order to reach the transpiration stream and get to the primary sites of metabolic activity in the shoot, ions have to reach the xylem, but the latter is dead tissue, so mineral ions have to be transported into it from the symplast. Different cells in plants, and different subcellular compartments, have different demands for mineral ions and, in the case of the large vacuole in many plant cells, act as a reservoir of them. For ions of many elements there is also significant redistribution via the phloem. Almost all of these ionic movements between compartments, cells, and tissues are controlled by the activity of transport proteins. One notable exception is revealed by the close relationship between concentrations of Ca and ^{90}Sr in soils and plants, which is thought to reflect significant apoplastic movement of Ca and ^{90}Sr. Thus, although there are instances in which transpirational flows of water control ion uptake, in general transport proteins regulate their uptake and distribution.

As in other organisms, transmembrane transport of ions in plants is driven by a proton-motive force (PMF) that is a product of electrical and pH differences across the membrane. In animal cells the membrane potential is generally around –100 mV, but in plants it is around –200 mV. The significant membrane potential in plants is generated by a unique H^+-ATPase. It has an ATP catalytic site through which energy from ATP is directly used to pump H^+ out across the plasma membrane. Its function is analogous to that of the Na^+/K^+ pump of animal cells. In plant cells there are also Ca^{2+}-ATPases, which are mainly used to pump Ca^{2+} into the vacuole, and heavy metal ATPases (HMAs), which enable efflux of a variety of divalent metals out of the root cells and into the xylem. There are eight HMAs in *Arabidopsis*, with two to four of these pumping Zn, Cd, Co, and Pb, five to eight pumping Cu, and one pumping Cu and Zn into chloroplasts. There are also many H^+-coupled transporters in plant cells in which the PMF drives H^+ flows that are coupled to the transport of other ions, either in the same direction as H^+ movement or in the opposite direction. In addition, there are numerous transporters that facilitate the movement of ions across membranes down electrochemical gradients. Thus transport is either actively driven against, or occurs down, electrochemical gradients. The direction and magnitude of these gradients can be calculated using the Nernst equation, and for a given ion they are opposite for influx and efflux across a membrane (Figure 7.3). In *Arabidopsis* there are about 1000 genes that encode proteins with transport activity. These include many that transport organic compounds, but a significant proportion of them make up the battery of transporters that are

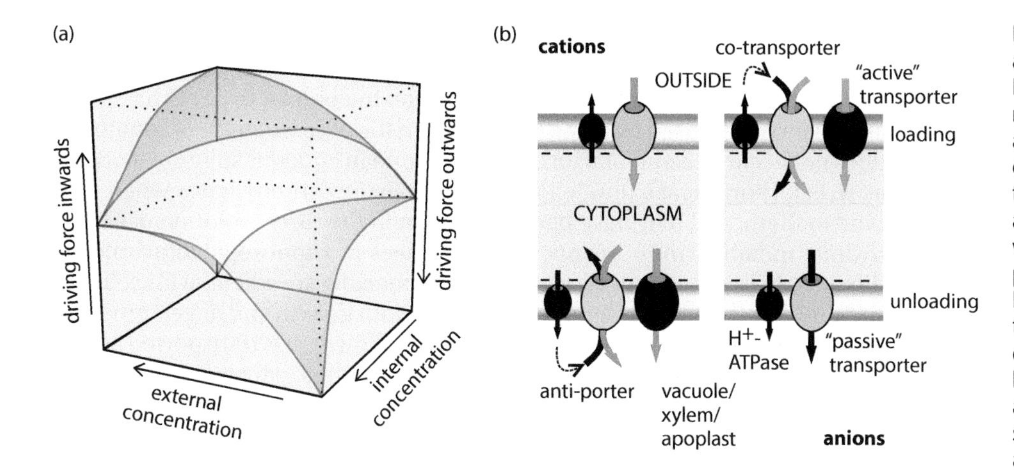

Figure 7.3. Transport proteins and the driving forces on ions. Most plant membranes have resting electrical potentials that are significant and negative (for example, −150 mV) relative to the outside of the cell. These are generated by the H^+-ATPase, which uses ATP to pump protons out of the cytoplasm. Membrane potentials contribute to electrochemical forces that drive the uptake of essential and beneficial elements, because almost all of them occur as ions in the soil solution. (a) At low internal and high external concentrations there are maximal inward driving forces on cations, with very high internal and very low external concentrations being necessary for there to be any outward forces on them (green surface). For anions, the forces are opposite in direction (gray surface). (b) Under most conditions, cations are transported into the cytoplasm through "passive" transporters (light green), but pumped out either by directly active transporters (black) or by anti-porters driven by the influx of protons down proton gradients. For anions the opposite process occurs.

distributed across plant membranes for mineral ions. Many of these are selective for specific ions, although there are many instances of interference between ions for transport. The selectivity mechanisms of many of the transporters are understood, in significant part because they have been fundamentally important during the evolution of life and thus have **homologs** in other organisms, many of which are more amenable to molecular investigation than multicellular plants.

Plants adjust to a variable supply of micronutrients by overexpressing homeostatic components

The availability of Fe, Zn, and Mn in soil is dramatically affected by pH, which can cause both **deficiency** and **toxicity** of each of these metals. Plants that inhabit soils of extreme pH have an array of adaptations to combat the effects of pH, and these adaptations therefore help to regulate the uptake of these micronutrients. Ni, Mo, Cu, and B are less affected by soil pH, and their concentrations in plants are, in general, regulated by a combination of transporter and ligand-binding activity. Plants require smaller amounts of Ni than of any other essential element. Ni is a co-factor in urease and therefore plays a role in N nutrition. It is taken up by several HMTs (Heavy Metal Transporters), and an Ni chaperone protein has been reported. Mo is a scarce element that occurs in the soil as MoO_4^{2-}, which is biologically inactive and has to bind to pterin in biological systems to act as a co-factor. MoO_4^{2-} is transported across plant membranes by the MOT family of transporters, which are similar to those that transport SO_4^{2-}. MOT transporters have been identified in various compartments, and aid the allocation of Mo to enzymes for which it acts as a co-factor, such as nitrate reductase and sulfite oxidase. Transporter and ligand-binding activity enables most plants to adapt to environmental variation in the supply of Ni and Mo.

Cu sulfides are only sparingly soluble, whereas Cu oxides are soluble. Cu began to be widely used by biological systems when the Earth became oxidizing, and it expanded the ancient functions of Fe, which has the opposite pattern of solubility. Cu is an essential co-factor for proteins in both photosynthesis and respiration, so plastids and mitochondria are its major sinks in plants. Cu evolved a vital role in controlling oxidative stress, especially in the superoxide dismutases that are commonly used by multicellular organisms. Transporters in the COPT family are widely used to transport Cu across plant membranes, and are closely related to the CT transporters found in yeasts and many plants. Together with chaperones such as CCH and HMA transporters that efflux Cu, COPT and CT transporters are used for Cu homeostasis in plants. SBL transcription factors have a role in orchestrating plant responses to variation in Cu supply. Cu homeostasis interacts with Fe homeostasis, reflecting their ancient linkage. In contrast to Cu, B has probably been widely used by living systems from the time when life originated. It does not occur in elemental form on Earth, and in aqueous solution it occurs as H_3BO_3 (boric acid). Borate ions stabilize many carbohydrates when they bind to them, and may have been important in the early evolution of carbohydrate metabolism. In plants, borate bridges in **rhamnogalacturonan-II** dimers, a key component of **pectic polysaccharide**, are essential for cell wall stability. B is therefore essential for normal plant growth and development—for example, deficiency quickly prevents root growth. B is transported across plant membranes by the BOR family of transporters, which are in the NIP family of aquaporins. Plants clearly adjust B transport to B supply, and in some cases have been shown to guttate B when there is an excess supply, as can occur in saline conditions. B nutrition interacts with water uptake, in part via common transporters but also because of its anatomical effects. Cu and B deficiencies in unmanaged ecosystems are rare, but they can be significant in some agricultural systems. In general in plants, however, homeostatic systems with a long history are able to counter environmental variation in Ni, No, Cu, and B levels.

Beneficial elements help many plant species to cope with a wide range of abiotic stresses

Several elements have been shown to be beneficial for some plants—that is, to promote growth. Availability of most of these elements can limit plant growth in the field, but Si in particular, although it is not essential, might have a major role in many plants. In the field, especially in unmanaged ecosystems, variability in the supply of beneficial elements may be more significant to plants than variation in some micronutrients. Co is essential for many microbes, but not for plants. However, for plants with N-fixing nodules, Co is beneficial because they have to supply it to microbial symbionts for them to synthesize nitrogenase. Most species that are **hyperaccumulators** of metals or metalloids grow best when these are available, even if the metal or metalloid is not an essential nutrient. Metal hyperaccumulation is often used as a defense against herbivores, so benefits in the field might be expected. However, even in hydroponic monocultures, hyperaccumulators of nutrients such as Ni, Zn, Cu, and Mn have been shown to grow optimally in much higher concentrations of these metals than are usual in plants. Interestingly, under such conditions, hyperaccumulators of non-nutrients such as Se, As, Al, and Se also show optimum growth when these are supplied in solution—the physiology of hyperaccumulators is attuned to growing in the presence of these elements even if they are not essential. Se is not essential for plants, but it is for animals, in which it is used to make selenoproteins with vital antioxidant activity. Se can substitute for S in cysteine (Se-Cys) and methionine (Se-Met). Plants have lost the ability to synthesize selenoproteins from Se-amino acids, but Se hyperaccumulators use Se-Cys and Se-Met to synthesize organic seleno compounds, especially methyl-SeCys. Se has also been shown to increase antioxidant capacity, as well as growth, in a number of non-Se hyperaccumulator species. With regard to As, only hyperaccumulator ferns benefit from this element being available. However, many plants that have adapted to the acid conditions in which Al is available, not just Al hyperaccumulators, benefit from Al supply. Many salt-adapted halophytes have been shown to have an essential requirement for Na, and many plants, especially crop plants with halophytic ancestry, have long been known to benefit from Na additions to soil. For C_4 and CAM plants, Na has an essential role in regenerating PEP from pyruvate, probably primarily by energizing the pumping of pyruvate into mesophyll chloroplasts. Overall, chronic differences in availability of these beneficial elements can have a significant ecological role, via effects on competitiveness on some soils, although Na, despite being essential for a considerable number of species, is generally regarded as never being deficient.

Si is essential for diatoms and for some horsetails (**Equisetaceae**), but not for any other terrestrial plants. However, there is increasing recognition that Si might have fundamentally important effects on the biology of plants and other organisms. It is the second most abundant element in most soils, and it occurs in soil solutions as silicic acid (SiH_4O_4), sometimes at concentrations approaching 1 mM. In the field almost all plants contain Si, with concentrations in the range 1–100 g kg^{-1}. In diatoms there are Si transporters that are driven by transmembrane Na^+ gradients, but in higher plants uptake occurs through aquaporins of the NIP family. Many species in the Poales and Cyperales accumulate particularly high concentrations of Si, probably because of extensive transport through NIPs, which can transport not only water and silicic acid but also a range of small, uncharged solutes, including boric acid, glycerol, and NH_3. Si is actively loaded into the xylem, in which it travels as silicic acid, and is then taken up into a range of cells, again via NIPs. Much of the Si that is taken up by plants is used to make silica ($SiO_2 \cdot H_2O$), which is, after calcium oxalate, the second most common biomineral synthesized by terrestrial plants, and can probably occur in all plant species. There are no reports of Si toxicity, despite the occurrence of very high concentrations (up to 25% of dry weight in some horsetails). Silica is biologically unreactive, and in many species it is made into phytoliths (structures composed of amorphous silica). Phytoliths can be widely distributed in the epidermis, hypodermis, and endodermis of plants. In Si-accumulating plants, much polysilicic acid can be present in the cell wall, and those that are members of the Poales and Equisetales have the hemicellulose (1-3,1-4)-β-D-glucan, which might aid the deposition of polysilicic acid in the cell wall. However, there is not only a physical role for silica in support, water regulation, and defense against herbivores and pathogens, but also a physiological role in fortification against stressors.

Oxidative stress is commonly associated with metal (Al, B, Cd, Mn, and Zn) and metalloid (As) toxicity, and is decreased by Si via enhanced antioxidant activity. Si also aids Fe uptake and homeostasis. In addition, pathogen attack can induce the activity of chitinases, peroxidases, and phytoalexins, an effect that is also enhanced by Si. In rice, which can take up significant quantities of Si, fertilization with Si can be used to enhance protection against **rice blast**—one of the most significant plant pathogens in agriculture. Interestingly, this fortification based on Si has no biomass production penalties, and its physiological basis is unclear. In the oceans, Si availability has clear effects on CO_2 fixation because diatoms fix at least 40% of the C in marine photosynthesis. On land, Si-accumulating plants, mostly grasses, carry out 55% of C fixation. It has therefore been suggested that biological cycling of Si has a major role in controlling C fixation on land. It has also been suggested that the contrasting environmental physiology of Al and Si might provide an insight into their importance. Al is biologically very reactive, was not available in large amounts early in the history of life, and is not essential to any organism. Si is biologically unreactive and has always been available. Aluminosilicates are common in the regolith, and have been found in several plants. It is possible that Si has a major role in counteracting the availability of Al, a reactive toxic ion that early life did not have to contend with, and that only became available later in evolution. The availability of Al is also increasing as a result of human activity. If the roles of Si in C cycling and Al detoxification are established, then Si, although not technically essential, will exemplify the potential major roles of uptake of beneficial elements by plants.

Sub-optimal sulfur availability can inhibit the synthesis of ecophysiologically important compounds

Many soils have enough available S to meet plant demands, primarily because of the weathering of mineral sulfates and, to a lesser extent, mineral sulfides. The decomposition of organic matter also releases sulfides, sometimes in significant amounts. In addition, there can be aerial inputs of S, such as SO_4^{2-} from seawater spray and SO_2 from combustion. Under aerobic conditions, bacteria in the genus *Thiobacillus* oxidize sulfides to sulfates, with the species involved depending on pH. The oxidation of sulfides tends to occur over a wider range of conditions than does the nitrification of NH_3, so in almost any aerobic soil the predominant form of S is SO_4^{2-}, for which all plants have a significant uptake capacity. Under anaerobic conditions, due to the action of bacteria such as *Desulfovibrio* species, which reduce SO_4^{2-}, sulfide is the predominant available form. With the exception of some H_2S, sulfides are not taken up by plants and can be toxic. The oxidation of sulfides releases H^+, but in many aerobic soils pH-buffering capacity is generally sufficient for soil pH not to be altered. However, not only do the sulfides of anaerobic soils produce significant S acquisition challenges for plants, but also if such soils are drained the sulfides are rapidly oxidized, which can release sufficient H^+ to dramatically reduce soil pH. Seawater is high in SO_4^{2-} and, due to bacterial reduction capacity, the anaerobic sediments beneath it contain high concentrations of sulfide. In many coastal regions there are significant areas of **acid sulfate soils** that are produced when sea-level changes expose previously anaerobic sediments to the air. This occurs as a result of long-term sea-level changes, but also has often been a significant challenge during land drainage. For example, during the formation of the Dutch **polders**, significant inputs of lime were often necessary to control soil pH.

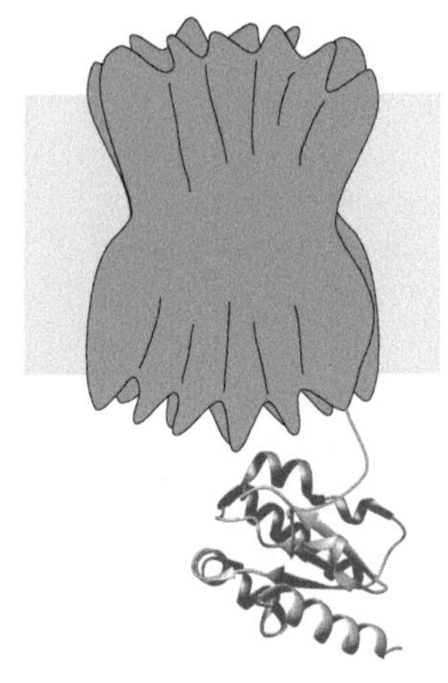

Figure 7.4. **Sulfate transporters in plants.** Different groups of SULTR transporters (shown as as the large green structure in the figure) have 12 membrane-spanning domains and control the flux of SO_4^{2-} across plant membranes. In contrast to other nutrient transporters in plants, SULTR has a cytoplasmic C-terminal STAS domain (shown as the ribbon structure in the figure), which is found in many other proteins and whose structure in other organisms is well known. STAS domains have been shown to connect proteins to a variety of signaling pathways. (STAS domain drawn in Chimera from PDB 4DGF; 12 generic α-helices shown to represent SULTR.)

Although plant S requirements are often met by SO_4^{2-} from mineral weathering, there are many aerobic soils in which the SO_4^{2-} supply is insufficient. Such soils often occur where intensive leaching has taken place over many millennia. For example, the soils of the extensive **Cerrado** ecoregion in Brazil are generally S deficient, as are several major soil types in Africa and Australasia. In the last few decades, S deficiency has become increasingly common in crop plants, particularly in countries that enacted clean air legislation in the last decades of the twentieth century. There is much evidence that, for example, across much of North America and Western Europe, this deficiency has increased in part because of a decline in SO_2 deposition from pollutants, which was contributing to S nutrition of many crops. Other contributors to this phenomenon were probably the decline in use of fertilizers that contained S, such as superphosphates and ammonium sulfate, and the increasing S demands of new varieties or species of crops, and greater supply of N and P. For the first time in several decades, fertilizer recommendations at the start of the twenty-first century for many crops in much of North America and Western Europe included the use of S. SO_4^{2-} is stable, and is readily taken up, transported, and stored in plants. Transmembrane transport of SO_4^{2-} is mediated by SULTR proteins, which are H^+/SO_4^{2-} cotransporters energized by the PMF. In plants, four different SULTR groups are known, which catalyze transport across the plasma membrane, the tonoplast, the xylem parenchymal membranes, and the chloroplast membrane. SULTR proteins have 10–12 membrane-spanning helices and a C-terminal STAS domain (Figure 7.4). STAS domains not only activate several different anion transporters, but are also known to link many other proteins to a variety of cellular signaling pathways. SULTR expression changes with S deficiency, and is responsible not only for increased uptake capacity but also for mobilization from vacuolar stores, and allocation to deficient tissues.

S assimilation in plants takes place via the action of *O*-acetylserine lyase on acetyl serine and S^-, which produces the S-containing amino acid cysteine. Reduction of SO_4^{2-} to S^- for incorporation into amino acids occurs primarily in plastids via sulfite (SO_3^{2-}), which is also used to produce **sulfoquinovose** from glucose in the first step of **sulfolipid** synthesis (Figure 7.5). During SO_4^{2-} reduction, SO_3^{2-} is produced from adenosine phosphosulfate (APS) by APS reductase, but SO_3^{2-} can also be produced during the **catabolism** of S-containing amino acids. Sulfite oxidase (SO) is a peroxisomal enzyme widely distributed in the kingdoms of life that oxidizes SO_3^{2-} to SO_4^{2-}. SO has a key role in controlling the concentration of SO_3^{2-}, which is toxic, in the plant cell. SO_2 entry from the atmosphere produces SO_3^{2-} in plant cells. If the amount produced is within the capacity of S-assimilatory enzymes, and the acidifying effects of SO_2 dissolution can be neutralized, SO_2 can contribute to plant S. Exposure to SO_2 can therefore increase both thiol-containing amino acids and SO_4^{2-} in plant cells because of the central role of SO_3^{2-} in S metabolism. Plants with altered expression of SO have significantly altered sensitivities to fumigation with SO_2, including the induction of classic SO_2 toxicity symptoms in SO knockout plants. APS is not only reduced by APS reductase, but can also be phosphorylated by APS kinase. The action of APS kinase produces phosphoadenosine phosphosulfate (PAPS). PAPS provides the sulfur for sulfonation during the production of numerous secondary compounds, and in *Arabidopsis* the kinetics of APS reductase and APS kinase determine the partitioning of S between primary and secondary metabolism. Changes in the production of secondary metabolites are frequently noted during S deficiency.

Cysteine is central to both S metabolism and the stress response of plant cells. Cysteine and methionine are the two S-containing amino acids essential to all life, which in terrestrial ecosystems are synthesized in plants. The thiol (S-H) groups of cysteine residues not only provide the disulfide bridges between amino acids that help to determine protein structure, but also have a key role in antioxidant and detoxification pathways. Cysteine is used in the synthesis of glutathione (GSH), a key water-soluble reductant in plant cells with thiols as the active group. Many environmental stressors increase the concentration of reactive oxygen species (ROS) in plant cells, and GSH is at the center of the pathways that help to control cellular redox when this occurs. Cysteine is also used in the synthesis of protein thiols such as **thioredoxins** and **glutaredoxins** that contribute to cellular control of redox. Some plants have the ability to chelate significant quantities of some reactive metal ions using **phytochelatins**, which are oligomers of GSH. Methionine is used, via SAM and ACC, to produce ethylene—a compound that often has a role in plant stress responses. SAM is also used in the synthesis of

Figure 7.5. **Sulfur assimilation and sulfur metabolites.** S assimilation (shown in black type) occurs in the chloroplast (shaded green) using reducing power from ferredoxin. Many S metabolites (shown in green type) of both primary and secondary metabolism are involved in the plant response to abiotic and biotic stressors in the environment. APS = adenosine phosphosulfate; GSH = glutathione; PAPS = phosphoadenosine phosphosulfate. OAS=o-acetylserine, γGC=γ-glutamylcysteine.

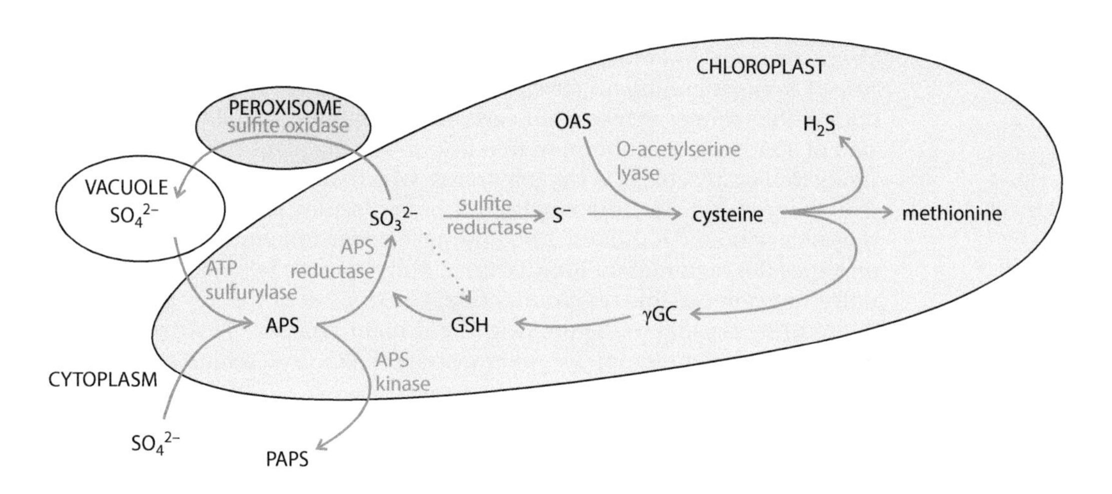

nicotianamine, which plays a key role in the plant response to Fe and Zn deficiency. Methionine is also used in the synthesis of **polyamines**, which are low-molecular-weight polyvalent cations often associated with stress tolerance in plants. Most species in the Brassicales produce **glucosinolates**, the synthesis of which usually begins with methionine. Cellular damage in the Brassicales causes the release of **myrosinase**, that acts on glucosinolates to produce a range of **isothiocyanates** ("mustard oils") that help to ward off herbivores and pathogens. Thus not only is variation in S supply sometimes a significant challenge for plants, but also S metabolism has a role in plant responses to both abiotic and biotic stressors in the environment. In general, biochemical and physiological adaptation probably enables plants to manage their S metabolism, as there are few other adaptations that have been shown to aid S capture.

Potassium can limit ecosystem production, but its use in fertilizer has a moderate environmental impact

After N and P, K is often the macronutrient added in the greatest amount in fertilizers, and the benefits of K addition in managed ecosystems are well established. K limitation is less often found in unmanaged ecosystems. The amount of total K in soils can be high, but there are generally only moderate concentrations (0.1–1.0 mM) in the soil solution. The ionic compounds of K are soluble, forming K^+ ions in the solution that adsorb to the cation-exchange sites common in most soils. This often provides a reservoir of exchangeable K^+ that can replace K^+ taken out of the solution by plants. In addition, some clay minerals can "fix" K^+, which then becomes available slowly (Figure 7.6). These processes mean that, in contrast to NO_3^- and except in sandy soils, leaching losses of K are generally moderate, and sustained fertilization with K can build up an accessible K reservoir in soils. K^+ is generally mobile in soils and readily supplied to plants by diffusion or mass flow. There are also few losses of K to the atmosphere, except in wind-eroded dusts. Most K fertilizers are made from a finite resource of K minerals, of which there are generally estimated to be several centuries of supply left. Thus, in general, the focus of K nutrition studies tends to be on understanding the consequences of deficiency, and on increasing the efficiency of K fertilizer use in managed ecosystems. This is in significant part because fertilization with N and P can produce K deficiency.

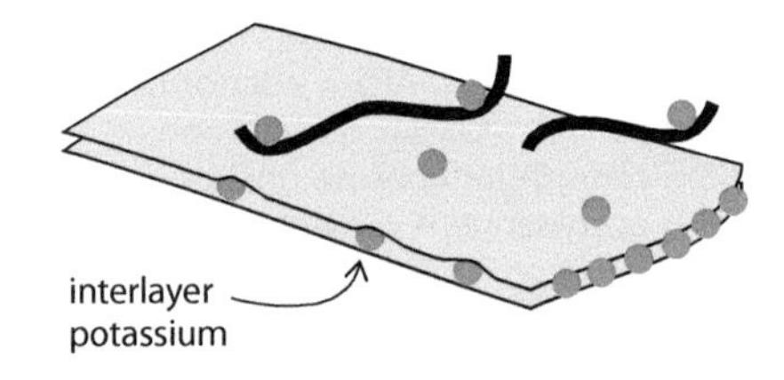

Figure 7.6. Potassium in soil. K availability in soils can be high because it is held in exchangeable forms. K^+ (shown as green dots) is attracted to negatively charged sites on organic matter (black lines) and on the surface and edges of micelles (shown in gray) of most clay types. K^+ has a high hydrated surface charge density but does not tend to form complexes, so significant quantities of it can be held exchangeably on these cation-exchange sites. In addition, in some clay types, K^+ can penetrate between micelles and partially collapse them, providing a slowly available, but sometimes significant, reservoir of K^+.

In most plants, K^+ is the predominant free cation. In the cytoplasm its concentration is quite well regulated in the range 50–100 mM, whereas in vacuoles its concentration can vary much more significantly. K^+ is transported across membranes rapidly, and is thus the major osmoticum used in plant cells for water-driven cell expansion. Fluxes of K^+ are also used to drive the opening and closing of stomatal guard cells and the rapid movements of Venus fly traps and touch-sensitive mimosas. In mature cells, some of the role of K in generating turgor can be fulfilled by organic acids, sugars, and compatible solutes. Most plant enzymes function optimally in a K solution of about the concentration that is found in the plant cell cytoplasm. K^+ binds to biomolecules via oxygen atoms, which are often a source of negative charges that slow the reaction of biomolecules with each other—the presence of bound K thus enables optimal enzyme function. For many plant enzymes the requirement for K is strict, and it cannot be replaced by other univalent cations. This requirement applies to most of the key enzymes of photosynthesis, and to the enzymes for protein synthesis in ribosomes. K^+ is also the key counter ion for the transport of NO_3^- and amino acids in the xylem, and sugars in the phloem.

Numerous K-transport proteins have been described in plants. Uptake from the soil solution occurs mostly in epidermal and cortical cells. At

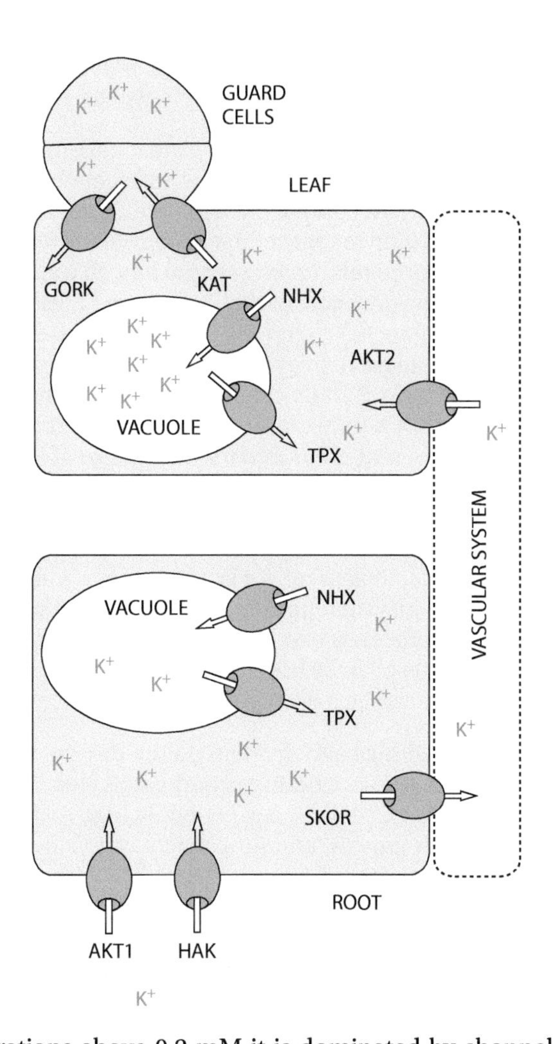

Figure 7.7. Potassium transporters in plants. K^+ fluxes across membranes are controlled by many transporters, and only those encoded by the main gene families are shown in the figure. K^+ influx through AKT1 transporters occurs down electrochemical gradients and is generally enhanced at low K concentrations by active HAK transporters. Efflux from the cytoplasm through SKOR and KAT transporters generally occurs down electrochemical gradients because of the relatively high cytoplasmic concentrations of K. Secondarily active transport through the K^+/H^+ exchanger NHX is used to drive significant accumulation of K^+ in vacuoles, whereas TPX transporters mediate efflux. In reality, many K transporters are found in a variety of organs and serve a variety of functions. K^+ levels in the vacuoles vary significantly and act as a reservoir. The detection of K deficiency results in changes in the expression of many of these transporters.

soil concentrations above 0.2 mM it is dominated by channels, often with similarity to Shaker channels in animals. In roots, AKT1 and its orthologs often dominate this low-affinity uptake, whereas in guard cells KAT1 does (Figure 7.7). K^+ influx through these channels occurs down electrochemical gradients, but necessitates efflux of H^+ if membrane polarization and electrochemical gradients are to be maintained. At soil concentrations below 0.2 mM, uptake from the soil is dominated by H^+/K^+ co-transporters, especially HAK transporters, but also HKT in monocots. The loading of K^+ into the vacuole occurs through the CHX K^+/H^+ antiporter. Loading of the xylem, which can almost always occur passively, occurs through the SKOR channel. These K-transport systems, of which there are probably many variants in the vast majority of terrestrial plant species that are as yet uninvestigated, respond to variation in external supply in order to maintain cytosolic K concentrations. It seems likely that AKT1, which can switch between low-affinity and high-affinity transport, is a K^+ receptor, initiating the events that help plants to cope with K deficiency. These include the use of classic signal pathways, based on Ca^{2+} and ROS, that results not only in adjustments to transporter activity but also to rapid decreases in rhizosphere pH that help to drive K^+ uptake and, in the long term, adjustments to morphology. Overall, in some unmanaged ecosystems the soil supply of K can be challenging for plants, which cope by means of a combination of physiological and morphological adjustments. In contrast to some other resources or stressors, no radical adaptations that are specific for K acquisition have been reported. K fertilization is widely effective in managed ecosystems with high levels of application of other nutrients, and is the focus of significant research aimed at improving K use efficiency, but with fewer environmental consequences than is the case for other fertilizer elements.

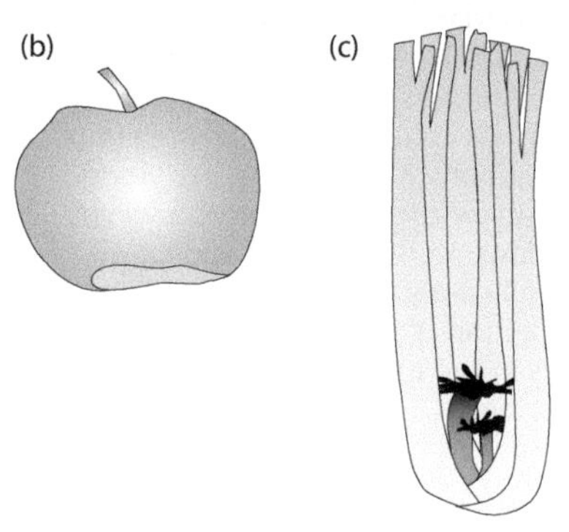

Figure 7.8. **Calcium deficiency in plants.** The most visible symptom of Ca deficiency is an inability to make new cell walls, so areas of meristematic activity become deformed. (a) In eudicots, the edges of leaves become deformed, producing "leaf-tip curl" in many species. A control leaf is shown behind a severely Ca-deficient leaf of *Ricinus communis* (castor oil plant). (b) In many fruits, Ca deficiency causes "tip burn," shown here in tomato with a black necrotic fruit tip. (c) In plants such as celery, shown here, "black-heart" regions are produced by stunted new leaves becoming black and necrotic.

Calcium deficiency can occur in a variety of plants, and magnesium deficiency in a variety of crops

After N and P, Ca and Mg generally occur in the next highest concentrations in plant shoots. Ca occurs at very low intracellular concentrations (*c.* 100 nM), but Mg has the second highest intracellular concentration after K. Ca:Mg ratios in plants tend to reflect those in soil, so extreme Ca:Mg ratios in soils are challenging for plants. Ca is essential for cell wall and membrane integrity, but its free concentration in the cytoplasm is very low because it would otherwise precipitate SO_4^{2-} and PO_4^{2-}. However, these properties do mean that it has evolved an extensive role in the calmodulin signaling system and, via dramatic fluctuations in its cellular concentration, as a secondary messenger. In plant cells, Ca influx to the cytoplasm occurs through a variety of transport channels, and efflux occurs via a range of Ca^{2+}ATPases and Ca^{2+}/H^+ antiporters. The CAX transporter that mediates fluxes into the vacuole is particularly well characterized. Much Ca transport and translocation occurs in the apoplast, and there is an important interdependence of Ca^{2+} and water fluxes in plants. In general, Ca is not phloem mobile, and it accumulates in tissues with high water flux. It is not remobilized significantly from vacuolar stores, so Ca deficiency occurs first in new growth (Figure 7.8). In some agricultural crops, especially leafy vegetables but also some fruit, such as tomato, Ca deficiency can be a problem.

Some plant species have long been divided, on the basis of their adaptation to low Ca or high Ca soils, into **calcifuges** and **calcicoles**, respectively. For calcifuges, rather than low Ca levels, it is generally high Al^{3+}, Fe^{2+}, and Mn^{2+} concentrations that limit growth, and for calcicoles it is low Fe^{3+} and PO_4^{2-} concentrations rather than high Ca levels that do so. Nevertheless, despite their Ca nutrition being adapted to scavenge Ca, many calcifuges have a high K:Ca **physiotype**. In contrast, calcicoles have a low K:Ca ratio in which Ca to a significant extent replaces K as an osmoticum—a "calciotrophic" physiotype. Many calcicole plants complex Ca with oxalate, often as an insoluble precipitate, which may help to keep free Ca at low concentrations. Ca biomineralization is quite common in terrestrial plants, either as carbonates in a variety of cell wall spaces, or as oxalates in the vacuole. Although such Ca minerals play a significant part in defense against herbivores, they probably also have a role in maintaining low concentrations of free Ca. Their capacity to do this is such that, in human nutrition, oxalate in food is regarded as an anti-nutrient for Ca. Thus Ca availability has a role in plant adaptation to some important soil types, and can be of importance to horticulture and human nutrition.

Mg is a common element in the Earth's crust, and it occurs in soils from weathering of its carbonates and sulfates. Mg^{2+} has a high charge density and thus attracts a large hydration shell. In the soil solution, its relatively low hydrated surface charge density means that although it is attracted to cation-exchange sites, it is exchangeable and mobile in soils—to a greater extent than are Ca, Mn, and K. When Mg concentrations are sufficient for plant growth, mass flow therefore supplies almost all Mg, but its mobility means that it is quite easily leached. If Mg is present in soils, plants can generally access it, but deficiency can occur because some rocks release little Mg, and in some soils it has been leached out. Furthermore, high concentrations of Ca, Al, and Mn in particular can inhibit its uptake. In soils on serpentine, Mg concentrations can be high enough to inhibit the growth of most plant species.

Mg is an essential element for all living cells. In plants, about 20% of Mg is found at the center of porphyrin rings in chlorophyll, with the rest being used as a co-factor for a wide variety of enzymes, including rubisco, and in complexes with nucleotide phosphates, especially ADP/ATP in the cytosol.

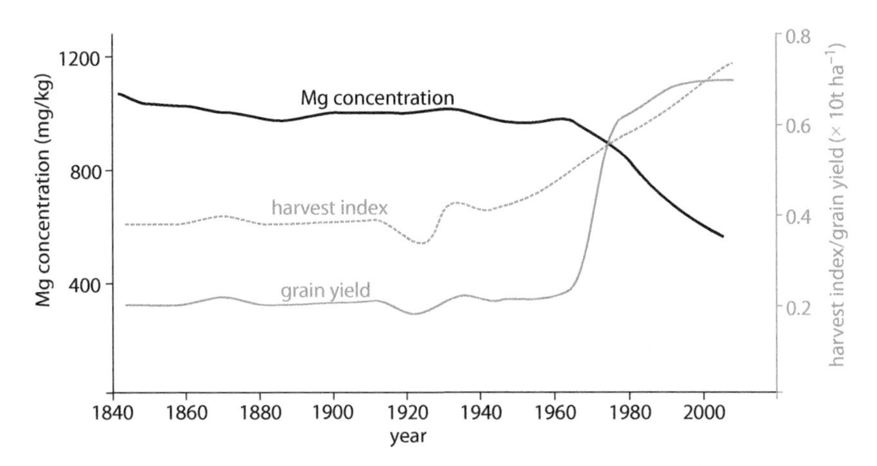

Figure 7.9. The decline in magnesium concentrations in crop products. The Broadbalk Wheat Experiment has been run continuously at Rothamsted Research Station in the UK since the mid-nineteenth century. In 1968, new short-straw cultivars were introduced which dramatically increased yield but decreased Mg concentrations. Similar decreases have probably occurred in harvested produce of many other crops. (Adapted from Rosanoff A [2013] *Plant Soil* 368:139–153. With permission from Springer.

Sub-optimal Mg concentrations probably occur quite widely in food and forage crops, which are the primary source of Mg in most human diets. Low Mg levels in humans are probably a significant contributing factor to hypertension, atherosclerosis, diabetes mellitus, and a variety of heart conditions. The use of high-yielding modern crop varieties has decreased Mg concentrations in many food products (Figure 7.9), and many modern diets are low in Mg. There is therefore significant interest in Mg deficiency in crops. In many species an early symptom of Mg deficiency is a build-up of carbohydrate in **source leaves** and a decrease in growth of **sink leaves.** Mg is phloem mobile, but is required to generate the PMF to actively load sucrose into the phloem via the SUT1 transporter. Mg concentrations in chloroplasts increase in the stroma during the day, which helps to activate rubisco, and the importance of Mg for key components of the circadian clock (especially the CCA, LHY, and TOC proteins) means that Mg deficiency decreases the ability of plant cells to maintain the circadian clock and hence the diurnal rhythms of photosynthesis. In more severe cases of deficiency there is an increase in chlorophyll a/b ratios, an increase in the proportion of PSI complexes, and a decrease in thylakoid organization. Mg deficiency-induced alterations in photosynthetic capacity are often associated with ROS production and hence an increase in the concentrations of ascorbate, glutathione, and ELIP proteins. Ultimately, Mg deficiency inhibits chlorophyll synthesis and produces the (often characteristic) deficiency symptom of inter-veinal chlorosis (Figure 7.10).

Some Mg has been shown to enter roots via **CNGC channels**, together with Ca and some other metal ions. However, research with model plants clearly shows that plants have specific Mg transport proteins that are homologous

Figure 7.10. Magnesium deficiency symptoms. Chlorosis is a clearly visible symptom of severe Mg deficiency. (a) In non-monocotyledonous plants, chlorosis induced by Mg deficiency is strongly inter-veinal, as shown here in *Ricinus communis.* A control Mg-sufficient leaf is shown behind an Mg-deficient leaf. (b) Close-up of Mg-deficient leaf of *Ricinus communis.* (c) In many species with palmate leaves it is the outer parts of the leaves that display chlorosis first.

Figure 7.11. Plant magnesium transporters. MRS transporters occur in many organisms, but plants have the greatest variety of MRS genes. MRS transporters (MRS2 is shown as a ribbon structure looking down into the transporter) are pentamers, with two transmembrane domains of each protein contributing to the central transport pore. Between the transmembrane domains there is a highly conserved Gly-Met-Asn (GMN) tripeptide that probably helps to confer the high specificity for Mg, as variations of this motif in other transporters confer specificity for other divalent cations. Mg^{2+} ions are transported in the unhydrated state, which probably helps to explain the transporter specificity. In a pentamer, individual MRS2 molecules can be coded for by different genes, several of which therefore have high levels of redundancy. (Drawn in Chimera using PDB 4EVF.)

with those in organisms of all domains of life, and that are responsive to Mg status. In prokaryotes and animals there is a single gene homolog of the MRS2 Mg-transport protein encoded by multiple genes in *Arabidopsis thaliana*. MRS2 transporters are involved in Mg uptake into the root, plastid, and vacuole, and into the cytoplasm from the xylem. MHX1 transporters have a role in transport in the tonoplast and the vascular tissue, and CNGCs have also been shown to have a role. MRS2 functions as a pentamer, with two transmembrane domains of each molecule helping to form the transport pore (Figure 7.11). High Mg concentrations help to decrease the toxic effects of Al, Cd, and NH_4^+ on plant cells. If MRS2 transporters maintain high Mg uptake, then high internal Mg concentrations (probably via interactions with Ca, transporters, and other proteins) help to reduce toxic effects. High Mg concentrations also help to decrease the toxic effects of salinity. Furthermore, it has been noted that heat stress makes Mg deficiency more severe. Overall, it is increasingly being suggested that Mg deficiency, especially in unmanaged but also in managed ecosystems, might more often be a limiting factor than is usually considered to be the case, certainly more so than K, and to an even greater extent than S.

Adaptations of root anatomy and morphology help plants to respond to chronic nutrient deficiency

Conditions in the rhizosphere can affect the development of the endodermis of plant roots, and hence the selectivity of element uptake. The SHR (short root hair) and SCR (scarecrow) TFs are key determinants of root form. They interact, for example, with miR165/166 to determine xylem development, but early on the activity of SHR in particular determines which cells will develop into endodermis. Depending on the environmental conditions, the Casparian strip starts to develop specifically in the medial walls of endodermal cells a few to many cells back from the root tip. At the Casparian strip the plasma membrane is attached to the cell wall, which prevents the movement of molecules within it. This enables the production of a **polarized epithelium** with different membrane properties on the inward- and outward-facing domains of the cell. The Casparian strip is initially formed by lignification of the cell wall, with **peroxidases** catalyzing the polymerization of **monolignol** molecules. This strip not only produces separate plasma membrane domains, but also initiates the binding together of adjoining cells, thus blocking the apoplast. Localized NADPH oxidase activity, produced in *Arabidopsis* by **RBOHf**, drives peroxidase activity. These processes occur on a scaffold of localized CASP proteins, with, for example, CASP1 being especially important in localizing peroxidase 64. Subsequently, there are often significant developments of the endodermis, particularly the deposition of suberin, with large suberin lamellae around mature endodermal cells (Figure 7.12). Mutants with altered endodermal development have been used to demonstrate that the development of the endodermis can significantly alter the flux of molecules into plants. For example, the movement of dyes such as **propidium iodide** can be used to visualize the barrier to penetration that is presented by the endodermis. For Ca in particular, endodermal mutants have less effect on fluxes, indicating that Ca probably enters roots near the tip before the endodermis develops. The endodermis clearly has a significant role in plant–water relations, defense against pathogens entering the root, and, in some species, the radial loss of O_2, but it is also affected by the composition of the soil solution. In particular, the presence of unusual elemental ratios or toxic elements often causes early and significant Casparian strip and endodermis development.

In addition to root anatomical adaptation, adaptation of root system architecture is caused by long-term variation in soil solution composition. Some soils have chronically low levels of essential and beneficial nutrients, but

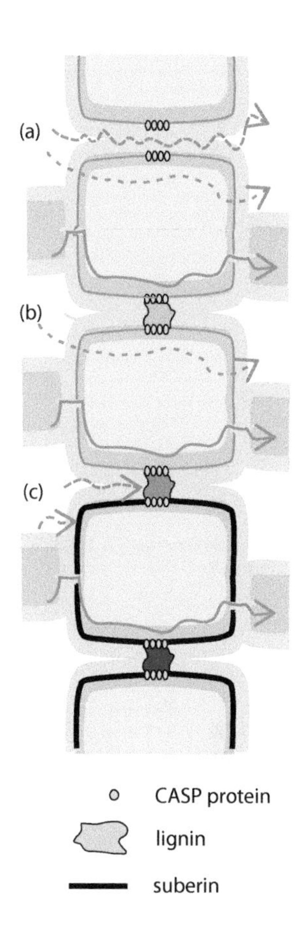

Figure 7.12. Development of the Casparian strip and endodermis. (a) CASP proteins in the plasma membrane initiate the development of the Casparian strip. At this point ions can pass through the apoplast (shown by broken lines with long dashes), be transported across an endodermal cell (shown by broken lines with short dashes), or move through the symplast (shown by solid line). (b) CASPs direct lignification that begins the process of producing a polarized epithelium and sealing the apoplast and stopping transport through it. (c). The endodermis is completed by the formation of suberin lamellae around the endodermal cells, blocking all but symplastic transport.

even in these soils there can be some short-term temporal variation induced by changes in weather and biological activity. Nutrient availability in soils is generally also spatially heterogeneous, often in patterns that change over time, not least because of the uptake activity of plants. Plant root systems are strikingly plastic in their development both temporally and spatially, and can be unusually asymmetric. The newest lateral roots often form ephemeral modules for resource acquisition, and are therefore, with respect to older roots, analogous to short-lived leaves on long-lived branches. This dual root functionality, which may differ between plants, enables them to respond to changes in soil conditions via the production and death of ephemeral roots. Although H_2O, N, and P are the primary determinants of root system architecture and its development, many other essential and beneficial elements also have significant effects, especially K and S. Mass flow of water can carry to the roots significant amounts of elements that occur in aqueous solution at high concentrations, such as K, S, Ca, Mg, Al, and Si. If root demand differs from supply, leading to either deficiency or toxicity, the root architecture can be adjusted, often in quantifiable ways that are specific to particular elements. Plant root systems therefore forage for nutrients and avoid toxicity in response to the balance between supply and demand. This flexibility underpins nutrient uptake efficiency and thus many key traits for understanding or managing ecosystems in the twenty-first century.

All root systems are initiated from the **radicle** of the embryo, which develops into the strongly **gravitropic** primary root. From this primary root, lateral roots emerge, mostly at an angle of 90°, but with significant gravitropism. Lateral root primordia form in the **pericycle**, mostly opposite the **protoxylem** points of the stele, and emerge through the endodermis, cortex, and epidermis. The production of subsequent lateral roots on them can, depending on the life span of the plant, produce a root system with second and third orders of roots (Figure 7.13). In dicotyledonous plants there can be significant **secondary growth** in higher-order (that is, older) roots. Especially toward the tips of the primary root and first-order laterals, there can be root hairs extending from epidermal cells. The growth of primary and lateral roots, the density of lateral roots, and the rate of production of higher orders of roots are all sensitive to nutrient variability, both temporally and spatially. Two key parameters that are altered by plastic root development are the volume of soil exploited and the surface area of root exposed to soil solution. The basic flexible pattern of development can also be overlaid by other significant developmental possibilities. Many plants produce roots from shoot

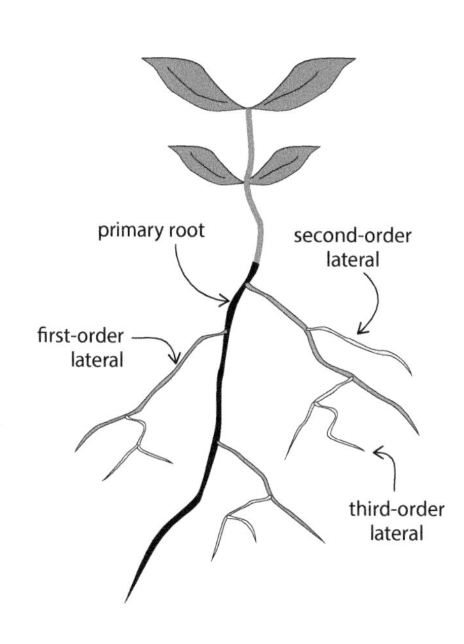

Figure 7.13. The orders of roots in non-monocotyledonous plants. After germination, a radicle emerges from the seed and develops into a primary root that persists throughout the life span of the plant. From the primary root the first-order lateral roots emerge, from these the second-order roots emerge, and from these the third-order roots emerge. In response to nutrient deficiencies, plants often develop not only different-sized root systems but also different ratios between roots of different orders.

nodes, the most important example being those produced by many monocotyledonous plants. In these plants, primary root development can be limited, and nodal roots from the base of the stem, which are only weakly gravitropic and tend to maintain a low diameter along their length, dominate the root system. In addition, many plants produce nodal roots from rhizomes, and many others produce nodal adventitious roots rather than constitutive nodal roots. Although adaptations analogous to N-fixing nodules or cluster roots, induced by lack of N and P, respectively, have not been found for other essential or beneficial nutrients, their availability can still induce significant changes in root system development.

K^+ is quite mobile in soils but, in general, chronic low K supply decreases total root length and surface area while increasing the proportion of lateral roots and the number of root hairs, perhaps reflecting an increased emphasis on, respectively, exploring new areas of soil and exploiting existing K. However, responses can vary widely within species, as exemplified by the significant differences between ecotypes of *Arabidopsis*, some of which decrease the production of lateral roots in response to low K. In most species that have been tested, a plentiful supply of K encourages root elongation, perhaps underpinning the capacity of plants to consume "luxury" amounts of K when it is available. Ca deficiency prevents root elongation, in general producing a stunted root system. S deficiency tends to increase the production of lateral roots, especially close to the root tip. In many species, root surface area responds to availability of Zn and Mn, mostly via changes in elongation rate. Fe deficiency tends to provoke a similar reaction in roots to P deficiency, namely an increase in lateral root, and root hair, production. These nutrient-induced changes in the root system, which can shape it over the course of days, weeks, or longer, are a vital complement to the physiological and homeostatic controls on nutrient concentration, and help to regulate nutrient supply in the long term. As with adaptions to N and P supply, auxin is the master hormone governing these responses. Cytokinins and ethylene play key roles, as do signaling molecules such as H_2O_2 and NO, and several TFs. An understanding of the mechanisms underpinning the responses is very useful for elucidating the nutrient dynamics of unmanaged ecosystems and increasing the efficiency of nutrient use in managed ecosystems.

Many plants use symbioses with fungi and changes in rhizosphere microflora to aid nutrient uptake

The **rhizosphere** is distinct from the bulk soil because it is influenced by the activity of roots, particularly on microbial populations, which are often attracted into it by exudates or emissions from the root (Figure 7.14). The arbuscular mycorrhizas (AM) that are formed by perhaps 80% of plants, and the ectomycorhizas (EcM) that are formed by about 10% of plants (mostly woody species) have well-established roles in P, H_2O, and N uptake by plants. There is also much evidence that they can contribute to plant uptake of other elements. In general, K uptake is much less affected by mycorrhizal status than is the uptake of P, H_2O, or N, although many studies have found effects, and its analog, radioactive ^{137}Cs, has been shown to have altered root-to-shoot translocation in mycorrhizal plants. Important interactions between K^+ uptake, which helps to adjust water potentials and aquaporin activity, and the effects of mycorrhizas on plant–water relations have also been reported. In addition, it is clear that mycorrhizas can alter S nutrition—for example, via changes in the expression of SULTR transporters. In general, however, these effects on S are only reported in P-sufficient plants. Effects of mycorrhizas on Mg and Ca nutrition have only occasionally been found, but effects on the micronutrients Zn and Cu are frequently reported and probably quite widespread. Plants control the extent of mycorrhization

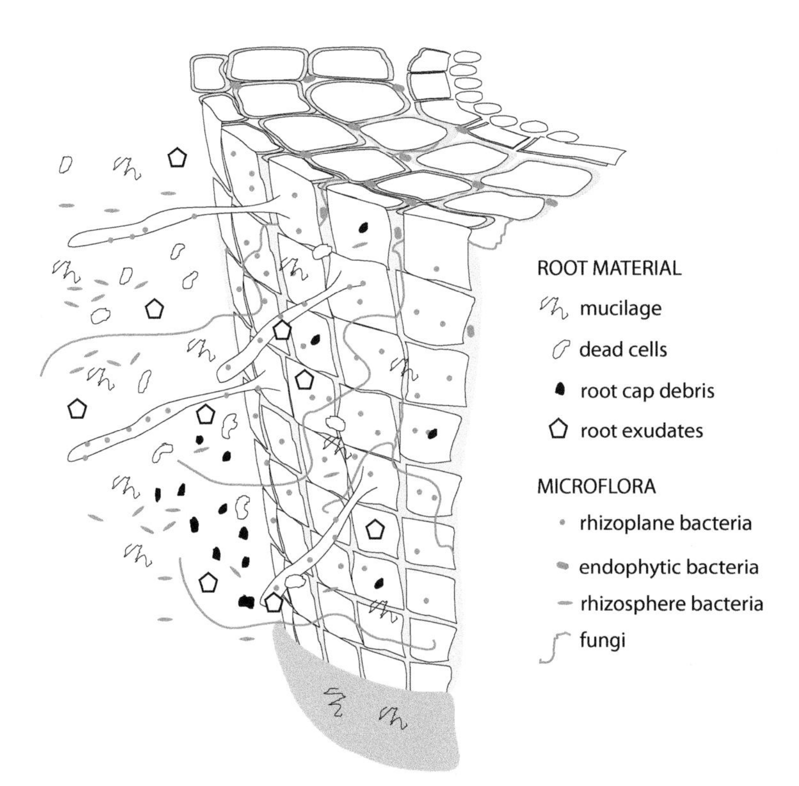

Figure 7.14. The rhizosphere. The soil surrounding a root is an attractive habitat for microorganisms because of the sloughing off of dead cells and root-cap debris, the production of carbohydrate mucus from the root tip, and the exudation of organic compounds from the root. The zone of soil in which there is a root-induced change in the population of bacteria, fungi, and other organisms is called the rhizosphere. The root surface often has a characteristic microflora, and is sometimes called the rhizoplane. In addition, the endophytic bacteria and fungi of the root apoplast are sometimes referred to as the endorhizosphere. Plant–microbe interactions in the rhizosphere have significant effects on the availability of many essential and beneficial elements.

in response to nutrient status, especially P levels. However, many low-P soils are deficient in other minerals, and mycorrhization also helps with their uptake. The frequency of the symbiosis is so high that, overall, it is an adaptation that has a significant effect on the form and function of plant roots, including the uptake of essential elements.

The rhizosphere also contains significant populations of microorganisms other than fungi, and it is clear that the nutritional status of a plant, including its K status, affects both the amount and type of root exudate. In many species the mixture of compounds that is exuded into the rhizosphere is unique to deficiency of particular nutrients, at least in part because it helps to encourage microbes that aid plant uptake of nutrients. Many plant growth-promoting bacteria, in particular diazotrophic species, have been shown to increase N supply to plants, but many also produce siderophores and can affect Fe and Zn availability. There are also some studies which have shown that K deficiency can be alleviated by rhizosphere microflora. Many rhizosphere and some endophytic microbes produce, or affect the production of, plant hormones with growth effects that alter many aspects of nutrition. Thus, although many plant–microbe interactions in the rhizosphere are generally most important for P, H_2O, and N, they can also have significant effects on the nutrition of other elements. These symbioses and interactions are particularly important for plants challenged by chronically low nutrient supply, but their management in agriculture has long been of considerable importance, and there is much interest in optimizing plant–microbe interactions for many aspects of crop nutrition.

Ionomics

C:N:P ratios in plants have long been useful for understanding ecosystems processes, including terrestrial ecosystems. Plant assimilation of these elements is subject to significant stoichiometric homeostasis, and environments that make this especially challenging promote the evolution of species adapted to particular C, N, and P regimes. Relationships between Ca and Mg concentrations in plants have also long been known, and the tendency to particular K:[Ca + Mg] ratios is well established in many plants. In addition, Fe:Zn and Fe:Mo ratios in plants are often related. Thus plants often have elemental signatures or profiles, including those for many of their essential and beneficial nutrients, prompted by stoichiometric homeostasis or adaptation when this is not possible. There are both physiological and environmental constraints on elemental ratios, so elemental profiles are ultimately a product of gene × environment interactions. The elemental composition of plants varies with developmental stage, but overall it seems clear that there are important constraints on variation in elemental composition. The elemental composition of a plant is its ionome. Ionomics attempts to describe the interrelated variation in concentrations of the elements, and to map it to allelic variation. This requires a refinement of Liebig's "law of the minimum," which tacitly assumes that demand for each element varies independently. It is perhaps as often the ratios of suites of elements that limit plant growth in terrestrial ecosystems as it is the availability of a single element.

Under different fertilizer regimes, willow genotypes with contrasting characteristics have elemental concentrations with not only a K:Ca:Mg group, but also an N:P:S group and an Fe:B:Zn:Al group, all of which are constrained within differences in availability. In brassicas grown in a single set of conditions, two- to threefold intraspecific differences in Ca and Mg have been found, but they are interrelated. In comparisons between domesticated and wild plants, multivariant analysis using element concentrations reveals groupings that change with domestication (Figure 7.15). Our understanding of these constraints on variation in plant ionomes will increase significantly as the number of species with appropriate genomic information to which to map these groupings increases, the statistical frameworks are developed, and data from high-throughput ionomic measurements are compiled. For some elements, variation can in part be explained by phylogeny. For example, despite many decades of contrasting fertilizer regime, in the Park Grass Experiment at Rothamsted Research Station in the UK there is significant interspecies variation in Ca, Mg, Zn, and Mn above the species level. There are also indications that phylogeny influences the concentrations of, for example, the alkaline earth metals (Figure 7.16). A phylogenetically informed gene × environment understanding of the ionome is likely to significantly improve our understanding of the way in which element availability affects plant growth. In unmanaged ecosystems this might be important for understanding adaptation to soils with unusual elemental ratios. For example, the soils of the serpentine zone of New Caledonia have a

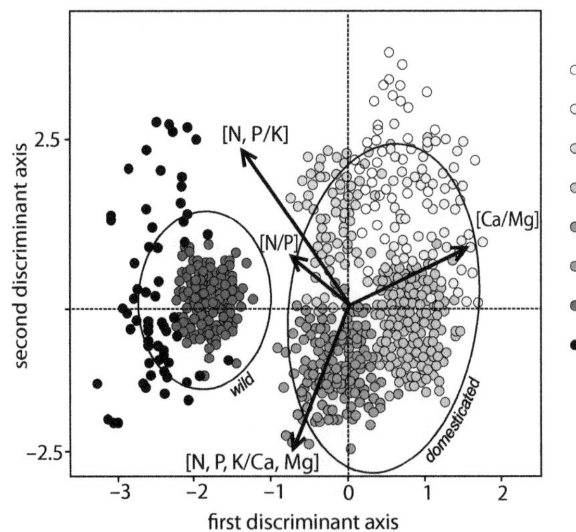

Figure 7.15. Ionomic variation in wild and domesticated fruits. Meta-analyses of ionomic variation between plant species reveal that, in different groups of plants, elements often vary in concert, but sometimes in different ways. In order to identify these patterns it is necessary to transform the raw data into a suitable form. In the example shown, isometric log-ratios of N, P, K, Ca, and Mg concentrations in some wild and domesticated fruits from various parts of the world were subjected to discriminant analysis. This shows that species can be distinguished on the basis of ratios of some key elements (square brackets), and that these differ between wild forms (shown in black/gray on the left) and domesticated forms (shown in green on the right). (From Parent S-E, Parent LE, Egozcue JJ et al [2013] *Front Plant Sci* 4, article 39. With permission from Frontiers Publishing.)

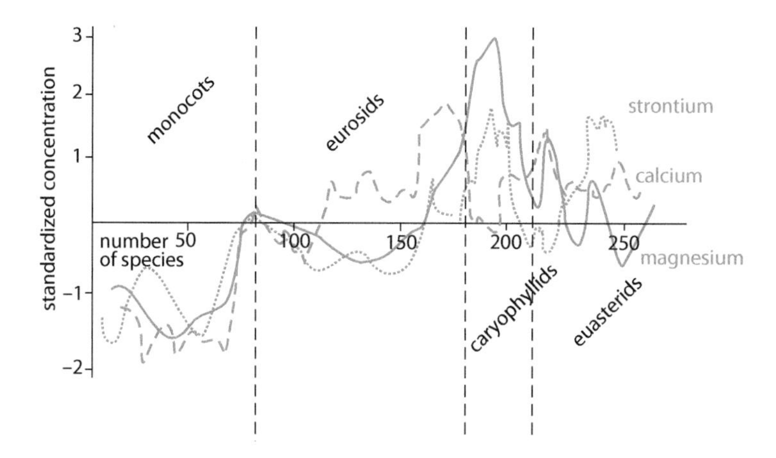

Figure 7.16. **Phylogeny and element variation in plants.** Studies that have collated data from plant species across many studies and mapped them onto phylogenies have begun to show that different clades of flowering plants can have, relative to each other, different element concentrations. The graph, which displays concentrations for the alkaline earth metals collated for about 250 species and standardized (that is, transformed to mean = 0 and standard deviation = 1), shows that major clades of angiosperms have lower or higher concentrations of alkaline earth metals than the average. (Redrawn to APG III from Willey NJ [2014] *J Environ Radioact* 133:31–34. With permission from Elsevier.)

disharmonic flora (that is, it does not have the taxonomic balance expected given the predicted colonization possibilities), which might be because of **exaptations** to the low Ca:Mg ratio found in serpentine in particular groups of colonizing plants. In managed ecosystems the control of nutrient concentrations, perhaps achieved by understanding the constraints on variation and using them to guide crop improvement, or understanding the long-term effects of fertilizer additions, is likely to be vital for food quality in the twenty-first century.

Summary

The mining of the soil by plants provides terrestrial ecosystems with a sustained supply of elements whose ancient availability made them a prerequisite for life. The molecules that evolved to bind and transport essential elements evolved into homeostatic systems that control element concentrations in plants by partitioning them between compartments and controlling their activity at physiologically important sites. Many species also have adaptations that aid the uptake of specific essential and beneficial nutrients (Figure 7.17). Many soils are deficient in K and S, which are widely used in fertilizers. The human impact on natural K cycles is modest compared with that on N, but anthropogenic S inputs to soil are environmentally significant. Although Ca is only occasionally deficient in crops, the development of many unmanaged ecosystems is fundamentally affected by Ca availability, while the important role that Mg might have in the health of terrestrial ecosystems is increasingly being recognized. Elements often have interrelated effects in plants—for example, those elements that are beneficial in the long term to plant growth often modify the effects of essential or toxic elements. The concentrations of elements in plants differ significantly even when their soil availabilities are similar, but there is no reason why this variation should only occur at the species level, so currently much research is aimed at understanding the ionomic relations of the elements in land plants.

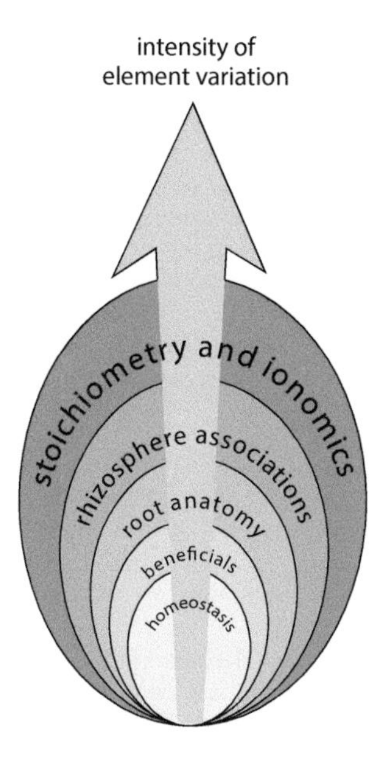

Figure 7.17. **The stress-response hierarchy for variation in availability of essential and beneficial elements.** When challenged with minor variation in the availability of elements, plants use homeostatic mechanisms to control element concentrations in key compartments. When more significant variation occurs, change in the overall fluxes of elements produced by significant changes in transporter activity supplement homeostatic control. With more profound variation in element availability, anatomical and morphological adaptations of roots and shoots are also used to increase control over element concentrations. Stoichiometric and ionomic analyses suggest that plant groups also adapt to chronic differences in the availability of elements, perhaps even on geological timescales.

Further reading

Evolution of element acquisition

Alassimone J, Roppolo D, Geldner N et al. (2012) The endodermis—development and differentiation of the plant's inner skin. *Protoplasma* 249:433–443.

Jones VAS & Dolan L (2012) The evolution of root hairs and rhizoids. *Ann Bot* 110:205–212.

Novoselov AA, Serrano P, Pacheco ML et al. (2013) From cytoplasm to environment: the inorganic ingredients for the origin of life. *Astrobiology* 13:294–302.

Stueeken EE, Anderson RE, Bowman JS et al. (2013) Did life originate from a global chemical reactor? *Geobiology* 11:101–126.

Effects of essential and beneficial elements

Brady KU, Kruckeberg AR & Bradshaw HD (2005) Evolutionary ecology of plant adaptation to serpentine soils. *Annu Rev Ecol Evol Syst* 36:243–266.

Clemens S, Aarts MGM, Thomine S et al. (2013) Plant science: the key to preventing slow cadmium poisoning. *Trends Plant Sci* 18:92–99.

Miller DD & Welch RM (2013) Food system strategies for preventing micronutrient malnutrition. *Food Policy* 42:115–128.

White PJ & Brown PH (2010) Plant nutrition for sustainable development and global health. *Ann Bot* 105:1073–1080.

Homeostasis of element concentrations

Alvarez-Fernandez A, Diaz-Benito P, Abadia A et al. (2014) Metal species involved in long distance metal transport in plants. *Front Plant Sci* 5:105.

Hedrich R (2012) Ion channels in plants. *Physiol Rev* 92:1777–1811.

Sinclair SA & Kraemer U (2012) The zinc homeostasis network of land plants. *Biochim Biophys Acta* 1823:1553–1567.

Zeng H, Wang G, Hu X et al. (2014) Role of microRNAs in plant responses to nutrient stress. *Plant Soil* 374:1005–1021.

Regulation of micronutrient uptake

Bittner F (2014) Molybdenum metabolism in plants and crosstalk to iron. *Front Plant Sci* 5:28.

Perea-Garcia A, Garcia-Molina A, Andres-Colas N et al. (2013) *Arabidopsis* copper transport protein COPT2 participates in the cross talk between iron deficiency responses and low-phosphate signaling. *Plant Physiol* 162:180–194.

Uraguchi S, Kato Y, Hanaoka H et al. (2014) Generation of boron-deficiency-tolerant tomato by overexpressing an *Arabidopsis thaliana* borate transporter *AtBOR1*. *Front Plant Sci* 5:125.

Wimmer MA & Eichert T (2013) Review: mechanisms for boron deficiency-mediated changes in plant water relations. *Plant Sci* 203:25–32.

Beneficial elements

Carey JC & Fulweiler RW (2012) The terrestrial silica pump. *PLoS One* 7:e52932.

Exley C (2009) Darwin, natural selection and the biological essentiality of aluminium and silicon. *Trends Biochem Sci* 34:589–593.

Gregoire C, Remus-Borel W, Vivancos J et al. (2012) Discovery of a multigene family of aquaporin silicon transporters in the primitive plant *Equisetum arvense*. *Plant J* 72:320–330.

Pilon-Smits EAH, Quinn CF, Tapken W et al. (2009) Physiological functions of beneficial elements. *Curr Opin Plant Biol* 12:267–274.

Sulfur

Gao Y, Tian Q & Zhang W-H (2014) Systemic regulation of sulfur homeostasis in *Medicago truncatula*. *Planta* 239:79–96.

Malcheska F, Honsel A, Wildhagen H et al. (2013) Differential expression of specific sulphate transporters underlies seasonal and spatial patterns of sulphate allocation in trees. *Plant Cell Environ* 36:1285–1295.

Varin S, Lemauviel-Lavenant S & Cliquet J-B (2013) Is white clover able to switch to atmospheric sulphur sources when sulphate availability decreases? *J Exp Bot* 64:2511–2521.

Zagorchev L, Seal CE, Kranner I et al. (2013) A central role for thiols in plant tolerance to abiotic stress. *Int J Mol Sci* 14:7405–7432.

Potassium

Cherel I, Lefoulon C, Boeglin M et al. (2014) Molecular mechanisms involved in plant adaptation to low K^+ availability. *J Exp Bot* 65:833–848.

Hafsi C, Debez A & Abdelly C (2014) Potassium deficiency in plants: effects and signaling cascades. *Acta Physiol Plant* 36:1055–1070.

Wang Y & Wu W-H (2013) Potassium transport and signaling in higher plants. *Annu Rev Plant Biol* 64:451–476.

White PJ (2013) Improving potassium acquisition and utilisation by crop plants. *J Plant Nutr Soil Sci* 176: 305–316.

Calcium and magnesium

Gilliham M, Dayod M, Hocking BJ et al. (2011) Calcium delivery and storage in plant leaves: exploring the link with water flow. *J Exp Bot* 62:2233–2250.

Hermans C, Conn SJ, Chen J et al. (2013) An update on magnesium homeostasis mechanisms in plants. *Metallomics* 5:1170–1183.

Lee J, Park I, Lee Z-W et al. (2013) Regulation of the major vacuolar Ca^{2+} transporter genes, by intercellular Ca^{2+} concentration and abiotic stresses, in tip-burn resistant *Brassica oleracea*. *Mol Biol Rep* 40:177–188.

Verbruggen N & Hermans C (2013) Physiological and molecular responses to magnesium nutritional imbalance in plants. *Plant Soil* 368:87–99.

Root anatomy and morphology

Geldner N (2013) The endodermis. *Annu Rev Plant Biol* 64:531–558.

Giehl RFH, Gruber BD & von Wiren N (2014) It's time to make changes: modulation of root system architecture by nutrient signals. *J Exp Bot* 65:769–778.

Kellermeier F, Chardon F & Amtmann A (2013) Natural variation of *Arabidopsis* root architecture reveals complementing adaptive strategies to potassium starvation. *Plant Physiol* 161:1421–1432.

Long Y, Kong D, Chen Z & Zeng H (2013) Variation of the linkage of root function with root branch order. *PLoS One* 8:e57153.

Effects of mycorrhizas and microbes

Carvalhais LC, Dennis PG, Fedoseyenko D et al. (2011) Root exudation of sugars, amino acids, and organic acids by maize as affected by nitrogen, phosphorus, potassium, and iron deficiency. *J Plant Nutr Soil Sci* 174:3–11.

El-Mesbahi MN, Azcon R, Manuel Ruiz-Lozano J et al. (2012) Plant potassium content modifies the effects of arbuscular mycorrhizal symbiosis on root hydraulic properties in maize plants. *Mycorrhiza* 22:555–564.

Miransari M (2011) Soil microbes and plant fertilization. *Appl Microbiol Biotechnol* 92:875–885.

Sieh D, Watanabe M, Devers EA et al. (2013) The arbuscular mycorrhizal symbiosis influences sulfur starvation responses of *Medicago truncatula*. *New Phytol* 197:606–616.

Ionomics

Agren GI & Weih M (2012) Plant stoichiometry at different scales: element concentration patterns reflect environment more than genotype. *New Phytol* 194:944–952.

Baxter I (2009) Ionomics: studying the social network of mineral nutrients. *Curr Opin Plant Biol* 12:381–386.

Parent S-E, Parent LE, Egozcue JJ et al. (2013) The plant ionome revisited by the nutrient balance concept. *Front Plant Sci* 4:39.

White PJ, Broadley MR, Thompson JA et al. (2012) Testing the distinctness of shoot ionomes of angiosperm families using the Rothamsted Park Grass Continuous Hay Experiment. *New Phytol* 196:101–109.

Chapter 8
Temperature

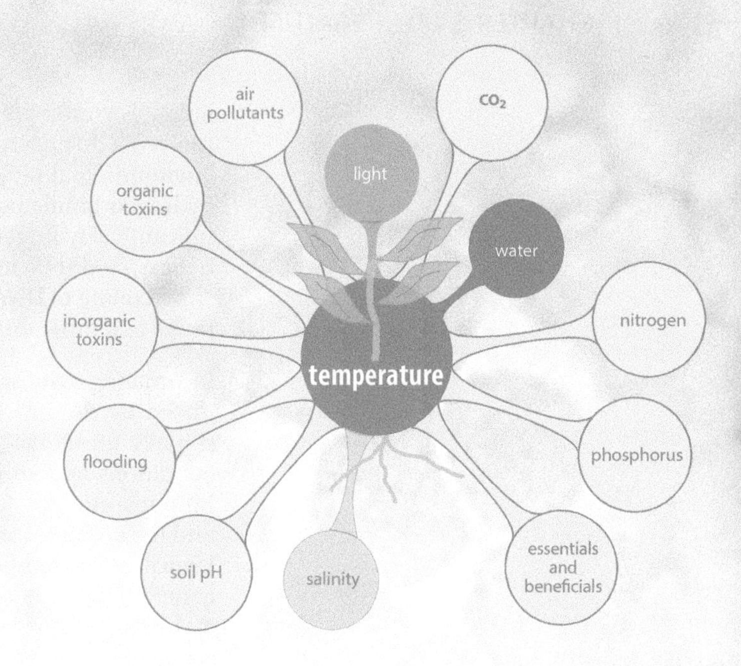

The challenge of variation in temperature interacts, in particular, with variation in water, light, and salinity levels.

Key concepts

- Plants are static poikilotherms that are adapted to significant variation in temperature.
- Plant growth and distribution are being affected by changes in the Earth's temperature.
- Plants detect temperature variation via physical changes in biomolecules.
- Temperature variation provokes canonical signaling pathways in plants.
- Some plants can acclimate physiologically to progressive changes in temperature.
- Modest temperature variation can initiate changes in membrane lipid composition.
- Some plants can use cryoprotectants and osmoprotectants at freezing temperatures.
- Some plants can reverse the denaturing of proteins that is caused by high temperatures.
- Morphological adaptations aid the growth of some plants at extreme temperatures.
- Some plants can adapt phenologically to chronic temperature variation.

Plants are static poikilotherms, so significant variation in temperature is a considerable challenge

Life depends on the interaction of molecules. Most abiotic variables that affect plant growth do so indirectly via changes in the chemistry of the environment that make it difficult to optimize internal chemical conditions for molecular interactions. In contrast, temperature changes affect organisms directly because they change the physical properties of molecules and hence their interactions. Although in the range of temperatures found in terrestrial environments some types of chemical bond, and therefore molecular properties, are affected more than others, temperature affects the properties of a wide range of molecules, including nucleic acids, proteins, and lipids. For example, the phosphate-ester bonds of the backbone of a DNA molecule do not change significantly in strength over the range of terrestrial environmental temperatures that are encountered, but the strength of the hydrogen bonds between the **pyrimidine** and **purine** bases is altered significantly. Hydrogen bonds between bases determine the geometry of DNA helices, and thus molecular **topology**. The torsional tension, bending, and supercoiling of DNA, which affect its function, change significantly over the range of temperatures found in terrestrial environments.

In proteins, **conformation** significantly affects function. Proteins are produced by the process of translation as linear sequences of amino acids. Initially these can form random coils, but a combination of temperature conditions and amino-acid sequence determines their final conformation. Proteins attain a functional conformation within an "energy landscape," and the amount of available energy is significantly affected by temperature. Hence, at the temperature range in which they originally evolved, proteins generally assume a functional conformation spontaneously, not necessarily by a set sequence of events, but because of minima in the energy landscape. Around such minima, minor variations in temperature can have minor effects on conformation and hence affect, for example, rate of action. However, there are also thresholds in the energy landscape at which there can be changes in conformation, such as aggregation or unfolding, which render the protein non-functional (Figure 8.1). Changes in the rate of action and in the conformation of most proteins occur over temperature ranges that are well within those experienced on the surface of the Earth. Minor conformational changes can be reversible, but dramatic changes such as aggregation have a low probability of reversal. For lipids, interactions between molecules determine membrane properties, which are consequently sensitive to temperature. Not only are membrane properties such as viscosity, fluidity (disordered vs. ordered organization), and permeability affected, and hence processes such as **cytosis**, but also the function of molecules that are embedded in membranes, mostly proteins. The physical properties of nucleic acids, proteins, and lipids are fundamental to life, and they are significantly affected by temperature.

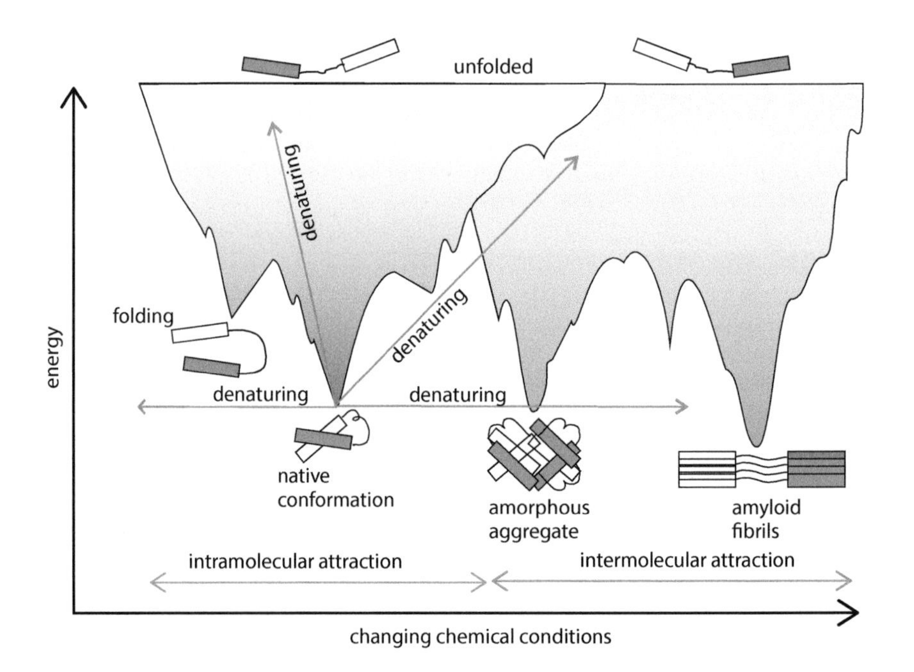

Figure 8.1. Conceptual energy landscape for protein conformation. Protein conformation depends on energy minima. Under optimal conditions a native, functional conformation is most likely to occur. Chemical conditions in the cytosol can affect the likely minima, and therefore conformation, but so can temperature because of its effects on available energy. Changes away from a functional conformation are described as "denaturation," and depend on changes in intra- and intermolecular attraction. Chemical conditions in the cytosol are subject to homeostatic control, so in plants changes in temperature can be a major cause of protein denaturation. (Adapted from Hartl FU & Mayer-Hartl M [2009] *Nat Struct Mol Biol* 16:574–581. With permission from Macmillan Publishers Ltd.)

Almost all organisms on Earth are **poikilotherms**, and tend to have their distributions restricted by where and when environmental temperatures confer appropriate properties on their biomolecules. **Thermogenesis**, which is common only in mammals and birds, which are **homeotherms**, is also found in some plants (Box 8.1), including those with the largest inflorescences, namely *Amorphophallus titanum* (Figure 8.2) and *Corypha umbraculifera*. However, thermogenic plants are exceptional, as almost all plants in terrestrial ecosystems are **ectotherms**. Most poikilothermic animals either inhabit environments in which temperatures do not vary greatly, or move if the temperature changes. Land plants are by far the largest poikilotherms, and because of their static habit and large surface area of interaction with the atmosphere, they face a particular challenge in either withstanding temperature changes or regulating them by gaining and losing heat.

Temperatures on the terrestrial surface of Earth vary dramatically both spatially and temporally. There are differences in mean temperature of almost 60°C with latitude and of 40°C with the altitude that occur on the Earth's surface. In some locations—for example, deserts at high altitude—diurnal temperature changes can be as much as 30°C (Figure 8.3a). Variation of this magnitude is well in excess of that which can have dramatic effects on the properties of biomolecules. Temperatures on the Earth's surface are affected by land mass and global climatic variables, so in the temperate latitudes there are pronounced differences in temperature regime between oceanic and continental climates, and in the tropics there are significant differences between wet and dry climates (Figure 8.3b). Near-surface air temperature tends to lag behind insolation rates, so in a daily cycle with maximum insolation occurring at 12 noon, maximum air temperatures can lag 3 h behind. Such lags are also characteristic of many seasonal climates (Figure 8.3b). For any given location on Earth, diurnal temperature can regularly change by 10–20°C (that is, up to 30% of the total variation in temperature found with latitude). Given the temperature sensitivity of biomolecules and the fact that plants are static ectotherms, temperature is a highly variable aspect of the environment, both within and between habitats, and is a significant challenge to plant life.

BOX 8.1. THERMOGENESIS IN PLANTS

Thermogenesis occurs in 14 families of plants (out of about 425 families of angiosperms, conifers, and cycads), but is restricted to cycad cones and primitive flowers, mainly in the Araceae. In most instances it is associated with attracting pollinators, either directly or by increasing the volatilization of chemical attractants. In species such as *Nelumbo nucifera*, *Philodendron selloum*, and *Symplocarpus foetidus*, thermogenesis is used for floral **thermoregulation**. It helps to control the floral environment for trapped pollinators, and perhaps decreases the effects of cold temperatures on the flowers. It also occurs in the world's largest flower, *Rafflesia arnoldii*. Thermogenic plants use the **cyanide-resistant respiratory pathway**, which is based on the **alternative oxidase**, to generate heat.

Changing global temperature regimes are affecting plant growth, development, and distribution

Plants have optimal temperature ranges for growth, which are often quite narrow. These optimal ranges can differ significantly, not only between species but also between subspecies, populations, and cultivars. Models of the effects of temperature on plant growth rate, often based on temperature differences from optima, have long been used to predict crop yields, and are now being used to predict the effects of changing temperature regimes on plants. In general, the negative effects of supra-optimal temperatures on growth rate accelerate more severely than those of sub-optimal temperatures, at

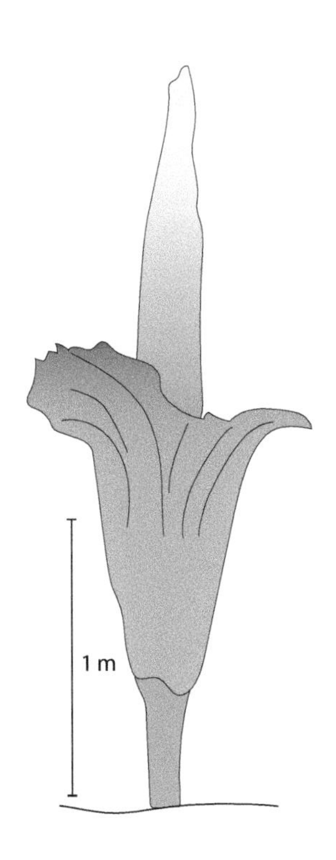

Figure 8.2. ***Amorphophallus titanum* (the titan arum).** The titan arum, native to Sumatra, is one of several species in the Araceae that are thermogenic. The massive inflorescence (shown in the figure) has a central spadix that emerges from a spathe. The flowers are arranged in rings at the bottom of the spadix, protected by the spathe. The spadix produces a carrion-like odor that attracts carrion-feeding insects, and it can reach a temperature of about 35°C. The high temperature of the spathe is thought to aid both volatilization of odiferous chemicals and the creation of the illusion of warm carrion. The massive inflorescence emerges from a corm that often weighs around 50 kg, and sometimes up to 100 kg—the largest known structure of this kind.

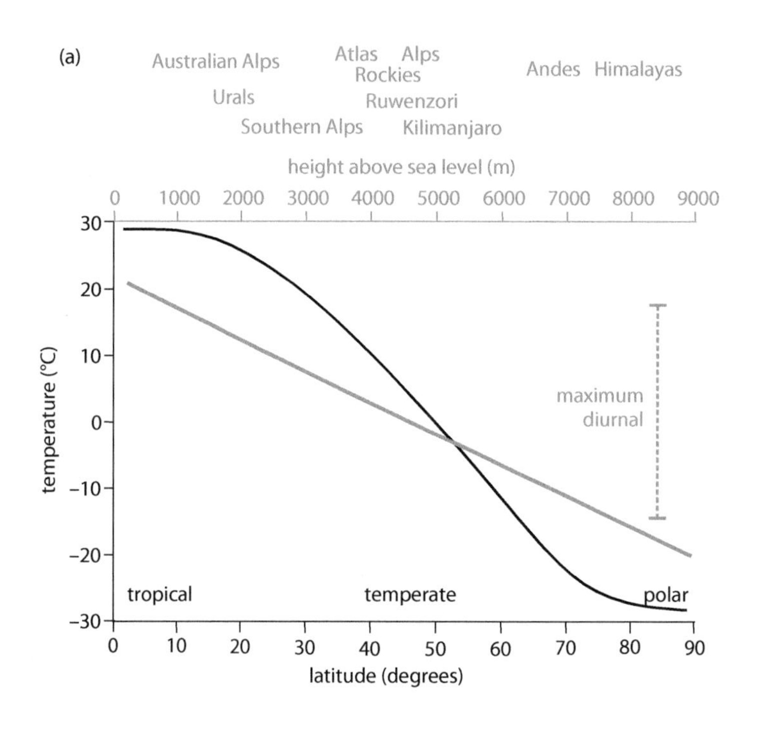

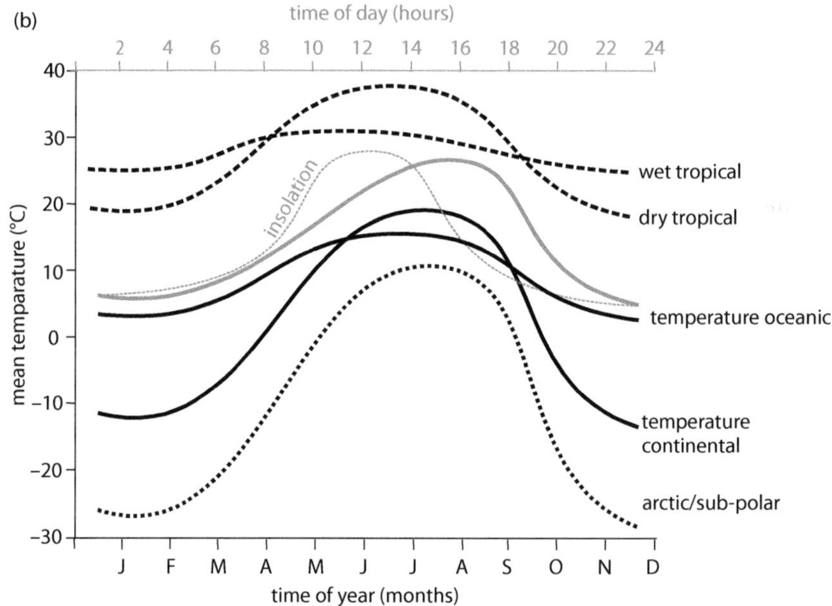

Figure 8.3. Temperature variation on the Earth's terrestrial surface. (a) At most latitudes, mean temperatures decrease by about 1°C for each 150 km (0.75°) towards the pole (black line). Mean temperatures decrease by about 1°C for each 220 m increase in altitude (green line) which, given the height of the mountain ranges shown, affects temperature almost as much as latitude does. Maximum diurnal temperature variation recorded in tropical alpine ecosystems is about 35°C (broken line). (b) The differences in the annual temperature regimes of different climatic zones on Earth. Diurnal temperature variation (shown in green for a summer day in the temperate zone) can be a significant proportion of the annual mean variation. Both annual and diurnal temperature variation lag behind insolation rate.

which the relationship is more sigmoidal (Figure 8.4). From such relationships calculation of, for example, growing degree units (GDU) can be made for days 1 to n as follows:

$$\mathrm{GDU} = \sum_{d=1}^{n} \left(\frac{T_{max} + T_{min}}{2} \right) - T_{base}$$

where T_{max} is the maximum ambient temperature, T_{min} is the minimum ambient temperature, and T_{base} is the lowest temperature at which the plant will grow. When days below T_{base} are counted as zero, this in effect calculates the accumulated heat. In many instances the use of GDU necessitates capping T_{max} at the beginning of the optimal range, and assuming that no temperature extremes occur and that relationships are linear. In ecosystems in which other factors are not limiting, including many managed ecosystems and some unmanaged ones, calculation of GDU can provide accurate estimates of growth. Even in ecosystems with other limiting factors, thresholds—such as the temperature at which growth ceases—can be a determinant of ecosystem processes. For example, in most places the treeline characteristic of high altitudes and high latitudes is primarily determined by temperature, which when low enough prevents the production of new tree growth sooner than that of herbaceous plants. In many climates, temperature is used as a developmental cue by plants. For example, **stratification** is essential for the germination of the seeds of many temperate species, and in many species flowering often occurs only after a particular number of GDU. These well-established phenomena demonstrate that temperature is a key determinant of plant growth and can have threshold effects on key life events in plants. They also indicate that extreme temperature events might have dramatic effects. Changing global temperatures and increasing frequency of extreme temperature events can therefore affect plant growth and the timing of life events (**phenology**) in plants, sometimes quite dramatically.

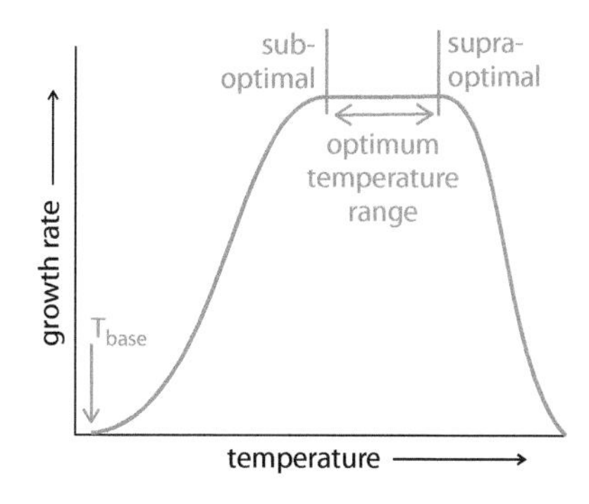

Figure 8.4. The relationship between plant growth rate and temperature. T_{base} is the minimum temperature necessary for growth to occur. The optimum temperature range for growth is sometimes quite narrow. Supra-optimal temperatures often decrease growth more rapidly than sub-optimal temperatures. For a particular species, subspecies, population, or cultivar, any or all parts of the curve may have different absolute values.

Using an estimate of anthropogenic increases in global temperature of 0.7°C, temperature-induced decreases in global yield of major crop plants were being calculated by the early years of the twenty-first century. There is now detailed evidence—for example, in rice—that increasing temperature is decreasing yield in some regions. Interactions between yield and temperature are complex, but in the case of rice increases in minimum temperature can decrease yields more than increases in maximum temperature. It has also been estimated that between 1980 and 2008, increasing global temperature decreased wheat yield by 5.5% and maize yield by 3.8%. Many of these effects are caused by changes not only in growth but also in events such as **anthesis**—with increasing temperature, growth rate may increase, but earlier anthesis results in less fixed C in total. Multi-site meta-analyses of data from the International Tundra Experiment (ITEX) indicate that, in many tundra ecosystems, increased temperature affects not only plant traits such as growth rate and phenology, but also, as a consequence, community composition and ecosystem processes. Many studies of alpine ecosystems are also reporting increases in the elevation of plant species. Overall, changes in mean temperatures are currently affecting both managed and unmanaged ecosystems.

If aspects of plant phenology reflect temperature then they can be used to monitor it. Using biological phenomena to monitor an environmental variable is complex, and is susceptible not only to variations in other influencing factors but also to extreme events. Nevertheless, it can provide a useful bio-indicator of variation. In North Dakota, on the great plains of the USA, first flowering dates (FFDs) of 178 species of prairie plant were recorded from 1910 to 1960, and have been compared with FFDs for the same species in the same location between 2007 and 2010. In Korea, FFDs have been recorded since 1922 in a phenological garden inside Seoul meteorological station. In Concord, Massachusetts, in the USA, Henry David Thoreau recorded phenological events for 473 species, starting in 1852, which have provided the baseline for modern comparisons (Box 8.2). In Geneva, Switzerland, records of the leafing date of chestnut go back to 1808. In England, phenological records, including many for plants, on the Marsham estate in Norfolk go back to 1736, and have contributed to a detailed 250-year phenological record. In Burgundy, France, grape harvest dates have been recorded since 1372, and in many other areas of Europe there are also long-term data for grape harvests. In Kyoto, Japan, the flowering date of the cherry *Prunus jamasakura* is known for 732 of the years since (and including) the oldest in 812 AD. The use of some of these records to reconstruct past temperatures can be contentious, mainly because of the high number of potentially confounding variables, but there is a strong indication of warming in recent decades.

BOX 8.2. PHYLOGENY AND CHANGES IN THE FLORA OF THOREAU'S WOODS

Henry David Thoreau was the author of *Walden*, a classic of self-sufficiency and self-reliance through which late-nineteenth-century American Romanticism continues to influence current environmental debate. The 2 years that Thoreau spent in a cabin near Walden Pond in Concorde, Massachusetts, initiated a detailed phenological record of 473 species in woods that he visited regularly between 1852 and 1858. Subsequent recording in the same woods during 1888–1902 and 2003–2007 has been used to describe changes to the woodland flora over a period of more than 150 years.

Around 25% of the species recorded by Thoreau are now locally extinct, and about one-third of them are threatened with extinction. The recorded temperature rise of 2.4°C at Concorde since the very early years of the twentieth century is associated with changed flowering patterns of extant species. The changing abundance of species is linked to evolutionary history (phylogeny)—that is, closely related species have been adversely affected or benefited similarly from changing temperatures (Figure 1). In particular, those families of plants that in general do not use temperature to initiate flowering have declined dramatically, and those that do use temperature in this way have continued to flower but at a slightly different time. Studies in the woods near Concorde first described in detail by Thoreau provide a unique insight into the effects that changing temperature can have on local ecosystems. They also serve as a reminder that not all species react similarly to changing temperatures, and that these responses can at least in part be predicted by phylogeny. It is significant that the adverse effects are occurring in many of the species that are regarded as characteristic of New England, and in woods that have an important role in provoking fundamental debate about human societies and their effect on the environment.

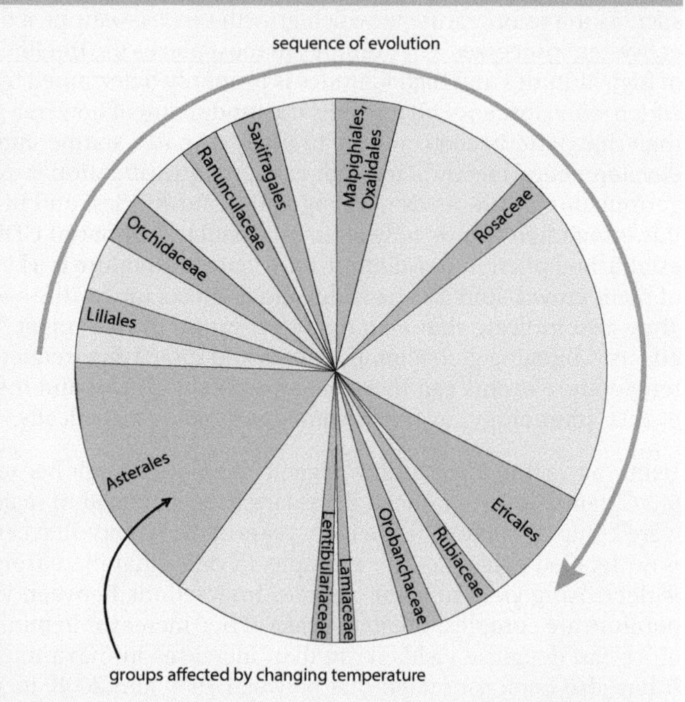

Figure 1. In Thoreau's woods some groups of plants have not been affected by increasing temperature (light green areas), but many other groups have been affected (darker green areas). These differences in response have significant consequences for ecosystem interactions.

Further reading: Willis CG, Ruhfel B, Primack RB et al. [2008] Phylogenetic patterns of species loss in Thoreau's woods are driven by climate change. *Proc Natl Acad Sci USA* 105:17029–17033.

Understanding the effects of temperature on plants is vital for predicting the effects of changing global temperatures on agricultural and natural ecosystems, but is also useful for interpreting long-term changes in plant behavior.

Plants detect temperature changes via physical changes in numerous biomolecules

The detection of temperature depends not on receptors for chemicals but on physical changes in particular biomolecules, so because the whole plant body tracks environmental temperature, a wide variety of molecules are potential thermosensors. In plants, classical experiments revealed that many responses to temperature shock can be induced by changes in the conformation of proteins, especially transcription factors (TFs). For example, temperature-induced changes in the conformation of CBF TFs can increase their binding activity as temperature decreases. Eukaryotic cells also constitutively express **heat stress transcription factors** (HSFs) that can bind to **heat stress-responsive elements** (HSEs) in the promoters of genes such as **heat stress-responsive proteins** (HSPs). High temperature affects the conformation of HSFs and hence their binding to each other, to other proteins, and to HSEs. Thus conformational changes in key TFs can change gene expression in response to temperature shock. Furthermore,

Table 8.1. Mechanisms of temperature sensing in plants

Biomolecule	Physical changes
Membrane lipids	Membrane state, membrane–protein interactions, activity of ion channels Rate of production of reducing/oxidizing metabolites in plastids
Chromatin	Binding of proteins to DNA
RNA	Folding of transcripts Spliceosome interaction with pre-mRNAs
Proteins	Protein stability Translation kinetics Enzyme activity Protein chaperone activity

(Data from McClung CR & Davis SJ [2010] *Curr Biol* 20:R1086–R1092.)

temperature-induced conformational changes in enzymes affect their activity differently, producing different **Q_{10} values**, so the ratio of metabolites changes with temperature. For example, the proportion of phosphorylated intermediates in photosynthesis can be affected in this way. However, significant temperature change, even shock, is necessary to trigger these changes. It has been suggested that physical changes in a variety of other biomolecules are also involved in temperature detection (Table 8.1), because plants can detect more subtle changes in temperature than those that trigger "temperature shock" responses.

RNA was one of the first biomolecules to evolve. Its structure, translation, and, in eukaryotes, splicing are all subject to temperature-dependent effects. For example, in *Listeria*, thermosensitive virulence is caused via 5′-untranslated regions of mRNA that form hairpins, which only dissociate at 37°C. Dissociation of hairpins allows entry to ribosomes and the translation of virulence genes during the infection of organisms with this body temperature. In eukaryotes there is evidence, including data from *Zea mays* and *Arabidopsis*, that **internal ribosomal entry segments** on mRNA act as heat-sensitive regulators of translation. In plants, patterns of **alternative splicing** of RNA are also affected by temperature. In *Synechocystis*, a cyanobacterium related to those that evolved into chloroplasts, *desB* encodes a key membrane lipid **desaturase** with strongly cold-induced expression. Cold increases the **supercoiling** of DNA in *Synechocystis*, which increases the transcription of *desB*, initiating pathways that change the saturation of membrane lipids. It was probably necessary for the early evolution of life that nucleic-acid-dependent processes were subject to temperature control, and they still have a role in temperature detection in plant cells via both nuclear and plastidic nucleic acids.

In eukaryotes, DNA occurs as **chromatin,** which has **nucleosomes** formed of DNA wrapped around **histone** proteins. In most eukaryotes, four histones occur (H2A, H2B, H3, and H4), and in *Arabidopsis*, as in a range of other organisms, the presence of the H2A.Z variant alters gene expression. H2A.Z is inserted into *Arabidopsis* DNA by the **SWR1 complex**. As the temperature increases so does the ratio of H2A to H2A.Z in the chromatin. Mutations in *ARP6*, which encodes part of the SWR1 complex, cause plants to have decreased H2A.Z at ambient temperature and the attributes of plants grown at warm temperatures. Over the 17–27°C range, the temperature-induced increase in H2A:H2A.Z ratio correlates with increasing expression of the heat shock protein HSP70, which has a long-established role in heat responses in plants. H2A.Z blocks transcription because it binds DNA more tightly in nucleosomes than does H2A. Temperature-induced increases in transcription increase the expression of the HSP70 promoter (Figure 8.5).

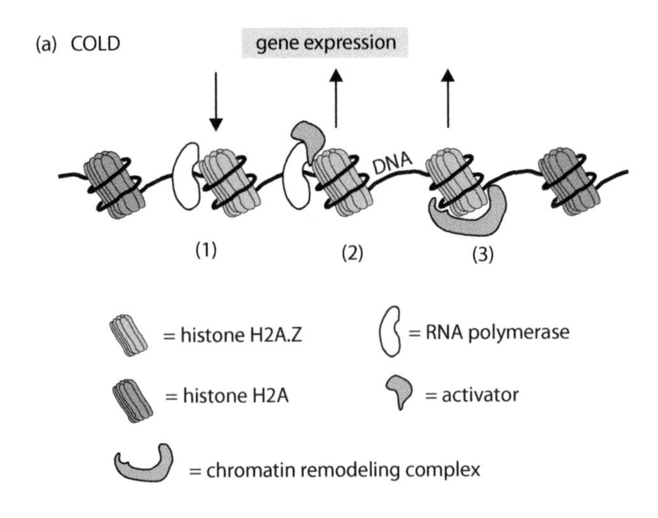

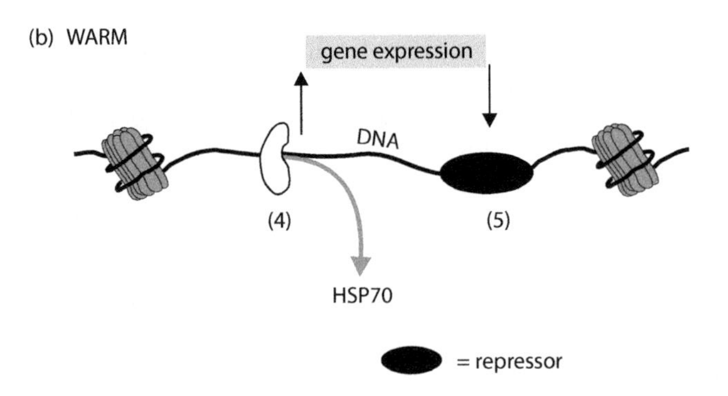

Figure 8.5. **A chromatin-based thermometer in plants.** DNA is wrapped around histone proteins to form the chromatin of the nucleus. (a) At low temperatures, H2A.Z histones occur at a relatively high ratio to H2A histones, but at higher temperatures they leave the chromatin. At low temperatures the action of RNA polymerase can be blocked by H2A.Z (1), the blocking can be relieved by activators (2), or a chromatin remodeling complex can allow access to genes for transcription despite the presence of H2A.Z (3). (b) As the temperature increases and H2A.Z leaves the chromosome, HSP70 and other genes are expressed (4), while the expression of other genes is decreased by the binding of repressors (5). These temperature-dependent, H2A.Z-associated changes in gene expression, especially of HSP70, can act as a "thermometer" for plants.

This chromatin detector of temperature also initiates changes in expression of flowering time repressor (*FLC*) and flowering locus T (*FT*) proteins, through which temperature effects on flowering time are mediated in *Arabidopsis*. In animal systems, HSF proteins ensure that changes to conformation only occur in histones associated with heat-induced genes, but none of the numerous HSFs in plants have yet been shown to have this role. Following heating, histones are changed in many ways that alter gene expression, including phosphorylation, glucosylation, acetylation, and methylation, but this seems unlikely to act as a primary mechanism of thermodetection.

The cell membrane has a fundamental role in separating the conditions for metabolism from those of the environment. It was perhaps the site of the first adaptations to environmental conditions. Membrane function, which necessitates both a separation from the environment and also simultaneously an interaction with it, is predicated on its fluidity. Some plant species have been shown to detect changes in temperature of only 1°C. Membranes are very sensitive to changes in environmental conditions, and this small change can affect membrane fluidity. In plants, membrane fluidizers and membrane rigidifiers initiate many of the physiological effects of heat and cold, respectively, as do genetic manipulations of membrane fluidity, which suggests that it is a key physical property for primary temperature sensing. Mutations in **fatty acid desaturase** (FAD) have revealed that cold-induced changes in the **phospholipase** C pathway are initiated by decreased fluidity. *fad* mutants also enhance the cold-induced proteolysis of a plasma-membrane-bound NAC transcription factor. One of the earliest responses in plant cells to either increased or decreased temperature is an influx of Ca^{2+} from the apoplast. Molecular manipulations of this influx show that it is a key part of the response to temperature. Experiments with the moss *Physcomitrella* indicate that changes in Ca^{2+} influx are among the earliest signals of temperature response. In animal cells, temperature-sensitive Ca channels are responsible for such influxes.

The plant **cytoskeleton** links cell architecture to intracellular signaling. Changes in temperature affect polymerization of the cytoskeleton and hence cellular architecture. Such changes, induced either by temperature or by chemical and molecular manipulation, alter the expression of a membrane-bound heat-activated MAP kinase (HAMK) that is linked to heat-signaling networks. In *Synechocystis* a membrane-bound histidine kinase (Hik33) is vital to cold responses. Artificial stabilization of the cytoskeleton prevents key parts of the heat response, such as the accumulation of HSP70. As plants are poikilotherms, changes in membrane and cytoskeleton properties are likely to be among the early cues of changing temperature. Signaling is likely to be initiated by membrane-bound components such as Ca channels and kinases. It is not yet clear, however, how Ca^{2+} influxes might specifically signal changes in temperature as opposed to other environmental variables, or how such temperature-related effects are integrated with those that occur in nucleosomes. The plant plasma membrane is spatially complex, as are its connections with the cytoskeleton. The mechanistic details of the interactions between the nucleic acid, plasma membrane, cytoskeleton, and protein thermosensors in plants have yet to be elucidated, but it seems clear that temperature is sensed by multiple mechanisms in plants—as might be expected of an environmental variable that affects the physical properties of so many biomolecules.

Chilling, freezing, and heat initiate changes in key components of different signaling pathways

Rapid changes in Ca concentrations in cells are integral to many signaling pathways, and can provide pathway-specific signals. They are involved in plant responses to chilling, freezing, and heat, but act as a common component of three distinct response pathways. In *Physcomitrella*, Ca^{2+} derived from internal stores does not initiate temperature responses, and neither do delayed influxes from the apoplast. Although it is not clear how this occurs, especially for chilling responses, in *Physcomitrella* at least, immediate cold-induced fluxes of Ca^{2+} from the apoplast include temperature-specific information. Once initiated, these changes in Ca concentration are transduced into distinct chilling and freezing response pathways. In higher plants responding to heat, the binding of HSFs to HSEs is affected by Ca concentration and, as in many other organisms, **calmodulins** (CaMs) probably have a role in interpreting heat-induced Ca fluxes. The expression of a number of CaMs is increased by heat, and the addition of CaM or of CaM antagonists affects the plant response to heat. CaM3 overexpression increases the expression of HSPs, and in *cam3* mutants their expression is decreased. In particular, CaM3 and CaM7 seem most likely to mediate signals in heat-induced Ca^{2+} influx. In *Arabidopsis* these calmodulins affect the phosphorylation of HSFA1a TFs and thus their binding to HSEs. Phosphorylation occurs via CaM-binding protein kinase 3 (AtCBK3), the manipulation of which also affects temperature responses. This suggests that, after the detection of heat, CaM-kinase pathways are involved in transducing the signal via phosphorylation of HSFs.

Other canonical components of signaling pathways in plants, in particular H_2O_2 and NO, are also involved in heat, chilling, and freezing responses. Membrane fluidizers and changes in temperature induce rapid changes in H_2O_2 concentrations in plant cells. The application of H_2O_2, or scavengers of it, affects the binding of HSFs to HSEs and thus the expression of some HSPs. *noa* mutants that produce reduced amounts of NO are less tolerant of increased temperature than WT plants, and show decreased accumulation of some HSPs and weaker binding of HSFs to HSEs. In particular, NO concentrations are related to CaM3 concentrations, and manipulation of the expression of the latter can affect the thermotolerance of *noa* mutants. Thus, although temperature detection in plants does not occur via a specific receptor, temperature changes are transduced into signaling pathways that are used by plants to adjust to many other changes in the environment.

In some plants, chilling temperatures can induce an acclimation response based on the CBF regulon

Many plant species, especially those of the temperate zone, can undergo cold **acclimation**, whereby a short period of moderate, sublethal low temperature increases their capacity to cope with subsequent cold, or even freezing, temperatures. Many plants that inhabit the tropical zone cannot acclimate to cold and are killed even by chilling (that is, exposure to temperatures below optimum but above freezing). The temperatures that trigger cold acclimation can vary significantly between species, and different parts of the same plant can acclimate differently. In those plants in which genome-wide studies are possible, cold exposure of even a few degrees below optimum growth temperature induces significant changes of the transcriptome. In *Arabidopsis,* an abrupt temperature drop of 10°C alters the expression of over 1000 genes in waves of expression over about 24 h. These changes help plants to respond to the immediate cold, but often also increase their tolerance of subsequent cold periods. Cold acclimation involves significant cold-stress-specific changes to membrane and cytoskeleton components. It also involves changes characteristic of other stresses—for example, a significant increase in the plant's capacity to deal with oxidative stress, including protection against photo-oxidative stress via the synthesis of protective pigments. There is also a notable increase in the concentration of polyamines in cold-acclimated plants, which probably plays a role not only in oxidative protection but also in controlling signaling pathways. Some of the genes with altered expression during cold (*COR* genes) also respond to other environmental changes—for example, some *COR* genes are up-regulated by abscisic acid (ABA) or drought (Figure 8.6). The evolution of the temperature acclimation response was probably driven by the progressive decreases

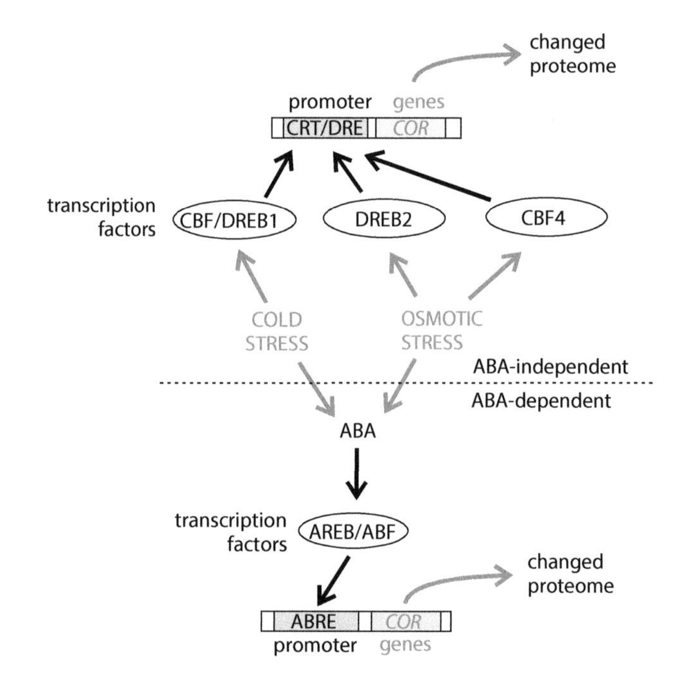

Figure 8.6. **Abscisic acid (ABA)-independent and ABA-dependent cold response pathways.** Many cold-regulated (*COR*) genes respond via ABA-independent pathways, but some are ABA dependent, and many genes have both CRT and ABRE regions in their promoters. Osmotic stress can also induce the expression of a proportion of *COR* genes via both ABA-dependent and ABA-independent pathways.

in temperature that characterize the approach of winter in the temperate zone. The freezing temperatures of winter can cause drought stress to plants, so the chilling-induced expression of genes involved in drought tolerance helps to prepare plants for freezing temperatures.

COR genes have been characterized in a wide range of plants. Three C-repeat binding factor TFs (CBF1, CBF2, and CBF3) that bind to the C-repeat motif, which has a 5-bp core sequence of CCGAC, in the promoters of many *COR* genes (Figure 8.7) show increased abundance shortly after exposure to cold. Together, these TFs and *COR* genes constitute the CBF regulon, which in *Arabidopsis* has about 100 genes. There are CBF homologs in a variety of wild and domesticated plants, including monocots and eudicots, annuals and perennials, woody and non-woody trees. The ectopic expression of *Arabidopsis* CBFs in other species, and vice versa, increases cold tolerance. A **quantitative trait locus** (QTL) that has one of the greatest effects on freezing tolerance in wheat and barley (Fr-A^{m}2/Fr-H2) includes *CBF* genes. Allelic variation in barley *CBF* genes across European accessions of cultivated and wild *Hordeum* helps to explain frost tolerance. The CBF regulon therefore plays an important role in, and is the best characterized molecular pathway of, plant response to cold. CBFs are a subfamily of the AP2/EREBP family of TFs, and are characterized by unique sequences that are highly conserved either side of their AP2 DNA-binding domain. These unique sequences confer binding specificity for the C-repeat motif. CBF1 and CBF2 activity increases within 1 h of cooling, whereas CBF2 activity takes about three times as long.

CBF3 expression is induced by a bHLH TF inducer of CBF expression (ICE), which binds to a promoter region upstream of *CBF3*. For *CBF2*, a number of promoter elements (CM1–7), some of which have similarity to ICE1, are necessary for expression. There is probably a network of promoter elements involved in the cold-induced expression of CBFs. The activity of ICE proteins is controlled via the SIZ1/HOS1 pathway. Sumoylation of ICE1 is

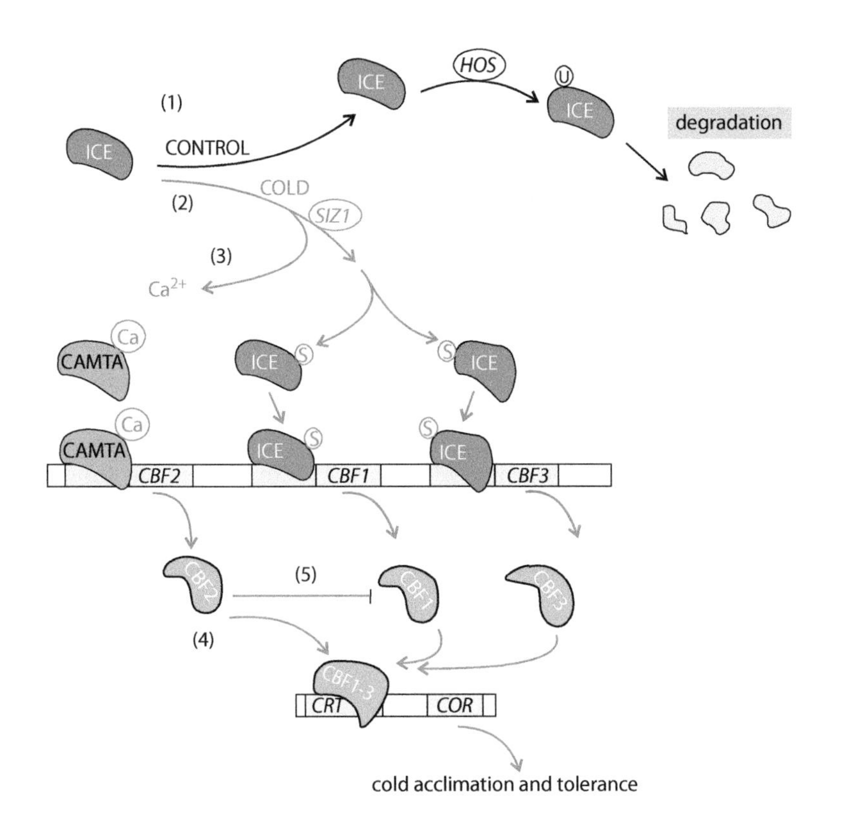

Figure 8.7. The CBF regulon. (1) In control conditions, HOS ubiquitinates ICE proteins and they are degraded. (2) In cold conditions, SIZ1 sumoylation of ICE proteins increases and they bind to ice box elements, increasing the production of CBF1 and CBF3 TFs and the expression of *COR* genes. (3) Cold also increases Ca^{2+} concentrations, activating CAMTA transcription factors for CBF2. (4) Together with CBF1 and CBF3, CBF2 can increase the expression of many *COR* genes. (5) In some circumstances, however, CBF2 blocks the activity of CBF1 and CBF3, attenuating the cold response. ICE = inducer of CBF expression.

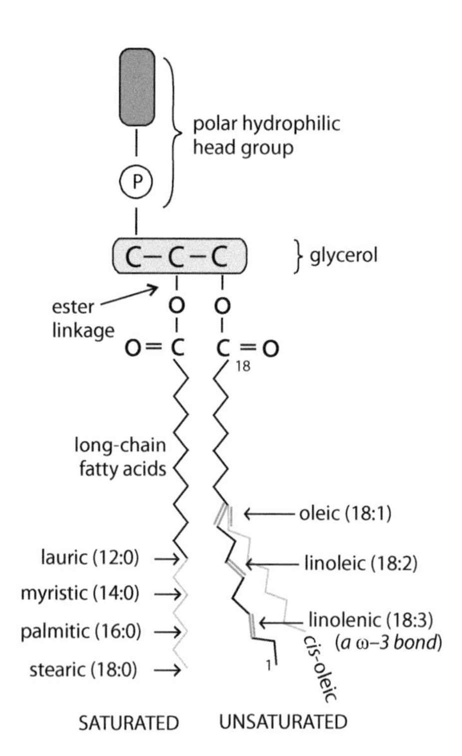

Figure 8.8. The glycerolipids of plant membranes. Phospholipids have different molecules (for example, choline or serine) attached via a phosphate group to form a polar hydrophilic head group. In the phospholipid shown in the figure, an attached glycerol is linked to two fatty acids to form a diacylglycerol. Galactolipids have one or two galactose molecules bonded directly to glycerol with no phosphate link, and the remaining sites are occupied by fatty acids. Sphingolipids have a polar hydrophilic head group, one fatty acid, and one sphingosine attached to glycerol. The fatty acids attached in glycerolipids can be saturated (that is, they contain no double bonds) or unsaturated (that is, they contain 1, 2, or 3 double bonds). Fatty acids with double bonds on the third C atom from the methyl end are ω-3 fatty acids. Some unsaturated fatty acids can have *cis-* and *trans-*double-bond isomers. Fatty acids are hydrophobic, but the head group is hydrophilic and helps to form lipid bilayers in the presence of water. Different fatty acids confer different membrane properties at different temperatures. (The numbers in parentheses indicate the number of C atoms and the number of double bonds in the chain.)

necessary for it to bind and activate *CBF3* expression. This is achieved under cold conditions (for example, 4°C) by SUMO E3 ligase (SIZ1). *HOS1* is constitutively expressed, and at optimum temperatures in *Arabidopsis* it targets ICE for ubiquitination and degradation. Sumoylation under cold conditions reverses the effects of ubiquitination and results in the expression of *CBF3*. In the promoter region for *CBF2* there are binding sites for calmodulin-binding transcription activator (CAMTA) proteins, which are necessary for expression and to confer cold tolerance. Decreases in temperature increase Ca^{2+} fluxes into cells, and hence increase the activity of CAMTA. Thus the CBF regulon allows some plants to adjust gene expression to fluctuations in temperature below the optimum level. In *Arabidopsis* and some other species, the *CBF* genes occur in sequence in a tandem repeat, and a number of *HO1S* and *CAMTA* genes or their close homologs are present. The details of the CBF regulon and its expression across all plants are likely to vary, but it is probably an important nexus of cold acclimation for plants that are capable of this. Across numerous *Arabidopsis* accessions, *CBF* expression only partly explains cold tolerance, and not all *COR* genes have CBF TFs, highlighting the importance of other pathways in the cold response. The well-known *esk1* (*eskimo*) mutant of *Arabidopsis* is freezing tolerant, and a number of *hos* mutants with altered cold tolerance are known, but their CBF regulons are all normal.

Adaptation to non-optimal temperature necessitates maintaining membranes in the liquid–crystal state

Plant membrane lipids, although varying significantly in composition between species, are primarily composed of two types of **glycerolipid**—the phospholipids and the **galactolipids**—together with some **sphingolipids** and a mixture of sterols (Figure 8.8). Glycerolipids have long-chain fatty acids ester-linked to a glycerol molecule. Triacylglycerides (TAGs) have fatty acids on all three of the C atoms in glycerol, and are extensively used for lipid storage in plants. Membrane lipids are diacylglycerides (DAGs), in which one C atom is linked not to a fatty acid but to a hydrophilic head group, in phospholipids via a PO_4^{2-} and in galactolipids directly to one or more galactose molecules. Fatty acids in liquid and solid form are commonly referred to as oils and fats, respectively. Saturated fatty acids contain no double bonds, whereas unsaturated fatty acids have double bonds at positions measured from the methyl end—hence ω-3 ("omega-3") oils have a double bond on the third C atom from the methyl end. In plant membranes, fatty acids in DAGs most commonly have 16 and 18 carbons, although fatty acids with 12, 14, and 20 carbons do occur at high concentrations in certain species, especially in TAGs in seeds. Cold-acclimated plants frequently have an increased proportion of phospholipids in their membranes relative to all other membrane constituents, including proteins, and these phospholipids contain an increased proportion of unsaturated fatty acids. These changes help to prevent the adverse effects of decreased temperature on membrane properties, and some of the *COR* genes help to bring them about. Fully functional membranes occur in the liquid–crystalline state, which allows movement of membrane rafts and of embedded molecules such as proteins. Below its transition temperature (T_c), a pure fatty acid solidifies and behaves as a gel rather than as a liquid–crystal. Due to the many molecules embedded in them, biological membranes rarely transition between states, but at low temperatures gel-like properties severely inhibit membrane function, including the movement of membrane rafts and the functioning of embedded molecules. This can have adverse effects on a significant proportion of all metabolic pathways, because many proteins and signaling molecules are membrane-bound.

Table 8.2. **Transition temperatures (T_c) of phosphatidylcholines (PCs)***

Biomolecule	Physical changes	T_c (°C)
Saturated	Distearoyl-PC (18:0, 18:0)	55
	Dipalmitoyl-PC (16:0,16:0)	41
	Dimyristoyl-PC (14:0, 14:0)	24
	Myristoyl, palmitoyl-PC (14:0, 16:0)	36
	Palmitoyl, myristoyl-PC (16:0, 14:0)	27
Unsaturated	Stearoyl, oleoyl-PC (18:0, 18:1)	6
	Stearoyl, linoleoyl-PC (18:0, 18:2)	–16
	Stearoyl, linolenoyl-PC (18:0, 18:3)	–13
	Palmitoyl, oleoyl-PC (16:0, 18:1)	–1
	Dioleoyl-PC (18:1, 18:1)	–19

* Values for pure artificial membranes (n:n denotes the number of C atoms in long-chain fatty acid, and the number of double bonds).

The degree of saturation of fatty acids in the phospholipids and galactolipids, and the position of their ester linkage, can significantly change T_c (Table 8.2). Importantly, significant changes in T_c are brought about not only by different fatty acids but also by *cis/trans* isomeric changes, and by the positions of the fatty acid in the DAG. In plants, FADs control the degree of saturation of fatty acids. Many of them are membrane-bound and activated by a temperature-induced change in membrane properties. In many plants, the range of FADs that control saturation of a range of membrane lipids is well known (Figure 8.9). In response to increased temperatures, decreased FAD activity can decrease the proportion of unsaturated membrane lipids, maintaining membrane properties. Mutants that are unable to synthesize FADs have increased cold sensitivity, as do plants that overexpress saturated fatty acids. A particularly strong correlation between changes in palmitic acid concentration and low temperature tolerance has been noted in some species, which is in accord with the change in fatty acid saturation that has the greatest impact on membrane structure. The activity of many FADs can change over short timescales and is strongly dependent on light, which helps to coordinate membrane properties with diurnal temperature fluctuations. Changes in temperature can also lead to an increase in oxidation of membrane lipids, mainly identifiable as an increase in malondialdehyde, which can alter membrane properties. There is some evidence that lipocalins—a ubiquitous and diverse family of proteins with lipid-binding capacity—help plants to prevent temperature-induced changes in the oxidation of membrane lipids.

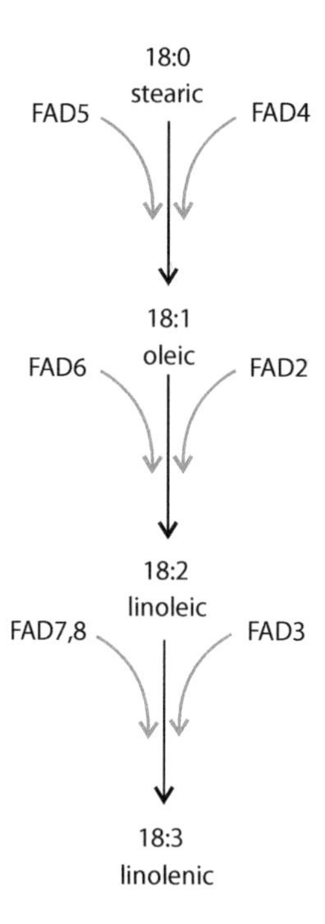

Figure 8.9. The production of unsaturated fatty acids by fatty acid desaturases (FADs). Different FADs are found in different cellular membrane systems, and they catalyze desaturation at different C atoms in long-chain fatty acids. The figure shows the FADs that desaturate stearic acid to linolenic acid. Plant species vary in their expression of different FADs, and hence in their ability to adjust their membrane fatty acid composition.

In several species, changes in the ratio of different classes of phospholipids have been shown to be important in temperature responses, with the ratio of phosphatidylcholine (PC) to phosphatidylethanolamine (PE) changing with temperature. In detailed analyses of *Arabidopsis*, many other classes of phospholipid, including phosphatidylserine and phosphatidylinositol, have been shown to have an important role in temperature tolerance. The replacement of fatty acids in phospholipids and changes in phospholipid class both require hydrolysis of phospholipids. Throughout the living world this process is catalyzed by **phospholipases** (PLs). A number of PLs that hydrolyze phospholipids at specific bonds to enable specific phospholipid changes are well characterized in plants (Figure 8.10). In *Arabidopsis*, the cold-enhanced activity of phospholipase D favours a decrease in PC relative to PE. PL activity is also necessary during changes in the ratio of 16C to 18C fatty acids. FAD/PL pathways are therefore intimately involved in

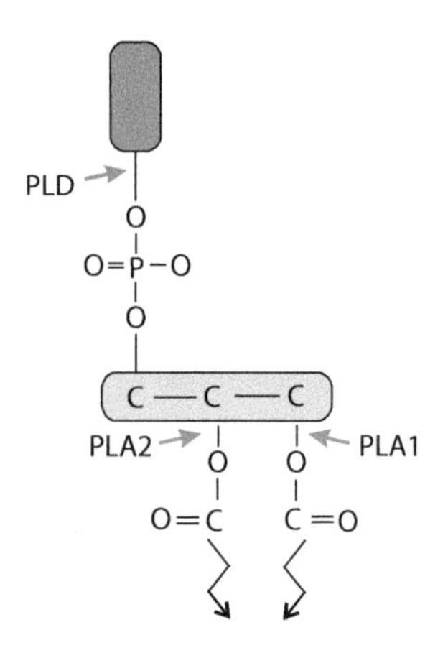

Figure 8.10. The action of phospholipases on membrane glycerolipids. The most common phospholipase (PL) in many plants is PLD, which is used to decouple the hydrophilic head of phospholipids. PLAs and PLBs can hydrolyze the ester bonds between fatty acids and glycerol. PLA1 and PLA2, like other PLs, act at specific bonds with glycerol. The action of PLs allows the fatty acid content of membrane lipids to be altered.

the management of membrane properties, and plants that can acclimate to temperature changes use them to optimize membrane properties. Probably because plants are poikilotherms and for much of the time are exposed to temperatures below those that many homeotherms maintain, their membranes are generally rich in unsaturated fatty acids and contain a wider variety of sphingolipids than do animal cells.

Increased concentrations of simple and complex sugars can also affect membrane properties. For example, in cold-acclimated plants the monosaccharide fructose and the disaccharide sucrose are found at increased concentrations within membranes, which help to maintain their fluidity. **Fructans**, which are polymers of fructose produced in the vacuole and used as a storage carbohydrate in many plants, are often produced in response to cold. In some plants, vesicle trafficking from the vacuole to the plasma membrane deposits fructan molecules between phospholipids, and this helps to maintain fluidity. In many plants, polyamines are synthesized from arginine and confer significant cold tolerance. Polyamines have multiple NH_2 groups and a significant positive charge. They can bind and help to stabilize the many negatively charged biomolecules in a cell, but their role in cold tolerance is not well known. In seed plants, lipid biosynthetic pathways were vital for the evolution of oil storage in seeds—a major factor in the success of seeds as a distribution strategy. The production and properties of seed oils are of great importance for the human diet and for biofuels. Furthermore, manipulation of plant lipid pathways that first evolved, in significant part, to counteract the effects of temperature fluctuation is increasingly important for the production of biofuels (Box 8.3).

Freezing-tolerant plants produce cryoprotectants and osmoprotectants

Temperatures below freezing occur at least at some time each year on about two-thirds of the Earth's surface, so plant responses to low temperatures are of significance to both managed and unmanaged ecosystems of the majority of the Earth's terrestrial ecosystems. In plants at sub-zero temperatures, extracellular apoplastic water freezes before intracellular water because it has a lower concentration of solutes. Apoplastic ice crystals can cause lethal physical damage if they lever cells apart or penetrate the cytoplasm. At sub-zero temperatures, ice often forms within freezing-tolerant plants, but its crystallization is controlled and often limited to extracellular spaces. In freezing-tolerant plants, for example, ice forms on their surface, in intercellular spaces, and in the xylem, but not in the cytoplasm. Between 0°C and -42°C water will only form ice if ice nucleators are present. If there are no ice nucleators, or if crystallization processes are inhibited, water can be supercooled in liquid form to -42°C. In a wide variety of freezing-tolerant organisms, ice nucleators are used to provide spatial control of freezing, and in freezing-tolerant plants they can be particularly important in restricting freezing to intercellular spaces. Ice nucleators are variable aggregates of proteins, lipids, and carbohydrates.

The **guttate** of freezing-tolerant plants does not freeze in the same manner as that of other plants, and it contains **antifreeze proteins** (AFPs). AFPs were first reported from Arctic fish, but occur in a wide variety of freezing-tolerant organisms, including plants. They are diverse, and evolved independently in different organisms, probably to a significant extent during the recent Quaternary glaciation (2.5 million to 10,000 years ago). The preceding Karoo glacial period was 360 to 270 million years ago, so many of the major extant groups of organisms, including flowering plants, evolved after it had ended, and have only recently been exposed to significant freezing stress. As the temperature decreases below zero, ice crystals form if nucleators are

BOX 8.3. CHANGING LIPID QUALITY FOR IMPROVED BIODIESEL AND FOR HUMAN HEALTH

Global production of vegetable oils increased sevenfold between 2000 and 2010. The production of these renewable high-calorific-value lipids, which have numerous industrial uses and are also important for human health, is likely to be vital to the development of environmentally benign economies. Vegetable oil crops for non-food use compete directly with food crops for high-quality agricultural land, and their fatty acid composition often restricts their industrial uses. There is therefore significant pressure to develop vegetable oil crops that can grow on marginal land and that have specific fatty-acid profiles.

Figure 1. *Jatropha curcas* (jatropha) produces seeds containing large quantities of oil that can be made into biodiesel. Several large-scale projects have used this product to run diesel engines and cars. The plant is native to the Americas and is a member of the Euphorbiaceae (spurge family).

Vegetable oils are predominantly triacylglycerides (TAGs) stored in lipid droplets derived from the endoplasmic reticulum. Fatty acids are primarily synthesized in plastids via polymerization of the 2-C acetyl CoA, and exported as oleic acid (18:1). Subsequent acyl editing produces the range of saturated and unsaturated fatty acids found in plants. The metabolic pathways underpinning these changes are quite well known, and have been manipulated to produce vegetable oils with specific properties. The calorific value of rapeseed oil has been manipulated to increase its combustion efficiency, and rapid increases in the quality of biodiesel produced from vegetable oils are likely to be possible in the early decades of the twenty-first century. Details of TAG production in alternative vegetable oil crops, such as crambe (*Crambe abyssinica*) and jatropha (*Jatropha curcas;* see Figure 1), that compete less for agricultural land are becoming much better understood. Metabolic engineering of fatty acids for human health is also likely to be of great importance. Most of the ω-3 fatty acids that are found in fish oils are very long chain (20- to 24-C), and these are not found in plants, but plants have been engineered to produce them. Some plants have a naturally high content of ω-3 fatty acids, primarily linolenic acid, but crops engineered to have enhanced production of a variety of these fatty acids are likely to become progressively more important in food production. Organisms probably initially evolved the ability to synthesize a range of fatty acids in order to maintain membrane fluidity over a range of temperatures. Multicellular organisms evolved to use these pathways to synthesize storage lipids. The biology of lipids, in current or alternative crops, is now likely to underpin the development of a number of renewable products that will be important both for industry and for human nutrition.

present. AFPs decrease the temperatures at which key events in ice crystallization occur, an effect that is measured by **thermal hysteresis** (TH) (Figure 8.11). The AFPs of some insects induce TH of –30°C, whereas those of Arctic fish induce much less TH (–3 to –5°C), but still sufficient to reduce the freezing point to below sea temperature. Plant AFPs produce very low TH values (–0.3 to –0.5°C). Different TH values reflect the diversity and function of proteins used as AFPs. Within the range of possible crystals that water can form, the most stable ice crystal thermodynamically has the lowest surface area. As ice crystals grow they continually recrystallize to achieve thermodynamic stability. It is recrystallization processes that allow the formation of large, damaging ice crystals. Some AFPs, rather than decreasing the temperatures at which crystallization events occur, inhibit recrystallization and prevent the formation of large ice crystals. In AFPs with low TH values, including those of plants, the inhibition of recrystallization (that is, crystal growth) is probably their primary function.

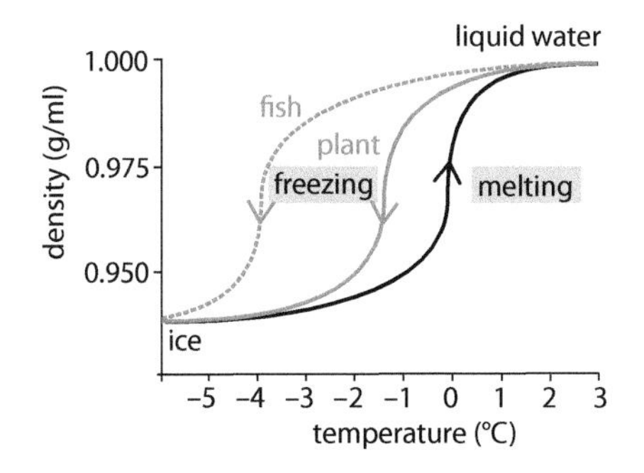

Figure 8.11. Thermal hysteresis and biological "antifreezes." Thermal hysteresis occurs when there is a difference between melting and heating curves. Fish antifreeze proteins (AFPs) depress the freezing point, as compared with the melting point, by a maximum of 4°C. Plant AFPs have a small thermal hysteresis effect—that is, they do not work entirely by depressing the freezing point, but rather by affecting recrystallization processes. Some insect AFPs depress the freezing point by up to 30°C.

Almost all AFPs have large planar hydrophobic surfaces that bind where water molecules in liquid form are incorporated into crystal lattices. Thus plant AFPs bind directly to the planar surfaces of ice, expelling liquid water and preventing crystal growth (Figure 8.12). The hydrophobicity of AFPs accounts for their effect on TH, indicating that AFPs act via non-colligative properties. Compounds such as ethylene glycol, more usually referred to as "antifreezes," act via **colligative properties**—that is, at high

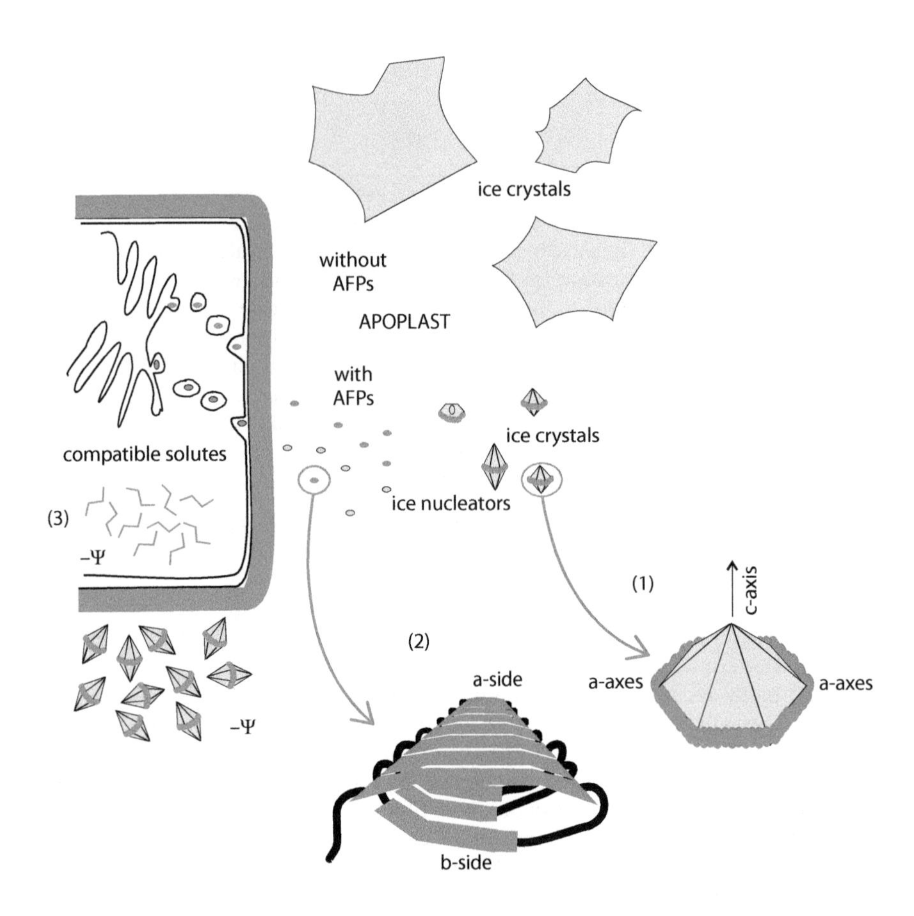

Figure 8.12. The action of plant cryoprotectants and compatible solutes. If a plant cannot produce cryoprotectants—that is, antifreeze proteins (AFPs)—freezing temperatures will induce the formation of large, planar, damaging ice crystals, starting in the apoplast where solution concentrations are lower than in the cytoplasm. If a plant can produce AFPs in response to cold or freezing temperatures, these proteins are exuded into the apoplast in conjunction with ice nucleators. (1) Plant AFPs then bind to planar surfaces on the a-axes of growing ice crystals, promoting limited pyramidal growth in the c-axis of many small ice crystals. (2) Plant AFPs are rolls of β-sheets whose "antifreeze" action is determined by the a-side, probably via its binding to the planar surfaces of ice. (3) Extracellular ice crystals generate a low ψ that is counteracted by the presence of intracellular compatible solutes. Plant AFPs therefore alter crystallization processes, rather than depressing the freezing point.

concentrations they change the ratio of solute and solvent molecules to depress freezing point—whereas the specific action of low concentrations of AFPs changes the freezing process. Plant AFPs are pathogenesis-related (PR) proteins characteristic of multicellular, terrestrial plants. It is likely that freezing-tolerant terrestrial plants, perhaps primarily during the recent Quaternary glaciation, have adapted to freezing conditions using proteins that first evolved to combat pathogens. Binding to pathogens, sometimes via planar surfaces, is important to many plant-pathogen recognition and inhibition strategies. A variety of plant AFPs have been identified, with individual species often producing particular proteins that function as oligomeric complexes. In plants, AFPs are often exuded into the apoplast together with ice nucleators. Freezing-tolerant plants therefore manage ice formation such that it occurs in particular extracellular spaces and forms only small non-damaging crystals.

Control of extracellular ice formation helps to minimize physical damage, but runs the risk of cytoplasmic dehydration because the ψ of ice is much lower than that of water. Freezing-tolerant plants are adapted not only to control extracellular freezing but also to limit the dehydration that it can cause. The synthesis of simple sugars such as sucrose increases significantly with cold stress. When the activity of water molecules decreases with decreasing temperature, many inorganic solutes in the cell tend to bind to membranes and proteins, inhibiting their function. The presence of high sugar concentrations in the cytoplasm slows the decrease in the activity of water to prevent the binding of solutes to important biomolecules, and helps to protect membranes. There is evidence that some sugars form a shell around some biomolecules as the temperature drops, trapping water around them to prevent dehydration-induced denaturation. Many of these sugars, plus other carbohydrates, also act as **compatible solutes**. During cold periods, compatible solutes probably help to maintain an aqueous environment that allows biomolecules to continue functioning, but under freezing conditions the low intracellular ψ that they produce also helps freezing-tolerant plants to avoid dehydration induced by extracellular ice (Figure 8.12). The *esk* 1 mutant of *Arabidopsis*, which is freezing-tolerant, has a very high intracellular concentration of proline. Other freezing-tolerant plants produce glycine-betaine in response to extreme cold. Cold-acclimated and freezing-tolerant plants also have characteristic proteomes at low temperatures. Some hydrophilic COR proteins probably have a similar role to compatible solutes, but many of the proteins isolated from a wide variety of freezing-tolerant plants are late embryogenesis abundant (LEA) proteins. LEA proteins are used across the kingdoms of life to confer extreme dehydration tolerance on other proteins. They are stable over a very wide range of temperatures, including boiling, and are extremely hydrophilic. These LEA proteins act as molecular shields, preventing proteins from becoming damaged during dehydration (see Chapter 4). In many plants their interaction with other molecules, such as trehalose, is important to their protective properties at freezing temperatures.

Heat-tolerant plants have protein curation mechanisms adapted to increase the rate of protein repair

At supra-optimal temperatures the metabolic limits of membrane fluidity and protein stability can easily be exceeded. In particular, proteins can become denatured and thus dysfunctional. Heat-tolerant poikilotherms, including plants, are adapted to this in significant part by the action of HSPs. One of the first eukaryotic genes to be cloned was HSP70, a protein named for the heat shock that induces its expression. In many organisms, including plants, a significant proportion (over 5%) of the protein in a heat-shocked cell can be HSP. Eukaryotic multicellular organisms use HSPs in numerous metabolic pathways, for responses to stresses other than heat, and, in mammalian cells, HSP70 can regulate apoptosis and is thus a significant target for anti-cancer drugs. However, despite their variety and their many metabolic roles, HSPs are clearly a central part of the response to supra-optimal temperatures in many organisms, including plants. Some cyanobacteria and yeast mutants that lack particular HSPs can have significantly altered thermotolerance, but in plants the effect in mutants or overexpressors is often less dramatic, indicating that HSPs only represent one aspect of the heat tolerance mechanisms in plants.

The native conformation of proteins occurs in a narrow region of an energy landscape, beyond which proteins can become denatured (Figure 8.1). The primary sequence of proteins promotes the formation of a secondary structure through hydrogen bonding of peptides. The most common secondary structures are the α-helix and β-sheet, but random coils that have no fixed structure are also quite common. Due to differences in hydrophobicity in the secondary structure, a tertiary structure can emerge and be stabilized by interactions such as disulfide bridges, salt bridges, or hydrogen bonding. The function of a protein is dependent on its structure, the grouping together of proteins to form subunits of a quaternary structure often being very important. Denaturation of a protein is the loss of quaternary, tertiary, or secondary structure. Aggregation of denatured proteins occurs when bonds form and hydrophobicity patterns emerge between monomers, often of different proteins. Protein aggregation is complex, and different proteins are susceptible to different degrees, but often the secondary structure, especially of the random coils, is important in determining susceptibility. Due to its role in a number of important neurodegenerative conditions and its promotion by prions, protein aggregation and the processes that control it are of great medical importance.

Mechanisms that curate proteins in order to control the probability of them occurring in native conformation have evolved for numerous reasons. As a protein is synthesized in a ribosome its final conformation cannot emerge from a partly made primary sequence—mechanisms are necessary to hold the emerging primary sequence until it reaches a size at which appropriate folding is possible. The functional conformation is also sometimes unlikely to emerge spontaneously in aqueous solution, so mechanisms to promote particular patterns of folding are necessary—for example, proteins targeted to membranes must attain their final conformation in a non-aqueous environment, but travel to it through the aqueous cytoplasm. Thus, from the earliest forms of life, organisms have had molecular chaperones for proteins that curate their conformation from when they emerge from the ribosome to their final site of action. Mutations can affect protein conformation, and mechanisms for dealing with proteins that do not have native conformation, preferably to mend them rather than destroy them (which would be costly to the plant), are advantageous. Many HSPs are molecular chaperones for proteins, and some have the capacity to restructure denatured proteins and disaggregate aggregated ones. The set of fundamental cellular stress-induced proteins that has been proposed to be the origin of stress responses in all organisms includes a number of HSPs. HSPs are therefore probably an early adaptation of the protein curation machinery to an environmental stressor.

Sets of HSPs, classified by their molecular weight, are used in heat-tolerant plants both to control the conformation of proteins and to disaggregate them. Small HSPs (sHSPs) with a low molecular mass (in the range 15–43 kDa) occur in prokaryotes and eukaryotes, but are particularly diverse in plants. As proteins undergo heat-induced denaturation, sHSPs bind to them, modifying, or in some instances eliminating, the aggregation process. sHSPs have no ATP binding site, and they frequently form large oligomers—with, for example, dodecameric oligomers being found in wheat. They have a characteristic **α-crystallin** domain of about 100 amino acids. This domain is similar to that of the α-crystallin found in the lens of the vertebrate eye, where it prevents aggregation and maintains transparency (which is lost when crystallin denatures to form a cataract). sHSPs bind as monomers or as large oligomers that can exchange monomers in a dynamic protective process. These processes probably affect both the availability of hydrophobic regions and the tendency to form hydrogen bonds, decreasing the tendency of a protein to undergo aggregation.

HSP100s have a molecular weight of about 100 kDa, and those with an **M-domain** can dissociate aggregated proteins. Those without an M-domain can unfold proteins but cannot resolubilize fully aggregated proteins. The yeast HSP104, whose function is known in detail, is homologous to plant HSP100s—for example, AtHSP101—and reciprocal expression has shown that AtHSP101 and HSP104 can be exchanged without loss of function. In plants, HSP100s are found in both the cytoplasm and the chloroplast. HSP100s are **AAA$^+$ proteins** that have a two-tiered hexameric form with a central pore through which proteins are fed during disaggregation. Each monomer has two characteristic **Walker-type nucleotide-binding domains** that bind ATP, the hydrolysis of which drives the process. ATP-driven conformational changes in the pore drive the protein through it. When the straightened protein emerges from the pore, the same processes that helped it to fold as it emerged from a ribosome are probably used to return it to its native conformation, reactivating it. HSP70s are frequently necessary for the disaggregation action of HSP100s, or at least increase their speed of action very significantly. At optimum temperatures when a protein emerges from a ribosome in eukaryotes, nascent polypeptide-associated complex (NAC) proteins hold it so that it can be folded or targeted to a particular region of the cell. HSP70s and TCP1-ring complex (TRiC) chaperonins then mediate folding of the protein. There are different models of how HSP70s and

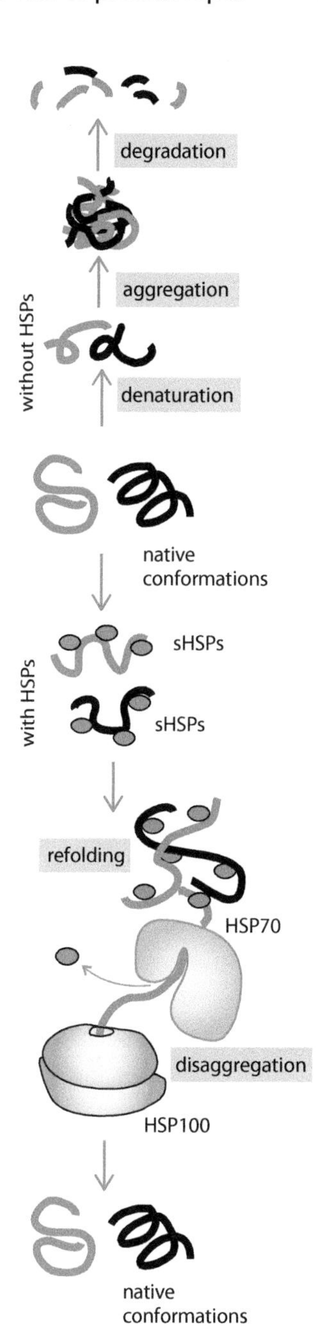

Figure 8.13. The roles of heat stress-responsive proteins (HSPs) in protein repair. Without HSPs at high temperatures, many proteins undergo denaturation and aggregation, and are degraded. With HSPs at high temperatures, small HSPs (sHSPs) help to minimize denaturation and aggregation, and promote the binding of HSP70. This allows proteins to be straightened through HSP100, breaking unwanted intramolecular bonds and enabling the proteins to refold into their native conformations.

HSP100s might interact in heat-tolerant eukaryotes, including plants, but it is most likely that HSP70s have been adapted to unfold aggregated proteins and feed them to HSP100. sHSPs increase the effectiveness with which HSP70s can unfold aggregated protein (Figure 8.13). HSP70s have one ATP-binding domain. The increased expression of HSPs in heat-tolerant plants, which enables them to deal with the increased rates of protein denaturation and aggregation induced by high temperature, is therefore an adaptation of the protein-curating process.

Heat shock factors (HSFs) activate gene expression by binding to HSEs of *HSP* genes. Plant HSP genes often have two or three HSEs, and plants seem to be particularly rich in HSFs—for example, there are 21 HSFs in *Arabidopsis*, compared with only 4 HSFs in many animals. Under optimum conditions, HSFs occur primarily in the cytoplasm, often bound to an HSP, which inactivates them. At supra-optimal temperatures they are released from the deactivating HSP, trimerize, and travel to the nucleus, activating the expression of HSPs. This regulon (Figure 8.14) is widely conserved across a range of organisms. In a number of plant species there is evidence that HSP70 controls the activity of HSFs. Some HSPs are much more important in prokaryotes, and thus in plastids, whereas others are important in the eukaryotic cytoplasm.

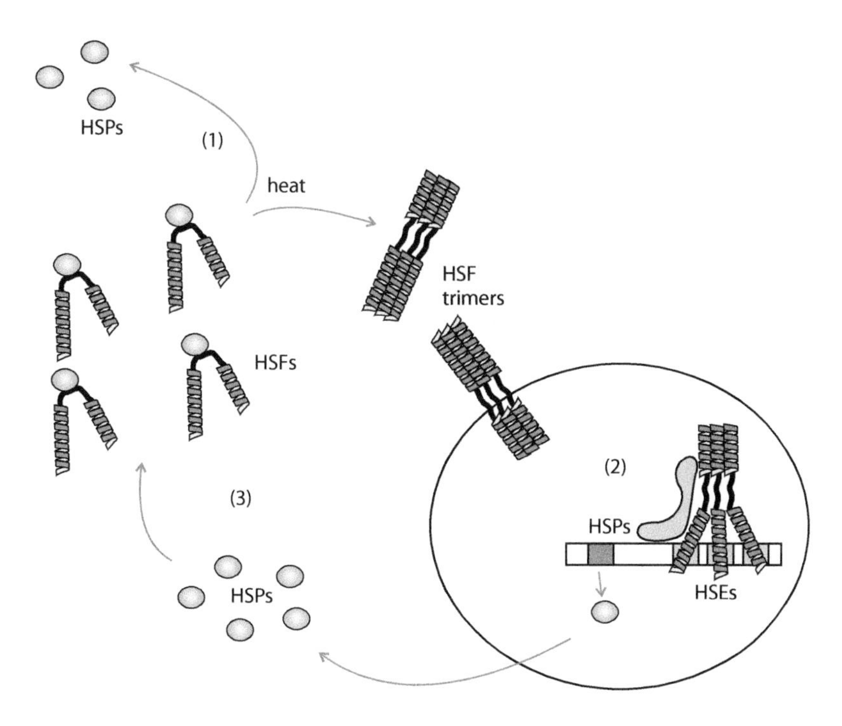

Figure 8.14. The role of heat stress-response factors in the plant response to heat. (1) Heat causes HSPs to dissociate from HSFs, releasing HSPs to meet demand from other heat-response pathways and initiating the trimerization of HSFs necessary for their entry into the nucleus. (2) The binding of HSFs to heat shock elements (HSEs) of promoters allows construction of the transcription complex and expression of HSPs. (3) The production of HSPs includes HSP70s, which help to attenuate the activation of HSPs.

In *Arabidopsis* there are 44 HSPs, and HSP genes are well characterized in a number of plant species. Plants vary in the proportion of different HSPs that are produced during heat stress, which probably reflects differences in their adaptation to heat. The very significant conservation across divergent organisms of the HSP mechanism for both protein curation and HSF activation suggests that it is an ancient pathway, probably active in all heat-tolerant terrestrial plants. Several other types of TF have also been implicated in heat stress responses. For example, manipulation of some WRKY TFs, which are unique to green plants and which help to control a wide variety of metabolic processes, changes the heat tolerance of some species.

Anatomical and morphological adaptations of leaves aid plant tolerance of prolonged cold and heat

Many of the most cold-tolerant plants on Earth are alpines that have not only physiological characteristics that help them to withstand extreme cold, but also anatomical and morphological adaptations. Ice formation on and within alpine plants is frequent, and some of these plants can freeze completely and then recover. The freezing of ice releases latent heat and, if cytoplasmic dehydration and any physical stresses can be withstood, the freezing of ice can be used to decrease the rate at which the plant cools. This effect is used when crops are sprayed with water before a frost—the freezing of external water during a cold night releases enough latent heat to prevent crop internal temperatures from falling below freezing (Figure 8.15). The formation of ice crystals, which occurs on both the outside and the inside of many alpine plants, provides some heat during decreases in external temperature, but the anatomy of many alpines is adapted to prevent the ice crystals from causing physical damage. In some alpines, the apoplastic spaces in which ice forms are structured to regulate the spread of freezing. **Infrared thermography** has been used to show that there are anatomical ice-formation barriers, including such barriers in the vasculature, which can impede the spread of ice and protect the particularly freezing-sensitive parts of some alpine plants (Figure 8.16).

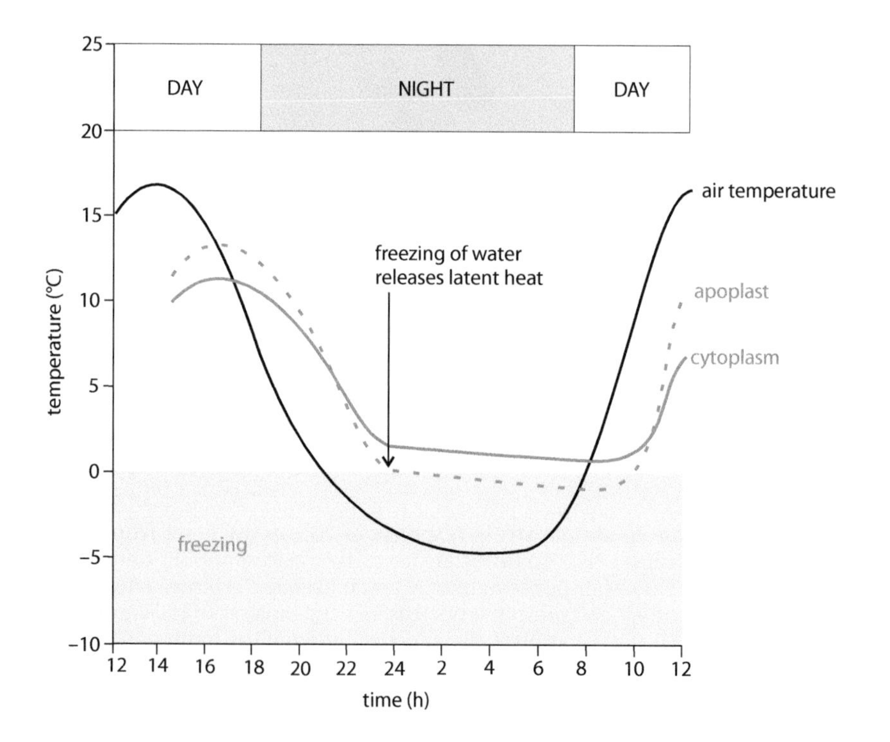

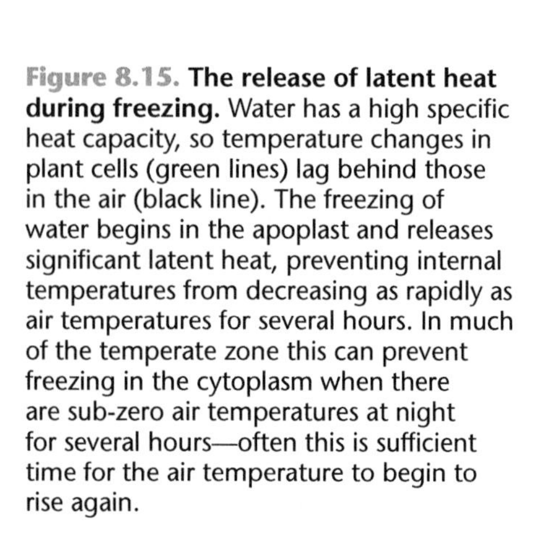

Figure 8.15. The release of latent heat during freezing. Water has a high specific heat capacity, so temperature changes in plant cells (green lines) lag behind those in the air (black line). The freezing of water begins in the apoplast and releases significant latent heat, preventing internal temperatures from decreasing as rapidly as air temperatures for several hours. In much of the temperate zone this can prevent freezing in the cytoplasm when there are sub-zero air temperatures at night for several hours—often this is sufficient time for the air temperature to begin to rise again.

Alpines, especially in the tropics, typically withstand significant diurnal temperature changes. Alpine communities often have plant species with a "cushion-like" or "hedgehog-like" morphology. For instance, the cushion plants of the puna of the Chilean Andes provide some notable examples, such as the yareta (*Azorella compacta*), which grows in compact cushions 2 or 3 m across, often above the treeline at an altitude of 3000–4000 m. The cushion-like habit means that these plants are less susceptible to desiccating winds and produce very little turbulence if there are airflows across them. It also means that if they are covered by snow this causes no physical damage. Snow is a significant insulator, and when the atmosphere is very cold the warmest place in a snow-covered ecosystem is the soil–snow interface, because the most significant heat source is the Earth. However, the cushion-like habit not only positions the plant in the most advantageous space in extreme cold, but also provides the morphology with the least internal temperature variation in the face of dramatic external temperature changes. Thus the morphology of alpine plants helps to regulate whole-body temperatures (Figure 8.17).

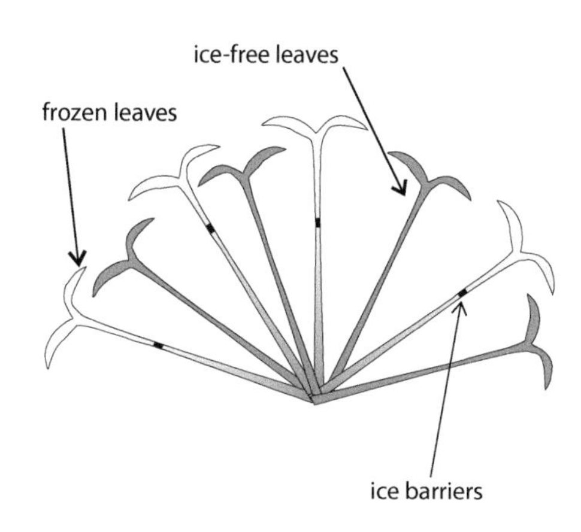

Figure 8.16. Freezing patterns in alpine plants. Thermography of herbaceous alpine plants has revealed that freezing of different parts of the plant, which is mostly initiated by ice nucleators, can occur independently, with certain parts freezing while other parts remain unfrozen. In some species, structural ice barriers help to control these freezing patterns, but in others warm inner parts block the spread of ice from one plant part to another thermally. During their early development, most organs of alpine plants are sensitive to cold, as are some organs throughout the plant's life span, so differences in freezing between the various parts of the plant can be important in limiting freezing damage.

Homeotherms generally find it easier to generate heat to avoid the effects of cold than to lose heat when temperatures are supra-optimal. For both poikilothermic and homeothermic organisms, excess heat can be a severe challenge because their physiological adaptations with regard to heat tolerance enable them to continue functioning at temperatures somewhat above optimal, but they do not help them to lose heat, which is often necessary at prolonged high temperatures. A plant leaf with a high water content that is adapted to intercept incoming solar radiation for photosynthesis has the potential to become very hot (temperatures in excess of 90°C have been demonstrated), so it is the plant anatomical and morphological adaptations that promote heat loss which keep internal temperatures in the range within which physiological adaptations can be useful. During insolation, plant leaves can rapidly reach temperatures significantly above that

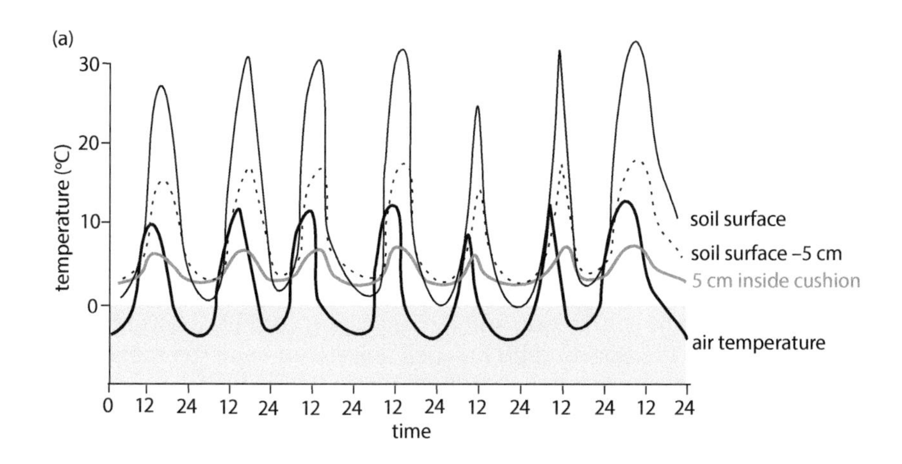

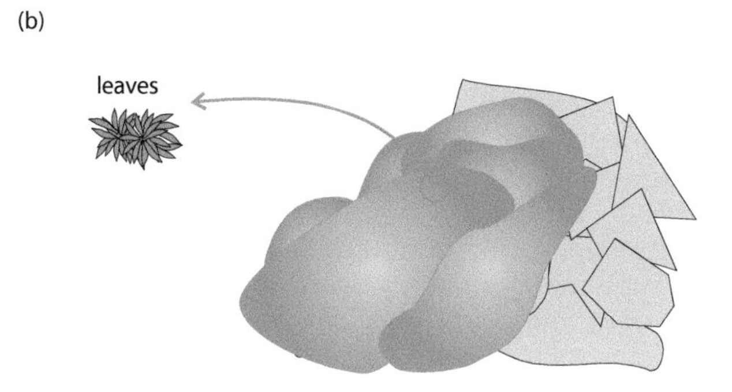

Figure 8.17. The temperature regimes of alpine cushion plants and their environment. (a) During insolation, soil temperatures (gray lines) can be significantly higher than air temperatures, and even maintain above-freezing conditions in the rhizosphere (–5 cm). During diurnal temperature fluctuations, temperatures within cushion plants (green line) are significantly buffered, which together with heat from the soil can prevent temperatures from falling below freezing despite sub-zero air temperatures. (b) The data shown in (a) are for the giant Andean cushion plant, *Azorella compacta*, which grows above the treeline in the Andes in the Chilean puna vegetation. (Data from Kleier C & Rundel P [2009] *Plant Biol* 11:351–358.)

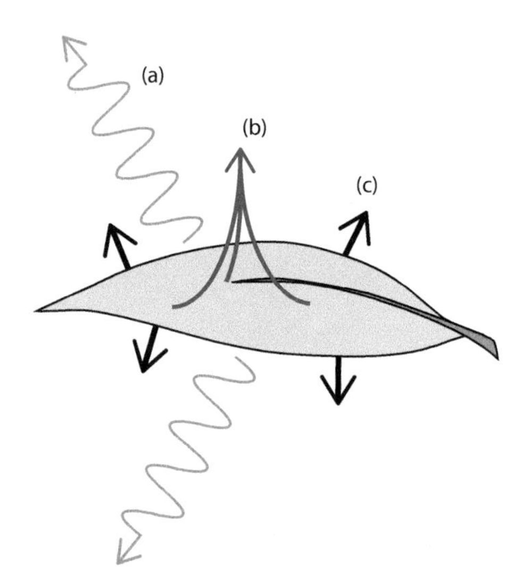

Figure 8.18. Mechanisms of heat loss from a plant leaf. Because their morphology is adapted to intercept incoming radiation for photosynthesis, leaves are at significant risk of overheating. (a) Long-wave radiation emitted from the leaf can contribute to heat reduction. (b) Heat used to drive evaporation from the leaf can have a significant cooling effect. (c) Sensible heat loss via conduction to the cooler surrounding air can contribute significantly to heat reduction.

of the surrounding air. There are then three possible mechanisms of heat loss—long-wave radiation, conduction, and latent heat of evaporation (Figure 8.18). Plant leaves can emit significant amounts of long-wave radiation, but in general they do not lose large amounts of heat in this way. The rate of heat loss by long-wave radiation can be significant during the night in some plants, but it does not counteract the heat gain that occurs in most plants during the day. Heat can also be conducted away from plant leaves via direct contact with cooler air—a process that can be very significant in leaf cooling. This is especially so if differential rates of conduction to the air warm it unequally and induce convection currents. These currents accelerate the movement of hot air away from the leaf, drawing cooler air across it, and maintaining maximal conduction of heat away from the leaf. Heat lost by conduction is often termed "sensible" heat. Transpiration of water from the leaf necessitates water undergoing a change of state from liquid to gas—a process that requires a significant input of energy and that can be very significant in leaf cooling. The ratio between sensible heat loss and evaporative heat loss is known as the **Bowen ratio**, and it varies significantly with leaf morphology.

In hot wet conditions, plant growth tends to be luxuriant, with most leaves located in a canopy in which wind speeds are very low. Such conditions promote the formation of boundary layers around leaves (Figure 8.19). These layers of still air, produced by frictional drag from the leaf which becomes proportionally more significant with decreasing wind speed, reduce the amount of heat that can be lost by conduction, because air is a poor conductor of heat. The Bowen ratio of tropical rain forest leaves is very low (values of 0.1–0.2 have been measured). In general, a large surface area to volume ratio promotes evaporative heat loss to a greater extent than it promotes sensible heat loss, so the leaves of many tropical rainforest plants are large and thin (Figure 8.19a)—a morphological adaptation to exploit the available water in order to maximize evaporative heat loss when conductive heat loss is limited. In uniform canopies, such as those of many crops, very significant boundary layers can produce Bowen ratios that approach zero—much irrigation water is used for heat loss by crops because evaporative cooling is their most significant available mechanism for heat loss. If evaporative heat loss is possible it has a very great cooling capacity, and leaves can generally be kept within a range that allows physiological adaptation to heat. In hot dry conditions the possibility of losing heat by evaporation is greatly reduced because of the lack of available water. The pinnate leaves of many plants can help to promote sensible heat loss. Direct conduction from the leaflets to the air is promoted because the morphology minimizes boundary layers while at the same time maximizing the surface area of contact, and it is inevitably unequal, setting up convection currents (Figure 8.19b). This promotes sensible heat loss by conduction and by convection currents drawing cooler air over the leaf. The Bowen ratio for such leaves can be in the range 5–10. However, such leaves can reach quite high temperatures because the capacity of conduction and convection for cooling is generally less than that of evaporation, so they tend to use physiological mechanisms of heat tolerance more than do plants that inhabit hot wet environments.

Under desert conditions, sensible heat loss, especially in those plants that eliminate transpiration during the day, can essentially be the only method of cooling, so the Bowen ratio approaches infinity. The general morphology of many cacti tends towards the lowest possible surface area to volume ratio (that is, a sphere), driven by the need to reduce water loss. However, this morphology decreases the potential for conduction because it minimizes the area of contact with the surrounding air. Many cacti use spines as significant conductors of heat, especially during the night when air temperatures are significantly lower than those of plants that have been heated the previous day (Figure 8.19c). Slow growth caused by lack of water also means that many cacti do not need to maximize the interception of sunlight, because it is not limiting to growth. The many hairs and spines on most cacti, together with significant wax layers on their surfaces, give them a high reflectance and a glaucous color, producing a high **albedo** (that is, most of the incoming radiation is reflected, which helps to minimize heat loss). The spines and hairs on cacti also aid the formation of a boundary layer, which to some extent

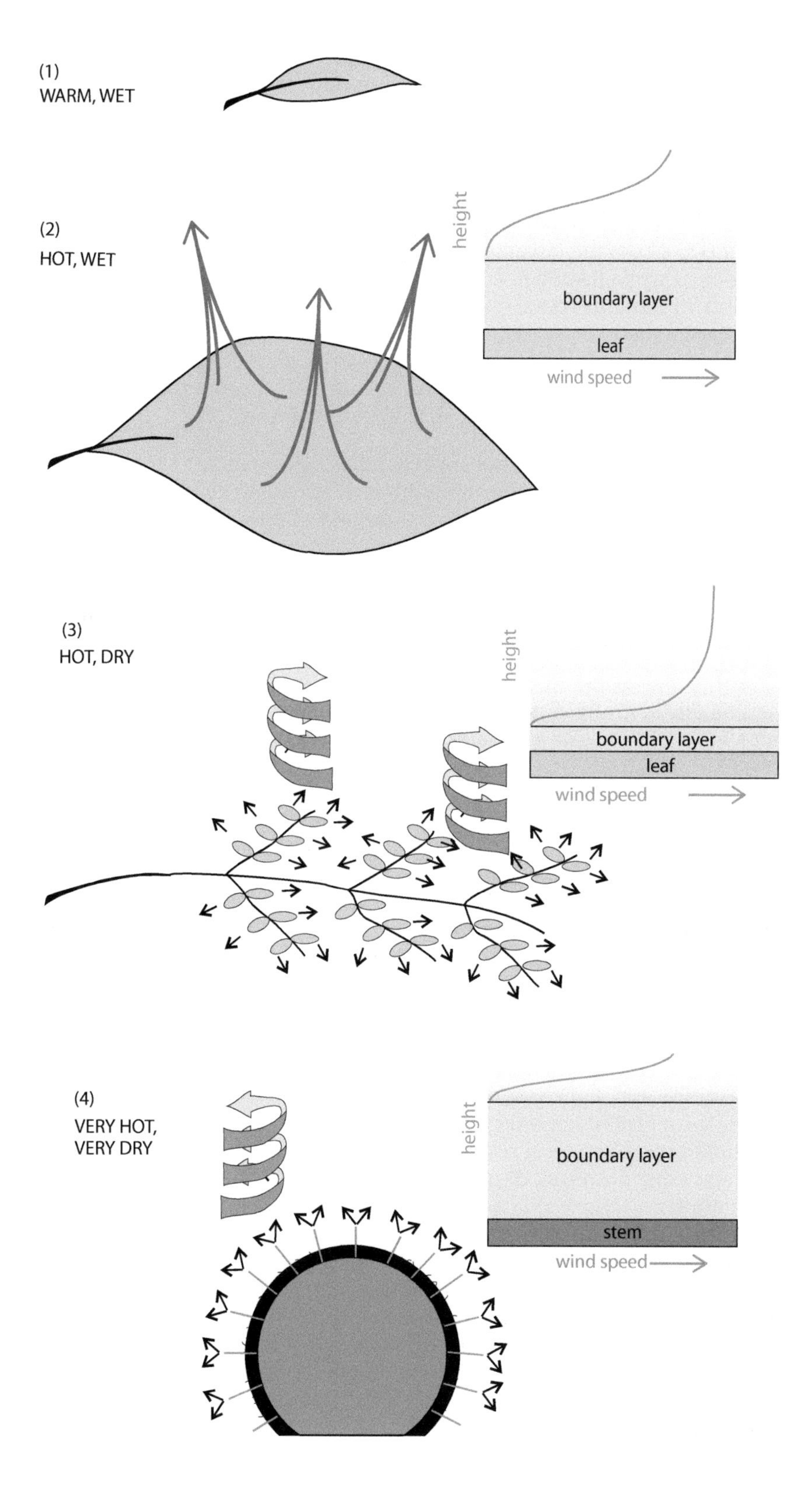

Figure 8.19. Heat loss and leaf morphology. (1) In warm wet conditions, leaf morphology can be optimized for photosynthesis. (2) In hot conditions, sensible heat loss is limited, but if water is available, evaporative heat loss is often maximized via thin leaves with a large surface area. Large leaves have a large boundary layer, but this is of no consequence because sensible heat loss is not important. (3) In hot dry conditions, sensible heat loss must be maximized, so differential conduction of heat to the surrounding air from, for example, pinnate leaves, can set up convection currents that carry heat away from the leaf and encourage the movement of cooler air over the leaf. Pinnate leaves minimize the depth of the boundary layer, which promotes the conduction of heat that sets up convection currents. (4) In xerophytes such as cacti, evaporative heat loss is almost zero. Morphologies that tend towards a spherical shape would minimize conductive heat loss if features such as leaves modified to form spines did not conduct heat, often by penetrating thick boundary layers, when ambient temperatures are lower than those of the plant. Hairs, spines, and a waxy cuticle produce a high albedo that reduces heat accumulation.

insulates the body of the plant from the changes in ambient air temperature. In general, however, the internal tissues of cacti can reach quite high temperatures, and physiological adaptation to heat is vital. Salinity-tolerant plants often have a similar xerophytic morphology, in particular cuticular waxes that produce a high albedo. This reflects the temperature regulation challenges that are encountered by halophytes.

Temperature-induced physiological changes trigger developmental and phenological responses

In a number of species the CBF regulon is **gated** by the circadian clock (that is, responses to the same cold shock vary at different times). In fact, a significant part of the cold response is regulated to occur daily during the day-night cycle. This gating is probably central to the interaction between the effects of daylength and temperature on many plants. This interaction is mediated by the circadian clock-associated 1 (CCA1)/late elongated hypocotyl (LHY) TFs of the circadian clock. The details of the molecular clock in *Arabidopsis* are well known, and many mutants of it are available. Studies of *cca1* and *lhy* mutants reveal that they both have a significantly impaired cold response and lack of circadian control over the CBF regulon. The direct link between the cold response and the circadian clock via CCA1/LHY is probably integrated with several other developmental responses to cold, helping to trigger the phenological effects of cold temperatures.

Gibberellic acid (GA) metabolism has long been known to interact with temperature. Classic experiments in the early twentieth century established interesting effects of varying temperature and of temperature differentials on stem growth. Cold acclimation in many species involves a reduction in stem growth, many of the genes that are involved in GA metabolism show altered expression with cold, and many GA mutants have an impaired cold response. GA responses in plants are regulated by DELLA proteins, which repress the GA response—targeted degradation of DELLA proteins eliminates this repression and allows growth responses such as an increase in stem length to occur. Cold induces a significant increase in the activity of DELLA proteins, stunting growth. Molecular dissection of this interaction has revealed that it is triggered by the CBF regulon. Temperature also affects auxin activity in plants, and interaction between auxin and GA during the development of roots and shoots is long established. Overall, the CBF regulon affects the activity of growth-regulating metabolism in plants, and thus helps to explain their morphological responses.

Vernalization is dependent on the interaction between GA metabolism and temperature. Cold treatment of plants that require vernalization increases the expression of GA-3 oxidase (GA3OX) and decreases the expression of GA-2 oxidase. This results in increased activity of biologically active GA. The expression of GA3OX is regulated by the SPATULA TF, a member of the phytochrome interacting factor (PIF) TF family, which therefore links temperature and light responses. In *Arabidopsis*, PIF4, which helps to control the morphological response of plants to low light levels, also controls the similar morphological response to high temperature. In addition, PIF4 promotes early flowering directly via the *FLOWERING LOCUS T* promoter. The complex network that underpins the transition from the vegetative to the reproductive phase in plants is also affected by cold. Although the interaction between temperature and flowering time is complex, with, for example, temperatures at particular times of the diurnal cycle being particularly important for certain species, the CBF regulon is linked to a number of TFs, and delays flowering time. These metabolic pathways that mediate temperature-induced changes in development and phenology are probably constitutively expressed in plants that are adapted to particular long-term temperature regimes.

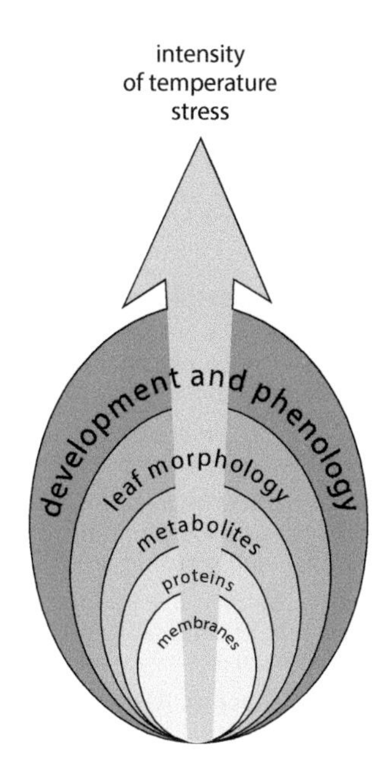

Figure 8.20. The stress-response hierarchy for variation in temperature. Relatively minor spatial and temporal variations in temperature cause minor stress. This can be detected by and counteracted via adjustments of the biochemical properties of membrane and sometimes cytoskeleton components. For tolerance of moderate temperature stress, a variety of physiological adjustments, often based on proteins and metabolites, are also necessary. Most plants that can tolerate quite severe temperature stress have morphological adaptations of their leaves, as well as biochemical and physiological adaptations. Plants that inhabit regions where extreme temperature stress occurs often have avoidance mechanisms based on developmental and phenological adjustments.

Summary

Interactions between biomolecules, especially lipids and proteins, are directly affected by temperature. Plants are poikilotherms and are particularly susceptible to changes in temperature because they are static. They adjust to minor fluctuations in temperature by means of adjustments of membrane and cytoskeleton properties, with plants from the temperature zone using such changes to acclimate to significant temperature variation. With more pronounced temperature variation, physiological adaptions (including cryoprotectants, osmoprotectants, and heat shock proteins) can help to maintain metabolic functioning. In plants that are adapted to chronically challenging temperature regimes, anatomical, morphological, and phenological adaptations complement biochemical and physiological adaptations (Figure 8.20). Understanding plant responses to temperature is currently particularly important because there is much evidence that climate changes are affecting plant growth in both unmanaged and managed ecosystems, upon which humans depend.

Further reading

Poikilothermy and plants

Guschina IA & Harwood JL (2006) Mechanisms of temperature adaptation in poikilotherms. *FEBS Lett* 580:5477–5483.

Hartl FU & Hayer-Hartl M (2009) Converging concepts of protein folding *in vitro* and *in vivo*. *Nat Struct Mol Biol* 16:574–581.

Seymour RS (2010) Scaling of heat production by thermogenic flowers: limits to floral size and maximum rate of respiration. *Plant Cell Environ* 33:1474-1485.

Effects of changing temperatures on plants

de Cortazar-Atauri IG, Daux V, Garnier E et al. (2010) Climate reconstructions from grape harvest dates: methodology and uncertainties. *Holocene* 20:599–608.

Lenoir J, Gegout JC, Marquet PA et al. (2008) A significant upward shift in plant species optimum elevation during the 20th century. *Science* 320:1768–1771.

Niu S, Luo Y, Li D et al. (2014) Plant growth and mortality under climatic extremes: an overview. *Environ Exp Bot* 98:13–19.

Rosenzweig C, Elliott J, Deryng D et al. (2014) Assessing agricultural risks of climate change in the 21st century in a global gridded crop model intercomparison. *Proc Natl Acad Sci USA* 111:3268–3273.

Temperature sensing in plants

Kumar SV & Wigge PA (2010) H2A.Z-containing nucleosomes mediate the thermosensory response in *Arabidopsis*. *Cell* 140:136–147.

Mittler R, Finka A & Goloubinoff P (2012) How do plants feel the heat? *Trends Biochem Sci* 37:118–125.

Penfield S (2008) Temperature perception and signal transduction in plants. *New Phytol* 179:615–628.

Saidi Y, Finka A & Goloubinoff P (2011) Heat perception and signalling in plants: a tortuous path to thermotolerance. *New Phytol* 190:556–565.

Temperature and signaling pathways

Franklin KA, Toledo-Ortiz G, Pyott DE et al. (2014) Interaction of light and temperature signalling. *J Exp Bot* 65:2859–2871.

Jung J-H, Park J-H, Lee S et al. (2013) The cold signaling attenuator high expression of osmotically responsive genei activates *flowering locus C* transcription via chromatin remodeling under short-term cold stress in *Arabidopsis*. *Plant Cell* 25:4378–4390.

Leivar P & Quail PH (2011) PIFs: pivotal components in a cellular signaling hub. *Trends Plant Sci* 16:19–28.

Miura K & Furumoto T (2013) Cold signaling and cold response in plants. *Int J Mol Sci* 14:5312–5337.

Acclimation to chilling

Dong C-H & Pei H (2014) Over-expression of *miR397* improves plant tolerance to cold stress in *Arabidopsis thaliana*. *J Plant Biol* 57:209-217.

Kang J, Zhang H, Sun T et al. (2013) Natural variation of *C-repeat-binding factor (CBFs)* genes is a major cause of divergence in freezing tolerance among a group of *Arabidopsis thaliana* populations along the Yangtze River in China. *New Phytol* 199:1069–1080.

Thomashow MF (2010) Molecular basis of plant cold acclimation: insights gained from studying the CBF cold response pathway. *Plant Physiol* 154:571–577.

Xu W, Zhang N, Jiao Y et al. (2014) The grapevine basic helix-loop-helix (bHLH) transcription factor positively modulates CBF-pathway and confers tolerance to cold-stress in *Arabidopsis. Mol Biol Rep* 41:5329–5342.

Membrane properties

Burgos A, Szymanski J, Seiwert B et al. (2011) Analysis of short-term changes in the *Arabidopsis thaliana* glycerolipidome in response to temperature and light. *Plant J* 66:656–668.

Chen G, Snyder CL, Greer MS et al. (2011) Biology and biochemistry of plant phospholipases. *Crit Rev Plant Sci* 30:239–258.

Liao P, Chen Q-F & Chye M-L (2014) Transgenic *Arabidopsis* flowers overexpressing Acyl-CoA-binding protein ACBP6 are freezing tolerant. *Plant Cell Physiol* 55:1055–1071.

Moellering ER, Muthan B & Benning C (2010) Freezing tolerance in plants requires lipid remodeling at the outer chloroplast membrane. *Science* 330:226–228.

Pham A, Lee J, Shannon JG et al. (2010) Mutant alleles of *FAD2-1A* and *FAD2-1B* combine to produce soybeans with the high oleic acid seed oil trait. *BMC Plant Biol* 10:195.

Osmoprotectants, cryoprotectants, and freezing

Hand SC, Menze MA, Toner M et al. (2011) LEA proteins during water stress: not just for plants anymore. *Annu Rev Physiol* 73:115–134.

Middleton AJ, Brown AM, Davies PL et al. (2009) Identification of the ice-binding face of a plant antifreeze protein. *FEBS Lett* 583:815–819.

Preston JC & Sandve SR (2013) Adaptation to seasonality and the winter freeze. *Front Plant Sci* 4:167.

Sasaki K, Christov NK, Tsuda S et al. (2014) Identification of a novel LEA protein involved in freezing tolerance in wheat. *Plant Cell Physiol* 55:136–147.

Wisniewski M, Gusta L & Neuner G (2014) Adaptive mechanisms of freeze avoidance in plants: a brief update. *Environ Exp Bot* 99:133–140.

Protein repair mechanisms

Bokszczanin KL & Fragkostefanakis S (2013) Perspectives on deciphering mechanisms underlying plant heat stress response and thermotolerance. *Front Plant Sci* 4:315.

Hasanuzzaman M, Nahar K, Alam MM et al. (2013) Physiological, biochemical, and molecular mechanisms of heat stress tolerance in plants. *Int J Mol Sci* 14:9643–9684.

Richter K, Haslbeck M and Buchner J (2010) The heat shock response: life on the verge of death. *Mol Cell* 40:253–266.

Waters ER (2013) The evolution, function, structure, and expression of the plant sHSPs. *J Exp Bot* 64:391–403.

Anatomy, morphology, and temperature

Dietrich L & Koerner C (2014) Thermal imaging reveals massive heat accumulation in flowers across a broad spectrum of alpine taxa. *Alp Bot* 124:27–35.

Koini MA, Alvey L, Allen T et al. (2009) High temperature-mediated adaptations in plant architecture require the bHLH transcription factor PIF4. *Curr Biol* 19:408–413.

Larcher W, Kainmuller C & Wagner J (2010) Survival types of high mountain plants under extreme temperatures. *Flora* 205:3–18.

Lujan R, Lledias F, Martinez LM et al. (2009) Small heat-shock proteins and leaf cooling capacity account for the unusual heat tolerance of the central spike leaves in *Agave tequilana* var. Weber. *Plant Cell Environ* 32:1791–1803.

Nicotra AB, Cosgrove MJ, Cowling A et al. (2008) Leaf shape linked to photosynthetic rates and temperature optima in South African *Pelargonium* species. *Oecologia* 154:625–635.

Developmental and phenological adaptations

Dong MA, Farre EM & Thomashow MF (2011) Circadian clock-associated 1 and late elongated hypocotyl regulate expression of the C-repeat binding factor (CBF) pathway in *Arabidopsis. Proc Natl Acad Sci USA* 108:7241–7246.

Sadras VO & Moran MA (2013) Nonlinear effects of elevated temperature on grapevine phenology. *Agricult Forest Meteorol* 173:107–115.

Salome PA, Weigel D & McClung CR (2010) The role of the *Arabidopsis* morning loop components CCA1, LHY, PRR7, and PRR9 in temperature compensation. *Plant Cell* 22:3650–3661.

Sarhadi E, Mahfoozi S, Hosseini SA et al. (2010) Cold acclimation proteome analysis reveals close link between the up-regulation of low-temperature associated proteins and vernalization fulfillment. *J Proteome Res* 9:5658–5667.

Chapter 9
Salinity

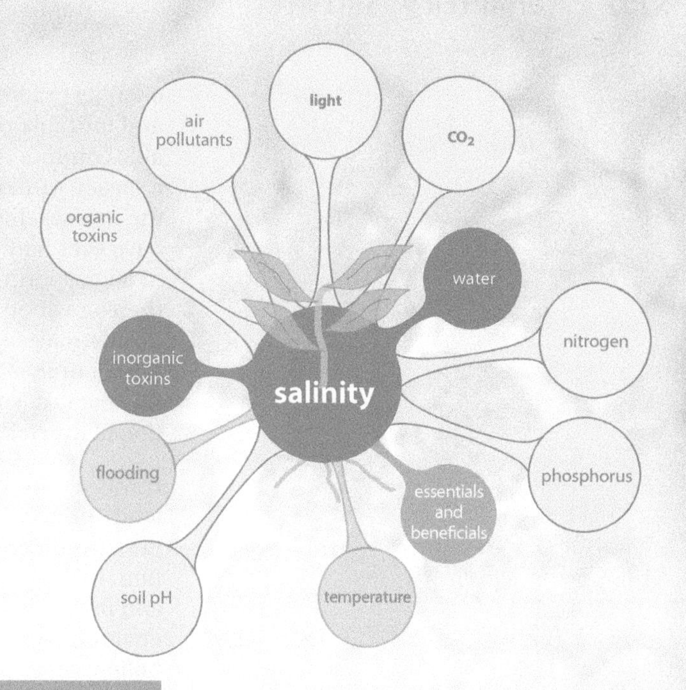

Salinity causes osmotic and ionic stress, often compounded by the stresses of waterlogging and temperature control.

Key concepts

- Terrestrial plants have freshwater ancestry and are rarely salt tolerant.
- Plant responses to salinity are globally significant with regard to irrigation and in shorefront communities.
- Salinity challenges plants with osmotic stress and ion toxicity, especially from Na^+ and Cl^-.
- Na^+ can enter plants through K^+-uptake systems, but plants have some capacity for Na^+ efflux.
- Salt-tolerant plants can control the uptake and distribution of Na^+ and Cl^-.
- Halophytes generate very negative water potentials by using specialized solutes.
- Physiological adaptations are the basis of moderate increases in salinity tolerance in crops.
- In chronically saline habitats, halophytes have distinct xerophytic morphologies.
- Some halophytes can excrete significant quantities of salt via specialized organs.
- In mangroves and salt-marsh plants the effects of salinity are compounded by waterlogging.

Terrestrial plants are descended from freshwater algae, so saline water is generally toxic to them

Life probably began more than 3 billion years ago in an aqueous ionic solution, and by about 2.5 billion years ago the majority of water on earth was saline. The great multicellular diversification of about 540 million years ago occurred in a saline ionic solution, so most extant phyla are diverse in marine environments, and most animal phyla first evolved in the sea. In terrestrial animals, including humans, Na^+ is the primary osmoticum, so their cells are bathed in saline solution. For example, the concentration of Na^+ in human blood plasma is maintained at about 140 mM. In animal cells, although K^+ is the most concentrated intracellular ion, Na^+ is an essential activator for numerous enzymes, and its electrochemical gradients are the primary energizer of ion transport. Na is therefore an essential element for animals, and for most of human history salt (NaCl) has been among the most valuable commodities traded. The essential roles of Na^+ in animal physiology are a legacy of marine ancestry. Almost no terrestrial plant phyla—including the mosses, liverworts, horsetails, ferns, and gymnosperms—have now, or have ever had, marine representatives. Among the 350,000 or more species of angiosperm that dominate the vegetation of many terrestrial ecosystems there are about 80 marine species. These are sea grasses in the Zosteraceae, Posidoniaceae, and Cymodoceaceae that inhabit shallow seas. These three families are all closely related and in the monocotyledonous order Alismatales, so terrestrial plants have been restricted to a single colonization of the marine environment.

It was about 475 million years ago that multicellular plants colonized the land surface. Before this the challenges of living on land, which included desiccation and, before the oxidation of the atmosphere, intense UV radiation, had probably only been met by lichens. The Earth's surface now has, and probably had then, a low but variable salinity, in contrast to the marine environment, which has had a constantly high salinity for at least the last billion years. The plant phyla that dominate the Earth's terrestrial vegetation are descended from a single origin in the freshwater algae, the Charales (Figure 9.1). K^+ is the primary osmoticum in the Charales, they have no enzymatic requirement for Na^+, and electrochemical gradients of H^+ are the primary energizer of ion transport. The evolution of a physiology that utilized K^+/H^+ rather than Na^+ was a vital step in the colonization of fresh water, and provided the physiological foundations for colonizing the Earth's surface. The descendants of the Charales (that is, all terrestrial plant phyla) inherited a physiology in which K^+ and H^+ fulfilled the functions that Na^+ has in organisms with marine ancestry. For species with C_4 and CAM photosynthetic pathways, small amounts of Na^+ are essential, and they benefit the growth of some other terrestrial plants, but in contrast to animals the majority of terrestrial plants have no requirement for Na. Alkali metal ions, such as Na^+ and K^+, are highly soluble and form the basis of osmotic and electrochemical gradients, but the animal and plant phyla that dominate terrestrial ecosystems have inherited physiologies with different alkali metal requirements.

Saline soils can be defined as those with concentrations of the chlorides or sulfates of Na, Ca, and Mg sufficient to produce an **electrical conductivity** in a soil extract (EC_e) of at least 4 dS m^{-1} and a sodium absorption ratio of less than 13 (Table 9.1). On the basis of this definition, about 7% of the Earth's surface has saline soil. This salinity is low compared with that of seawater (EC in the range 50–60 dS m^{-1}), but an EC of 4 dS m^{-1} or more causes severe problems for the growth of almost all terrestrial plants, and in many species adverse effects are detectable at a much lower conductivity (Figure 9.2).

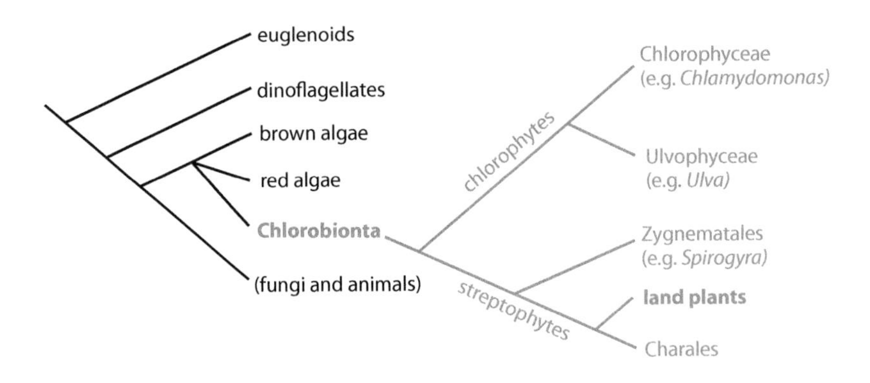

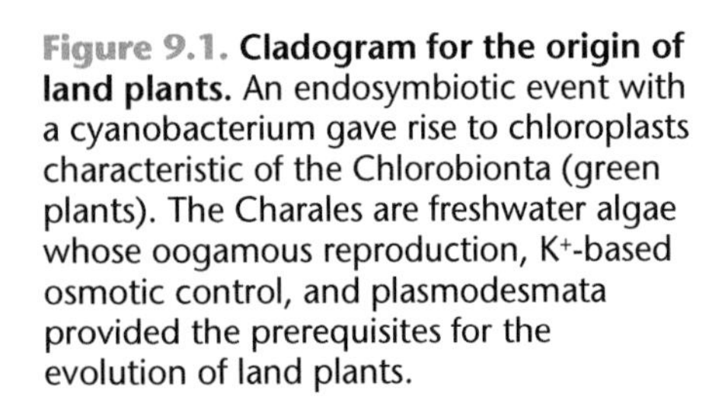
Figure 9.1. Cladogram for the origin of land plants. An endosymbiotic event with a cyanobacterium gave rise to chloroplasts characteristic of the Chlorobionta (green plants). The Charales are freshwater algae whose oogamous reproduction, K^+-based osmotic control, and plasmodesmata provided the prerequisites for the evolution of land plants.

Table 9.1. Definition of soil salinities and sodicity. Soil salinity is generally defined by its effects on plant growth, and is measured by electrical conductivity in a soil paste. The sodium absorption ratio (SAR), based on Na/(Ca+Mg) in soil, determines the availability of Na. Saline soils are high in NaCl, whereas sodic soils are high in $NaCO_3$

Soil type	EC range (dS m^{-1})	SAR	Effects on plants
Non-saline	0–2	< 13	Negligible effects on plant growth
Slightly saline	2–4	< 13	Some effects on sensitive plants, some benefits for halophytes
Saline	4–8	< 13	Severe effects on sensitive plants, optimal for halophytes
Strongly saline	8–16	< 13	Tolerated by few plants other than halophytes
Very strongly saline	> 16	< 13	Tolerated only by halophytes
Saline-sodic	> 4	> 13	Effects on Na-sensitive plants
Sodic	< 4	> 13	Poor structure and high pH prevent growth of many species

(Salinity definitions based on Abrol IP, Yadav JSP & Massoud FI [1988] Salt-affected soils and their management. FAO Soils Bulletin 39.)

Soils with an EC_e in the range 2–4 dS m^{-1} are defined as mildly saline, and this definition would extend the area covered by saline soils to about 15% of the land surface. Sodic soils have high $NaCO_3$ concentrations, and because of their low Ca+Mg concentrations they have particularly high sodium absorption ratios. They can cause problems for plant growth because of their poor structure and high pH as much as their Na concentration. Salt-sensitive species such as *Cicer arietinum* (chickpea) cannot survive in conditions of mild salinity. A few plant species are **halophytes** ("salt plants") that can grow at relatively high salinity, although their optimum growth almost always occurs at salinities below that of seawater.

Just over 2600 angiosperm species are known to tolerate an EC_e of 8 dS m^{-1} in their substrate, and of these about 350 species are able to complete their life

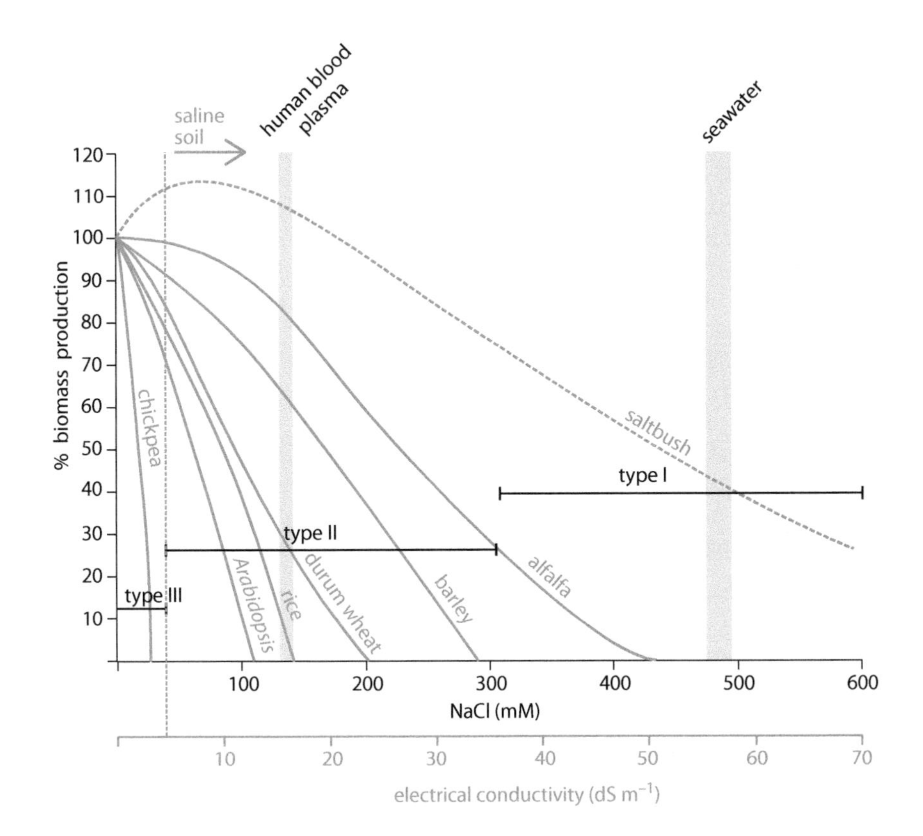

Figure 9.2. The sensitivity of terrestrial plants to salinity. Even halophytes such as saltbush with a "type I" response to salinity, which require some salt for optimal growth, show inhibited growth at seawater salinities. At the United States Department of Agriculture (USDA) threshold for a "saline" soil (EC of 4 dS m^{-1}), the growth of most plant species is significantly decreased in a "type II" response to salinity. The many salt-sensitive species with a "type III" response to soil salinity cannot grow at all in saline soils. (From Munns R & Tester M [2008] *Annu Rev Plant Biol* 59:651–681. With permission from Annual Reviews.)

cycle in 20 dS m^{-1}. The salinity tolerance of many species is not known, but these figures suggest that perhaps 1% of the world's plant species are tolerant of moderate salinity (8 dS m^{-1}), and 0.1–0.2% are tolerant of salinity that is about half that of seawater. A very few species (for example, *Arthrocnemum macrostachyum*) can survive at 500–1000 mM NaCl. Some salt-tolerant species (for example, *Atriplex nummularia*, *Suaeda maritima*, and *Disphyma australe*) do not just survive but have optimum growth at concentrations of 200 mM NaCl or more. Such species show a type IV response to salinity and are few in number. In other species (for example, *Distichlis spicata*), growth in increasing concentrations of NaCl is unimpaired until a threshold is reached, after which it declines—this is a type III response (Figure 9.2). Based on tolerance of 80 mM NaCl, halophytism has evolved independently in each of the major clades of angiosperms (Figure 9.3), but the Caryophyllales are particularly rich in halophytes, and have twice as many halophytic species as the next most halophyte-rich terrestrial order, the Malpighiales (Figure 9.4). The **PACMAD clade** of the Poaceae, which contains the C_4 grasses, also includes many halophytes.

Terrestrial plants almost all grow well only in fresh water, also known as "sweet" water, and are hence termed **glycophytes** (that is, "sweet plants"). Few crop species are tolerant of even mild salinity, and many of the world's staple food crops are notably salt-sensitive. Crops with halophytic ancestry, in particular older varieties of chard and beets, benefit significantly from Na addition even when sufficient K is present, and some crops gain a certain amount of benefit from Na (Table 9.2). These Na-induced increases in growth are potentially significant agriculturally, and might be quite widespread among higher plants, sometimes conferring on Na the status of a beneficial element. These benefits occur at concentrations of around 1 mM Na in the soil, which is well below the concentration that classifies a soil as saline. Thus Na is beneficial only at very low concentrations, and the majority of terrestrial plants have no essential requirement for it. For most plant species even mild salinity is highly toxic, and even many halophytes cannot survive in seawater.

Plant responses to salinity are important in irrigated agriculture and in salt marshes and mangrove swamps

Due to the salt sensitivity of most crops, irrigation water cannot be saline. Irrigated agricultural systems produce, by economic value, around 50% of all food from about 20% of agricultural land. Globally, rain-fed agricultural systems cannot now, and in many parts of the world never could, support significant human populations, so extensive irrigation currently underpins global food production. Of the land that is currently irrigated worldwide, about 25% is affected by salinization, and this percentage is increasing. In many parts of the world, salinization of irrigated land means that the total area under crop production is decreasing. Irrigated agricultural production has frequently proved to be unsustainable, and although soil waterlogging and silt accumulation can be problematic on irrigated land, the primary cause is soil salinization. Some of the earliest agricultural systems—for example, those in ancient Mesopotamia—became unsustainable

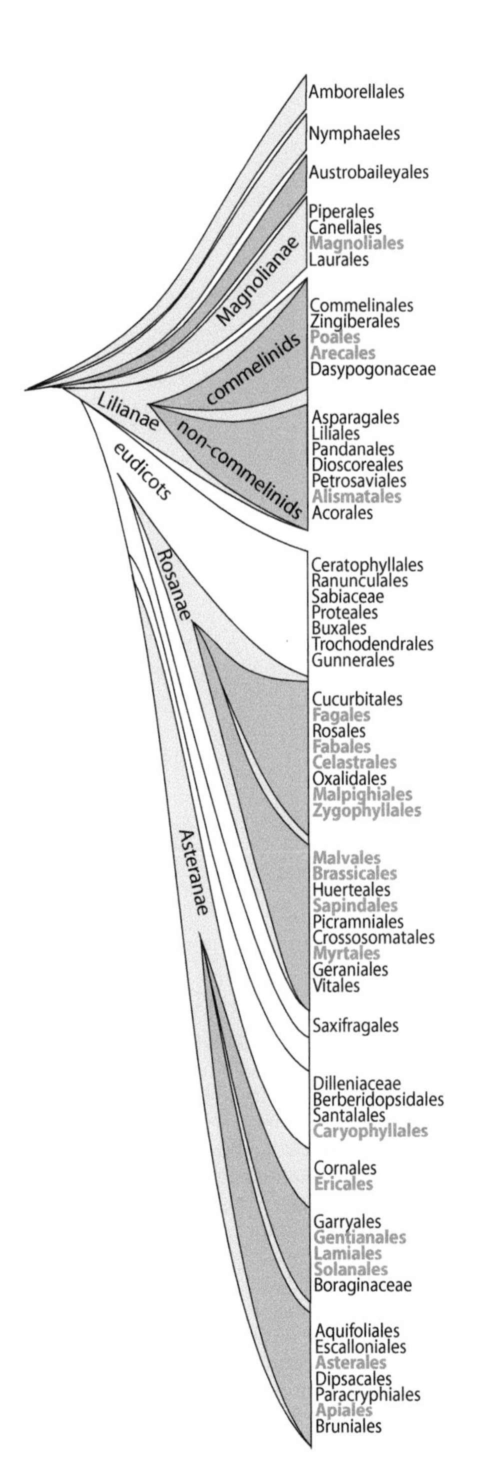

Figure 9.3. **The phylogenetic distribution of halophytes.** Halophytic angiosperms have evolved independently in several major clades. Those in the Alismatales are three families containing 80 species of "seagrass." "Mangroves" are woody halophytes found in several eudicot orders. Temperate salt-marsh and arid-zone halophytes are found in both monocot and eudicot orders. (Data from Flowers TJ, Galal HK & Bromham L [2010] *Funct Plant Biol* 37:604–612.)

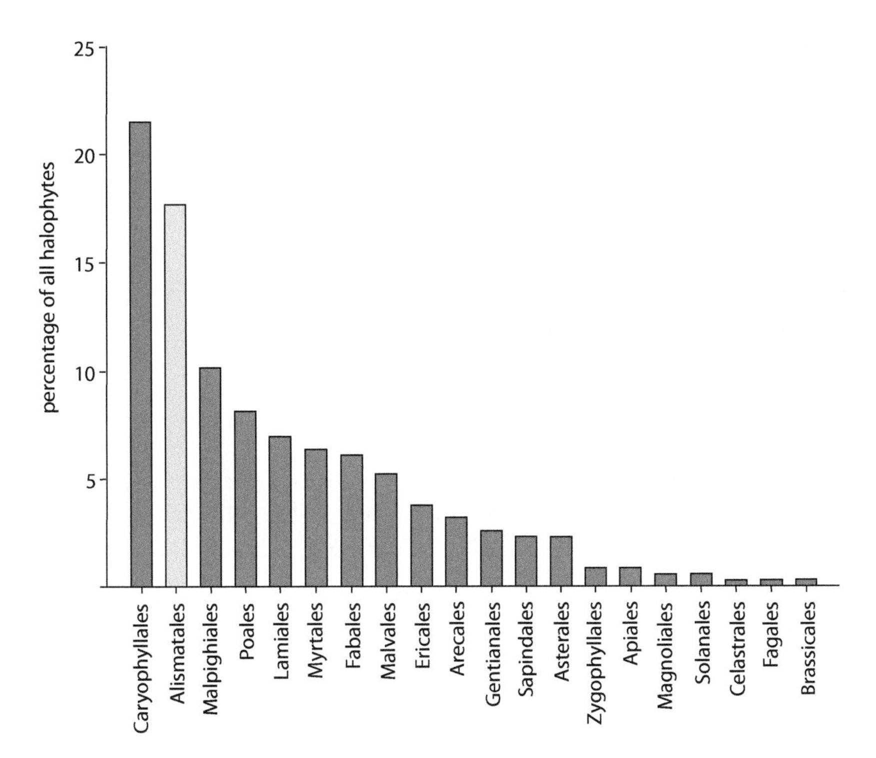

Figure 9.4. The taxonomic distribution of halophytes. There are about 320 true halophytes. The halophytes in the Alismatales are all marine sea grasses. There are more than twice as many halophytes in the Caryophyllales as there are in the next most halophyte-rich terrestrial plant order, the Malpighiales. (Data from Flowers TJ, Galal HK & Bromham L [2010] *Funct Plant Biol* 37:604–612)

Table 9.2. Crop plants that benefit from Na addition. Under conditions of K deficiency a considerable range of crops can benefit from Na addition. The table lists crops that, even when K-sufficient, derive some benefit from Na addition. Those that derive a large benefit from Na addition have halophytic ancestry

Slight to medium benefit	Large benefit
Cabbage	Celery
Celeriac	Mangel
Cassava	Sugar beet
Horseradish	Swiss chard
Kale	Red beet
Kohlrabi	Turnip
Mustard	
Radish	
Oilseed rape	

(From Subbarao GV, Ito O, Berry WL & Wheeler RM [2003] *CRC Crit Rev Plant Sci* 22:391–416. Taylor and Francis Group LLC.)

because of salinization. Wheat is more salt sensitive than barley, and the switch from one crop to the other as soils became saline is evident across ancient Mesopotamian civilizations. Fresh water contains numerous solutes (Table 9.3), so irrigation of crops with fresh water adds not only water but also solutes to the soil. High evaporation rates and/or high water tables decrease the **leaching** of solutes from soils and favor a net build-up of salts in the rooting zone (Figure 9.5). It has been suggested that differences in these processes between Mesopotamia and ancient Egypt account for the

Table 9.3. The solutes in fresh water. Seawater has a high concentration of salts, whereas fresh water is generally defined as having less than 0.5 g salt l^{-1}. However, the concentrations of solutes in fresh water vary significantly, with many having a characteristic composition. The ranges for fresh water listed in the table are for different sources and sampling times in the USA, together with average values for seawater

Solute	Fresh water (mg l^{-1})	Seawater (mg l^{-1})
Calcium	0.8–22	400
Magnesium	0.14–17	1350
Sodium	0.46–14	10,500
Potassium	0.11–0.5	380
Bicarbonate	4–129	28
Sulfate	1.3–36	185
Chloride	0.06–33	19,000
Silica	0.3–30	3

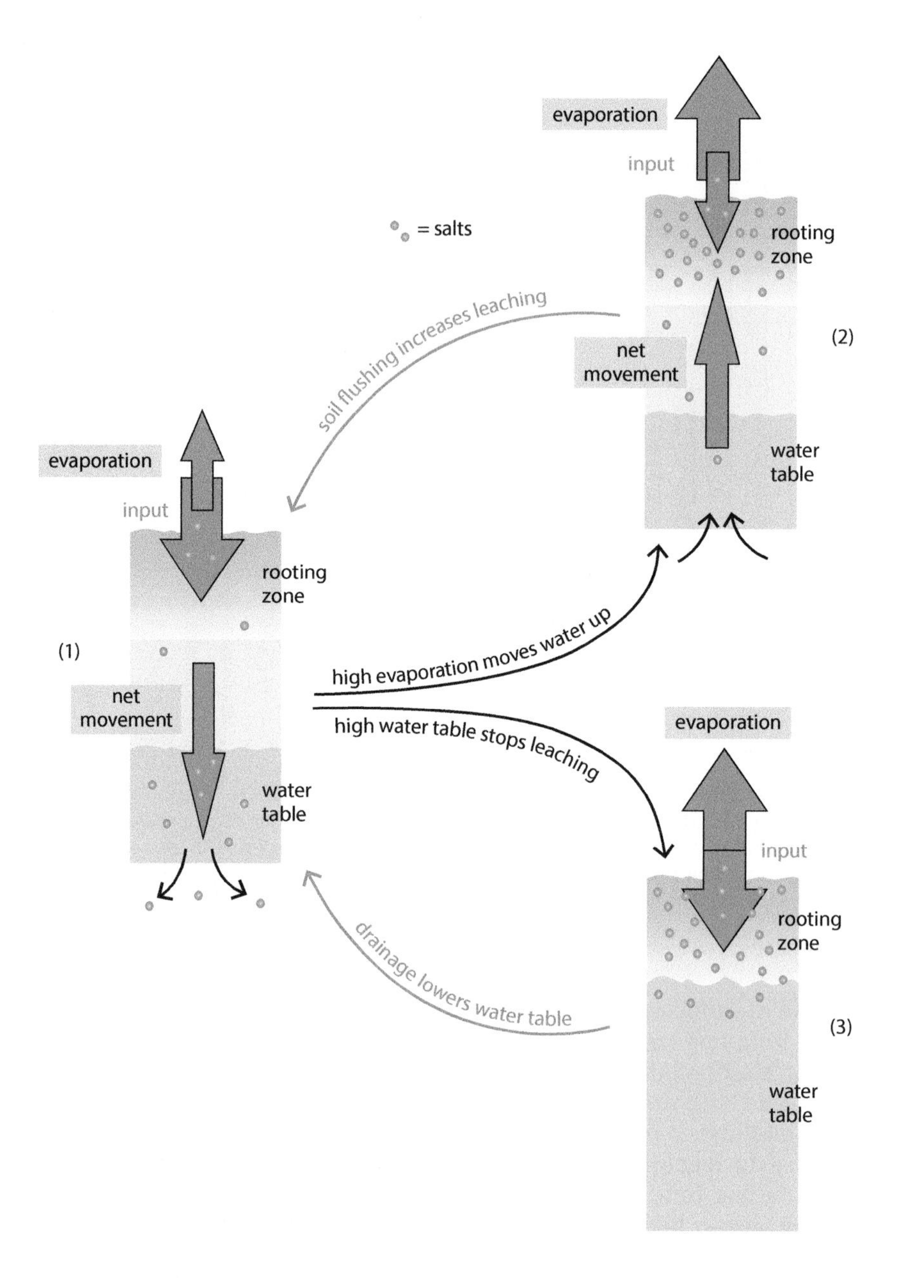

Figure 9.5. The salinization of soils. (1) In soils in which the water table is below the rooting zone, and input of water from irrigation and/or precipitation exceeds evaporation, there is a net movement of water down the soil profile, minimizing the potential for accumulation of salts around the roots. (2) In soils with a net upward movement of water, often because of high evapotranspiration rates, salts can accumulate in the rooting zone—a process that can be reversed by regular flushing of the soil with water. (3) High water tables can also inhibit the removal of salts from the rooting zone, and if evaporation rates are significant can lead to the accumulation of salts in the rooting zone—a process that can be alleviated by soil drainage.

differences in longevity of their civilizations (Box 9.1). These different fates illustrate what is necessary for sustainable irrigation systems—regular soil leaching down to a low water table, and minimization of evaporative build-up of solutes. It is possible to fulfill these criteria in many, although not all, irrigated systems, but often only with sustained effort, so sustainable irrigation practice—where it is possible—can be costly.

There are a number of reasons why plant responses to salinity might be even more important for agriculture in the foreseeable future than they have been in the past. First, salinity problems in irrigated soils can take many years to reach critical levels, and a large proportion of currently irrigated soils have been brought into production only in the last few decades—for example, the irrigated land area on Earth nearly tripled between 1950 and 2000. Second, it is likely that, unless forests are cleared, all of the land suitable for non-irrigated agriculture using current crops is already being used. If increased food production to meet the needs of a growing human population requires the use of more land area, which it might well do because continuing increases in yield from current agricultural land are now stalling, then much of the increase in area will have to be from irrigating marginal

BOX 9.1 ANCIENT MESOPOTAMIAN CIVILIZATIONS AND THE SUSTAINABILITY OF IRRIGATED AGRICULTURE

In ancient Mesopotamia, flood-irrigated systems of the wide, flat Tigris and Euphrates valleys had a high water table from January to April, caused by rainfall in uplands to the north and west (Figure 1). This was followed directly by high rates of evaporation during the hot season from May to August. Thus, when the rooting zone of soils in Mesopotamia was wet, it had a propensity for both evaporation because of the hot season, and decreased leaching because of the high water table. This promoted salinization of the soil. It has been suggested that, in addition to the most significant factor, namely drought, soil salinization probably contributed to the relatively rapid rise and fall of numerous civilizations in Mesopotamia, such as those of Assyria, Akkad, Babylon, and Sumer, because it contributed to the difficulties involved in long-term agricultural production.

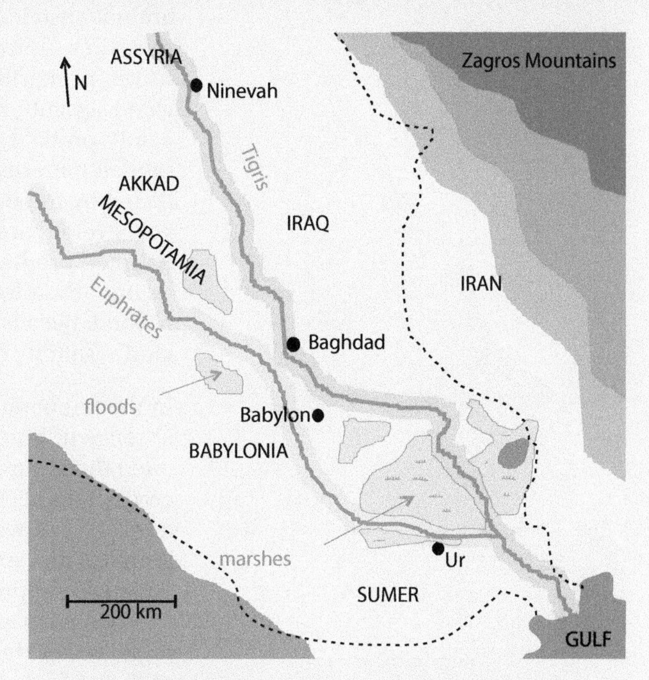

Figure 1. The rivers and marshes of Mesopotamia. The climate and hydrology of the Tigris and Euphrates plain have varied significantly during the last 10,000 years, but there has almost always been a significant flux of water down the rivers, often with associated flooding, in the early part of the year. Extensive canal and dam systems have long diverted this water for irrigation. At the southern end of the rivers there have usually been extensive marshes (light green). By the end of the twentieth century these had mostly been drained, leaving just small patches of marsh (dark green). Large-scale schemes are currently restoring these marshes, which might well have been the original "Garden of Eden."

In ancient Egypt, although there were significant drought periods, soils were probably much less prone to salinization. The flooding of the Nile is caused by rain in East Africa, which arrives in Egypt 9–10 months later (Figure 2). In ancient Egypt, the season of inundation along the Nile (from July to September) was after the period of maximum evaporation potential (April to June) and before the main period of irrigation (October to February). Thus flooding occurred when the water table was low, promoting the leaching of soils. The narrow Nile valley had deep soils, helping to promote leaching out of the rooting zone. Flood waters for irrigation were then available outside the period of maximum evaporation, minimizing the build-up of solutes. These hydrological patterns helped to minimize soil salinization and perhaps contributed to sustained periods of agricultural production in ancient Egypt. The erratic flooding of the Nile is now controlled by the Aswan Dam, which has enabled modern Egypt to avoid floods and droughts associated with variations in the degree of inundation, but it has increased the tendency to salinization along the Nile valley—soil salinization is thus more of a challenge in modern Egyptian agriculture than it was in ancient times. The actual significance of salinization in explaining the differences in the longevity of civilizations in Mesopotamia and ancient Egypt is difficult to gauge, but these differences serve as a reminder that irrigated agriculture is only sustainable in the long term if there is no build-up of solutes in the rooting zone, which usually entails regular soil flushing to low water tables.

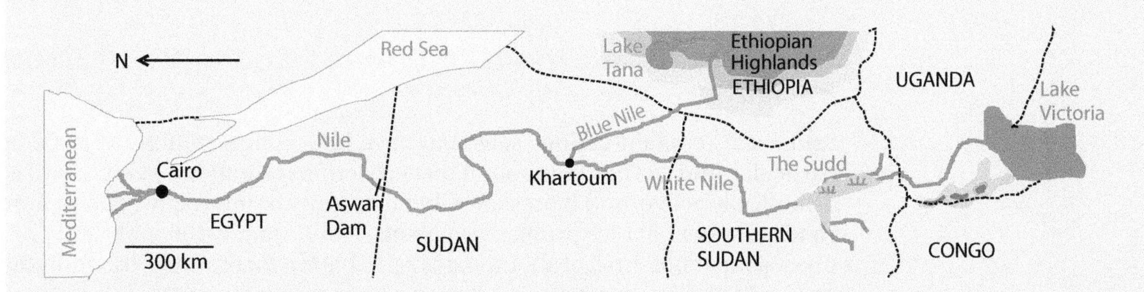

Figure 2. Egypt and the Nile valley. The Ethiopian and other East African highlands provide the water for the flooding of the Nile, but the length of the river is such that the waters do not reach Egypt until 9–10 months later. The properties of the White Nile are produced in part because of the time the water spends in one of the great wetlands of the world, the Sudd, which helps to provide its characteristic color. Before the Aswan Dam was built, when the waters arrived in Egypt the water table was low, so there was substantial leaching that washed solutes out of the soil. Irrigation also did not coincide with the period of maximum evaporation, minimizing the evaporative build-up of solutes.

land that may be susceptible to salinization. Third, irrigated agriculture currently uses about 70% of all the fresh water extracted by humans from rivers, lakes, and aquifers. The current water extraction rates from these sources are not sustainable, and human societies are now entering what, by most measures for most people, is likely to be a prolonged period of water crisis driven primarily by demand for irrigation water. In fact, the spread of irrigated agriculture at the beginning of the twenty-first century slowed significantly, probably limited by the lack of available fresh water. Finally, changing rainfall patterns on a warmer Earth, even in the least severe scenario predicted by the Intergovernmental Panel on Climate Change (IPCC) in 2014, will, overall, probably enhance both evaporation rates and the demand for irrigated production. Crop responses to salinity in irrigated agriculture will be important for human habitation of Earth in the twenty-first century and beyond, because they are vital to confronting what has been called humankind's "salinity dilemma."

In the temperate zones a significant proportion of the shorefront is, in the absence of human destruction, **salt marsh**. In the tropics, **mangroves** are often the natural vegetation of the shorefront, and they probably once covered around 200,000 km^2. Salt marshes and mangrove swamps experience a mix of salt water and fresh water, a mix of terrestrial and aquatic challenges to life, and have a high **allochthonous** nutrient content, giving rise to unusual and productive ecosystems. In comparison with other shorefront habitats, such as beaches or cliffs, salt marshes and mangrove swamps have high species densities, and there are many species that spend at least part of their life cycle in these habitats. A significant proportion of the Earth's halophytes inhabit salt marshes and mangrove swamps, and these habitats often provide a physical buffer between the sea and the land. Although they occupy a greater land area than most constructed sea defenses, they can attenuate storm surges and high tides. In many parts of the world, "flood defense" strategies are being replaced by "flood risk management" strategies in which salt marshes and mangroves have a positive role. In some regions, 30–80% of mangrove swamps have been destroyed, and they are disappearing at a rate of 1–2% a year. This is having a variety of environmental effects, but the link with exacerbated storm surges and tsunamis has been much discussed since the Asian tsunami of December 2004. In addition, if the sea level rises, the response of salt marshes and mangrove swamps will have consequences for the surrounding low-lying areas, in which there are often significant human populations. The halophytes of salt marshes and mangrove swamps can not only provide physiological insights on which global food security might depend, but they also produce the biomass on which numerous species depend, and that forms a significant buffer between the land and the sea.

Exposure to salt induces osmotic and ionic stresses in plants

Saline soils contain various salts, but have high concentrations of NaCl in particular, and it is this compound that predominantly affects plants. NaCl is osmotically active, and if present either externally or internally adds significantly to water balance problems in plants—for instance, 100 mM NaCl produces a water potential of about –0.4 MPa. In barley, for example, the addition of 75 mM NaCl to hydroponic solution causes a cessation of leaf elongation within minutes. Similar but less sensitive effects have been shown for root growth. Almost simultaneously there is a rapid increase in Ca concentration in the cytosol. These effects are also inducible by other osmolytes, and plants rapidly but partially recover to grow at a reduced rate that depends on ψ. Within hours, such stress initiates hormonally regulated changes in stomatal conductance and root hydraulic conductivity that reduce water uptake, mainly by ABA-mediated pathways. These reductions in growth and water uptake are associated with increased production of reactive oxygen species (ROS) in chloroplasts, peroxisomes, and mitochondria (Figure 9.6). In chloroplasts, to which CO_2 supply is reduced by decreased stomatal conductance, the effects are primarily on the water–water cycle at PSI and on singlet oxygen production at PSII, because electron transfers become directed towards O_2. In peroxisomes, antioxidants are consumed, and in mitochondria, manganese superoxide dismutases (Mn-SODs) are consumed and alternative oxidase activity increases. These initial responses are essentially the same as those induced by drought. In *Arabidopsis*, microarrays have demonstrated that more than 2000 genes show altered expression after 3 h of salt exposure, and more than 3000 genes do so after 24 h. About 75% of these genes show altered expression in response to other abiotic and biotic stressors, with a significant number of these changes being associated with ABA- or Ca-mediated responses and oxidative stress. Acute salt exposure thus produces an osmotic shock which is similar to that induced by drought, and which provokes responses that overlap with those to other stressors.

In plants, osmotic shock induced by salt exposure initiates ABA-dependent and ABA-independent responses that postpone osmotic stress. The trigger for an increase in ABA concentration is probably a change in turgor pressure detected via the cytoskeleton. The first response to increased ABA levels is

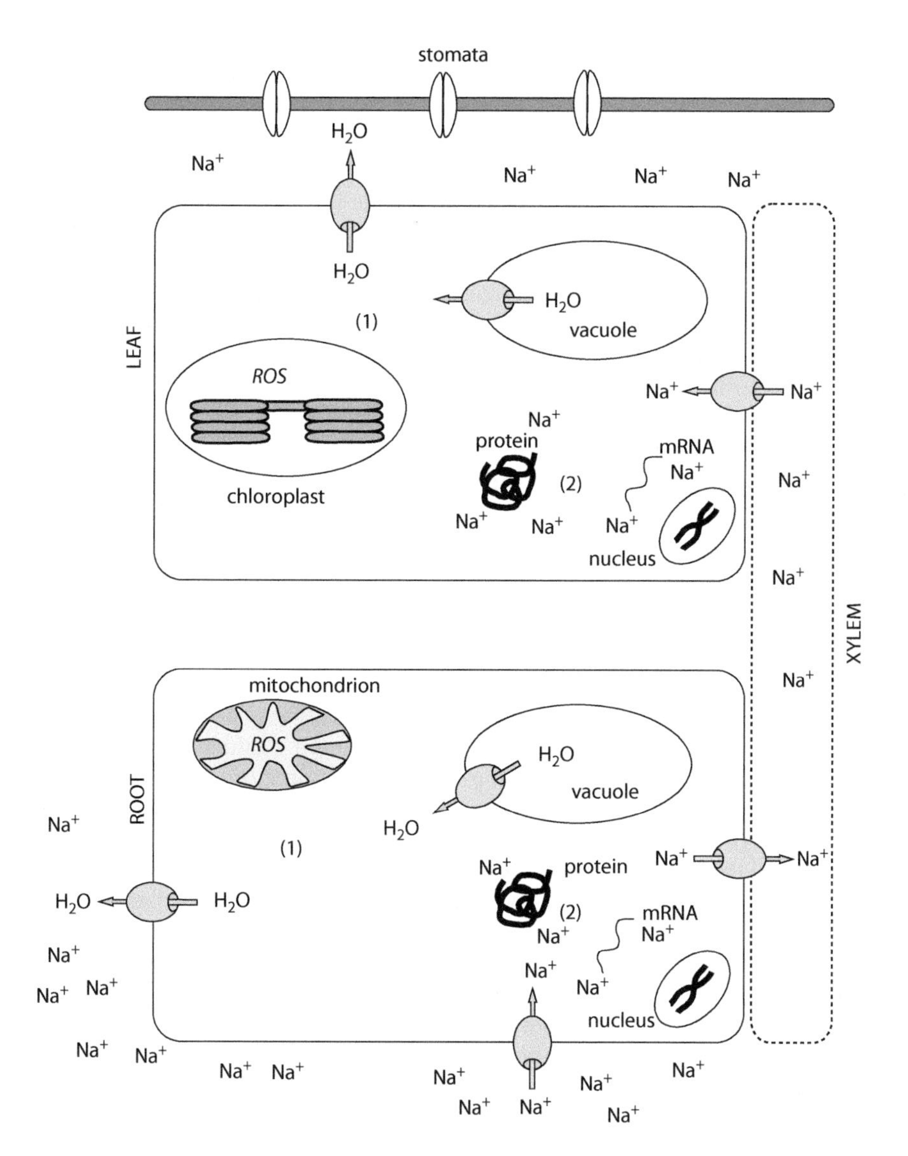

Figure 9.6. Sodium-induced toxicity in plant cells. (1) External and apoplastic Na^+ quickly causes water to be drawn out of plant cells, causing osmotic stress, closing the stomata, and increasing the production of reactive oxygen species (ROS) in plastids of the root and shoot. (2) Na^+ can enter plant cells via alkali metal transporters, and gradually accumulates inside the cells, causing toxicity effects—particularly to proteins, which lose their conformation as Na^+ concentrations increase, but also to other molecules, such as mRNA, which become unstable. (Organelles not drawn to scale.)

a decrease in stomatal aperture that reduces transpiration, causing a rapid rise in leaf temperature. In guard cells, receptors of at least three types in both the plasma membrane and the cytoplasm bind to ABA, initiating responses mediated through phospholipase D, inositol 1,4,5-trisphosphate (IP_3), and NO, and increasing Ca^{2+} concentrations in the cytoplasm to trigger K^+ and Cl^- efflux and decrease guard cell turgor. The stomatal response is coordinated via transcription factors and miRNAs that have almost all been shown to respond to salt-induced osmotic stress. ABA-dependent pathways directly increase the activity of bZIP transcription factors that bind to ABA-response element (ABRE) promoter regions of many genes associated with salt stress, but also increase the synthesis of MYB and NAC transcription factors. ABA-independent pathways result in increased activity of transcription factors (TFs) that bind to drought-responsive elements in promoter regions. Numerous TFs in these families that are activated by salt have been identified, as have numerous miRNAs that help to coordinate responses. The molecular responses to salt-induced osmotic stress are also, to a significant extent, characteristic of heat stress. Thus, interestingly, prior exposure to heat can prepare plants to deal with some of the short-term effects of salt stress.

With NaCl exposure for longer than a few hours, the Na and Cl concentrations in non-tolerant plants start to increase, often interfering significantly with their mineral nutrition, particularly that of K. Uptake of NaCl results in high NaCl fluxes in the transpiration stream, and salt build-up in leaves with high transpiration rates (mostly older leaves). This causes the dehydration and visible necrosis of older leaves ("scorch") that is characteristic of salt injury, and results in leaf death within days or weeks. Cytoplasmic Na concentrations of 10 mM start to affect protein structure and function, with increases to 100 mM causing major metabolic problems. Group 1A metals (alkali metals such as Na and K) have a strong tendency to hydration and high exchange rates when they coordinate with organic ligands. They do not bond to organic macromolecules, but interact via strong electrostatic effects. In the aqueous solution of the cytoplasm these interactions are crucial to the structure and function of proteins. In addition, Group 1A metals act as co-factors or allosteric effectors for specific enzymes. In plants, K is the preferred metal for these interactions, and increasing Na concentrations interfere significantly with protein conformation (Figure 9.6). After 2–3 weeks of low to moderate salt exposure, the difference in growth between salt-sensitive and salt-tolerant crop varieties becomes significant because of the difference in their ability to deal with ionic effects. Depending on the NaCl dose and the salt sensitivity of the species, osmotic and ionic stresses result in effects ranging from reduced growth (with concomitant changes in many other phenological parameters) to the death of leaves or entire plants.

For many woody and leguminous species in saline soils, the effects of Cl^- ions are more significant than those of Na^+ ions. Cl^- is an essential nutrient which, as the hydrated anion, regulates protein conformation and acts as a counter ion to cations during membrane transport (for example, to K^+ during regulation of stomatal aperture), and is an essential co-factor in hydrolytic proteins during photosynthesis. At supra-optimal Cl^- concentrations all of these functions are disrupted, with the primary visible symptom of necrosis occurring in tissues with high transpiration rates, especially leaf margins and older leaves. Plant uptake of NO_3^- and SO_4^{2-} is also affected by Cl^- concentrations. For sensitive species, Cl^- toxicity occurs at tissue concentrations of 100–175 mM. Agricultural genera in which Cl^--induced effects are important to salinity responses include *Vitis* (grapevine), *Citrus*, *Medicago*, and *Trifolium*. Although the ionic effects of salinity are primarily due to Na^+ and Cl^-, other elements, in particular boron (B), can occur at significant concentrations and cause toxicity.

Sodium can enter plants via symplastic and apoplastic pathways, but can be removed from the cytoplasm

Plant cells generally have cytoplasmic concentrations of 1–5 mM Na, so even in the least saline soil-plant systems the electrochemical driving force on Na^+ ions is strongly inward across the plasma membrane. Na^+ can therefore enter the cytoplasm through passive transport systems, but efflux has to be active. The influx of Na^+ into plant cells is generally not saturable at the concentrations that are found in saline soils. Non-selective cation channels (NSCCs) are numerous in plant membranes. There are classes that are voltage sensitive or insensitive, and that are gated by cyclic nucleotides, glutamate, and ROS. Ordinarily they probably contribute significantly to the transport of divalent ions, especially Ca^{2+}. In plant cells, exposure to Na^+ does not affect membrane polarization, but **patch-clamp** studies of voltage-insensitive (VI)-NSCCs have shown Na^+ transport through them. In published selectivity series, selection against Na^+ in favor of K^+ is generally significant and is greater than selectivity against Cs^+, an ion with less chemical similarity to K^+. Primary Na^+ influx to some plants has been shown to have pharmacological properties characteristic of VI-NSCC, with known blockers of these channels such as lanthanum (La^{3+}) and gadolinium (Gd^{3+}) reducing Na^+ influx. However, Na^+ influx through NSCCs is saturable at 20–40 mM Na^+, and decreased by Ca^{2+} concentrations in the range found in most saline soils. Given their diversity and ubiquity, NSCCs are probably a route for the entry of some Na^+, but a particular class of them, or even NSCCs as a whole, seem unlikely to be the only route for influx of Na^+.

The chemical similarity of Na^+ to K^+ means that Na^+ can be transported by many of the systems that evolved for K^+ uptake. Studies using model plants have shown that KUP/HAK/KT and AKT families of transporters are probably the dominant systems for K^+ uptake from soil. Numerous members of these families transport Na^+ at high rates, and starving plants of K^+, which increases expression of these systems, increases Na^+ influx. K^+ flux through KUP/HAK/KT systems is sensitive to NH_4^+, as is Na^+ influx, with numerous studies having shown increased Na^+ influx in NH_4^+-grown plants. However, altered expression of KUP/HAK/KT and AKT transporters does not always alter Na^+ influx, and the latter often decreases their expression. HKT1 was the first high-affinity K^+ transporter to be isolated from plants, but it is expressed mainly in the root cortex and in mesophyll cells around the leaf vasculature, although HKT2 transporters in roots can enable Na^+ influx in some species. Cation-chloride co-transporters (CCCs) that have well-established roles in osmoregulation in animal systems also occur in plants, and there is increasing evidence that they might be significant in Na^+ and Cl^- influx. Overall, there is not a particular set of transport proteins through which primary Na^+ influx occurs, but a variety of entry routes in different taxa and ionic scenarios. It seems that after evolving a physiology focused on K^+, plant entry of Na^+ is dependent on it "leaking" through systems that evolved for transport of other ions. However, the substantial inward driving forces on Na^+ do mean that influxes of this ion through transporters can be rapid and significant.

The suberized Casparian strip in the endodermis that surrounds the vascular strand of roots seldom provides a complete barrier to apoplastic movement of ions. Developing root tips where much ion uptake occurs take time to differentiate an endodermis, lateral roots disrupt it, and environmental conditions do not always favor its development. Apoplastic bypass is significant for the influx of many ions, including Na^+, for which influx correlates with the flow of apoplastic dyes such as **PTS** and radiotracers of apoplastic flows such as radiolanthanum. Apoplastic bypass is the major route of Na^+ entry in rice, and is significant in many species because their endodermal

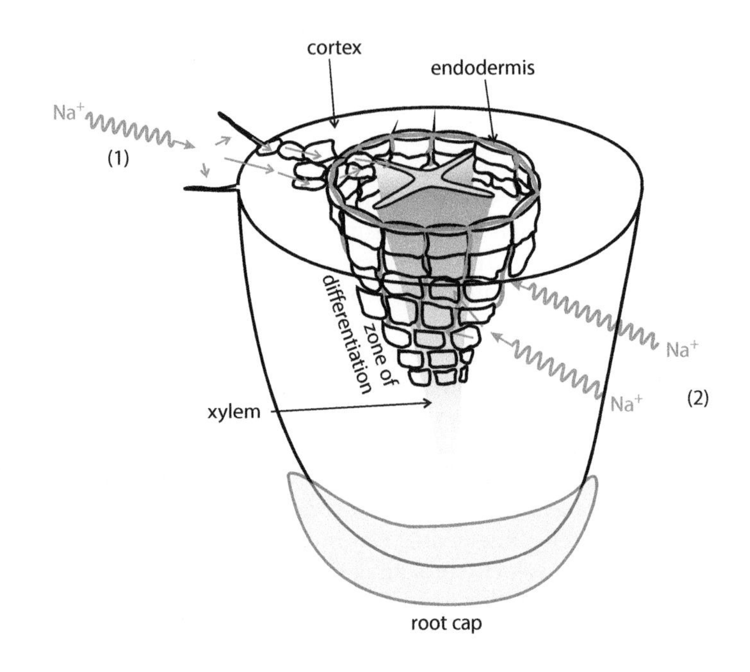

Figure 9.7. Pathways of influx of Na^+ to plant roots. (1) Na^+ can enter plants through K^+ transporters in the root hairs and epidermis, and can bypass the endodermis by traveling through the symplast to the vascular tissue. (2) In many non-halophytes, often in the zone of differentiation in the root, there is also significant apoplastic bypass of the endodermis by Na^+, which then enters the vascular tissue directly.

mutants have altered Na^+ influx. Thus Na^+ influx to plants can occur through both symplastic and apoplastic pathways (Figure 9.7). The contribution of apoplastic influx is highlighted by root cortical adaptations that decrease Na^+ influx—for example, salt-tolerant brassicas have **phi cells**, some grasses have silica deposits, and mangroves can have hydrophobic root barriers.

Arabidopsis can grow at up to 50 mM NaCl because significant Na^+ efflux minimizes Na^+ accumulation, but its overly salt-sensitive mutants (*sos*) are very sensitive to millimolar concentrations of Na^+, and deficient in Na^+ efflux. The salt overly sensitive (SOS) pathway is related to the salt-inducible kinase (SIK) network in mammals (which includes the Li^+-sensitive phosphatases that are a target for lithium therapy for bipolar disorder). In plants, Ca^{2+} that has moved into the cytoplasm following salt exposure binds to SOS3 via EF hands with sequences characteristic of calcineurins, promoting dimerization (Figure 9.8). SOS3 also has a β-subunit similar to those with phosphatase activity in calcineurins, but has no phosphatase activity, and requires *N*-myristoylation in order to function. In its dimeric form SOS3 binds to SOS2, which is a serine/threonine kinase (that is, it phosphorylates the OH group in serines and threonines). When SOS2 is not bound to SOS3, an FISL motif inhibits its kinase activity. Dimeric SOS3 fits into the cleft formed by the FISL motif, exposing the kinase.

The primary target of the Ca-SOS3–SOS2 kinase complex is the Na^+-efflux pump SOS1 (Figure 9.8). The catalytic region of SOS2 is similar to those of the SNF kinases in yeast and the AMP-activated kinases (AMPKs) in animals, but SOS2 has a different regulatory domain, an auto-inhibitory FISL motif, and a region with phosphatase activity (PPI). The PPI region can remove phosphate ions from SOS1, thereby activating it. SOS1 is an Na^+-efflux pump. It is an Na^+/H^+ antiporter, energized by the efflux of H^+ via the plasma-membrane H^+-ATPase that creates an electrochemical gradient for proton influx through SOS1—and for each proton traveling inwards there is efflux of an Na^+ ion. SOS1 shows much similarity to other Na^+/H^+ antiporters, all of which are of fundamental biological importance, as in the majority of organisms they are the primary regulators of cellular Na^+ concentration, cellular pH, and cell volume. *Arabidopsis* plants that have been engineered to overexpress SOS1 have increased salt tolerance, and silencing

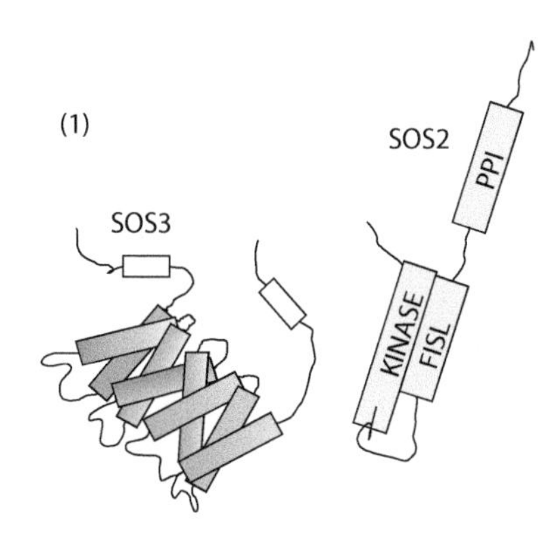

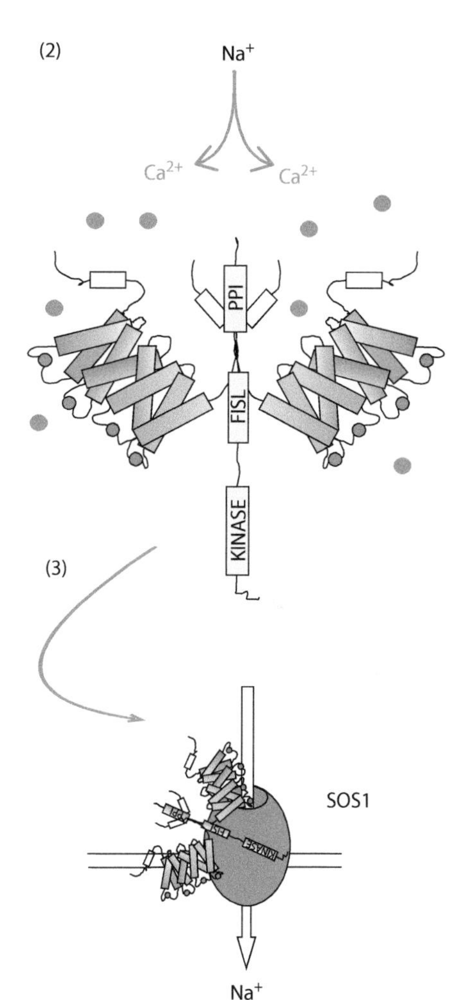

Figure 9.8. The salt overly sensitive (SOS) pathway for efflux. (1) SOS3 and SOS2 occur as monomers in the cytoplasm. SOS2 has a FISL motif that inhibits its own kinase activity. (2) Na^+ in the cytoplasm triggers an influx of Ca^{2+}, which binds to SOS3, causing it to dimerize and bind to SOS2, exposing its kinase. (3) The SOS3–SOS2 complex binds to SOS1, phosphorylating it and triggering efflux of Na^+. SOS1 has much similarity with other Na^+/H^+ antiporters in the living world. PPI, protein phosphatase interaction.

its expression increases Na sensitivity in salt-tolerant tomato lines. SOS1 has a C-terminal tail that is different to that of other Na^+/H^+ pumps and is likely to be involved in Na activation. The *SOS1* transcript is only stable in the presence of Na^+. In addition, SOS1 interacts with the GIGANTEA TF that controls flowering (which probably explains the decrease in flowering that can occur under salinity stress), and is also involved in triggering changes in root architecture.

SOS1 is highly expressed not only in root tip epidermis but also in cells surrounding the vascular system, especially the xylem. Root tip cells are poorly vacuolated and cannot remove Na^+ from the cytoplasm into the vacuole, thus increasing the importance of Na^+ efflux. In terrestrial plants the SOS pathway seems likely to be the first line of defense against increased Na^+ levels in soils. This Na^+ efflux system is likely to have persisted in terrestrial plants because of exposure to low concentrations of Na. Na^+-ATPases are not expressed in higher plants, but their use by some mosses suggests that they were lost during adaptation to low-Na terrestrial environments. The SOS pathway probably links Na^+ efflux in plants to Ca-signaling networks affected by salt stress, including ABA-mediated effects. However, the efflux of Na^+ from root cells is not sufficient to explain the tolerance of prolonged or intense salt stress that is observed in many species, because Na^+ efflux has limited capacity, Na^+ accumulation in the apoplast (which is where SOS1 enables Na^+ efflux) is a prominent part of salt stress, and many salt-tolerant plants in fact have high internal Na^+ concentrations.

Salt-tolerant plants compartmentalize sodium, and halophytes also control potassium:sodium ratios

During salt exposure, the more salt-tolerant varieties of most cereal crops accumulate lower leaf concentrations of Na^+ than do salt-sensitive varieties. Cereal salt tolerance is a quantitative trait, and crosses between salt-tolerant and salt-sensitive lines have enabled genetic loci associated with the trait to be investigated. These loci have genes encoding HKT1 transporters expressed in vascular tissue that take up Na^+ from xylem sap, decreasing concentrations in the transpiration stream. Na^+ transport into root cell vacuoles then controls root cytoplasmic Na^+ concentrations. For example, in the salt-tolerant *indica* rice variety Nona Bokra, the locus *SCK1* accounts for about 40% of salt tolerance, and includes genes that encode HKT1. Durum wheat (*Triticum turgidum* subsp. *durum*) is generally more salt sensitive than bread wheat (*T. aestivum*), but in salt-tolerant lines of *T. turgidum* the locus of salt tolerance is associated with an HKT transporter, HKT8. The *Kna1* salt-tolerance locus of *T. aestivum* is a homolog of *Nax2*. There are multiple HKT transporters in most monocot genomes, perhaps eight or more, but fewer in eudicot genomes, which suggests that retrieval of Na from the transpiration stream might be a particularly important mechanism of moderate salt tolerance in monocots. The net effect of Na^+ retrieval from the transpiration stream, at a given salinity, is to increase the K^+/Na^+ ratio in the leaves of tolerant plants compared with those of sensitive plants. *HKT* genes of salt-tolerant lines encode a serine residue in the loop of the first pore, which probably makes them highly selective for Na^+. The association of transport

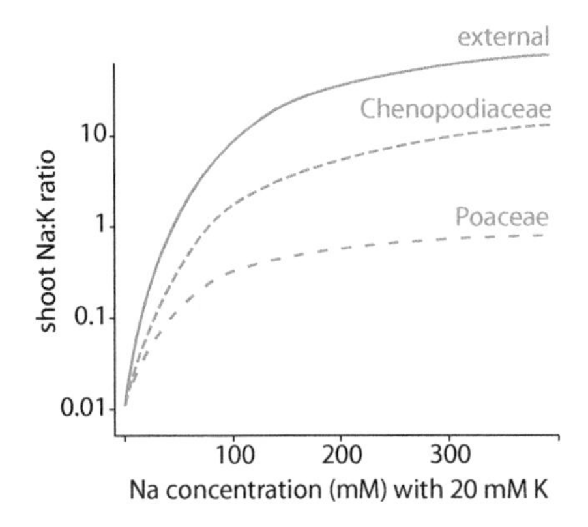

Figure 9.9. Halophytism and Na:K discrimination. Salt-tolerant plants, and in particular halophytes, discriminate against Na during both uptake from soil and transfer to shoots. In the scenario shown in the figure there is 20 mM external K but an increasing external Na concentration. The Na:K ratios in shoots of halophytic Chenopodiaceae are closest to the external ratios, but still represent up to 10-fold discrimination against Na. Halophytic Poaceae have the highest shoot discrimination against Na, and proportionally the lowest leaf Na^+ concentrations. For Chenopodiaceae, $S_{K:Na}$ = 4 (*n* = 16 species), 6 (*n* = 31), and 9 (*n* = 39) at low, medium, and high Na concentrations (20, 200, and 400 mM), respectively, and for Poaceae, $S_{K:Na}$ = 22 (*n* = 5), 42 (*n* = 7), and 60 (*n* = 11), respectively. (Based on $S_{K:Na}$ from Flowers TJ & Colmer TD [2008] *New Phytol* 179:945–963.)

genes that aid cellular compartmentation with salt tolerance in some major crop plants is clear, and is of great importance in breeding salt-tolerant crops, but does not fully explain either their salt tolerance or the more extreme salinity tolerance of halophytes.

Saline terrestrial ecosystems are rich in mineral ions, and halophytes have unusual concentrations and ratios of ions. Before the Industrial Revolution, plants were an important source of alkaline ash for a variety of processes, with halophytes being combusted to produce **soda ash** (mostly $NaCO_3$), and glycophytes being combusted for **potash** (mostly KCO_3). Glycophytes, in which Na^+ influx can occur through K^+ systems, can be overwhelmed by Na during salt exposure if its efflux does not occur at a sufficient rate, whereas in halophytes there is significant discrimination against Na^+ during uptake. Soil-to-leaf flux calculations show that halophytes minimize the Na:K ratio in leaves by preventing over 90% of the Na that arrives at the root surface from reaching the leaves. In monocotyledonous halophytes, discrimination against Na is particularly strong (Figure 9.9), but even in dicotyledonous plants in the Chenopodiaceae (which have the highest concentrations of Na, and were formerly used for making soda ash in many parts of the world) the change in Na:K ratio during uptake reflects up to 10-fold discrimination against Na. Thus, although halophytes have high Na concentrations, these are not as high as they would be if these plants did not discriminate against Na during uptake. The ability to minimize Na uptake is vital to salt tolerance and, for example, in salt marshes dominated by *Spartina,* significantly enhances Na concentrations in the root zone.

In saline substrates it is difficult for glycophytes to achieve K homeostasis. One symptom of salinity stress is K deficiency, because Na can be taken up instead of K, but cannot fulfill its physiological roles. Na competes with K for transport proteins, but its influx also depolarizes the plasma membrane, decreasing the driving forces for K uptake. In the halophyte *Suaeda maritima,* K uptake rates twice as high as those in cereals have been measured, and, in the absence of Na, halophytes can accumulate very high K concentrations. In glycophytes exposed to Na, K^+ efflux rates increase, and in some crops the ability to decrease K^+ efflux can be related to salt tolerance. The halophyte *Thellungiella halophila* (salt cress) is significantly more selective for K during uptake than its close relative *A. thaliana.* Such characteristics suggest that plant salt tolerance is underpinned by the ability to maintain K^+ influx from substrates with high Na:K ratios. For many species, high Ca concentrations moderate the effects of salinity, and the addition of Ca-containing compounds to soils has long been part of agronomic strategies for dealing with mild salinity. High Na concentrations interfere with Ca uptake, and Ca deficiency is often reported in salt-affected crops. Low Ca concentrations are associated with symptoms such as a decrease in membrane integrity, but plant responses to salinity are also mediated by Ca fluxes in the cytoplasm that are disrupted by salinity-induced Ca deficiency.

At high concentrations, Na^+ has toxic effects in the cytosol, but these can be avoided by removal of the ion into the vacuole. In all plants that have been investigated to date, Na^+ transport into the vacuole occurs almost entirely via NHX—an Na^+/H^+ antiporter in the tonoplast. NHX transporters are ubiquitously expressed and have roles in numerous functions, but their expression is affected via the SOS pathway, which probably serves to increase their expression during salt exposure (Figure 9.10). NHX expression is activated by MYB transcription factors. Numerous forward and reverse genetic approaches have been used to link NHX expression directly with salt tolerance. Vacuolar H^+-translocating pyrophosphatases have a role in driving Na^+ transport through NHX. In particular, the expression of *Arabidopsis* vacuolar H^+-pyrophosphatase (AVP) has been shown to increase Na^+ influx to the

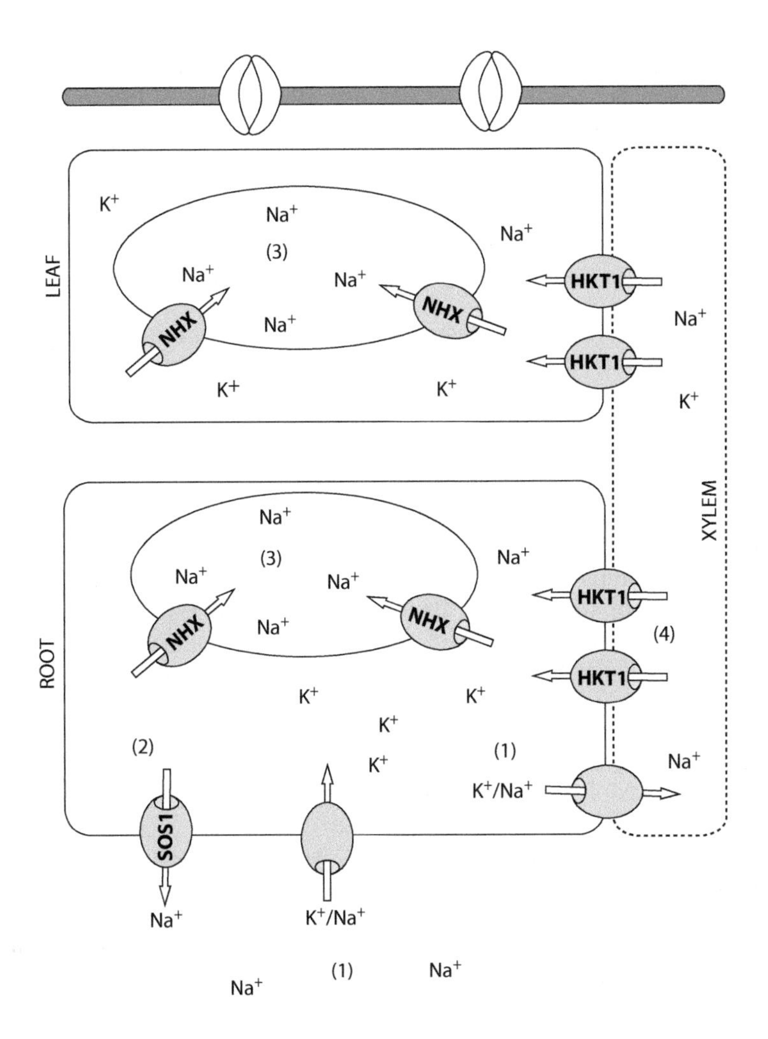

Figure 9.10. Physiological mechanisms for controlling internal distribution of Na. (1) Halophytes can discriminate against Na^+ during uptake, decreasing its rate of entry into the root and transfer to the xylem. (2) The efflux of Na^+ through SOS1 can help to decrease its concentration in the cytoplasm. (3) The transport of Na^+ into the vacuole via NHX transporters can help to decrease its concentration in the cytosol. (4) The retrieval of Na^+ from the transpiration stream by HKT1 can help to decrease the rate of Na^+ arrival at the shoot.

vacuole—its activity produces the electrochemical gradient of H^+ across the tonoplast that provides the driving force for Na^+ influx. The high concentrations of Na in halophytes are almost entirely in the vacuole as a result of NHX activity. Thus increasingly dramatic control over the internal distribution of Na, together with adaptations to maintain a physiologically workable K:Na ratio, helps to explain the increasing degree of salinity tolerance from crops to halophytes.

At high salinity, halophytes synthesize specialized metabolites in order to adapt to osmotic challenges

The environments that halophytes inhabit provide a chronic osmotic challenge. If a glycophyte is grown on a saline soil with an EC of 4 dS m^{-1} or more in the medium to long term, it faces problems of dehydration. Most glycophytes generate a maximum total water potential (ψ) in their roots of -2 MPa, but in saline soils ψ can be significantly lower than this, and the ψ of halophyte roots can be as low as -10 MPa. Direct measurements of halophytic and glycophytic root cells have found no differences in ψ_p and ψ_m that might account for the large negative ψ_{root} in halophytes. However, they have found low ψ_s. In saline soils in almost all halophytes low ψ is generated by elevated Na tissue concentrations that produce a low ψ_s. However, almost all plant enzymes are sensitive to high Na concentrations, and there is little evidence that key enzymes in halophytes are less sensitive to Na than those in glycophytes. In halophytes under intense salt stress, as well as in glycophytes with moderate salt tolerance, Na^+ is sequestered at high concentration in the vacuole. At the concentrations attained in halophytes (for example, 200–400 mM), Na^+ acts as a significant osmoticum. This concentration in the vacuole risks dehydrating the surrounding cytosol. Adaptation to high salinity requires not only significant removal of Na^+ from the cytosol to the vacuole, but also reductions in ψ in both root and shoot cells via mechanisms that do not dehydrate the cytosol, and that minimize Na^+ toxicity. Like marine algae, halophytes use high concentrations of inorganic and organic ions to help to generate low ψ and moderate the effects of Na^+ toxicity.

Many halophytes synthesize organic solutes, sometimes targeted at the cytosol, which decrease ψ_s. These solutes are uncharged at cytosolic pH, osmotically active, and not toxic at high concentrations—that is, they are compatible with metabolic functioning. Halophytes thus utilize the large amounts of osmotically active Na^+ available to them, together with the properties of compatible solutes, to generate low ψ_{root} (Figure 9.11). The ratio of inorganic to organic solutes used to generate ψ varies between species, as does the identity of the organic compounds. Halophytes in the Chenopodiaceae have higher Na^+/K^+ ratios than do species in the Poaceae, and synthesize glycine-betaine as a compatible solute. The Na^+ concentration is probably higher in the vacuole than it is in the cytosol, although the presence of glycine-betaine and other molecules in the cytosol reduces the activity of Na^+ significantly, which means that it can be present at elevated concentrations in the cytosol without causing toxicity. The Chenopodiaceae contain a higher number of halophytes (over 380) than any other family (Figure 9.4), perhaps reflecting the effectiveness of Na^+/glycine-betaine adjustments in balancing ψ. On a whole-tissue basis, the concentration of glycine-betaine is not sufficient to contribute to ψ, but if glycine-betaine is located primarily in the cytosol it can contribute to a localized reduction in ψ. Although measurements of the cellular distribution of compatible solutes are difficult to obtain, glycine-betaine concentrations of 320 mM and 0.24 mM have been measured in the cytosol and vacuole, respectively, of the chenopod *Atriplex gmelini*. Glycine-betaine production has been detected in halophytes from numerous families. Its biosynthesis involves the dehydrogenation of choline, which can be derived from various amino acids, but especially **alanine** and **proline** (Figure 9.12). Betaine synthesis is known to be initiated in halophytes in

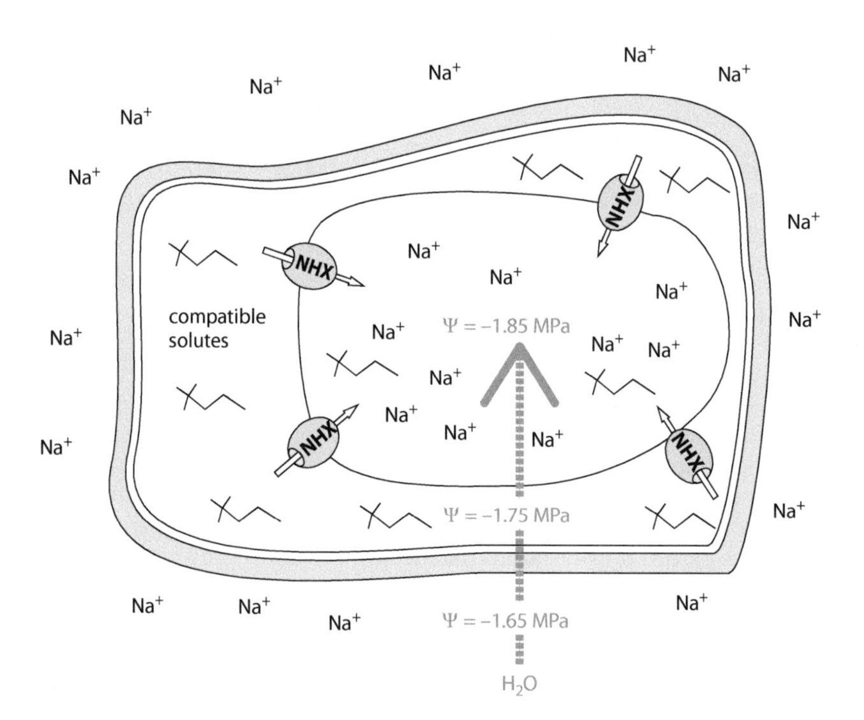

Figure 9.11. Compatible solutes and control of ψ in plant cells. In halophytes, Na^+ concentrations in the cytosol are kept low, with all of the Na^+ being accumulated in the vacuole. In many halophytes, compatible solutes in the cytosol also keep its ψ low, preventing it from dehydrating, but in addition maintaining a water potential gradient for water influx.

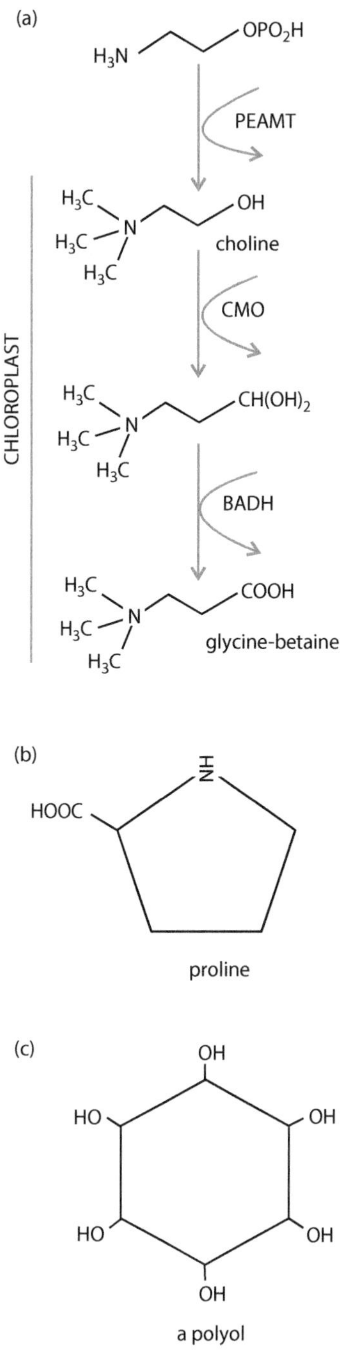

Figure 9.12. Compatible solutes and their synthesis in terrestrial halophytes. (a) Betaines are quaternary ammonium compounds synthesized in the chloroplast from choline, one of the most common being glycine-betaine. (b) Proline is a common compatible solute, and (c) numerous polyols, which have multiple OH groups, can also act as compatible solutes. PEAMT = phosphoethanolamine *N*-methyltransferase; CMO = choline monooxygenase; BADH = betaine aldehyde dehydrogenase.

response to either salt stress or water stress, and up-regulation of the synthesis of betaine in plants has sometimes successfully increased their salt tolerance, as has exogenous application of glycine-betaine to crops.

Halophytes of the Plumbaginaceae (60 species) synthesize betaines not only of glycine but also of alanine, proline, and hydroxyproline, as well as **choline** sulfate, which is formed in a single step via the replacement of CH_2OH on choline with SO_4^{2-}. Choline sulfate occurs in some marine algae as an osmoticum, and might also help plants to render metabolically inert the high concentrations of SO_4^{2-} in saline waters. Many other halophytes have high concentrations of proline, and some produce **polyols** (sugar alcohols with multiple hydroxyl groups) and polyamines. Certain polyols are produced in response to stress by many plants, and can lower ψ in cellular compartments. **Inositol** and **pinitol** are polyols produced by some halophytes, but polyols such as **mannitol** that are produced in response to other stresses are not often associated with salt stress. In algae, dimethylsulfoniopropionate (DMSP), which is synthesized from the sulfur-containing amino acid methionine, is an important compound in osmotic adjustment. The synthesis of DMSP from methionine evolved independently in some halophytes, in particular *Spartina* species (the cord grasses that dominate many salt marshes in the temperate zone).

A wide variety of organisms, ranging from archaea to animals, produce osmoprotectants that not only contribute to ψ, but also during dehydration have antioxidant activity and help to stabilize proteins and membranes. The compounds that act as osmoprotectants in other organisms are often those which occur as compatible solutes in halophytes. For example, glycine-betaine is a significant osmoprotectant in some bacteria, this role in *E. coli* being well established. There is less evidence for compatible solutes acting as general osmoprotectants in halophytes than in some other organisms, but it would be surprising if they did not have a significant osmoprotective role, and they are often referred to as such. The synthesis of compatible solutes/osmoprotectants is metabolically costly to the plant, compromising growth rates, but it is a common biological response across all the kingdoms of life in environments that are highly stressful osmotically.

Salt tolerance in crops has been increased by manipulating biochemical and physiological traits

The biochemistry and physiology that confer salt tolerance upon plants are multigenic, quantitative traits. It is variation in these traits—both intra- and interspecific—that provides the raw material for developing salt-tolerant crops. For example, in cotton, which is quite salt tolerant, breeding has provided lines with enhanced tolerance. In India, the Central Soil Salinity Research Institute has successfully released to cultivation CSR10, a salt-tolerant rice cultivar, and has developed KRL1-4, a salt-tolerant wheat cultivar. In rice, well-known cultivars for salt-affected soils include Pobbeli (Indonesia) and IR2151 (Sri Lanka). The identification of quantitative trait loci (QTLs) for salt tolerance together with much improved genetic markers in major crop plants is increasing the speed at which intraspecific breeding can be performed. Thus the identification of the *Nax* loci of salt tolerance in old landraces of durum wheat has enabled breeding using molecular markers (for example, marker *gwm*312 for *Nax1*) to produce durum wheat with enhanced salt tolerance. However, intraspecific breeding has not so far matched the great success that the approach has had in developing other desirable traits in crop plants, primarily because of limited variation in salt tolerance within individual crop species.

For cereal crops there are related species that have high salt tolerance. There are about 150 halophytes in the Poaceae, some of which are in the same tribe or even genus as major cereals. For example, *Triticum monococcum* (wild emmer wheat), which occurs in the Near East close to where wheat was first domesticated, and which can be bred with both durum and bread wheat, is genetically diverse for salt tolerance, with some accessions growing well on 175 mM NaCl and producing grain on 250 mM NaCl. The progenitor of barley, *Hordeum spontaneum*, which occurs from the Mediterranean region to Iran, has genotypes that can ripen when grown in 350 mM NaCl. *Hordeum marinum* (sea barley), native to Southern and Western Europe but now widely distributed on most continents, has fully salt-tolerant ecotypes that inhabit salt marshes. Such species represent a significant resource for interspecific breeding. Species in genera related to cereals, such as *Elytrigia elongata* (in the Triticeae) and *Porteresia coarctata* (in the Oryzeae), offer possibilities for intergeneric breeding. Advances in techniques for transferring traits and fusing genetic material from different species are increasing the possibility that salt-tolerant relatives will contribute to the development of salt-tolerant cereals. However, for non-cereal crops, in particular the legumes, there are fewer halophytic relatives.

The elucidation of some of the molecular mechanisms underpinning salt tolerance has prompted the development of transgenic salt-tolerant plants. One of the first examples of altered expression of proteins involved in Na^+ transport conferring salt tolerance was *AtNHX1* in *Arabidopsis*. Transformation of tomato plants with *AtNHX1* enhanced salt tolerance and enabled growth in 200 mM NaCl. Transport of Na^+ into vacuoles through NHX1 can clearly contribute to salt tolerance, and in tomatoes resulted in only fruit rather than leaves with elevated Na^+ concentrations. Salt-tolerant cell lines of a number of species with increased expression of Na^+/K^+ antiporters have now been developed. Overexpression of AVP1 in *Arabidopsis* and in wheat to drive Na^+ influx to the vacuole increases salt tolerance. However, NHX1-transformed plants have high internal concentrations of Na—often undesirable in crop plants—and are not necessarily more tolerant of osmotic stresses. Altered expression of TFs that control the expression of multiple salt tolerance genes has also been effective. *DREB*-like genes overexpressed in various non-host species have been shown to increase salt tolerance. The NAC and MYB/MYC TFs have been used to similar effect. The *HVA1* gene from the aleurone layer in barley encodes a promoter that has been used in conjunction with salt-responsive promoters. miRNAs have now been shown to affect the translation of mRNAs synthesized during salt stress, and may also provide a method of enhancing salt tolerance in crops.

Enhanced concentrations of osmoprotectants were an early target of attempts to engineer salt tolerance. The manipulation of *CMO* and *BADH* expression (Figure 9.12) in a variety of plant species revealed that the synthesis of choline is the limiting factor in glycine-betaine synthesis, but increased expression of these enzymes did not increase the glycine-betaine concentration. A complication was that the dehydrogenation steps catalyzed by these enzymes occur in the chloroplast, which is therefore where choline concentrations have to be increased. Such barriers have now been overcome to produce plants with enhanced glycine-betaine levels. Increased proline concentrations via enhanced expression of PCS (Figure 9.12) and increased expression of bacterial choline dehydrogenase in plants to increase glycine-betaine concentrations have also been achieved. Transgenic plants with increased concentrations of compatible solutes have enhanced stress tolerance, especially in short-term experiments, and in particular they have enhanced tolerance of osmotic stress. However, such plants often suffer Na^+ and Cl^- toxicity, and can have significantly impaired function in the long term because of the metabolic cost of synthesizing compatible solutes. Refinements in the expression of compatible solutes to minimize metabolic costs are likely to be possible, as similar outcomes have been achieved for other biosynthetic pathways.

Few of the plants developed with enhanced salt tolerance have resulted in varieties of crops that are widely used, primarily because their yield is lower. Most crops show different sensitivity to salt at different stages of development. Enhancing salt tolerance via cell-specific gene expression and at the appropriate developmental stages is a significant challenge. However, given the time that most crop developments take to get from laboratory to market, the use of marker-assisted selection and genetic engineering for developing salt tolerance is still in its infancy. Nevertheless, most crops with enhanced salinity tolerance show modest increases in tolerance and modest yields, and there may be a limit of moderate salinity tolerance that can be achieved. This may equip crops to grow in soils in which agriculture is currently restricted, but high salinity tolerance necessitates more than biochemical and physiological adaptations. It is for this reason that there is a long history of investigation of the possibility of using halophytes directly for food and forage (Box 9.2).

Halophytes that face severe osmotic stresses have morphological and physiological adaptations

Soils with a high NaCl content have soil water potentials as low as those of desert soils, so in response many halophytes are xeromorphic. Xeromorphy provides a type of cellular organization that enhances plant responses to the osmotic and ionic challenges of salinity. As it does in desert plants, it reduces water requirements, but in halophytes it also enables increased control over the compartmentation of NaCl. Halophyte morphology is variable, but succulence of varying degrees is almost universal. In halophyte leaves a dense photosynthetic palisade layer lies directly inside the epidermis and surrounds extensive, large-celled parenchyma cells with few chloroplasts and massive vacuoles. Many halophytes have Kranz anatomy with distinct cellular sheaths around the vascular bundles, and C_4 photosynthesis. Often the abaxial and adaxial leaf surfaces are similar. Extreme halophytes such as *Salicornia* species and *Arthrocnemum* species are stem succulents with no true leaves. In contrast to glycophyte stems, which have a central pith surrounded by a ring of vascular bundles, halophytes have stems with a prominent central stele surrounded by extensive water-storing parenchyma and a strip of chlorophyllous palisade parenchyma cells (Figure 9.13). The extensive vacuolated parenchyma in various organs of xeromorphic halophytes provides significant water and NaCl storage capacity.

Stomatal densities and apertures of halophytes are low—a few tens of stomata per mm^2 with apertures less than 10 μm in diameter, compared with hundreds of stomata per mm^2 with apertures over 15 μm in diameter in many glycophytes. Halophyte stomata are often located in crypts below the leaf surface and/or surrounded by papillae or trichomes (Figure 9.13). The low surface area:volume ratio of succulent organs, stomatal organization, C_4 photosynthesis, and deep cuticles produce low transpiration rates per unit biomass. This helps halophytes to meet the osmotic challenges of salinity, but reduces evaporative heat loss. Many halophytes are therefore glaucous with a high **albedo**. Light reflection is achieved from thick cuticles and trichomes, and by salt deposition. Low substrate ψ ensures that only small differentials in ψ between soil and leaf are possible, limiting the height of halophytes. Mangroves are among the tallest halophytes, but are rarely more than 5 or 6 m in height. The endodermis in halophyte roots is often extensively suberized, providing a tight Casparian strip. A hypodermis with a secondary Casparian strip to decrease apoplastic bypass is often reported (Figure 9.13). In true halophytes, xeromorphy therefore helps to enhance the response to the ionic and osmotic challenges of salinity.

BOX 9.2 CAN HALOPHYTES BE USED FOR FOOD AND FORAGE PRODUCTION?

It has frequently been suggested that halophytes might be used to produce food and forage on salinized soils. Either domestication of halophytes or their direct use in mixed systems might be possible. In addition, the competition for fresh water between irrigation and other uses—a competition that is intensifying—might be alleviated if saline water could be used for irrigated agriculture, which might be possible if halophytes could be used agriculturally. There are additional incentives for utilizing halophytes for food and forage production—water drawn from many aquifers is increasing in salinity, saline waste waters are a significant but currently under-utilized resource, and landscape gardening (often even in desert regions) frequently uses glycophytes with high demand for fresh water, but could utilize saline water if halophytes were used horticulturally. In Europe, since Roman times some salt-tolerant plants, such as *Smyrnium olusatrum* (Alexanders), have been used as vegetables, as have some true halophytes, such as *Salicornia europaea* (marsh samphire) (Figure 1). Seaweeds and the salt-tolerant scurvy grasses (*Cochlearia* species) have also long been used for food. In several regions, saltmarsh-grazed lamb and beef have long been among the most highly prized forms of meat, which suggests that halophytes could be suitable for use as forage plants.

Figure 1. ***Salicornia europaea.*** This halophyte grows around the coast of Europe and has long been used for food. For example, it is sold in the UK as "marsh samphire." However, its primary use was to produce soda ash for glass making, and this gave rise to its most common English name—glasswort.

Halophytic traits result in high concentrations of minerals, especially Na^+, in shoot biomass, and tend to be associated with low seed production. In staple human foods, high concentrations of Na^+ are undesirable, so although, for example, the Cocopah peoples of the Colorado river valley utilized the halophytic grass *Distichlis palmeri* for grain, there has been little success with grain production from halophytes. The qualities necessary for forage—essentially enough digestible energy from non-toxic biomass—are less stringent. As far back as the 1950s and 1960s, large-scale trials with saline irrigation of *Salicornia bigelovii* established its potential in mixed forage, as have trials with *Atriplex* and *Kochia* species. In addition, in desert environments some halophytes have been used for soil stabilization while also providing forage. In Australia, the most successful use of halophytes in intensive forage has been in mixtures—for example, *Atriplex nummularia,* a bushy halophyte native to Australia but now found in dry environments in the Americas and Africa, grown with herbaceous species such as *Medicago polymorpha.* In Pakistan, a system using the salt accumulator *Suaeda fruticosa* grown around *Panicum turgidum,* which benefits from salt accumulation in *S. fruticosa* and (like most Poaceae) tolerates salt by limiting Na^+ transport to the shoot, has been used to produce palatable *P. turgidum* forage under saline conditions. *Atriplex lentiformis* (quailbush), a halophyte native to the western USA, which is widely used for forage and for revegetation schemes, grows well on the saline concentrate from reverse-osmosis desalination plants. *Salicornia bigelovii* has been used to produce vegetable oil, and the use of halophytes to produce biofuels on saline soil or with saline water is possible. However, agriculture involving the production of halophytes has emphasized the specific agronomic demands of growing them—they need specific water, cropping, planting, and processing regimes, and large-scale management of saline waters is challenging. Pest, disease, and weed control presents further challenges.

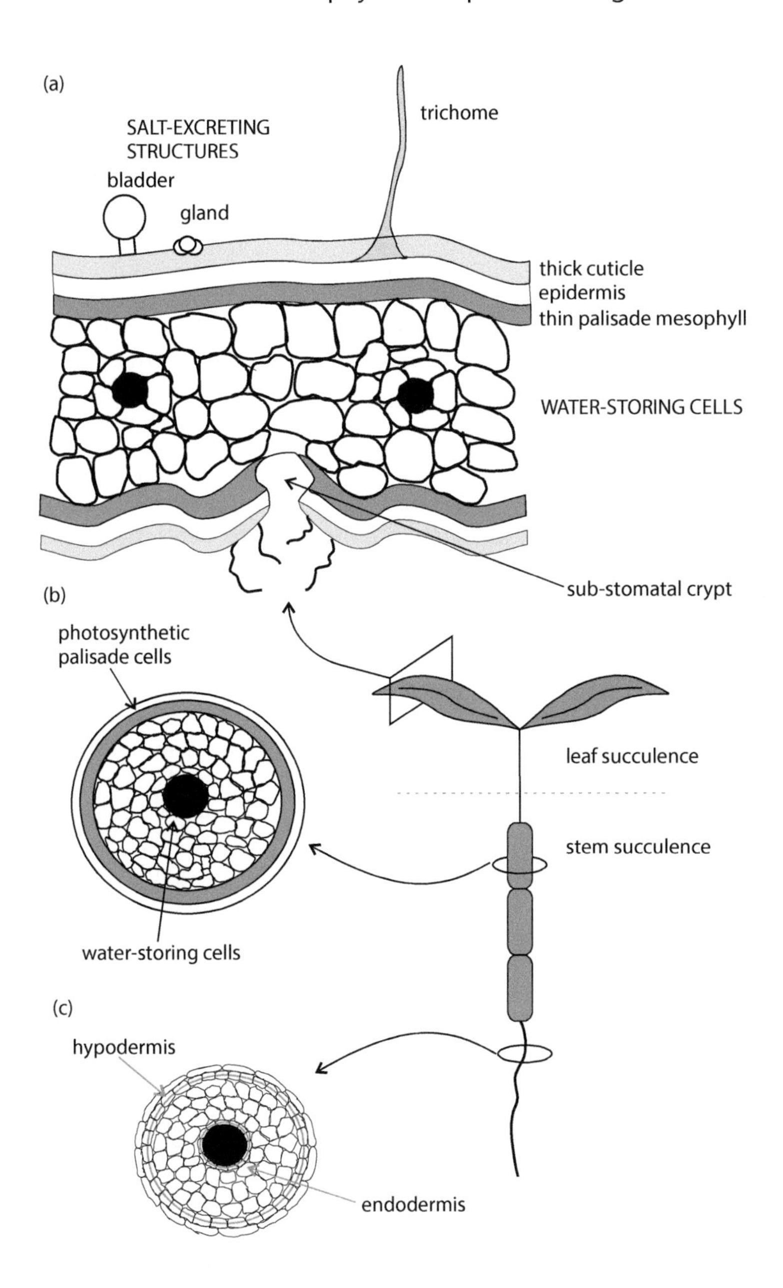

Figure 9.13. The xerophytic characteristics of halophytes. (a) In many halophytes the leaf is succulent and consists mainly of water-storing cells with vacuoles that have a high NaCl content. Waxes and trichomes help to slow down transpiration, but this decreases evaporative heat loss. Waxes and trichomes also reflect incident radiation, minimizing heat gain. (b) The stems of many halophytes are also succulent, and can be important sites of photosynthesis. (c) In roots, the presence of a hypodermis with a Casparian strip can help to decrease apoplastic bypass of Na^+.

Some halophytes use specialized organs to excrete sodium chloride from their leaves

Some halophytes are able to excrete NaCl, mostly onto the surface via either salt bladders or salt glands. Excretion is more pronounced at high NaCl concentrations, having been observed at seawater concentrations or above in some species. At high salt concentrations, such plants can excrete onto leaf surfaces a significant proportion of all the Na^+ in the transpiration stream. In the Chenopodiaceae (the family with the largest number of halophytes), many species have salt bladders developed from leaf hairs. These consist of a stalk cell surmounted by a bladder cell, with a cuticle covering the whole structure (Figure 9.14). The stalk cell has no vacuole, has strengthened cell walls waterproofed by cutin, and is rich in endoplasmic reticulum. NaCl is actively loaded into the stalk cell, from which vesicles loaded with NaCl move into the bladder cell and fuse with its vacuole. The bladder cell can reach 200 μm across and eventually ruptures, releasing salt onto the leaf surface. In young *Atriplex* leaves, 80% of the NaCl in the leaf can be in the

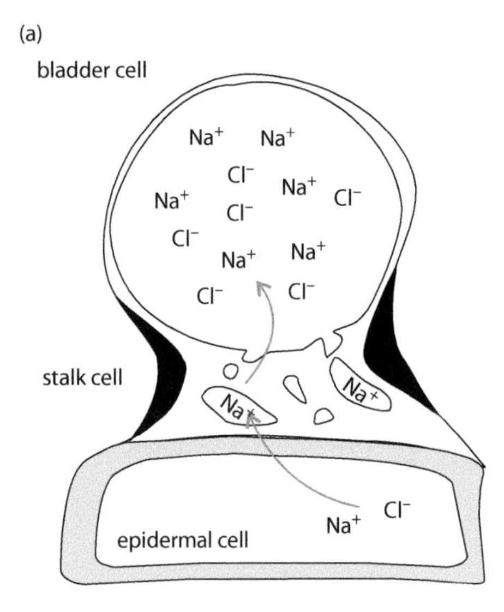

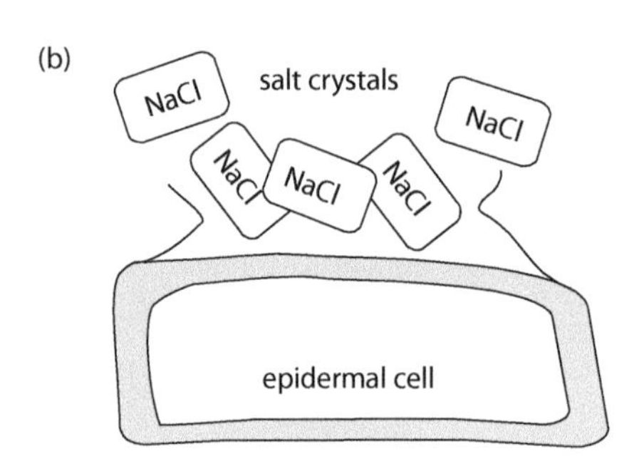

Figure 9.14. The salt bladders of the Chenopodiaceae. (a) Salt is transported into the bladder cell by pinocytosis. (b) When salt becomes so concentrated that it precipitates, the bladder bursts, leaving salt crystals on the surface. These can be washed off, carrying salt away, but they often persist, probably helping to reflect incident radiation and thus decreasing heat gain.

bladder cells. Salt bladder cells are reported to become less frequent and less important as the leaves get older, with the ratio of NaCl in mesophyll cells to that in salt bladders progressively increasing. Salt on the surface of the leaves decreases the albedo of the leaf, which is likely to be useful in decreasing leaf insolation at high salinities.

Halophytic NaCl-excreting grasses, all of which are in the subfamily Chloridoideae of the Poaceae, have bicellular salt glands and a significant capacity to excrete NaCl. Bicellular salt glands are developed from protodermal cells. In *Aeluropus littoralis*, for example, they are distributed on the adaxial and abaxial leaf surfaces at densities of 5000 cm^{-2} and 7000 cm^{-2}, respectively (about 15–20% of the density of stomata). Excreted salt crystals are almost pure NaCl with a characteristic cuboid structure, indicating a significant selectivity against K^+ when measured against the xylem sap or external solutions. In some species, the bicellular salt glands significantly increase in number and/or diameter in response to external salt. In *Spartina alterniflora*, which often dominates salt marshes on the Atlantic coast of the Americas, excretion rates of up to 32 $\mu mol\ g^{-1}\ h^{-1}$ leaf fresh weight have been measured at seawater concentrations of 600 mM NaCl. In *Aeluropus* and *Spartina*, 50% of absorbed NaCl can be excreted via salt glands, with the proportion increasing with salinity. In *Spartina alterniflora*, monthly excretion can represent up to 3% of the NaCl in sediment. If NaCl crystals are washed off leaves, salt excretion is detectable again within hours, and for *Sporobolus* excretion has been suggested to occur mainly at night.

Bicellular salt glands have no direct connection to vascular tissue, and consist of a large basal cell surmounted by a smaller cap cell (Figure 9.15). The basal cell has no vacuole and its plasma membrane is strongly invaginated from the cap cell side, providing a large surface area of contact between the basal and cap cells. There are few chloroplasts in basal cells, but there are numerous mitochondria and extensive microtubule development. Water and NaCl enter the basal cell via the plasma membrane. They are then transferred to the cap cell via the apoplast and symplast. The neck of the basal cell is lignified, which not only keeps apoplastic fluid excreted from the basal cell in the folds of the plasma membrane, but also helps to direct NaCl through the plasmodesmata and into the cap cell. From the cap cell, NaCl is excreted into a collecting chamber formed by a porous cuticular dome. Pressure in the cells of the salt gland pushes the NaCl-laden fluid from the collecting chamber out onto the surface of the cuticle (Figure 9.15). Metabolic inhibitors have been used to show that salt excretion through bicellular glands is an energy-requiring process. For loading into the basal cell, Cl^- transport has

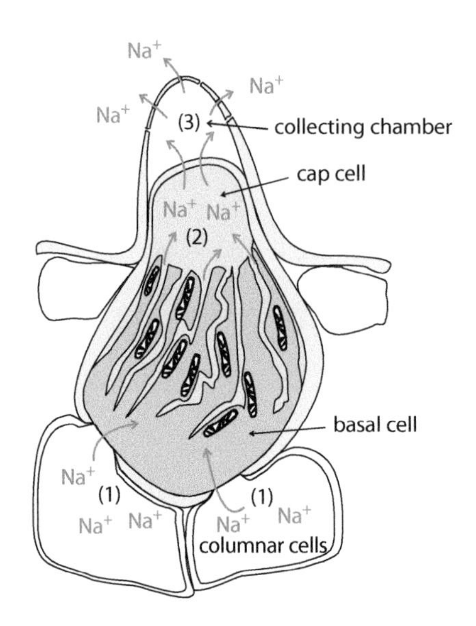

Figure 9.15. The bicellular salt glands of the Poaceae. (1) NaCl is transported into the salt gland from cells with large vacuoles that are often found in columns beneath it, and that probably help to conduct salt towards the gland. (2) NaCl solution is transported into the cap cell under pressure, probably via active processes in the mitochondria-rich basal cell. (3) Salt solution is squeezed from a collecting chamber through pores in the cuticle onto the leaf surface, where it can either crystallize or be washed away. Bicellular salt glands have been found in halophytic grasses in the genera *Aeluropus, Chloris, Cynodon, Distichlis, Odyssea, Spartina, Sporobolus,* and *Zoysia.*

to be active and, given the high concentrations in the basal cell, a significant active component of Na^+ transport is likely, the energetics being reversed at loading from the cap cell into the collecting chamber. It has been suggested that the numerous mitochondria and microtubules in the basal cell might be associated with a physical pumping of fluid via pulsing mechanical contractions. All of the genera of grasses with bicellular salt glands have been shown to have C_4 photosynthesis, and in some species the salt glands are associated with Kranz anatomy. For example, in *Distichlis* the salt glands are associated with large, water-storing, chlorophyllous parenchyma cells with large vacuoles located between the photosynthetic mesophyll cells. Similar cells are frequently associated with the hinging of grass leaves in response to water deficit. Thin regions of cell wall in which the cytoplasm and vacuole are separated from the apoplast that feeds into the basal cell by only a thin layer of membrane have been observed. Such cells have a role in salt transport, and might act as collecting cells for the salt glands.

There are a variety of multicellular salt glands developed from protodermal cells in a diverse range of eudicot halophytes. *Avicennia* (in the Lamiales) has salt glands that consist of basal collecting cells with a stalk cell surmounted by a number of secretory cells (Figure 9.16). The gland is covered with cuticle, especially around the stalk cell, which is linked to the basal cells by numerous plasmodesmata, the gland being 20–40 μm in diameter. Salt moves towards the collecting cells in the apoplast, enters the symplast of the basal cell, and is then transported into the secretory cells. From the secretory cells, NaCl is loaded into the subcuticular space, which is covered by a porous dome of cuticle. The cutinized walls of the stalk cell prevent apoplastic backflow, and probably contribute to a positive pressure in the secretory cells and subcuticular chamber, which aids NaCl excretion. The secretory cells are rich in cytoskeletal actin-like filaments, especially towards the periphery, which may have a role in squeezing salt from the salt gland. In *Aegiceras corniculatum* (in the Ericales) up to 90% of the salt that reaches the leaves can be excreted via salt glands. Many halophytes in the Plumbaginaceae also have characteristic multicellular salt glands. In *Limonium bicolor* the density and functioning of these salt glands have been shown to be dependent on Ca^{2+} supply, and their functioning is inhibited by Na_3VO_3, suggesting a role for Ca/plasma-membrane H^+-ATPase-dependent pathways in their development and excretion activities. Plants in the Frankeniaceae have salt glands that are structurally and functionally similar.

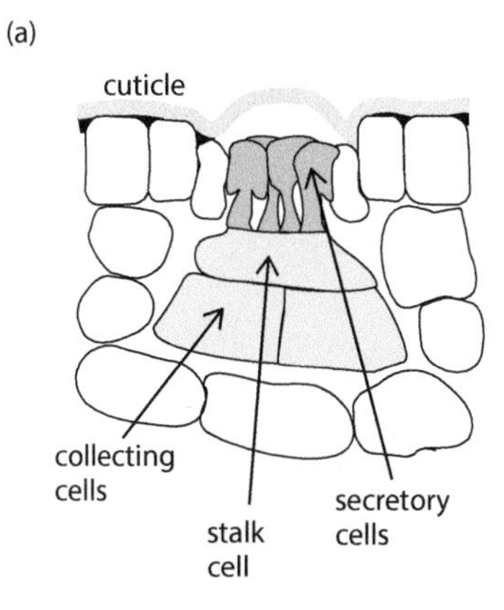

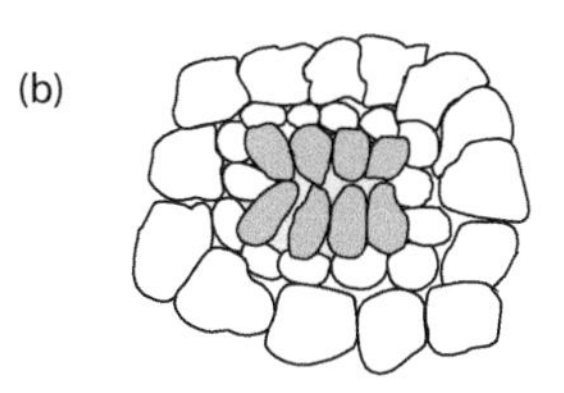

Figure 9.16. The multicellular salt glands of mangroves. (a) Longitudinal and (b) top view of the multicellular salt gland of the mangrove *Avicennia officinalis*. The secretory cells show cytoskeletal and mitochondrial specialization that helps to actively excrete salt into a subcuticular cavity from which it seeps out onto the leaf. The collecting cells are not linked via plasmodesmata to the mesophyll cells, but collect salt excreted into the apoplast by them. With the exception of the Chenopodiaceae, halophytes in dicotyledonous families have characteristic multicellular salt glands. In *Avicennia*, the salt glands on the adaxial surface tend to be sunken below the leaf surface (shown in the figure), sometimes significantly so, whereas those on the abaxial surface tend to be surrounded by trichomes.

Mangrove and salt-marsh plants tolerate waterlogging and salinity

For the few crop species in which it has been studied, waterlogging decreases the ability of plants to restrict Na^+ and Cl^- transport into the leaves, thus exacerbating sensitivity to salinity. Decreased O_2 supply to the roots inhibits their functioning and is the primary stress in waterlogged soils. Loss of discrimination during uptake and transfer to shoots probably explains the increased salt sensitivity of crops during waterlogging. Many halophytes inhabit waterlogged soils in which root O_2 demand cannot be met from the soil, particularly salt-marsh and mangrove species. The stresses produced in waterlogged soils are well known, as are numerous plant adaptations, but few of them have been extensively investigated in halophytes that inhabit wet saline soils.

High internal root porosity has been reported in some halophytes of saline wetlands (for example, *Spartina anglica*, *S. patens*), and the ability to produce **aerenchyma** in response to O_2 deprivation in roots seems to be similar in wetland halophytes and glycophytes. Salinity is not directly related to induction of aerenchyma in wetland halophytes, and can reduce it—perhaps

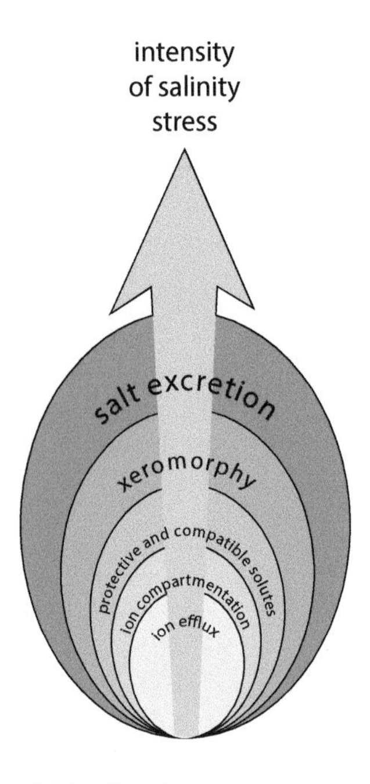

Figure 9.17. The stress-response hierarchy for salinity. Most terrestrial plants have Na^+ efflux systems that maintain very low internal Na^+ concentrations and confer tolerance of mild salinity. Tolerance of moderate salinity also involves compartmentation of Na^+ that enters plants via systems of transporters. In addition, salt-tolerant plants contain solutes that protect physiologically important molecules and generate low ψ while also being compatible with physiological functioning. True halophytes are also mostly xeromorphic, and almost universally have salt excretion systems.

reflecting a salinity-induced slowing of growth and of O_2 demand. The effect of other root adaptations that enhance the flow of O_2 to root tips in glycophytic wetland plants has seldom been investigated in halophytes. In general, these adaptations (which include suberized hypodermal layers, a decrease in lateral roots, and an increase in surface adventitious roots) decrease both radial O_2 loss and apoplastic influx of toxic ions. It is likely that such adaptations, which have been reported in some wetland halophytes, will decrease the apoplastic bypass of NaCl into plants. There is some evidence of S^{2-} tolerance and of fermentation metabolism in the roots of wetland halophytes. Increased alcohol dehydrogenase (ADH) activity has been shown in some halophytes, but other end products of fermentation metabolism, such as lactate and alanine, have seldom been investigated. Some studies suggest that tolerance of S^{2-} and the induction of fermentation metabolism are decreased under conditions of increased salinity, but the interactions between salinity and waterlogging are not well understood. Increased flooding from the sea due to rising sea level and the use of saline water for irrigation are two important instances in which an understanding of the interactions between salinity and waterlogging tolerance would be useful.

Summary

Temporal variations in substrate salinity occur over long time scales, generally over months or years, but there is significant spatial variation in soil salinity. Salt stress is therefore often chronic, and in the relatively small number of salt-tolerant angiosperms there is a hierarchy of responses to increasing intensity of salinity stress (Figure 9.17). Many plant species have an Na^+ efflux system which is probably a vestige of the Na^+ control mechanisms that are found in all organisms, and which acts as a first line of defense against low concentrations of Na^+ in soils. At moderate salinity, internal compartmentation of salt and physiological adjustments are also necessary to combat general osmotic and specific ion toxicity threats. Significant exclusion/compartmentation of $Na^+$and Cl^- combats ion toxicity, but the high fluxes of Na^+ in the transpiration stream mean that complete exclusion of salt during moderate stress does not seem to have been possible, and might not be desirable because of the increase in rhizosphere Na^+ levels that it would promote. At high salt intensities, further physiological and also morphological and anatomical adjustments are common, including adaptations that are used to reduce transpiration and to produce significant excretion of salt. Overall, salinity tolerance is unusual in angiosperms, infrequent in crop plants, and difficult to alter significantly even by advanced molecular breeding methods. The direct use of halophytes in agriculture has been promoted as a potential solution to the dilemma of how to produce more crops from land that is becoming increasingly salinized. The ecology of halophytes, which in some species includes tolerance of drought and heat, and in others includes tolerance of waterlogging, is likely to be of increasing importance to solving the salinity dilemma that faces humankind.

Further reading

Freshwater ancestry and salt sensitivity

Bromham L & Bennett TH (2014) Salt tolerance evolves more frequently in C_4 grass lineages. *J Evol Biol* 27:653–659.

McCourt RM, Delwiche CF & Karol KG (2004) Charophyte algae and land plant origins. *Trends Ecol Evol* 19:661–666.

Munns R & Tester M (2008) Mechanisms of salinity tolerance. *Annu Rev Plant Biol* 59:651–681.

Page MJ & Di Cera E (2006) Role of Na^+ and K^+ in enzyme function. *Physiol Rev* 86:1049–1092.

Irrigation, salinity, and coastal communities

Duarte CM, Losada IJ, Hendriks IE et al. (2013) The role of coastal plant communities for climate change mitigation and adaptation. *Nature Clim Change* 3:961–968.

Hillel D & Vlek P (2005) The sustainability of irrigation. *Adv Agron* 87:55–84.

Rengasamy P (2010) Soil processes affecting crop production in salt-affected soils. *Funct Plant Biol* 37:613–620.

Tanaka N (2009) Vegetation bioshields for tsunami mitigation: review of effectiveness, limitations, construction, and sustainable management. *Landscape Ecol Eng* 5:71–79.

Salinity and toxicity

Miller G, Suzuki N, Cifti-Yilmaz S et al. (2010) Reactive oxygen species homeostasis and signalling during drought and salinity stresses. *Plant Cell Environ* 33:453–467.

Riadh K, Wided M, Hans-Werner K et al. (2010) Responses of halophytes to environmental stresses with special emphasis on salinity. *Adv Bot Res* 53:117–145.

Rivero RM, Mestre TC, Mittler R et al. (2014) The combined effect of salinity and heat reveals a specific physiological, biochemical and molecular response in tomato plants. *Plant Cell Environ* 37:1059–1073.

Vialaret J, Di Pietro M, Hem S et al. (2014) Phosphorylation dynamics of membrane proteins from *Arabidopsis* roots submitted to salt stress. *Proteomics* 14:1058–1070.

Sodium influx and efflux in plants

Adams E & Shin R (2014) Transport, signaling, and homeostasis of potassium and sodium in plants. *J Integr Plant Biol* 56:231–249.

Ji H, Pardo JM, Batelli G et al. (2013) The Salt Overly Sensitive (SOS) pathway: established and emerging roles. *Mol Plant* 6:275–286.

Waters S, Gilliham M & Hrmova M (2013) Plant high-affinity potassium (HKT) transporters involved in salinity tolerance: structural insights to probe differences in ion selectivity. *Int J Mol Sci* 14:7660–7680.

Zhang J-L, Flowers TJ & Wang S-M (2010) Mechanisms of sodium uptake by roots of higher plants. *Plant Soil* 326:45–60.

Compartmentation of sodium

Deinlein U, Stephan AB, Horie T et al. (2014) Plant salt-tolerance mechanisms. *Trends Plant Sci* 19:371–379.

Jha D, Shirley N, Tester M et al. (2010) Variation in salinity tolerance and shoot sodium accumulation in *Arabidopsis* ecotypes linked to differences in the natural expression levels of transporters involved in sodium transport. *Plant Cell Environ* 33:793–804.

Sottosanto JB, Saranga Y & Blumwald E (2007) Impact of AtNHX1, a vacuolar Na^+/H^+antiporter, upon gene expression during short-term and long-term salt stress in *Arabidopsis thaliana*. *BMC Plant Biol* 7:18.

Yao X, Horie T, Xue S et al. (2010) Differential sodium and potassium transport selectivities of the rice OsHKT2;1 and OsHKT2;2 transporters in plant cells. *Plant Physiol* 152:341–355.

Halophytes and osmotic adjustments

Ashraf M & Foolad MR (2007) Roles of glycine betaine and proline in improving plant abiotic stress resistance. *Environ Exp Bot* 59:206–216.

Flowers TJ & Colmer TD (2008) Salinity tolerance in halophytes. *New Phytol* 179:945–963.

Liu D, He S, Zhai H et al. (2014) Overexpression of *IbP5CR* enhances salt tolerance in transgenic sweetpotato. *Plant Cell Tiss Organ Cult* 117:1–16.

Wani SH, Singh NB, Haribhushan A et al. (2013) Compatible solute engineering in plants for abiotic stress tolerance - role of glycine betaine. *Curr Genomics* 14:157–165.

Salt-tolerant crops

Nevo E & Chen G (2010) Drought and salt tolerances in wild relatives for wheat and barley improvement. *Plant Cell Environ* 33:670–685.

Qadir M, Oster JD, Schubert S et al. (2007) Phytoremediation of sodic and saline-sodic soils. *Adv Agron* 96:197–247.

Roy SJ, Negrao S & Tester M (2014) Salt resistant crop plants. *Curr Opin Biotechnol* 26:115–124.

Ruan CJ, da Silva JAT, Mopper S et al. (2010) Halophyte improvement for a salinized world. *Crit Rev Plant Sci* 29:329–359.

Xeromorphy in halophytes

de Vos AC, Broekman R, Groot MP et al. (2010) Ecophysiological response of *Crambe maritima* to airborne and soil-borne salinity. *Ann Bot* 105:925–937.

Debez A, Saadaoui D, Slama I et al. (2010) Responses of *Batis maritima* plants challenged with up to two-fold seawater NaCl salinity. *J Plant Nutr Soil Sci* 173:291–299.

Flowers TJ, Hajibagheri MA & Clipson NJW (1986) Halophytes. *Q Rev Biol* 61:313–337.

Redondo-Gomez S, Mateos-Naranjo E, Parra R et al. (2010) Modular response to salinity in the annual halophyte, *Salicornia ramosissima*. *Photosynthetica* 48:157–160.

Salt excretion

Barhoumi Z, Djebali W, Abdelly C et al. (2008) Ultrastructure of *Aeluropus littoralis* leaf salt glands under NaCl stress. *Protoplasma* 33:195–202.

Oi T, Taniguchi M & Miyake H (2012) Morphology and ultrastructure of the salt glands on the leaf surface of Rhodes grass (*Chloris gayana* Kunth). *Int J Plant Sci* 173:454–463.

Oi T, Miyake H & Taniguchi M (2014) Salt excretion through the cuticle without disintegration of fine structures in the salt glands of Rhodes grass (*Chloris gayana* Kunth). *Flora* 209:185–190.

Tan W, Lim T & Loh C (2010) A simple, rapid method to isolate salt glands for three-dimensional visualization, fluorescence imaging and cytological studies. *Plant Methods* 6:24.

Salt marshes and mangroves

Adam P (1990) *Saltmarsh Ecology*. Cambridge University Press.

Feller IC, Lovelock CE, Berger U et al. (2010) Biocomplexity in mangrove ecosystems. *Annu Rev Mar Sci* 2:395–417.

Kathiresan K & Bingham BL (2001) Biology of mangroves and mangrove ecosystems. *Adv Mar Biol* 40:81–251.

Parida AK & Jha B (2010) Salt tolerance mechanisms in mangroves: a review. *Trees* 24:199–217.

Chapter 10
Soil pH

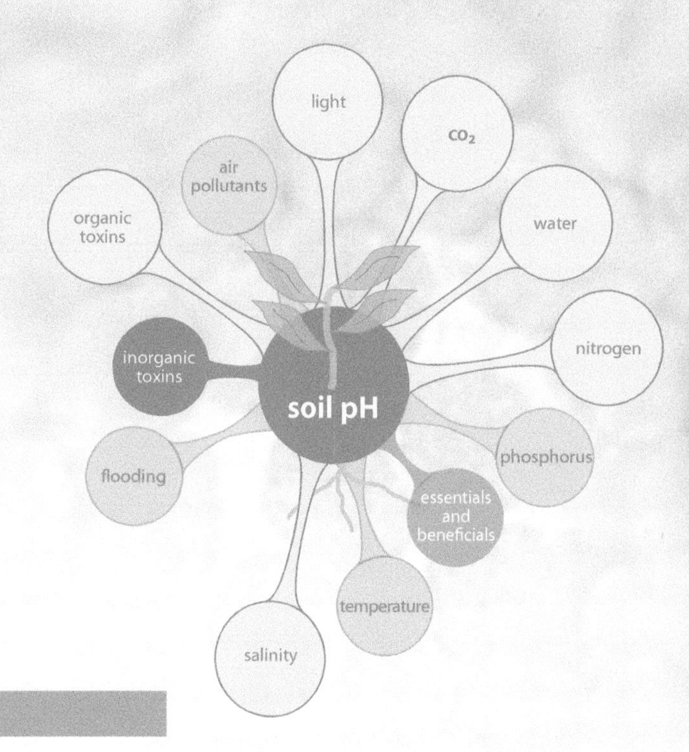

The effects of soil pH are exacerbated by its influence on ion availability, and complicated by flooding and temperature variation.

Key concepts

- Soil pH affects plant growth in both unmanaged and managed ecosystems.
- Human activity is changing soil pH on a global scale.
- Cytosolic pH-stats provide some buffering against variation in soil pH.
- As soil pH decreases, ionic toxicity in plant roots increases.
- Some plants resist the effects of moderate acidity by excluding toxic ions.
- Mycorrhizas significantly increase the ability of some plants to tolerate acid soils.
- Al-tolerant plants on severely acidic soils can take up and detoxify Al^{3+}.
- Low availability of Fe, Zn, and Mn in basic soils induces deficiency symptoms in plants.
- On basic soils, some plants scavenge Fe from the rhizosphere by reducing it.
- Nicotianamine aids Fe homeostasis and underpins Fe^{3+} acquisition in some plants.
- Extending the soil pH range on which crops can grow is important for food security.

Soil pH affects the growth of both wild and domesticated plants

It has long been recognized that some substances are "sour" and that other substances can counteract "sourness." Pliny the Elder, writing in *Naturalis Historia* (AD 77), described the liming of fields to counteract the sourness of soil, and by AD 1000, Islamic scholars had an advanced understanding of such properties in, for example, plant ash. The word "acid" is derived from the Latin *acidus* (meaning "sour"), and the word "alkali" is derived from the Arabic *al-qaly* (meaning "the ashes"). Characteristic assemblages of plants occur on soils with extreme pH values, and many early experiments showed that transplants from one extreme to the other suffered toxicity symptoms. Many plants, including crops, have soil pH preferences (Figure 10.1), which are often used to categorize them (Box 10.1). Species richness–soil pH

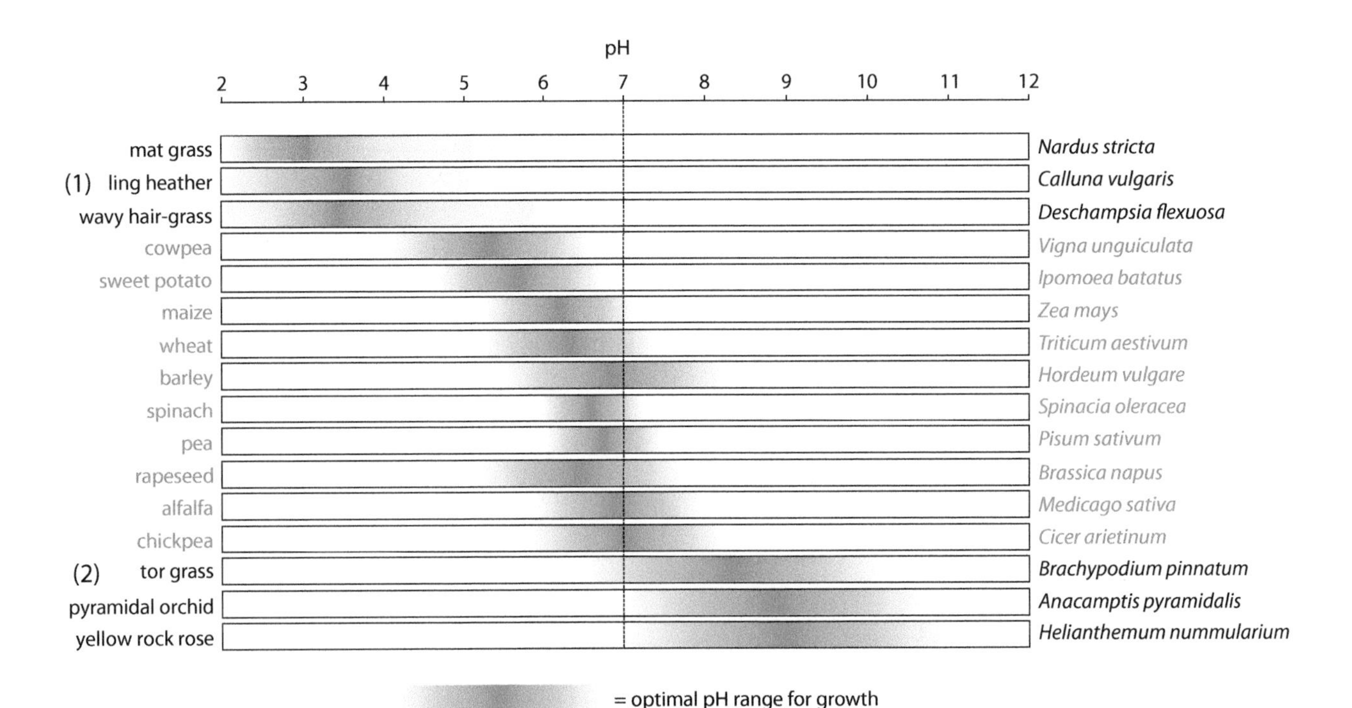

Figure 10.1. Optimum pH range for growth of selected species. Most cultivated species (shown in green) grow optimally at mildly acidic to neutral pH. Basifuges (1) grow best at acidic pH, whereas basicoles (2) grow best at alkaline pH. The pH ranges within which crop plants grow can vary significantly with the variety. Some species can inhabit a wider range of soil pH than others, but no plants can inhabit both truly acidic and truly basic soils. (Common names are shown on the left, and scientific names on the right.)

curves within biomes often show a decrease in species richness at extreme pH values, especially with acidity (Figure 10.2), although globally some of the most species-rich habitats occur on acidic tropical soils. The NO_3^-/NH_4^+ preferences of plant species correlate with their soil pH preferences. Due to such phenomena, soil pH is of great significance in both unmanaged and managed ecosystems.

Acidic soils cover approximately 30% of the ice-free land surface, primarily in two bands, one in the wet tropics and the other in the boreal zone, and corresponding to a total area of about 4000 million hectares (Figure 10.3). Acid mineral soil covers over 40% of the topics. At high latitudes, spodosols with deep organic surface horizons predominate and entirely organic soils (histosols) also occur. Histosols formed by the remains of *Sphagnum* moss can be particularly acidic. In the tropics, the predominant natural vegetation on acid soils is savannah or rainforest, but at high latitude it mainly consists of coniferous forests. Acid soils therefore underpin the most extensive near-pristine ecosystems on Earth. Many early civilizations developed, and the majority of humans still live, between these two bands of acidic soil. Many

BOX 10.1. DESCRIBING PLANT pH PREFERENCES

Some plant species have distributions that are clearly confined to either low-Ca or high-Ca soils. They thus became described as **calcifuges** ("calcium haters") and **calcicoles** ("calcium lovers"). The term "base" came into use when alkalis served as the base for making useful salts. There are some soils that are basic but low in Ca, so the terms **basifuge** and **basicole** are often used to describe the plants that inhabit acidic and basic soils, respectively.

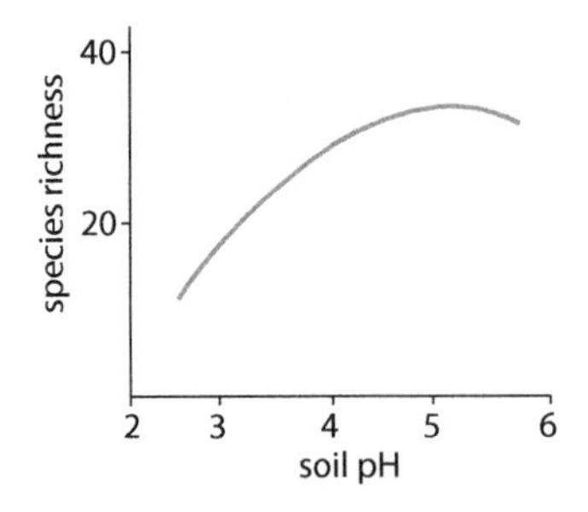

Figure 10.2. Species richness–soil pH curve. Data sets such as the one shown in the figure, measured in Germany in deciduous forests across their range of mineral soil types, demonstrate that key ecosystem characteristics, such as species richness, are affected by soil pH. (From Peppler-Lisbach C & Kleyer M [2009] *J Veg Sci* 20:984–995. With permission from John Wiley and Sons.)

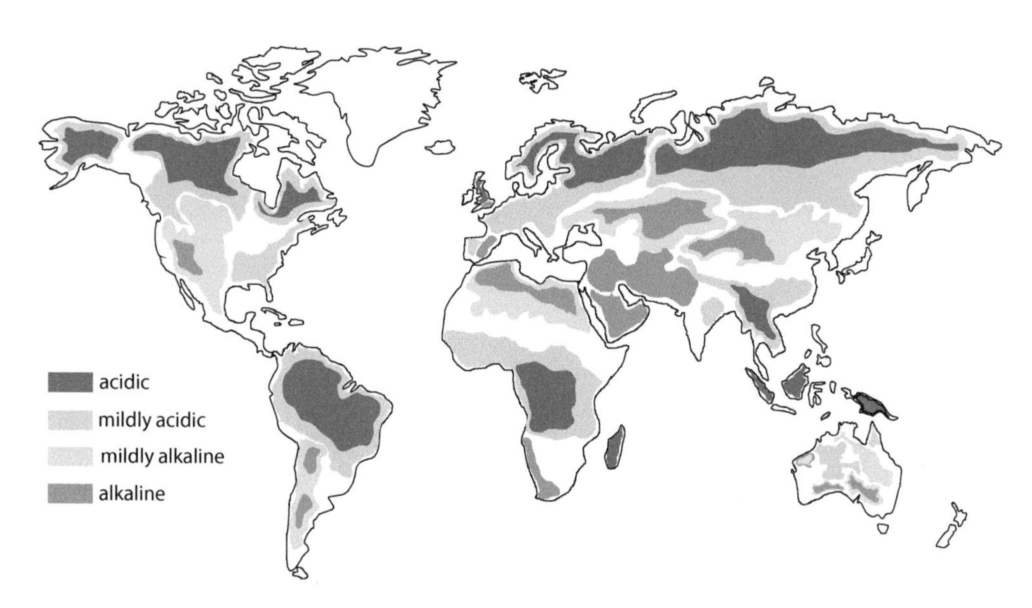

Figure 10.3. The pH of soils on Earth. Acid soils are more common at high and low latitudes, where they are predominantly organic and mineral, respectively. Alkaline soils occur over calcareous rocks anywhere on Earth, but they are most common in areas of high evaporation.

domesticated crop plants grow best at around neutral soil pH, especially the cereals that produce the majority of calories now consumed by humans. In significant part because of the expansion of agriculture onto new land in the twentieth century, around 40–50% of arable land is now at least mildly acidic. This is a significant limitation to the productivity of agricultural ecosystems. Only about 5% of acid soils are cropped, and there is much pressure to expand agricultural production on these soils, but the regulation of soil pH by the addition of "lime" and other soil amendments is already a significant expense for global agriculture. Some of the few acid-tolerant crops include cassava (*Manihot esculenta*), cowpea (*Vigna unguiculata*), pigeon pea (*Cajanus cajan*), peanut (*Arachis hypogaea*), and tea (*Camellia sinensis*), but the crops that are grown on the greatest area by humans are not acid tolerant.

Calcareous soils have a pH buffered by $CaCO_3$, so their pH ranges up to about 8.5 (that is, they are basic). Calcareous soils cover over 30% of the world's terrestrial surface (Figure 10.3). They are particularly important over limestone, but are also widespread in semi-arid and arid climates where accumulation of basic ions in the upper layers of soil can occur. In addition, saline and sodic soils, which cover over 400 million hectares, are frequently alkaline. Basic soils are generally challenging for plant and microbial life, and ecosystems that have developed on them rely on a distinct set of rhizosphere interactions. Almost one-third of agricultural production now occurs on basic soils, but crops grown on these soils can be deficient in micronutrients, affecting the health of over 2 billion people.

Soil pH is operationally defined and human activities are affecting it on a global scale

pH is generally defined as $-\log_{10}[H^+]$, although an H^+ ion (that is, a proton) does not exist in isolation in an aqueous solution, but associates strongly with one or more water molecules to form H_3O^+ or $H_9O_4^+$, and for soil is measured in an extract (Box 10.2). In pure water, at any one time few H_2O molecules are dissociated into H^+ and OH^- ions. At 25°C the concentration of H^+ and OH^- ions is 1.0×10^{-7} M. Therefore in pure water the pH is $-\log_{10}(1.0 \times 10^{-7})$, which is equal to 7. However, water in the environment, which includes that in the soil, has substances dissolved in it, namely minerals from the geosphere or gases from the atmosphere. The Arrhenius definition of an acid or a base is a substance that ionizes in water to give H^+ or OH^-, respectively, examples being HCl and NaOH. In soils, an important

BOX 10.2. MEASURING SOIL pH

Soil pH is measured using a soil extract, the commonest extractants being H_2O and $CaCl_2$. Thus "soil pH" is in practical terms $-\log_{10}[H^+]$ in a soil extract. In general, a 1:5 soil:solution extraction with $CaCl_2$ is thought to approximate the pH experienced by plant roots. Dilution of a soil solution with water tends to increase H^+ bound to cation-exchange sites, but $CaCl_2$ tends to prevent this—$CaCl_2$-extracted pH is almost always lower than H_2O-extracted pH.

Arrhenius acid is carbonic acid. Carbonic acid (H_2CO_3) is formed when CO_2 dissolves in water. This is important in soils, because they are in contact with the atmosphere, inhabited by respiring organisms that produce CO_2, and recharged with rainwater that has significant quantities of CO_2 dissolved in it. Carbonic acid dissociates in water to increase the H^+ concentration, and at atmospheric CO_2 concentration (pCO_2) produces a pH of about 5.7:

$$CO_2 + H_2O \leftrightarrow H_2CO_3 \leftrightarrow HCO_3^- + H^+$$

Soil scientists use a wider, "operational" definition of acid and base with regard to soil pH. The operational definition of an acid or base includes any substance that changes the H^+ or OH^- concentration in the soil solution, whether it dissolves to produce H^+ or OH^- itself or not. For example, $CaCO_3$, which is a significant component of agricultural lime, acts as a base in soils because it decreases the H^+ concentration in the soil solution:

$$CaCO_3 \leftrightarrow Ca^{2+} + CO_3^{2-} \leftrightarrow CO_3^{2-} + H^+ \leftrightarrow HCO_3^-$$

Soil pH is spatially correlated and well buffered temporally. Semivariograms for soil pH differ but, whatever their scale, generally have nuggets of up to 2 pH units and ranges of large distances—that is, there are differences in soil pH of up to 2 units in most places, but greater differences only between significantly different locations. pH has been measured in soils at timescales ranging from hours to decades, but at particular locations changes of more than 2 units over time seem to be rare. Overall, soil pH is quite variable at a given location, but differs more between well-separated locations that tend not to change greatly over time (Figure 10.4).

Humans have altered the soil pH on a local scale for millennia. For example, drainage of previously saturated soils frequently decreases their pH. However, in the twentieth century the Earth's soils were not only drained but also eutrophied, polluted, and degraded to an unprecedented extent. These processes have had adverse environmental consequences across the globe, in significant part because they change the soil pH. The United Nations Environment Program (UNEP) has identified acidifying cycles as an important aspect of land degradation, and suggests that the pH of up to 25% of soils on Earth is being changed by human activity. Fertilizer application to soil can change its pH. The longest experiment describing such changes is the Park Grass Continuous Hay Experiment at Rothamsted Research Station, in the UK (Box 10.3). Many air pollutants also acidify soil. In Europe and North America, soil acidification over extensive areas from the 1950s to the 1980s caused many of the problems of "acid rain." The UNEP estimates that over much of Europe and eastern North America the soil pH has decreased by 0.5–1.5 units, primarily due to air pollution. By the year 2000, acidified soils in much of Europe and North America were recovering, but in some places models suggest that decreases of a further pH unit are still possible. Across Asia the challenge of soil acidification, particularly that caused by air pollution, is ongoing, with many arable soils in China and South-East Asia being affected. Degraded soils with decreasing pH are also a continuing consequence of deforestation, because the latter increases leaching, particularly of basic ions, often presenting a challenge to agricultural production. Salinization following irrigation can increase soil pH to 10 or more, primarily because of the accumulation of $NaCO_3$. Taken together, these human-induced effects amount to global-scale changes in soil pH that are significantly more pronounced than those which occur naturally.

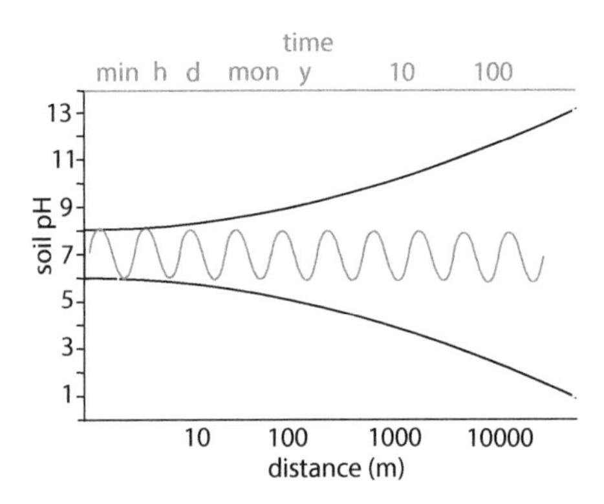

Figure 10.4. **The variability of soil pH.** Spatial variation in soil pH (shown in black) has a nugget of about 2 pH units, but over distances of hundreds of meters or more can vary across most of the possible pH range. On timescales of up to 100 years (shown in green), at a particular place soil pH varies by no more than 2 units. Anthropogenic changes of 1 unit or more produced within a few years represent substantial changes in soil pH.

In addition, organic acid soils play a key role in the carbon cycle. They contain the largest reservoir of labile carbon, and are significant producers of methane (CH_4), a potent greenhouse gas (Box 10.4). Anthropogenic climate change might have a major impact on water regimes of acid organic soils

BOX 10.3. SOIL pH EFFECTS IN THE WORLD'S OLDEST CONTINUOUS GRASSLAND EXPERIMENT

In 1856 at Rothamsted Manor in England in the UK, John B. Lawes and Joseph H. Gilbert initiated an experiment to test the efficacy of fertilizer additions in improving yields of hay from the Park Grass meadow. At the time most fertilizers were complex organic materials such as farmyard manure (FYM) or guano, and little was known about the effects of individual mineral elements on yield. With some additional treatments, the basic experiment of Lawes and Gilbert has continued to the present day. Among the many interesting findings that have been revealed by this experiment, now the oldest continuous one conducted on grassland, are some effects on soil pH.

The Park Grass Continuous Hay Experiment was not designed as it would be now, as the statistics for analyzing the designs that might be chosen today had yet to be invented (in the early twentieth century, researchers at what is now Rothamsted Research would play a role in inventing them). Nitrogen was added at different levels as NO_3^- and NH_4^+, as were numerous other minerals (Figure 1). The differences in pH between N treatments rapidly became sufficiently great that by 1905 half of each plot was limed. In 1965 the plots were subdivided and limed to pH 5, 6, and 7, with plot d remaining unlimed. Plots 3 and 12 have had no fertilizer additions since 1856. There are now dramatic differences between the plots and subplots in yield, biodiversity of flora and fauna, and numerous soil parameters, including pH. Medium additions of $(NH_4)_2SO_4$ without lime—for example, plots 10*d*, 9*d*, and 4*d*—have produced significant changes in soil pH that contrast with those produced by NO_3^- on plot 14 (Figure 2). These differences are emphasized by the different amounts of lime necessary to maintain the pH values on *a*, *b*, and *c* subplots of these treatments. Perhaps most interestingly, the pH of the plots with no additions has varied significantly over time. The lowest pH values were recorded in the 1980s, and are related to the N and S pollution deposition rates at around this time. These changes in pH have been related to various changes in the plots, including changes in biodiversity.

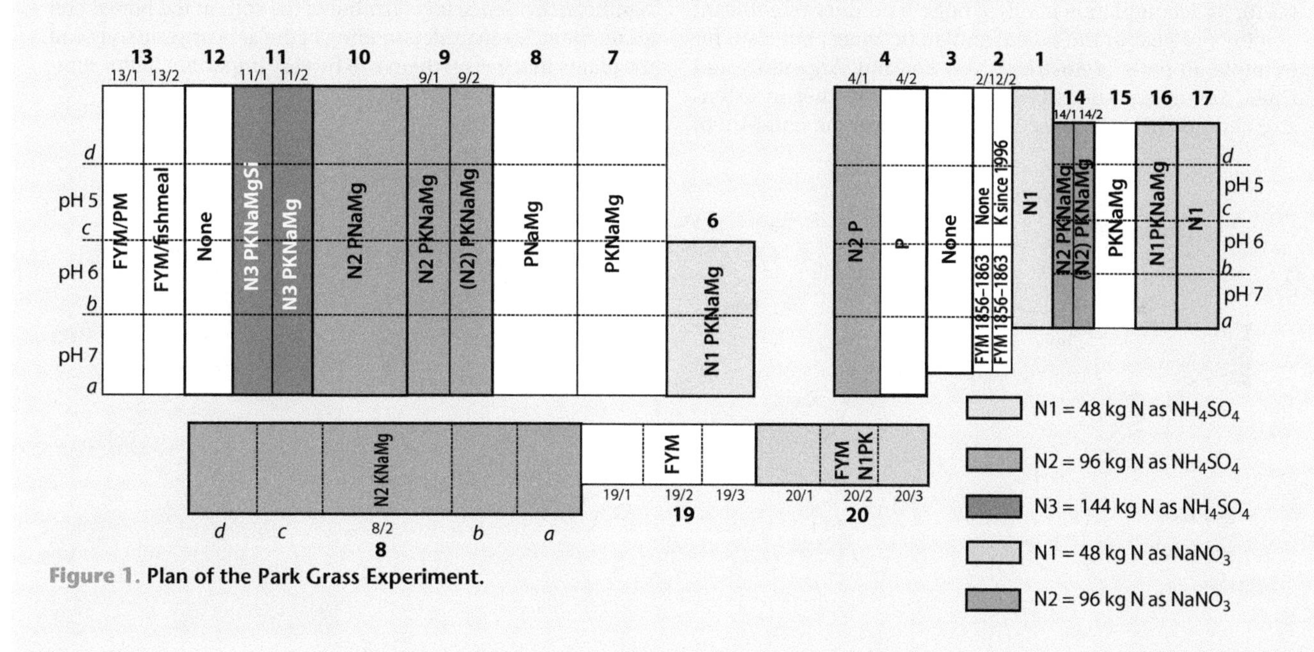

Figure 1. Plan of the Park Grass Experiment.

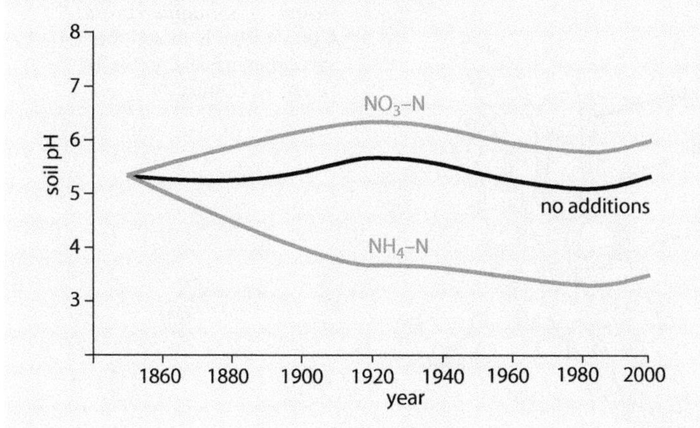

Figure 2. Soil pH in the Park Grass Experiment.

BOX 10.4. CLIMATE CHANGE AND ACID ORGANIC SOILS

The soils of boreal and arctic ecosystems are acidic and contain huge amounts of organic matter in the soil profile (Figure 1). According to the USDA soil taxonomy, gelisols are soils that have permafrost near the surface, and horizon processes dominated by cryoturbation. In general, they only have an A horizon, and this contains significant quantities of organic matter because of the slow decomposition rates. Spodosols have significant surface O horizons—that is, build-up of organic matter—which overlie leached mineral horizons. Histosols have horizons that are primarily organic ("peat"), mainly because excess water inhibits decomposition. There are some boreal and arctic areas where histosols are dominant (Figure 1), but in many localities anywhere in the world where water input greatly exceeds evaporation there can be extensive peat accumulation. Together, the acid organic soils of high latitudes hold a significant proportion of the labile carbon reserves on Earth.

In many instances where peat accumulates, *Sphagnum* moss plays a key role. *Sphagnum* is in significant part responsible for peat accumulation in blanket bogs, and entirely responsible for its accumulation in raised bogs. It produces significant amounts of peat in the boreal and arctic zones, but also, for example, in parts of Australia, New Zealand, Argentina, and Chile. *Sphagnum* stems (Figure 2) grow together in extensive clumps that retain water and encourage the build-up of peat. Water retention can be so significant that the *Sphagnum* and its peat can retain their own water body, well above the surrounding water table (a "raised bog"). *Sphagnum* encourages the build-up of peat because of its water-retaining and decomposition-inhibiting properties. The leaf-like **microphylls** of *Sphagnum* contain large achlorophyllous hyaline cells (Figure 3), which fill with water, retaining many times the plant's own weight. *Sphagnum* exudes significant quantities of protons and organic acids, such that the water it retains is acidic. This can contribute significantly to the acidity of blanket and raised bog systems. It also inhibits microbial growth, thus encouraging the accumulation of peat.

If peat dries out and decomposes it releases very large quantities of CO_2. Some studies suggest that at the end of the Quaternary ice ages the release of massive quantities of CH_4 and CO_2 from organic soils increased global CO_2 concentrations significantly, and hastened the warming of the global climate. If acid organic soils dehydrate and begin to decompose they could release CH_4 and CO_2 in quantities sufficient to be of great significance for the global climate. There is already significant evidence for warming of the soils in the boreal and arctic zones, so an understanding of the acid organic soils and the plants that live on them will be very important in the coming decades.

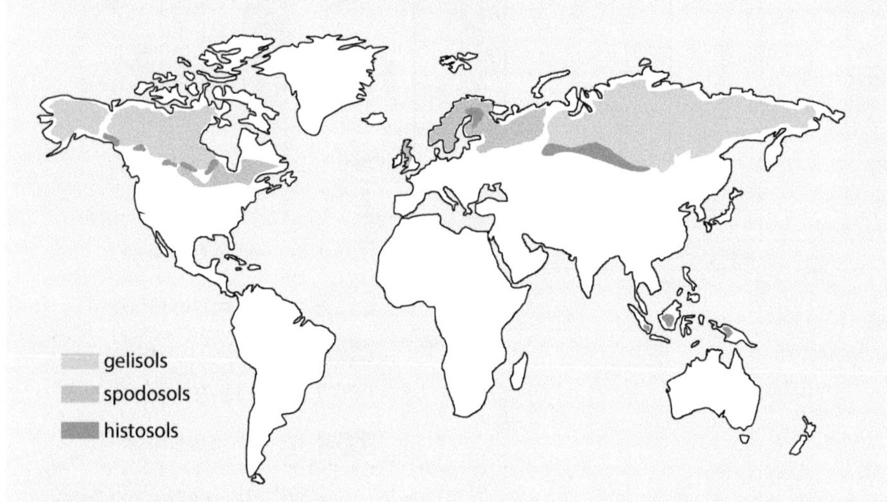

Figure 1. Acid organic soils at high latitudes.

Figure 2. *Sphagnum* moss. A single stem is shown, but *Sphagnum* usually grows in extensive clumps.

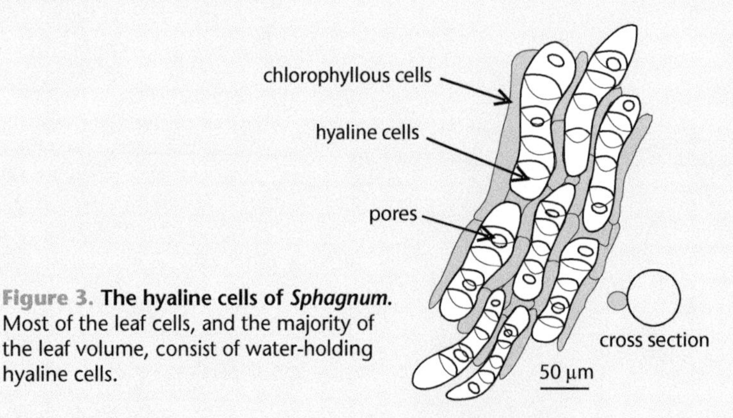

Figure 3. The hyaline cells of *Sphagnum*. Most of the leaf cells, and the majority of the leaf volume, consist of water-holding hyaline cells.

and hence on the carbon cycle. Inorganic acid soils in the tropics support most of the rainforest on Earth, which has a key role in uptake of CO_2 from the atmosphere. Understanding how plants function on soils with extreme pH values is therefore important not only to the conservation of unmanaged ecosystems, attainment of global food security, and understanding of soil degradation, but also to the prediction of global climate.

Plant cells have multiple mechanisms for buffering cytosolic pH

In plant cells, cytosolic pH is generally in the range 7.2–7.5 (32–63 nM H^+), and vacuolar pH is in the range 4.5–5.5. In general, a change of 1 unit in external pH only causes a change of 0.1 unit in the cytosolic pH, indicating significant pH-buffering capacity. Animal cells often have much less buffering capacity, but are bathed in plasma, the pH of which is closely controlled. Plant cells can generally buffer pH over a range of about 1.5–2 units (approximately the nugget of soil pH variograms), but plants that are not adapted to extreme pH values are not able to counteract the more severe buffering challenges that these cause. When buffering capacity is overwhelmed there are direct effects of pH on plant cells that are separate to, although they overlap with, indirect effects caused by a change in the chemistry of the environment.

In plant cells, slight increases in cytosolic H^+ or HCO_3^- concentrations can be counteracted by efflux of H^+ or HCO_3^- to the apoplast or the vacuole. In addition, the cytosolic buffering capacity (due to organic acids, bicarbonate ions, and phosphate compounds) is around 20–100 mEq per pH unit. These processes constitute a "biophysical pH-stat," which can buffer cytosolic pH against changes of 1 unit or more in external pH (Figure 10.5). However, with more significant changes in pH, H^+/HCO_3^- efflux decreases because of the limited buffering capacity of the apoplast (5 mEq per pH unit) and the

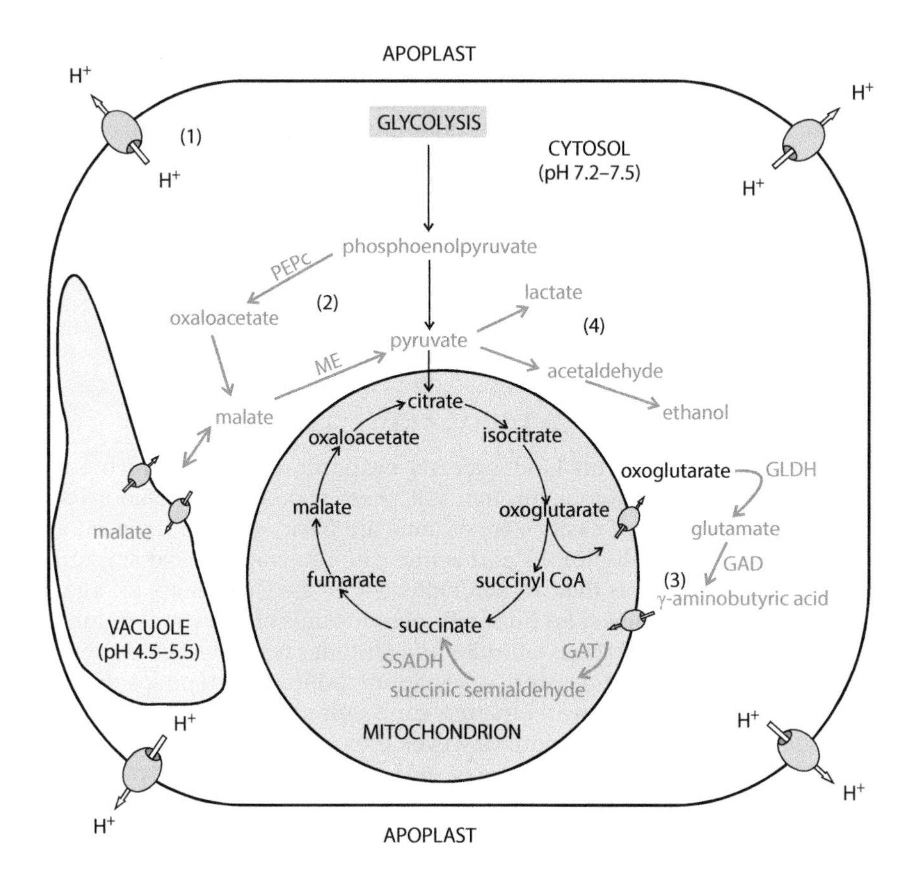

Figure 10.5. The regulation of cytosolic pH in plant cells. A range of biochemical and physiological mechanisms help to regulate cytosolic pH. (1) The contents of the cytoplasm act as a pH buffer, but there is also significant capacity for exchanging protons with the apoplast. (2) The pH-dependent enzymes phosphoenolpyruvate carboxylase (PEPc) and malic enzyme (ME) regulate the exchange of malate/malic acid with the vacuole, helping to control cytosolic pH. (3) The γ-aminobutyric acid (GABA) shunt from the tricarboxylic acid (TCA) cycle can be used to control the concentration of glutamate/glutamic acid, helping to regulate pH. (4) Particularly under hypoxic and anoxic conditions, many plants can produce ethanol rather than lactate/lactic acid, helping to prevent acidosis. GLDH, glutamate dehydrogenase; GAD, glutamate decarboxylase; GAT, GABAaminotransferase; SSADH, succinate semialdehyde dehydrogenase.

limited volume of the vacuole. In plant cells, following glycolysis a proportion of phosphoenolpyruvate (PEP) can be converted into malate via oxaloacetate (OAA). PEP carboxylase (PEPc), which catalyzes the conversion of PEP to OAA, is a pH-sensitive enzyme with increasing activity above pH 6.0, and its action on PEP and HCO_3^- increases the cytosolic H^+ concentration. Conversely, malic enzyme (ME), which catalyzes the conversion of OAA to malate, has increasing activity at decreasing pH, and consumes H^+ ions. The activity of PEPc kinase (PPCK), which activates PEPc, increases in the cytosol with increased pH, and the activity of PEP carboxykinase, which activates the conversion of OAA back to PEP, decreases (Figure 10.5). The contribution that this PEPc/ME "biochemical pH-stat" makes to pH homeostasis depends on how much oxoglutarate is being extracted from the TCA cycle for use in biosynthetic pathways. If the proportion is significant, a large amount of malate can be used anaplerotically—that is, to replenish the TCA cycle as oxoglutarate is extracted—thereby decreasing its contribution to pH control. Nitrogen assimilation in plants occurs via the action of glutamine synthetase on glutamate and NH_3, so nitrogen metabolism affects the contribution of the PEPc/ME pH-stat to pH homeostasis.

In plant cells, glutamate can enter the γ-aminobutyric acid (GABA) shunt. The *stop 1* (*sensitivity to proton rhizotoxicity*) *Arabidopsis* mutant suffers cytosolic **acidosis** at acid pH, and in comparison with acidosis-resistant plants has decreased expression not only of ME but also of GLN, GAD1, and GABA-T from the GABA shunt (Figure 10.5). The activity of the GABA shunt is affected by numerous stressors (both abiotic and biotic), and is frequently reported to aid pH homeostasis. Many stressors rapidly affect glycolytic activity, which influences proton production and then the activity of the biochemical pH-stat and the GABA shunt. In addition, glutamic acid decarboxylase (GAD), which catalyzes the conversion of glutamate to GABA, is a calmodulin-activated enzyme, and increases in cytosolic Ca levels are frequently produced in stressed plant cells. The alternative oxidase (AOX) in plant mitochondria may also aid pH control. Under aerobic conditions, **complex II** oxidizes succinate from the TCA cycle, providing electrons to the ubiquinone pool, which in plants is accessed by AOX as well as **complexes III** and **IV**. AOX catalyzes the production of H_2O and, in contrast to the reactions of complexes III and IV, does not generate protons. Anaerobic conditions, such as waterlogging, produce cytosolic acidosis, and anaerobic metabolic pathways also have a role in counteracting this (Figure 10.5). All plants probably have to use some, perhaps all, of these mechanisms to counteract minor changes in soil pH, but they have also been adapted to enable plants to inhabit soils with pH values that are somewhat different to that of the plant cell cytoplasm.

Acid soils contain high solution concentrations of ions that are toxic to plant cells

Wherever precipitation on land exceeds evapotranspiration, which can promote the leaching of ions through soil, there is potential for soil acidification. The majority of easily leached ions are basic, and are replaced at ion-exchange sites with protons and acidic cations. Tropical acid soils are mostly ultisols or oxisols that were not affected by the Quaternary ice ages, and that have been leached for hundreds of thousands of years, and in some cases probably longer. At high latitudes and altitudes where wet cold conditions prevail, decomposition of organic matter is inhibited, so spodosols and histosols are common. Many temperate-zone soils also have higher rates of precipitation than of evaporation. However, they are relatively young due to the impact of the Quaternary ice ages, and are mildly acidic alfisols. Soils with a pH above 7 mostly develop either over Ca-containing rocks or where evapotranspiration exceeds precipitation.

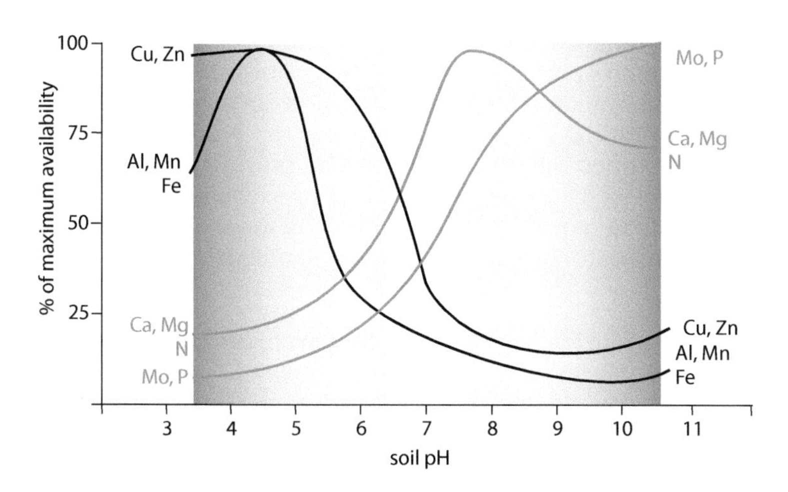

Figure 10.6. The availability of important ions in the soil solution across the pH range. The solubility, and hence plant availability, of nutrient ions (Cu, Zn, Mn, Fe, Ca, Mg, N, Mo, and P) and potentially toxic ions (Cu, Zn, Al, Mn, Fe, and Mo) varies significantly with soil pH. Due to varying plant demand for particular ions, soil acidity is particularly associated with deficiency of P, Mo, and Mg, and with toxicity from Al, Zn, and Cu. Basic soils are especially prone to Fe, Mn, and Zn deficiency, but generally contain high levels of Mo, Mg, and N. Interactions, such as those between Al and P in acid soils, and between Ca and P in basic soils, can also significantly affect availability.

Acid soils have low levels of basic ions, especially Ca^{2+}, Mg^{2+}, and MoO_4^-, but high levels of acidic ions, such as Al^{3+}, Mn^{2+}, and Fe^{2+} (Figure 10.6). Aluminum, the third most common element in the Earth's crust, is not susceptible to leaching, and occurs at particularly high concentrations in acid soils. Cu^{2+} and Zn^{2+} are much more available in acid soils than in basic soils. PO_4^{3-}, which is precipitated as $AlPO_4$ under the acidic conditions that increase Al^{3+}, has decreased availability in acid mineral soils. In addition, the availability of K and S decreases in acid conditions. Fe in particular, but also Cu, Zn, and Mn, have low availability in high pH soils, and B has minimum availability at pH 7–8. Therefore in both acidic and basic soils, certain macronutrients have low availability, but in acid soils there is the additional problem of high concentrations of Al^{3+}. The speciation of Al in soil solution is affected by pH (Figure 10.7). The $Al(OH)_3$ that dominates soil Al at pH 6–8 is insoluble, but Al^{3+} is very soluble and available to plants below pH 5. Al^{3+} is toxic, and is the most important factor inhibiting plant growth on acid mineral soils. Al^{3+} occurs as the octahedral hexahydrate $Al(H_2O)_6^{3+}$ in aqueous solution. In organic acid soils the Al^{3+} concentration is low, but the acid conditions present pH homeostasis problems.

Al is a hard **Lewis acid**—that is, a strong electron-pair acceptor, especially from oxygen in **carbonyl** (C=O), PO_4^{3-}, and SO_4^{2-} groups. The pectate of plant cell walls has a high proportion of D-galacturonic acid with carbonyl functional groups. Root growth is inhibited within minutes at concentrations as low as 1 μM Al, because Al^{3+} inhibits Ca-pectate formation. The transition zone between the mitotic meristem and the zone of elongation is an early site of Al toxicity because of its high demand for Ca-pectate for the production of new cell wall. Border cell growth in the transition zone can be particularly inhibited, leading to rupture of the root as the central portions continue

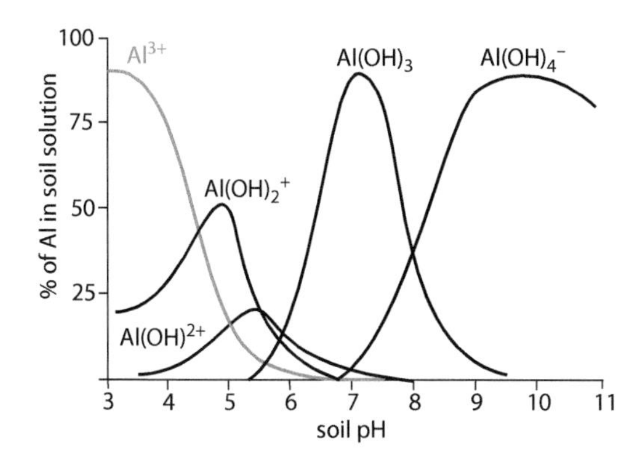

Figure 10.7. The speciation of Al in soil solution across the range of soil pH. $Al(OH)_3$ is a solid and dominates soil Al in the mid-range of pH, rendering it insoluble. $Al(OH)_2^+$ and $Al(OH)^{2+}$ are sparingly soluble, and Al^{3+} is very soluble, dominating soil Al (shown in green) below pH 5.

to grow. Lack of Ca-pectate also affects cell wall flexibility and expansion, so with sustained exposure to Al across the root zones, cell-cycle progression and differentiation are also inhibited, and the quiescent center of the root meristem can be lost. **Callose** synthesis can also be triggered in the cell wall, further decreasing its flexibility and providing a commonly used indicator of toxicity. Disrupted root growth causes changes in root morphology, which lead to root malfunction, inhibiting water and nutrient uptake. Overall, in roots there are morphological similarities between Al toxicity and Ca deficiency.

Al^{3+} also binds to the PO_4^{3-} in phospholipids, decreasing membrane integrity and elasticity. Changes in these membrane properties initiate ROS production, primarily via lipid peroxidation, because membrane-bound electron carriers are affected (Figure 10.8). Membrane changes affect transporter function, but Ca^{2+} transport can also be blocked directly by Al^{3+}. Some Al^{3+} also enters the symplast, where it can affect Ca homeostasis, the cytoskeleton, and DNA. Al^{3+} in the cytosol outcompetes Ca for binding sites on Ca-binding proteins, interfering with Ca homeostasis and Ca-mediated signaling (Figure 10.8). In Al-tolerant plants, Ca^{2+} spikes in the cytoplasm are an important early signal of the presence of Al, inhibited by cytosolic Al in Al-sensitive plants. The microtubules of the cytoskeleton are composed of tubulin bound with guanosine triphosphate (GTP)—an interaction that is adversely affected by Al^{3+}, which can bind directly to GTP, decreasing cytoskeleton integrity. Al^{3+} that reaches the nucleus can bind to the PO_4^{3-} in the backbone of DNA, affecting its structure and thus gene expression. Callose formation in plasmodesmata and at sieve plates also occurs, isolating cells from their neighbors. The numerous carbonyl, PO_4^{3-}, and SO_4^{2-} groups on many other molecules in plant cells, together with the many other Ca- and Mg-bound organic molecules, provide a plethora of additional sites of potential Al toxicity.

Excess Mn^{2+} in acid soils not only aggravates Al^{3+} toxicity symptoms by interfering with Ca uptake and physiology, but also affects shoot functioning. A decrease in Mn transport to the shoots plays a significant role in Mn tolerance, because excess Mn interferes with electron transport in the thylakoids and hence with numerous aspects of photosynthetic performance. There are many Mn-activated enzymes, including some decarboxylases and dehydrogenases in the TCA cycle, but the greatest demand for Mn in plants is from Mn-dependent proteins in the oxygen-evolving complex (OEC) of PSII. Excess Mn in plant leaves affects the fluorescence of chlorophyll *a* and decreases the maximum quantum yield of photosynthesis (F_v/F_m). These effects are produced by an increase in Q_A in the reduced state, which

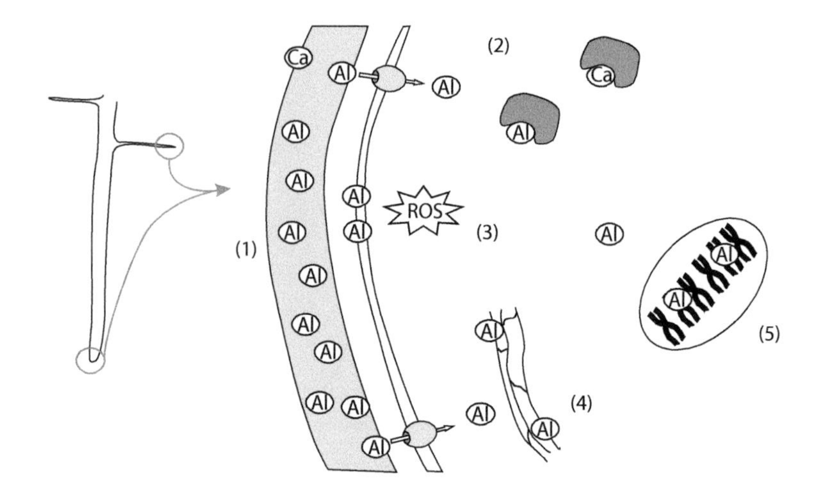

Figure 10.8. Mechanisms of Al toxicity in plant cells. There are numerous mechanisms of Al toxicity in plant cells. (1) Al inhibits the formation of Ca-pectate in the cell wall, especially at growing root tips. (2) Al is transported into plant cells, inhibiting both the uptake of Ca and the binding of Ca to Ca-binding proteins. (3) Al binds to phospholipids in the plasma membrane, initiating the production of reactive oxygen species (ROS). (4) Al binds to the cytoskeleton and disrupts it. (5) Al binds to DNA in the nucleus, affecting genome stability and gene expression.

reflects a decreased capacity to supply electrons from the OEC. Ultimately, a decrease in the supply of reductants from PSI will decrease carbon assimilation. Excess Mn^{2+} can also affect the activity of rubisco, which probably contributes to decreased photosynthetic performance. Not only Fe^{2+}, but also Mn^{2+}, Zn^{2+}, and Cu^{2+} participate in Fenton reactions to produce OH^-. Thus Mn^{2+} toxicity and tolerance are associated with changes in oxidative stress metabolites and antioxidant activity. The effects of excess Fe^{2+}, Zn^{2+}, and Cu^{2+} on root border cells, cell walls, and antioxidants are similar to those of Mn^{2+}, and probably contribute to mineral stresses in plant roots on acid soils.

Some plants resist the effects of moderate soil acidity by excluding aluminum from the cytoplasm

Plants that are adapted to moderately acid soils resist the toxicity effects that are produced in non-adapted plants. One function of the root cap is to protect the root meristem. The mucilage that is produced from the root cap is rich in polysaccharides, especially **uronic acids**, and includes over 2000 proteins. This mucilage lubricates the root and has numerous chelation sites for multivalent cations, including Al^{3+}, altering their concentration at the root tip. Early studies showed that the production of root mucilage was greater in Al-resistant than in Al-sensitive plants. However, recent studies have noted that the primary site of Al^{3+} toxicity (the transition zone) is distal of the root cap, that removing the root cap does not necessarily increase Al sensitivity, and that mucilage can have a limited total Al adsorption capacity. In some species, mucilage contributes to the formation of a rhizosheath—a distinct cylinder of soil and mucilage bound around the root that has been shown to increase Al^{3+} resistance in wheat. In rice, root Fe-plaques formed from the precipitation of Fe around the root probably have a similar effect. Although Al^{3+} chelation in mucilage is perhaps only sometimes a contributor to Al resistance in plants, investigations of it provided an early indication that root exudates can have a role in disarming Al^{3+}.

Wheat (*Triticum aestivum*) is a generally Al-sensitive species—in many wheat cultivars, root growth inhibition is detected from about 10 μM Al^{3+} upwards. However, there is significant variation in Al resistance, which was used to breed cultivars that can be grown on moderately acid soils. The most important locus of this resistance is on the long arm of chromosome 4D. The origin of resistance alleles at this locus can be traced back to land races of hexaploid wheat in the Old World, the progeny of which were taken to Brazil, where the preponderance of acid mineral soils meant that selecting for Al^{3+} resistance was vital to breeding programs. Many modern wheat cultivars with resistance to moderate Al exposure are based on cultivars bred in Brazil in the early twentieth century, one of the most well known being Atlas 66.

In 2004 it was shown that the *TaALMT1* gene at the resistance locus in wheat encodes an **organic anion transferase**. TaALMT1 (*Triticum aestivum* aluminum-activated malate transferase 1) promotes the efflux of malate particularly effectively, and is activated to do so by Al^{3+}. TaALMT1 is expressed primarily at the root apex, and the exuded malate ions form complexes with Al^{3+} outside the root. These complexes reduce root apex exposure to Al^{3+}, and are not taken up by root transporters (Figure 10.9). Variation in Al resistance between some cultivars is due to tandem repeats in *cis*-regulatory regions of *TaALMT1*, rather than variation in the gene itself. These repeats have not been detected in the progenitor of the D genome, *Aegilops tauschii*, indicating that acid resistance in wheat has developed in the last 10,000 years, perhaps via regular selection for it during the domestication of *Triticum aestivum*. Across numerous wheat cultivars and other species, resistance to moderate Al concentrations in the soil solution correlates with malate efflux.

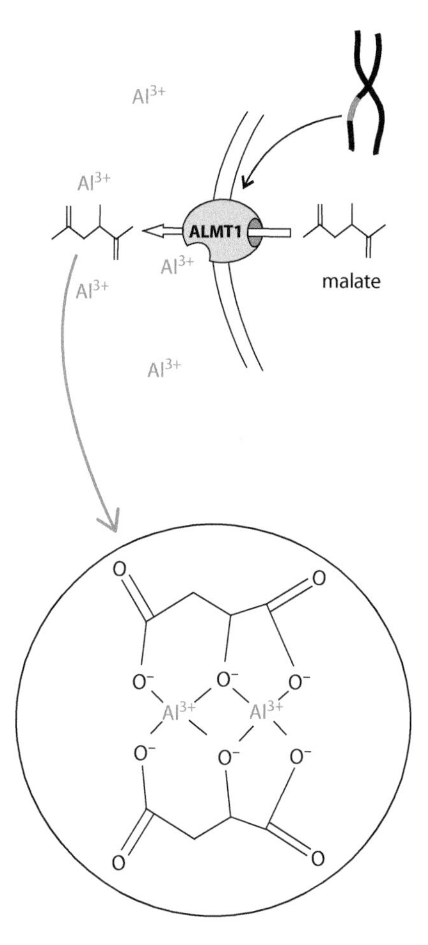

Figure 10.9. The ALMT system for malate exudation from roots. In some plants that can respond to the presence of Al^{3+}, there are Al-activated malate transport (ALMT) proteins which allow efflux of malate when Al^{3+} binds to them. In many plant species the expression of ALMT genes is up-regulated by the presence of Al^{3+}. In general, two malate anions can together chelate two Al^{3+} ions, detoxifying them.

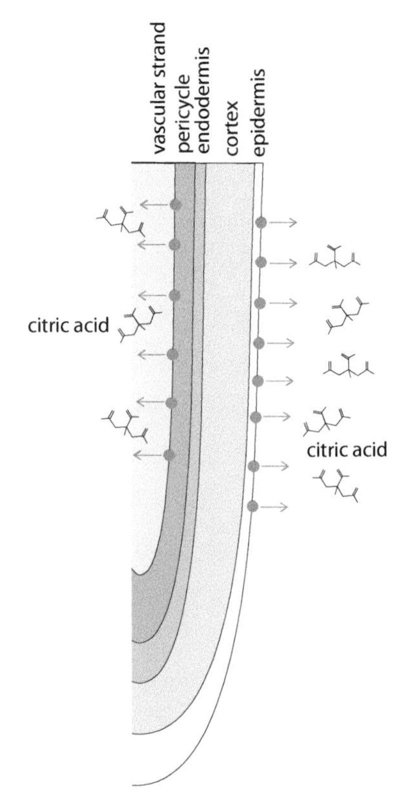

Figure 10.10. The MATE system for citrate transport in plant roots. Plants are very rich in multidrug and toxic compound extrusion (MATE) proteins, which are involved in the transport of metabolites across the kingdoms of life. In plant roots, MATE transporters (green dots) are used to allow citrate efflux, especially in the pericycle and epidermis. Pericycle MATE transporters load the xylem with citrate, where it is used for transport to the shoot, in particular to complex Fe^{2+}. In the epidermis, in acid-tolerant plants MATE transporters allow the efflux of citrate ions, which form complexes with Al^{3+} in the apoplast and rhizosphere, preventing it from reaching sites of toxicity in the root.

The isolation of the *TaALMT1* gene enabled genetic analysis of other species, and *ALMT* genes are now known to contribute to Al resistance in many species. Numerous genetic transformations have confirmed the contribution of *ALMT1* to Al resistance—for example, acid-sensitive barley cultivars transformed with *TaALMT1* have increased Al resistance, and this translates into increased yields over a growing season on moderately acid soils.

Before the identification of *ALMT* genes it was known that, in response to Al^{3+}, some species exude citrate from their roots, and that Al resistance in sorghum and barley—for example, at the well-known Alt_{SB} resistance locus in sorghum—did not map to loci with *ALMT* genes. In sorghum, barley, *Arabidopsis*, and several other species, citrate is exuded through a family of organic anion H^+-driven antiporters encoded by *MATE* (multidrug and toxic compound extrusion) genes (Figure 10.10). MATE antiport proteins are ubiquitous in the kingdoms of life, but are particularly common in plants. MATE transporters are involved in the internal transport of numerous toxins, such as nicotine, and in the transport of flavonoids, pyrrolidines, and organic anions such as citrate (Figure 10.11). In bacteria they are often involved in antibiotic resistance because they can transport antibiotics that are organic anions. The family was designated as containing MATE proteins a number of years before a role in Al resistance was discovered. Citrate exudation produces Al resistance in numerous cereals, legumes, and citrus species, with *MATE* gene expression in root tip regions often having been demonstrated. For some species—for example, rye, which is in general the most Al-resistant cereal—both *ALMT* and *MATE* genes are involved in Al resistance. *Arabidopsis,* which is less Al resistant than rye, also uses both systems. The ubiquity of citrate biochemistry enabled pioneering attempts to engineer Al resistance in plants (Box 10.5). It is often noted that plant species can differ in the timing of release of organic acids from the roots, probably because of different patterns of molecular control. In "pattern I" species, organic acid release occurs almost immediately upon exposure to Al^{3+} (the "pattern I" response), whereas in "pattern II" species it occurs after a few days.

Oxalate is the simplest dicarboxylate (Figure 10.12), and is accumulated to concentrations of 5–15% in a variety of oxalate accumulator species. In some of these species—for example, *Fagopyron esculentum* (buckwheat), *Hydrangea* species, *Amaranthus* species, and *Spinacia oleracea* (spinach)—the presence of micromolar concentrations of Al^{3+} induces rapid efflux of oxalate anions. In *Camellia sinensis,* oxalate is the key Al chelate in roots. Oxalate anions have been shown to complex Al^{3+} in the rhizosphere and apopolast of a number of species, and are likely to help to reduce exposure to Al^{3+}. However, the biochemistry of oxalate production and the molecular biology of oxalate permeases in roots are much less well known than those of malate or citrate, so the contribution of oxalate anions to Al resistance is not as clear. Given the preponderance of oxalate in many plant species, it may be significant. It is often noted that in forest ecosystems with acidic

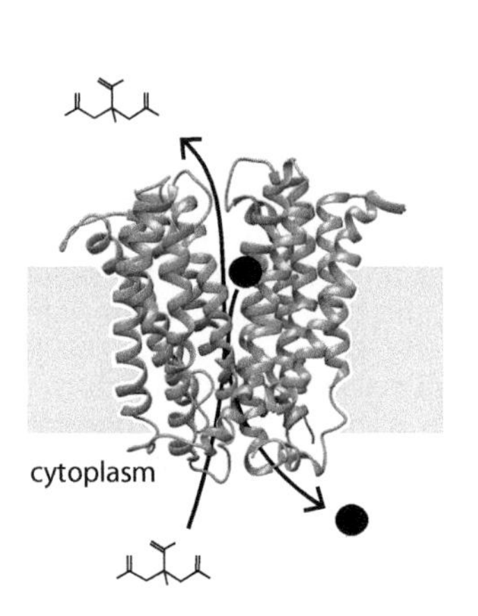

Figure 10.11. The operation of MATE transporters. MATE transporters have 12 membrane-spanning domains, and are antiporters driven by fluxes of univalent cations (black dots) down concentration gradients. In the externally open conformation (shown in the figure), the univalent cation binds to a site within a cleft formed from two groups of 6 membrane-spanning helices, which changes the conformation of the protein such that an organic anion binds to the internal surface, driving the antiport of the cation and anion. The transporter shown is from *Vibrio cholerae*, which is driven by Na^+. Plant root MATE transporters are similar in structure and function, but are driven by H^+ gradients. (Drawn in Chimera from PDB 3MKU: He X, Szewczyk P, Karyakin A et al [2010] *Nature* 467:991–994)

BOX 10.5. EARLY ATTEMPTS TO ENGINEER ALUMINUM TOLERANCE IN PLANTS

In one of the first attempts to engineer Al tolerance in plants, the citrate synthase (CS) gene from the bacterium *Pseudomonas aeruginosa* was expressed in tobacco plants, which increased citrate levels in root cells and increased citrate efflux from roots. These plants had increased resistance to the effects of Al above 50 μM, much better root hair development than non-transformed plants, and much reduced Al penetration into the roots. *Carica papaya* similarly transformed was able to grow normally in up to 300 μM Al, whereas root development failed in non-transformed plants in 50 μM Al. However, plant transformation with CS from *P. aeruginosa* does not always produce Al resistance, so transformation with CS genes, often from plant mitochondria, has also been used to produce Al-resistant plants. Given the ubiquity of citrate metabolism, it is likely to have a role in moderate Al resistance in many plant species.

Further reading: De la Fuente JM, Ramirez-Rodriguez V, Cabrera-Ponce JL et al. (1997) Aluminium tolerance in transgenic plants by alteration of citrate synthesis. *Science* 276:1566–1568.

mineral soils there are many plants that produce large quantities of oxalic acid (for example, *Oxalis acetosella*). It also seems possible that other carboxylates, many of which are known to occur in plants, might be used to reduce exposure to Al. Organic anion-based systems that function to reduce plant exposure to Al^{3+} have been most extensively investigated in annual crops. Demonstration of their use by an increasing number of wild species indicates that they probably also play a significant role in allowing many wild species to inhabit moderately acid soils. In addition, in a number of species, phenolic compounds—including **coumarins**, and in *Eucalyptus* some **tannins**—have been shown to chelate Al^{3+} in the rhizosphere, indicating that across the plant kingdom a range of compounds probably function to exclude Al^{3+} from roots.

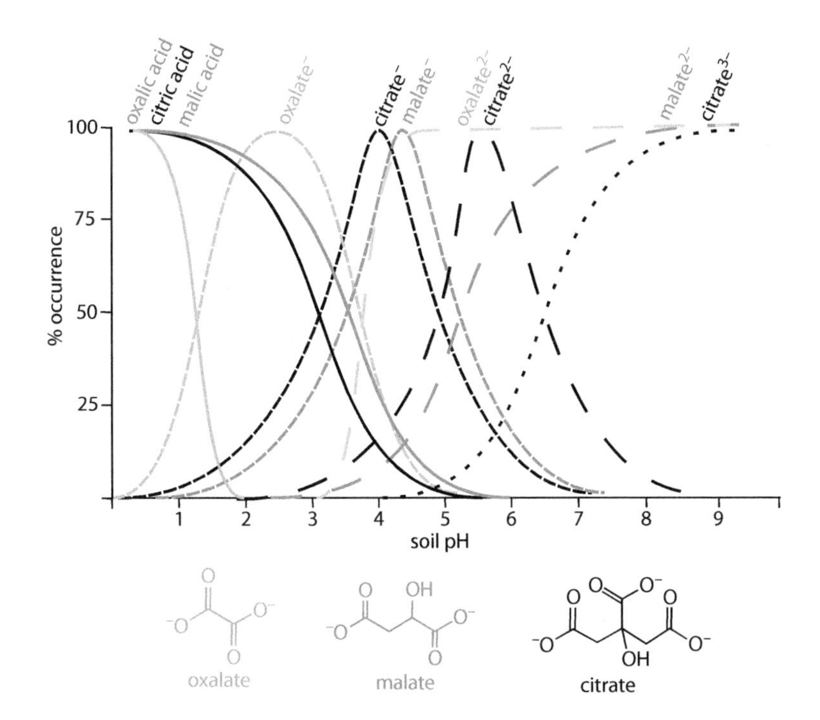

Figure 10.12. The dissociation of carboxylate ions that are excreted from plant roots. The conjugate bases of carboxylic acids are anions, and are the form in which they are excreted from roots. Those most often excreted from plant roots are dicarboxylates (oxalate and malate) or tricarboxylates (citrate) that have multiple pKa values and dissociate to form different anions in the pH range that occurs in acid soils (pH 2–6). The role of malic acid and citric acid in chelating toxic ions such as Al^{3+} is known in most detail, but oxalate ions can also have a significant role.

The capacity for Al^{3+} exclusion via external chelation is probably only sufficient to reduce Al exposure on moderately acid soils, and even in wheat—with its strong link to one Al resistance locus—genomic and proteomic investigations are a reminder of the importance to Al resistance of other processes. In maize, Al resistance is much less strongly linked to a single locus, and about twice as many genes are associated with Al resistance as in wheat. In rice, which is quite Al resistant, organic acid efflux often does not correlate with resistance, and there is no single QTL for resistance. In resistant plants much Al is located in the apoplast, and thickening of the cell wall, methylation of pectins, and changes in UDP-glucose production (which affects cellulose production) have been shown to contribute to resistance in some species. Decreases in the proportion of phospholipids in membranes can also contribute to the resistance of plants to the toxic effects of Al^{3+}, and in rice CDT proteins bind to the membrane, perhaps helping to stabilize it. Al-resistant rice cultivars, and several other species, also disarm Al^{3+} in roots not only externally but also internally. In such cultivars, Al^{3+}, perhaps when the maximum external capacity for chelation is reached, enters the cell through NRAT transporters, and undergoes efflux into the vacuole, where chelation with organic acid can occur, through ALS transporters of the ABC transporter family. In addition, antioxidant pathways can be up-regulated and the production of ROS decreased by increased use of AOX in mitochondria. ART1 and STOP1 are homologous transcription factors that help to control many of these processes in different species. A number of miRNAs and the WRKY46 transcription factor help to control expression of many resistance-associated genes. The identity of the sensor of Al^{3+} toxicity that provokes these responses is not yet clear, but Ca^{2+} and NO have established roles in transduction. Thus a battery of adaptations has been found in plants, often expressed to a different extent in different species or cultivars, that help them to resist the effects of Al^{3+} on roots by restricting its interaction with the root cell wall and cytoplasm.

For many plants on acid soils, mycorrhizal associations increase aluminum resistance

Many species in the order Ericales are strong calcifuges (Box 10.6). Ericoid mycorrhizas, unique to this order, have long been known to confer increased tolerance of toxic metals in soils, and they do so for Al in particular. There is much evidence that Al is transported into the fungi and chelated internally, reducing the exposure of plant cells to Al^{3+}, so ericoid mycorrhizal associations are in large part responsible for the ability of many ericaceous plants to inhabit acid soils. The geographical center of diversity for *Erica*, the most diverse genus in the order, is the acid heathlands of Southern Africa.

Fungi of ecomycorrhizas (EcM) can decrease root exposure to Al^{3+}. This has been demonstrated many times for pines and related conifers that frequently inhabit acidic soils. The Pinaceae, in which individuals with EcM have been shown to be more Al^{3+} tolerant than non-mycorrhizal individuals, include loblolly pine (*Pinus taeda*), pitch pine (*Pinus rigida*), eastern white pine (*Pinus strobus*), and Norway spruce (*Picea abies*). In general, EcM plant roots are tolerant of higher concentrations of Al^{3+} than Al^{3+}-tolerant non-mycorrhizal varieties—that is, species that are dependent on root exudates. Staining with fluorescent probes that bind Al^{3+}, such as **lumogallion**, has shown that in lateral roots Al^{3+} is accumulated and chelated in the fungal hyphae, producing low Al^{3+} concentrations in plant cells. The formation of complexes with organic acids exuded from the fungi and with organic acids and phosphates within hyphae has been reported in various fungal partners. There is also evidence that Al^{3+} is deposited in fungal cell walls. Overall, EcM reduce root cell exposure to Al^{3+} and contribute to plant growth over significant areas of acid soil. This is possible because Al is kept away from

BOX 10.6. THE ERICALES AND ACID SOILS

The order Ericales contains families that include rhododendrons, heathers, bilberries, and blueberries, many of which have a strong preference for acid soils. Heathers and their relatives are often a particularly important component of acid heathlands and moorlands. However, some heathers, such as *Erica vagans* (Figure 1), which occurs in heathlands around the coast of North-West Europe, tolerate alkaline soils and provide, within a single genus, an interesting example of the basifuge/basicole dichotomy.

Figure 1. *Erica vagans.* This heather is found on the coasts of Ireland, Cornwall, France, and Spain.

the meristem of the growing roots, as it is found preferentially in cells surrounding the meristem but not in the meristem itself. The improved nutrition of trees with EcM, especially with regard to P, contributes significantly to their ability to withstand high concentrations of Al^{3+}. Such effects enable EcM trees to dominate many forests over acidic soils, especially in the boreal zone. *Eucalpytus* species are important in ecosystems on many acidic soils in Australia, and have been widely planted on acid soils around the world. *Eucalyptus* plants with EcM are more Al^{3+} tolerant than non-EcM plants, and such associations seem likely to contribute significantly to the competitiveness of *Eucalyptus* on acid soils. However, EcM often do not infect primary root tips, which are an important site of Al toxicity. *Pinus taeda* primary root tips with no mycorrhizal partner have the ability to withstand Al concentrations of up to 500 μM, so other mechanisms of cellular resistance in the plant partner can also be significant.

In many acidic and moderately acidic soils, arbuscular mycorrhizal (AM) fungi are common and frequently form associations with plant roots. Such associations increase plant uptake of water and nutrients, which helps to provide protection against toxic metals. Seedling survival and plant growth are enhanced by AM in plants growing on soils containing toxic metals. *Eucalyptus globulus* plants with AM are more tolerant of high soil Al concentrations than are plants without AM, an effect which has been reported in numerous plant species that form associations with AM fungi, including acid-tolerant grasses such as *Andropogon virginicus*. However, AM fungi also affect the dynamics of organic acid exudation into the rhizosphere (both by contributing themselves and by promoting exudation from their host plant), form complexes with significant amounts of Al^{3+} in their cell walls, and accumulate Al in their vacuoles in the form of polyphosphates. AM fungi also exude **glomalins** into the rhizosphere. Sometimes these are the commonest proteins in a soil, and they can contribute significantly to the chelation of Al^{3+}. In some species, AM plants have enhanced concentrations of Al in their shoots, so AM associations not only control exposure to Al but can also affect Al distribution.

On very acidic soils, some plants take up and compartmentalize aluminum

In the 1990s, Al-sensitive ("*als*") *Arabidopsis* mutants were identified. *Arabidopsis* is not particularly resistant to Al^{3+}, but these mutants were especially sensitive to very low concentrations. The mutations were identified as lying within a gene encoding an ABC transporter. In *Arabidopsis*, *ALS1* and *ALS3* are expressed throughout the plant vasculature, including the root tip and hydathodes. ABC transporters have been shown to transport Fe–S clusters and other metal chelates, and thus it seems likely that ALS proteins are involved in loading Al complexes, perhaps Al–citrate complexes, into the phloem for distribution to less sensitive parts of the plant. They are involved in distributing Al to vacuoles and perhaps also in exuding Al from hydathodes. The fact that these processes are occurring in a species that is not particularly Al resistant suggests that many species probably have such mechanisms to help them to manage the Al that they take up during the short-term and limited-magnitude variation in Al^{3+} that occurs both spatially and temporally in most soils. However, it also provides a clue as to the mechanism whereby plants tolerate high internal concentrations of Al^{3+}.

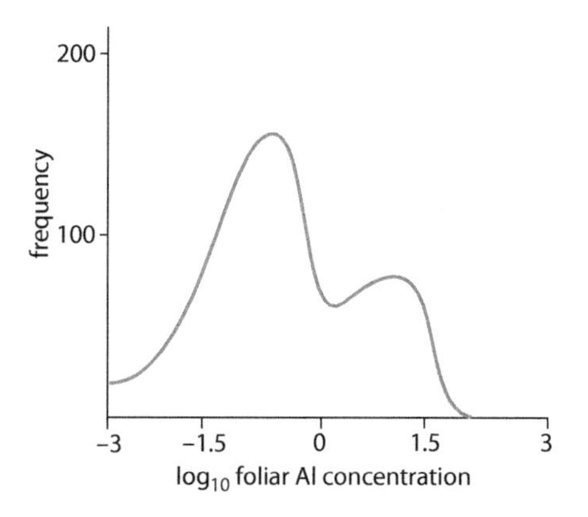

Figure 10.13. Aluminum-accumulating plants. In large collations of Al concentrations in plant leaves (mg Al g^{-1} dry biomass), most leaves have low Al concentrations, but some have distinctly elevated Al concentrations and are said to hyperaccumulate Al. The data shown are for 815 angiosperm species. (Data from Metali F, Salim KA & Burslem DFRP [2012] *New Phytol* 193:637–649.)

Many species, mostly woody ones on the basal lineages of rosids and asterids, have been designated as hyperaccumulators of Al (over 1000 ppm Al dry weight), a feature that might characterize at least 18 families of plants (Figure 10.13). Well-investigated examples include *Camellia sinensis* (tea), *Hydrangea macrophylla* (hydrangea), *Melastoma malabathricum*

(Singapore rhododendron), and *Fagopyron esculentum* (buckwheat). Concentrations of up to 30 mg Al g^{-1} are found in mature tea leaves, primarily bound to **catechins** in epidermal cells, which can result in up to 600 mg Al g^{-1} in young leaves. In *Hydrangea* leaves, in which concentrations of up to 1 mM Al have been reported, Al is found in 1:1 complexes with citrate, and in *Fagopyron* and *Melastoma* leaves, it is found in complexes with oxalate. Some investigations have also found Al bound to phenols, and it seems likely that other chelators could play a role. In blue petals of *Hydrangea macrophylla* and some other *Hydrangea* species an Al-delphinidin complex has been found (Box 10.7).

Al accumulator plants compartmentalize chelated Al in particular leaf vacuoles to prevent it from becoming toxic. For such compartmentalization to occur, Al transport through complex tissues has to be coordinated while toxicity is avoided. In accumulator plants, Al in the xylem sap is bound to citrate, other organic anion complexes seldom being reported, whereas oxalate ions are the primary root exudates. *Fagopyron* takes up Al-oxalate complexes from the rhizosphere and transports them across the root to the vasculature, where oxalate is exchanged with citrate for transport to the shoot. In the shoot of at least some Al accumulators the citrate must be exchanged for other chelators and the complexes targeted to certain cells. The proteins that mediate these processes and the mechanisms that control them are not well characterized, but analogous systems exist for other metals. In *Melastoma malabathricum*, an Al accumulator that grows on acid sulfate soils, the root mucilage has pronounced Al-chelation capacity and aids Al accumulation because its organic acids displace Al chelated in the soil and make it available for uptake. It is likely that high concentrations of toxic metals in plant tissue act as a herbivore deterrent. In severely acid soils the total amount of available Al^{3+} probably reduces the effectiveness of exclusion as a resistance strategy, so some plants have evolved to tolerate, and perhaps even to exploit, high internal Al^{3+} concentrations.

BOX 10.7. SOIL pH AND THE FLOWER COLOR OF HYDRANGEAS

Many hydrangeas (Figure 1) are Al accumulators on acid soils. Those with white flowers (Figure 2) do not change flower color if the soil pH varies, but those with pink/blue flowers do so. On acid soils, Al is accumulated and in petals is bound in a complex with delphinidin 3-glucoside and 3-caffeoylquinic acid. This complex gives a blue color to the flowers, and it decreases in concentration as the pH increases, turning the flowers pink.

Further reading: Hotta H, Wang Q, Fukuda M et al. (2008) Identification of aluminum species in an aluminum-accumulating plant, hydrangea (Hydrangea macrophylla), by electrospray ionization mass spectrometry. *Analyt Sci* 24:795–798.

Figure 1. *Hydrangea macrophylla.*

Figure 2. White flowers of *Hydrangea*.

Basic soils are low in important nutrients and induce characteristic symptoms in plants

Calcareous soils vary greatly in their organic matter content, and in the many such soils in which it is low, water-holding capacity can be very poor. Many calcareous soils also have a relatively low clay content, which in conjunction with the low organic matter content can produce a poorly developed soil macrostructure. This means that compaction can easily occur, with its attendant problems of poor aeration and poor drainage. In many calcareous soils, exposure of the underlying bedrock occurs. The temperature of these very thin soils can vary widely, and this can affect many aspects of the growth of higher plants. These physical properties pose significant challenges to plant growth on many soils, but a particularly challenging feature of calcareous soil is a chemical one, namely low Fe availability. Iron, the second most common metal in the Earth's crust, is the commonest metal to occur in two redox states (Fe^{3+} and Fe^{2+}) in aqueous solution at around cellular pH, and is used in numerous redox reactions by all organisms. Electron transport in respiration uses cytochrome, which has a heme group with an Fe atom. In plants, photosynthetic electron transport utilizes cytochrome b_6f. In addition, redox enzymes such as ferredoxin, catalase, and peroxidase require Fe, and the biosynthesis of chlorophyll is Fe-dependent. Animals are also dependent on other heme-based molecules, such as hemoglobin. In terrestrial ecosystems almost all Fe enters food chains via plants that extract it from the soil, including those upon which humans depend for food. An understanding of the impact of low Fe availability in basic soils is important not only for plants but also for all organisms in terrestrial systems.

The compounds of Fe^{3+} that form in aerobic aqueous solutions are only sparingly soluble, with a concentration at pH 7 of 10^{-11} to 10^{-10} M. In the range of conditions that are found in soils, either Fe^{3+} (oxidized, ferric) or Fe^{2+} (reduced, ferrous) can occur because the transformation between Fe^{3+} and Fe^{2+} is affected by pH and E_h, both of which can change markedly both spatially and temporally in soils (Figure 10.14). Below pH 5 under aerobic conditions both Fe^{2+} and Fe^{3+} are formed in soils and are soluble. However, at pH 5–8 in aerobic conditions Fe^{2+} is not formed in soils, and $Fe(OH)_3$ is the dominant species. $Fe(OH)_3$ is sparingly soluble, and available Fe^{3+} concentrations drop dramatically between pH 4 and pH 8. At pH 9 or higher,

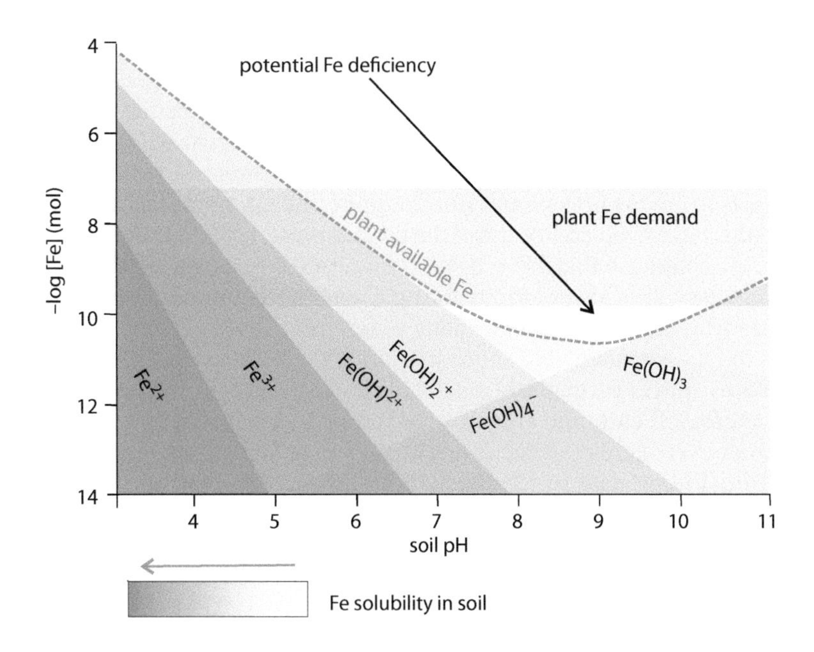

Figure 10.14. The effects of soil pH on Fe solubility. In soil solution, Fe forms a series of free ions and hydroxides. Plants can take up Fe^{3+} and Fe^{2+}, which are present at minimum concentration in the soil solution in the range of pH values that occur in many basic soils (pH 8–10). At their minimum availability there is generally insufficient Fe to meet plant demand, so there are characteristic plant species on basic soils which often have adaptations that enable them to mobilize Fe in soil.

available Fe concentrations can increase due to the formation of soluble $Fe(OH)_4^-$, and in waterlogged soils of various pH values Fe^{2+} can occur. A significant proportion of aerobic soil has a moderately to strongly basic pH and therefore has very low concentrations of available Fe. Fe availability is frequently insufficient to meet plant demand, and lack of Fe limits plant growth on a global scale in both unmanaged and managed ecosystems. In addition, microorganisms face the same challenges of obtaining Fe on basic soils, and compete strongly with plants for this nutrient.

The availability of nutrients other than Fe is also affected at alkaline pH values. The speciation of Zn is pH-dependent, with increasing pH decreasing the amount of available Zn^{2+} in soil (Figure 10.6), an effect that can induce Zn deficiency in crops growing on calcareous soils. Rice paddies formed from clay soils that have been puddled and flooded frequently have an alkaline pH, and Zn deficiency is common in rice cultivation. In many calcareous soils the plant-available concentration of P is also very low, often due to the precipitation of any PO_4^{2-}/HPO_4^- as insoluble $CaPO_4$. Chernozem soils are an exception to this—they have a slightly alkaline pH and high Ca content, but also a high organic matter content that helps to maintain P availability.

The classic symptom of non-adapted plants in soils with a high lime ($CaCO_3$/CaO) content is "lime-induced chlorosis," long used by agriculturalists and horticulturalists to diagnose Fe deficiency. Chlorosis of the leaves, which turn from green to yellow/white in color, is due to inhibition of chlorophyll synthesis. Chlorosis induced by alkaline soil pH is frequent in crop plants and problematic in hydroponic systems, in which it can be challenging to maintain sufficient soluble Fe. Lime-induced chlorosis is a significant factor in, for example, European viticulture (Box 10.8) and olive production. It begins between the veins of new leaves and can spread to an even chlorosis, but in general the old leaves remain green. In plants, as in many other organisms, much Fe is stored in **ferritin**. Old leaves in Fe-deficient plants remain green because Fe is released relatively slowly from ferritin, and translocation of Fe is generally insufficient to meet demand from new leaves. In monocotyledonous plants with parallel veins, Fe deficiency produces a characteristic "yellow-stripe" pattern. Chlorophyll and heme both have a tetrapyrrole ring ("a porphyrin"), but with central Mg and Fe atoms, respectively. The synthesis of porphyrin starts with the transformation of glutamic acid into 5-aminolevulinic acid. This rate-limiting step in chlorophyll biosynthesis is catalyzed by an Fe-dependent enzyme, glutamyl-tRNA reductase. Fe deficiency inhibits the action of this enzyme and thus the biosynthesis of chlorophyll.

In addition to low Fe, calcicoles face the particular problem of very high Ca and HCO_3^- availability, and low PO_4^{2-} availability. Cytosolic concentrations of free Ca^{2+} have to be kept very low, primarily because the formation of insoluble calcium phosphates would severely interfere with metabolic processes involving phosphates (for example, phosphorylation of enzymes). Calcicoles have evolved to control the concentration of free Ca^{2+} in the cytoplasm. Adaptations that have been shown to be involved in this include transporters with a lower affinity for Ca^{2+}, sequestration in the vacuole, and chelation by organic acids. In many plants the production of Ca-oxalate seems to be associated with sequestration of Ca^{2+}. Ca-oxalate clearly deters herbivores, and is particularly noticeable in many calcicoles. In some of the latter, increased carbonic anhydrase activity has been reported, which can help to convert excess HCO_3^- into CO_2 and H_2O. Calcicoles often mobilize PO_4^{2-} from the vacuole, which can help not only with P depletion in the cytoplasm but also with the biochemical pH-stat. High pH also produces metabolic and oxidative challenges for plants because many parts of the TCA cycle and antioxidant systems are known to be affected by it.

BOX 10.8. MANAGING FE-DEFICIENCY-INDUCED CHLOROSIS IN GRAPEVINES

In many crops that are grown on calcareous soils a lack of chlorophyll ("chlorosis") induced by Fe deficiency is quite common. Chlorosis can severely inhibit plant growth and thus substantially reduce yield. Fe-deficiency-induced chlorosis is particularly problematic in grapevines (Figure 1), because of a combination of the soils on which the grapes for many desirable wines are produced, and the rootstocks that are used for a significant proportion of production. The remedies for chlorosis in grapevines, which commonly focus on adding Fe supplements, are often expensive and sometimes environmentally undesirable. The choice of rootstock for particular soils and the generation of rootstocks that have good Fe uptake are vital to many vineyards.

The majority of grapes are produced from domesticated *Vitis vinifera*, a species indigenous to Eurasia. There are *Vitis* species in both the Old World and the New World, but it is the fruit of *V. vinifera* that are the mainstay of global grape production. The many calcareous soils of Europe and the Mediterranean basin have long been important in wine production, in part because of the wines they produce, but also because they are more suitable for *V. vinifera* than for many other crops. Grapes have been produced in the Mediterranean basin for several thousand years, and were probably among the earliest fruit to be domesticated by humans, but in the nineteenth century European wine production was severely jeopardized by epidemics of phylloxera. By the second half of the nineteenth century, several *Vitis* species from North America had been introduced to Europe, not only for ornamental purposes but also in an attempt to improve grape production. These introductions brought with them from North America phylloxera (*Daktulosphaira vitifoliae*), a sap-sucking insect related to aphids and scale insects. North American vines have some natural resistance to phylloxera, but *V. vinifera* does not, and by the late nineteenth century phylloxera had devastated grape production across much of Europe. It continues to be a significant problem in grape production in most parts of the world, including North America, because of the importance of *V. vinifera*. Phylloxera nymphs feed on the phloem sap in the roots. Severe infestations can inhibit root function very significantly, and they also allow soil-borne diseases to infect the root.

By the end of the nineteenth century, in an attempt to increase phylloxera resistance, *V. vinifera* hybrids with several North American species were produced. However, phylloxera-resistant hybrids have only ever found minority use because they often, although not always, produce lower-quality grapes. It was the grafting of *V. vinifera* **scions** onto phylloxera-resistant North American rootstocks that enabled European grape production to recover, and that is the mainstay of grape production to this day (Figure 2). There are now only a few "old-vine" vineyards that do not use phylloxera-resistant rootstocks. Phylloxera has spread to most grape-producing areas, the only exceptions being some islands and parts of South America and Australasia. Phylloxera-resistant rootstocks tend to be more susceptible to Fe-deficiency-induced chlorosis than those of *V. vinifera*, because the North American *Vitis* species are less well adapted to living on calcareous soils. Chlorosis resistance of rootstocks is now a major factor in grape production, and the remedy of Fe deficiency is often a significant cost. Traditional breeding of grapevines takes many years, so there is much interest in using advanced techniques to enhance the chlorosis resistance of rootstocks. For example, QTLs have been identified and might form the basis of marker-assisted selection to provide rootstocks in which the management of chlorosis can be less economically and environmentally costly.

Further reading: Bert P-F, Bordenave L, Donnart M et al. (2013) Mapping genetic loci for tolerance to lime-induced iron deficiency chlorosis in grapevine rootstocks (*Vitis* sp.). *Theor Appl Genet* 126:451–473.

Bavaresco L, Goncalves MIVDB, Civardi S et al. (2010) Effects of traditional and new methods on overcoming lime-induced chlorosis of grapevine. *Am J Enol Vitic* 61:186–190.

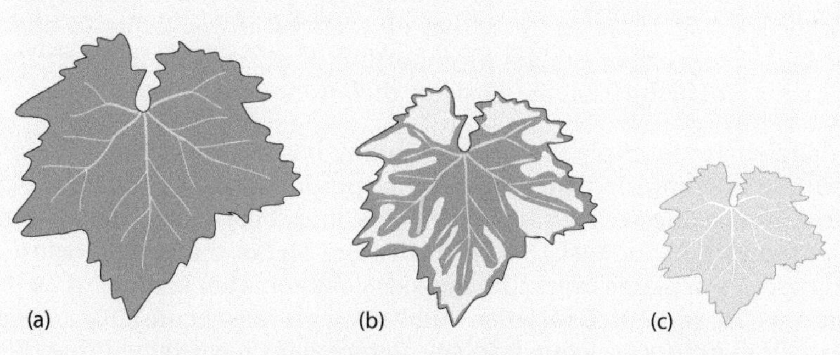

Figure 1. Symptoms of Fe-deficiency-induced chlorosis in grape. As Fe is relatively immobile in plants, deficiency symptoms are most severe in new leaves. With mild Fe deficiency, new leaves have interveinal areas lacking chlorophyll ("chlorosis") (b) and are smaller than Fe-replete leaves (a). With severe Fe deficiency (c) the leaves are small and widely chlorosed.

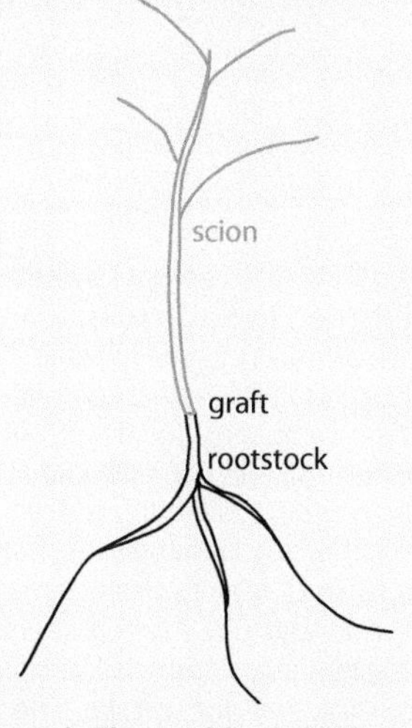

Figure 2. The grafting of *Vitis vinifera*. Most grapevines that are used in commercial production consist of scions of *V. vinifera*, which provides the best grapes for wine production and for eating, grafted onto a phylloxera-resistant rootstock.

Some plants have adapted to scavenge iron, zinc, and manganese from basic soils

In soils with a pH in the range 6–7, the amount of plant-available Fe is less than the requirements (10^{-9} to 10^{-8} M) of most plants. For a wide variety of plant species the most common optimum bulk soil pH for growth is about 6.5, but the rhizosphere pH of most plants is generally 1 or more pH units less than that of the bulk soil, due to the presence of root exudates. Protons in particular have to be exuded because they are produced in excess by numerous processes in plant cells. For plants, a bulk soil pH of 6.5 therefore means a pH of 5.5 or less in the rhizosphere—the range at which Fe^{2+} becomes available, but before Al^{3+} becomes toxic. A soil pH of about 8.5—that is, the pH of calcareous soil—is especially difficult for plants because the rhizosphere pH of about 7.5 results in minimal Fe availability.

In *Arabidopsis*, increasing rhizosphere pH induces an increase in efflux of protons through ATPases encoded by *AHA* genes. Such increases in H^+ efflux have now been shown in many species, and are probably a widespread response that produces increases in rhizosphere pH and hence increases the availability of Fe^{2+}, Zn^{2+}, and Mn^{2+}. In addition, in many species, enzymes of the ferric reductase/oxidase (FRO) family are produced during Fe deficiency, some of which are targeted at the root epidermis. For example, *FRO2* in *Arabidopsis* encodes an Fe-deficiency-inducible FRO targeted at the outside of the root. FROs have been identified in a number of species, and their induction explains the increase in Fe-reduction capability detected in the rhizosphere of many plants in response to alkaline pH. Plant FROs are closely related to ferric reductases in yeast, and to NADPH oxidases such as the $gp91^{phox}$ involved in the oxidative burst in human **neutrophils**. FROs have a large cytoplasmic loop with FAD and NADPH binding sites, and are flavocytochromes with heme moieties. FAD and NADPH act as the source of electrons transported across the membrane by two intra-membrane heme groups. In $gp91^{phox}$, O_2 is reduced by the addition of electrons, but with FROs in plant roots Fe^{3+} is reduced to Fe^{2+}. Thus, by increasing the efflux of protons and electrons, plants can increase the concentration of soluble Fe^{2+} in the rhizosphere. Fe^{2+} is taken up by plants via an iron-regulated transporter (IRT1) which functions as a passive divalent cation transporter with high specificity for Fe^{2+}. Increases in the activity of H^+-ATPases, FROs, and IRT1 work in concert to reduce and take up Fe in what is termed a "strategy I" response to low Fe availability (Figure 10.15). As part of this strategy of reducing Fe^{3+} to Fe^{2+}, several plant species have also been shown to increase the exudation of phenolics (especially coumarins) and flavins in response to Fe deficiency, and in the case of carob, quinic acid. In rice the transcription factor IDEF1 helps to control strategy I responses, and has homologs in other species. It seems likely that differences in the ability to operate strategy I components explain a proportion of differences in tolerance of basic soils and of the basifuge/basicole dichotomy.

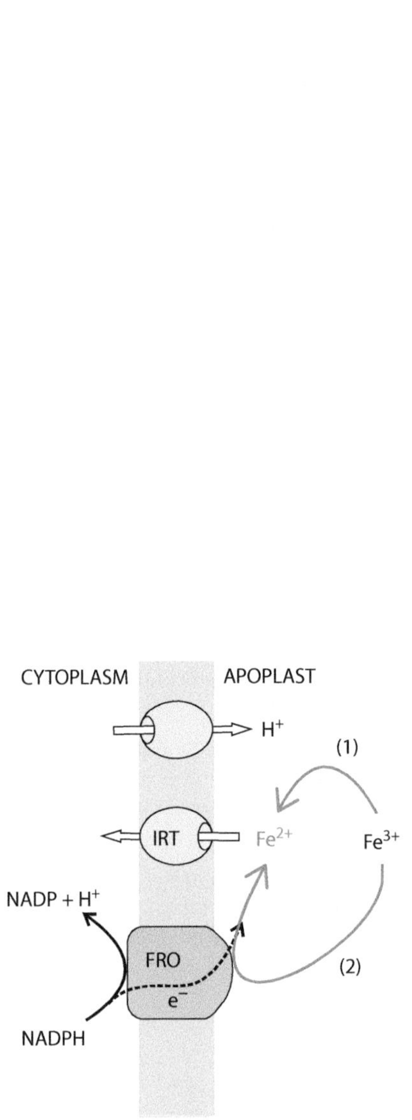

Figure 10.15. The strategy I system for Fe acquisition in basic soils. Many basicoles are adapted to mobilize Fe^{2+} and take it up. (1) Many of these plants have significant capacity for decreasing the rhizosphere pH via the efflux of H^+, thus increasing the availability of Fe^{2+} for uptake. (2) Many basicoles also have ferric reductases (FROs) in the plasma membrane that can directly reduce Fe^{3+} to Fe^{2+} using electrons from NADPH, again mobilizing Fe^{2+} for uptake. The IRT transporters are in the ZIP family of divalent metal cation transporters, which occur widely in eukaryotes, and can also transport Zn and Mn. The mobilization of Fe^{2+} and uptake through IRT is generally referred to as "strategy I" for Fe uptake from basic soils. It is the primary strategy for non-grass species, but is also found in some grasses.

Zn deficiencies limit plant growth on many basic soils, and are the most important micronutrient limitation in rice production. Zn availability in the rhizosphere is affected by carboxylic acids, and there are numerous transporter families, including CDFs, NRAMPs, and MHXs, that can transport Zn. Many species of plants from unmanaged ecosystems are tolerant of Zn deficiency, as are some rice varieties. Analyses with rice **recombinant inbred lines** (RILs) produced using low-Zn-tolerant lines have shown that differences in tolerance are not associated with differences in transporter activity, but with the exudation of organic anions. Zn-isotope fractionation studies also indicate that complexation in the rhizosphere helps to explain the tolerance of low Zn soils by rice. The mobilization of Zn, in part via malate exuded from the roots, followed by uptake of Zn chelates or Zn mobilized from chelates in the apoplast, helps to explain how some plants inhabit low-Zn soils.

Mycorrhizas are abundant in many calcareous ecosystems, and these infections may also be important in Zn acquisition.

In biological systems, Mn deficiency is generally thought to be rare because Mn is the second most abundant transition metal and, although it is an essential nutrient, demand for it is low. However, in aerobic calcareous soils, Mn^{2+} availability is particularly low and Mn deficiency can limit plant growth. IRT can efficiently transport Mn^{2+}, and changes in its expression or in Fe^{2+} supply affect Mn^{2+} uptake into plants. In many organisms, including plants, NRAMP transporters control intracellular concentrations of Mn^{2+}. In plants they help to control Mn supply to chloroplasts, and enhanced expression of these transporters is associated with increased efficiency of Mn use. NRAMP1 in *Arabidopsis* has been shown to mediate Mn^{2+} uptake in roots, and can supply Mn in the absence of Mn^{2+} transport via IRT. It is likely that Mn^{2+} uptake by many plants occurs through IRT and NRAMP proteins, and that differences in tolerance of Mn deficiency might be produced by differences in the expression of these proteins.

Nicotianamine aids iron homeostasis, and in grasses evolved into root exudates that chelate iron

Fe^{2+} can be toxic because it catalyzes Fenton reactions that produce hydroxyl radicals, and it is Fe^{3+} that is needed by enzymes, so not only concentrations but also redox states of Fe in plants have to be controlled. Nicotianamine (NA) is a key molecule in plant Fe homeostasis and transport. It is synthesized from methionine by a one-step condensation of three molecules of *S*-adenyl-methionine (SAM) catalyzed by NA synthase (NAS). NA has three amine and three carboxyl groups in an **azetidin** ring, and at cytosolic pH can form very stable chelates with metal ions, including both Fe^{2+} and Fe^{3+} (Figure 10.16). Plants with mutations in NAS (for example, the *chl* mutant of tomato) or overexpression of NAS have altered Fe homeostasis and transport. Disruption of NAS leads to chlorosis even under conditions of adequate Fe supply, but so can overexpression of NAS because of overchelation of Fe. Altered NAS expression affects the expression of numerous genes associated with the FRO/IRT Fe uptake system, and affects plant response to Fe deficiency, with plants that have increased NA generally coping better. Disruption of NA affects plant development, most importantly the development of the reproductive parts. In addition, it affects the transport of other metals, including Zn, Mn, and Ni, indicating that NA also plays a key role in distributing them.

NA–Fe complexes have to be transported across membranes, and this process is probably the most important mechanism of intercellular transport of Fe in plants. The YSL ("yellow-stripe like") proteins are known to have numerous members in a range of plant species, and to transport NA–Fe complexes. They are members of the oligopeptide transporter (OPT) group, which occurs only in plants, fungi, bacteria, and archaea. The 8 YSL transporters of *Arabidopsis* and the 16 YSL transporters of rice are expressed differently in different tissues, with some being expressed primarily in vascular tissue, developing endosperm, or senescing tissue. It seems very likely that YSL transporters have a key role in the intercellular allocation of Fe in plants. RNAi experiments that have altered YSL expression have shown altered expression of *IRT* and *NAS* genes, and of transcription factors (TFs) associated with Fe deficiency. The manipulation of *NAS* and *YSL* genes has been used to increase the Fe concentration in seeds in a number of plant species. The FRO/IRT/NAS/YSL pathways of Fe homeostasis are fundamental not only to plant survival on low Fe soils, but also to managing the supply of Fe in human food chains (Figure 10.17).

(a)
COOH COOH COOH
N N NH2
H

(b)

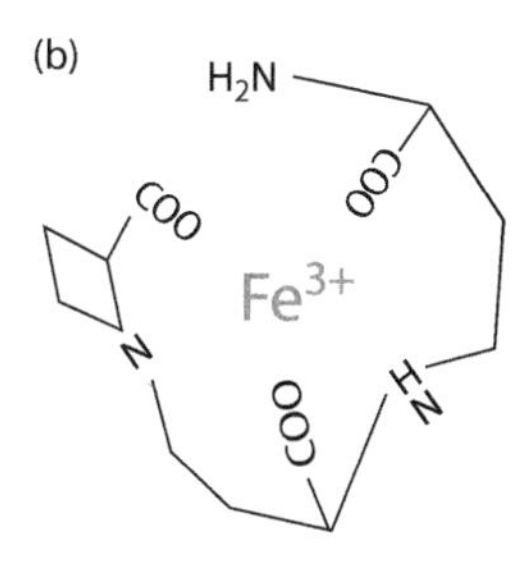

Figure 10.16. Nicotianamine (NA) and the chelation of metals. Nicotianamine (a) is synthesized from methionine, and is a vital chelator of Fe^{3+} (b) and Fe^{2+} in plant cells. Chelation of Fe with NA aids the control of cytosolic Fe concentrations. NA–Fe chelates are also transported across many membranes and thus have a vital role in the compartmentation of Fe.

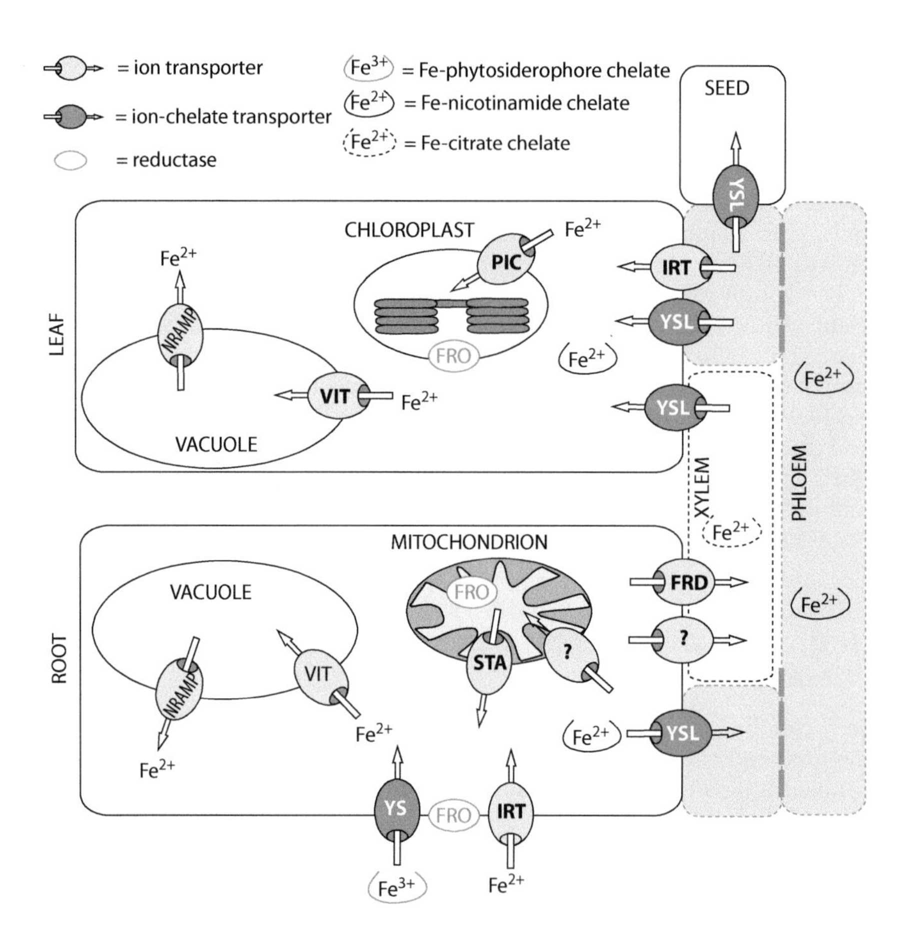

Figure 10.17. Fe homeostasis in plants. Plants have transporters for Fe ions and for Fe ion–chelates, and also have reductases that can control Fe speciation. In concert these proteins control the transport of Fe into and out of plant cells and cell compartments. They also control the loading of Fe into the vascular tissue in the root, and hence control root-to-shoot transport. IRT, iron transporter; YS, yellow stripe; YSL, YS-like; VIT, vacuolar iron transporter; NRAMP, natural resistance-associated macrophage protein; PIC, permease in chloroplast; FRD, ferric chelate reductase; FRO, ferric reductase/oxidase; STA, STARIK1 ABC transporter; ?, unknown transporters.

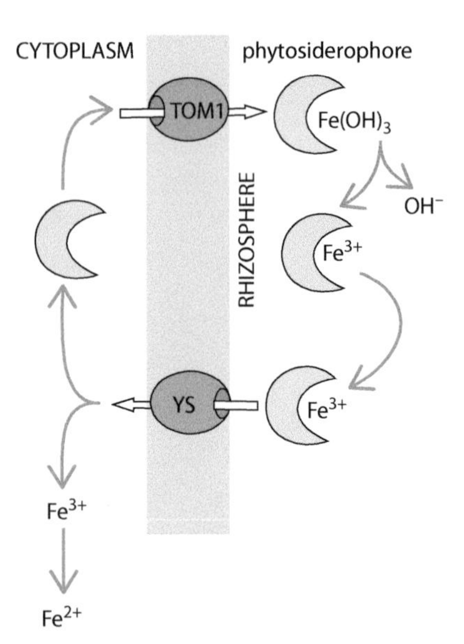

Figure 10.18. The strategy II system for Fe uptake in basicole grasses. Many basicole grasses have the ability to synthesize, and then exude into the rhizosphere, NA derivatives called phytosiderophores (PSs). These are mugineic acids that have the ability to chelate Fe^{3+}. Much of the Fe3+ that they chelate is derived from $Fe(OH)_3$, but they can also chelate other multivalent cations, including Zn^{2+} and Mn^{2+}. PS–Fe complexes are transported across the membrane by yellow-stripe (YS) transporters (named for the yellow striped leaves that grasses lacking them develop under conditions of Fe deficiency). Fe^{3+} is released in the cytoplasm and reduced to Fe^{2+}, and the PS is exuded by as yet unknown transporters.

During Fe deficiency, many grasses, and in particular those adapted to basic soils, excrete phytosiderophores (PSs) into the rhizosphere. These are iron-chelating ("siderophore") mugineic acids that are synthesized from NA, undergo efflux through the TOM1 transporter, and chelate Fe^{3+} in the rhizosphere (Figure 10.18). Many PSs have a diurnal pattern of exudation. PS-Fe^{3+}chelates are taken up from the rhizosphere into roots by YS (yellow-stripe) transporters named for the chlorotic, yellow-striped *ys1* maize mutant that was first used to identify them. PSs and enzymes for their synthesis have so far only been found in the Poaceae (Figure 10.19), but the YS1 transporter that is found in the root epidermis was the basis for identifying the YSL transporters involved in NA–Fe transport in all plants. It seems very likely that the PS system for adapting to low-Fe soils, which is termed "Strategy II," is a graminoid adaptation that evolved from the Fe-homeostasis mechanism common to all plants. The range of siderophores produced by bacteria is well characterized. Their production reflects the challenges that microbes face in obtaining enough Fe in aerobic aqueous environments. Siderophores were probably important in the evolution of microbial pathogenicity to multicellular animals, because the low concentrations of free Fe^{3+} in animals can be a challenge for microbes—**virulence factors** in some pathogenic bacteria are siderophores. The PSs that have been identified to date are evolutionarily distinct from microbial siderophores, although both use high-affinity hexadentate ligands to chelate Fe^{3+}—that is, they have six donor atoms. The independent evolution of these systems was probably driven by low bioavailability of Fe^{3+} on the oxidized Earth.

In rice, which is generally sensitive to low soil Fe, strategy II for iron uptake is supplemented by the action of IRT1 transporters. In barley, which is generally the cereal that is most tolerant of low Fe, much larger amounts of not only 2′-deoxymugineic acid (DMA) but also mugineic acid (MA) and

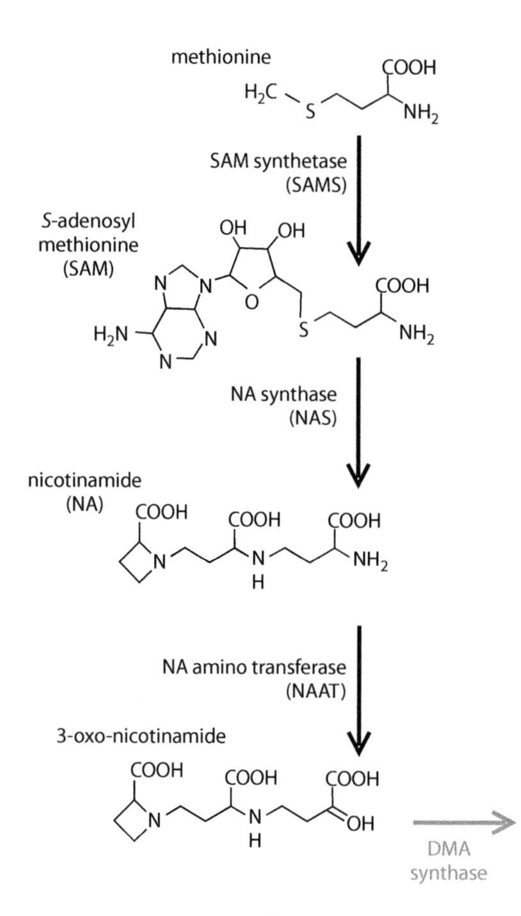

Figure 10.19. The synthesis of phytosiderophores. The nicotianamine (NA) precursor of phytosiderophores is synthesized from methionine. Deoxymugineic acid synthesized from NA is the precursor for mugineic acids and avenic acid, all of which can chelate Fe^{3+} and other multivalent cations.

epimugeneic acid (epiMA) are exuded. Hydroxy and epihydroxy MAs have also been found in some root exudates. Tolerance of low Fe in grasses correlates with PS exudation, and manipulation of *NAAT* and *DMAS* expression affects PS production and Fe-uptake efficiency. In rice, a decrease in MA synthesis in *NAAT* mutants increases the uptake of Fe^{2+}. YS1 transporters are expressed in the root epidermis, and in other parts in some species. YS1 transporters have been shown to transport PS–Fe chelates in shoot tissues in a number of species, indicating that intercellular transport of Fe might occur not just as NA–Fe complexes. Avenic acid in root exudates of *Avena sativa* has been shown to chelate Fe^{3+}, and it is likely that other Fe^{3+}-chelating compounds have evolved in plants. PSs also chelate other metals, including Cu, Zn, and Cd. Cu uptake efficiency is increased and Cd toxicity is moderated by PSs. Some mycorrhizal fungi have been shown to release siderophores, and may make a significant contribution to Fe mobilization in the rhizosphere of unmanaged ecosystems.

Ecologically important iron and zinc deficiency responses are finding important agricultural uses

In the strategy I plant *Arabidopsis*, FIT TFs are essential for regulating the response to Fe deficiency. FIT TFs are homologs of the IRO TFs known in rice. Both are bHLH TFs and affect expression of numerous Fe nutrition genes. FRO gene expression is regulated transcriptionally by FIT via *cis*-regulatory sequences, whereas IRT is controlled post-translationally. In rice, IRO TFs affect the expression of genes that are part of the strategy II response as well as the strategy I response, and in addition the *cis*-acting NAC TFs IDEF1 and IDEF2 regulate numerous Fe-responsive genes. It is also clear that miRNAs regulate the expression of many Fe-responsive genes. As not only systems for Fe uptake and distribution but also those for the production of many Fe compounds intersect with Fe homeostasis, the pathways of Fe homeostasis must be complex and vital to all plants, but especially basicoles. Crops with naturally efficient Fe mobilization and uptake strategies are already agronomically useful, and it is likely that crops engineered to have these characteristics will find increasing use in the twenty-first century. The biofortification of crops with increased Fe and Zn concentrations is a particular focus of research efforts.

A number of established eudicot/Poaceae intercrops have been shown to have increased efficiency of Fe and Zn uptake. Combinations involving strategy I, and especially strategy II, plants have been shown to increase Fe/Zn uptake in peanut/maize, chickpea/wheat, guava/maize, and guava/sorghum intercrops. Other combinations, such as beans and ryegrass, have also been noted to produce increased Fe uptake, and grasses have been used to increase Fe availability to grapevines. In particular, the increased availability of Fe^{3+} due to the presence of PSs in the rhizosphere from the gramineous partner appears to increase Fe uptake in the eudicot partner. The overlaps and complementarity of the two Fe acquisition strategies suggest that genetic transformation, breeding, and agronomic strategies can all play a role in increasing Fe uptake by crop plants. They are also a reminder that in Fe-limited ecosystems, rhizosphere processes that affect Fe nutrition are of great ecological significance.

Summary

Plant root cells have pH-stats that can buffer cytosolic pH against variation in soil pH. Chelation of Al in the apoplast and protoplast is sufficient to prevent toxicity from Al^{3+} exposure induced by slightly acid soil. In acid soil, exclusion of Al and the formation of mycorrhizas are common adaptations in resistant plants. In very acidic soils, Al accumulation and compartmentation underpin Al tolerance. These systems for dealing with progressively more intense exposure are probably important not only for Al but also for other toxic ions. Plants that inhabit basic soils have evolved systems for scavenging Fe from low concentrations. These systems are based on methionine-derived molecules widely used in biology for this purpose, and scavenge not only Fe but also other nutrients, such as Zn and Mn. Anatomical and morphological adaptations to extreme soil pH values are minor—the strong chelates that are formed with key ions are sufficient basis for the adaptations (Figure 10.20). Soils of extreme pH are widespread, underlie some important natural ecosystems, and are occasionally used for agricultural production. Wheat, rice, and maize do not grow well on acidic soils, so much of the research into acid toxicity has involved these plant species, but there is growing awareness of the global importance of natural ecosystems with

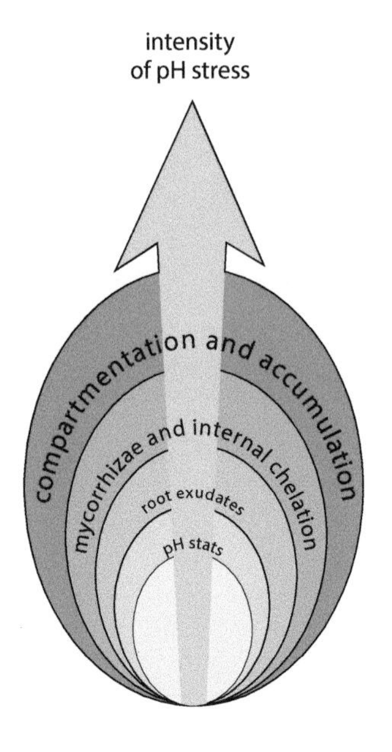

Figure 10.20. Plant adaptation to extreme soil pH. Plant cells have biochemical buffers, metabolic pathways, and transport systems that, during exposure to routine variations in soil pH, act as pH-stats to control cytosolic pH. In response to the additional challenges of ion availability that moderately acidic or basic soils present, root exudates are also extensively used to control the uptake in particular of Al and Fe, respectively. Plant species that are characteristic of truly acidic or basic soils often also have well-developed mycorrhizas that help to control ion availability to plants and/or use chelates to control internal concentrations of ions. Particularly in very acid mineral soils, some plant species are Al accumulators that survive by compartmentalizing the large quantities of Al to which they are exposed. In general in plants, morphological and phenological adaptations are not important for adapting to the challenges of extreme soil pH.

acid soils, especially acid organic soils. Increased agricultural production, especially from acid soils, is very likely to be important in achieving global food security. Many ecosystems with alkaline soils are sensitive to pollution because it frequently decreases soil pH. Overall, there is a great incentive to understand plant responses to soils of extreme pH for both conservation management and agricultural production.

Further reading

Soil pH and plant growth

Bartelheimer M & Poschlod P (2014) The response of grassland species to nitrate versus ammonium coincides with their pH optima. *J Veg Sci* 25:760–770.

Colombo C, Palumbo G, He J-Z et al. (2014) Review on iron availability in soil: interaction of Fe minerals, plants, and microbes. *J Soils Sediments* 14:538–548.

Peppler-Lisbach C & Kleyer M (2009) Patterns of species richness and turnover along the pH gradient in deciduous forests: testing the continuum hypothesis. *J Veg Sci* 20:984–995.

Sumner ME (2005) Food production on acid soils in the developing world: problems and solutions. *Soil Sci Plant Nutr* 51:621–624.

Soil pH variation and environmental change

Azevedo LB, van Zelm R, Hendriks AJ et al. (2013) Global assessment of the effects of terrestrial acidification on plant species richness. *Environ Pollut* 174:10–15.

Guo JH, Liu XJ, Zhang Y et al. (2010) Significant acidification in major Chinese croplands. *Science* 327:1008–1010.

Kim TK, Silk WK & Cheer AY (1999) A mathematical model for pH patterns in the rhizospheres of growth zones. *Plant Cell Environ* 22:1527–1538.

Richter DD (2007) Humanity's transformation of earth's soil: pedology's new frontier. *Soil Sci* 172:957–967.

Control of plant cell pH

Britto DT & Kronzucker HJ (2005) Nitrogen acquisition, PEP carboxylase, and cellular pH homeostasis: new views on old paradigms. *Plant Cell Environ* 28:1396–1409.

Chen Z, Jenkins GI & Nimmo HG (2008) pH and carbon supply control the expression of phosphoenolpyruvate carboxylase kinase genes in *Arabidopsis thaliana*. *Plant Cell Environ* 31:1844–1850.

Fait A, Fromm H, Walter D et al. (2008) Highway or byway: the metabolic role of the GABA shunt in plants. *Trends Plant Sci* 13:14–19.

Iuchi S, Koyama H, Iuchi A et al. (2007) Zinc finger protein STOP1 is critical for proton tolerance in *Arabidopsis* and coregulates a key gene in aluminum tolerance. *Proc Natl Acad Sci USA* 104:9900–9905.

Ion availability and toxicity in acid soils

Bojorquez-Quintal JEA, Sanchez-Cach LA, Ku-Gonzalez A et al. (2014) Differential effects of aluminum on *in vitro* primary root growth, nutrient content and phospholipase C activity in coffee seedlings (*Coffea arabica*). *J Inorg Biochem* 134:39–48.

Hiradate S (2004) Speciation of aluminum in soil environments. *Soil Sci Plant Nutr* 50:303–314.

Li Q, Chen L, Jiang H et al. (2010) Effects of manganese-excess on CO_2 assimilation, ribulose-1,5-bisphosphate carboxylase/oxygenase, carbohydrates and photosynthetic electron transport of leaves, and antioxidant systems of leaves and roots in *Citrus grandis* seedlings. *BMC Plant Biol* 10:42.

Panda BB & Achary VMM (2014) Mitogen-activated protein kinase signal transduction and DNA repair network are involved in aluminum-induced DNA damage and adaptive response in root cells of *Allium cepa* L. *Front Plant Sci* 5:256.

Aluminum exclusion from cell walls and cytoplasm

Liu J, Pineros MA & Kochian LV (2014) The role of aluminum sensing and signaling in plant aluminum resistance. *J Integr Plant Biol* 56:221–230.

Ma JF, Chen ZC & Shen RF (2014) Molecular mechanisms of Al tolerance in gramineous plants. *Plant Soil* 381:1–12.

Nunes-Nesi A, Brito DS, Inostroza-Blancheteau C et al. (2014) The complex role of mitochondrial metabolism in plant aluminum resistance. *Trends Plant Sci* 19:399–407.

Tahara K, Hashida K, Otsuka Y et al. (2014) Identification of a hydrolyzable tannin, oenothein B, as an aluminum-detoxifying ligand in a highly aluminum-resistant tree, *Eucalyptus camaldulensis*. *Plant Physiol* 164:683–693.

Mycorrhizas and aluminum

Aguilera P, Cornejo P, Borie F et al. (2014) Diversity of arbuscular mycorrhizal fungi associated with *Triticum aestivum* L. plants growing in an Andosol with high aluminum level. *Agr Ecosyst Environ* 186:178–184.

Arriagada CA, Herrera MA, Borie F & Ocampo JA (2007) Contribution of arbuscular mycorrhizal and saprobe fungi to the aluminum resistance of *Eucalyptus globulus*. *Water Air Soil Pollut* 182:383–394.

Klugh-Stewart K & Cumming JR (2009) Organic acid exudation by mycorrhizal *Andropogon virginicus* L. (broomsedge) roots in response to aluminum. *Soil Biol Biochem* 41:367–373.

Seguel A, Cumming JR, Klugh-Stewart K et al. (2013) The role of arbuscular mycorrhizas in decreasing aluminium phytotoxicity in acidic soils: a review. *Mycorrhiza* 23:167–183.

Aluminum accumulation by plants

Brunner I & Sperisen C (2013) Aluminum exclusion and aluminum tolerance in woody plants. *Front Plant Sci* 4:172.

Hajiboland R, Rad SB, Barcelo J et al. (2013) Mechanisms of aluminum-induced growth stimulation in tea (*Camellia sinensis*). *J Plant Nutr Soil Sci* 176:616–625.

Klug B & Horst WJ (2010) Spatial characteristics of aluminum uptake and translocation in roots of buckwheat (*Fagopyrum esculentum*). *Physiol Plant* 139:181–191.

Watanabe T, Misawa S, Hiradate S & Osaki M (2008) Characterization of root mucilage from *Melastoma malabathricum*, with emphasis on its roles in aluminum accumulation. *New Phytol* 178:581–589.

Properties and effects of basic soils

Babuin MF, Campestre MP, Rocco R et al. (2014) Response to long-term $NaHCO_3$-derived alkalinity in model *Lotus japonicus* ecotypes Gifu B-129 and Miyakojima MG-20: transcriptomic profiling and physiological characterization. *PLoS One* 9:e97106.

Canasveras JC, Sanchez-Rodriguez AR, Carmen del Campillo M et al. (2014) Lowering iron chlorosis of olive by soil application of iron sulfate or siderite. *Agron Sustain Dev* 34:677–684.

Colombo C, Palumbo G, He J-Z et al. (2014) Review on iron availability in soil: interaction of Fe minerals, plants, and microbes. *J Soils Seds* 14:538–548.

Ma JF & Ling H-Q (2009) Iron for plants and humans. *Plant Soil* 325:1–3.

Fe^{3+} reduction and Fe^{2+} uptake

Barberon M, Dubeaux G, Kolb C et al. (2014) Polarization of IRON-REGULATED TRANSPORTER 1 (IRT1) to the plant-soil interface plays crucial role in metal homeostasis. *Proc Natl Acad Sci USA* 111:8293–8298.

Kobayashi T & Nishizawa NK (2012) Iron uptake, translocation, and regulation in higher plants. *Annu Rev Plant Biol* 63:131–152.

Socha AL & Guerinot ML (2014) Mn-euvering manganese: the role of transporter gene family members in manganese uptake and mobilization in plants. *Front Plant Sci* 5:106.

Vasconcelos MW, Clemente TE & Grusak MA (2014) Evaluation of constitutive iron reductase (AtFRO2) expression on mineral accumulation and distribution in soybean (*Glycine max* L). *Front Plant Sci* 5:112.

Iron homeostasis and strategy II

Clemens S, Deinlein U, Ahmadi H et al. (2013) Nicotianamine is a major player in plant Zn homeostasis. *Biometals* 26:623–632.

Kobayashi T & Nishizawa NK (2014) Iron sensors and signals in response to iron deficiency. *Plant Sci* 224:36–43.

Oburger E, Gruber B, Schindlegger Y et al. (2014) Root exudation of phytosiderophores from soil-grown wheat. *New Phytol* 203:1161–1174.

Ricachenevsky FK & Sperotto RA (2014) There and back again, or always there? The evolution of rice combined strategy for Fe uptake. *Front Plant Sci* 5:189.

Agronomy and iron/zinc nutrition

Ding H, Duan LH, Li J et al. (2010) Cloning and functional analysis of the peanut iron transporter *AhIRT1* during iron deficiency stress and intercropping with maize. *J Plant Physiol* 167:996–1002.

Omondi EC, Ridenour M, Ridenour C & Smith R (2010) The effect of intercropping annual ryegrass with pinto beans in mitigating iron deficiency in calcareous soils. *J Sustain Agr* 34:244–257.

Prasad R, Shivay YS & Kumar D (2014) Agronomic biofortification of cereal grains with iron and zinc. *Adv Agron* 125:55–91.

Xiong H, Shen H, Zhang L et al. (2013) Comparative proteomic analysis for assessment of the ecological significance of maize and peanut intercropping. *J Proteomics* 78:447–460.

Chapter 11
Flooding

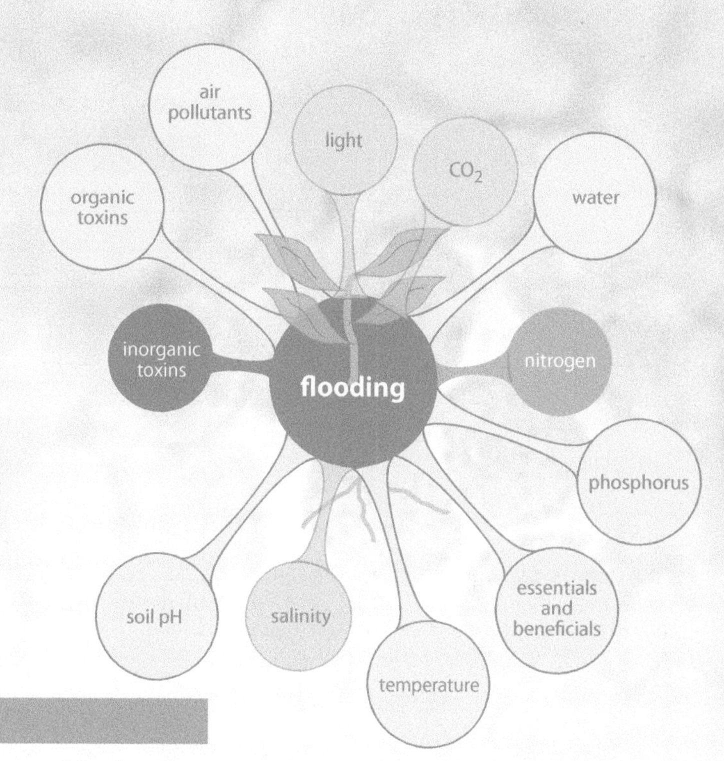

Waterlogging affects the chemistry of the soil solution, and submergence limits the availability of light and CO_2.

Key concepts

- Wetland systems provide important ecosystem services and foodstuffs.
- Human activity is destroying wetlands and increasing the incidence of flooding.
- Excess water alters soil chemistry by decreasing O_2 availability.
- Many plant species show flexibility of anaerobic metabolic capacity.
- Ethylene-induced pathways underpin many plant responses to excess water.
- Many plants can form lysigenous aerenchyma to adapt to O_2 demand and supply.
- In some species, constitutive aerenchyma enables significant internal flows of O_2.
- Mangroves have multiple adaptations to a highly variable water regime.
- Wetland plants are often adapted to oxygenate their rhizosphere.
- Some plants can adapt to periods of submergence.
- Emergent aquatic macrophytes can force O_2 into their submerged organs.
- Submerged aquatic macrophytes face challenging O_2, CO_2, and light regimes.

Flooding is a significant variable in both unmanaged and managed terrestrial ecosystems

The atmosphere has a high O_2-supplying capacity because it consists of 21% O_2, and O_2 diffuses rapidly in air. During the evolution of terrestrial multicellular organisms it was the delivery of O_2 to cells within tissues that was a challenge, rather than O_2 supply from the environment. In response, animals evolved internal gas delivery systems, and terrestrial plants evolved branching surfaces for external gas exchange. However, O_2 diffuses slowly in water, so in saturated soils microbial demand for O_2 is generally much greater than the rate at which it can diffuse in from the atmosphere. This produces reducing "waterlogged" soils that cannot supply O_2 to roots. Submergence of shoots decreases O_2 supply to plants still further. Therefore if flooding waterlogs the soil matrix, and especially if it also submerges the plant shoots, it presents a significant challenge to terrestrial plants.

Table 11.1. **Some ecosystem services provided by wetlands**

Type	Service	Examples
Provisioning	Fresh water	Agricultural, domestic, and industrial water
	Food	Fish, crops, game
	Fiber and fuel	Papyrus, biofuels, fodder
Regulating	Water	Hydrologic flows and groundwater recharge
	Water purification	Recovery of nutrients and removal of pollutants
	Erosion	Retention of soils and sediments
	Climate	Regulation of greenhouse gases, precipitation, and temperature
	Flood waters	Flood control and storm protection
Supporting	Soil formation	Sediment and organic matter retention
	Nutrient cycling	Storage, recycling, and processing of nutrients
Cultural		Recreational, spiritual, and aesthetic

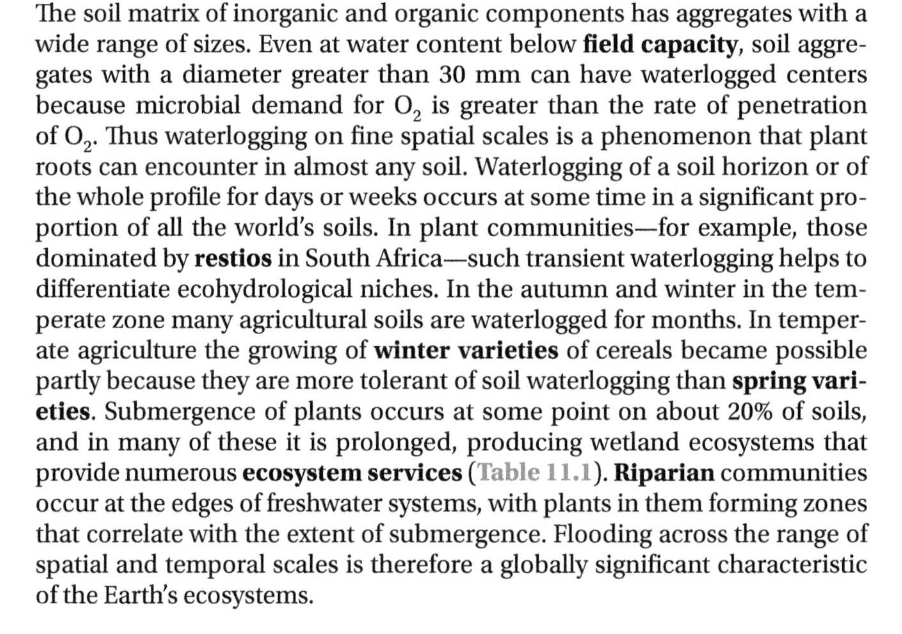

The soil matrix of inorganic and organic components has aggregates with a wide range of sizes. Even at water content below **field capacity**, soil aggregates with a diameter greater than 30 mm can have waterlogged centers because microbial demand for O_2 is greater than the rate of penetration of O_2. Thus waterlogging on fine spatial scales is a phenomenon that plant roots can encounter in almost any soil. Waterlogging of a soil horizon or of the whole profile for days or weeks occurs at some time in a significant proportion of all the world's soils. In plant communities—for example, those dominated by **restios** in South Africa—such transient waterlogging helps to differentiate ecohydrological niches. In the autumn and winter in the temperate zone many agricultural soils are waterlogged for months. In temperate agriculture the growing of **winter varieties** of cereals became possible partly because they are more tolerant of soil waterlogging than **spring varieties**. Submergence of plants occurs at some point on about 20% of soils, and in many of these it is prolonged, producing wetland ecosystems that provide numerous **ecosystem services** (Table 11.1). **Riparian** communities occur at the edges of freshwater systems, with plants in them forming zones that correlate with the extent of submergence. Flooding across the range of spatial and temporal scales is therefore a globally significant characteristic of the Earth's ecosystems.

Figure 11.1. **Bunded rice paddies.** Much rice is grown in small flooded fields, known as "paddies," in which water flows are carefully managed to provide appropriate conditions at particular developmental stages. The paddy bottoms consist of clay that is "puddled" by compaction to provide a sealed bottom that does not allow water infiltration. The figure shows paddies being prepared for rice planting in South India. There is a bund between two rice paddies of different heights, with water flowing from one to the other.

Rice is currently the staple crop of about 3 billion people, providing about 20% of all the calories and 15% of all the protein consumed by humans. About 700 million tonnes of rice, with a value of over US$ 200 billion, are produced each year from about 160 million hectares of land. Rice prices significantly affect the calculation of global poverty and the economic well-being of many nations. There has been a linear increase in global rice production since 1960, from about 160 million tonnes, which shows no sign of slowing down. Paddy rice, in which flooded conditions are maintained in **bunded** fields (Figure 11.1), accounts for about 50% of rice area and 75% of production. Rain-fed rice accounts for about 20% of rice production, and about 10 million hectares of rice, mostly in South-East Asia, are grown in deep-water conditions in which plants can be submerged by meters of water. *Corchorus capsularis* (jute) accounts for about 3 million of the 25 million tonnes of natural fibers produced globally. It is primarily produced under flooded conditions in the Ganges Delta from India to Bangladesh. Historically, jute was of great importance for the economy of Bengal, and although the development of artificial fibers reduced its importance in the twentieth century, a resurgence of interest in natural fibers is now contributing to its increasing importance. *Colocasia esculenta* (taro, eddo) (Figure 11.2) is widely grown across the Pacific region and Asia, often in moist or flooded conditions, for its high-carbohydrate tubers. Although some crop plants grow on flooded soils, the majority cannot do so, and flooding is therefore a significant risk in managed systems that provide food for humankind. Thus the challenges presented by excess soil water and plant responses to these underpin many important unmanaged and managed ecosystems on Earth.

(a)

(b)

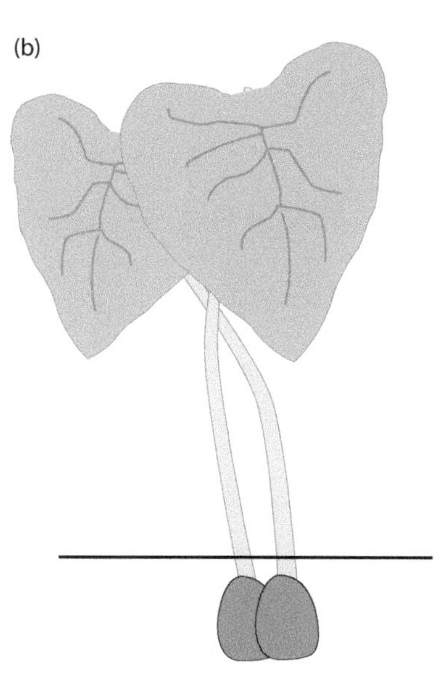

Figure 11.2. ***Colocasia esculenta.*** Taro or eddo is widely grown on the Indian subcontinent and in South-East Asia in moist, even wet soils. (a) It has very large cordate leaves (underside shown), and is thus commonly known as "elephant ear." (b) It has an underground corm (a swollen stem base) that is used as a source of starch, and the leaves are also sometimes eaten.

Human activity is adversely affecting wetlands and increasing the incidence of flooding

Many of the earliest human civilizations developed along flooded river valleys—for example, the Tigris and Euphrates in the Middle East, the Indus in the Indian subcontinent, the Nile in Africa, and the Yangtze in East Asia. The ecosystem services that wetlands provide perhaps played a key role in enabling these civilizations to develop. Only from the Yangtze valley has a significant flood-tolerant crop—rice—entered modern agriculture, but many flood-tolerant plants, such as *Cyperus papyrus* (papyrus), were vital to early civilizations (Figure 11.3). Awareness of the value of wetlands provoked the 1971 Convention on Wetlands (the "Ramsar Convention"), and was a focus of the **Millennium Ecosystem Assessment** (MEA). The Ramsar Convention and the MEA clearly demonstrate not only that wetlands are among the most important ecosystems on Earth, but also that they have been damaged more by human activity than by any other factor. They include ecosystems of great conservation importance, such as the **Sudd**, the **Okavango Delta**, the **Everglades**, the **Sunderbans**, and the flooded forests of the Amazon. In many parts of the world at least 50% of wetlands have been entirely lost. In many others, water extraction, overharvesting, nutrient and pollutant loading, and alien species are very significant and increasing threats. Wetland restoration schemes are among the most significant conservation efforts, and constructed wetlands are increasingly being used for water treatment.

Flooding has always been among the most destructive natural disasters for humankind. Crop losses are frequently one of the most far-reaching effects of flooding, and flood tolerance is among the most useful attributes of many crop varieties. There is significant evidence that, due to environmental changes, both the incidence and impact of flooding are increasing. These changes include altered watershed characteristics, hydrological flows, and climate. The MEA concluded that global climate change is also expected to exacerbate the decline or loss of wetlands in the next few decades. In addition, about 50% of methane emitted to the atmosphere is from wetlands, and an increasing amount of the remainder is from rice paddies. Methane is a potent greenhouse gas that has significant effects on climate. An understanding of plant function in wetlands and in flooded agricultural soils is thus potentially very useful for understanding some key aspects of environmental change arising from human activity.

Waterlogged soils are low in oxygen and some nutrients, but high in toxins

Fick's first law quantifies diffusion rate at equilibrium as a function of concentration gradient and a conductivity constant, the diffusion coefficient (D), which reflects the net "random walk" of atoms or molecules down the gradient. In homogeneous unstirred media, D_o can be measured for O_2 as 0.205 $cm^2\ s^{-1}$ in air and $2.267 \times 10^{-5}\ cm^2\ s^{-1}$ in water. Therefore, for O_2 in water, D_o is about 9000 times slower than in air, because liquid water slows down the movement of O_2 molecules as compared with air, but over quite short time periods O_2 can still diffuse tens of centimeters into unstirred water.

(a)

(b)

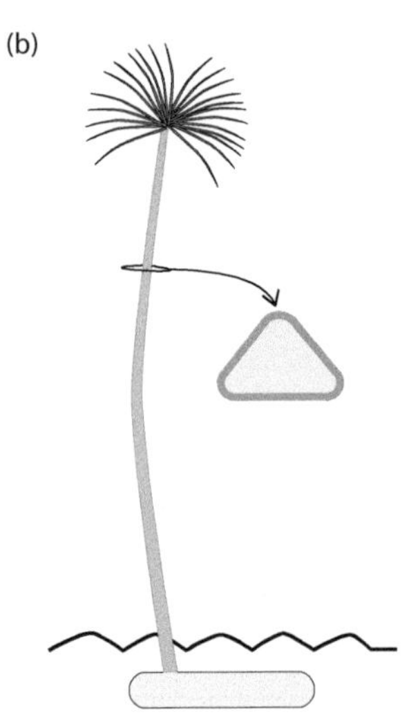

Figure 11.3. ***Cyperus papyrus.*** This species flourished along the Nile, especially in the Sudd and the Nile Delta, but is much less common than it was formerly. It was cultivated in ancient Egypt and its pith was compacted to form the paper-like papyrus (the origin, via the Greek, of the English word "paper"). It is a member of the sedge family (Cyperaceae), and is one of the many plants in this family that inhabit waterlogged soils. (a) The end of the main stem bears a spray of branches that produce flowers. (b) The stem emerges from a rhizome and can reach a height of several meters. The triangular stems have greatly enlarged pith.

However, soil is a heterogeneous medium with pores that are tortuous and discontinuous, so the effective D (D_e) in soil water is significantly less than D_o. In most wet soils, D_e is 0.5–10 × 10^{-6} $cm^2\ s^{-1}$, so in a sterile flooded soil O_2 diffuses down to about 5–10 cm and will reach below that depth only in vanishingly low concentrations. However, in non-sterile soils, microorganisms consume O_2 while it diffuses down the concentration gradient from the atmosphere. In most wet soils, low D_e together with O_2 consumption ensures that O_2 penetrates down no further than 1 cm, and often just a few millimeters (Figure 11.4). Thus waterlogging makes the rooting zone in soils anaerobic. In most soils this occurs within 24–36 h. Plant roots have a significant requirement for O_2, which, in the absence of excess water, in most species is met from the soil. During transient waterlogging, plant roots must withstand anaerobic conditions for a brief period of time, but during prolonged waterlogging plants face a significant challenge in supplying O_2 to roots.

During aerobic respiration, electrons are transferred from organic molecules to O_2, the most widely available eager electron acceptor in the terrestrial environment. Respiration is initiated by glycolysis, which produces pyruvate together with some ATP, and under aerobic conditions continues via an electron transfer chain. This produces the proton motive force for further production of ATP by transferring four electrons to the terminal electron acceptor, O_2. In flooded soils, aerobic microbes disappear when O_2 is used up, and are replaced by anaerobes that either ferment pyruvate or carry out anaerobic respiration using terminal electron acceptors other than O_2. The populations of anaerobes that are found in waterlogged conditions are determined by the propensity of redox couples to accept electrons. The **redox potential** (E_h) is commonly measured relative to a standard hydrogen redox couple (H_2/$H^+ + e^-$), and for O_2 it is +820 mV. At 4% O_2 in soil, which produces an E_h of about +350 mV, the O_2 supply to aerobes becomes limiting and E_h decreases because the capacity of the soil constituents to accept electrons is diminished. The E_h for NO_3^- is +400 mV, so bacteria such as *Paracoccus denitrificans*, which can use NO_3^- as a terminal electron acceptor, flourish because they possess the most energetically favorable respiratory pathways at that E_h. The denitrification of NO_3^- produces gaseous N_2O or N_2, and is responsible for N losses from anaerobic microsites in soil aggregates on a global scale. In flooded soils, mineralization of organic matter continues to produce NH_4^+, but in the absence of O_2 it cannot be nitrified. Thus within a period of days excess water rapidly changes soils from aerobic, NO_3^--dominated environments to anaerobic, NH_4^+-dominated environments (Figure 11.5).

In most soils, E_h down to −200 mV is dominated by redox couples of Fe and Mn. Numerous facultative and obligate anaerobes can use Fe^{3+} or Mn^{4+} as terminal electron acceptors, and most mineral soils contain large quantities of Fe^{3+} hydroxides and Mn^{4+} oxides. In many soils the capacity of Fe and Mn

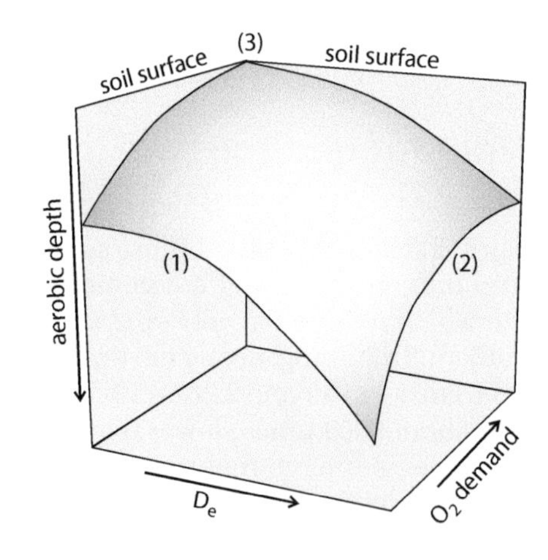

Figure 11. 4. The control of aerobic depth in waterlogged soil. If a soil becomes waterlogged, the depth of the aerobic layer is a function of the effective diffusion coefficient (D_e) of O_2 and O_2 demand from microorganisms. (1) In a sterile soil with no O_2 demand, O_2 penetration is dependent on D_e, which varies with, for example, the tortuosity of soil pores. (2) In a non-sterile soil, the greater the O_2 demand from microorganisms, the shallower the depth to which O_2 penetrates the soil. (3) The minimum depth of penetration, often only a few millimeters, occurs in soils with low D_e and high O_2 demand.

redox couples is sufficient to prevent E_h from dropping below -200 mV, but this produces significant quantities of Fe^{2+} and Mn^{2+}, which are soluble and toxic to plants (Figure 11.6). Below about -200 mV, SO_4^{2-} reduction to sulfide (the most important plant toxin in many flooded soils) by bacteria such as *Desulfovibrio* is possible, and the production of physiologically active and potentially toxic H_2S can also be significant. At E_h below -250 mV, **methanogens** produce CH_4, using as a terminal electron acceptor CO_2 from other microbes that have fermentation capacity (Figure 11.5). In addition, changes in E_h affect the chemistry of numerous other nutrient and pollutant ions in soil solution. In general, reducing conditions decrease the availability of Zn, Ni, Cu, Se, Pb, and Cd, but increase the availability of SiO_2, PO_4^-, As, Hg, and Sb. Concomitant with redox changes, fermentation of organic matter in soil produces not only CO_2 but also carboxylic acids such as butyric, propionic, and fumaric acid.

Varying redox potential, and in particular a changing oxic/anoxic boundary, can have profound effects on other aspects of soil chemistry. Redox affects H^+ concentrations, with low E_h favoring neutral pH. When an alkaline soil is waterlogged the pH tends to decrease, whereas in an acid soil it increases. The organic matter content of soil affects microbially mediated changes in redox because it is the source of electrons, and low redox reduces the formation of Fe and Mn hydroxides/oxides that are important for soil structure. In environments such as transient wetlands and paddy fields, where soils have a varying water regime, the interplay between soil constituents and redox can be complex but significant. In wetlands, water regimes affect the ratio between the mobilization of ions from soil into the water column and immobilization in the soil. Natural wetlands and constructed wetlands, in which waterborne pollutants often collect, can—depending on the water regime—be significant sources or sinks of PO_4^{2-}, metals, and arsenic. In paddy rice, pH and redox conditions for most of the growth season can produce Zn deficiency and Fe toxicity but good supplies of PO_4^{2-} and SiO_4, whereas draining of paddies before harvest changes the soil chemistry and can be important in determining, for example, Cd influx to rice. If flood water is saline, its ions present further toxicity challenges. The chemistry of a particular flooded soil depends on its composition, and interactions between micoorganisms, minerals, and plants can produce rhizospheres with a very different chemistry to that of the bulk soil. Despite these challenges, flooded ecosystems—both unmanaged and managed—are among the most sustainably productive on Earth, because plants that are adapted to them are able to take advantage of favorable soil regimes of organic matter, nitrogen, pH, and micronutrients.

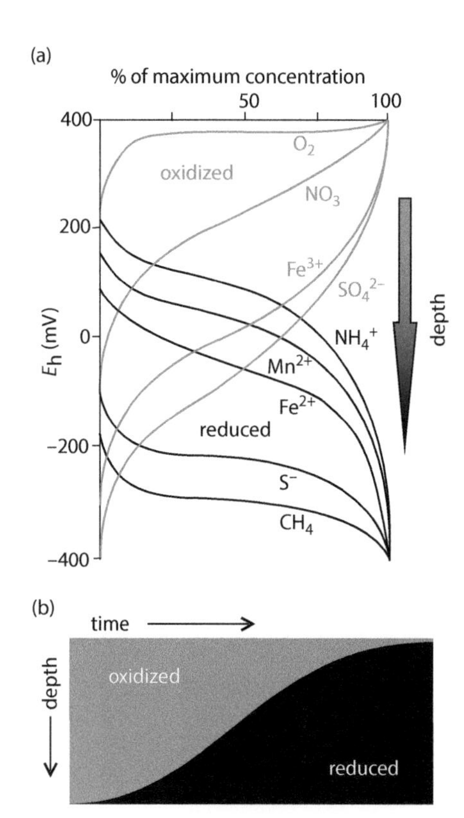

Figure 11.5. Soil redox (E_h) and ionic species in waterlogged soil. (a) With restricted O_2 influx to waterlogged soil, microorganisms use a series of progressively less eager electron acceptors (shown in green), producing a progressively lower E_h with increasing depth. In reducing soils there are increasing concentrations of reduced forms of the electron acceptors (shown in black), which are often toxic to plants, with increasing depth. (b) If an oxidized soil is flooded, because of microbial action the profile becomes progressively more reduced over time (often over just a few days), with the final depth of the aerobic layer (often very shallow) depending on the D_e of O_2 in the soil and on O_2 demand.

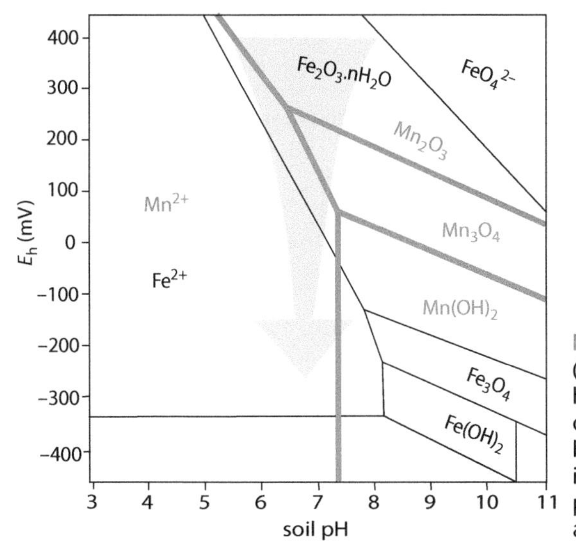

Figure 11.6. A Pourbaix diagram for Fe and Mn. The ionic species of Fe (shown in black) and Mn (shown in green) vary with both E_h and pH. At high E_h and pH, the oxidized ionic species (Fe^{3+} and Mn^{3+}) form a series of oxides. As E_h and pH decrease, reduced ionic species (Fe^{2+} and Mn^{2+}) form hydroxides and, when E_h and/or pH are low enough, free Fe^{2+} and Mn^{2+} ions. The large green arrow represents the effect of flooding on soil E_h and pH, and therefore Fe and Mn species—E_h decreases and the pH tends to approach neutral.

Soil waterlogging rapidly induces hypoxia, cellular acidosis, and decreased water uptake

Cellular **hypoxia** induced by inadequate O_2 supply results in an energy crisis caused by decreased ATP production in mitochondria. In plant roots, hypoxia can occur within a day of soil waterlogging. In waterlogging-sensitive plants, during the onset of hypoxia there is decreased flow of electrons to O_2, but glycolysis still occurs, using ADP and **NAD^+** to produce some ATP and NADH. Without replenishment of NAD^+, glycolysis is inhibited and all metabolic activity ceases. Most organisms, including plants, can replenish NAD^+ for a short time during hypoxia using lactate dehydrogenase, an NADH-dependent **fermentation** enzyme that converts pyruvate into lactic acid (Figure 11.7). Glycolysis supported by anaerobic lactic acid generation produces only 2 molecules of ATP per glucose molecule, and occurs in the cytosol, which normally has a pH of about 7.5. Hypoxia in flood-sensitive plants rapidly leads to a shortage of ATP and a drop in cytosolic pH that inhibits many metabolic pathways. Cytosolic acidosis is initiated by lactic acid production, but the cellular pH-stat is dependent on ATPases so, for example, the transport of H^+ ions into the vacuole is slowed down, equalizing the cytosolic pH with the acid pH of 5.5 in the vacuole. Lactate dehydrogenase has a limited capacity to replenish NAD^+ because its activity is inhibited by acidity. Many plants, in contrast to animals, also have a significant capacity to produce NAD^+ via the production of ethanol from pyruvate, which does not produce H^+ and therefore slows down acidosis (Figure 11.7). The effects of hypoxia on glycolysis have consequences for the TCA cycle and carbon metabolism. Overall, many plants have some capacity to use fermentation pathways to delay the effects of hypoxia.

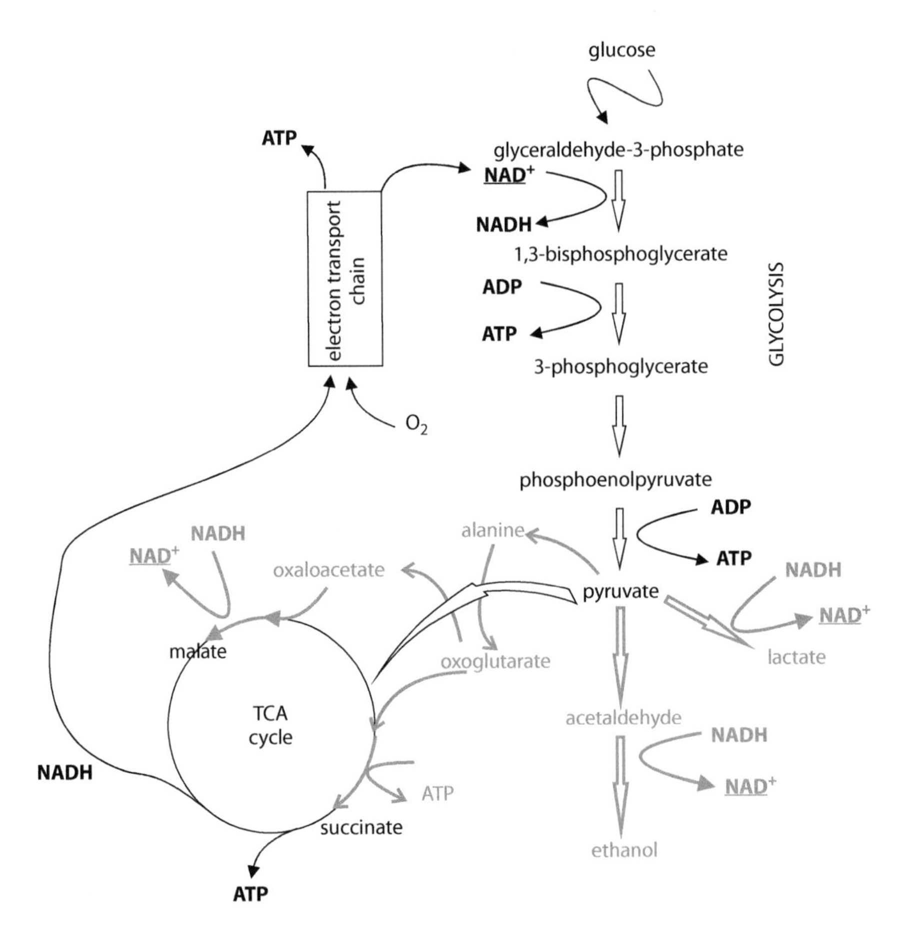

Figure 11.7. Anaerobic respiratory pathways in plants. In aerobic conditions, NAD^+ for glycolysis is regenerated, using electron transport chains for oxidative phosphorylation, from NADH produced from the TCA cycle (shown in black). In anaerobic conditions, plants use other pathways (shown in green) to regenerate NAD^+. Like most organisms, plants can use lactate production to regenerate NAD^+, but they also often use fermentative ethanol production and several adaptations of the TCA cycle.

Another early effect of excess water on plants is a decrease in water uptake. Water transport into the symplasm is mediated by aquaporins. PIP2 aquaporins are gated by protonation during cytosolic acidosis (Figure 11.8), which has been shown to account for the decrease in water uptake induced in plants by flooding. Aquaporins have six transmembrane domains and cytosolic N and C termini, giving three external loops (A, C, and E) and two cytosolic loops (B and D). Plasma membrane intrinsic proteins (PIPs) have a characteristic D loop that gates the channel. Serine 115 when unphosphorylated allows the attachment of loop D to the N terminus, closing the channel. Phosphorylation of Ser 115 breaks this attachment and opens the channel. Protonation of His 193 closes the channel by allowing it to attach to Asp 28 near the N-terminus. In water-sufficient plants, PIP2s are phosphorylated and open, whereas in hypoxia- and drought-stressed plants, dephosphorylation, protonation, and bound Ca^{2+} close the aquaporins. Under waterlogging conditions, the reduction in water uptake is associated with the closing of stomata and a decrease in C fixation, but is probably helpful in reducing the mass flow of toxic ions to roots, especially Fe^{2+} and Mn^{2+}. The uptake of nutrients is an active process in growing roots, so any decline in ATP availability, root growth, and ion-uptake capacity will cause a decline in nutrient uptake. For example, for K and Na the discrimination against Na that characterizes plant uptake declines in hypoxic roots, resulting in a decreased K:Na ratio in the shoot. Xylem loading with cations from stelar parenchyma cells is active because of the approximately –100 mV membrane potential (E_m). E_m is maintained by active transport, and has been shown to decline significantly in hypoxic roots. The greater the decline in E_m, the more dominant the transport of cations such as Na via non-selective cation channels becomes. Many non-adapted plants suffer ion toxicity in waterlogged soils, which is probably exacerbated by a reduction in their ability to select ions during uptake to the stele. In soils that are waterlogged by saline water, the ability to maintain discrimination against Na during prolonged waterlogging, and thus maintain a high shoot K:Na ratio, is a significant factor in salinity tolerance. Many of the physiological effects of flooded soils result in an increase in oxidative damage to membranes and other cellular components. The vast majority of terrestrial plants can only transiently withstand the effects of flooding, but probably have to do so in many soils.

Waterlogging-induced changes quickly affect root growth. Mild O_2 deficiency at the root surface can produce severe hypoxia in the central **stele** of roots with a diameter of only 1 mm. Measurements of O_2 concentrations in roots and mathematical modeling using realistic resistances to O_2 movement show that hypoxia of the stele occurs rapidly. A key function of the root is to supply water and nutrients to the shoot via the vascular tissue, which requires energy expenditure in the stele, especially for xylem loading of cations. Root function is impaired from the inside out during sustained hypoxia. Reductions in root development and function cause a decline in shoot function. The concentration of photosynthetic products in shoots generally increases during hypoxia, which suggests that, initially at least, photosynthesis declines less rapidly than growth.

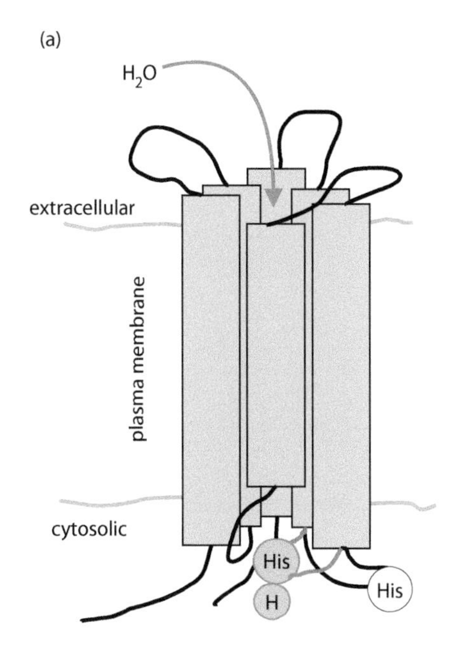

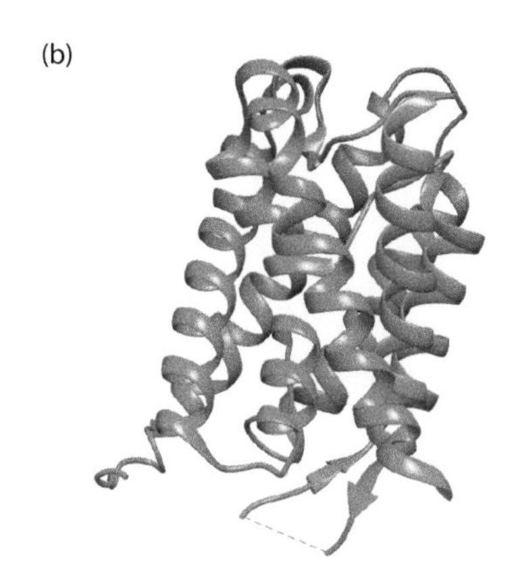

Figure 11.8. The effect of cytosolic acidosis on aquaporin function. (a) Aquaporin monomers have 6 membrane-spanning domains that form a pore for the passage of water. Cytosolic acidosis leads to protonation of histones (His) on a cytosolic loop, bending it inward to close the pore and block the passage of water (shown in green). (b) Ribbon diagram of a spinach aquaporin (PIP2;1) monomer. PIP2;1 occurs as tetramers in the plasma membrane.

Physiological adjustments enable some plants to withstand soil waterlogging for short periods

In flood-tolerant species, hypoxic conditions give rise to characteristic physiological changes that acclimate the plants to flooding. These changes provide an indirect sensor of O_2 concentrations, but plants also use a direct sensor based on post-translational degradation of VII ERF transcription factors (TFs). These have a constitutively exposed Cys residue on their N-terminus which, when oxidized, targets them for degradation by proteasomes. As O_2 concentrations decrease they persist and activate, especially in flooding-tolerant plants, hypoxia-responsive genes. Other mechanisms decrease the expression of particular genes. For example, in addition to the degradation of unnecessary mRNAs, the oligouridylate-binding protein UBP1 binds and stabilizes specific mRNAs during hypoxia, preventing their translation. An important effect of these changes in gene expression is an altered ability to manage ATP production and consumption. Transporters for lactate efflux are expressed during hypoxia in some species, which might help to delay cytosolic acidosis and sustain NAD^+ production. Hypoxia-tolerant plant species have a significant capacity for production of ethanol, which is not toxic to plants and can diffuse out of cells. Hypoxia-tolerant plants also use TCA-cycle components in unusual pathways. In such species, NAD^+ can be generated by the action of malate dehydrogenase on oxaloacetate, and alanine and succinate accumulate whereas aspartate and glutamate levels decline. Alanine aminotransferase, an enzyme that has high activity in many plants during hypoxia, uses pyruvate and glutamate to produce 2-oxoglutarate directly, in contrast to the multiple steps necessary to form this compound in the TCA cycle. Conversion of 2-oxoglutarate to succinate uses NAD^+ but generates ATP. These pathways enable hypoxia-tolerant plants to generate anaerobically four ATP molecules for each glucose molecule. *Arabidopsis pdc* and *adh* mutants have increased sensitivity to hypoxia, and enhanced pyruvate decarboxylase (PDC) and alcohol dehydrogenase (ADH) activity is widely reported in hypoxia-tolerant plants. ADH expression is controlled by ROP-GTPases via the effect of O_2 deficiency on the GTP/GDP ratio (Figure 11.9).

During hypoxia, significant increases in the flux of glucose into anaerobic pathways are necessary to generate sufficient ATP for metabolism. Increased breakdown of stored starch to sucrose is common, as is a shift to the sucrose synthase-dependent pathways for sucrose degradation that require half as much ATP as the invertase-dependent pathways. In hypoxia-tolerant plants, ATP demand in hypoxic cells is adjusted in conjunction with ATP production. For example, ATP-dependent enzymes such as phosphofructokinase (PFK) of glycolysis are replaced with pyrophosphate-dependent enzymes—in the case of PFK, with pyrophosphate fructose-6-phosphate phosphotransferase. Genomic and proteomic studies have reported decreases in ATP-requiring protein synthesis, cell wall production, and numerous other metabolic processes. In a variety of species, and in many tissues that experience prolonged hypoxia, a quiescent state in which very little ATP is used can be reached.

Flood-tolerant plants also adjust their nitrogen metabolism. During hypoxia, GABA levels can increase, probably due to low pH activated glutamic acid decarboxylase (GAD), which catalyzes the production of GABA from glutamate. The concentration of **polyamines** (especially putrescine, spermidine, and spermine), from which GABA is also synthesized, can increase. GABA helps to control cellular pH, the TCA cycle and, together with polyamines, the plant response to oxidative stress. During hypoxia an NH_4^+-N source leads to a decrease in nitrite reductase (NiR) activity, which means that any NO_2^-, which is toxic, must be dealt with by other means. Increases in hemoglobin concentration have been noted in many flood-tolerant plants. This helps to bind NO, which in flooded conditions is produced in increased quantities by the action of nitrate reductase (NR) and, in mitochondria, cytochrome c

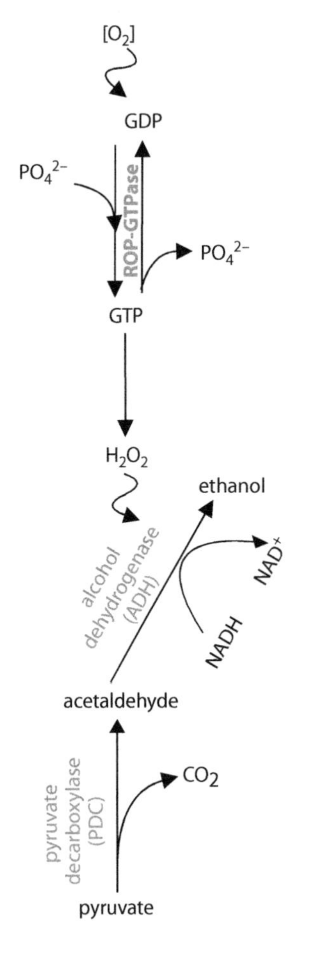

Figure 11.9. Hypoxia-induced up-regulation of alcohol dehydrogenase activity. The guanosine triphosphate (GTP) molecule is a component of numerous signaling pathways in organisms. RAS proteins are widespread in eukaryotes, are involved in the development of several cancers in humans, and are GTPases (RAS-GTPases). One type of RAS-GTPase (RHO-GTPase) is involved in the generation of reactive oxygen species (ROS). In plants these are referred to as "RHO of plant origin"-GTPases, or "ROP"-GTPases. ROP-GTPases help to control the production of H_2O_2 in response to changes in O_2 concentration, thereby helping to control the activity of ADH.

oxidase (COX) on nitrite. COX utilization of NO_2^- can produce significant amounts of ATP under hypoxia ("NO:nitrite respiration"), but hemoglobin is necessary to help to control the concentration of NO that it produces.

Flood-tolerant species are also resistant to oxidative damage caused by reoxygenation as the water recedes. Antioxidant activity can be particularly important during this phase, but additional mechanisms include reoxygenation-dependent disintegration of UBP1 cytoplasmic stress granules in which mRNAs are stabilized during hypoxia, which allows translation of normoxic mRNAs to resume. In some species, epinastic shoot growth can be used as an indicator of flooding sensitivity. During **epinasty** the adaxial (upper) side of the leaf petiole and leaf surface grows more quickly than the abaxial (lower) surface, leading to downward curvature of the leaves (Figure 11.10). Epinastic growth is induced by high concentrations of ethylene in the shoot. In order to synthesize ethylene from its precursor, 1-aminocyclopropane-1-carboxylic acid (ACC), plants require O_2. In hypoxic roots, ACC therefore increases in concentration and enters the transpiration stream, which delivers it to shoots where O_2 is available for ethylene synthesis. Epinasty was first reported in tomato, and is often thought to be characteristic of solanaceous plants. However, epinastic responses in, for example, crossed *Quercus robur* lines have been correlated with their waterlogging tolerance. Epinasty is seldom reported in flood-tolerant plants, which suggests that their physiological adjustments help them to avoid it, and that they respond differently to ethylene.

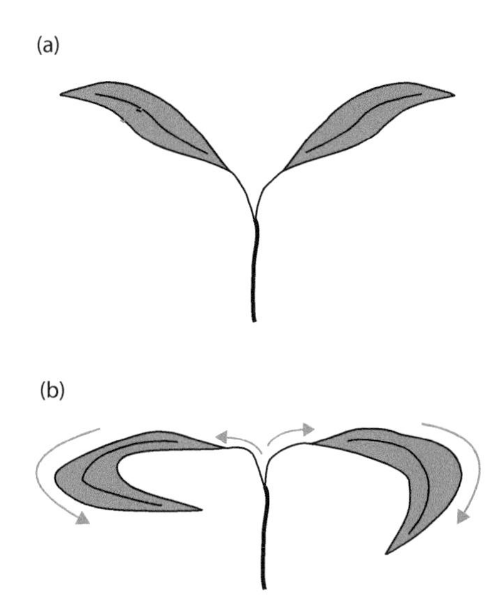

Figure 11.10. **Epinasty in plants growing on waterlogged soils.** In non-adapted plants, soil waterlogging can produce epinastic growth. Nastic growth is a growth response to a non-directional stimulus, such as environmental conditions. Epinasty is excess growth of the upper surfaces of leaves and petioles; (a) and (b) in the figure show, respectively, shoots before and after epinastic growth. With complete submergence, some species can show hyponasty, in which there is excess growth of the lower surfaces, directing the leaves up to the water surface. Both of these responses are mediated by ethylene, which accumulates in plant tissues during waterlogging and flooding, and to which different tissues can react in different ways.

Ethylene signaling is central to plant responses to excess water

Ethylene is constitutively synthesized in plants from methionine. Ethylene concentrations increase rapidly in roots in waterlogged soils and, if they are submerged, in whole plants. This is not only because of decreased diffusion out of the roots and shoots (which is perhaps why ethylene is used to transduce responses to waterlogging), but also because of increased ethylene synthesis. Ethylene receptors in plants are located in the endoplasmic reticulum, and the first such receptor to be isolated was ETR1 from *Arabidopsis* (Figure 11.11). There are multiple similar ethylene receptors in all plants

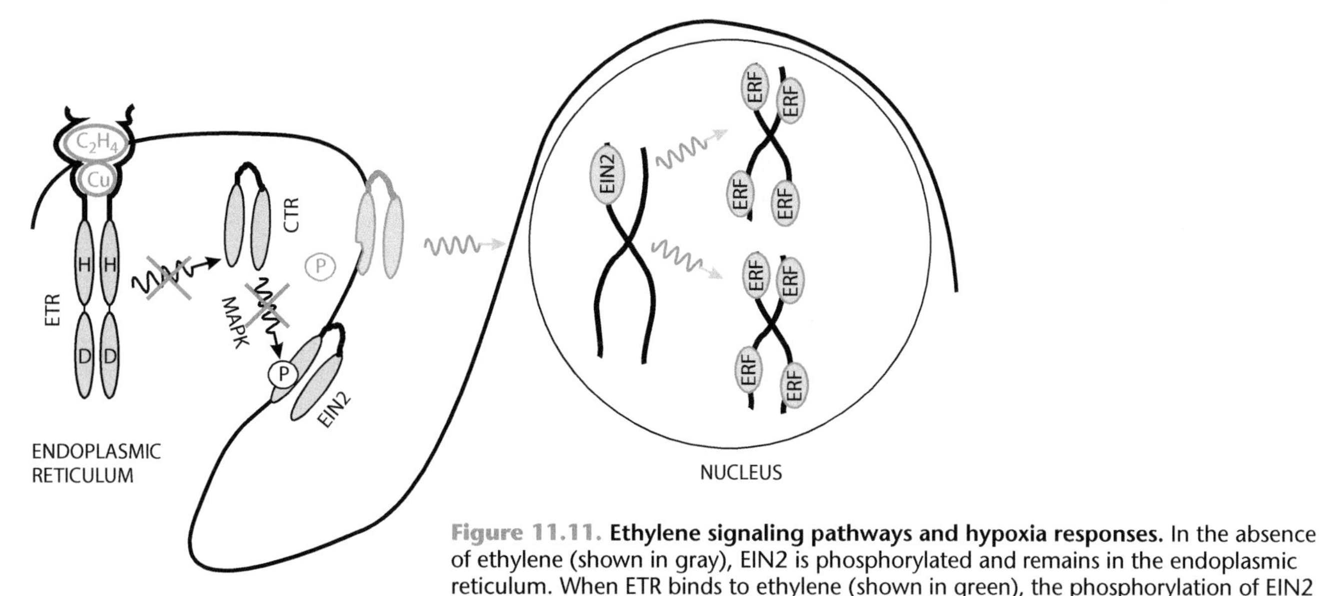

Figure 11.11. **Ethylene signaling pathways and hypoxia responses.** In the absence of ethylene (shown in gray), EIN2 is phosphorylated and remains in the endoplasmic reticulum. When ETR binds to ethylene (shown in green), the phosphorylation of EIN2 is prevented and it travels to the nucleus, setting in train the expression of ethylene response factors (ERFs) that control ethylene-induced gene expression. Hypoxia induces the accumulation and sometimes *de novo* synthesis of ethylene, resulting in increased ERF activity. MAPK = mitogen-activated protein kinase.

that have been investigated so far. During waterlogging, spatial and temporal differences not only in ethylene concentrations but also in ethylene receptors and their activity are likely to help to explain the different responses of tissues to ethylene. Ethylene receptors are homodimers, and ethylene binds to them via a Cu co-factor. Alkenes such as ethylene bind strongly to monovalent transition metals, and Ag^+ can be used to interfere with responses to ethylene. It is possible that the increased concentrations of metals that occur in waterlogged soils are phytotoxic partly because of their effects on co-factors for ethylene binding. Mutations in RAN, a Cu transporter, affect ethylene binding, and competition for such transporters may be significant.

After binding ethylene, ETR1 autophosphorylates on a histidine residue, with the PO_4^{2-} being rapidly transferred to an aspartate residue on the C-terminus of the protein. ETR1 and other ethylene receptors therefore have much similarity with two-component receptors in prokaryotes because the binding to the receptor towards the C-terminus results in histidine kinase activity, which ultimately phosphorylates a receptor domain to activate physiological effects. Ethylene receptors probably originally evolved from transport proteins in prokaryotes, and there are several of them in most plant species—mutagenizing individual receptors often has little effect on ethylene response, or merely decreases its intensity. The receptor domain interacts with constitutive triple response 1 (CTR1) serine/threonine kinase on its cytoplasmic side. CTR1 is related to serine/threonine protein kinases that amplify signals in response pathways, probably in all eukaryotes. At very low background concentrations of ethylene, CTR1 is bound to the ethylene receptor and its kinase activity is activated. Together with proteasome degradation induced by the F-box proteins ETB1 and ETB2, this maintains a low concentration of EIN2 —a key protein in the control of ethylene-dependent gene expression. When ethylene binds to its receptor, the kinase activity of CTR1 is inhibited, the activity of ETB1 and ETB2 decreases, and the concentration of EIN2 increases. EIN2 activates EIN3, a primary TF located in the nucleus that initiates a transcriptional cascade primarily by increasing the expression of ethylene response factor (ERF) TFs. ERFs are a subfamily of the AP2 TF family, and bind via a GCC box to the ethylene response elements of numerous genes (Figure 11.11). G proteins clearly also have a role in plant responses to ethylene. Many plants are rich in both small GTPases and GTPase-activating proteins. Rop GTPases are involved in ROS/H_2O_2 signaling during ethanolic fermentation.

Plant ethylene signal transduction pathways have similarity to those in prokaryotes and other eukaryotes, and are likely to be important in most plants that are adapted to flooded environments. Manipulation of ethylene signaling has changed crop responses to flooding, and differences in ethylene responses are likely to underpin differences in plant ability to withstand flooding in unmanaged ecosystems. The control of differential tissue responses to short-term hypoxia and the constitutive development of morphologies that are characteristically induced by ethylene are especially notable in flood-tolerant plants. Thus ethylene-mediated pathways are important not only to physiological responses but also to anatomical and morphological responses to hypoxia. Elucidation of the interactions of these pathways with ethylene responses is likely to provide major insights because EIN3, AP-like transcription factors, and G proteins are also involved in plant responses to other environmental factors via other hormones.

In many plants, waterlogging-induced hypoxia induces changes in root anatomy

If plant shoots are in an O_2-rich environment while the roots are in the anoxic environment of a waterlogged soil, there will be a significant internal concentration gradient of O_2. In the majority of plants, which are waterlogging intolerant, internal resistance to shoot-to-root diffusion of O_2 down this gradient is high, whereas in those plants that can tolerate periods of waterlogging of a few days or longer, internal resistance to diffusion often decreases rapidly. In the few centimeters behind the tip of wheat roots the cells are closely hexagonally packed, resulting in porosity as low as 1% and a high resistance to O_2 diffusion. In rice roots that are adapted to wet soils, the cells behind the root tip are loosely cubically packed, and on flooding the porosity increases, approaching 10%. In wetland species, porosities of 40–50% have been reported. The anatomy of the root–shoot junction is complex and can inhibit long-distance flow of O_2, but in plants that are adapted to waterlogged soils it too can show an increase in porosity.

By 1900, the anatomy of porous roots and shoots of waterlogging-tolerant plants had been described in detail, and its tissue had been named "aerenchyma." The cortical parenchyma between the endodermis and epidermis of plant roots is often quite loosely packed with cells, but in aerenchyma there are **lacunae** that can occupy a significant proportion of the cortical volume and extend longitudinally for a significant distance. In some species these lacunae can be continuous between the root and the shoot. Much aerenchyma is formed, in response to waterlogging, by the targeted death and dissolution of cortical cells, a process that is termed **lysigeny** (Figure 11.12). This process has similarities with **programmed cell death** (PCD) in animal cells, but without the phagocytic activity or removal of debris via a circulatory system, and instead with the digestion of cellulose cell walls. Lysigenous aerenchyma, although it can be somewhat disordered, develops in particular cells and does not occur in some parts of the root—for example, where lateral roots emerge. In the Cyperaceae, such as *Papyrus*, it proceeds tangentially to produce a regular aerenchyma. Lysigenous aerenchyma is induced by waterlogging in some tolerant species, including varieties of crops such as maize, but can be constitutive in others, such as rice. During constitutive development, cells are formed from the root meristem, but even in oxic conditions they undergo lysis at a set stage in development. Soil compaction (which decreases aeration), nutrient deficiency, and in some cases exposure to the organic acids found in waterlogged soils can induce the development of lysigenous aerenchyma, and in all of these cases ethylene inhibitors prevent its development. In waterlogged soils, ethylene concentrations increase in plant tissues, providing the trigger for lysigenous aerenchyma development, and in some species hypoxia increases ethylene synthesis. The development of aerenchyma increases the porosity of roots and shoots such that diffusion of O_2 from the atmosphere down to the roots is possible. It also reduces the demand for O_2 in the roots, and decreases the size of the hypoxic core of the root by decreasing resistance to radial O_2 flows. Low resistance to gas flow through the root and shoot also means that efflux of gases from the substrate to the atmosphere can occur, which for methane is important on a global scale (Box 11.1).

Hypoxia or treatment with ethylene induces rapid increases in Ca^{2+} concentration in root cells, and chemicals that alter intracellular Ca^{2+} levels affect the development of lysigenous aerenchyma. Microarray analyses in hypoxic roots have revealed changes in genes for many Ca-signaling molecules, including calcineurin- and calmodulin-like proteins. Together with changes in Ca-dependent kinases and Ca transporters, these molecules increase cytoplasmic Ca^{2+} concentrations, probably using apoplastic and mitochondrial pools. An important target of increased Ca^{2+} levels in maize is **RBOH**, an NADPH oxidase that is homologous to the $gp91^{phox}$ which is involved in mammalian apoptosis. Ca^{2+} stimulates RBOH to make H_2O_2, which is a trigger for cell lysis. Genes encoding scavengers of ROS are downregulated in root cells targeted for lysis, which suggests that the induction

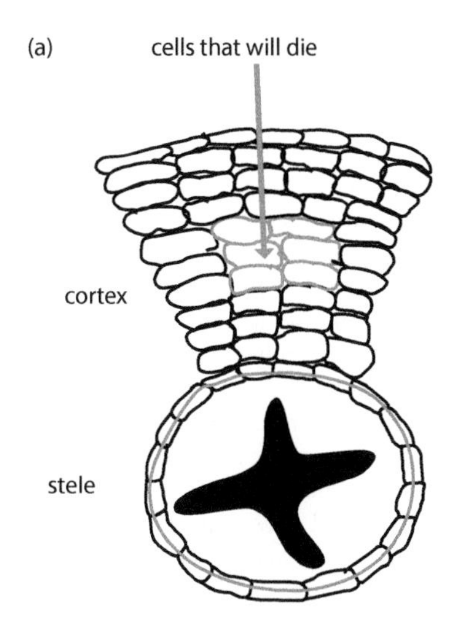

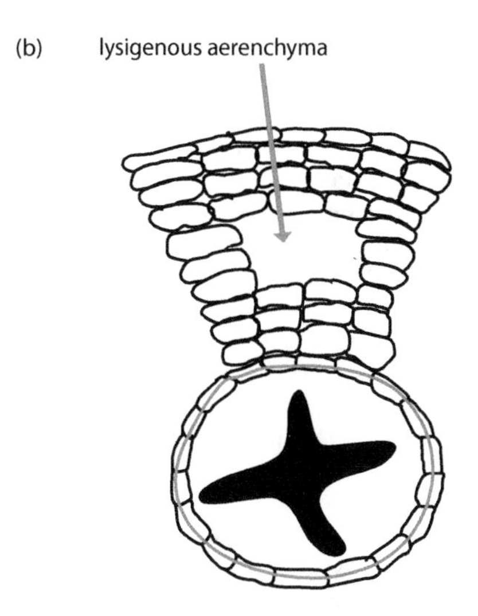

Figure 11.12. The formation of lysigenous aerenchyma. In many plant species, when waterlogging occurs particular cells can undergo a process akin to programmed cell death, known as lysigeny, to form aerenchyma—that is, tissue containing lacunae (shown before and after development in (a) and (b) respectively). Lysigenous aerenchyma can consist of quite extensive contiguous lacunae, but also occurs in discontinuous patches. It can be constitutive, but is more often induced. In the figure the development of root lysigenous aerenchyma is shown in cross section.

BOX 11.1. MANAGING METHANE EMISSIONS FROM WETLANDS

Methanogenic archaea flourish in substrates with very low redox potential and a high organic matter content—conditions that are often found in wetlands. About one-third of the methane that is released into the atmosphere each year is from wetlands. Up to two-thirds of this is from rice production. Methane is a very active greenhouse gas that contributes significantly to the greenhouse effect, so emissions from wetlands, and especially rice, are of global importance. The Intergovernmental Panel on Climate Change (IPCC) has issued guidelines for the management of methane emissions from rice to aid countries that are trying to meet emission targets. Landfill sites that have been used for disposal of biological waste are also associated with significant production of methane. Methane that is produced within waterlogged substrates diffuses out into the atmosphere slowly because of the high diffusive resistance. However, wetland plants greatly reduce this resistance, and methane as an uncharged molecule can diffuse into and up through plants (Figure 1). There are also methanotrophic bacteria that in anaerobic conditions, but especially in aerobic conditions, use methane as a terminal electron acceptor, producing a range of reduced carbon compounds, some of which are toxic to plants. In wetlands, the balance of methanogenic and methanotrophic microbial activity together with the venting of methane through plants determines methane production rates that are of global significance to climate. Many management systems for rice have been developed to decrease methane emissions and to manage the production of reduced organic compounds by methanotrophs. These include decreasing the organic matter content of soils and controlling the activity of aerobic methanotrophs. Managing rhizosphere oxidation and K/Cu concentrations, allowing growth of aerating weeds, and timing the addition of organic matter and fertilizers have all been found to have significant effects. Methanotrophic bacteria from wetlands also contribute to bioremediation protocols, because they can often oxidize a range of small carbon compounds, including some organic pollutants.

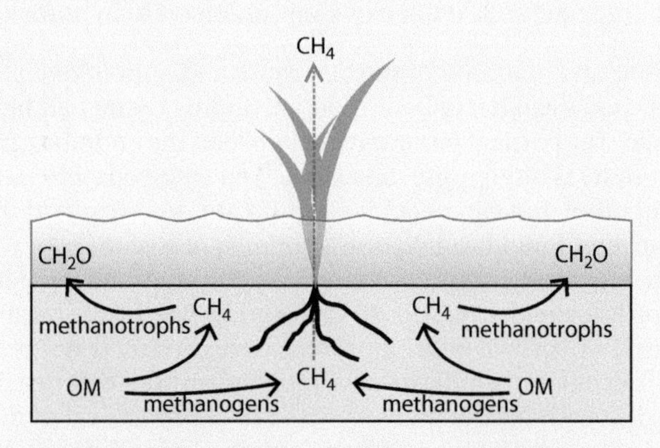

Figure 1. Methane in anoxic soils. In waterlogged anoxic soils, anaerobic methanogens produce methane (CH_4) from organic matter (OM). The CH_4 can be used by methanotrophs to produce formaldehyde (CH_2O), especially at anoxic–oxic boundaries. CH_2O usually enters methanotrophic metabolism, but can result in the production of aldehydes and other organic compounds.

of lysigenous aerenchyma is dependent on an oxidative burst. The formation of membrane-bound parcels of organelles occurs in PCD, and has also been noted during lysigeny. Genomic and proteomic investigations have revealed changes in numerous genes and proteins during lysigeny that show clear similarities with PCD in other cells. The disintegration of chromatin that can occur during lysigeny is characteristic of PCD, and changes in histone methylation and acetylation have been associated with both PCD and aerenchyma formation. In addition, increases in enzymes that loosen and degrade cell walls have frequently been reported in the later stages of lysigeny, as have decreases in enzymes associated with cell wall production. This includes the activity of expansins that break cellulose and hemicellulose apart, and of xyloglycan-endo-transglycosylase (XET), cellulase (CEL), xylanase, and pectinase that break down plant cell wall components (Figure 11.13). These processes of cell death and lysis release small molecules and ions into the apoplast, from where they are absorbed by the remaining cortical cells. The mechanism of cell targeting, including the apparent resistance of the remaining cortical cells to hydrolytic enzymes released by lysing cells, is obscure, but many details of the process of cell lysis are well known. The development of lysigenous aerenchyma is dependent on ethylene, but excessive ethylene concentrations can induce necrosis in almost all plant cell types. Aerenchyma probably functions not only to increase O_2 flow to the roots, but also to prevent the build-up of ethylene to toxic concentrations by allowing it to vent through the shoot.

The root tip is particularly sensitive to hypoxia, and is the tissue that is furthest from the atmospheric O_2 source surrounding the shoot. Furthermore,

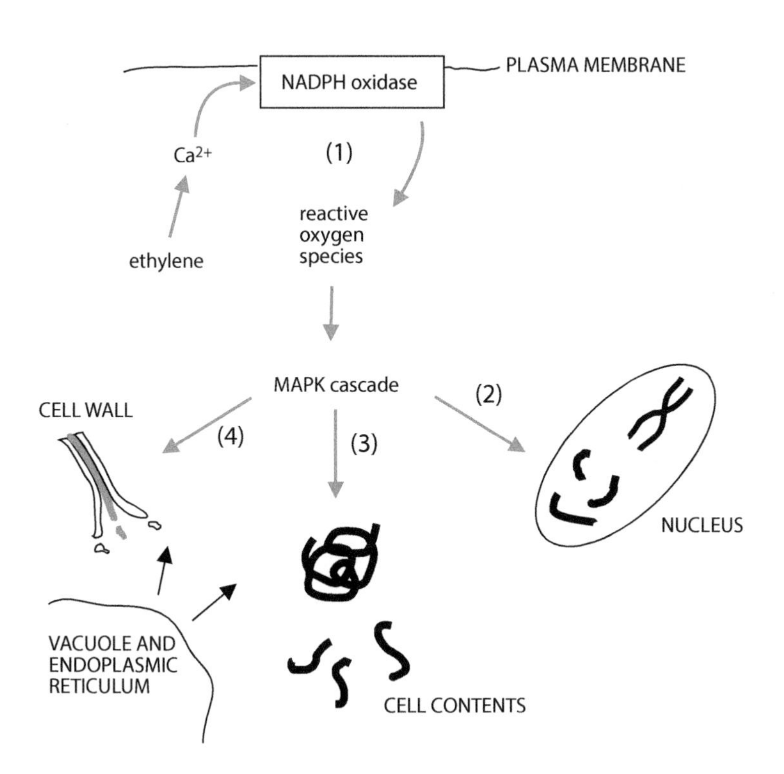

Figure 11.13. The process of lysigeny in plant cells. Lysigeny is a process akin to programmed cell death and dissolution, whereby aerenchyma is formed in plants. Waterlogging leads to the accumulation of ethylene, changes in Ca concentration, and a burst of ROS produced by NADPH oxidase (1). This initiates mitogen-activated protein kinase (MAPK) cascades that lead to cell death and dissolution, including chromatin condensation and oligonucleosomal DNA fragmentation (2), hydrolysis and proteolysis of cell contents by enzymes released from the vacuole and internal membrane systems (3), and loosening of the middle lamella and degradation of the cell wall (4).

there is a steep O_2 concentration gradient from the aerenchymal lacunae to the soil, which drives diffusive radial oxygen loss (ROL) out of the root. Thus as O_2 diffuses down the root aerenchyma, O_2 consumption by root cells and ROL can decrease the amount of O_2 that penetrates to the root tip. Measurements in many species have shown that, in response to waterlogging, ROL from roots is significantly reduced, particularly in the basal zones (Figure 11.14). A barrier to ROL promotes O_2 penetration through aerenchyma to the root tip in many hypoxia-tolerant species, and many hypoxia-sensitive species cannot prevent ROL from roots. The barrier to ROL occurs because of anatomical adjustments in the outer layers of roots. Many wetland plant roots have a hypodermis—that is, a layer of densely and often hexagonally packed cells beneath the epidermis. **Sclerified** fibers have been reported in the hypodermis from species such as rice. Lignification and suberization to form an apoplastic Casparian strip and true exodermis have frequently been noted (Figure 11.15). These hypodermal characteristics not only prevent ROL, because O_2 is slow to diffuse through the aqueous

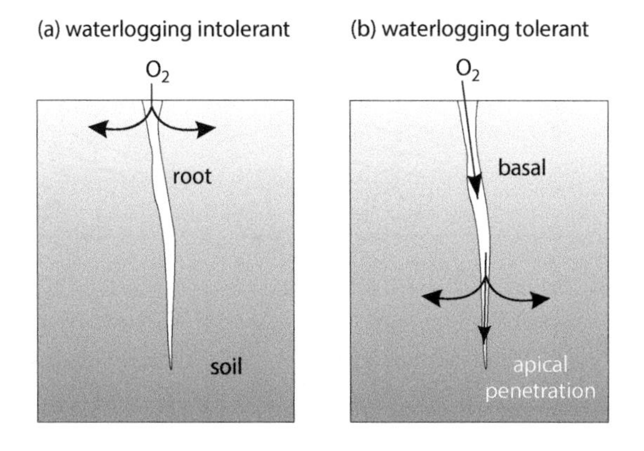

Figure 11.14. O_2 penetration down into the roots during waterlogging. In waterlogged soils there is a very low O_2 concentration (green areas), which results in the creation of a steep concentration gradient between air and soil. Plant roots can provide a conduit for O_2 down this concentration gradient. (a) In the roots of waterlogging-intolerant plants, O_2 moving down this concentration gradient does not penetrate far into the root, resulting in hypoxia in the peripheral roots, because it moves out into the anaerobic soil and is also consumed by basal root cells. (b) In waterlogging-tolerant plants, O_2 penetrates much further because the root anatomy minimizes efflux (for example, by means of a suberized hypodermis) and consumption (for example, by the formation of aerenchyma), and thus the apices of peripheral roots are oxygenated.

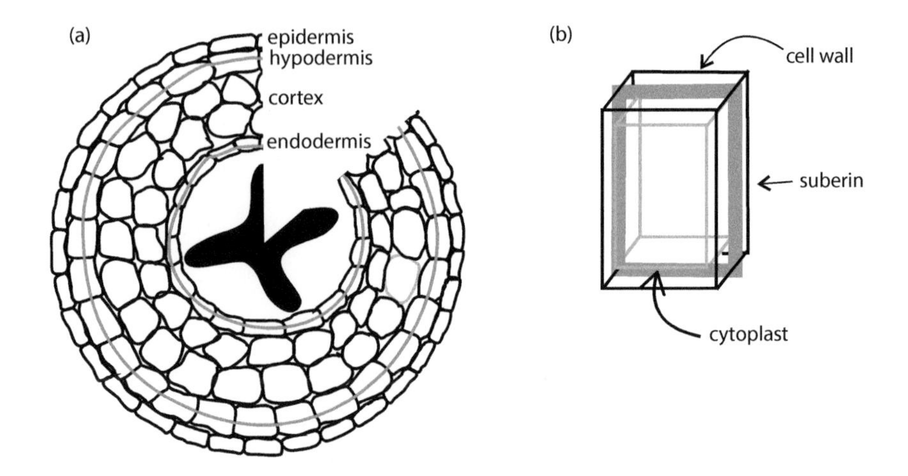

Figure 11.15. Generalized root anatomy in hypoxia-tolerant plants. (a) The development of a hypodermis with cell walls impregnated with suberin (shown in green) often occurs in hypoxia-tolerant plants. The hypodermis helps to minimize the efflux of O_2 and the influx of inorganic toxins. (b) Impregnation of the cell wall with suberin seals the apoplast and prevents the movement of molecules through it.

symplast of hypodermal layers, but also provide some structural support for a root with aerenchyma in the cortex. The stele at the center of roots adapted to hypoxia also often has a small diameter. In wheat, the stele occupies about 15% of the cross-sectional area near the root tip, whereas in hypoxia-adapted rice roots it is nearer 5%, which minimizes the O_2 demand at the center of the root. The root anatomy of waterlogging-tolerant plants is therefore adapted to maximize both the supply of O_2 and the efficiency of its use. Variation in soil hypoxic intensity occurs quite frequently at fine to medium temporal and spatial resolution—that is, up to a few days and a few tens of centimeters—and the ability to adapt the root anatomy probably helps to explain why some plants are able to survive such variation.

Wetland plants form extensive constitutive aerenchyma and adapt morphologically to flooding

Differences in growth between layers of cells can produce significant mechanical stresses in complex expanding tissues. In many wetland plants such stresses produce large lacunae in shoots and roots as development proceeds. These lacunae are produced by **schizogeny**—that is, cells separating at the middle lamella of cell walls. Mathematical models of plant growth have shown that the development of such lacunae is consistent with stresses that build up in roots and shoots during development. Schizogeny probably involves enzymatic weakening of the cell wall combined with stress differentials (Figure 11.16). Both the build-up of stresses and cell wall weakening are under close developmental control, as schizogenous aerenchmya in wetland plants has a highly ordered architecture that is often species-specific.

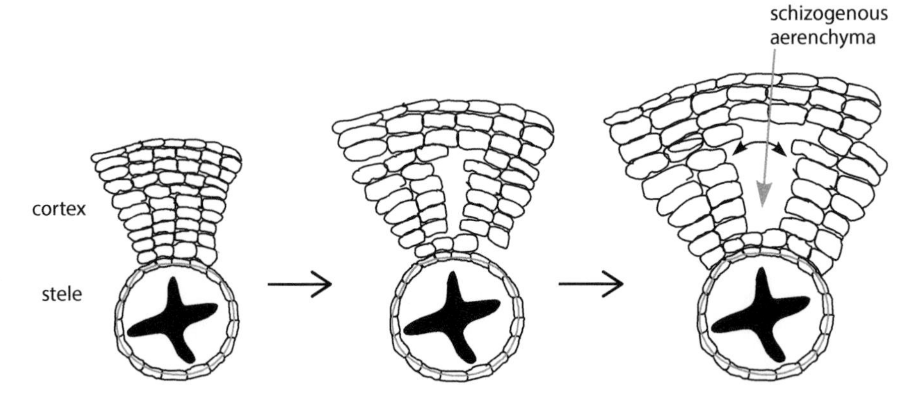

Figure 11.16. The development of schizogenous aerenchyma. In plant species that are adapted to prolonged waterlogging, during the development of particular tissues, such as the root cortex (shown in the figure), cells are pulled apart at zones of weakness to produce lacunae. These lacunae are mostly continuous and constitutive, and can be architecturally complex.

In some species, cell separation is followed at an early developmental stage by cell expansion and division, resulting in the production of a honeycomb-like tissue termed expansigenous aerenchyma. Constitutively developed aerenchyma has a wide variety of lacunal patterns, ranging from numerous fine intercellular spaces across the entire cortex in *Carex pseudocyperus* to the few massive lacunae in petioles of *Nelumbo nucifera*, and can include septa and other architectural features. The development of these constitutive aerenchymal tissues often combines minimization of the resistance to internal flow of gases with maximization of mechanical support.

Nypa fruticans is the only palm that can survive in flooded conditions, and it commonly inhabits mangroves. The apical meristem of its short stem is often entirely submerged. Its leaves can be many meters long, and often abscise above the waterline to leave leaf bases with extensive cortical aerenchyma, which helps to supply O_2 to the stem and root tissue. Many, but not all, wetland species in which aerenchyma develops constitutively are herbaceous. Woody dicotyledonous wetland plants face problems in producing airflows via cortical aerenchyma because the cortex can be squashed by secondary growth in both roots and shoots. During secondary growth the epidermis is replaced by the **periderm**, in which a layer of **phellogen** produces **phellem** towards the outside of the shoot or root. Periderm is a protective layer, and phellem is often highly suberized, inhibiting the diffusion of O_2. In aboveground tissues, the periderm generally inhibits O_2 entry and hence O_2 supply to the aerenchyma. However, in for example, *Melilotus siculus*, aerenchymatous phellem in the hypocotyl has been shown to be a significant entry point for O_2, which is then supplied to the roots through aerenchyma in the phellem (Figure 11.17). In the stem periderm, **lenticels** allow the influx and efflux of gases, and some waterlogging-tolerant woody plants have been shown to be particularly rich in lenticels above the soil surface, and to use them to supply O_2 to the phellem aerenchyma. The rich variety of constitutively expressed aerenchyma in wetland plants confirms its importance for flood tolerance.

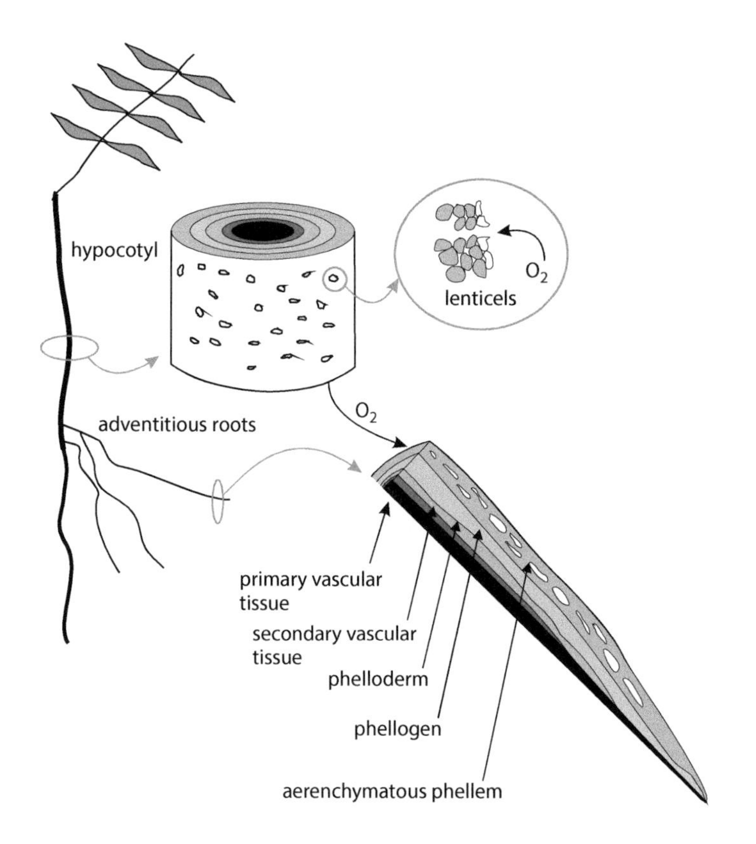

Figure 11.17. Peridermal aerenchyma in *Melilotus siculus*. In dicotyledonous plants with secondary development, the development of periderm squashes the cortex, eliminating it as a possible location of aerenchyma. Some waterlogging-tolerant species, such as *M. siculus* (shown in the figure), develop aerenchyma in the phellem of adventitious roots, allowing the penetration of O_2. The periderm seals the shoot, which O_2 therefore has to enter through lenticels.

As flooding increases in intensity or duration, the morphology of many hypoxia-tolerant species changes. In many species, standing water provokes **hyponasty**, in which the leaves grow upward, away from the water. In plants that withstand prolonged flooding for part of the year, shoot morphology can change during flooding to adjust leaf area, leaf longevity, and stomatal number. The production of adventitious roots combined with the death of old roots can reconfigure root systems to focus on more oxic soil layers and to draw in O_2 from oxygenated water. In rice and other species, aquatic adventitious roots that develop into flood water have extensive aerenchyma and poorly developed Casparian strips, which lower the resistance to O_2 entry and fluxes. Targeted cell death occurs, mediated by RBOH, to aid the eruption through the cortex of adventitious roots induced by flooding. In mangroves, secondary growth in adventitious roots from the stem forms stilt roots, which not only provide support in an unstable environment, but are also often rich in lenticels and aerenchyma to aid the flow of O_2. Many wetland plants grow clonally, with individual **ramets** linked by rhizomes or stolons. In several species the removal of ramets in shallow water affects the growth of those in deep water, which suggests that clonal growth can be important not only for withstanding an unstable environment, but also for the exchange of O_2, and perhaps other ions and molecules. Thus, for example, in *Alternanthera philoxeroides* (alligator weed) the anatomy and morphology of the plant allow a physiological integration that aids survival in deep water. Overall, plants of habitats that experience prolonged and chronic flooding use constitutive anatomical adaptations and induced morphological adaptations to expedite O_2 flow into the roots. These adaptations depend on a variety and combination of developmental processes, regulated primarily by ethylene. Perhaps the most extensive constitutive aerenchyma develops in an expansigenous manner in the Nymphaeales, Acorales, and Alismatales (some of the most primitive orders of flowering plants), and it has been suggested that it is a primitive angiosperm feature. Certainly some of the earliest flowering plants inhabited flooded environments. Prolonged waterlogging tolerance is generally more common in primitive and monocotyledonous plants, and although it is also found in plants that evolved more recently, it depends on adaptations that many less primitive plants have lost.

In some flooded soils, pneumatophores help woody plants to aerate their roots

Mangroves, which have terrestrial ancestors, inhabit environments that flood regularly. The secondary growth of mangrove trees, which is crucial to the development of the prop roots that provide support against tidal flows in many species, inhibits the influx of O_2 to the cortex. Many mangrove trees, such as *Avicennia marina*, use aerating roots called **pneumatophores** to enhance O_2 influx. The gymnospermous swamp cypress (*Taxodium distichum*) also produces pneumatophores. Mangrove root systems are complex, and include pneumatophores, lateral roots, stilt roots, and feeding roots. In contrast to most roots, pneumatophores are negatively geotropic, growing vertically upward from laterals. They have highly developed aerenchyma that runs, within a strengthened cortex, in long continuous columns down into the root system. Pneumatophores are often profuse, many thousands per plant having been recorded in some species, and in many cases they are a conduit for much of the O_2 that reaches the root system. The covering of pneumatophores with sediment or pollutants such as oil can kill mangroves, and H_2S toxicity in plants with blocked pneumatophores has often been reported.

Mature pneumatophores have a well-developed phelloderm that includes layers of sclerified cells which provide structural support (Figure 11.18). The phellem is bark-like and is a significant barrier to the entry of O_2. However, pneumatophores have extensive development of lenticels through which

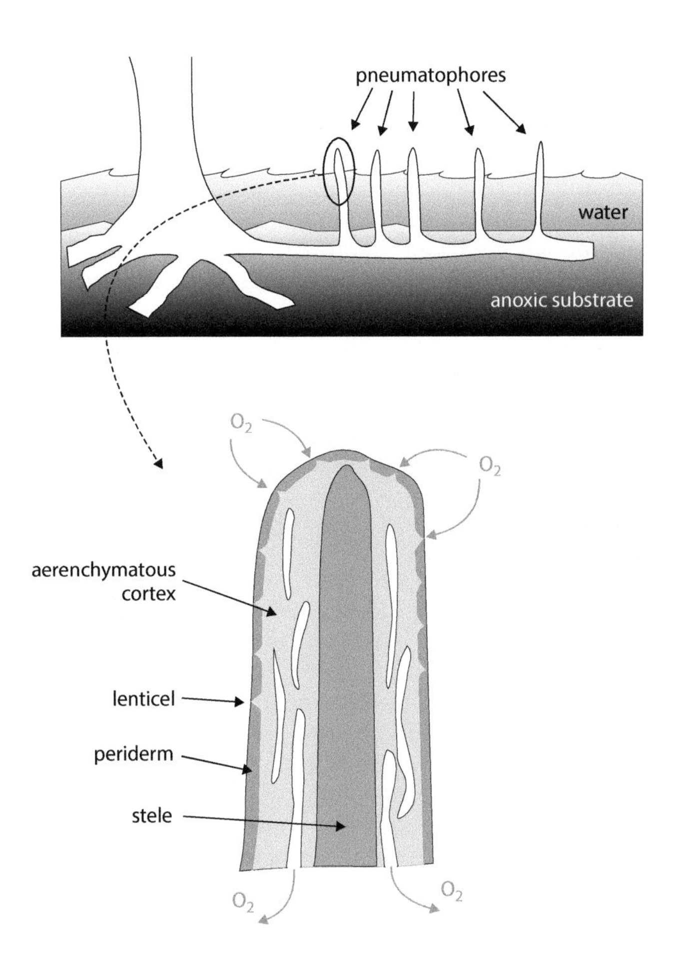

Figure 11.18. The pneumatophores of *Avicennia*. Black and gray mangroves (*A. germinans* and *A. marina*, respectively, of the Acanthaceae) have pneumatophores that allow the entry of O_2 above the water surface and its percolation down into the root system in an anoxic substrate. Similar structures are formed by *Sonneratia alba* (mangrove apple) and *Taxodium distichum* (swamp cypress). Red mangroves (*Rhizophora mangle* of the Rhizophoraceae) and related species form prop and stilt roots that lift them above the water. *Bruguiera* species (Rhizophoraceae) form long roots with "knees" that arch up above the waterlogged substrate and act as pneumatophores.

O_2 can enter, especially at the distal end which protrudes above the water. Lenticels, which normally develop in shoot tissue, are loose accumulations of cells that provide gaps in the phelloderm to permit the flow of gases. Thus pneumatophores are organs with secondary growth but that are adapted to allow significant flow of gases into the root system. Some pneumatophores have lenticels that include cutinized and suberized cells which are hydrophobic and that might serve to repel water, thereby keeping the lenticels water-free in an environment in which they are frequently splashed or inundated with water. Other roots and the stems of many mangrove trees are also very rich in lenticels, allowing influx of O_2 to the aerenchyma. In addition, the leaf morphology of mangroves varies with flooding intensity, thus helping to manage O_2 supply and demand. Pneumatophores can also have colonies of cyanobacteria that fix up to 25% of the N entering the mangrove system. In the low-N anaerobic sediments of mangrove swamps this may be a significant function of pneumatophores. Thus, in addition to their physiological capacities, mangroves have profound anatomical and morphological adaptations in their roots and shoots that enable them to aerate their roots and obtain nutrients while rooted in a substrate that is regularly flooded.

The adaptations of wetland plants often produce oxidized rhizospheres

A plaque forms in the rhizosphere of many wetland plants, including mangroves and seagrasses. This plaque formation has been correlated with many aspects of waterlogging tolerance—for example, in rice. Fe^{2+} toxicity in plants produces leaf bronzing, and in rice decreased Fe^{2+}-induced bronzing correlates with root plaque formation. The plaque is primarily composed of Fe(III) oxyhydroxides, primarily ferrihydrate, a mixture of hydrated forms of Fe_2O_3 and FeOOH. It has a complex crystalline structure and is a precursor to the precipitation of hematite (Fe_2O_3) and goethite (FeOOH). Root plaques also often contain significant quantities of Mn(III/IV) oxyhydroxides. Fe and Mn therefore occur in an oxidized state in the rhizosphere plaque, in contrast to their usual reduced state in waterlogged soil. Root plaques are generally only a few millimeters wide, and vary in thickness along roots. The formation of plaques is dependent on the O_2 supply from the roots, but the production of ferrihydrate is often microbially mediated. Plaques do not develop so extensively or quickly under sterile conditions, and Fe-oxidizing bacteria such as *Sideroxydans paludicola* can be used to enhance plaque formation under controlled conditions. Many flooding-tolerant plants therefore decrease exposure to toxic Fe^{2+} and Mn^{2+} by rhizosphere oxygenation, which encourages the build-up of a microbial-enhanced plaque. In most plants, and especially those that inhabit alkaline soils, roots have a reducing effect on the rhizosphere, so the effects of wetland plant roots are unusual.

With continuous soil anoxia, rhizosphere oxidation may be necessary to avoid ion toxicity, but it also has other important effects. Plaque development significantly increases Al tolerance in plants and, together with the pH increase often produced by soil waterlogging, helps to explain why many varieties of rice will grow well on quite acid soils—Al^{3+} is bound to oxyhydroxides in the plaque and therefore reaches the root in decreased amounts. Rhizosphere oxygenation also affects PO_4^{2-} uptake, mostly enhancing it by reducing the concentration of Fe^{2+} and hence the formation of insoluble $FePO_4$. In As-containing soils, rhizosphere oxygenation affects As uptake. Arsenate (AsO_4^{3-}), which is produced in oxidized soils and has to compete with PO_4^{2-} for uptake, is immobilized in plaques, increasing the proportion of As available as arsenite (As_2O_3), with which PO_4^{2-} does not compete for uptake sites. Thus waterlogging can increase As uptake even by plants that can oxygenate the rhizosphere, increasing the risk of As toxicity in plants and ecosystems. The presence of Si moderates the uptake of As and PO_4^{2-} in reducing conditions, but this effect is decreased by rhizosphere oxidation. Zn has decreased availability in oxygenated rhizospheres, and in mangroves high Zn concentrations have been shown to induce plaque formation that limits Zn influx. Oxygenation of the rhizosphere by wetland plants therefore has an important influence on the biogeochemistry of ecosystems with anaerobic soils. Rhizosphere oxidation is also a vital property of plants in constructed wetlands that are used for water treatment, as it changes and increases the population of microorganisms, favoring many of those that carry out biodegradation.

A combination of physiological, anatomical, and morphological characteristics helps to determine the capacity for internal O_2 supply to and efflux from roots. In many wetland species, anatomical adjustments minimize ROL for much of the root, forcing O_2 down to the root tip. This ensures not only that the root extremities are supplied with O_2, but also that the most active and sensitive part of the root forms a plaque, which is laid down as the root grows. However, the barrier to ROL formed in many wetland plants is not absolute, and does not occur in all parts of the root system. Some species only have a partial barrier to ROL and, even with tight barriers, O_2 loss—although a small proportion of the total O_2 in the root—might be sufficient to have an effect on rhizosphere redox potential. Lateral root formation penetrates the hypodermis and can provide a route for O_2 efflux. O_2 penetration through lateral roots, even those near the root base, is greater in wetland plants. Laterals on adventitious roots can also be a significant source of oxygenation of the rhizosphere. If significant amounts of O_2 can be supplied via aerenchyma, the gradient for diffusion into the soil is steep, so root tips and the gaps in the barrier to ROL are sufficient to allow significant efflux of O_2 into the rhizosphere. In wetland plants the development of anatomical and morphological features therefore optimizes the O_2 supply both to the root tips and to the rhizosphere, providing the roots with aerobic rhizospheres in otherwise anoxic soils.

Some plants can adapt to submergence of their shoots

Sometimes flooding can submerge not only the roots but also the shoots. This presents a significant challenge not only for O_2 supply but also for CO_2 uptake and light harvesting. A spectacular example of submergence occurs each year in the floodplain forests of the Amazon, where an area of around 300,000 km^2 can be flooded to a depth of many meters for many months. Plants that survive submergence have morphological adaptations, but they also often become physiologically quiescent. Most varieties of lowland rice, although grown in up to 25 cm of standing water, are not very tolerant of submergence, which causes significant yield losses. However, some varieties become quiescent and are the basis of improved submergence-tolerant varieties (Box 11.2). There is a major QTL (*SUB1*) for flood tolerance on chromosome 9 of the FR13A cultivar. The QTL includes genes for three ethylene-response factors (ERFs). Almost all rice varieties have *SUB1B* and *SUB1C* at this location, but a subset of *indica* varieties, including FR13A, also have *SUB1A*, which arose by duplication of *SUB1B* after the origin of *indica* rice. A serine residue at amino acid 186 instead of a proline residue is now known to confer submergence tolerance.

The build-up of ethylene during submergence induces expression of *SUB1A*, especially in tissues involved in elongation, such as nodes. This decreases the consumption of soluble sugars and starch. It also restricts the action of gibberellic acid (GA) in promoting stem growth, because it helps to maintain levels of SLR proteins that repress GA signaling (Figure 11.19). Many

BOX 11.2. THE DEVELOPMENT OF SWARNA-SUB1 RICE

The majority of rice is grown in flooded paddy soils, which for much of the growing season have water depths of 10–20 cm. Almost without exception, the most important traditional varieties of paddy rice are not tolerant of submergence. For example, in India in significant rice-growing areas such as Tamil Nadu, up to 5% of the rice crop is affected by floods each year. In India as a whole, several million tonnes of rice are lost each year because of flooding. In many of the wet, rice-growing areas of Asia, especially South-East Asia, there are similar significant losses of rice production due to flooding. Much rice is grown in wet environments, and there is a very real risk of there being too much water.

Some varieties of rice are tolerant of submergence and can enter a quiescent state for 2 weeks or longer under water. This ability is often controlled by the *sub1a* gene, and many experiments have shown that the presence of this gene enables rice plants to tolerate a period of submergence (Figure 1). Most varieties of paddy rice do not possess *sub1a*, but have properties that make them desirable for agronomic or dietary reasons. The Swarna variety of rice is one of the main, and most popular, varieties grown in India. It is sensitive to submergence, and accounts for a significant proportion of the rice lost each year to flooding. The *sub1a* gene was identified at the International Rice Research Institute, and by 2009 had been bred into Swarna rice and released through the Central Rice Research Institute of India as "scuba rice," Swarna-sub1a. Since then the use of Swarna-sub1a has spread rapidly and became integral to the Bill & Melinda Gates Foundation's "Stress-Tolerant Rice for Africa and South Asia" programme. Swarna-sub1a was produced through a marker-assisted backcross, so has avoided the issues that are sometimes associated with the introduction of GM varieties. It is capable of surviving at least 15 days of submergence, and maintains the desirable properties of the Swarna variety.

Further reading: Bailey-Serres J, Fukao T, Ronald P et al. (2010) Submergence tolerant rice: *SUB1*'s journey from landrace to modern cultivar. *Rice* 3:138–147.

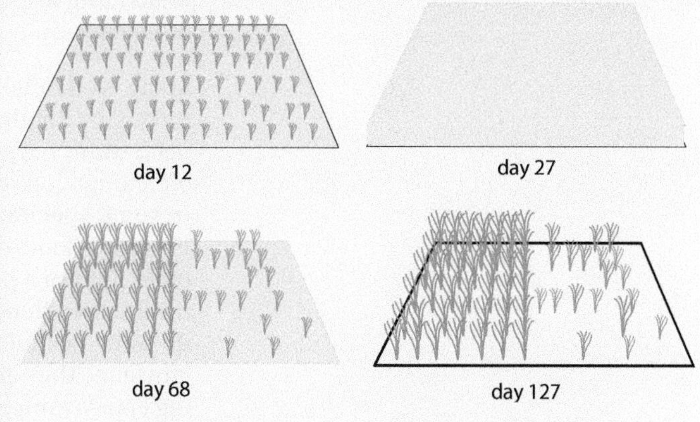

Figure 1. The effect of *sub1a* on flooding tolerance. Most varieties of rice are planted out as seedlings into standing water (day 12). Total submergence for several days after this (day 27) results in significant loss of the crop without *sub1a* (right-hand side in the figure), whereas the crop with *sub1a* fully recovers (left-hand side).

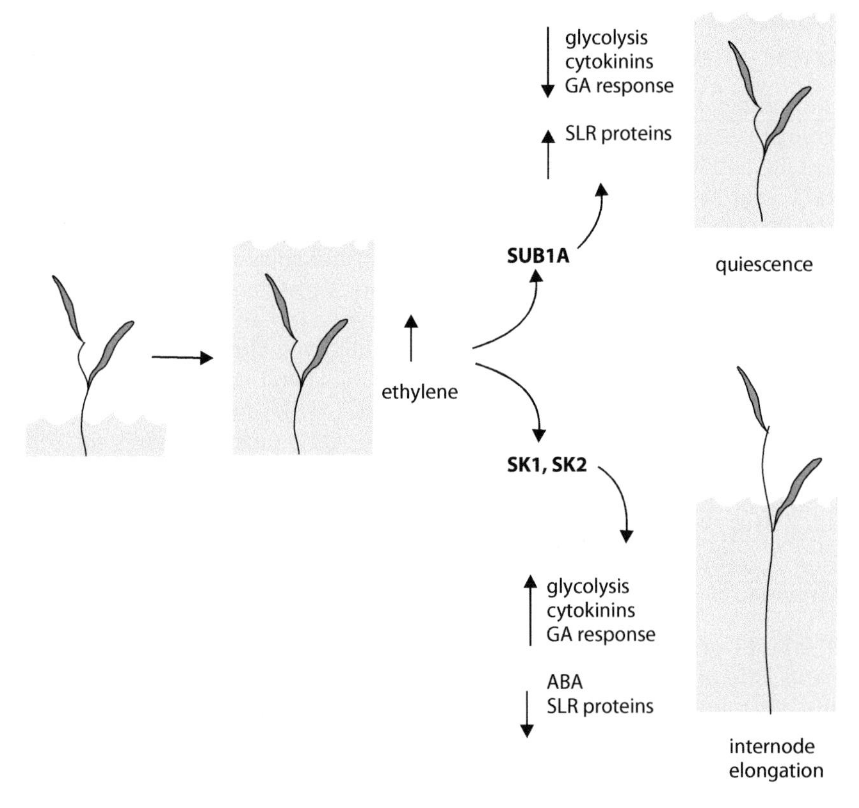

Figure 11.19. The responses of submergence-tolerant plants. In tolerant plants, the "submergence" (SUB) and "snorkel" (SK) ethylene-induced TF families mediate the response to flooding. In some species, SUB TFs induce a quiescent state that enables survival of a period of submergence. SK TFs induce hyponastic growth and internode elongation, enabling some plants to grow up and out of the water

other processes necessary for stem elongation, such as cell wall loosening and cytokinin production, are muted in *SUB1A*. Like many other submergence-tolerant plants, *SUB1A* rice maintains its chlorophyll content to a much greater extent than non-tolerant plants, which tend to become etiolated when submerged. Submergence-tolerant varieties are also much more tolerant of the oxidative burst that can occur after submergence. The leaf morphology of many submergence-tolerant plants promotes the formation of a gas film around the leaf during submergence. This aids the continuation of respiration and photosynthesis, with significant influx of O_2 into the leaf, and significant cycling of O_2 through it during the day–night cycle, having been recorded. The molecular details of the quiescence strategy in plants other than rice is less well known, but the frequent involvement of ethylene in flooding responses suggests that similar pathways of quiescence occur in other species.

Many wetlands have prolonged, often predictable, floods that are many meters in depth. Some plant species that are adapted to submergence in deep water have an escape strategy and avoid anoxia by internode or petiole elongation, which rapidly reconnects the leaves with the atmosphere. In some species, elongation rates of 25 cm a day have been recorded, and in some "flood-rice" varieties, plants 6 or 7 m tall are produced. Elongation often follows a period of hyponasty that orientates the shoot or petiole vertically after flooding-induced flattening. Species that avoid submergence do so by becoming, in contrast to quiescent species, especially sensitive to ethylene. They also down-regulate abscisic acid (ABA) synthesis by inhibiting epoxycarotenoid dioxygenase (which is involved in ABA synthesis), and they activate ABA breakdown. In *Rumex palustris* this promotes increasing concentrations of GA_1. This increased GA sensitivity and production induces cell wall acidification and loosening, often with the expression of expansins and xyloglycan hydrolases. In *Nymphoides peltata*, following increases in GA expression, proteins that promote cell division have been detected. Such changes are accompanied by GA-induced changes in carbohydrate metabolism that fuel growth. Deep-water varieties of rice can grow in several meters of water. This trait is significantly associated with a QTL on chromosome 12 where two genes, *SNORKEL1 (SK1)* and *SNORKEL2 (SK2)*, are located. SK1 and SK2 are ERF-type transcription factors. Transgenic non-deep-water rice plants that express SK1 and SK2 display rapid internode elongation in response to submergence. Following ethylene-induced changes, the response has much molecular overlap with the shade avoidance response, and is in part serving a similar purpose. In *R. palustris* there are populations that show either the quiescence response or the avoidance response to submergence. These have been used to analyze the costs of each strategy, and to show that quiescence can be the optimal strategy for short-duration "flash" floods, whereas avoidance via growth is favored if there is prolonged deep flooding.

Emergent aquatic macrophytes can force oxygen down through organs buried deep in anoxic mud

Emergent **aquatic macrophytes** grow in permanently deep water, but have shoots that emerge onto or above the water surface. Many of these species, such as water lilies, have leaves that float on the water surface, with an abaxial leaf surface and petiole in contact with water, and the other organs buried deep in anoxic mud (Figure 11.20). Diffusion does not provide a sufficient O_2 supply to the roots of such plants, because of the limited shoot area available for O_2 entry. Pressure differentials between different parts of these plants have been shown to produce internal flows of air that move gases from the atmosphere down into the submerged organs in quantities that add significantly to those provided by diffusion. Chambers with small pores can have internal gas pressures that are significantly higher than those of the surrounding air if there are humidity differentials. If the air inside a chamber is humid and a large volume of the surrounding air is much less humid, the internal pressure increases because the presence of H_2O decreases the partial pressure of, for example, O_2 and N_2, promoting their influx by diffusion and increasing the overall internal air pressure. An internal pressure is only maintained if the humidity inside the chamber continues to be higher than that outside, which is only possible if the pores are less than a few μm in diameter. Higher temperatures increase the internal pressure that can be attained.

The floating leaves of a number of aquatic macrophytes have anatomies and morphologies that promote humidity-induced positive internal pressure. The conditions necessary for such internal pressurization occur in young leaves of water lilies such as *Nuphar lutea* because the stomata are sufficiently small in diameter, evaporation from leaf parenchyma cells is sufficient

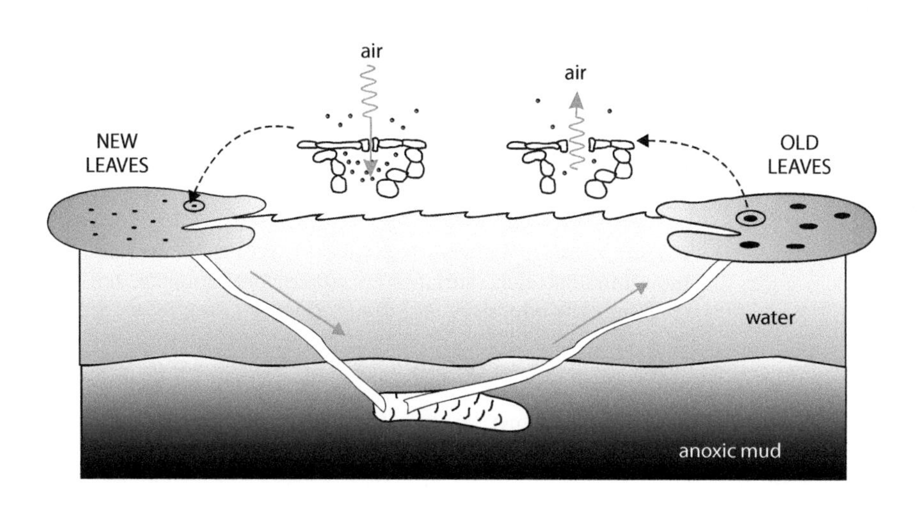

Figure 11.20. **Humidity-induced airflow in water lilies.** If conditions are more humid inside than outside the leaf, warming causes pressure to build up, especially in new leaves, which have small stomata. Atmospheric gases have a lower partial pressure in humid air in the sub-stomatal cavity than outside, so they diffuse inward from the surrounding air, contributing to a rise in pressure. If the stomata are small, there is significant resistance to airflow outward, so quite high pressures can be attained. In old leaves with larger stomata there is less resistance to pressure flows, which therefore vent through them, setting up an airflow through the buried rhizome. This airflow can supply O_2 to the rhizome and vent gases such as C_2H_4 and CH_4.

to maintain internal humidity, and leaf temperatures are above ambient, either because of insolation during the day, or at night because of heating from water that is above air temperature. Aerenchyma provides a path of least resistance to internal gas flows, but these can only be maintained if air exits the plant. Venting occurs from old leaves in *Nuphar* and from the central button of the leaves of *Nelumbo nucifera* (sacred lotus). Floating leaves of aquatic macrophytes therefore provide a system of humidity-induced pressurized gas flows through submerged organs that adds significantly to the provision of O_2 (Figure 11.20). The leaves of *Nelumbo nucifera* and other water lilies are also superhydrophobic, ensuring that water does not rest on the leaf surface but rolls off, taking any dust and dirt with it. In plants that must exchange gases through the leaf surface it is vital that the stomata are not blocked by water, dust, or dirt. The nanostructure of *Nelumbo* waxes provided the basis for superhydrophobic and self-cleaning materials in a very successful example of **biomimetic** design (Box 11.3).

BOX 11.3. THE "LOTUS EFFECT" AND SELF-CLEANING MATERIALS

Superhydrophobicity is a very useful property not only for waterproofing materials but also for making them self-cleaning, because water runs off them, taking dirt particles with it. In common with some other plants, the leaves of *Nelumbo nucifera* (sacred lotus) have been known for thousands of years to have self-cleaning properties. Given that *N. nucifera* inhabits anoxic mud, the emergence of its clean leaves perhaps helps to explain why it has for so long been a symbol of purity. In contrast to the unrelated water lilies, it has leaves that stand above the water surface rather than floating on it (Figure 1). As the leaves emerge from the mud and rise up out of the water, any mud on the leaves is removed by water droplets running off their surface.

Figure 1. *Nelumbo nucifera* (sacred lotus). (a) Leaves standing above the water surface. (b) The characteristic peltate leaf (circular in shape with central petiole attachment).

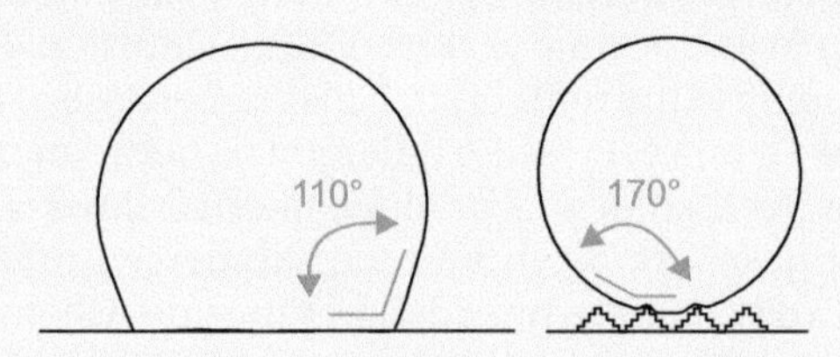

Figure 2. Contact angles of water droplets. Water droplets have low contact angles with smooth surfaces, but sculpted surfaces, such as the waxes of a lotus leaf, produce contact angles that approach 180°C.

The hydrophobicity of a surface is measured using the contact angle between a water droplet and the surface. If a water droplet rests on a surface with a contact angle of, for example, 110° or less, there is a strong attraction between the water and the surface, so water tends to stick to it. As the contact angle approaches 180° the attraction to the surface diminishes until there is not much attraction to the surface and the water runs off—often even if the surface is flat, because of air movements (Figure 2). Water droplet contact angles of 160–170° have now been measured on the leaves of lotus and other plant species. By the 1990s, the complex nanostructure underpinning the self-cleaning "lotus effect" was being revealed. In most instances, leaf waxes have a hierarchy of structures of two different scales that make small contact angles unlikely. The "lotus effect" has now been used in the production of many self-cleaning materials in a very successful example of biomimetic design. For lotus flowers it is an adaptation that keeps their leaves clean, which is essential for generating the influx of air to the leaves to supply O_2 to the rhizome buried in the anoxic mud below the water.

Further reading: Bhushan B & Jung YC (2011) Natural and biomimetic artificial surfaces for superhydrophobicity, self-cleaning, low adhesion, and drag reduction. *Prog Mater Sci* 56:1–108.

In *Phragmites australis* (common reed) leaf sheaths above the water surface, the stomata have diameters that produce humidity-induced internal positive pressure. This can contribute significantly to air movement down through the aerenchyma and into rhizomes buried in anoxic mud. The leaves of some mangroves have stomata with large sub-stomatal cavities and intimate connections to the aerenchyma of the leaf, and from there to the root aerenchyma. These allow humidity-induced pressurized flows of O_2 into the mangrove through the leaves. Many tropical wetland plants have humidity-induced gas flows, and in a variety of ecosystems the depth of water that emergent plants can withstand is related to their ability to generate such flows. *P. australis* has also been shown to have internal airflows produced by the **Venturi effect**, in which airflow across dead, broken culms produces decreased internal pressures that induce airflow from younger leaves through the aerenchyma. This mechanism of internal airflow makes a small but perhaps significant contribution to airflow in *P. australis*, and might also do so in other emergent wetland plants. The highest humidity-induced pressurized flows, of 10 cm s^{-1}, giving total air fluxes of 120 cm^3 min^{-1}, have been measured in *Equisetum telmateia*. Different *Equisetum* species have similar morphologies but differing anatomies. Studies of these have shown that humidity-induced internal airflows correlate with continuity of aerenchymal lacunae and internal resistances produced by anatomical features. The Equisetaceae are an ancient family of plants, and are closely related to the Calamitaceae that were a significant component of the Carboniferous swamps that produced the coal measures. The anatomical features of fossil calamites suggest that they too used significant humidity-induced airflows to aerate their roots, helping to produce the organic matter that has formed many fossil fuels.

Some aquatic macrophytes are adapted to living permanently submerged

Many aquatic macrophytes grow permanently submerged in water with no direct access to atmospheric O_2. Their only sources of O_2 are photosynthesis and the water that surrounds them. In order to drive photosynthesis, CO_2, which diffuses slowly in water, and light, which can be attenuated significantly on its passage through water, are both necessary. Many aquatic macrophytes have leaves with a very large surface area, almost no cuticle or stomata, a very thin epidermis, and an extensive aerenchyma (Figure 11.21). In such leaves there is a pronounced diurnal change of gas regime, with significant build-up of O_2 and depletion of CO_2 during the day, whereas at night the opposite occurs. Bifurcating leaves such as those of *Ranunculus aquatilis* maximize the surface area for uptake of CO_2 from water without significant escape of O_2, the loss of which is inhibited by the surrounding water. Such leaves can be formed because they do not have to control water loss or provide structural support. Aerenchyma helps submerged aquatic plants to distribute O_2, which is mainly derived from photosynthesis during the day. Many submerged aquatic plants use carbon-concentrating mechanisms to increase the flow of CO_2 to rubisco, and to minimize photorespiration.

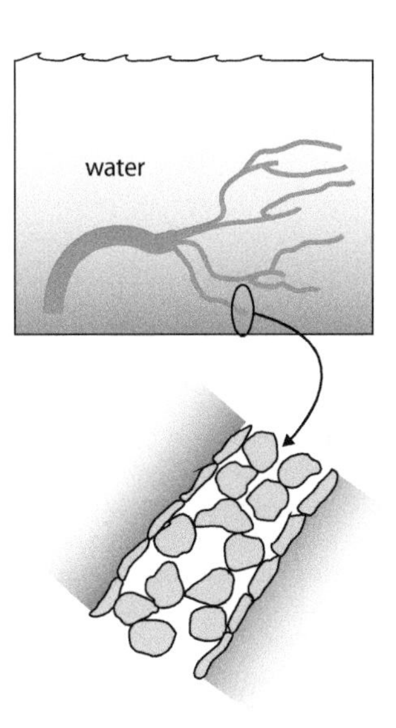

Figure 11.21. **The leaves of submerged aquatic macrophytes.** Gases diffuse through water very slowly, and in rivers and some lakes water movement can be significant, so the leaves of aquatic macrophytes often tend to be bifurcating. This increases the surface area available for gas exchange as well as decreasing mechanical stresses due to resistance to water flow. The leaves also often have a poorly developed epidermis and loosely packed chlorenchyma cells. Aquatic macrophytes commonly use PEP carboxylase to fix HCO_3^- to supplement fixation dependent on the diffusion of CO_2. During exposure to light, O_2 concentrations in the leaf increase because O_2 cannot escape quickly, thus helping to overcome hypoxia. In the dark, CO_2 concentrations increase and can contribute to subsequent carbon fixation.

Summary

Excess water inhibits the diffusion of O_2 into plant tissues, causing an energy crisis and cellular acidosis. Physiological flexibility allows most plants to withstand the short-term waterlogging that is a frequent occurrence in many soils. Some physiological pathways have been amplified in plant species that can withstand more sustained waterlogging. Progressively more profound anatomical and morphological adaptations are necessary to maintain the O_2 supply during prolonged flooding, and to cope with the increasing occurrence of toxic ions. Plants that can survive permanent

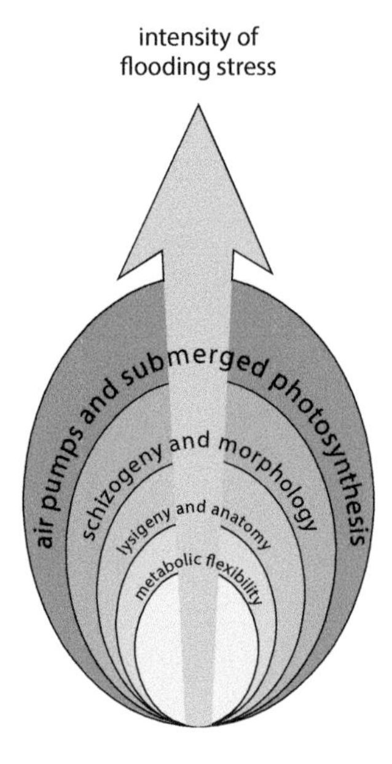

Figure 11.22. Plant adaptation to waterlogging and submergence. Almost all plants have some biochemical flexibility that enables them to withstand a certain degree of variation in O_2 supply. Plants that can tolerate transient waterlogging often have an amplified anaerobic metabolic capacity. With prolonged waterlogging, only plants that also have anatomical adaptations that enhance internal O_2 flows (aerenchyma) and oxygenation of the rhizosphere (hypodermis) can survive. Some plants adapt morphologically to submergence and permanent waterlogging by elongation growth, or by producing unique permanent structures such as pneumatophores.

waterlogging or complete submergence have an array of physiological, anatomical, and morphological adaptations (Figure 11.22). Hypoxia and ethylene are key signals for the development of adaptation to excess water, and ethylene-responsive transcription factors are probably the nexus for interaction with the numerous pathways underpinning the ability to inhabit what is a challenging environment for a terrestrial plant. Excess water is ecologically and agriculturally important on a global scale, and an understanding of how plants function in the face of this challenge will be useful for sustainable human habitation of Earth.

Further reading

The importance of flooding

Bailey-Serres J & Voesenek LACJ (2008) Flooding stress: acclimations and genetic diversity. *Annu Rev Plant Biol* 59:313–339.

Carmichael MJ, Bernhardt ES, Braeuer SL et al. (2014) The role of vegetation in methane flux to the atmosphere: should vegetation be included as a distinct category in the global methane budget? *Biogeochemistry* 119:1–24.

Jackson MB, Ishizawa K & Ito O (2009) Evolution and mechanisms of plant tolerance to flooding stress. *Ann Bot* 103:137–142.

Sorrell BK, Tanner CC & Brix H (2012) Regression analysis of growth responses to water depth in three wetland plant species. *AoB Plants* 4:pls043.

Human impacts on wetlands and flooding

Ballantine K & Schneider R (2009) Fifty-five years of soil development in restored freshwater depressional wetlands. *Ecol Appl* 19:1467–1480.

Cai ZC, Kang GD, Tsuruta H et al. (2005) Estimate of CH_4 emissions from year-round flooded rice fields during rice growing season in China. *Pedosphere* 15:66–71.

Millennium Ecosystem Assessment (2005) *Ecosystems and Human Well-Being:* Wetlands and Water. *Synthesis.* World Resources Institute.

Šíma J, Diáková K & Holcová V (2007) Redox processes of sulfur and manganese in a constructed wetland. *Chem Biodivers* 4:2900–2912.

Flooding and soil chemistry

Cook FJ, Knight JH & Kelliher FM (2013) Modelling oxygen transport in soil with plant root and microbial oxygen consumption: depth of oxygen penetration. *Soil Res* 51:539–553.

Fageria NK, Carvalho GD, Santos AB et al. (2011) Chemistry of lowland rice soils and nutrient availability. *Commun Soil Sci Plant Anal* 42:1913–1933.

Lamers LP, Govers LL, Janssen IC et al. (2013) Sulfide as a soil phytotoxin—a review. *Front Plant Sci* 4:268.

Shaheen SM, Rinklebe J, Rupp H et al. (2014) Temporal dynamics of pore water concentrations of Cd, Co, Cu, Ni, and Zn and their controlling factors in a contaminated floodplain soil assessed by undisturbed groundwater lysimeters. *Environ Pollut* 191:223–231.

Adverse effects of flooding on plants

Frick A, Jarva M & Tornroth-Horsefield S (2013) Structural basis for pH gating of plant aquaporins. *FEBS Lett* 587:989–993.

Herrera A (2013) Responses to flooding of plant water relations and leaf gas exchange in tropical tolerant trees of a black-water wetland. *Front Plant Sci* 4:106.

Rocha M, Licausi F, Araujo WL et al. (2010) Glycolysis and the tricarboxylic acid cycle are linked by alanine aminotransferase during hypoxia induced by waterlogging of *Lotus japonicus*. *Plant Physiol* 152:1501–1513.

Shabala S (2011) Physiological and cellular aspects of phytotoxicity tolerance in plants: the role of membrane transporters and implications for crop breeding for waterlogging tolerance. *New Phytol* 190:289–298.

Physiological adjustments to soil flooding

Limami AM, Diab H & Lothier J (2014) Nitrogen metabolism in plants under low oxygen stress. *Planta* 239:531–541.

Qi B, Yang Y, Yin Y et al. (2014) *De novo* sequencing, assembly, and analysis of the *Taxodium* 'Zhongshansa' roots and shoots transcriptome in response to short-term waterlogging. *BMC Plant Biol* 14:201.

Sorenson R & Bailey-Serres J (2014) Selective mRNA sequestration by oligouridylate-binding protein 1 contributes to translational control during hypoxia in *Arabidopsis*. *Proc Natl Acad Sci USA* 111:2373–2378.

Voesenek LACJ & Sasidharan R (2013) Ethylene – and oxygen signalling – drive plant survival during flooding. *Plant Biol* 15:426–435.

Flooding and ethylene signaling

Liu Q, Xu C & Wen C (2010) Genetic and transformation studies reveal negative regulation of ERS1 ethylene receptor signaling in Arabidopsis. *BMC Plant Biol* 10:60.

Shakeel SN, Wang X, Binder BM et al. (2013) Mechanisms of signal transduction by ethylene: overlapping and non-overlapping signalling roles in a receptor family. *AoB Plants* 5:plt010.

Steffens B & Sauter M (2010) G proteins as regulators in ethylene-mediated hypoxia signaling. *Plant Signal Behav* 5:375–378.

Yang C-Y (2014) Hydrogen peroxide controls transcriptional responses of ERF73/HRE1 and ADH1 via modulation of ethylene signaling during hypoxic stress. *Planta* 239:877–888.

Anatomical adaptations to waterlogging

Joshi R & Kumar P (2012) Lysigenous aerenchyma formation involves non-apoptotic programmed cell death in rice (*Oryza sativa* L.) roots. *Physiol Mol Biol Plants* 18:1–9.

Teakle NL, Armstrong J, Barrett-Lennard EG et al. (2011) Aerenchymatous phellem in hypocotyl and roots enables O_2 transport in *Melilotus siculus*. *New Phytol* 190:340–350.

Watanabe K, Nishiuchi S, Kulichikhin K et al. (2013) Does suberin accumulation in plant roots contribute to waterlogging tolerance? *Front Plant Sci* 4:178.

Yin D, Chen S, Chen F et al. (2013) Ethylene promotes induction of aerenchyma formation and ethanolic fermentation in waterlogged roots of *Dendranthema* spp. *Mol Biol Rep* 40:4581–4590.

Constitutive and morphological adaptations to waterlogging

Chomicki G, Bidel LPR, Baker WJ et al. (2014) Palm snorkelling: leaf bases as aeration structures in the mangrove palm (*Nypa fruticans*). *Bot J Linn Soc* 174:257–270.

Luo F-L, Chen Y, Huang L et al. (2014) Shifting effects of physiological integration on performance of a clonal plant during submergence and de-submergence. *Ann Bot* 113:1265–1274.

Sauter M (2013) Root responses to flooding. *Curr Opin Plant Biol* 16:282–286.

Rich SM, Ludwig M & Colmer TD (2012) Aquatic adventitious root development in partially and completely submerged wetland plants *Cotula coronopifolia* and *Meionectes brownii*. *Ann Bot* 110:405–414.

Mangroves and pneumatophores

Arrivabene HP, Souza I, Oliveira Co WL et al. (2014) Functional traits of selected mangrove species in Brazil as biological indicators of different environmental conditions. *Sci Total Environ* 476:496–504.

Evans LS & Bromberg A (2010) Characterization of cork warts and aerenchyma in leaves of *Rhizophora mangle* and *Rhizophora racemosa*. *J Torrey Bot Soc* 137:30–38.

Evans LS, Okawa Y & Searcy DG (2005) Anatomy and morphology of red mangrove (*Rhizophora mangle*) plants in relation to internal airflow. *J Torrey Bot Soc* 132:537–550.

Janarthine SRS & Eganathan P (2012) Plant growth promoting of endophytic *Sporosarcina aquimarina* SjAM16103 isolated from the pneumatophores of *Avicennia marina* L. *Int J Microbiol* 2012:532060.

Induced redox changes in the rhizosphere

Gutierrez J, Atulba SL, Kim G et al. (2014) Importance of rice root oxidation potential as a regulator of CH_4 production under waterlogged conditions. *Biol Fert Soils* 50:861–868.

Jiang FY, Chen X & Luo AC (2009) Iron plaque formation on wetland plants and its influence on phosphorus, calcium and metal uptake. *Aquat Ecol* 43:879–890.

Mei X-Q, Yang Y, Tam NF-Y et al. (2014) Roles of root porosity, radial oxygen loss, Fe plaque formation on nutrient removal and tolerance of wetland plants to domestic wastewater. *Water Res* 50:147–159.

Weiss JV, Emerson D & Megonigal JP (2005) Rhizosphere iron(III) deposition and reduction in a *Juncus effusus* L.-dominated wetland. *Soil Sci Soc Am J* 69:1861–1870.

Adaptations to submergence

Bailey-Serres J, Lee SC & Brinton E (2012) Waterproofing crops: effective flooding survival strategies. *Plant Physiol* 160:1698–1709.

Nagai K, Kondo Y, Kitaoka T et al. (2014) QTL analysis of internode elongation in response to gibberellin in deepwater rice. *AoB Plants* 6:plu028.

Parolin P, Waldhoff D & Zerm M (2010) Photochemical capacity after submersion in darkness: how Amazonian floodplain trees cope with extreme flooding. *Aquat Bot* 93:83–88.

van Veen H, Mustroph A, Barding GA et al. (2013) Two *Rumex* species from contrasting hydrological niches regulate flooding tolerance through distinct mechanisms. *Plant Cell* 25:4691–4707.

Humidity-induced gas flows

Koch K & Barthlott W (2009) Superhydrophobic and superhydrophilic plant surfaces: an inspiration for biomimetic materials. *Philos Trans A Math Phys Eng Sci* 367:1487–1509.

Konnerup D, Sorrell BK & Brix H (2011) Do tropical wetland plants possess convective gas flow mechanisms? *New Phytol* 190:379–386.

Richards JH, Kuhn DN & Bishop K (2012) Interrelationships of petiolar air canal architecture, water depth, and convective airflow in *Nymphaea odorata* (Nymphaeaceae). *Am J Bot* 99:1903–1909.

Sorrell BK & Hawes I (2010) Convective gas flow development and the maximum depths achieved by helophyte vegetation in lakes. *Ann Bot* 105:165–174.

Submerged aquatic macrophytes

Colmer TD, Winkel A & Pedersen O (2011) A perspective on underwater photosynthesis in submerged terrestrial wetland plants. *AoB Plants* 3:plr030.

Pedersen O, Pulido C, Rich SM et al. (2011) *In situ* O_2 dynamics in submerged *Isoetes australis*: varied leaf gas permeability influences underwater photosynthesis and internal O_2. *J Exp Bot* 62:4691–4700.

Pedersen O, Rich SM, Pulido C et al. (2011) Crassulacean acid metabolism enhances underwater photosynthesis and diminishes photorespiration in the aquatic plant *Isoetes australis*. *New Phytol* 190:332–339.

Pedersen O, Colmer TD & Sand-Jensen K (2013) Underwater photosynthesis of submerged plants – recent advances and methods. *Front Plant Sci* 4:140.

Chapter 12

Inorganic Toxins

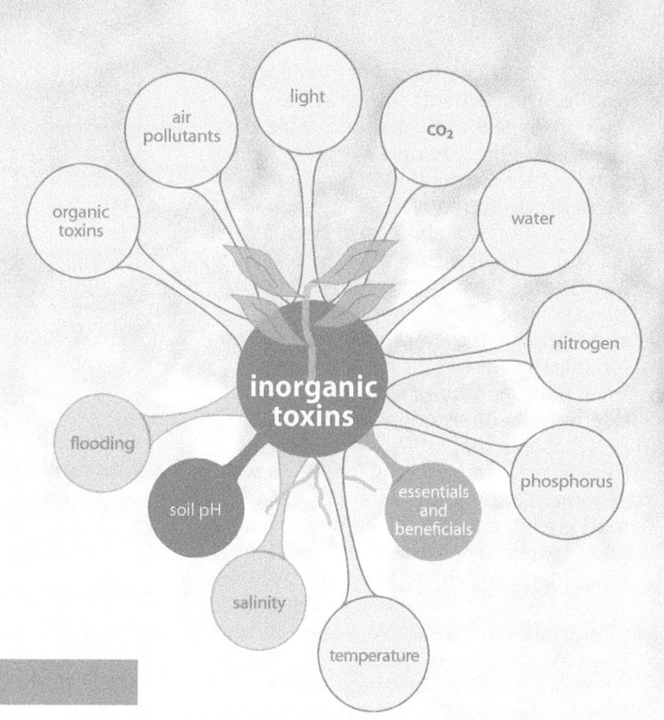

Exposure to inorganic toxins is affected by soil pH, salinity, and flooding, and can be moderated by beneficial elements.

Key concepts

- Inorganic toxins affect plants and thus heterotrophs in many ecosystems.
- Human mobilization of reactive elements is altering the biogeochemistry of Earth.
- Homeostatic mechanisms control internal concentrations of reactive elements in plants.
- Toxins can overwhelm homeostasis, with adverse physiological and genetic effects.
- Toxin-tolerant plants have amplified homeostatic mechanisms in their roots.
- Exclusion and compartmentation of toxins enhance metal-tolerant physiologies.
- Mycorrhization with toxin-tolerant fungi is a key exclusion mechanism.
- Inorganic toxins can be hyperaccumulated using amplified homeostatic mechanisms.
- Chronic exposure of plants to toxins provides ecological and evolutionary insights.
- The management of plant–toxin interactions is vital for food quality and phytoremediation.

A few reactive elements are essential, but they and many non-essential elements can also be toxic

The autotrophic nutrition of plants is based on 17 elements. Plants use C, H, O, N, P, and S to build the carbohydrate, protein, and lipid molecules that they need, and that non-autotrophs rely on them to synthesize. They use a further 11 elements to build essential biomolecules and to aid enzymatic control of metabolism. These 11 elements are also essential for most heterotrophs, which obtain them in significant part from plants. In addition, most animals require a further seven elements. There are 92 naturally occurring elements, but the physiology of organisms is based on a few relatively light elements, including numerous transition metals (Figure 12.1). The physico-chemical properties of the elements and the geochemistry of Earth

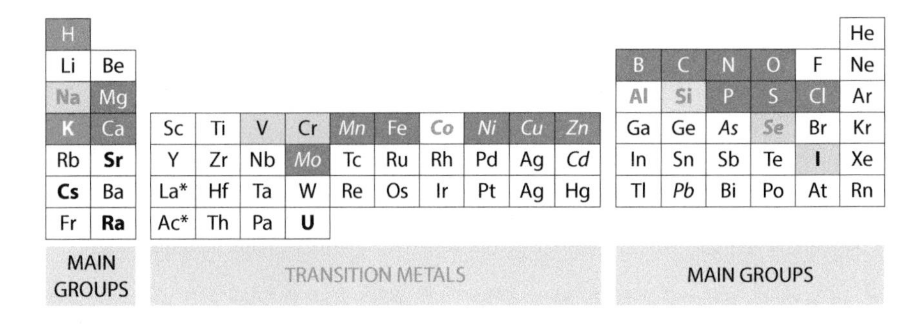

Figure 12.1. The essential elements for life. The elements shown in the dark green cells in the periodic table are essential for life. Essential elements tend to be lighter elements, with main group elements being required in large amounts and transition metals in small amounts. Na and Si are essential for some species of plants (green cells) but Na is essential for all organisms other than plants. V, Cr, Se, and I (shown in the gray cells) are essential for animals but not for plants. A very small number of organisms have been discovered with essential requirements for Br, Cd, Sn, Sr, Ba, or W. Some elements (green text) are beneficial for the growth of some plants. Some elements (italic text) are hyperaccumulated by some plant species. The commonest causes of inorganic toxicity to plants are high concentrations of transition metals and high concentrations of Na, Al, As, Se, and Pb. Radioisotopes of Cs, Sr, I, Ra, and U (shown in bold) are common contributors to enhanced plant exposure to ionizing radiation. *Elements of the lanthanide series (La) and actinide series (Ac) are not shown.

during the evolution of life help to explain which elements became essential. Studies of the **ionome** of archaea and bacteria suggest that Mg, Fe, Co, and Ni were probably the first metals to be chelated by biomolecules and incorporated into metabolism. The evolution of eukaryotes was associated with significant changes in the concentrations of Na, Ca, K, Mg, and Fe, but perhaps the biggest change occurred with the evolution of aerobic respiration—the concentrations of Zn, Cu, Mn, Ni, and Fe changed significantly in aerobes, probably due to the evolution of enzymes that regulate oxidative stress or deal with decreased Fe availability. With the exception of Zn, transition metals have electrons not only in *s*- and *p*-orbitals but also in *d*-orbitals, from which they are easily lost. This is reflected in the high **electronegativity** and multiple **valence** states of many transition metals. In contrast to the Group I and Group II essential metals, which tend to form ionic compounds, the essential transition metals form covalent compounds and complexes, often incorporating neutral molecules such as water. Life evolved to utilize these properties, but at high concentrations the reactivity of essential transition metals can cause toxicity. Thus the reactions of life evolved to make use of the properties of particular elements, but variation in their availability has been, and still is, a challenge for organisms.

There are many non-essential elements that are chemically similar to essential elements and that can cause toxicity if taken up by plants. For example, Cd enters plants and causes toxicity because it is chemically similar to Zn. The metalloid As and the non-metal Se frequently occur in the environment as $H_2AsO_4^{2-}$ and SeO_4^{2-} ions, respectively, which have chemical similarity to HPO_4^{2-} and SO_4^{2-}, so they can enter plants because there is insufficient discrimination against them during uptake. Some elements, such as Al, are not analogs of nutrients, but can accumulate in plants until they reach toxic concentrations. In general, the toxic effects of non-essential ions become apparent at much lower concentrations than do those of essential ions (Figure 12.2). Some essential elements also have **primordial radionuclides** (for example, ^{40}K), and naturally occurring decay series have some chemical analogs of essential elements that occur at high concentrations in some soils (for example, ^{222}Ra, which is chemically similar to Ca). Thus a variety of naturally occurring metals, metalloids, non-metals, and radionuclides can enter plants and reach concentrations that are toxic. Plants provide the major conduit for elements from the soil into organisms of terrestrial ecosystems.

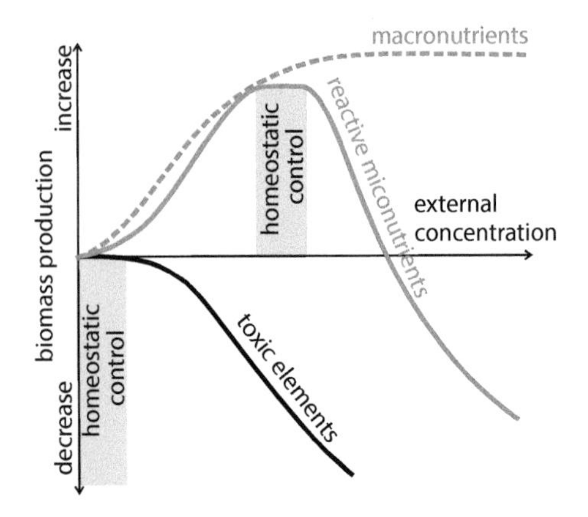

Figure 12.2. The effects of inorganic toxins on biomass production. In general, increasing external concentrations of macronutrients produce growth maxima but not toxicity unless the concentrations are extremely high and have osmotic effects (green broken line). There is homeostatic control of reactive micronutrient concentrations in plants to maximize growth, but this can be overwhelmed at high concentrations, causing toxicity and reduced biomass production (green solid line). The influx of non-nutrient toxic elements in plants can be controlled at very low external concentrations, but increased external concentrations quickly cause toxicity and decreased growth (black line).

For example, with the exception of Na and I, humans rely on food chains in which plants are the ultimate conduit for mineral elements from the soil—even those that are not essential for plants themselves. Plants have evolved to discriminate against non-essential elements, but this discrimination is never complete, so inorganic toxins are taken up by plants and can adversely affect not only the plants but also the organisms that rely on them for food. Inorganic toxins can therefore significantly affect biomass quality and production in both unmanaged and managed ecosystems.

Plant ionomes in part reflect the elemental composition of the soils that the plants inhabit, but the same plant species growing on different soils can maintain a characteristic ionome. Molecular phylogenies are helping to clarify the contribution of fundamental phylogenetic patterns to plant ionomes. In general, for many elements monocotyledonous plants accumulate relatively low concentrations in shoots, whereas the shoots of other clades can be mineral-rich (Figure 12.3). These fundamental phylogenetic patterns provide the background variation upon which adaptations to allow tolerance and accumulation of toxins are superimposed, which produces significant biodiversity in plant responses to inorganic toxins. It is likely that the natural selective force of inorganic toxins has contributed to the evolution of fundamental differences in the ionomes found in angiosperms.

Human activity is significantly increasing the concentrations of inorganic toxins in the Earth's ecosystems

From at least the time when humans learned to make bronze from copper and tin, the development of human civilizations has been linked to the exploitation of metals. In the twenty-first century humans are as dependent on the extraction of metals from the Earth's crust as they have ever been. The extraction of metals, and their ultimate release into the environment, represent among the most significant alterations to biogeochemical cycles

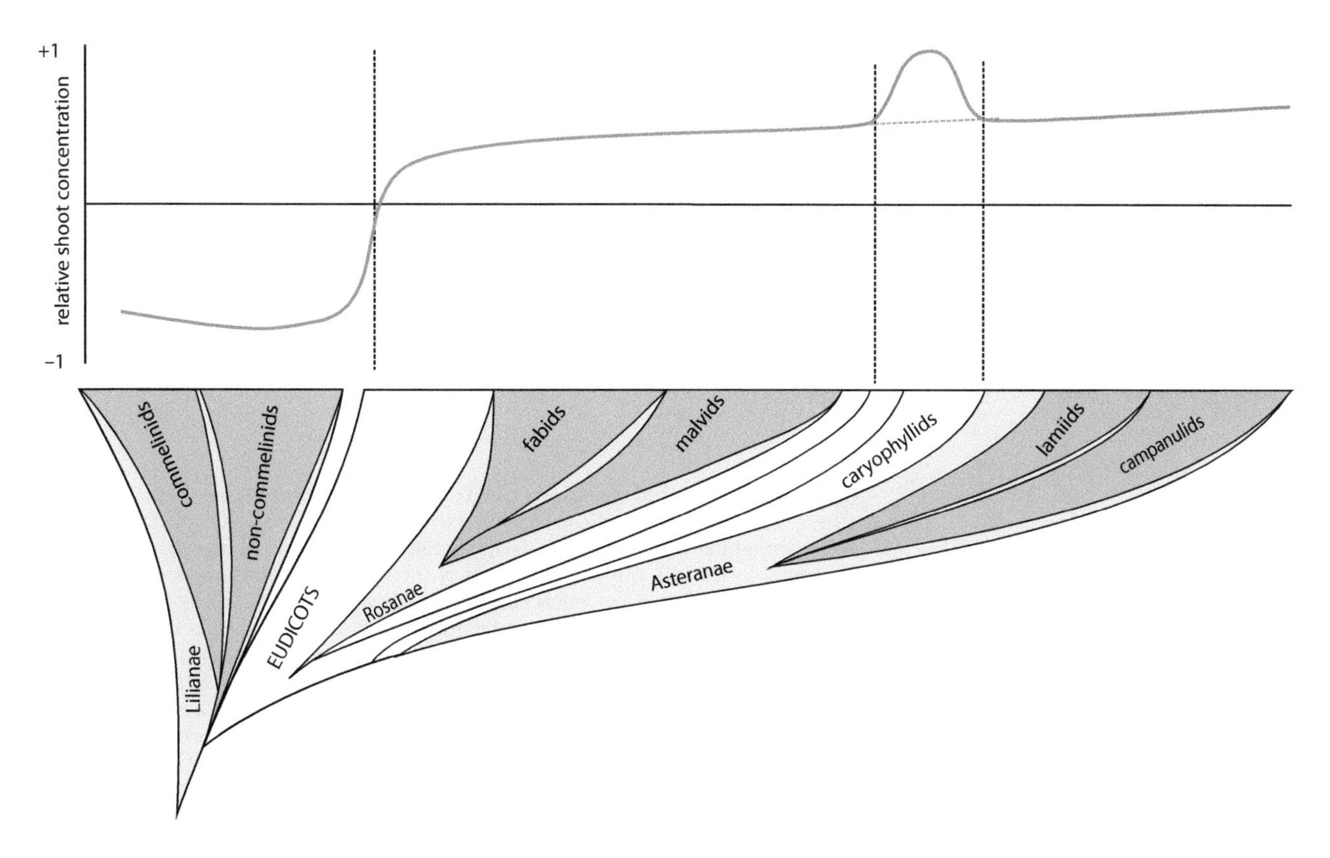

Figure 12.3. Phylogenetic effects on element concentrations in plants. Meta-analyses and experimental evidence suggest that there are phylogenetic constraints on the concentrations of elements, including many toxins, in plants. Detailed descriptions are not available for any elements, but current data suggest that the Lilianae, and the Poaceae and relatives in particular, have relatively low shoot concentrations of many elements compared with the Rosanae and Asteranae. For some elements, including Co, Ni, and Cs, the caryophyllids have particularly high concentrations.

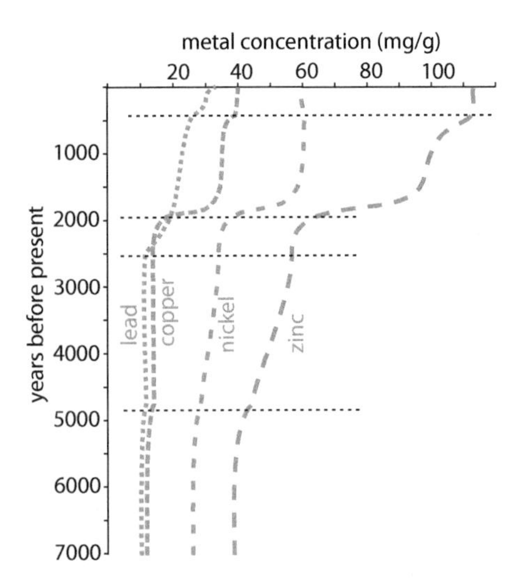

Figure 12.4. The historical mobilization of metals into the environment in China. Metal concentrations over a period of 7000 years are recorded in sediments of Liangzhi Lake in the Jianghan Plain of the middle reaches of the Yangtze River, Hubei, China. Initial increases in Pb, Cu, Ni, and Zn concentrations coincide with the Bronze Age civilization in this part of China, and further increases coincide with subsequent periods of mining and smelting activity. (From Lee CSL, Qi SH, Zhang G et al. [2008] *Environ Sci Technol* 42:4732–4738.)

caused by human activity. Each year, human activity is responsible for mobilization of at least as much of many metals and metalloids as are natural processes. This anthropogenic global mobilization of metals has contributed significantly to suggestions that the Earth is now in the "Anthropocene" epoch. Metal contamination from past civilizations is detectable in many datable lake and peat deposits. In Europe, peaks of lead, zinc, and cadmium from Roman times are easily detectable in many deposits—around 100,000 tonnes of lead were mined and processed each year when Roman civilization was at its height. Lake deposits in Central China provide a record not only of the ancient beginnings of human mobilization of metals, but also of the significant acceleration of metal release in the last few centuries (Figure 12.4). Estimates of the quantity of metals released each year show dramatic increases during the twentieth century in particular (Figure 12.5).

By the 1980s, the effects on the environment of the release of toxic metals were being regarded as unsustainable because they probably amounted to a slow poisoning of the biosphere. For example, in 1973, Japan was the largest smelter of zinc in the world, and mining for copper and zinc had caused copper, zinc, and cadmium contamination of the environment for most of the previous century. By the 1970s this had rendered perhaps 10% of Japan's rice paddies unfit for rice production. The smelting of metals in numerous others parts of the world was often associated with similar adverse effects. Legislation to prevent emissions has now been largely successful across the developed world, and culminated with controls on one of the most significant inputs of metals to the environment, namely lead in fuels. By the early twenty-first century, more than 3,000,000 tonnes of lead were being produced from ore each year, but uncontrolled emissions to the environment were declining significantly compared with levels in the 1980s, in part because of the recovery of at least as much lead from waste. In the annual reports of the Blacksmith Institute and Green Cross International on Earth's most polluted sites, the majority of these sites involve metals. By 2010 there was also significant evidence that total global emissions of inorganic toxins to the environment were increasing again, especially in the rapidly growing economies of China, India, and Brazil. Some estimates suggest that human activity has degraded 2 billion hectares of land (that is, an area greater than that used for agriculture), a significant proportion of this being due to contamination with toxins. In particular, post-industrial and "brownfield" sites are frequently contaminated with inorganic toxins. These toxins contribute significantly to the pollution threat to the lives of 200 million people, and to the pollution-related deaths of several million people each year.

The release of inorganic toxins into the environment is often associated with emissions from mining and industry, but other human activities also increase their availability in soils. For example, the reduced conditions of paddy fields mobilize significant quantities of Zn and Mn, phosphate fertilizers

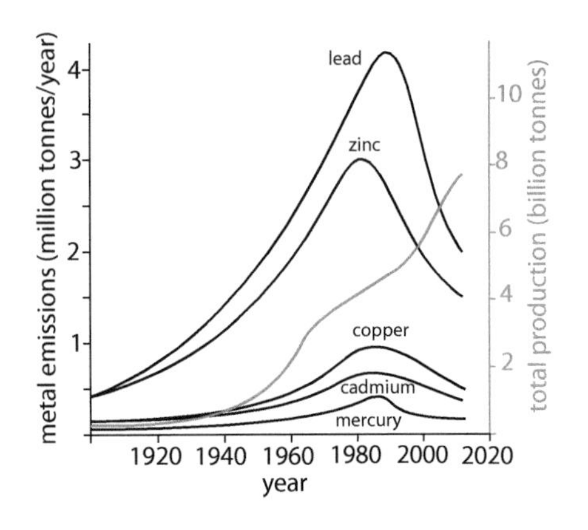

Figure 12.5. Global emissions of toxic metals. In the late twentieth century there was the most significant increase in toxic metal emissions into the environment in human history. Environmental legislation decreased emissions overall (black lines), but in many parts of the world emissions are accelerating again, and are above the levels considered to produce adverse effects on human and environmental health. The extraction of ores and minerals continues to accelerate (green line). Given that emissions overall are decreasing, this indicates that there are increasing amounts of toxic metals in use and in wastes that have yet to be dealt with. (Data from Nriagu JO [1996] *Science* 272:223–234, and Harmens H, Norris D, Mills G, and the participants of the moss survey [2013]. Heavy metals and nitrogen in mosses: spatial patterns in 2010/2011 and long-term temporal trends in Europe. ICP Vegetation Programme Coordination Centre, Centre for Ecology and Hydrology, Bangor, UK, 63 pp.)

routinely added to most crops can contain significant amounts of Cd impurities, and many of the substances that are added to soils, including many manures and sewage sludges, can contain inorganic toxins. The importance of human impacts on the subsurface for the mobilization of inorganic toxins has also become apparent in recent decades. Around 150 million people worldwide have to consume drinking water with an arsenic concentration above the maximum safe limit recommended by World Health Organization (WHO) guidelines, in most cases because it is mobilized from a contaminated **aquifer** (Box 12.1). In the western USA, the accumulation of selenium

BOX 12.1. THE WORST MASS POISONING IN HISTORY?

The Bengal Delta, formed by the confluence of the Ganges and Brahmaputra Rivers, is one of the largest, most fertile, and most densely populated deltas in the world. It has been a major food-producing region for centuries. Rice, the staple food crop, and jute, the primary cash crop, are both mostly grown in flooded soils. The low-lying topography and massive influx of water during the monsoon season aid these flood-requiring crops, but have also contributed to high levels of waterborne disease. The delta is still very active, and has mostly built up during the **Holocene**. Its sediments store huge quantities of recently trapped water in aquifers under anaerobic conditions. Wells bored into these aquifers provide clean, disease-free water. From the 1970s there was a massive increase in the number of wells that drew water from the aquifers, resulting from internationally funded programs and local availability of the required technology. These wells contributed significantly to reductions in disease, but by the early 1990s it was clear that water drawn from them often contained high concentrations of arsenic (As). The drawing of water from the aquifers has probably contributed significantly to the mobilization of As via increases in oxidation of As.

The WHO's recommended upper limit of 0.01 mg/l As has been much discussed, but in large areas of the Bengal Delta the As concentrations in water exceed this limit by so much (Figure 1), and the symptoms of As poisoning are so clear, that effects on human health on a massive scale are indisputable. About 100 million people live in As-affected areas, with about 30 million people being affected by As poisoning in Bangladesh and 6 million in the Indian state of West Bengal. In most areas, As ingested in food is at least as significant a problem as As consumed in drinking water. In the areas affected by As, up to 70% of calories in the diet can come from rice grown in flooded conditions. The plough pans formed during the puddling of paddy fields mean that contaminants in water can build up in the upper layers of soil because there is limited soil flushing. In the anaerobic conditions of flooded soil, As occurs primarily as arsenite (H_3AsO_3), which can be taken up by plants through NIP aquaporins that transport H_4SiO_4. The Poaceae, of which rice is a member, take up particularly high concentrations of Si, and rice takes up the highest concentration of any cereal. Paddy rice can therefore become contaminated with As at high concentrations and contribute significantly to As poisoning. In plants, As(III) compounds can be converted into As(V) compounds and methyl arsenic acids. Both As(II) and As(V) compounds are more toxic than methyl arsenic acids.

Efforts to minimize As contamination of food include numerous agronomic and agricultural practices. There are significant differences in As uptake between rice varieties, and several varieties with low As uptake have been bred with changes to their transporters, In addition, the formation of Fe plaques on roots, and the use of phosphate to minimize uptake are now part of a battery of techniques available. These are useful not only in the Bengal Delta, but also in the many other areas where mining or industrial activities have contaminated the soil with As.

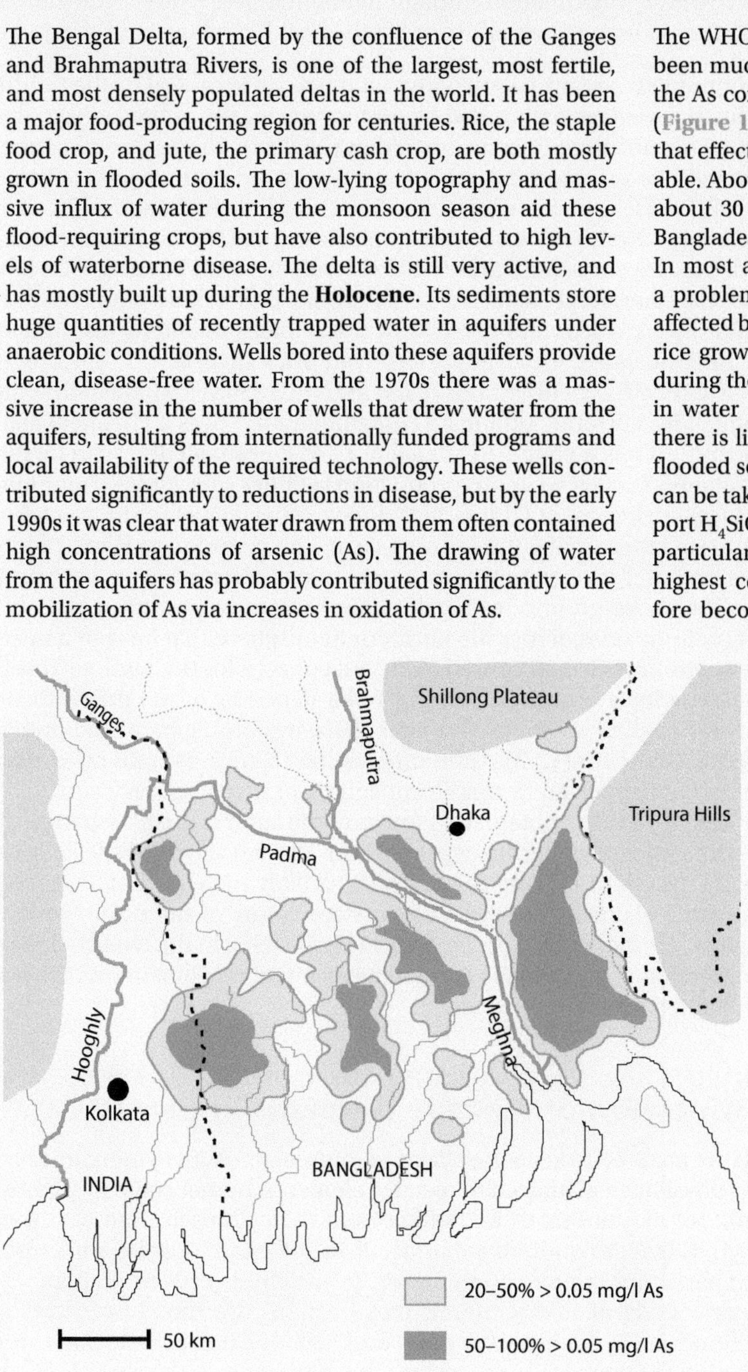

Figure 1. The Bengal Delta. The percentage of water samples in which arsenic (As) concentrations exceed 0.05 mg/l in contaminated regions.

by some plants has caused cases of livestock poisoning since the early twentieth century, but by the late twentieth century it was apparent that drainage water, especially from irrigated agriculture that used water from aquifers, contained high concentrations of mobilized selenium. The WHO estimates that each year there are several million deaths which are directly attributable to exposure to chemicals, a proportion of which are metals and metalloids. Most estimates indicate that the combined toxic effect of metals and metalloids is greater than that of organic toxins and radionuclides. By the late twentieth century it was being suggested that elevated concentrations of mobile metals in the environment were exposing around 1 billion people to sublethal doses of inorganic toxins, with unknown effects—and this number may increase further in the first half of the twenty-first century. Effects on human health are predominantly caused by lead, cadmium, mercury, and arsenic, with contamination through food often being very significant, but many other inorganic toxins also affect the health of plants and ecosystems.

Environmental contamination with radionuclides is caused by the products of industrial nuclear fission and by anthropogenically enhanced activity of naturally occurring radioactive materials (NORM). The causes of contamination with fission products include accidents, nuclear weapons development, and nuclear power generation. The East Urals Radioactive Trace produced by the "Kyshtym" waste-tank explosion at the Mayak complex in 1957 includes an area of several thousand km^2 with elevated ^{90}Sr and ^{137}Cs levels, as does the Chernobyl Exclusion Zone. At Fukushima there is an area of several hundred km^2 contaminated with ^{137}Cs. These incidents also released other radionuclides, especially short-lived ones such as ^{131}I, but the total activity and half-lives of ^{90}Sr (29 years) and ^{137}Cs (30 years) mean that these radionuclides currently dominate contamination. ^{90}Sr is a chemical analog of Ca, and ^{137}Cs is a chemical analog of K, so if available they can be taken up by plants. Nuclear weapons production facilities sometimes contaminate their local environment with radioisotopes, especially of uranium and plutonium, so, for example, the ongoing clean-up operation at Hanford in the USA is one of the largest environmental remediation projects ever undertaken. The above-ground testing of nuclear weapons produced global fall-out of radionuclides across the northern hemisphere that has left a trace in many environments and caused sufficient concern for the Limited Test Ban Treaty, banning atmospheric testing, to be signed by most nations in 1963. At several nuclear weapons test sites there are problems of radionuclide contamination (Box 12.2). Uranium-mining tailings containing U-decay-series radionuclides are a significant source of increased levels of NORM, but one of the most problematic sources of the latter is the phosphogypsum waste from phosphate fertilizer production. Human activity, especially from the early twentieth century onwards, has greatly increased the availability of inorganic toxins in the soil–plant system. Some of the most significant challenges to unmanaged and managed ecosystems in the twenty-first century necessitate being able to predict the resultant movements of toxins and understand their consequences.

Homeostatic mechanisms control the uptake and translocation of reactive elements in plants

Plants are adapted to keep the effects of routine exposure to inorganic toxins within physiological limits. For reactive elements, homeostatic mechanisms control not only influx to and efflux from the cytoplasm, but also transit through it. Under conditions that occur in soils and in cells, most metals, and in particular the transition metals, form soluble cations. In plant cells, the major sinks of micronutrient metal cations are the chloroplasts and mitochondria (in shoots, 80% of Fe and 50% of Cu can be located in the chloroplasts), whereas the major repository for toxins is the vacuole. Transit mechanisms that carry potentially toxic ions in unreactive states through the cytoplasm to organelles are therefore vital to homeostasis. In plant cells, the influx of divalent (for example, Zn^{2+}, Mn^{2+}, Ni^{2+}, Co^{2+}, Cd^{2+}, Pb^{2+}) and monovalent (for example, Cu^{+}) metal cations occurs down electrochemical

BOX 12.2. FALLOUT FROM THE NUCLEAR AGE

By 1990, almost 18% of the world's electricity was being generated in nuclear power stations (Figure 1), and this figure has subsequently declined to about 15%. In 2010, France produced over 70% of its electricity from nuclear power, and about 20 countries used it to produce 10–50% of their electricity. By 2012, there were about 430 nuclear power stations, 240 nuclear research reactors, and about 50 reactors that were decommissioning. The first generation of nuclear power reactors were developed from those designed to make plutonium for nuclear weapons, but almost all reactors are now of a much more modern design. Despite only one approval for a nuclear-waste repository (in Sweden), and some high-profile accidents, the increasing demand for energy in the twenty-first century is such that nuclear power generation looks likely to be a significant source of electricity in the years to come.

The detonation of the first nuclear weapons at Hiroshima and Nagasaki in 1945 was followed in the 1950s by rapid development of nuclear weapons, which were mostly tested above ground. Above-ground tests at the Nevada Test Site (by the USA), Maralinga/Christmas Island (by the UK), Reggane (by France), and Semipalatinsk (by the former Soviet Union) culminated in the H-bomb explosions at Bikini (USA) and Novaya Zemlya (former Soviet Union) with 15 Mt and 50 Mt equivalents, respectively. The Hiroshima and Nagasaki blasts had been about 15 kt and 20 kt, respectively. These above-ground nuclear tests significantly contaminated the atmospheric circulation of the northern hemisphere—in fact much of what was then known about atmospheric circulation derived from studies of the contamination. ^{90}Sr and ^{137}Cs were the most significant and biologically available components of fallout. Concern about them contaminating food chains around the world through fallout resulted in the Limited Test Ban Treaty, signed by most nuclear nations in 1963, which drove nuclear testing underground. Dealing with radioactive contamination at some test sites is an ongoing issue. In some instances, ground contamination was minor, as most of the radioactivity was dispersed in the atmosphere. In others, contaminated soil was buried, and at Bikini potassium was used to stop plants taking up ^{137}Cs. Much of the fallout from nuclear weapons testing has now decayed, so the test sites and weapons production facilities have the major legacy of contaminated land. Its area is relatively small, but capping or burial *in situ* is the primary means of dealing with it, as decontamination is too difficult.

The three major accidents of the nuclear age, at Kyshtym, Chernobyl, and Fukushima, each resulted in widespread contamination and exclusion zones in which habitation and agriculture were discouraged. At present there is no real solution to the problem of radioactively contaminated land in these exclusion zones—most crops would become contaminated, alternative uses of the land are hampered by the need for personnel, and there is no method of decontaminating such large volumes of soil. At nuclear power station sites that are being decommissioned there can also be small volumes of radioactively contaminated soil. There is thus a significant legacy of contaminated soil from the nuclear age, and much of the environmental hazard that it presents is due to the fact that it results in contamination of plants that grow on it.

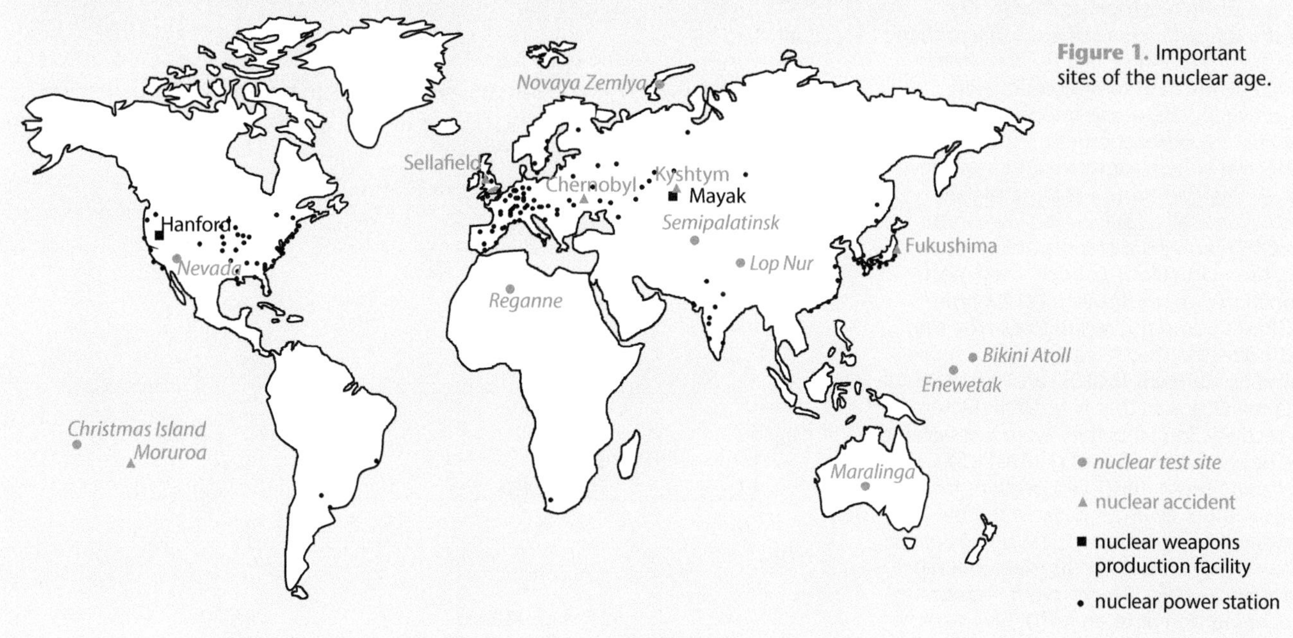

Figure 1. Important sites of the nuclear age.

gradients through a wide variety of transporters. Some specific transporter families dominate uptake for particular metals, and most non-essential metals are inadvertently transported as nutrient analogs, but generally at reduced rates. Members of the zinc/iron permease (ZIP) family of metal transporters are particularly important for influx of the micronutrients Fe^{2+}, Zn^{2+}, and Mn^{2+}, but Cd^{2+} also enters plants through these transporters. They have 8 membrane-spanning domains and short external N- and C-terminal domains. Natural-resistance-associated macrophage protein (NRAMP) transporters are metal-H^+ antiporters in which divalent metal uptake is coupled to H^+ efflux. NRAMPs have been shown to transport Fe^{2+}, Zn^{2+}, Mn^{2+}, and Mo^{2+}. Numerous ZIP and NRAMP transporters are expressed in particular plant parts and are specific for divalent metals. The Cu transporter (COPT) mediates Cu^+ uptake in plants (Figure 12.6).

P_{1B}-type ATPases and cation diffusion facilitator (CDF) proteins mediate efflux of divalent metals, including Zn, Fe, Cu, Mn, Cd, Pb, Co, and Ni, from the cytoplasm (Figure 12.6). Efflux to the xylem is vital for translocation of many metals from root to shoot, whereas efflux from the cytoplasm to the vacuole or organelles is vital for metal homeostasis. Efflux of divalent metals occurs against electrochemical gradients and uses energy directly from ATP (P_{1B}-type ATPases) or from cation influx (CDFs). The plant P_{1B}-type ATPases that transport divalent metals, mostly denoted in metal-tolerant plants as heavy-metal ATPases (HMAs), are involved in numerous metal homeostatic processes and in the distribution of contaminants such as Cd^{2+}, Pd^{2+}, and Hg^{2+}. They have not only an ATP-binding site but also a cytoplasmic heavy-metal-binding domain. In bacteria, a loading platform at the center of the protein is associated with metal discrimination and can allow specific transport of, for example, Cu^+. In plants, HMAs have frequently been shown to be involved in xylem loading. P_{1B}-type ATPases are central to metal

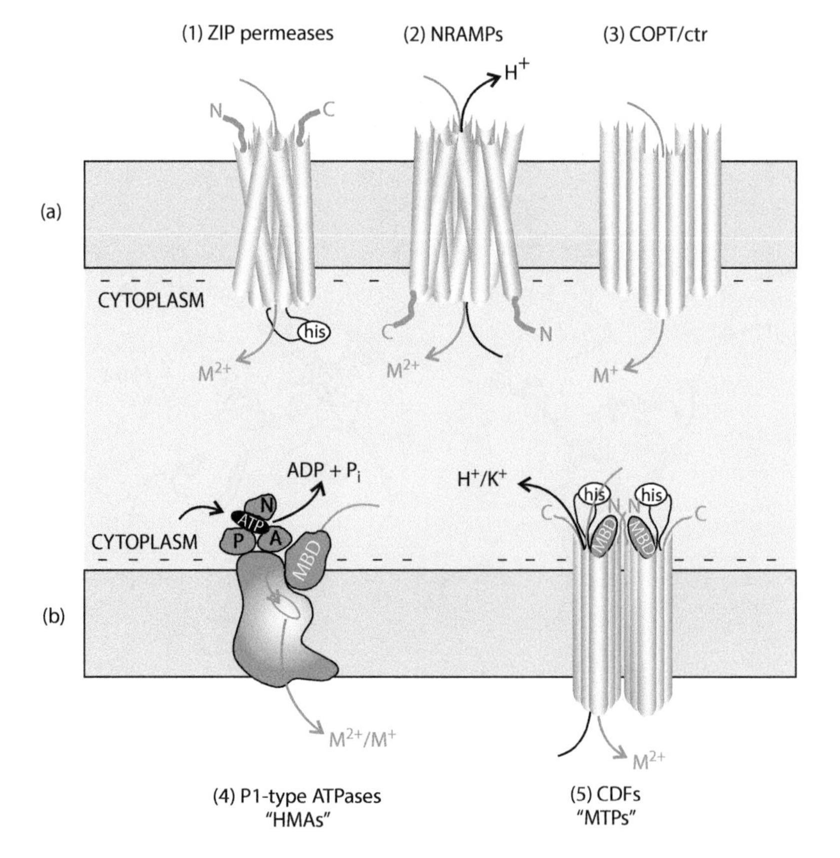

Figure 12.6. Important metal-transporter families in plants. The transporters described here evolved primarily for metal homeostasis, although some have unusual expression patterns in metal-tolerant plants. (a) Some of them enable movement down electrochemical gradients, whereas (b) others actively pump metals out of the cytoplasm. There are many variants of these transporters, and several other transporters known to have a role in metal transport in plants. (1) Zinc/iron permease (ZIP) proteins have 8 membrane-spanning domains and a His-rich cytoplasmic loop. (2) Natural-resistance-associated macrophage proteins (NRAMPs), first isolated from bacteria resistant to antibiotics with reactive metals, act as antiporters with H^+. Numerous other channels, especially non-selective voltage-sensitive and insensitive ones, can also transport toxic metals into plant cells. (3) Copper transporter (COPT) proteins transport univalent metals, particularly copper. They mostly function as homotrimers. (4) P1-type ATPases pump metals using energy from ATP directly. They have complex metal-binding domains (MBDs) and pass metals via binding sites that help to regulate selectivity. In plants they were first isolated as heavy-metal ATPases (HMAs). (5) Cation diffusion facilitator (CDF) proteins facilitate metal efflux by antiport powered by gradients of monovalent cations. In plants they were first isolated as metal tolerance proteins (MTPs). They have 6 membrane-spanning domains, an MBD, and work as homodimers.

homeostasis across the domains of life, and impairment of their function is associated with metal toxicity or deficiency in many organisms, including humans. CDF transporters have 6 membrane-spanning domains, cytosolic N- and C-terminal domains, and a histidine-rich cytoplasmic loop. They are antiporters that couple metal efflux primarily to H^+ influx, and probably function as homodimers. In plants they were first denoted as metal tolerance proteins (MTPs). MTPs transport Zn^{2+}, Cd^{2+}, Ni^{2+}, Mn^{2+}, Co^{2+}, and Pb^{2+} from the cytoplasm, mainly into vacuoles. The differential expression and discrimination of the many efflux proteins for metals in plants have a central role not only in metal homeostasis but also in determining when toxicity occurs from particular inorganic ions, both essential and non-essential. However, there are some other proteins with metal efflux capability (for example, members of the CAX, ABC, and MRP families of transporters) that can also contribute to metal homeostasis and that can have a role in toxicity and tolerance in plants.

The non-metal Se, and the metalloids As and Sb, can all cause toxicity in plants. Under most soil and cell conditions they form a series of oxyanions and are taken up by plants as neutral or anionic molecules. In the environment, As can occur as arsenite (As(III)) oxyanions, especially H_3AsO_3, which enters plants via silicic acid pathways. Si, in those plants for which it is beneficial, is taken up as silicic acid (H_4SiO_4) primarily through NIP IIIs. NIPs are a group of aquaporins unique to plants, which catalyze the diffusion of small neutral molecules through membranes. Rice can take up Si to a higher concentration (up to 10%) than many macronutrients, and when grown in paddy production systems that can have reduced soils containing As(III) ions, can produce grain contaminated with As via NIPs. Antimony, although much less often found at toxic concentrations, also enters plants through aquaporins as H_3SbO_3. In oxidized conditions, As(V) ions, primarily $H_2AsO_4^{2-}$, enter plants through phosphate transporters, in particular PHT and PHO transporters, and Se(VI), particularly in the form of SeO_4^{2-}, can enter plants through SO_4^{2-} transporters.

The transport of metal ions through the cytoplasm is often carried out by proteins that act as **metallochaperones**. At least one-third of all proteins are **metalloproteins**, and metallochaperones play a role in all organisms in delivering metal ions to them by holding the ions in an unreactive state during transit through the cytoplasm. In plant genomes there are numerous metallochaperone genes, which often encode heavy-metal-associated **isoprenylated plant proteins** (HIPPs), a family of proteins unique to plants. **Metallothioneins** can act as metallochaperones, but are also used to control toxic ion concentrations in many cells, including those of plants. In most cells there are also molecules that routinely detoxify or chelate reactive metals. Glutathione and amino/carboxylic acids, respectively, carry out these functions in plant cells. Although some proteins and secondary metabolites may have a role in transport in vascular tissue, for most metals the commonest ligands are organic acids. The expression of components of metal homeostatic systems is signaled in particular by ROS and gene expression subject to both translational and post-translational control, including miRNA and epigenetic effects. In general, these homeostatic mechanisms function at low external concentrations of inorganic ions, and maintain extremely low concentrations of free ions in cells—some calculations suggest that there are just one or two free ions per cell. Toxicity occurs when they are overwhelmed by unusually high concentrations of potentially toxic nutrient or non-nutrient ions, and the concentrations of free ions increase (Figure 12.7). For many toxic metals there are well-established hormetic effects—that is, a very low level of exposure increases growth, probably because it activates many of the molecules necessary for defense against biotic stresses. Some plants with little homeostatic control are hypersensitive to inorganic toxins, but

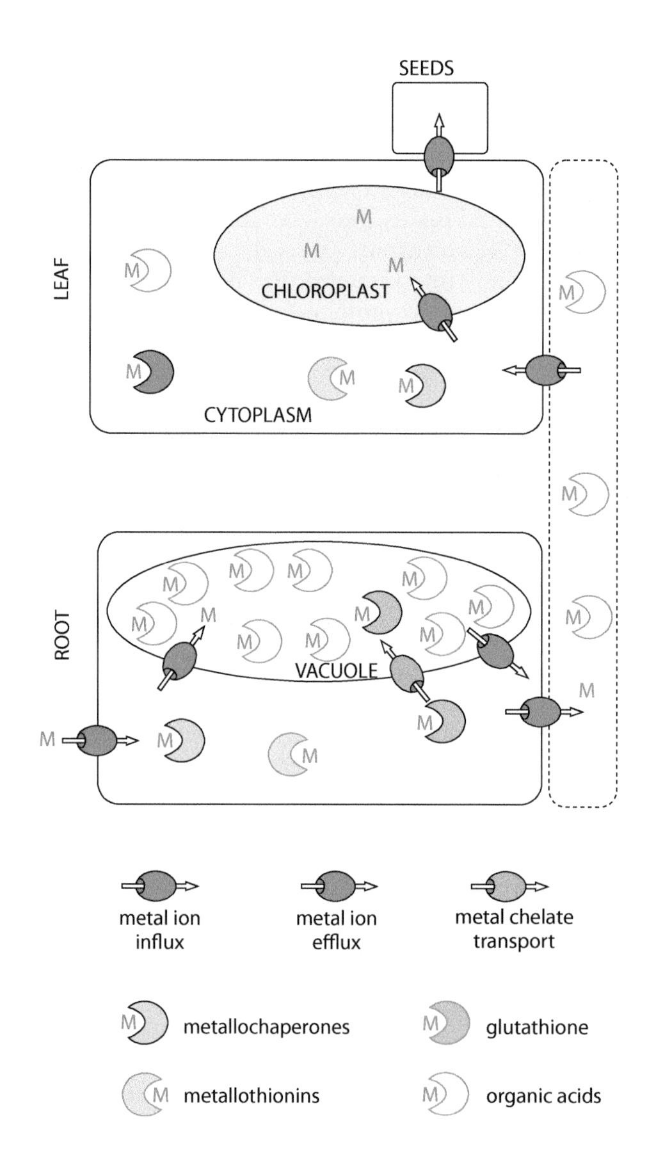

Figure 12.7. Mechanisms of metal homeostasis in plants. Plants control influx and efflux of metals to and from the cytoplasm to minimize free concentrations, but metals must reach the regions where they are needed, especially chloroplasts in shoots, by moving through the cytoplasm. Metallochaperones and metallothioneins transport metals through the cytoplasm in unreactive forms. Organic acids are used to store metals in vacuoles or to transport them through the vascular system in unreactive forms. Glutathione plays a role in directly detoxifying reactive metals, and also in transporting them into vacuoles.

homeostatic mechanisms have also provided the basis for the adaptations that enable some plants to inhabit areas that contain particularly high concentrations of inorganic toxins.

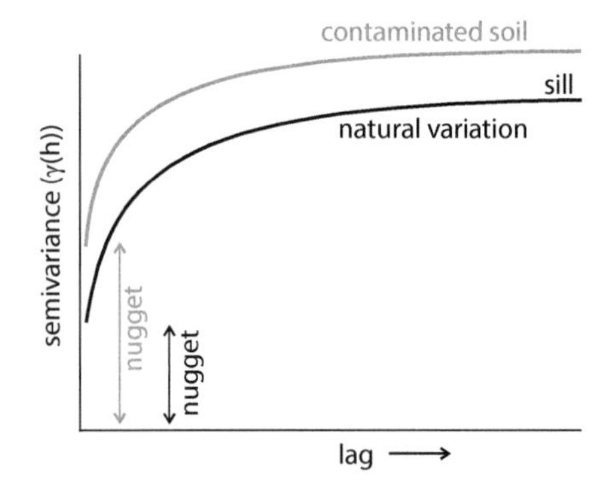

Figure 12.8. Semivariograms of metal concentrations in soils. The concentration of naturally occurring metals often varies significantly over short distances, producing a significant nugget in the semivariogram. Soil contamination can increase variation overall, and significantly increases the nugget of variation.

Exposure to inorganic toxins decreases growth and reproduction via physiological and genetic effects

In general, for micronutrients the difference between plant optimum concentration and average soil availability can be much greater than for macronutrients, so terrestrial plants have evolved to scavenge metal micronutrients from the soil. If environmental concentrations are unusually high, the scavenging capacity for essential metals overwhelms homeostatic mechanisms and renders plants susceptible to toxicity. In most soils, the inorganic elements that can be toxic have generally low but highly variable concentrations. On semivariograms for metals in natural soils, the nugget of variation can be 30% or more. Not only the root systems of individual plants but also even single roots are exposed to significant variation in the concentration of inorganic toxins. Analyses of variation in contaminated soils reveal large increases in short-range variation in toxin concentration (Figure 12.8). Natural long-range variation is associated with changes in soil chemical properties related to geology or hydrology, but is also exacerbated by contamination. For plants the toxicity threshold is generally considered to be a 10% reduction in growth, which occurs at much lower concentrations for non-essential elements than for essential ones. The soil concentration at which phytotoxicity occurs depends on soil properties, whereas the plant concentration at which it occurs depends on the plant species, and in both cases there is significant variation. Meta-analyses of the effects of toxins on plants suggest that biomass production is usually affected more quickly than reproduction. The central role of metal-dependent photosynthesis and respiration in the physiology of plants means that early symptoms of toxicity often relate to these processes. Usually there are also effects on stomatal conductivity, root morphology, membrane permeability, and protein synthesis, and in many cases chlorosis, anthocyanin levels, and senescence all increase. Detailed analyses of particular species have shown the many subtle effects of low concentrations of metals on vegetative and reproductive structures in plants (Figure 12.9). They have also revealed changes in the physiological, anatomical, and morphological properties of leaves, with a tendency to become slightly xeromorphic and to increase the production of certain secondary metabolites. In crop systems these effects can significantly reduce yields, and symptoms that can be used to diagnose specific toxicities in specific crops are well known.

The three most important mechanisms of inorganic toxicity in plants are ionic interference, oxidative stress, and genotypic damage. There are at least 1500 metalloenzymes in plants that require Zn^{2+}, several hundred that require Fe^{2+}, 40–50 that require Cu^{2+}, and several that require Mn^{2+}. Interference with homeostasis of these ions contributes significantly to phytotoxicity. For example, Zn^{2+} is found in the active sites of hydrolases, lyases, and superoxide dismutases (SODs), helps to control the activity of kinases and phosphatases, and helps to structure the zinc finger domains that regulate many protein–protein and protein–nucleic acid interactions. Zn^{2+} and Cd^{2+} are chemically similar, so Cd^{2+} in plants interferes significantly with Zn^{2+} homeostasis, adversely affecting physiology and thus growth and development. It also affects Fe^{2+}, Cu^{2+}, and Mn^{2+} homeostasis and hence ferredoxin, Fe-S clusters, cytochromes, plastocyanins, and the oxygen-evolving complex, all of which are essential to the light-dependent reactions

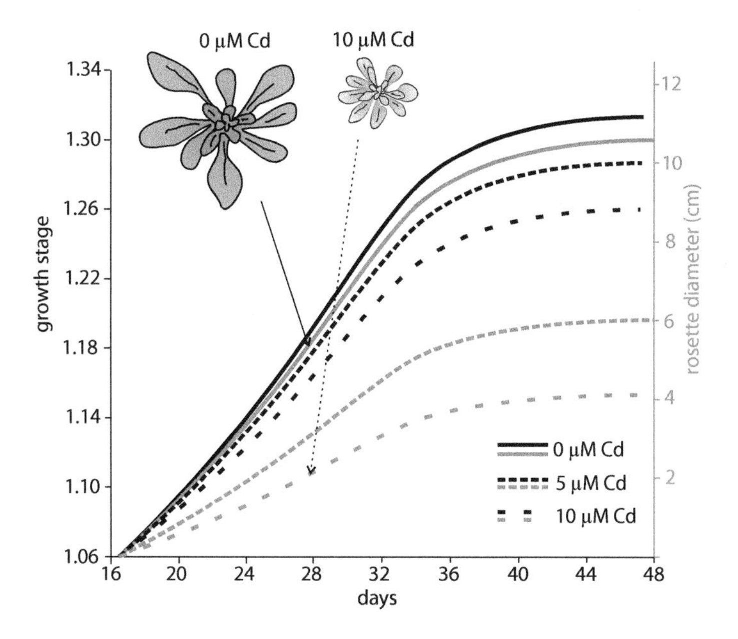

Figure 12.9. The effects of cadmium (Cd) on leaf morphology. For *Arabidopsis*, Boyes et al. (2001) defined growth stage 1 as leaf development. For the data shown in the figure, Cd exposure occurred when the plants had six leaves (growth stage 1.06), stages reached over the next 32 days are shown in black, and the rosette diameter is shown in green. For plants grown in sand with Hoagland's solution, 10 μM Cd clearly stunted development and caused chlorosis of the leaves. Plants are depicted from day 28. (Adapted from Keunena E, Truyensa S, Bruckers L et al [2011] *Plant Physiol Biochem* 49:1084–1091. With permission from Elsevier.)

of photosynthesis. Cd^{2+} also has a similar hydrated size to the redox-unreactive Ca^{2+}, so can interfere with Ca^{2+} signaling pathways and with cell wall biosynthesis, which is especially noticeable in roots. The full cascade of effects of the presence of Cd^{2+} in plants cells has yet to be described. However, in significant part because of interference with the numerous roles of Zn, effects on active auxin concentrations (and hence growth and lignification), the circadian clock, and the cellular pH-stat have been reported. Hg^{2+} toxicity occurs because of similar patterns of interference. Pb^{2+} interferes in particular with Ca^{2+}-based cellular signaling mechanisms. Arsenate in cells can interfere with PO_4^{2-} biochemistry, causing disruption of energy flows via effects on ATP formation. Se can replace S in cysteine and methionine, producing non-proteinogenic amino acids and hence phytotoxicity. In many instances, ionic interference by non-essential inorganic ions can be moderated by increasing the concentrations of their nutrient analogs.

All aerobic cells are susceptible to oxidative stress because, if the properties of the terminal sink for electrons in mitochondrial transport chains weaken, electrons can leave early at complexes I and III and form superoxide ions ($O_2^{\bullet-}$), most of which undergo **dismutation** to H_2O_2. In photosynthetic cells, the electron transport chain of the light reactions in chloroplasts is an additional and often even greater source of reactive oxygen species (ROS) if its electron sink properties weaken. Aerobic cells have evolved antioxidant systems that control cell redox potential if ROS are produced during respiration or photosynthesis. The enzymatic components of this system include SODs, catalases, and peroxidases, and in plant cells there are also significant non-enzymatic components, including glutathione (GSH), ascorbate, and α-tocopherol. Non-redox-active As, Cd, Co, Hg, Ni, and Pb can all, primarily via ionic interference, disrupt the components both of the electron transport chains (increasing ROS production) and of the antioxidant systems (decreasing the cell's ability to deal with ROS), presenting a significant oxidative challenge to plant cells. In addition, the presence of free ions of redox-active metals such as Cr, Cu, and Fe can produce ROS directly via Haber-Weiss and Fenton reactions (Figure 12.10).

If the cell loses redox control, the effects of excess ROS include oxidation and degradation of lipids, changes to redox-sensitive proteins (which, for example, can affect mRNA translation), and protein aggregation. All of these effects, if sustained, can lead to apoptosis and cell death. In plant cells, GSH is the hub of antioxidant capacity (Figure 12.11). It can directly reduce H_2O_2 to produce H_2O. Two oxidized GSH molecules react to produce GSSH (reduced glutathione), and NADPH-dependent glutathione reductase (GR) then regenerates GSH. Generally in plant cells there is high constitutive activity of GR, which keeps 90% of the GSH/GSSH as reduced GSH and maintains a reducing redox potential. Toxic concentrations of inorganic ions in cells are frequently associated with a rapid decline in GSH and increased oxidative challenge to the cell. This is because GSH is used both to directly

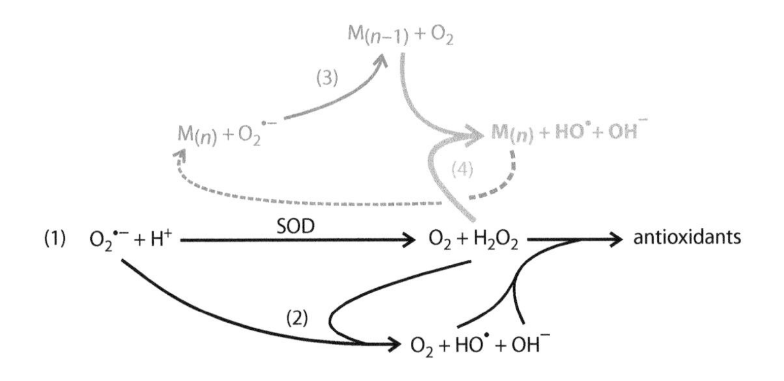

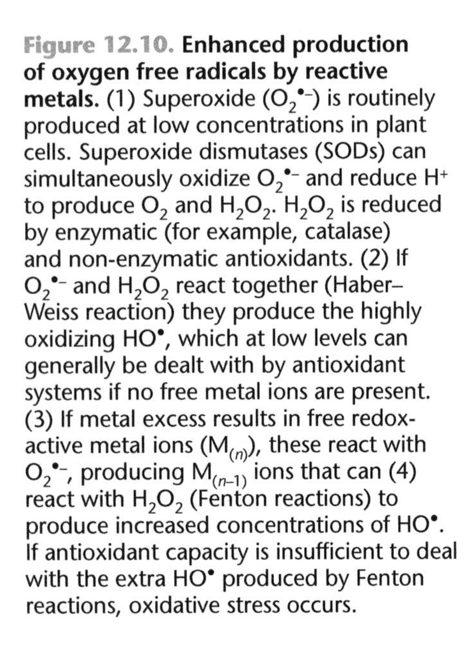

Figure 12.10. **Enhanced production of oxygen free radicals by reactive metals.** (1) Superoxide ($O_2^{\bullet-}$) is routinely produced at low concentrations in plant cells. Superoxide dismutases (SODs) can simultaneously oxidize $O_2^{\bullet-}$ and reduce H^+ to produce O_2 and H_2O_2. H_2O_2 is reduced by enzymatic (for example, catalase) and non-enzymatic antioxidants. (2) If $O_2^{\bullet-}$ and H_2O_2 react together (Haber–Weiss reaction) they produce the highly oxidizing $HO^{\bullet}$, which at low levels can generally be dealt with by antioxidant systems if no free metal ions are present. (3) If metal excess results in free redox-active metal ions ($M_{(n)}$), these react with $O_2^{\bullet-}$, producing $M_{(n-1)}$ ions that can (4) react with H_2O_2 (Fenton reactions) to produce increased concentrations of $HO^{\bullet}$. If antioxidant capacity is insufficient to deal with the extra $HO^{\bullet}$ produced by Fenton reactions, oxidative stress occurs.

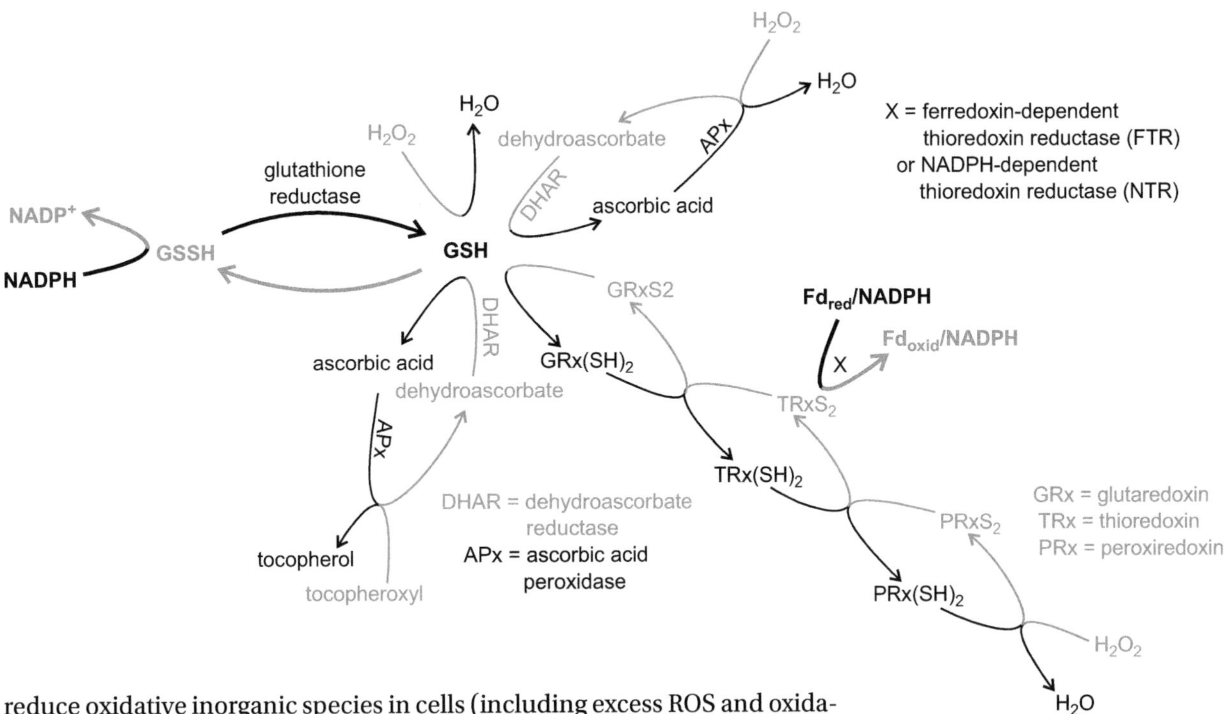

Figure 12.11. The primary antioxidant system in plant cells. Reducing power from NADPH, which is ultimately derived from photosynthesis, passed on via glutathione (GSH) is the nexus of antioxidant capacity in plant cells. Ascorbic acid is present at high concentrations in cells, and is particularly important for reducing H_2O_2. Tocopherol reduces lipid hydroperoxides, helping to minimize lipid peroxidation and producing the tocopheroxyl radical, which can be reduced back to tocopherol using ascorbic acid. Excess concentrations of inorganic toxins increase the production of oxidants, and ultimately H_2O_2, which in unadapted cells overwhelms the antioxidant system, resulting in oxidative stress. (Reduced species are shown in black, and oxidized species are shown in green.) GSSH= oxidized GSH.

reduce oxidative inorganic species in cells (including excess ROS and oxidative metal/metalloid species) and to reduce the non-enzymatic antioxidants ascorbate and α-tocopherol. Oxidative effects of essentially all inorganic toxins have been widely reported in plants. Lipid peroxidation and the activity of some antioxidants are sometimes used as biomarkers of toxicity. Ionizing radiation, by definition, produces ROS during radiolysis of water. In contrast to cellular effects at very high dose rates (Gy/h), at the dose rates found even in radioactively contaminated environments the levels of ROS produced by radiolysis are very low compared with the antioxidant capacities of plant cells, and the direct effects on cellular redox are negligible. However, uranium at the concentrations in which it is found in contaminated environments, such as mining spoil, does have some direct chemical effects on cellular redox.

Genotoxicity of inorganic toxins can be induced by the effects of ROS on DNA. The commonest genotoxic effect of ROS, probably in plants as well as in other organisms, is the oxidation of guanine to oxoguanine. This can alter G-C pairing to G-T. ROS from inorganic toxins have also been shown to induce single-stranded breaks in DNA. Some metals have been shown to bind to DNA in plants, but most do so more weakly than to other ligands in the cell, and the most significant genotoxic effects, especially for Cd and Cu, occur via disruption of the DNA repair machinery. DNA damage occurs routinely during the life of a cell, but it can generally be repaired. There are multiple mechanisms of DNA repair in plant cells (Figure 12.12), most of which depend on proteins with DNA-binding sites containing cysteine residues with bound Zn (for example, zinc finger groups). Inhibition of DNA repair leads to genome instability. Studies of animal cells have provided most evidence for the inhibition of DNA repair by Cd, but investigations in plant cells and for other metals indicate that similar mechanisms of toxicity are likely to occur.

In animal cells, excess Cd prevents the functioning of proteins involved in nucleotide excision repair (NER), and disrupts the expression of the well-known tumor suppressor gene *p53* that contributes to NER. Base excision repair (BER) is most commonly needed because of oxoguanine, with the commonest enzyme involved being oxoguanine-DNA glycosylase 1 (OGG1), the action of which is inhibited by Cd. In yeast and animal cells

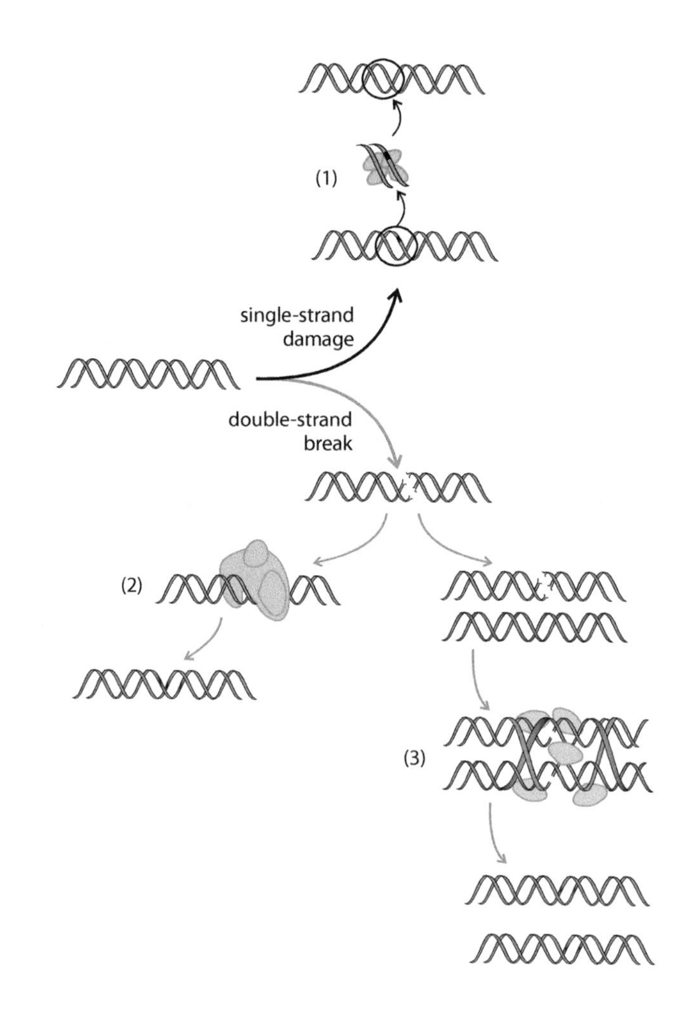

Figure 12.12. DNA repair mechanisms. (1) Inorganic toxins can cause damage—sometimes directly, but mostly via production of ROS—to one DNA strand. If this causes mismatches, these can be repaired during DNA replication by the process of "mismatch repair." If there is damage to or changes in bases of a single strand, which causes disruption to DNA structure, it can be repaired by excision and replacement of bases or nucleotides, using the complementary strand as a template. Base excision repair and nucleotide excision repair underpin routine maintenance of the genome, but can also occur during transcription of genes. (2) Inorganic toxins and especially ionizing radiation can cause double-stranded breaks in DNA, the broken ends of which can be simply joined back together by non-homologous end joining. This can result in an altered sequence if nucleotides are lost during breakage. (3) The repair of double-stranded breaks by homologous recombination uses the sequence of a chromosome's pair as a template for the remade sequence. The processes of DNA repair, which mainly rely on templates from other strands of DNA, probably help to explain why genetic information is stored in double-stranded DNA in pairs of chromosomes.

at least, Cd also inhibits mismatch repair (MMR) by interfering with initial protein binding to DNA. Double-stranded breaks in DNA are repaired by two separate protein complexes that accomplish either non-homologous end joining (NHEJ) or homologous recombination (HR). There is some evidence that metals interfere with NHEJ and HR, but their primary relevance to inorganic toxicity is their increased occurrence during exposure to ionizing radiation, a phenomenon that has been shown to occur not only in the laboratory but also at radioactively contaminated sites. In addition, a number of toxic metals have significant epigenetic effects, which have also been reported in plant cells. **Chromatin** structure affects gene expression—genes in condensed chromatin have decreased expression, whereas those in extended chromatin have increased expression. The state of chromatin is affected by DNA methylation patterns and by histone modifications, both of which are mitotically and meiotically heritable. In a range of cell types there is good evidence that Ni, Pb, As, Cd, Cu, and Cr have epigenetic effects on gene expression. Overall, at the low levels of exposure to inorganic toxins that occur in many environments, toxic effects are negated by high concentrations of nutrient ions, by the antioxidant systems of cells, and by DNA repair mechanisms. Significant exposure to toxins overwhelms these systems, resulting in increases in the rate of damage.

Amplified homeostatic mechanisms in the roots of some species produce a metal-tolerant physiology

The characteristic floras of naturally occurring mineral outcrops have long suggested that some plants are more tolerant of toxic metals than others.

By the late twentieth century, the evolution of metal-tolerant plant ecotypes that could colonize mining spoil containing high concentrations of toxic metals had become an exemplar of natural selection in action. In general, it was established that only those non-metallophyte species which have natural populations with a few tolerant individuals can colonize spoil heaps, that the degree of metal tolerance that has evolved in a population correlates with metal availability, that tolerance of different metals is often independent, and that tolerant × non-tolerant crosses indicate that tolerance is often controlled by relatively few genes. Studies also showed that there are concentrations of metals at which even tolerant plants cannot survive, and that exposure to multiple toxic metals is particularly challenging for plants. Metal-tolerant plants, especially the many grass species that have evolved tolerance, generally have a specific balance of influx, efflux, and transfer of metals. For example, although tolerant plants sometimes take up less metal in total than do non-tolerant ecotypes of the same species, in the majority of cases they take up as much metal as non-tolerant plants, but sequester it in the roots, mostly in the vacuoles. In some instances they have increased concentrations of metal-binding molecules, such GSH or phytochelatins (PCs). The enzymes of metal-tolerant plants are often just as susceptible to metal toxicity as those of non-tolerant plants. Interestingly, exposure to metals can result in genotypic diversity, probably via genotoxicity, so the evolution of metal tolerance has provided insights into phenotypic **plasticity** and **canalization**.

Patterns of uptake and allocation of inorganic toxins have now been linked to tolerance for many toxins and for many plant species, ranging from metallophyte species to crop cultivars. The role of transporters in mediating uptake and allocation has been confirmed many times by the manipulation of specific transporters. Tolerance of several metals can be increased in *Arabidopsis thaliana* by increased expression of HMA4 (located in the tonoplast), and differences in Cd tolerance of rice cultivars arise from differences in HMA3 activity (located in the plasma membrane in xylem parenchyma). Some studies have noted increased efflux of metals from the roots in tolerant plants, and increased efflux of organic ligands that bind toxic metal ions. In a range of species (for example, Mn-tolerant grape cultivars, Cd-tolerant rice cultivars, and As-tolerant *Rubus ulmifolius*), root-to-shoot transport of toxins is minimized, whereas in metal-tolerant coffee cultivars the toxins are directed to the shoots, or in *Evodiopanax innovans* to the bark. In general, metals are directed to less metabolically active tissues in the roots or leaves, such as older roots or epidermal cells, and even to senescing tissues that in deciduous plants can provide a route for voiding toxins. Other aspects of transport capacity can also differ in tolerant plants. For example, some As-tolerant plants have decreased expression of phosphate transporters that operate at low concentrations, and Cd tolerance in *Arabidopsis thaliana* is linked to the activity of an NRT1 nitrate transporter that retrieves NO_3^- from the xylem in order to increase root NO_3^- concentrations. Genomic and proteomic comparisons of tolerant and non-tolerant plants have emphasized the importance of transporter expression patterns to tolerance. However, the manipulation of transport proteins alone has not generally entirely replicated tolerance.

The production of GSH and PCs is often constitutively high, or expression is quickly increased, in metal-tolerant plants, especially for Cd and As. GR activity is also often enhanced. The tripeptide GSH is used to make PCs (Figure 12.13), so although its expression is increased, the levels of GSH can be lower in tolerant plants that make a large amount of PCs. Arsenate (As(V)) that enters plant cells is quickly reduced to arsenite (As(III)). Cd^{2+}, As(III), and Hg^{2+} are strongly attracted to thiols on GSH and especially on PCs. The stability constants for Cd with thiols increases in the order

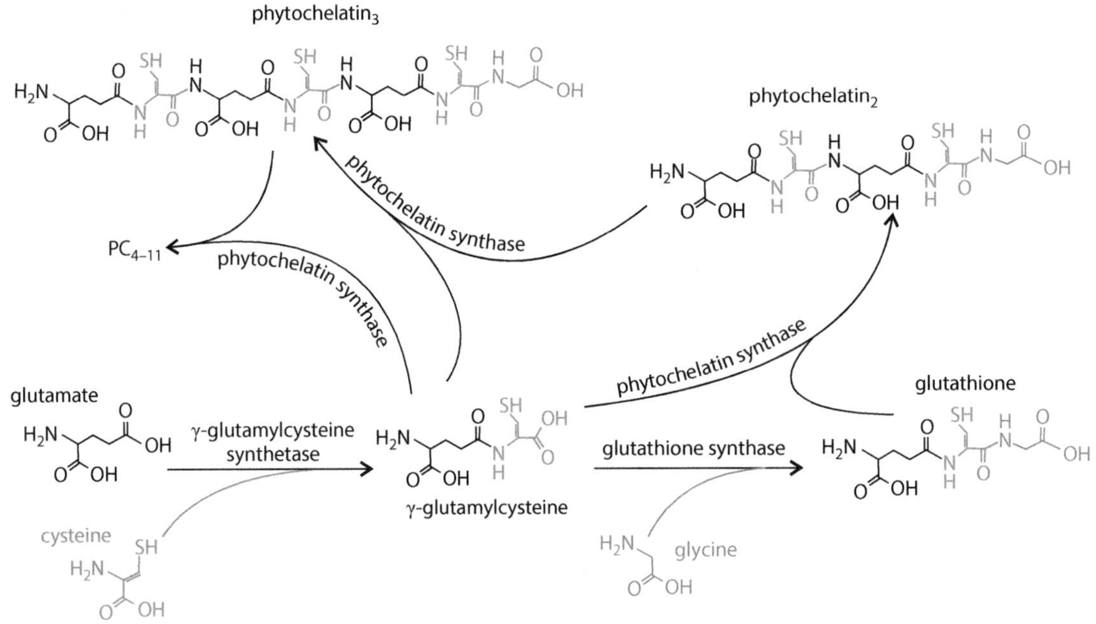

Figure 12.13. Glutathione and phytochelatin synthesis. Glutathione (GSH) is a tripeptide that is synthesized enzymatically in the cytoplasm. From GSH and γ-glutamylcysteine, phytochelatins (PCs) can also be synthesized enzymatically in the cytoplasm. PC_2 to PC_{11}, each with an additional γ-glutamylcysteine residue, have been reported, but PC_2, PC_3, PC_4, and PC_5 are most important in detoxification of inorganic toxins, especially Cd and As. For most divalent metals, four thiol (SH) groups bind each metal ion, so multiple GSH or PC molecules provide the ligands for detoxification. In some plants, amino acids other than glycine (for example, serine, alanine, glutamine) have been reported in PCs. Synthesis of the enzymes in the GSH/PC pathways is activated by the presence of inorganic toxins, and the activity of PC synthase is activated by metal binding to it.

$GSH < PC_2 < PC_3 = PC_4 = PC_5$, with As(III)-$PC_4$ and As(III)-PC_3 being the predominant As complexes. In some tolerant plants the majority of Cd and As has been shown to be bound to PCs, and ATP-dependent transporters that allow efflux of As/Cd–PC complexes into the vacuole have been described in yeast and plants. PCs occur not only in plants but also in some fungi and animals. It is likely that they evolved in plants as part of a metal homeostasis system, most probably for Zn. It is often suggested that their role in removing reactive elements other than Cd and As to the vacuole has been neglected, but with regard to other elements, such as Ni and Cu, they seem to play only a minor role. Tolerance of and sensitivity to Cd and As have been manipulated in a number of plant species by altering PC expression, which affects tolerance but does not fully replicate it.

Metal-tolerant plants often also have increased antioxidant capacity. In laboratory experiments and in the field, in plants of unmanaged ecosystems and in crop plants, increased activity of almost all of the components of plant antioxidant systems has been reported in metal-tolerant plants. This probably helps to explain why the engineering of transporter or GSH/PC activity in non-tolerant plants frequently does not achieve full tolerance—if cells contain increased levels of inorganic toxins, there are frequently also increased levels of oxidants that have to be neutralized. Antioxidant capacity is triggered and regulated via numerous mechanisms, many of which have been shown to be involved in metal tolerance. Most divalent metals have been shown to interfere with mitogen-activated protein (MAP) kinase cascades in certain species, and Cd in particular can affect calmodulin-regulated pathways, so in toxin-tolerant plants these pathways are probably robust to toxin exposure and help to trigger downstream responses.

Nitric oxide (NO) signaling clearly plays a role in metal tolerance in many plants, probably by helping to trigger antioxidant responses, and in some but not all plants the exogenous application of NO can increase metal tolerance. Many molecules that are involved in either classical signaling pathways (for example, CDPK and IP_3) or the generation of oxidative bursts (for example, NADPH oxidase) have also been implicated in metal tolerance. However, such molecules are not just involved in the signaling pathways and regulation of antioxidant capacity in metal-tolerant plants. Differences in physiology controlled by almost all plant hormones, including ethylene, auxin, ABA, jasmonic acid, and salicylic acid, have been reported between toxin-sensitive and toxin-tolerant plants. Numerous *cis*-elements and transcription factors, which are almost always also involved in responses mediated by these hormones, have a role in regulating the response to metals. Intriguingly, there are some organisms which tolerate toxic metals that are non-essential for all other organisms by incorporating them into their physiology. In the upper oceans, where Zn can be deficient but Cd is available, some diatoms can use Cd instead of Zn in carbonic anhydrases. A receptor-like protein kinase has been proposed as a sensor for extracellular divalent metals, but in general the details of the receptor and signaling pathways involved in responses to metals, and that might differ in metal-tolerant plants, have yet to be elucidated. Overall, in plants with metal-tolerant physiologies, various aspects of metal homeostasis are amplified so that, in susceptible cell compartments or organelles, toxic metal activities are maintained within physiological limits.

Some plants have the capacity to minimize the uptake of toxins from high external concentrations

Some plants have not only a physiology that can redistribute inorganic toxins, but also physiological, anatomical, and morphological mechanisms for excluding them. Plant roots exude a variety of compounds, but particularly carboxylic acids. Such exudates are used by plants to actively influence the availability of ions in the soil solution around the root. They have a long-established role not only in helping plants to solubilize some micronutrients when they are poorly available, but also in reducing exposure to Al under acid conditions. It has become increasingly evident that some plants can use these exudates to exclude inorganic toxins. The tolerance of Pb in rice varieties correlates with their exudation of oxalate, and exposure of tolerant sorghum and maize varieties to Cd correlates with increased exudation of malate and citrate. Ionic speciation modeling shows that the concentration of carboxylic acids in such exudates is sufficient to bind the majority of metal ions to which the roots are exposed. In some experiments, exogenous application of carboxylic acids at rates equivalent to tolerant-cultivar exudation confers tolerance on sensitive cultivars. The **isoflavonoids** that are frequently exuded from the roots of members of the Fabaceae have been shown to form complexes with a range of metals and with Se.

In some plant species there is significant movement of ions via the apoplast into the xylem in individuals not previously exposed to toxic ions, but a decrease in flow through the apoplast after exposure. These changes correlate with accelerated development of the Casparian strip resulting in a more complete barrier to apoplastic flow (Figure 12.14). In some plants a hypodermis, or even periderm (which is more usually associated with secondary growth), develops in response to exposure to metals. It is likely that, as with Na^+ uptake into salinity-tolerant plants, differences in root anatomy during exposure to metals can increase the proportion of available metal that is excluded. In experiments in which root systems are differentially exposed to metals, effects of exposure on root morphology have been observed. Where patches of toxic metals occur, it is often reported that the roots of many species proliferate in patches with low concentrations of toxins, and avoid patches with high concentrations. Such morphological adjustments can help to prevent uptake of toxic ions.

The most important mechanism for exclusion of toxic ions from plants is probably mycorrhizal association. Numerous microorganisms have mechanisms that increase or decrease the concentrations of free metals in their environment. In general, microorganisms, including fungi, are thought to be more tolerant of high concentrations of inorganic toxins than are plants.

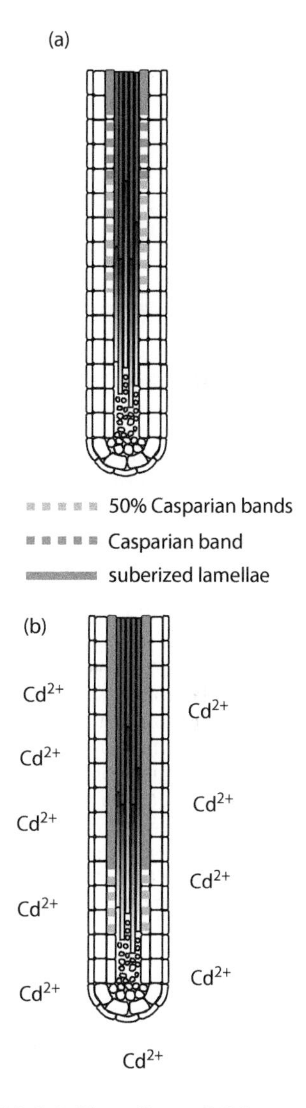

Figure 12.14. The effect of Cd on endodermal development. (a) In maize, the Casparian strip of the endodermis usually starts to develop several cm back from the root tip. It starts with the formation of Casparian bands and culminates in suberization of the cell wall lamellae. (b) In roots exposed to Cd, the Casparian strip develops much closer to the root tip, probably helping to minimize the apoplastic movement of Cd into the vascular tissue at the center of the root. (Data from Lux A, Martinka M, Vaculík M, White PJ [2011] *J Exp Bot* 62:21–37.)

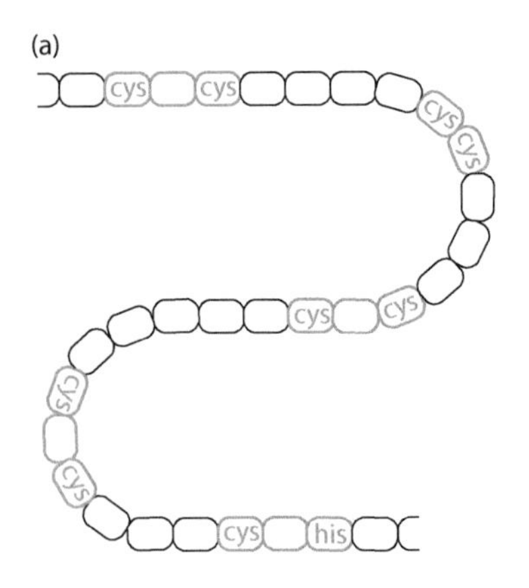

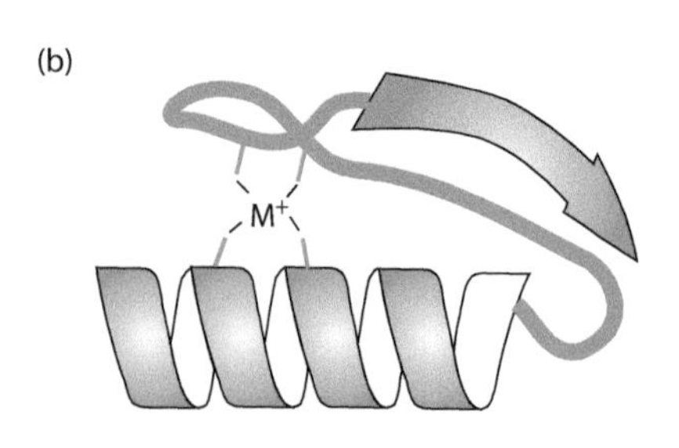

Figure 12.15. Metallothioneins bind metals in plants and mycorrhizal fungi. (a) Metallothioneins are low-molecular-weight polypeptides with repeat Cys-X-Cys/Cys-Cys and sometimes repeat Cys-X-His motifs. They have evolved independently several times, and different classes of metallothioneins have different molecular weights and structure. They help to regulate free concentrations of both micronutrient and toxic ions in cells. They are important in mycorrhizal fungi that contribute to metal tolerance in plants. (b) In all of the diverse metallothioneins, thiol groups on cysteines bind metal ions. (Data from Keunen E, Truyens S, Bruckers L et al. [2011] *Plant Physiol Biochem* 49:1084–1091.)

Fungi, including those that form mycorrhizal associations with plant roots, have mechanisms of metal homeostasis that share many features in common with those in plants, and that are accentuated in metal-tolerant fungi. At both naturally occurring and contaminated sites with high metal concentrations, numerous strains of metal-tolerant fungi have been reported. In an analogous situation to that in plants, particular fungi most often develop metal-tolerant strains. Metal-tolerant fungi have a significant capacity to chelate metals from the cytoplasm. In particular, metallothioneins (MTs) (Figure 12.15) have been implicated in decreasing the free concentration of Cu, Co, Cd, Zn, Ni, and Pb in fungi. MTs in fungi, like those in plants, are encoded in the genome, but are often shorter than plant MTs, with around 24 amino acids, while still containing characteristic Cys-Cys or Cys-X-Cys sequences. PCs also occur in many metal-tolerant fungi, and transporters for the flow of MT–metal and PC–metal complexes into fungal vacuoles have been isolated. In yeast, Zn concentrations in the vacuole of up to 100 mM have been reported, demonstrating the efficient compartmentalization of metals that can be achieved in fungi. Many organic acids, as well as amino acids, have been shown to chelate metals in vacuoles. In fungi, the cell wall has a significant adsorption capacity for metals, with chitin, glomalin, and other molecules binding significant quantities of metal. The capacity of the cell wall is finite, but this **biosorption** makes a contribution to tolerance at least in the initial phases of exposure. Many metal-tolerant fungi have been shown to have significant antioxidant capacity and even, for example, to exude SODs into the soil, probably to reduce metals and their availability.

Arbuscular mycorrhizas (AM) are beneficial to plant health in many species via uptake of P, micronutrients, and water, which can aid stress tolerance, but they can also specifically decrease the transfer of toxins to shoots. During As exposure in tolerant plant–fungal AM associations, As accumulates disproportionately in fungal components. This can occur via changes in direct P uptake by the plant and the fungus, and by decreased translocation of As to plant–fungus exchange structures. PCs expressed in AM fungi in response to a range of metals have been shown to reduce plant exposure to metals in several species. The zinc violets of Europe are subspecies of *Viola lutea* that tolerate high Zn levels in calamine soils. AM increase their tolerance of Zn, and some studies suggest that if the AM are absent, zinc violets are not Zn tolerant. Changes in S metabolism of AM fungi upon exposure to metals, which could affect both GSH and PC synthesis, have also been reported. In ectomycorrhizal (EcM) associations in many species, including important tree species, and in ericoid mycorrhizal (ERM) associations, similar mechanisms that decrease metal transfer to plant shoots in particular have been noted. These include the exudation of organic chelates and antioxidants into the soil solution.

Some plants can hyperaccumulate inorganic toxins in their shoots

Basal metal tolerance generally results in shoot:root ratios of metals that are significantly less than 1. In contrast, some plant species can hyperaccumulate metals or metalloids in their shoots, producing shoot:root ratios of 10–100 or even higher. Not only can these plants hyperaccumulate elements in their shoots, but also their shoot:root metal ratios are often 100–1000 times higher than those of other plants growing on the same soil, and their absolute shoot metal concentrations can be a few per cent of their dry weight. They are not only tolerant of internal toxin concentrations that would normally be lethal to other plants, but also often grow best on them. There are about 500 species of hyperaccumulators known, from a variety of plant families and able to hyperaccumulate a variety of metals and metalloids. This is a phenotype that has evolved independently numerous times for a variety of inorganic toxins (Figure 12.16), and it results in the highest concentrations of some metals in any biological system. The distribution of hyperaccumulators suggests that natural outcrops of **calamine, serpentine,** and **chalcocite** have driven the evolution of Zn/Cd/Pb, Ni/Mn, and Cu/Co hyperaccumulators, respectively. Similarly, As and Se hyperaccumulators are often associated with naturally occurring deposits of high concentrations of these elements.

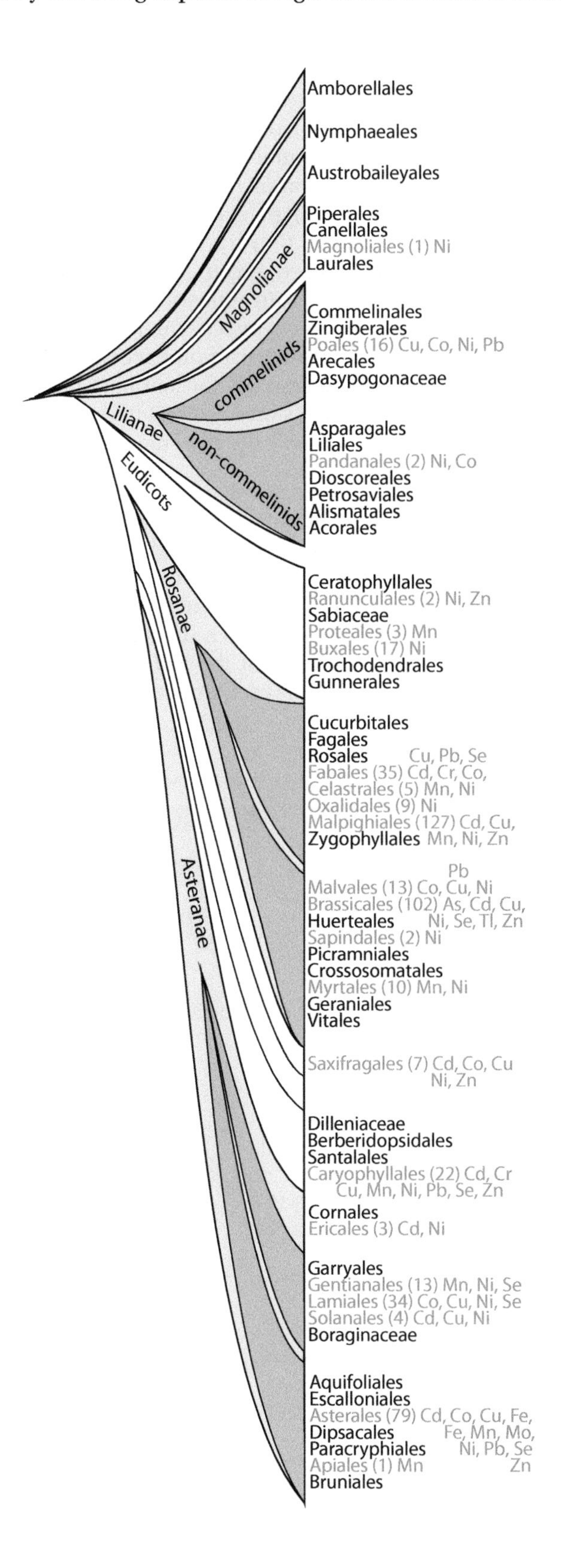

Figure 12.16. The phylogenetic distribution of hyperaccumulator species. Hyperaccumulation has evolved independently in different clades for different elements. Number of species is shown in parentheses, followed by the elements hyperaccumulated. (Data from Cappa JJ & Pilon-Smits H [2014] *Planta* 239:267–275.)

Several of the metal-tolerant plants that were used in geobotanical prospecting to locate mineral-rich ores (from perhaps the earliest mining activities until well into the twentieth century) are hyperaccumulators.

The molecular mechanisms of Zn/Cd/Pb hyperaccumulation are known in most detail because the Zn/Cd hyperaccumulators *Arabidopsis halleri* and *Noccaea caerulescens* are brassicas related to *A. thaliana*. The evolution of these Zn/Cd hyperaccumulator species is not explained by the evolution of new genes or even gene variants, and there are few significant anatomical or morphological differences between hyperaccumulators and the non-metal-tolerant *A. thaliana*. Hyperaccumulation and hypertolerance result from greatly enhanced activity of metal homeostasis transporters to produce dramatically altered internal metal allocation (Figure 12.17). In those instances in which transporters involved in hyperaccumulation have been directly compared with those in *A. thaliana*, they have been found to be essentially identical. In both *A. halleri* and *N. caerulescens*, dramatic increases in *HMA4* transcript abundance are associated with gene triplication and quadruplication, respectively. In addition, differences in *cis*-elements upstream of *HMA4* genes increase expression by increasing promoter activity. Increased HMA4 activity dramatically increases the capacity for loading Zn and Cd into the xylem from root cells. Ectopic expression of HMA4 in *A. thaliana* increases tolerance but does not entirely replicate the hyperaccumulation phenotype.

In Zn/Cd hyperaccumulators, MTPs that load Zn/Cd into vacuoles, especially in shoots, are also essentially identical to those in non-tolerant *A. thaliana*, but have greatly increased expression. ZIP transporters such as ZNT1, and NRAMPs that take up Zn/Cd from the xylem stream into shoot cells, are also highly expressed. RNAi experiments have shown that nicotianamine (NA) is required for Zn/Cd to reach the xylem, and that hyperaccumulators have increased expression of the YSL transporters that efflux NA–metal complexes from the cytoplasm (NA is probably used to increase

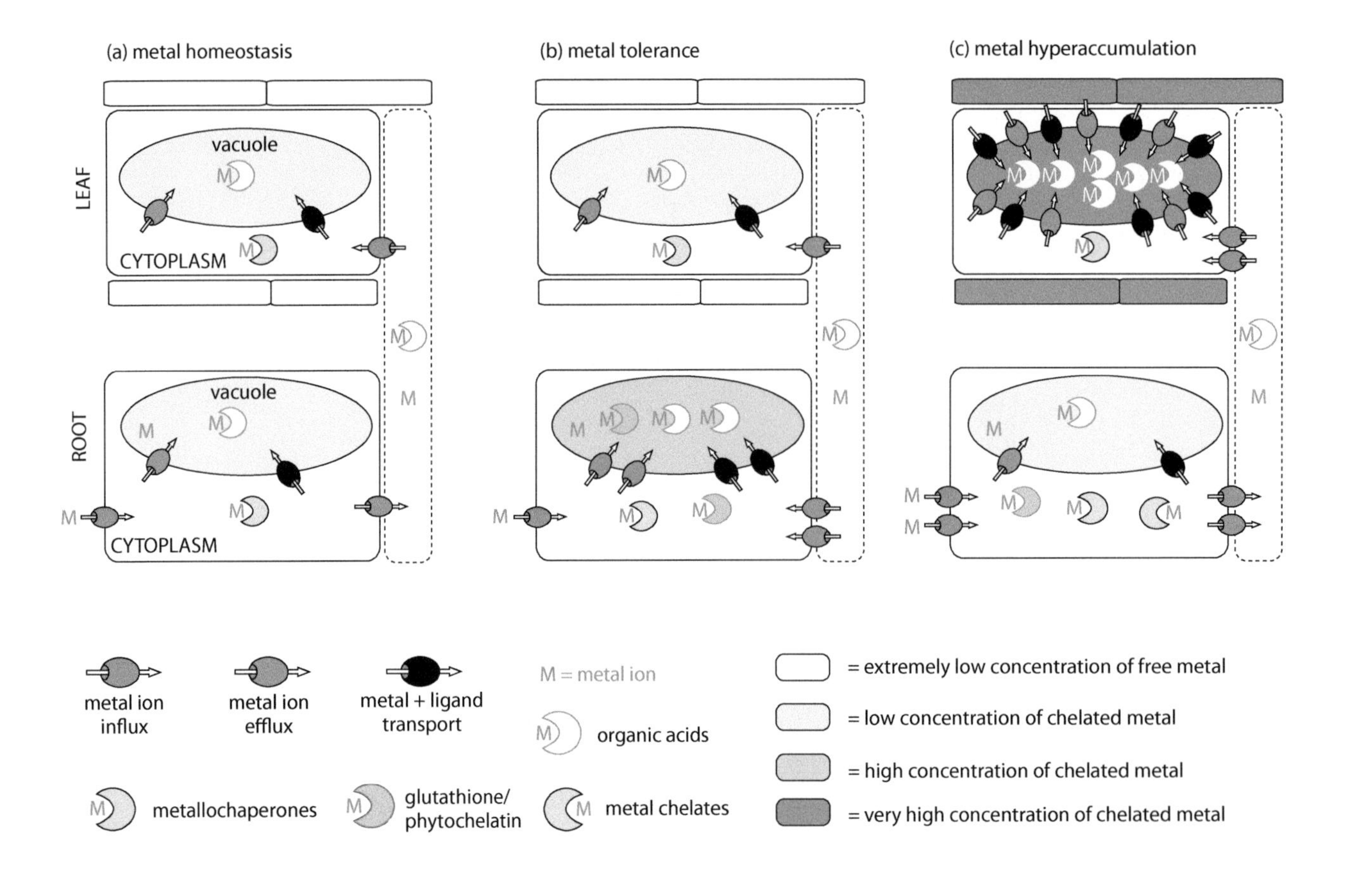

Figure 12.17. Mechanisms of plant metal tolerance and hyperaccumulation. In plants with metal-tolerant physiologies, mechanisms that transport metals into root vacuoles and chelate them are more highly expressed (b) than during homeostasis in non-tolerant plants (a), together with mechanisms that retrieve metals from the xylem to prevent them from reaching the shoot. Glutathione polypeptides (phytochelatins) often play a role in binding metals, and are transported into the vacuole. In hyperaccumulators, metals are moved across the root in significant quantities, and are then transported up through the xylem, sometimes as free ions, but also bound in a variety of organic acids. In the shoot they are pumped into vacuoles, often specifically in epidermal cells, and chelated with organic acids (c). As and Se are transported to and accumulate in the vacuole, but in free and organic forms, respectively. In general, the same transport proteins and chelators are involved in homeostasis, tolerance, and hyperaccumulation, but they are expressed differently.

the safe symplasmic transit of Zn/Cd). The timing of the *A. halleri/A. thaliana* speciation event coincides with the triplication of *HMA4*, and many of the genes with different expression in microarray studies of hyperaccumulators relate to Zn or Fe deficiency. Massive efflux of Zn/Cd into the xylem has probably driven the evolution of sequestration events in the shoot, and the evolution of efficient Zn/Fe uptake, including symplastic transit, in roots. The evolution of the Zn/Cd hyperaccumulation phenotype, and hence new species, was probably initiated by a few changes in gene expression.

In *N. caerulescens*, and in the only non-brassicaceous Cd hyperaccumulator, *Sedum alfredii*, leaf epidermal vacuoles have the highest Zn/Cd concentrations, whereas in *A. halleri* these occur in the mesophyll cell vacuoles. In vacuoles, which tend to be acidic, Zn and Cd are associated with malate and citrate in particular. **δ^{66}Zn** decreases significantly from root to shoot in hyperaccumulators, indicating that in roots Zn is also complexed with organic ligands. Zn and Cd can move as free ions in the xylem (although citrate has also been suggested to bind Zn during such movement), and inorganic phosphate in root cell walls and phytate in tissues such as seeds might also be important ligands at particular life stages. There is some evidence that rhizosphere microflora affect metal uptake by hyperaccumulators. There are 15 Zn hyperaccumulators, of which about five can hyperaccumulate Cd, and one (*T. caerulescens*) also has Ni-hyperaccumulating populations. There are also about 15 species that can hyperaccumulate Pb, many but not all of these being members of the Brassicaceae, which probably use similar mechanisms to Zn/Cd hyperaccumulators. The extent of Zn/Cd/Pb hyperaccumulation can vary significantly from one individual to another, and in some species hyperaccumulation is universal, whereas in others it only occurs in particular populations.

Although the molecular physiology of Ni/Mn and Co/Co hyperaccumulators is less well understood than that of Zn/Cd hyperaccumulators, there are many more of them—almost 400—for Ni. These occur in serpentine floras on all continents except Antarctica. About 25% are in the Brassicaceae and 25% in the Euphorbiaceae, but the fact that they occur in many other families as well strongly suggests numerous independent origins. Ni hyperaccumulators direct Ni into leaf vacuoles, often in the epidermis, but also sometimes out into the cuticle. Vacuolar loading occurs via enhanced MTP transporter activity in the tonoplast. Stem epidermal cells in some species have high Ni concentrations—for example, in the woody *Sebertia acuminata*, latex in the bark can have spectacular metal concentrations of over 20% of dry weight, and in many species in the Euphorbiaceae the latex has the highest Ni concentration. Some Ni hyperaccumulators have numerous leaf trichomes but, in contrast to a number of other elements, only have high Ni concentrations in the trichome pedicel. Ni uptake into hyperaccumulators is active, NA seems to play a role in symplastic transit, and xylem loading is significant. In hydroponically grown *Alyssum bertolonii*, Ni is bound to histidine in the xylem, and manipulation of histidine synthesis in some species affects tolerance. In other Ni hyperaccumulators, and in the field, Ni has been shown to be bound to numerous other ligands or to occur as the free ion in the xylem sap. In vacuoles, Ni is almost certainly bound primarily to organic acids, particularly citrate and malate, but in the xylem a variety of ligands bind Ni, perhaps depending on N status. In lactifers of stems and bark a wide range of compounds have been implicated in Ni binding. About 20 hyperaccumulators of Mn are known, the majority of them being woody plants of the Western Pacific region from Australia to Japan. Their subcellular distribution of Mn can include high concentrations in photosynthetic cells, but Mn is predominantly complexed with carboxylic acids in vacuoles. Around 25 hyperaccumulators of Co and 35 hyperaccumulators of Cu have been identified, mostly herbaceous plants of the Central African copper belt.

BOX 12.3. PHYTOREMEDIATION USING ARSENIC HYPERACCUMULATORS

Pteris vittata, a fern in the Pteridaceae, has ecotypes that hyperaccumulate arsenic (As). *P. vittata* is found in a variety of habitats, and can hyperaccumulate As from relatively low soil concentrations. Its use in phytoremediation has been patented, and a cultivar of it, the "edenfern™," is commercially available for phytoremediation from the Edenspace Systems Corporation. The As is accumulated in the vacuoles of epidermal cells and trichomes of fronds as free As(III). In their natural habitats, *P. vittata* and other As hyperaccumulators probably primarily take up As(V) through phosphate transporters in rhizoids. However, in the latter, As(V) is reduced to As(III), which is the predominant translocated and stored form. As(III) is generally transported across biological membranes through NIPs, but As(III) transport in *P. vittata* involves additional, perhaps active, transporters in the roots. In vacuoles, ACR3 transporters allow the efflux of As(III) from the cytoplasm into the vacuole. ACR3s from yeast enable the efflux of As, and can be replaced by PvACR3, and their loss from angiosperms might help to explain their lack of As hyperaccumulators, whereas *ACR3* duplication might have helped to drive the evolution of As hyperaccumulation in the Pteridaceae. In plants with basal As tolerance, As is frequently bound in the roots with thiol groups of PCs, but no significant role of PCs has been found in As hyperaccumulators, which often have lower PC concentrations in their roots than As-tolerant plants.

A number of fern species in the Pteridaceae can hyperaccumulate As, and have provided a unique example of the use of a hyperaccumulator in phytoremediation (Box 12.3).

In non-tolerant plants, exposure to Se results in selenate accumulation in cells, leading to protein malfunction, but Se hyperaccumulators sequester methyl-selenocysteine (methyl-SeCys) in the vacuoles of epidermal cells and leaf hairs. Methyl-SeCys is non-proteinogenic, and is produced by enhanced activity of SeCys methyltransferase (SMT). Se hyperaccumulators have enhanced ubiquitin activity, which probably helps to rid the cells of malformed proteins. In non-hyperaccumulators, selenomethionine is the primary organic form, which can produce low rates of volatilization to dimethyl selenide, whereas in hyperaccumulators SeCys can volatilize to dimethyl diselenide at significant rates. Se uptake and incorporation occur via SO_4^{2-} transport and assimilation pathways.

Chronic exposure to toxins in metalliferous ecosystems provides some unique biological insights

The ecosystems that have developed over calamine, serpentine, and chalcocite have long been recognized as unique. Outcrops of these minerals tend to occur as patches with distinct flora and fauna. These "islands" have high rates of metal-tolerant plants, some hyperaccumulators, and distinct biomass production dynamics. Evolution is in significant part driven by local adaptation (that is, habitat specialization). In plants, as noted by Wallace in 1858, in some of the earliest writing about natural selection, good examples of habitat specialization include adaptation to soil conditions. Metalliferous ecosystems provide unique examples of such adaptation, and have been used to study not only the advantages of metal tolerance and accumulation, but also many micro- and macro-evolutionary processes. Most experiments that have involved growing non-metal-tolerant plants in metalliferous soils have clearly demonstrated their toxicity, whereas reciprocal experiments have shown that although metal-tolerant plants can grow on other soils they are generally outcompeted by other plants—in other words, tolerance has a cost but it enables plants to exploit a stressful niche. The explanation of metal hyperaccumulation is more complex. The selective advantages for which there is greatest empirical support are herbivore and pathogen deterrence. However, there are some insects that flourish specifically with hyperaccumulators, and there is also good evidence, especially for Se, of elemental allelopathic effects. The importance of interactions with microorganisms, and in particular mycorrhizal symbioses, to tolerance and ecosystem dynamics has been emphasized by a number of field studies.

On calamine deposits and spoil heaps of Eurasia, short grassland with numerous metal-tolerant grasses and herbs occurs. Not only hyperaccumulating brassicas but also tolerant plants such as zinc violets and leadwort (*Minuartia verna*) occur. In zinc violets the numerous ecotypes have been used to describe the evolution of barriers to reproduction—a vital microevolutionary step in speciation that is often difficult to demonstrate. Rocks that contain over 70% ferromagnesian (mafic) minerals are classed as ultramafic, the commonest of these generally being serpentine. Serpentine floras generally have low productivity, a high rate of **endemism**, and a distinct vegetation that is of lower height and more xeromorphic and sclerophyllous than the surrounding non-serpentine vegetation. Serpentine soils generally have a very low Ca:Mg ratio, low clay content and water-holding capacity, and high metal concentrations. The vegetation of these soils is generally tolerant of all these characteristics, and has a high number of endemics and hyperaccumulators. Serpentine outcrops occur throughout the world.

The plant genus *Leucocroton*, which inhabits serpentine soils in the Caribbean region, provides an excellent example of sympatric speciation. Studies of macro-evolutionary processes in the serpentine floras of California have revealed that colonization of serpentine entails a decrease in diversity, that variation in tolerance promotes the evolution of endemics but can also cause the loss of tolerance, and that pre-adaptation to some of the conditions of serpentine is necessary for colonizing it. The southern end of the main island of New Caledonia in the Western Pacific has large outcrops of serpentine with extremely high rates of endemism. The island's long history of isolation since it split from Gondwana explains the existence of numerous non-angiosperm endemics, but not that of the endemics of the very Ni-rich serpentines (New Caledonia has perhaps 25% of the world's Ni reserves). The Celastrales, Oxalidales, and Malpighiales are particularly well represented among the Ni-tolerant and Ni-hyperaccumulating angiosperms of New Caledonia. Overall, the New Caledonian serpentine endemics evolved not only because of the dispersal properties of certain species, but also because some of these must have been pre-adapted to some of the demands of living on serpentine. Mineral outcrops that are rich in Cu are relatively rare, with significant deposits only in Arizona, Peru, Chile, Australia, and Central Africa. On the particularly rich deposits of Katanga, in Central Africa, a distinctive short grassland occurs with more than 30 endemic species forming a now critically endangered ecosystem. All of these metalliferous ecosystems are of high conservation value because of their rarity, high rate of endemism, and unique provision of evolutionary insights.

Control of soil-to-plant transfer of inorganic toxins is useful in agriculture and phytoremediation

Cd and As in particular are often food contaminants at concentrations that are hazardous to human health. An understanding of the mechanisms that control the influx of these elements to plants, and that tolerant plants use to prevent their movement into the shoots, is likely to contribute significantly to minimizing the entry of these toxins to the human food chain. Prevention of As contamination and Zn toxicity in rice production systems is an important aim for the agronomy of this crop. Rock phosphate can contain Cd impurities, especially the lower-grade ore that is now more frequently being used, so control of Cd addition to food crops via P fertilizer will be important to food production in the coming decades. The mechanisms that plants use to tolerate toxins when they occur naturally are proving to be very useful to efforts to minimize the contamination of the human food chain. In addition to efforts to control the concentrations of other inorganic toxins, an understanding of the mechanisms that control concentrations of essential elements, such as Zn, Fe, and Se, is useful for the development of **biofortification**.

The restoration of contaminated and degraded land is a key challenge facing humankind, and the use of metal-tolerant plants in restoration ecology has a long history and widespread application. The pressure from urban development on natural and agricultural ecosystems can be relieved—some estimates suggest very significantly—by restoring post-industrial sites to use. In most industrial and post-industrial nations there are very significant numbers of contaminated sites. Most current methods of decontamination involve removal or washing of topsoil. For some end purposes these methods are effective, but they disrupt soil ecology and render the sites unfit for many uses. The existence of so many undeveloped contaminated sites of potentially high value suggests that current methods of decontamination are not always suitable. For radioactively contaminated soils there is simply not the capacity in approved nuclear-waste repositories to remove contaminated soils, so they mostly remain *in situ*.

Phytoremediation uses plants, often those with specific toxin tolerance, to stabilize contaminated soil and either minimize or maximize contaminant extraction. Some contaminants can be removed from soils very rapidly by plants. For example, $^{99}TcO_4^-$, a common constituent of nuclear waste, is highly mobile in aerobic soils and can be removed from soil, even from below the rooting zone, via the transpiration stream, after one or two harvests. Contaminants have infrequently been successfully phytoextracted from sites to meet regulatory guidelines—phytoextraction from most sites tends to be slow and incomplete. Thus hyperaccumulators have attracted much attention, as have soil amendments and microbiological manipulations that can increase availability. There is also much interest in trying to use advanced breeding methods and transgenetic approaches to produce high-uptake plants to remove contaminants whose availability has been artificially enhanced. What would in effect be solar-powered contaminant extraction accords with what are likely to be many twenty-first-century priorities, and might develop increasing efficacy and application during this century.

Many decontamination projects now incorporate phytoremediation as one aspect of decontamination regimes. For example, electrodics has been quite widely used to manage contaminant availability in soils, and might be combined with the growing of plants. Alternatively, soils and water can be moved to decontamination lysimeters or circulation tanks and managed to optimize plant extraction before being replaced mostly intact biologically. Water treatment by reed beds has been established for decades, and plants that tolerate inorganic toxins have sometimes been incorporated in them. It is becoming an exemplar of how semi-engineered phytoremediation that exploits ecosystem properties associated with particular plants can work, even in individual buildings in cities with a high population density. For example, the headquarters of the San Francisco Public Utilities Commission were designed to incorporate water treatment using coastal wetland plants and their associated rhizoflora. The necessary new methods of decontamination and new ecodesigns of the coming decades may well utilize plants that possess the mechanisms which enable them to grow in the presence of inorganic toxins.

Phytoremediation projects frequently raise the question of what can be done with the contaminated plants. Ashing reduces volumes significantly, but it increases costs and produces a concentrated waste. Plants have at least removed the toxins from the environment, but the plant waste problem has raised the possibility of recovering useful amounts of metal from this waste. Not all of the inorganic toxins that can be taken up by plants are commercially useful, but many of them are. Biomining using bacteria to mobilize metals is an established technique, with a significant proportion of Cu being mined in this way. Given the continuing increase in human demand for metals, the decrease in quality of remaining ore reserves for some of them but their ubiquity at low concentrations in many soils, and the increasing environmental pressures on surface mining, phytomining of metals from soils just might find a role in the future. It also exemplifies what might be the key to the more widespread use of phytoremediation—finding economic value in the biomass produced. The many uses of biomass and the possibility of engineering even more uses suggest that such value might be realized.

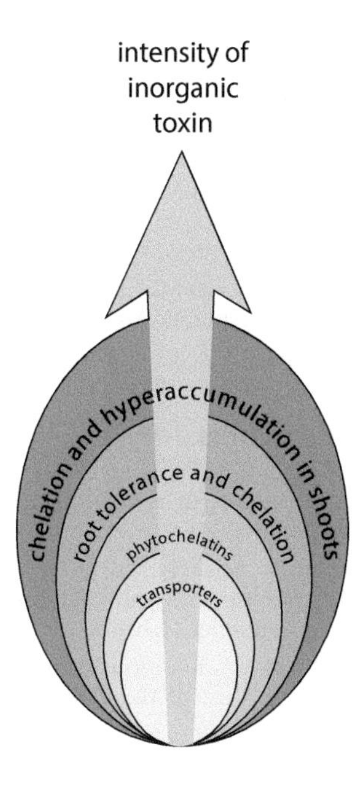

Figure 12.18. Plant responses to increasing levels of exposure to inorganic toxins. Plants have homeostatic mechanisms based on transporters and chelators that help to control the concentration of inorganic toxins in the cytoplasm. Some plants are able to withstand higher than usual concentrations of inorganic toxins by producing phytochelatins. If the concentrations are too high to control by chelation, some plants tolerate them by root accumulation. At high concentrations some plants have adapted to hyperaccumulate inorganic toxins, which not only enables them to cope with high levels of exposure, but is also an advantageous herbivory deterrent.

Summary

Plants evolved to select light, often reactive, trace elements for essential physiological functions. They scavenge trace elements from soils because these elements tend to be poorly available, but accumulation of concentrations that are too high can cause toxicity. Plants have finely tuned homeostatic mechanisms that favor the entry of essential trace elements and supply them to active sites while minimizing their free concentrations. Homeostatic mechanisms cope with significant variation in the availability of potentially toxic inorganic ions, but in most plants are overwhelmed at high concentrations. Those plants that have evolved basal tolerance or that can hyperaccumulate toxic ions do not have new genes, but rather they have amplified expression of particular components of homeostatic mechanisms (Figure 12.18). The significant differences in tolerance and accumulation that arise from adaptations of homeostatic pathways are an exemplar of adaptation to the local environment, and are extremely helpful for the manipulation of metal tolerance and accumulation in plants. This might be very useful for minimizing contamination of crops, for biofortification of crop plants with trace elements, and for developing phytoremediation systems, all of which are likely to become increasingly important during the twenty-first century.

Further reading

Reactive elements and plant evolution

Conn S & Gilliham M (2010) Comparative physiology of elemental distributions in plants. *Ann Bot* 10:1081–1102.

Frausto da Silva JJR & Williams RJP (2001) The Biological Chemistry of the Elements, 2nd ed. Oxford University Press.

Solymosi K & Bertrand M (2012) Soil metals, chloroplasts, and secure crop production: a review. *Agron Sustain Dev* 32:245–272.

White PJ, Broadley MR, Thompson JA et al. (2012) Testing the distinctness of shoot ionomes of angiosperm families using the Rothamsted Park Grass Continuous Hay Experiment. *New Phytol* 196:101–109.

Impact of heavy metals in the environment

Clemens S, Aarts MGM, Thomine S et al. (2013) Plant science: the key to preventing slow cadmium poisoning. *Trends Plant Sci* 18:92–99.

Nagajyoti PC, Lee KD & Sreekanth TVM (2010) Heavy metals, occurrence and toxicity for plants: a review. *Environ Chem Lett* 8:199–216.

Sharma AK, Tjell JC, Sloth JJ et al. (2014) Review of arsenic contamination, exposure through water and food and low cost mitigation options for rural areas. *Appl Geochem* 41:11–33.

Singh BR, Gupta SK, Azaizeh H et al. (2011) Safety of food crops on land contaminated with trace elements. *J Sci Food Agric* 91:1349–1366.

Mechanisms of metal homeostasis

Alvarez-Fernandez A, Diaz-Benito P, Abadia A et al. (2014) Metal species involved in long distance metal transport in plants. *Front Plant Sci* 5:105.

Leszczyszyn OI, Imam HT & Blindauer CA (2013) Diversity and distribution of plant metallothioneins: a review of structure, properties and functions. *Metallomics* 5:1146–1169.

Palmer CM & Guerinot ML (2009) Facing the challenges of Cu, Fe and Zn homeostasis in plants. *Nat Chem Biol* 5:333–340.

Smith AT, Smith KP & Rosenzweig AC (2014) Diversity of the metal-transporting P_{1B}-type ATPases. *J Biol Inorg Chem* 19:947–960.

Mechanisms of toxicity

Boyes DC, Zayed AM, Ascenzi R et al (2001). Growth stage-based phenotypic analysis of *Arabidopsis*: a model for high throughput functional genomics in plants. *Plant Cell* 13:1499–1510.

Dubey S, Shri M, Misra P et al. (2014) Heavy metals induce oxidative stress and genome-wide modulation in transcriptome of rice root. *Funct Integr Genomics* 14:401–417.

Fragou D, Fragou A, Kouidou S et al. (2011) Epigenetic mechanisms in metal toxicity. *Toxicol Mech Methods* 21:343–352.

Lin A, Zhang X, Zhu Y et al. (2008) Arsenate-induced toxicity: effects on antioxidative enzymes and DNA damage in *Vicia faba*. *Environ Toxicol Chem* 27:413–419.

Sobrino-Plata J, Meyssen D, Cuypers A et al. (2014) Glutathione is a key antioxidant metabolite to cope with mercury and cadmium stress. *Plant Soil* 377:369–381.

Mechanisms of tolerance

Batista BL, Nigar M, Mestrot A et al. (2014) Identification and quantification of phytochelatins in roots of rice to long-term exposure: evidence of individual role on arsenic accumulation and translocation. *J Exp Bot* 65: 1467–1479.

Choppala G, Saifullah, Bolan N et al. (2014) Cellular mechanisms in higher plants governing tolerance to cadmium toxicity. *Crit Rev Plant Sci* 33:374–391.

Fischer S, Kuehnlenz T, Thieme M et al. (2014) Analysis of plant Pb tolerance at realistic submicromolar concentrations demonstrates the role of phytochelatin synthesis for Pb detoxification. *Environ Sci Technol* 48:7552–7559.

Sobrino-Plata J, Meyssen D, Cuypers A et al. (2014) Glutathione is a key antioxidant metabolite to cope with mercury and cadmium stress. *Plant Soil* 377:369–381.

The exclusion of toxins from roots

Amir H, Lagrange A, Hassaine N et al. (2013) Arbuscular mycorrhizal fungi from New Caledonian ultramafic soils improve tolerance to nickel of endemic plant species. *Mycorrhiza* 23:585–595.

Leonhardt T, Sacky J, Simek P et al. (2014) Metallothionein-like peptides involved in sequestration of Zn in the Zn-accumulating ectomycorrhizal fungus *Russula atropurpurea*. *Metallomics* 6:1693–1701.

Pinto AP, Simoes I & Mota AM (2008) Cadmium impact on root exudates of sorghum and maize plants: a speciation study. *J Plant Nutr* 31:1746–1755.

Siemianowski O, Barabasz A, Kendziorek M et al. (2014) HMA4 expression in tobacco reduces Cd accumulation due to the induction of the apoplastic barrier. *J Exp Bot* 65:1125–1139.

Plant hyperaccumulation of toxins

El Mehdawi AF & Pilon-Smits EAH (2012) Ecological aspects of plant selenium hyperaccumulation. *Plant Biol (Stuttg)* 14:1–10.

Kazemi-Dinan A, Thomaschky S, Stein RJ et al. (2014) Zinc and cadmium hyperaccumulation act as deterrents towards specialist herbivores and impede the performance of a generalist herbivore. *New Phytol* 202:628–639.

Kraemer U (2010) Metal hyperaccumulation in plants. *Annu Rev Plant Biol* 61:517–534.

Park W & Ahn S-J (2014) How do heavy metal ATPases contribute to hyperaccumulation? *J Plant Nutr Soil Sci* 177:121–127.

Visioli G & Marmiroli N (2013) The proteomics of heavy metal hyperaccumulation by plants. *J Proteomics* 79:133–145.

Evolutionary ecology of metallophytes

Anacker BL, Whittall JB, Goldberg EE et al. (2011) Origins and consequences of serpentine endemism in the California flora. *Evolution* 65:365–376.

Boyd RS (2013) Exploring tradeoffs in hyperaccumulator ecology and evolution. *New Phytol* 199:871–872.

Jestrow B, Amaro JG & Francisco-Ortega J (2012) Islands within islands: a molecular phylogenetic study of the *Leucocroton* alliance (Euphorbiaceae) across the Caribbean Islands and within the serpentinite archipelago of Cuba. *J Biogeogr* 39:452–464.

Kolar F, Dortova M, Leps J et al. (2014) Serpentine ecotypic differentiation in a polyploid plant complex: shared tolerance to Mg and Ni stress among di- and tetraploid serpentine populations of *Knautia arvensis* (Dipsacaceae). *Plant Soil* 374:435–447.

Kuta E, Bohdanowicz J, Slomka A et al. (2012) Floral structure and pollen morphology of two zinc violets (*Viola lutea* ssp. *calaminaria* and *V. lutea* ssp. *westfalica*) indicate their taxonomic affinity to *Viola lutea*. *Plant Syst Evol* 298:445–455.

Biofortification and phytoremediation

Meier S, Borie F, Bolan N et al. (2012) Phytoremediation of metal-polluted soils by arbuscular mycorrhizal fungi. *Crit Rev Environ Sci Technol* 42:741–775.

Pinto E, Aguiar AARM & Ferreira IMPLVO (2014) Influence of soil chemistry and plant physiology in the phytoremediation of Cu, Mn, and Zn. *Crit Rev Plant Sci* 33:351–373.

Sebastian A & Prasad MNV (2014) Cadmium minimization in rice. A review. *Agron Sustain Dev* 34:155–173.

White PJ & Broadley MR (2009) Biofortification of crops with seven mineral elements often lacking in human diets—iron, zinc, copper, calcium, magnesium, selenium and iodine. *New Phytol* 182:49–84.

Zhang C, Sale PWG, Doronila AI et al. (2014) Australian native plant species *Carpobrotus rossii* (Haw.) Schwantes shows the potential of cadmium phytoremediation. *Environ Sci Pollut Res Int* 21:9843–9851.

Chapter 13
Organic Toxins

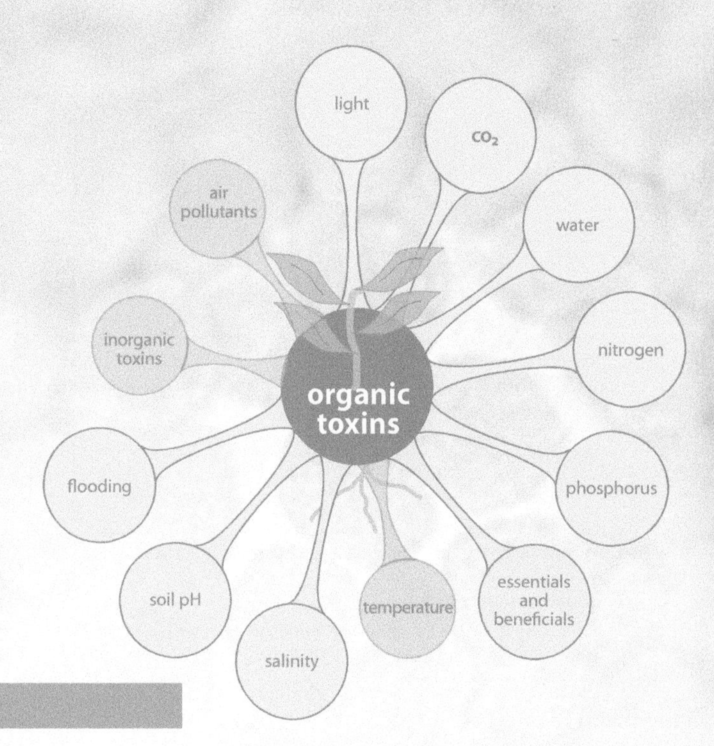

Plant responses to organic toxins interact with their responses to other pollutants, and can be affected by temperature.

Key concepts

- Plants synthesize, and can detoxify, biologically active organic compounds.
- Synthetic organic toxins contaminate water, food, and ecosystems.
- Plant uptake of organic toxins is passive and depends on chemical properties.
- Organic toxins can elicit general and specific stress effects in plants.
- At low levels of exposure, many organic toxins can be transformed enzymatically in plants.
- Some plants can conjugate and detoxify organic toxins in significant amounts.
- Organic toxin conjugates can be allocated to metabolically inactive compartments.
- Some plants have evolved general resistance to herbicides.
- After chronic exposure to particular herbicides, complete specific tolerance can evolve.
- Phytoremediation can enhance the environmental attenuation of organic contaminants.
- Manipulation of plant–toxin interactions is ecologically and agriculturally useful.

Plants can control the reactivity of many organic functional groups

Many of the **phytochemicals** of plant secondary metabolism have evolved to be biologically active (for example, to attract pollinators) or toxic (that is, to deter herbivores and pathogens) (Figure 13.1). The biosynthesis and use of these phytochemicals are underpinned by mechanisms that control their reactivity *in planta*. Using mechanisms that probably evolved from those used in the synthesis of reactive organic molecules, plants can also disarm chemical attacks by pests and pathogens, with disease-resistance genes often encoding specific detoxification mechanisms. In addition, many plants are exposed to allelopathic chemicals produced by other plants (Box 13.1), an interaction that is of great importance not only ecologically

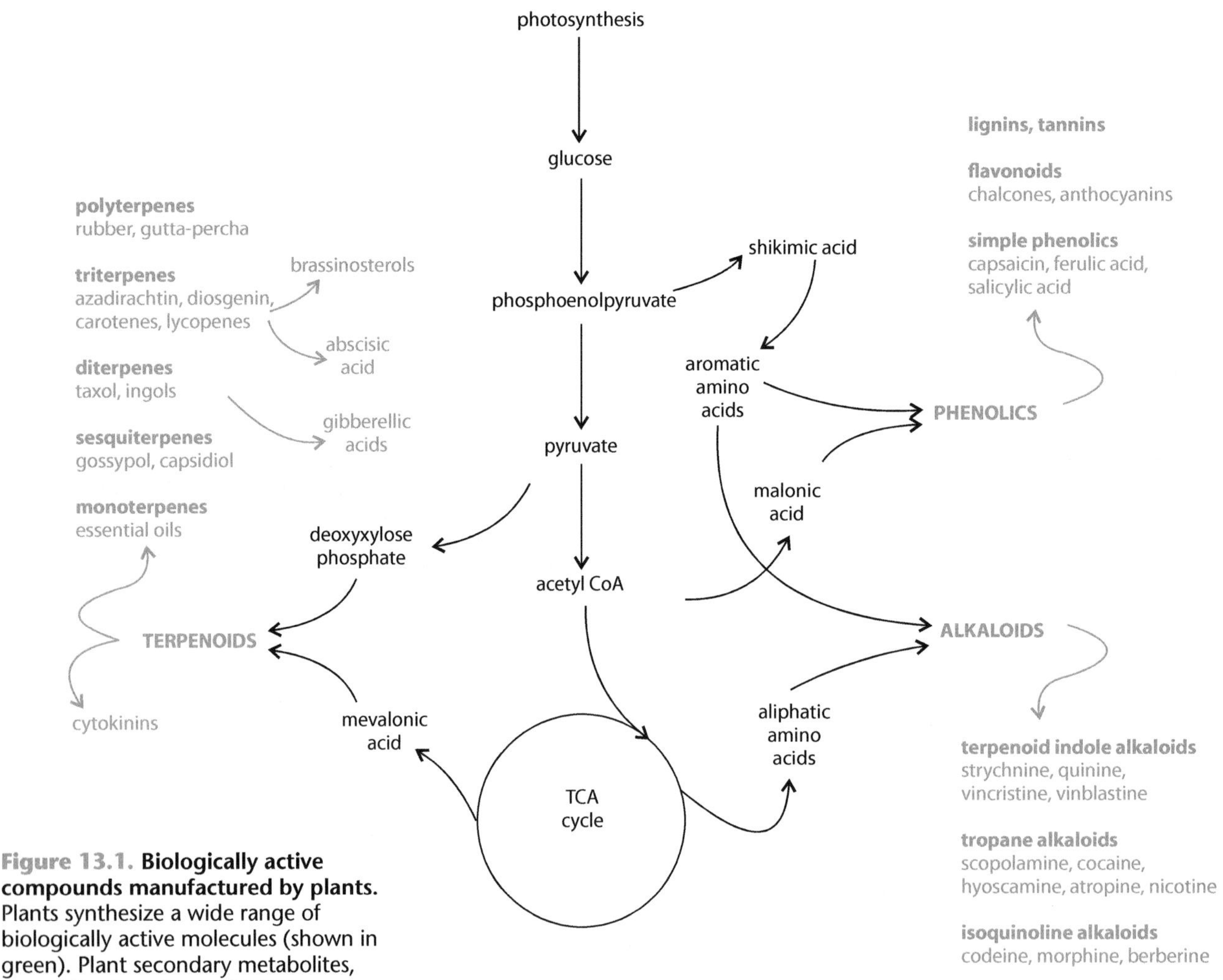

Figure 13.1. Biologically active compounds manufactured by plants. Plants synthesize a wide range of biologically active molecules (shown in green). Plant secondary metabolites, especially phenolics (*c.* 10,000), terpenoids (*c.* 25,000), and alkaloids (*c.* 12,000), include many toxins that act as herbivore deterrents, phytoalexins, or allelochemicals.

but also in the management of weeds and invasive plants. Thus plants have evolved mechanisms for controlling the reactivity of, and detoxifying, a wide range of organic compounds. Many synthetic organic compounds have the same functional groups as reactive phytochemicals, so plants can have a significant capacity to detoxify organic toxins, even many that are synthetic and to which they have only been exposed in recent decades. Overall, as a result of the complex chemistry that has been vital to their evolution, plants have a complex response to organic toxins, including the ability to resist and detoxify these compounds.

Heterotrophic microbes have the most diverse catabolic metabolism of all organisms, and microbes have been found with the capacity to mineralize many of the organic xenobiotics that humans release into the environment. Plants harvest energy from sunlight and so have not evolved a complex energy-producing catabolic metabolism. However, they have the most diverse metabolic pathways of **anabolism** of all organisms on Earth, and synthesize at least 200,000 and perhaps up to 1 million different compounds, including, for example, around 20,000 **triterpenes**, most of which cannot be chemically synthesized. The synthesis of such a diverse array of phytochemicals necessitates both the production of reactive functional groups, and then the control of their reactivity during the construction of complex organic skeletons. Many of the oxygenases, reductases, and conjugating enzymes that plants use in such syntheses have close phylogenetic relationships with those in many other organisms, but they are often particularly diverse in

BOX 13.1. MANAGING ALLELOPATHY FOR ECOLOGY AND AGRICULTURE

Plants synthesize a wide array of toxic organic compounds. When exuded by plants, these can act as allelochemicals that decrease growth (Figure 1), which can have a significant role in key ecological processes. It helps to explain why some plants have the capacity to cope with exposure to certain synthetic organic compounds, but it might also provide a way of replacing them in order to reduce their environmental impact.

Almost 2000 years ago it had been noted that some plant species could affect the growth of others. By the early twentieth century there were many reports of this phenomenon. Chemicals that affect interactions with other organisms are, in their widest sense, said to be allelopathic, but the term is often restricted to those that have negative effects. The significance of allelopathic effects was at first difficult to demonstrate, but invasive plant species and investigations of intercrops have shown that these effects can be significant. In the last 250 years, progressively increasing numbers of plant species have been transferred to non-native habitats. Many of these species do not become invasive, but those that do cause huge problems to the native flora and to crops. It is likely that the problems caused by, and the expense of dealing with, invasive species will increase as the twenty-first century progresses. A powerful, but seldom comprehensive, explanation of why some species are so invasive is a lack of natural herbivores and pathogens. However, invasiveness is frequently explained by the "new weapons" that invasive species can possess. These new weapons are often allelochemicals against which plants in an invaded ecosystem have no defense. For many invasive weeds, allelochemicals are now held responsible, at least in significant part, for the existence of monospecies stands of invaders that have severe adverse ecological consequences which are very expensive to address.

Many traditional intercrops have been shown to work in part by allelopathic suppression of weeds. Monospecies stands of invasive weeds that are dependent on allelopathic chemicals have also long suggested that allelopathy might be useful in crop monocultures. Many rice cultivars produce a range of allelopathic chemicals that help to suppress weeds, and allelopathic reactions with some serious weeds, such as *Echinochloa crus-galli* (in the Poaceae), have long been recognized as a means of controlling them. In general, many allelopathic grasses produce benzoxazinoids, which are terpenoids. Benzoxazinoids not only from rice but also from crops such as rye are increasingly being produced, isolated, and added to crops, and also used as templates for herbicide development. Many other species produce phenolic allelochemicals, of which one from sorghum—sorgoleone—has attracted particular interest (Figure 2). Plant production of and responses to toxic organic allelochemicals are therefore of major ecological and agricultural interest. Understanding and managing plant reactions to these toxic organic compounds may be useful not only for developing sustainable plant management, but also for finding new mechanisms of action for biodegradable agrochemicals.

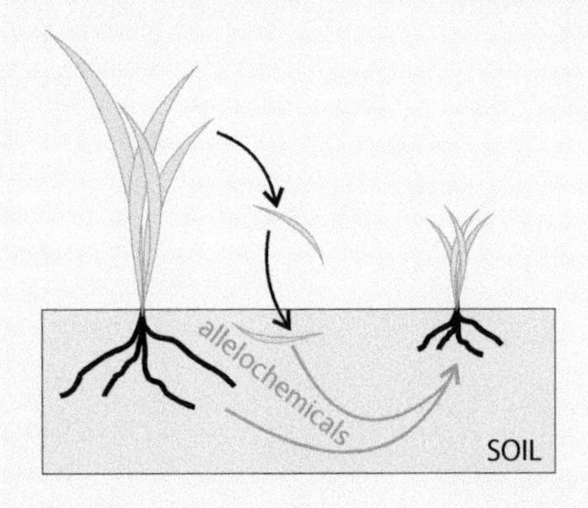

Figure 1. Allelopathy. Many plant species exude allelopathic chemicals from their roots, or deposit them with their leaves at leaf fall, to decrease the growth of competitors. In some instances this can have ecologically and agriculturally significant effects on the growth of other plants.

(a)

(b)

Figure 2. *Sorghum bicolor.* (a) Sorghum is a cereal, native to Africa, that produces numerous small round grains. (b) There are many varieties of *S. bicolor*, including some that produce large quantities of sorgoleone, which has allelopathic properties.

plants. Plants have adapted some key enzymes from anabolic pathways for use in detoxification systems. However, the anabolically focused metabolism of plants is not well equipped to mineralize reactive organic compounds. Many microbes can completely mineralize organic xenobiotics to simple inorganic molecules such as CO_2, but this ability is very rare in plants. In many animals the liver detoxifies reactive organic species, including organic xenobiotics such as pharmaceuticals, and releases the products, primarily in conjugated form, into excretion systems. Plants have only crude excretion systems (root exudates and leaf drop partially fulfill this function), but instead have significant non-metabolic vacuolar and apoplastic spaces in which the products of detoxification are compartmentalized. In summary, plants have the capacity to detoxify reactive organic compounds and, rather than then mineralizing or excreting them, they compartmentalize the products. They can therefore function as "green livers" that excrete detoxification products to non-metabolic compartments rather than to the external environment. This capacity helps to explain the fate of organic toxins in many plants and ecosystems.

Synthetic organic compounds underpin modern life but can have a significant environmental impact

In the late nineteenth century the synthesis of artificial dyes and then of explosives such as TNT initiated synthesis of organic compounds on an industrial scale. There are now more than 50 million registered synthetic compounds, some of which are made in vast quantities. These compounds underpin many aspects of human life. Without them, food production systems, healthcare systems, energy production, and manufacturing industries would be profoundly different. Synthetic chemicals are now found in most environmental media on Earth, often in significant amounts, and almost always in detectable amounts. By the 1980s, significant concern about the toxicological effects of a number of synthetic chemicals—chlorofluorocarbons (CFCs) and bioaccumulating organochlorines in particular—had resulted in environmental legislation that is still being developed today. For example, in Europe the Registration, Evaluation, Authorisation and Restriction of Chemicals (REACH) legislation of 2007 and the Water Framework Directive of 2015 have significantly increased the pressure to manage the environmental risk posed by chemicals and the level of water contamination, respectively.

In many parts of the world, environmental protection legislation has significantly reduced the release of synthetic organic compounds into the environment, but even in regions with some of the strictest environmental legislation there is often debate about the release of particular chemicals or the allowable limits for others. For instance, currently there is debate about the significance of emerging contaminants, including, for example, many pharmaceutical and personal care products, and of the breakdown products of, for example, persistent organic pollutants (POPs). There is also a significant legacy of contaminated sites. Since the 1980s, initiatives such as the US Environmental Protection Agency's Superfund program have focused on cleaning up the many hazardous waste sites identified in the USA. However, in the USA, Europe, and elsewhere there are tens or perhaps even hundreds of thousands of additional contaminated sites. In many instances there is much pressure to decontaminate these sites, and soil decontamination has been part of site development for many decades. At hazardous-waste sites and contaminated sites, organic toxins are an important (and often the most important) type of contaminant. In many parts of the world, environmental legislation is more recent, and contamination with organic xenobiotics is a source of increasing problems. For example, China is now the second largest (after the USA) producer and consumer of pesticides in the world, but attempts to control environmental contamination have much more recently been enacted. In general, the production of synthetic compounds has shifted significantly to countries in which environmental legislation is less stringent. Many organic xenobiotics can ultimately be mineralized in the environment, but for some this takes decades and release rates often exceed mineralization rates, resulting in long-term contamination problems. Thus current societies are dependent on the production of synthetic chemicals on an industrial scale, but there are significant environmental issues involved in dealing with a legacy of contamination, understanding the impacts of new chemicals, understanding environmental impacts in regions where environmental legislation is less stringent or is not enforced, understanding the impacts of accidental releases, and predicting the effects of environmental change on their mobilization. Because of the importance of plants both to food production systems and to natural ecosystems, a knowledge of the uptake and fate of synthetic organic chemicals by plants is vital for understanding their environmental impacts.

The extremely wide range of organic xenobiotics can be categorized by their environmental fate, their use, their origin, or their physico-chemical properties. POPs resist mineralization in the environment and can have adverse effects on health and the environment. They now circulate widely in the biosphere, hydrosphere, and geosphere, and at high concentrations have a significant impact. Herbicides are a diverse group of chemicals used for weed control, and represent one of the most significant releases of organic toxins to the environment. Many of them degrade quite quickly, but those that do not, or that reach natural ecosystems in significant amounts, can have significant effects on primary producers, the most important of these being the impact on phytoplankton and aquatic plants. For some, such as atrazine, there is long-standing concern about their impact on organisms other than plants, resulting in bans in some places, while very significant use continues elsewhere. There are significant challenges in global agriculture from herbicide-tolerant weeds that are prompting changes in the amounts and types of herbicides used. Compounds such as trinitrotoluene (TNT) and cyclotrimethylenetrinitramine (RDX) are still used to make most explosives, are widespread in the environment, and are established toxins. Petroleum hydrocarbons (PHCs) originate from oil and are widely dispersed, with many sites significantly contaminated. They include alkanes (for example, methane, propane), aromatics (for example, benzene, toluene, ethylbenzene, and xylene, also known as "BTEX"), and polycyclic aromatic hydrocarbons (PAHs). PAHs contain numerous aromatic rings fused together, and some of them occur naturally, but their main source is combustion of fossil fuels. Some of them are established carcinogens. Volatile organic compounds (VOCs) are defined by their vapor pressure, and have a wide variety of origins. Human societies now benefit enormously from synthetic chemicals, which as mentioned previously have enabled the development of fundamental aspects of civilization ranging from healthcare systems to food production systems, but they can and do have adverse environmental consequences. An understanding of the environmental behavior of organic toxins, including their behavior in primary producers, is necessary both for minimizing their impacts on health and the environment, and for cleaning essential water and soil resources, making it a key global challenge for the twenty-first century.

The entry of organic toxins into plants depends on soil, plant, and chemical properties

The entry of organic toxins into plants is predominantly passive. Roots evolved to interface with a large volume of soil, and shoots to interface with a large volume of atmosphere, so the uptake of organic toxins from either soil or atmosphere, although passive, can be substantial. There are circumstances in which apoplastic bypass can allow water to flow almost directly into the xylem, but in general root anatomy prevents this and necessitates movement across membranes for substances to enter the root cells and beyond. Plant roots evolved to take up inorganic ions, so the entry of organic toxins is coincidental and depends on their chemical properties. Those that are non-polar can pass quickly through plant membranes into the root cells with essentially no active control (Figure 13.2). Those that are non-polar but lipophilic tend to be incorporated into the membranes rather than passing through them (Figure 13.2). Some organic toxins are **zwitterionic** but generally behave as non-polar molecules during uptake by plants. For those that are weak acids, at most environmental pH values their pK_a reflects a significant proportion of uncharged acid in solution which can pass through root membranes. In many plant cell compartments the pH is higher than in the environment, so weakly acidic organic toxins dissociate in them to give high concentrations of charged conjugate base, which cannot pass across membranes and thus becomes trapped in the cell. The transport across membranes of organic toxins that ionize in aqueous solution is slow and dependent on lipophilicity and/or the presence of transporters.

The transpiration stream is driven passively by water potentials and, when it occurs, is generally much faster than the actively driven flow of solutes in the phloem. In general, non-polar organic toxins that pass quickly through membranes can be carried in the transpiration stream, and leak out of it relatively slowly because xylem has lignified cell walls. Such substances leak out of the membrane-bound phloem very rapidly, generally into the xylem, and are thus only transported to areas that are reached by the transpiration stream. In contrast, ionic compounds that can enter plants tend not to leak out of the phloem easily, and are carried in it throughout the plant, as well as in the transpiration stream. Many systemic herbicides, such as glyphosate, are ionic compounds that can enter plants and be transported in the phloem and xylem (Figure 13.2). Many organic toxins are semi-volatile or volatile, and in contaminated environments occur in significant concentrations in the gaseous state. Around plant roots, these compounds volatilize into soil

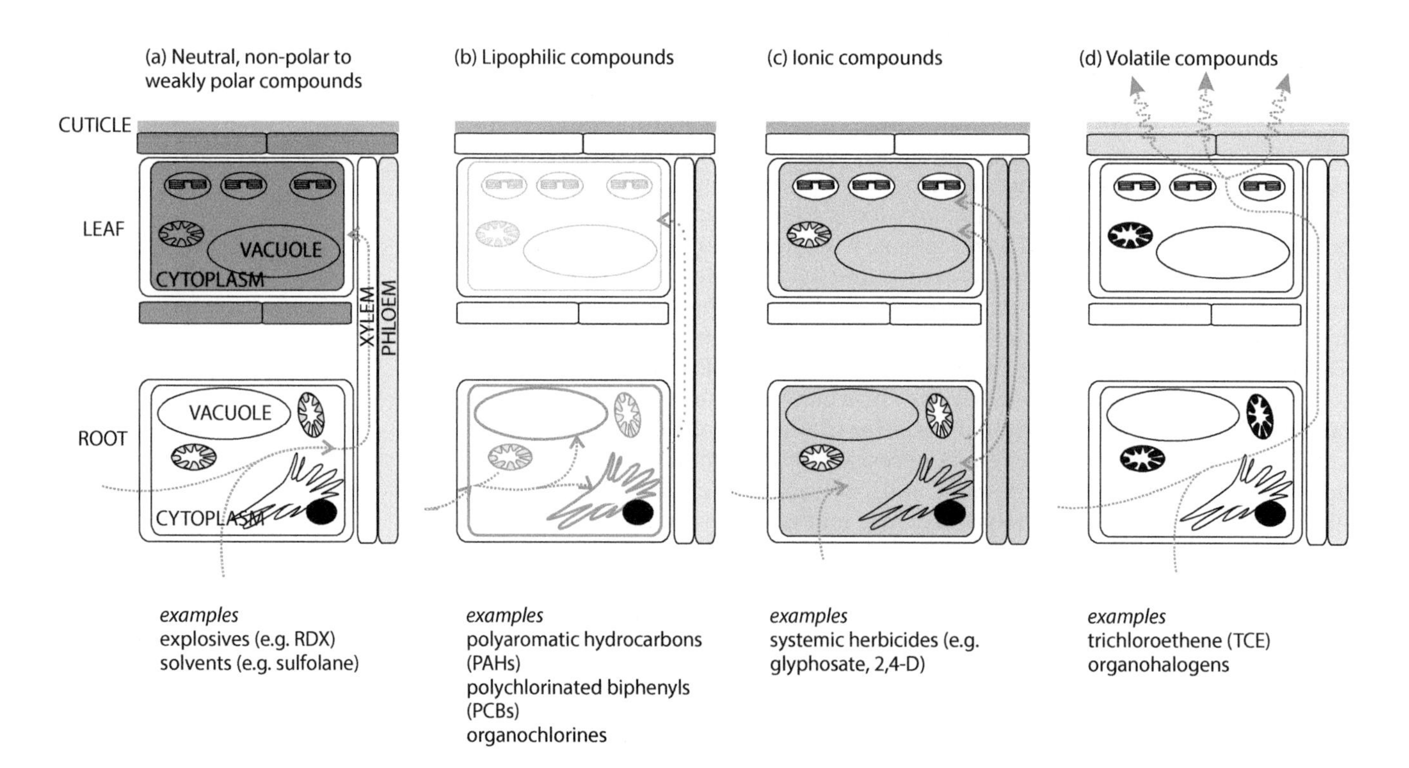

Figure 13.2. The general fate of organic xenobiotics after root exposure. In general, after root exposure the fate of organic xenobiotics is affected by their ability to penetrate membranes, their lipophilicity, and their volatility. (a) Soil-to-shoot bioconcentration factors (BCFs) are highest for neutral, non-polar, and weakly polar compounds because they pass through membranes and are transported in the transpiration stream, but cannot exit the shoot if they are not volatile. (b) Lipophilic compounds enter plants but remain in the membrane systems of roots with very low transpiration stream concentration factors (TSCF). (c) If ionic compounds can enter plants they are widely distributed because they can be transported in the phloem. (d) Volatile compounds tend to exit quite rapidly though the shoot unless temperatures are low. Extremely lipophilic compounds tend not to enter the roots because they remain outside the mass flow of water that drives uptake, but they can accumulate in root cell walls.

air spaces and enter plants primarily as gases, their ability to cross root membranes being the controlling factor (Figure 13.2). Volatile compounds in the atmosphere can enter plants through open stomata and then do not face the endodermal obstacle that they would encounter in roots, but can move significant distances by diffusion in the apoplastic space, and cross membranes if they are able to do so. Hydrophobic volatile compounds can cross the cuticle that covers plant shoots and thus enter plants, sometimes at substantial rates. In many instances, especially in the field, plants can also be contaminated by direct deposition on external surfaces, especially when organic toxins are adsorbed to particles that are deposited on plants. The many organic toxins that are polar, or that can ionize, adsorb to particles in significant quantities, especially to soil organic matter and charged minerals. Wind and rain splash can deposit significant quantities of soil material on the external surfaces of plant shoots. Organic toxins can diffuse into and through the cuticle from these particles or become embedded such that they are significant contributors to heterotrophic diets if whole plants are eaten. Overall, the entry of organic toxins into plants depends on the properties of the chemical and the plant, and these can vary significantly such that some organic toxins barely enter plants, whereas others accumulate to high concentrations.

Experiments on the entry of organic toxins into plants can be conducted using sterile culture media so that the confounding effects of soil availability and microbial action are eliminated. From these experiments, and much field data, some general principles concerning the entry of the myriad organic toxins have emerged. From the soil, persistent non-polar non-volatile organic toxins can be taken up rapidly, enter the transpiration stream, reach the leaves in large quantities, and have the highest bioconcentration factor (BCF), which for plants is equal to $C_{plant}/C_{external}$ (where C_{plant} is the concentration in the plant and $C_{external}$ is the concentration in the growth medium). Volatile organic chemicals, many of which are non-polar, can enter plants but exit rapidly, mostly through the leaves, and do not remain in plants at high concentrations. For volatile compounds, the rate of passage into the cuticle in response to gradients of vapor pressure can be generally quantified by the octanol–air coefficient, K_{oa}, which is equal to $1 - C_{octanol}/C_{air}$ (that is, a specific instance of **Henry's law**). Lipophilic compounds do not generally reach the transpiration stream, but can accumulate in significant amounts in roots (Figure 13.2). Some ionic compounds, and compounds that mimic those which occur naturally in the plant (for example, plant hormones), can enter plants and be circulated throughout them. When a root is exposed to an organic toxin, the root cell walls are the first potential point of contact. Polar compounds dissolve in water and generally do not interact significantly with the cell wall. If they are only weakly polar they can pass through membranes and enter the transpiration stream. Very polar and lipophilic compounds tend to be incorporated into lipid membranes in the roots, which inhibits their entry to the transpiration stream. Extremely lipophilic compounds may not even reach the membrane because they tend to separate out onto the root cell walls from apoplast solution.

In general, the properties underpinning these entry patterns can be approximated by the octanol–water coefficient, K_{ow}, which is equal to $1- C_{octanol}/C_{water}$. At very low K_{ow} values, entry of the compounds into plants is slow because they cannot penetrate any membranes. As K_{ow} increases so does the entry rate, because passage through membranes is possible. At very high K_{ow} values, entry—especially to the leaves—decreases because the compounds tend not to reach the root membranes, and if they do, they are incorporated into them and thus do not reach the transpiration stream (**Figure 13.3**). These entry processes are ultimately driven by the transpiration stream and the chemical activity of the compounds, but there is some evidence that a

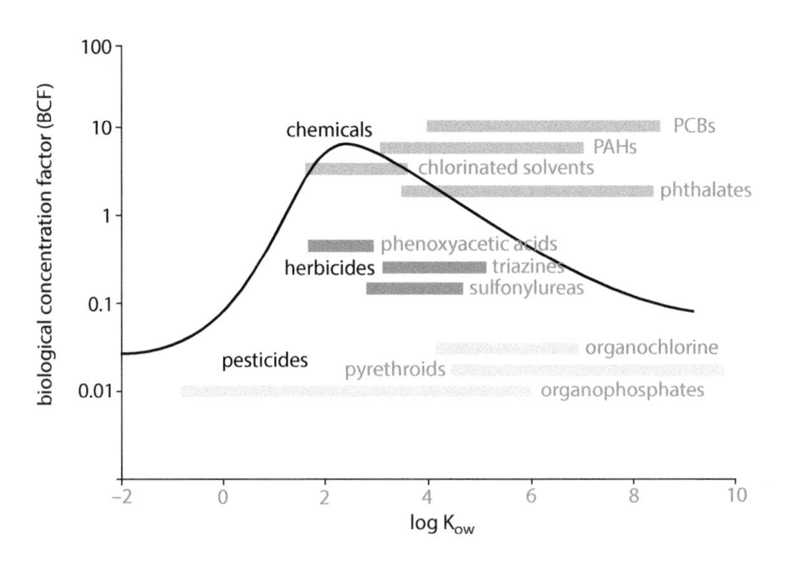

Figure 13.3. The relationship between bioconcentration factor (BCF) and K_{ow}. When K_{ow} is between –1 and 3, BCF increases linearly with increasing K_{ow}. At very low K_{ow} values, below –1, the BCF is often reported to be unexpectedly high. This may be because many compounds have some uptake via the shoot, or because of apoplastic bypass, both of which produce higher BCFs than would be expected from very low K_{ow} values. There are many reports of BCF decreasing at high K_{ow} values, although this is not always the case, and some compounds with very high K_{ow} values produce a very high BCF. Many widely used herbicides are systemic and are taken up and transported around plants rapidly, often because these compounds ionize. Many pesticides and chemical contaminants have a low BCF, but some have high BCFs, and those with low BCFs can accumulate significant concentrations in roots. PCB = polychlorinated biphenyl; PAH = polycyclic aromatic hydrocarbon.

certain amount of active uptake might also be involved, even for some non-ionic compounds. Plant membranes have transporters that actively transport a variety of organic compounds, including amino acids, sugars, plant hormones, and secondary metabolites. Experiments in axenic culture with, for example, the PAH phenanthrene show electrophysiological and pharmacological behaviors that are consistent with a proportion of uptake occurring actively. Roots have often been considered to actively transport only inorganic ions, but it is now clear that they also actively transport amino acids and organic chelates such as phytosiderophores. Active uptake processes might therefore contribute to the entry of a range of organic toxins.

Modeling the contamination of plants with organic toxins is very important for estimating levels of human exposure to these toxins. Any such model has to take into account wide variation, because the plant, soil, and environmental factors that affect uptake all vary significantly. If many plant species are simultaneously exposed to the same organic contaminant under the same conditions, species differ significantly in their BCF and generally produce a log-normal distribution of BCFs (Figure 13.4). Such a distribution occurs with many contaminants and many organisms, and demonstrates that much variation in BCFs is due to plant factors. In general, non-ionic organic contaminants bind to organic matter in the soil, thus reducing their availability to plants. Organic matter content and constituents can vary significantly between soils, producing differences in availability and BCF that are as great as the differences in uptake between plant species under one set of conditions. The distribution coefficient K_d, which is equal to $C_{soil}/C_{solution}$, describes soil adsorption and can be related directly to organic matter content for many organic contaminants. K_{ow} and K_d have been used in numerous regression models to make general predictions of BCFs. They have been extensively used in dose assessments by environmental protection agencies because of their simplicity and the large data sets available for a variety of compounds. One of the limitations of this approach to modeling entry is that it does not take account of plant metabolism and is not dynamic. Pharmacokinetic analyses relate the properties of compounds to adsorption, distribution, metabolism, and excretion (ADME) behavior. For organic compounds, the quantitative structure–activity relationship (QSAR) can be modeled for many variables and related to ADME behavior. QSAR methodologies can be used to obtain improved predictions of BCFs for organic contaminant entry into plants. Mass-balance models that quantify compartmentalization in the soil–plant system have also been constructed, as well as some dynamic models. The dynamic models provide a reminder that environmental variables have significant effects on both passively driven

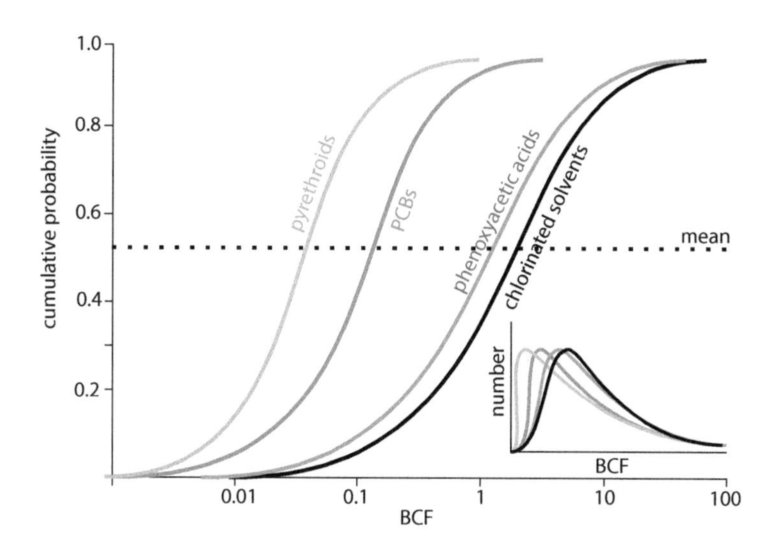

Figure 13.4. Variation in BCFs for organic xenobiotics. Many factors affect the BCF for different types of organic xenobiotic, but for each chemical there is also significent variation in BCF. This is almost always log-normally distributed, as reflected in the cumulative probability plots shown in the figure. For each chemical, a range of plant species or soil conditions tends to produce many values in the lower part of the range (inset), and fewer higher values. For this reason, geometric means are frequently calculated for BCFs. PCB = polychlorinated biphenyl.

entry processes and active detoxification in plants. For example, the entry of volatile compounds into plant leaves is temperature dependent. In the temperate zone, in spring and autumn semi-volatile organic compounds diffuse into the leaves via the cuticle, but at higher temperatures in summer they diffuse out into the atmosphere.

Organic toxins elicit reactive and perhaps also proactive stress responses in plants

The myriad organic toxins in the environment have a significant range of toxicities. For many of these compounds, toxicity has been extensively investigated in animal systems, and adverse effects have prompted stringent controls on their release into the environment. There are fewer data on the toxicological effects of organic toxins on plants, but the established cytotoxic, genotoxic, **teratogenic**, and carcinogenic effects on animals often occur via pathways that are common to all eukaryotes, which suggests that there are adverse effects on plants. In general, controlled experiments have shown significant effects on plant growth, biomass production, and reproduction for a range of organic toxins, including chlorinated benzene, trichloroethylene (TCE), polychlorinated biphenyls (PCBs), PAHs, and organochlorines. Toxic effects in plants can often be related to the same chemical characteristics that produce toxicity in animals—primarily the existence of **electrophilic** sites so that, for example, the degree of chlorination of organochlorines correlates with their toxicity. Overall, for many chemicals, toxicological studies involving plants, ranging from cellular to population levels, suggest that they are generally more tolerant of organic toxins than are animals. However, there are significant genotypic effects, with some plants being particularly sensitive to or tolerant of particular chemicals. Interestingly, **hormetic** effects of organic toxins have relatively frequently been reported in plants (Figure 13.5). In complex systems such as soils or cells it is difficult to determine whether xenobiotics are really promoting growth, or preventing it from being inhibited by other factors, but true hormesis is probably not uncommon in plants that are exposed to organic toxins. Organic toxins also interact with each other and with inorganic toxins to produce toxic effects. There are many detailed examples of antagonistic and synergistic effects between organic and inorganic toxins, but as yet few principles that can be used to predict them. This is of great importance because many sites are polluted with more than one contaminant.

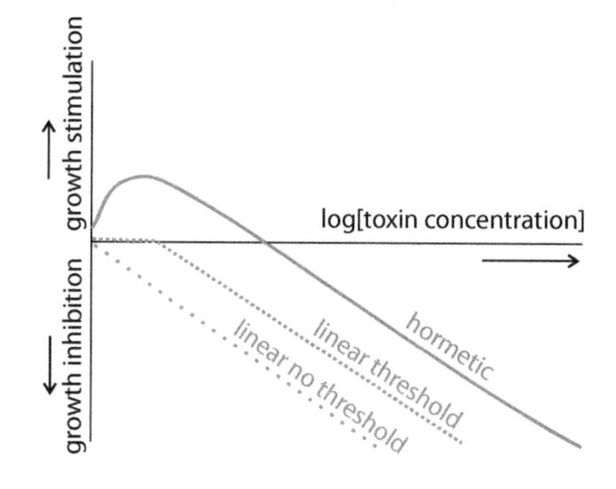

Figure 13.5. Theoretical dose–response curves for the effects of organic xenobiotic toxins on plants. Various aspects of plant growth can be used as end points in toxicity assessments. Linear no-threshold (LNT) responses are often assumed in toxicology, and have been described for many toxins, including some organic toxins in plants. They suggest that even extremely low concentrations of toxins will have some adverse effects, even if these are minor. Linear-threshold (LT) responses have also been described for many toxins, which suggests that organisms can deal with a certain concentration of toxin before any effects are manifested. Hormetic responses occur if a low concentration of toxin actually stimulates growth. Some data for plants and organic toxins seem to follow this pattern, and have contributed to debates in toxicology about how widespread hormetic responses to toxins may be.

When toxicity from organic compounds occurs in plants it is frequently associated with membrane damage, oxidative stress, and DNA damage.

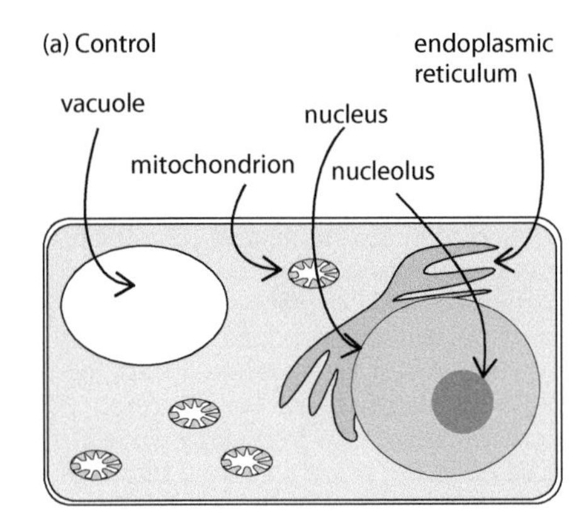

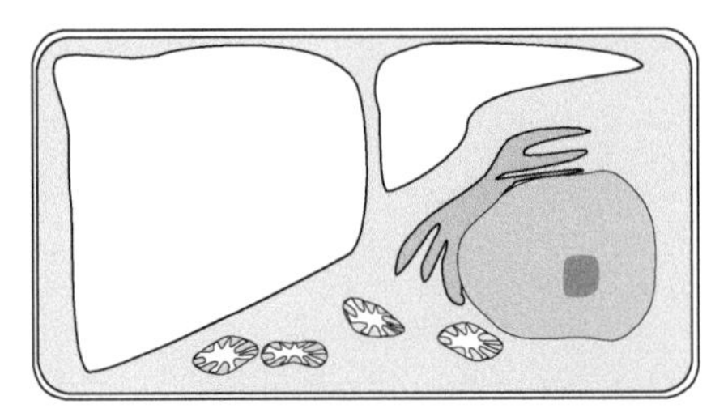

Figure 13.6. **Ultrastructural changes induced by phenanthrene.** The polycyclic aromatic hydrocarbon (PAH) phenanthrene has three rings, and, compared with controls (a), 300 μM phenanthrene in a hydroponic solution for 21 days induces significant ultrastructural changes in tomato root tip cells (b). (Data from Ahammeda GJ, Gaoa C-J, Ogweno JO et al [2012] *Ecotoxicol Environ Saf* 80:28–36.)

Lipophilic xenobiotics accumulate in plastids in the roots and shoots, where alteration of membranes causes leakage of reactive oxygen species (ROS). Membrane damage is also frequently associated with changes in membrane transport and electrolyte leakage. **Malondialdehyde** is produced by the action of ROS on unsaturated lipids, and is frequently observed during toxicity caused by organic toxins. For certain chemicals there are effects on soluble protein content, some of which might result from direct interaction with proteins. As with abiotic stressors, many of the cytotoxic effects culminate in oxidative stress. Meta-analyses of global genomic and proteomic responses to organic toxins have revealed significant overlap with abiotic and biotic stress responses. For some organic toxins there is a certain degree of specificity of response. However, although much genomic and proteomic data has been produced from acute high-dose exposure of plants, it seems likely that the effects of organic toxins can best be understood within the framework of the stress-response pathways of plants. For example, a number of studies have noted overlap with the heat-shock (HS) response in plants. HS responses serve to stabilize membranes, and may at least in part be initiated by changes in the membrane, which might be particularly significant for lipophilic toxins. Many ultrastructural changes associated with organic xenobiotic toxicity have been described, often initiated by changes in membrane properties (Figure 13.6).

Not only can organic toxins initiate stress responses indirectly, but also, because of their effects, it seems increasingly likely that plant cells are able to directly detect at least some of them. Plant cells contain numerous complex organic compounds, the reactivity of which has to be controlled, and many plants have to combat allelopathic chemicals. Organisms that both contain and are surrounded externally by bioactive organic compounds might gain a significant advantage from being able to detect them. In animals there are molecules that detect particular organic groups, acting as receptors for them. The well-characterized toll-like receptors (TLRs) of the innate immune system have evolved to recognize specific chemicals produced by microbes. The aryl hydrocarbon receptor (AhR) in vertebrates is a cytosolic transcription factor that is activated when organic toxins bind to it. This activates the transcription of genes, often by binding to xenobiotic response elements (XREs), so called because of the importance of the genes in responses to xenobiotics. Plant genomes are rich in receptor kinases. The leucine-rich repeat receptor-like kinases (LRR-RLKs), which many plants contain in abundance, have a similar role in pathogen interactions to that of TLRs in vertebrates. In many microbes, histidine-receptor kinases that operate in two-component signaling systems are used not only in stress signaling but also to detect complex organic compounds during, for example, **chemotaxis.** Two-component signaling systems occur in plants, and might also be used to aid the detection of organic compounds. Anthropogenic xenobiotics do have many unique characteristics, but they also have much in common with, for example, allelopathic chemicals. It seems likely that plants cells can detect at least some organic toxins, and that plant responses to these might be proactive as well as reactive (Figure 13.7).

Some organic toxins affect DNA, with mutagenic, carcinogenic, and teratogenic effects that are well established in animal systems but less well investigated in plant cells. Such effects can arise from either direct or indirect impact of organic toxins on DNA. For example, alkylating agents directly damage DNA, causing mutations. Furthermore, the metabolism of organic toxins can produce intermediates that can induce mutations. Such mutations can affect any of the numerous pathways associated with teratogenesis or cancer susceptibility. Some organic toxins affect DNA-repair mechanisms, resulting in increased mutation rates, whereas others affect factors such as methylation that control gene expression, resulting in epigenetic effects. Given that,

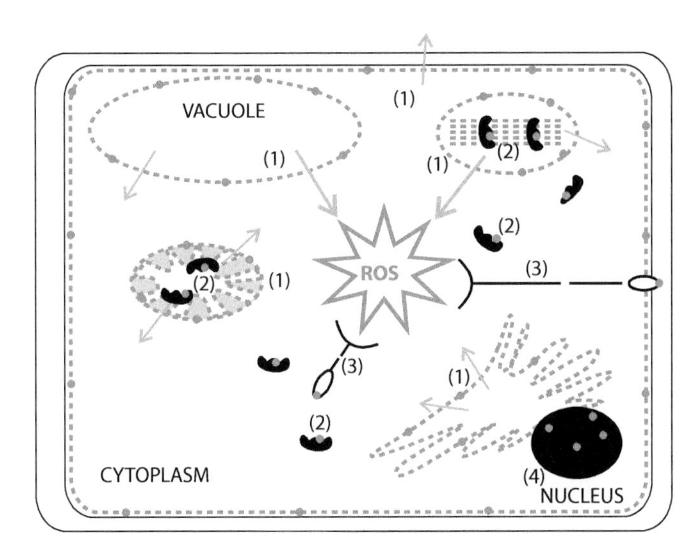

Figure 13.7. The effects of toxic organic xenobiotics on plant cells. Many but not all organic xenobiotics (shown as green dots in the figure) are toxic to plant cells. Many of them, especially those that are lipophilic, affect membrane properties (1), often resulting in leakage of ions and metabolites from membrane-bound compartments. Mitochondrial and chloroplast function depends on membrane integrity, so these organelles release significant quantities of reactive oxygen species (ROS) if it breaks down. Some organic xenobiotics bind directly to proteins, disrupting metabolism, and this results in an increased concentration of ROS (2), especially in photosynthesizing leaves. There is some evidence that plants may have receptors that detect the presence of organic xenobiotics (external and/or internal) that are used to trigger, for example, antioxidant production which contributes to the capacity to tolerate xenobiotically induced ROS (3). Some xenobiotics have genotoxic effects that can result in deleterious mutations (4).

in plants, inorganic toxins such as Cd have established effects on DNA-repair mechanisms and induce epigenetic effects, it is likely that organic xenobiotics also induce significant effects via these mechanisms.

In plant cells, many organic toxins can be transformed enzymatically

The persistence of organic toxins in an active form within an organism can be measured using their **biological half-life**, which for many organic toxins in many plants is relatively short because plants have the capacity to transform these toxins enzymatically. Although the primary substrate of peroxidases is probably the O–O bond in H_2O_2, there are many peroxidases that can oxidize organic substrates with O–O bonds if appropriate electron donors are available. The electron donors can include cytochrome or glutathione. Horseradish peroxidase (Figure 13.8) is perhaps the best characterized of these peroxidases, and has been shown to decrease the biological half-life of a variety of polyphenols. Peroxidases probably also contribute significantly to the ability of some root exudates to degrade certain organic toxins. Polyphenol oxidase activity has also been reported from a range of plant species. Laccases oxidize phenols, and probably evolved to catalyze the synthesis of lignin, a polyphenol that is the second most abundant biopolymer

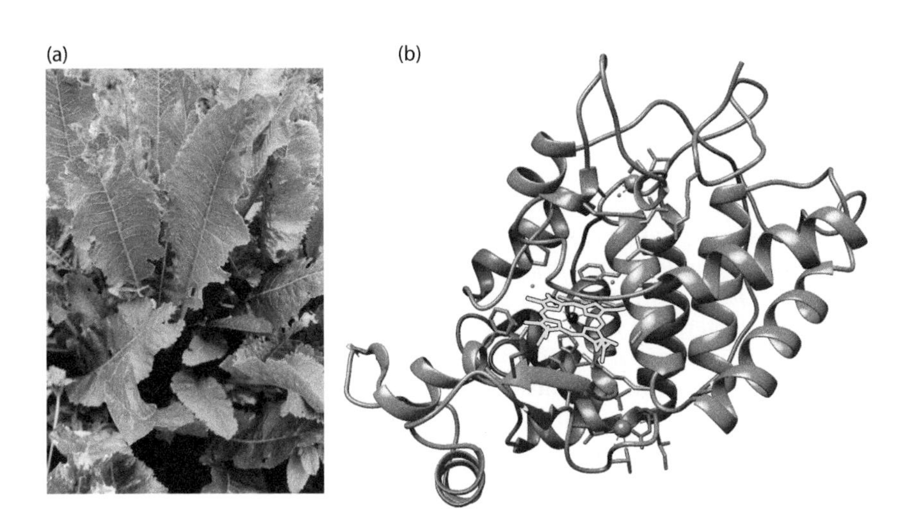

Figure 13.8. Horseradish peroxidase. (a) *Armoracia rusticana* (horseradish) is a member of the Brassicaceae and is native to Western Asia. Like other brassicas, it produces glucosinolates that give it a distinctive flavor, but it also produces an easily extractable, rapidly acting, and powerful peroxidase. (b) Using H_2O_2, horseradish peroxidase (HRP) utilizes a heme co-factor (shown in light green on the structure in the figure) to oxidize a wide range of organic compounds. It is commonly used in molecular biology with conjugates that, for example, are oxidized and change color or luminesce due to oxidation in the presence of their substrate. Under appropriate conditions, HRP can oxidize a variety of compounds. For this reason it has been widely investigated in phytoremediation systems, and it has often been found to have a significant capacity to degrade phenols and chlorophenols.

on Earth. Laccase was first isolated from the latex of *Toxicodendron vernicifluum* (black lacquer tree). A number of white-rot fungi have evolved to use laccase to completely break down wood (which is essentially lignified cellulose), and are therefore among the most important decomposer organisms on Earth. Laccases exist in plants and can contribute to the breakdown of polyphenols. The genetic modification of, for example, tobacco with fungal laccases that are excreted into the rhizosphere has produced plants with the ability to degrade phenols. Olive-mill waste waters have notably high levels of phenols, so laccases, perhaps from genetically modified plants, might have a significant role in treating them. Esterases have been shown to hydrolyze some organic toxins, phosphatases can attack organophosphates, and amidases can hydrolyze some N-containing organic toxins. In some plant species these reactions can provide the primary transformation capacity, but in most species for most organic toxins the primary transformation capacity is provided by oxygenation using cytochrome P450 (CYP) enzymes.

The capacity for monooxygenation (that is, the addition of a single O atom to an organic compound) is vital for a very wide range of anabolic and catabolic reactions. In all organisms, CYPs are the primary monooxygenases. They are the most diverse group of enzymes known, and are particularly diverse in higher plants, in which more than 7000 CYPs have been described. Although the protein folds and active site of CYPs are so highly conserved that it seems that all CYPs evolved from a single gene, there is sufficient diversity of these enzymes that, given their fundamental role in metabolism, they can be used to understand key macro-evolutionary events. Eukaryotes have significant microsomal CYP activity, which is located primarily on the cytosolic side of the endoplasmic reticulum (ER). These CYPs are particularly diverse in plants, with over 300 in *Arabidopsis* (in which they are the third most numerous gene family) and over 350 in *Oryza*. In prokaryotes, CYP capacity is primarily cytosolic, and the migration to the ER might have been important to the changing metabolic capacity of eukaryotes. Certainly in plants the diversity of CYPs on the ER seems to be related to their very wide metabolic diversity. In all eukaryotes, CYPs are necessary for the synthesis of sterols, lipids, and hormones, but in plants they are also necessary for the synthesis of, for example, the terpenoids, alkaloids, polyphenols, and glucosinolates of secondary metabolism. When organic toxins enter plant cells they can be transformed by monooxygenation using CYPs. This can bring about, for example, dehalogenation, dealkylation, hydroxylation, demethylation, epoxidation, and deamination. In most instances this transformation increases hydrophobicity, and in many (although not all) it reduces electrophilicity and hence toxicity. It seems likely that this capacity of CYPs to detoxify organic toxins was a selective advantage, with their diversity in static plants being consistent with the need to detoxify many organic compounds that enter plants. It also means that CYPs have great biotechnological potential.

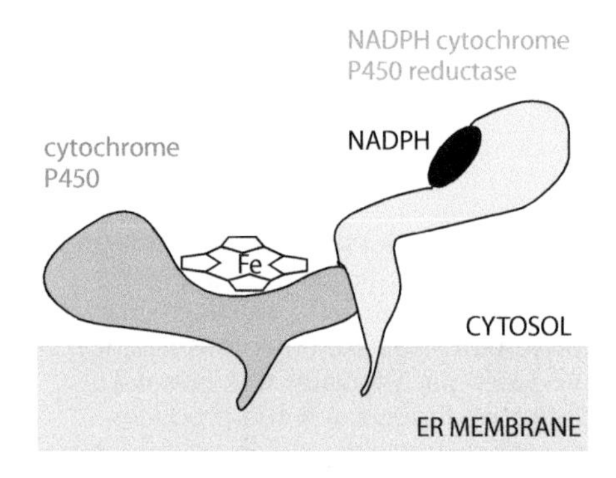

Figure 13.9. Plant cytochrome P450s (CYPs). In plants, CYPs occur primarily on the cytoplasmic side of the endoplasmic reticulum (ER). They mainly use NADPH as a source of electrons. When an organic compound binds to CYP, conformational changes lead to the transfer of an electron from NADPH to the Fe in its heme group, initiating reactions that result in the insertion of an O atom into the organic compound. There are several hundred CYPs in plants, and many of these enzymes are found in all other organisms. They have varying specificities and electron sources.

CYPs need an electron donor. For most plant CYPs this is NADPH P450 reductase, which transfers electrons from NADPH to the heme group in the cytochrome (Figure 13.9), although there are many CYPs that use other electron donors. The reaction of atomic O with organic molecules is **spin-forbidden**, so O has to be activated. The binding of an organic molecule with a CYP changes its conformation so that an electron can be added to the Fe^{3+} in heme, producing Fe^{2+}, and allowing molecular O_2 to bind to Fe^{2+}. Addition of a further electron and two protons results in the production of an oxygenated organic molecule (Figure 13.10). Many CYPs have evolved a high degree of specificity, but others can act on quite a wide range of substrates. In addition to decreasing hydrophobicity, oxygenation often increases the exposure of functional groups, preparing organic toxins for further reactions. Flavin monooxygenases and Cu monooxygenases are also used for oxygenation in many microbes, but although the genes for flavin

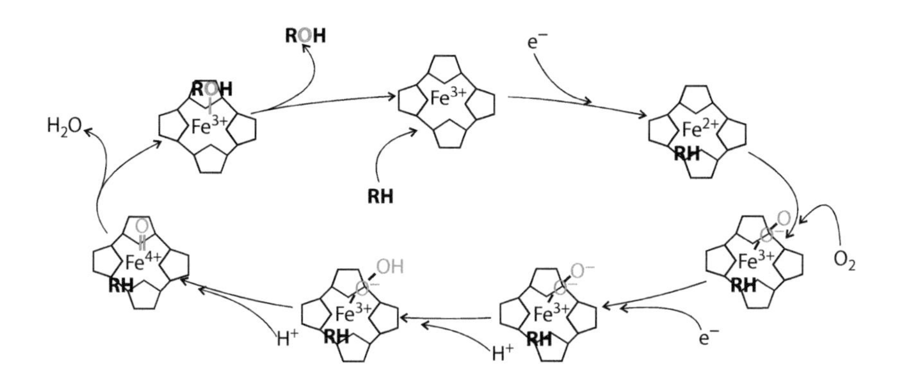

Figure 13.10. The catalytic cycle of cytochrome P450 (CYP). Organic molecules, including xenobiotics (RH), bind to CYP, changing its configuration so that electrons can be passed to Fe^{3+}, reducing it to Fe^{2+}. This reacts with O_2 and then two protons to produce the reactive Fe(IV)O group, which reacts with the bound RH to produce ROH and return the heme to its initial state.

monooxygenases exist in plants, their role in detoxification reactions is less clear. Not only can oxidases, peroxidases, and CYPs initiate detoxification of many organic toxins, but also their combined action can be important. For example, rhubarb (Figure 13.11) can detoxify sulfonated anthroquinones. These compounds are the basis of many persistent dyes and pigments, and the combined action of detoxifying enzymes in rhubarb is able to detoxify them in significant amounts. In general, the reactions that transform organic toxins are termed phase I of detoxification. Their products are often still quite toxic, sometimes more so than the original contaminants, but their hydrophobicity and exposed functional groups make them amenable to further detoxification. If polar compounds, which are not transferred readily to plant shoots, are degraded, phase I detoxification in the roots is generally responsible.

Figure 13.11. Rhubarb (*Rheum rhabarbarum*). The genus *Rheum* is in the Polygonaceae. A number of *Rheum* species have significant peroxidase and monooxygenase activity, and can break down (among other organic xenobiotics) sulfonated anthroquinones. *Rheum* species have been used in Chinese herbal remedies for centuries, and rhubarb roots (primarily for treating stomach disorders) were one of the most valuable commodities of the spice trade. European culinary rhubarb (shown in the figure) was a by-product of efforts to grow rhubarb for medicinal use. (See Page V & Schwitzguébel J. [2009] *Environ Sci Pollut Res Int* 16:805–816..)

In some plants, organic toxins and their transformation products can be deactivated by conjugation

During phase II of detoxification, organic toxins are conjugated, which can be especially important for the detoxification of non-polar compounds. For example, most cells, including those of plants, have the capacity to glucosylate many lipophilic compounds using cytosolic UDP-glucosyltransferases (GTs), which is ultimately important for controlling the composition of membranes. On exposure to organic toxins, many plants have been shown to increase the expression of GTs and to use them in glucosylation, which clearly plays a role in detoxification. Some studies have suggested that GT1s in particular can catalyze the majority of conjugation. In glucosylated form, the reactive sites of organic toxins are not exposed and the conjugates are often **amphipathic**. One of the most important **nucleophilic** compounds in plant cells is glutathione (GSH). GSH clearly has an ancient and vital role as a reducing agent in cells, but its ability as a nucleophile to conjugate electrophilic molecules is also important. A very large number of metabolic intermediates, especially in pathways that synthesize biologically active compounds, are strongly electrophilic. GSH-electrophile conjugates have been demonstrated to play important roles in controlling the activity of electrophilic metabolic intermediates. For example, GSH binds to intermediates in the biosynthetic pathway of jasmonic acid, allowing them to be transported between compartments. During the synthesis of **phytoalexins** such as camalexin, GSH helps to protect the plant from the reactive sites that evolved to be toxic to pathogens. Although there is no direct evidence for such a role in the syntheses of all the toxic compounds that plants can produce, it is likely that GSH conjugation plays a role in many of them. GSH is also important in the synthesis of S-containing metabolites such as glucosinolates, which in the Brassicaceae are the source of reduced S. The nucleophilic properties of

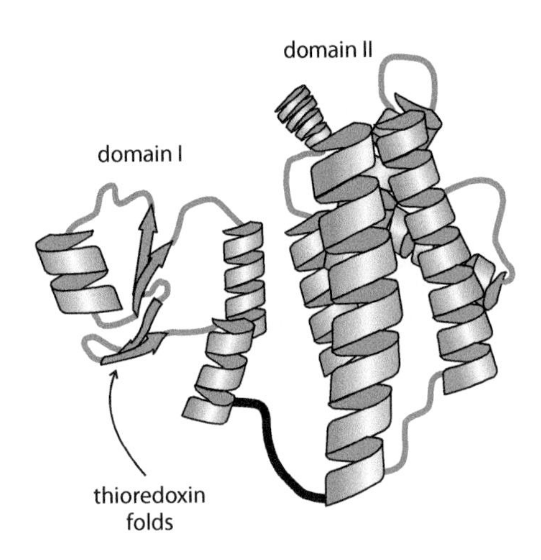

Figure 13.12. The structure and function of glutathione transferases (GSTs). There are numerous GSTs in plants and other organisms, but they are monophyletic and have a similar structure and function. The N-terminal thioredoxin folds in domain I are composed of four β-sheets, one of which is antiparallel, that form the G-site and bind GSH with great specificity. The C-terminal domain has six α-helices that can vary significantly. They form an H-site, with different variants binding different molecules. Induced-fit changes of conformation bring the molecule bound to the H-site into close proximity to the thiol on GSH bound to the G-site.

GSH mean that if electrophilic organic toxins enter the cytoplasm, whether they are natural or synthetic, they can be liable to conjugation. This conjugation can render them less reactive and more amphipathic.

There are numerous organic toxins that can enter the cytoplasm but which are only conjugated slowly. A vital role of the CYPs is to increase the electrophilicity of molecules and their probability of being conjugated. Despite this, spontaneous conjugation of many organic toxins is slow. All organisms have glutathione transferases (GSTs) that catalyze the conjugation of GSH to electrophilic molecules. In some instances, GSTs contribute up to 10% of cytosolic protein, and in a number of plant species there are known to be more than 50 genes encoding them. GSTs are monophyletic and have a well-established role in the detoxification of organic toxins, including many pharmaceuticals, in animal cells. In plant cells, GSTs are involved in many stress responses, in the synthesis of secondary metabolites, and in the detoxification of natural and synthetic organic toxins. For example, the breakdown products of chlorophyll are potentially toxic because they can trap light and produce ROS. GSTs have a role in detoxifying them. Many GSTs can also inadvertently confer the ability to detoxify synthetic toxins—best demonstrated by the many examples of herbicide resistance due to the action of particular GSTs. Some GSTs have a broad specificity for toxins, whereas others are highly specific, this property being determined by their structure.

GSTs have two domains—an N-terminal domain containing a highly conserved G-site with high specificity for GSH, and a C-terminal domain with a variable H-site that binds to hydrophobic xenobiotics (Figure 13.12). GSTs function as a dimer with the active site in a central cavity. The G-site has evolved from thioredoxin, having its characteristic four-stranded β-sheets and three α-helices. It binds GSH very specifically and deprotonates the thiol group on its cysteine residue, producing a thiolate anion that is nucleophilic. There is significant variation in the α-helices of domain II, but most of them seem likely to allow induced-fit mechanisms to bind electrophilic xenobiotics and bring them close to the thiolate anion bound on the G-site. GSTs are classified into seven classes (Table 13.1), including two (the φ- and τ-GSTs) that are unique to and particularly numerous in plants. They are functionally distinct in having a serine rather than a tyrosine residue in the active site. In some important leguminous crops (for example, soya), homoglutathione

Table 13.1. Classes of glutathione transferases in *Arabidopsis**

GST class	Number of genes	Occurrence	Active site	Function
τ-GST	28	Plants	Serine	Conjugation, high affinity for fatty-acid derivatives
φ-GST	14	Plants	Serine	Conjugation of a wide range of electrophilic molecules
DHAR	4	Plants	Cysteine	Conversion of dehydroascorbate (DHA) to ascorbate
θ-GST	3	Eukaryotes	Serine	Conjugation, but also GSH-dependent peroxidase (G-POX) activity
ζ-GST	2	Eukaryotes	Serine	Isomerization of *cis*-maleylacetoacetate to *trans*-fumarylacetoacetate
λ-GST	3	Eukaryotes	Cysteine	Conjugation of small molecules
TCHQD	1	Prokaryotes and eukaryotes		Tetrachlorohydroquinone dehalogenase (TCHQD) can conjugate chlorinated xenobiotics
Microsomal	1	Prokaryotes and eukaryotes		Membrane-bound GSH-dependent transferase

*The range of GSTs in many other species, although variable, is probably similar. There are plant-specific classes of GSTs that, although similar to other GSTs, have serine at active sites. In contrast, animals mostly have tyrosine. Many GSTs are involved in protecting the cell from electrophilic sites in natural and synthetic xenobiotics, but some, such as DHARs and ζ-GST, have other specific roles.

(γ-Glut-Cys-Ala) rather than GSH occurs, and in some cereals (for example, wheat and rice) hydroxymethylglutathione (γ-Glut-Cys-Ser) is an additional thiol—there are GSTs that specifically conjugate organic toxins to these alternative thiols. Plants often have most GST activity in their roots, and in general the GST activity in crops is up to 20-fold higher than that in native weeds. This is believed to be the result of selection for a trait unrelated to detoxification.

In some plants, conjugated organic toxins can be allocated to metabolically inactive compartments

If glucosylated and thiolated conjugates of organic toxins remain in the cytoplasm, they can still be toxic, and they make equilibria unfavorable for binding further toxins. In resistant plants, conjugated toxins are therefore mostly rapidly transported out of the cytoplasm, primarily to the vacuole. This compartmentation was apparent in some early studies of herbicides, and can be visualized with various fluorescent xenobiotics (Figure 13.13). ATP-binding cassette (ABC) transporters in the tonoplast actively transport numerous organic metabolites that are found in the cytoplasm into the vacuole. They function mostly as homodimers, with a highly conserved cytoplasmic domain that binds the nucleotides of ATP, and a variable transmembrane domain that can catalyze the transport of a wide range of molecules. Most of the metabolites in the cytoplasm are hydrophilic and thus do not readily diffuse across lipid membranes. Many conjugates are anions under cytosolic conditions. ABC transporters have a central pore that allows hydrophilic and anionic molecules to be transported across membranes. In plants there are several ABC transporter family C (ABCC) transporters, with ABCC1 often transporting organic toxin conjugates into the vacuole. Although the ABC transporters were first isolated using their xenobiotic-conjugate-transporting activity, the wide range of molecules that they are now known to transport strongly suggests that this was not their original function. For example, GSH itself is transported by ABCC1 after its synthesis in plastids.

Many xenobiotics do not persist as conjugates in the vacuole, but are subjected to phase III detoxification. In many instances the GSH has Gly and Glu residues removed to leave a Cys-conjugate. In some species a carboxypeptidase removes glycine first, whereas in others γ-glutamyltranspeptidase (GGT) removes glutamic acid first. In *Arabidopsis* roots, and perhaps in other species, phytochelatin synthase (PCS) in the cytoplasm has GGT activity and there is a carboxypeptidase in the plasma membrane providing degradative capacity in the cytoplasm. Cys-conjugates have a variety of fates—for example, acetylation, malonylation, methylation, and carboxylation have all been reported. In some instances a second round of glutathionylation in the cytoplasm has been documented. In some plants, volatile degradation products can be formed, but for some xenobiotics little degradation is possible, whereas for others a variety of intermediates can be formed. A significant proportion of Cys-conjugated xenobiotics and their degradation products can be exported from the vacuole and from the cell into the apoplast. Here they can be incorporated irreversibly into the cell wall together with lignin and cellulose, thereby completely removing the threat that they pose to plant metabolism. In the cell wall they can still be toxic to consumers of plant material, and they can enter soil organic matter during decomposition. These processes of conjugation, compartmentation, and removal all probably evolved to control the activity of reactive metabolic intermediates, but, having been useful since primordial times for detoxification, they provide many plants with a significant capacity to tolerate many organic toxins (Figure 13.14). However, there are many organic toxins that most plants have difficulty detoxifying, and some plants are particularly susceptible to the toxic effects of certain chemicals.

(a) time 0 min

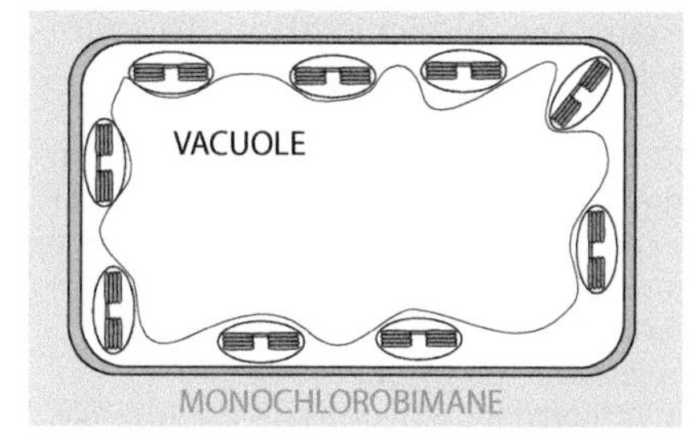

(b) time 30 min

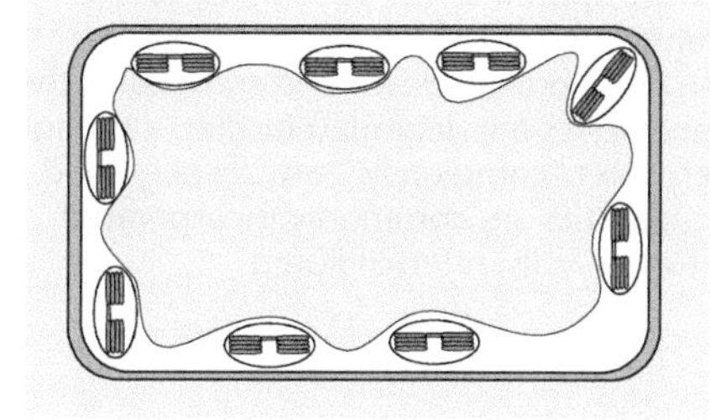

(c) time 180 min

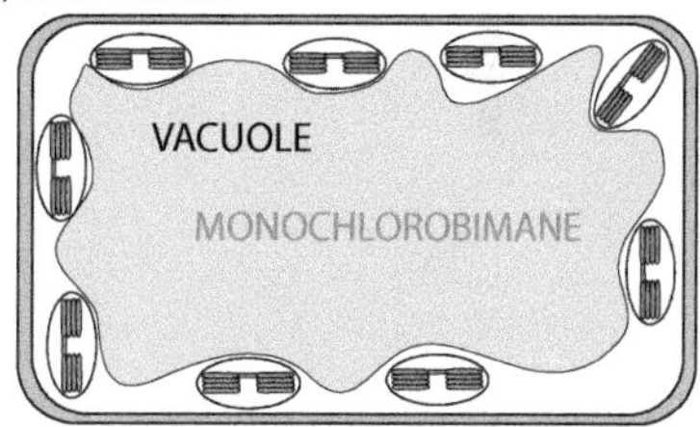

Figure 13.13. Compartmentalization of electrophilic organic compounds in the vacuole. Monochlorobimane (MCB) fluoresces on conjugation with glutathione, and is used as a model electrophilic xenobiotic. MCB studies reveal the rapid compartmentalization of electrophilic organic toxins.

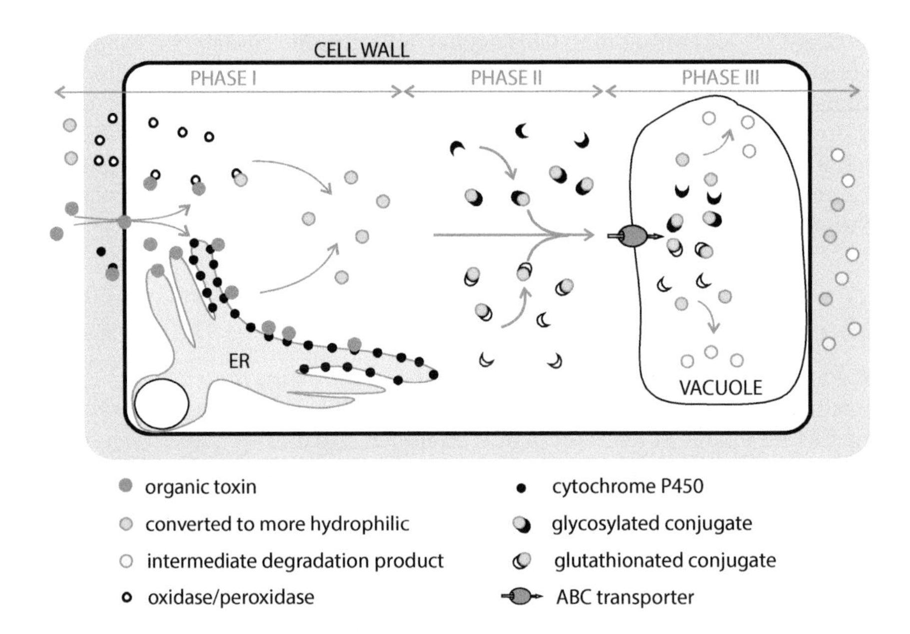

Figure 13.14. The detoxification of organic xenobiotics in plants. Organic xenobiotics are oxidized during phase I of detoxification, in part in the apoplast, but for lipophilic xenobiotics in particular primarily by cytochrome P450s (CYPs) in the endoplasmic reticulum (ER). Oxidation tends to increase hydrophilicity and reactivity. In phase II of detoxification, conjugation by glycosylation and glutathionylation neutralizes electrophilic sites on xenobiotics and prepares them for transport into the vacuole by ABC transporters. In the vacuole, conjugates are often broken down and sometimes the xenobiotics are degraded further, although very rarely completely. Partially degraded xenobiotics are commonly incorporated into cell walls, often in lignin.

Non-target-site herbicide resistance can evolve from xenobiotic detoxification mechanisms

Synthetic herbicides are organic toxins that have been selected for their toxicity to plants. The many available synthetic herbicides have a range of modes of action (Figure 13.15). Synthetic herbicides started to enter widespread commercial use in the 1950s, and until about 1990 there were regular introductions of herbicides with new modes of action. However, after the early 1990s no herbicides with new modes of action were introduced (in contrast to the development of new pesticides and fungicides). This has been attributed in significant part to the introduction of herbicide-resistant genetically modified (GM) crops, which changed the market for herbicides dramatically, because new modes of action are (at least in theory) possible. Weeds are now known that have developed resistance to herbicides with all the modes of action currently available. There are also numerous weeds that are resistant not only to a single type of herbicide but also to multiple types of herbicide. Currently available herbicides essentially all have specific targets in plant metabolism. Much herbicide resistance occurs because of target-specific changes, but non-target-specific herbicide resistance makes a significant contribution and is based on plant detoxification mechanisms.

For some herbicides, such as acetolactate synthase (ALS) inhibitors, resistance was reported within 3 to 4 years of their introduction and is widespread across numerous species, but for 20 years after the introduction of glyphosate there were no reports of glyphosate-resistant weeds. It was suggested that the mechanisms whereby glyphosate target-site-specific resistance could occur in plants would inevitably decrease the efficiency of its metabolically essential enzyme target, preventing the evolution of resistance. The introduction of glyphosate-resistant GM crops in 1996 resulted in what was, from both an agronomic and an ecotoxicological perspective, a widely successful herbicide control regime. However, it did lead to a large increase in single-tactic weed control, based only on glyphosate, and there has subsequently been an increasing list of weeds with significant glyphosate resistance. It has quickly led to the development of weeds with a resistance once thought to be very unlikely to occur. Since 1990, food production systems have therefore become particularly entwined with the phenomenon of herbicide resistance in plants. It is clear that resistance to the primary global weed control strategy, namely herbicides, can evolve rapidly, that this is

CELL METABOLISM

A. inhibitors of lipid synthesis (ACCase)
target: acetyl CoA carboxylase, enzyme of lipid synthesis
application: post-emergence for phloem translocation
location: meristems, where lipid synthesis is active
susceptibility: especially Poaceae, dicots less susceptible
examples: cyclohexanediones (DIMs), aroxylphenoxypropionates (FOPs), phenylpyrazoline

N. other inhibitors of lipid synthesis
target: cell development
application: pre-emergence for xylem translocation
location: mostly meristems
susceptibility: monocots and dicots
examples: thiocarbamates (organosulfurs), chlorocarbonic acids, benzofuranes

B. inhibitors of ALS
target: acetolactate synthase (acetohydroxy acid synthase), key plant enzyme for synthesis of branched-chain amino acids
application: pre- or post-emergence, xylem and phloem
location: often mostly shoots
susceptibility: broad spectrum
examples: sulfonylureas, imidazolinones

G. inhibitor of EPSPS
target: enolpyruvylshikimate-3-phosphate synthase (EPSPS), key plant enzyme for synthesis of aromatic amino acids
application: post-emergence, phloem translocation
location: roots and shoots
susceptibility: broad spectrum
example: glyphosate

H. inhibitor of GS
target: glutamine synthetase, key plant enzyme for assimilation of ammonia
application: post-emergence, phloem translocation
location: shoots
susceptibility: broad spectrum
example: glufosinate

CELL DIVISION AND GROWTH

O. synthetic auxins
target: auxin action sites
application: post-emergence, xylem and phloem
location: meristems, auxin-dependent growth sites
susceptibility: broadleaves especially
examples: phenoxy-carboxylic acids, benzoic acids

K1. inhibitors of microtubule assembly
target: cytoskeleton
application: pre-emergence for xylem translocation
location: mostly meristems
susceptibility: mostly monocots but some dicots
examples: dinitroanilines, pyridines

K2. inhibitors of microtubule polymerization
target: cytoskeleton
application: pre-emergence for xylem transport
location: mostly meristem
susceptibility: mostly monocots but some dicots
examples: carbamates

K3. inhibitors of cell division
target: long-chain fatty acid synthesis, especially at sites of cell division
application: pre-emergence, xylem transport
location: roots and shoots
susceptibility: mostly monocots but some dicots
examples: choloracetamides, oxyacetamides, acetamides

L. inhibitors of cellulose synthesis
target: developing cell walls
application: pre-emergence, xylem transport
location: roots and shoots
susceptibility: mostly dicots
examples: benzamide, alkylazines

PHOTOSYNTHESIS AND PIGMENTS

C. inhibitors of photosynthesis (PSII)
target: plastoquinine binding site on D1 polypeptide
application: pre- and post-emergence, xylem translocation
location: photosystem II complexes in thylakoids
susceptibility: broadleaves more susceptible
examples: triazines, triazinones, phenylureas, uracils, nitriles, amides

E. inhibitors of PPO
target: protoporphyrinogen oxidase (PPO) in the chloroplast, which is essential for biosynthesis of chlorophyll
application: post-emergence for contact or xylem transport
location: shoots
susceptibility: mostly dicot plants
examples: diphenylethers, thiadiazoles, triazolinones

D. inhibitors of PSI
target: ferredoxin of photosystem I
application: post-emergence, contact or xylem transport
location: shoots
susceptibility: broad spectrum
examples: bipyridyliums

F. inhibitors of carotenoid synthesis
target: enzymes of carotenoid biosynthesis or their co-factors, lycopene cyclase
application: often pre-emergence, xylem transport
location: shoots
susceptibility: monocots and dicots
examples: dimethylethers, isoxazolidinones, triazoles

Figure 13.15. Mechanisms of action of herbicides. Herbicides have a number of different modes of action, but few new modes of action have been discovered recently. Some herbicides are moved to all parts of the plant rapidly through the xylem and phloem, some only move with mass flow of water through the xylem, while others act only on contact. This helps to explain the strategies developed for applying herbicides, which can be used from pre-emergence for weeds through to defoliation of root crops for harvest. Some herbicides work best on monocots such as grasses, whereas others work best on broadleaved dicots. The widely used Herbicide Resistance Action Committee (HRAC) classification categorizes herbicides from A to Z according to their mode of action. Some important classes of herbicide, organized according to the general aspect of plant metabolism on which they act, are shown in the figure.

significantly complicating weed control, and that not only does the currently most successful weed control strategy depend on dramatic modification of herbicide resistance in crops, but also it is probably effective at producing this in weeds. The most widely considered solutions to future weed control challenges also focus on herbicide resistance. GM crops that are resistant to herbicides other than glyphosate are gaining a larger market share, there is evidence that the market opportunities offered by herbicides with a new mode of action are encouraging their development, and there is also significant optimism about the use of crops with "stacked" herbicide resistance to two or more herbicides. Herbicide resistance in weeds is mostly, but not always, associated with fitness costs—there are few "superweeds"—but it has agronomic costs, and the changes in herbicide use that this entails can have environmental and financial implications. Some herbicide-resistant weeds also cause ecological problems. The many agronomic systems that have been developed for weed control that do not rely exclusively on synthetic herbicides are not only finding more widespread use, but are also the focus of more detailed scientific investigation. Underpinning many aspects of these systems are the production of and resistance to organic toxins.

Some early studies of weeds that evolved herbicide resistance found metabolites which suggested that herbicides were being broken down by CYPs. Numerous *in-vitro* studies using microsomes have shown that plant CYPs

Table 13.2. The most problematic herbicide-tolerant weeds*

Species	Family
Lolium rigidum	Poaceae
Avena fatua	Poaceae
Setaria viridis	Poaceae
Echinochloa crus-galli	Poaceae
Eleusine indica	Poaceae
Amaranthus retroflexus	Amaranthaceae
Amaranthus hybridus	Amaranthaceae
Kochia scoparia	Chenopodiaceae
Conyza canadensis	Asteraceae

*For many weeds there are numerous populations with different herbicide-resistance profiles, but the species listed here are regarded as the most problematic herbicide-resistant weed species by the Weed Science Society of America.

can detoxify a variety of herbicides, that herbicide-resistant weeds often have CYPs that can detoxify herbicides, and that particular plant CYPs expressed in, for example, yeast can detoxify herbicides. CYP81A6 from a sulfonylurea-resistant rice population, when engineered into *Arabidopsis* and tobacco, confers sufficient resistance for it to have been suggested as a marker for transformation. Inhibitors of CYPs, such as malathion, when applied to herbicide-resistant weeds can be used to distinguish non-target-site resistance due to changes in CYPs from target-site resistance. By the late 1990s it was well established that changes in the capacity of CYPs could contribute significantly to the evolution of herbicide resistance in the field. Thus in 2012 a survey of herbicide resistance in wild oats, one of the world's most problematic herbicide-tolerant weeds (Table 13.2), and economically the most important in the Canadian prairies, found it to be due in large part to CYP-associated resistance to acetyl co-enzyme A carboxylase (ACC) and ALS inhibitors. *Lolium rigidum* in its native Eurasia, and particularly in Australia where it is economically the most important weed, has populations that are resistant to numerous herbicides, and CYPs make a significant contribution to much of this resistance. Screening in crops such as sunflowers has revealed sufficient resistance from CYPs for these to have been suggested as the basis for developing herbicide-resistant crops. The genetics of CYP-induced herbicide resistance are complex and variable, especially in a fecund outbreeder such as *L. rigidum*. There are instances of specific CYPs detoxifying specific herbicides, and also of CYPs with broad-spectrum activity. Resistance due to CYP activity is particularly challenging to address because it can evolve quickly, it can confer resistance to multiple herbicides, and it is unpredictable. In human cells, increased CYP activity is often due to gene duplication, but in plants it is likely that changing expression patterns, gene duplications, and epigenetic effects all contribute. CYPs make a particularly significant contribution to herbicide resistance in grasses.

Shortly after the introduction of triazine herbicides in the 1950s, it became apparent that many varieties of maize were relatively resistant to them. By 1970 this had been linked to the high activity of GSTs. There are now numerous examples of evolved herbicide resistance in weeds caused by the activity of particular GSTs. In wheat and oil crops in Europe, especially in the UK, herbicide-resistant *Alopecurus myosuroides* (black grass) has been a significant problem for more than 25 years. Populations with resistance to herbicides operating via different mechanisms of action (phenylurea inhibitors of PSII and ACC inhibitors) can detoxify these herbicides using GSTs. Interestingly, direct glutathione peroxidase (G-POX) activity is at least as important in detoxification as conjugation activity. There is significant resistance to protoporphyrinogen oxidase (PPO) inhibitors in soya bean and other legumes that has been ascribed to G-POX activity of τ-GSTs. Resistance to thiocarbamates in several weeds has been ascribed to φ-GSTs. The molecular details of GST binding of several herbicides are known, site-directed mutagenesis has been used to alter specificity, and some of the associated genetics has been unravelled. The activity of GSTs can therefore help to explain the evolution of non-target-specific herbicide resistance. GST activity complements CYP activity, and often seems not only to contribute to detoxification of a specific herbicide but also to help plants to resist the stresses, especially oxidative ones, that herbicides produce. Some important metabolites of glyphosate detoxification can be identified in plant cells, and are often found in glyphosate-resistant weeds. However, they do not generally explain a significant proportion of non-target-specific glyphosate tolerance. Several glyphosate-resistant weeds have enhanced vacuolar sequestration of glyphosate, and ABC efflux transporters can be used to engineer glyphosate-resistant *E. coli*. Phases I, II, and III of detoxification can therefore play a role in herbicide resistance.

Some herbicide safeners were discovered during the early development of herbicides. These are compounds that, when applied to crops, make herbicides safe to use on them, by in effect inducing herbicide resistance. Several herbicide safeners have been widely used for decades (Table 13.3). They work almost exclusively on cereal crops. Herbicide safeners have now been shown to work by inducing detoxification phase I-III mechanisms, especially the GSTs that are widely active in the Poaceae—that is, they prepare cereals to tolerate toxin attack. The pathways via which safeners induce defense responses are complex, but include oxylipins (the products of lipid oxidation) and several families of transcription factors. However, induced defense responses cannot explain much of the non-target-site-specific herbicide resistance in, for example, glyphosate-resistant weeds. In naturally glyphosate-resistant plants such as *Mucuna pruriens* (velvet bean), cuticle waxes limit glyphosate uptake, and glyphosate translocation is greatly reduced. In several important weeds that have evolved glyphosate resistance, including *Sorghum halepense*, *Lolium rigidum*, *Conyza canadensis*, and *Digitalis insularis*, it is their limited translocation of glyphosate in the phloem that contributes most to non-target-site resistance (Figure 13.16). Genetic analyses have indicated that this is a single gene trait that seems likely to be associated with a transport protein. Thus essentially all aspects of the organic toxin detoxification mechanisms of plants have contributed to the development of herbicide-tolerant weeds.

Table 13.3. Some widely used herbicide safeners

Safener	Crop safened
Dichlormid	*Zea mays*
Benoxacor	*Zea mays*
Fenclorim	*Oryza sativa*
Cloquintocet-mexyl	*Triticum, Hordeum*
Fluorazole	*Sorghum vulgare*

Target-site resistance helps plants to adapt to catastrophic exposure to herbicides

Much of the herbicide resistance in weeds can be directly attributed to changes in herbicide targets. In population terms, plant exposure to herbicides is a catastrophic event. Together with non-target-site resistance, the molecular details of target-site resistance provide not only insights of profound practical importance, but also a better understanding of the capacity of plant populations to deal with catastrophic exposure to organic toxins. Target-site resistance to glyphosate makes a significant contribution to resistance in the many weeds that have evolved this capacity. It is due to changes in the expression and/or properties of 5-enolpyruvylshikimate-3-phosphate (EPSP) synthase. This enzyme catalyzes the synthesis of EPSP from shikimate-3-phosphate and phosphoenolpyruvate (PEP). EPSP synthase is used in the synthesis of aromatic amino acids, including tryptophan, phenylalanine, and histidine. Glyphosate binds to EPSP synthase, blocking the PEP-binding site, inhibiting the synthesis of aromatic amino acids, and fatally disrupting metabolism. In glyphosate-resistant *Amaranthus palmeri*,

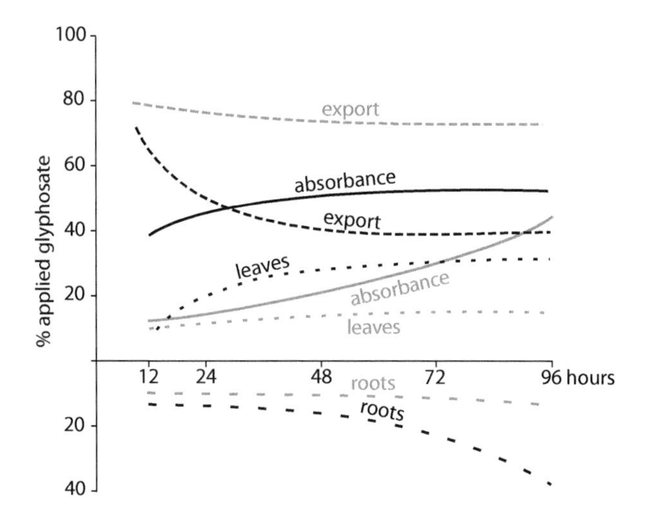

Figure 13.16. Glyphosate translocation in susceptible and resistant populations of *Digitaria insularis*. In Brazil, *D. insularis* is an important weed, and significant glyphosate-resistant populations have been identified. Compared with susceptible populations (black lines), resistant populations (green lines) absorb glyphosate slowly and export it less from areas of application to leaves and roots, helping to confer resistance. (Data from de Carvalho LB, Alves PL, González-Torralva F et al [2012] J *Agric Food Chem* 60:615–622.)

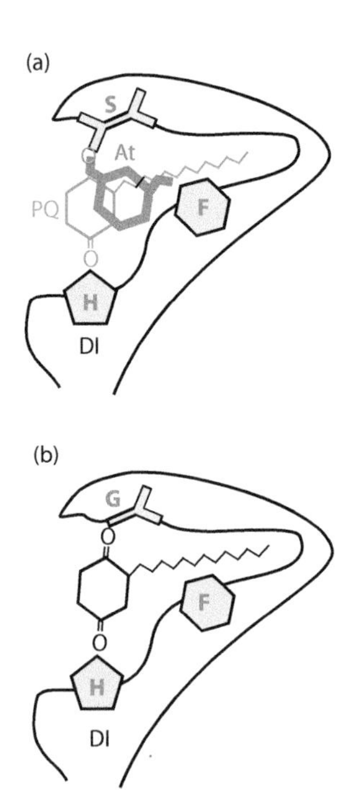

Figure 13.17. Site-specific triazine resistance. (a) Triazine herbicides such as atrazine (At) block the plastoquinone (PQ)-binding site of the D1 polypeptide of photosystem II. Usually a combination of binding to histidine-215 (H), phenylalanine-225 (F), and serine-264 (S) means that At can compete with PQ for binding. (b) Several mutations at serine-264 are known, but the commonest one in many species is serine to glycine at 264, which favors PQ binding. Mutations at serine-264 explain a high proportion of the widespread triazine resistance in plants, and they have occurred many times. (Data from Powles SB & Yu Q [2010] *Annu Rev Plant Biol* 61:317–347)

gene amplification (up to 150-fold) is associated with large increases in the concentration of active EPSP synthase and with increased glyphosate resistance. In many other glyphosate-resistant species there have been substitutions at proline 106 of EPSP synthase that change the spatial surrounds of the active site, decreasing the binding of glyphosate and allowing greater access of the PEP in the presence of glyphosate. These changes only confer moderate glyphosate resistance by themselves, but when combined with other mechanisms of resistance they demonstrate the gamut of mechanisms that can be quickly selected for when plants are subjected to catastrophic insult with organic toxins. Many glyphosate resistance changes have consequences for plant fitness, and are therefore currently problematic agriculturally rather than ecologically, but given the speed with which glyphosate resistance—once thought to be unlikely—has developed, the evolution of traits that confer resistance but do not have fitness consequences is probably to be expected.

Triazines have been among the most important herbicides for more than half a century. They bind to the binding site for plastoquinone (PQ) on the D1 polypeptide of the photosystem II complex of thylakoid membranes. This eliminates the quenching of PSII, producing fatal amounts of ROS. Triazines block the PQ-binding site but do not bind to the same amino acids as PQ does. A widespread substitution of serine by glycine at amino acid 264 in D1 still allows PQ to bind, but not triazines (Figure 13.17). This substitution also confers resistance to many urea herbicides that bind to this site. Several other substitutions for Ser-264 have been described, but Ser-264-Gly accounts in significant part for resistance to triazines in a wide variety of weed populations around the world. There is also widespread resistance to ALS-inhibitor herbicides, caused by substitution at proline 197 in the acetohydroxyacid synthase (AHAS) that catalyzes the first step in the synthesis of branched-chain amino acids, such as valine, leucine, and isoleucine. Numerous substitutions are known that prevent the action of ALS inhibitors but allow enzyme function. ALS inhibitors do not bind to the active site of AHAS, but block the entry channel. Substitutions that prevent herbicide binding but do not prevent function have evolved relatively easily, and can be specific to particular ALS-inhibitor herbicides. For other classes of herbicides there are also sequence variants in key proteins that have conferred herbicide resistance. Plant populations therefore respond to the massive selection pressure generated by herbicide use by relatively quickly evolving resistance in what is now an exemplar of forced evolution. There are many parallels with the evolution of antibiotic-resistant microbes as a result of the widespread use of antibiotics. Herbicide-resistant plants that have evolved in response to the widespread use of herbicides are an increasingly expensive threat to world agriculture, in significant part because more herbicides are needed in order to control them, with consequent toxicological implications.

Plants enhance the bioremediation of water and soils contaminated with organic xenobiotics

Due to the global pressure on freshwater resources, methods for treating the large proportion of fresh water that becomes contaminated are vital. Water is such a valuable resource that for point-source, high-concentration contamination, energy-intensive, centralized "pump-and-treat" methods are economically viable. Water contaminated at low concentration via non-point sources is more challenging to treat. Wetland systems in which plants have a vital role have been developed for decentralized treatment of such water, and provide significant insight into the potential of plants for the bioremediation of environmental media contaminated with organic toxins. Constructed wetlands have proved useful for treating sewage over several decades. In these wetlands, organic matter in sewage is broken down by

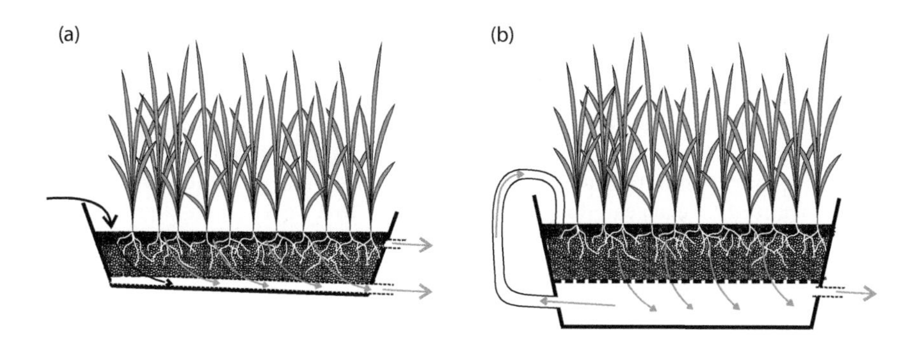

Figure 13.18. Constructed wetlands for water treatment. (a) Most constructed wetlands have an impermeable layer or membrane that is filled in with substrate (often inert) and drainage channels. Inlet and outlet pipes can be set at different heights to encourage vertical flow or horizontal flow, which can change the proportion of surface, light, and volatilization processes. (b) Reservoirs and recirculation of water are possible, as is sedimentation in holding tanks before passage through the constructed wetland. Hydrological flows, plant species, and physico-chemical properties can all be varied for specific uses.

plant-enhanced activity of microorganisms. In natural wetlands, the concentration of organic toxins in water exiting the system can be lower than that in water entering it. This finding, and the ability of many plants to tolerate organic toxins, inspired constructed wetlands designed for bioremediation of organic toxins. For contaminated run-off—for example, from roads or fields—buffer strips of plants are quite widely used to decrease the input of contaminants into water courses. More elaborately designed constructed wetlands can be aimed at specific organic toxins and more complete treatment (Figure 13.18).

Many plants that are grown in tissue culture or hydroponic systems can degrade organic toxins either after taking them up or via root exudates. Floating aquatic macrophytes can thus help to remove organic toxins directly from water (Box 13.2). In constructed wetlands, in almost all instances the existence of plants enhances degradation of contaminants. For some organic toxins in some wetlands this is because of plant uptake and detoxification, but in most it is not. In most constructed wetlands, organic toxin degradation occurs predominantly as a "rhizosphere effect." In general, wetland plants with controlled O_2 flow in their roots, including species in the genera *Phragmites*, *Typha*, *Scirpus*, and *Juncus*, are most effective in promoting the rhizosphere effect. The rhizosphere of such plants includes a high population of heterotrophic microorganisms with diverse metabolic activity because of the steep redox gradients and the diversity of organic compounds in root exudates. Many organic toxins are strongly adsorbed to particulate matter, and the plants in constructed wetlands have a significant role in promoting their accumulation and enhancing the mobilization of toxins from them, mostly because of redox processes. For volatile organic compounds, phytovolatilization—both from phytodegradation products and directly through aerenchyma—can be significant and sometimes even environmentally problematic. Thus plants are a vital component of constructed wetland bioremediation systems for organic toxins, primarily because they can tolerate organic toxins while injecting the oxygen and exudates necessary for rhizosphere formation, but they can also alter the flow of particulate matter and water to favor degradation, and can contribute to volatilization and detoxification.

Organic toxin concentrations in soils naturally attenuate because microorganisms degrade them. This process is often too slow to remediate soils that are contaminated with widely dispersed organic contaminants at low concentrations. At high-value sites, or for relatively small volumes contaminated at high concentrations, there are long-established soil-washing methods for decontamination. There are also established bioremediation techniques based on injection of microbial populations into soils. Soil washing in particular disrupts the soil, and neither technique exploits the rhizosphere effect, so more benign phytoremediation techniques that do so have been developed. Soil contamination with petroleum hydrocarbons (PHCs) is widespread, and natural attenuation is relatively slow. The BTEX,

BOX 13.2. FLOATING AQUATIC MACROPHYTES AND WATER TREATMENT

Floating aquatic macrophytes take up nutrients from the water on which they float. Many of the freshwater systems that these plants inhabit have been contaminated with toxic organic compounds. Some aquatic macrophytes have proved to be quite resistant to the effects of xenobiotic organic contaminants, perhaps because of the range of toxic compounds that can be found in some of the water systems they inhabit. One such species is the water hyacinth, *Eichhornia crassipes* (Figure 1).

E. crassipes is indigenous to Amazonia, but during the twentieth century it spread to the freshwater systems of most tropical and subtropical regions of the world, most commonly being transported by humans. It is now the world's most problematic aquatic weed. It primarily inhabits **lentic** freshwater systems, mostly lakes. In many parts of the world at various times in the twentieth century *E. crassipes* caused severe problems in even medium-sized lake systems. By the 1990s it was a severe problem in the world's second largest freshwater lake, Lake Victoria in Central Africa, which it had probably reached after being introduced to Burundi (Figure 2). In the late 1990s it covered almost 80 square miles near the north and northeastern shores of Lake Victoria.

E. crassipes floats on the water surface, supported by swollen stems containing a honeycomb of air pockets (Figure 1). It also produces a dense root system that absorbs nutrients from the water on which it floats. It can form extensive rhizome systems and spreads by means of stolons, forming extensive, interconnected mats. Under non-limiting conditions these mats can double in area every 1 to 2 weeks. *E. crassipes* can be harvested from mats manually and mechanically, but this often expensive process has had mixed success in controlling this weed. The mats of *E. crassipes* when formed in harbors and fishing areas not only alter the water quality and biota beneath them, but also physically prevent the passage of small and even medium-sized boats. On Lake Victoria there have been wide variations in the extent of infestation with *E. crassipes*, with maximal infestation often related to the influx of sewage and nutrient-rich waters. *E. crassipes* has proved to be tolerant of quite high concentrations of a wide range of organic toxins. Its roots have a high capacity for absorbing molecules from water, and, in conjunction with their microflora, can detoxify and degrade a range of organic toxins. The rapid growth of *E. crassipes* and its ability to absorb organic xenobiotics have led to its use in the development of a range of phytoremediation systems for water contaminated with toxic organic compounds.

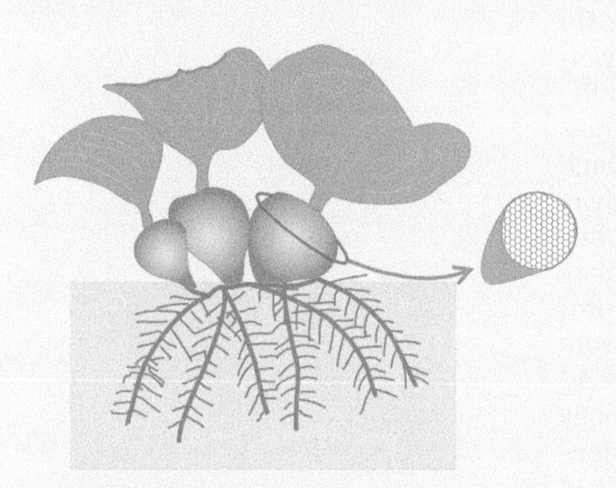

Figure 1. *Eichhornia crassipes* (water hyacinth). This species is native to Amazonia, but is now found across the tropics and subtropics.

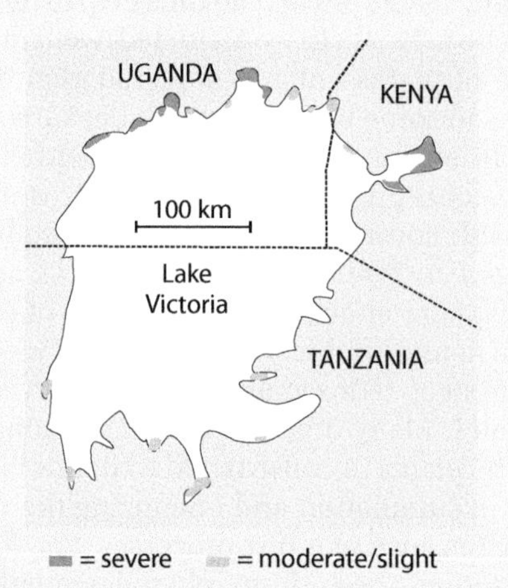

Figure 2. Distribution of *Eichhornia crassipes* on Lake Victoria. *E. crassipes* reached Lake Victoria in the late twentieth century and spread rapidly. The impact of the infestation has varied since then, but overall the increased difficulties of access to ports and fishing grounds, together with the costs of control measures, have been very significant.

alkanes, and PAHs contained in many PHCs are tolerated and can be taken up and detoxified by many plant species. In the field, plants grown on PHC-contaminated soils enhance degradation significantly. In most instances this is because of the rhizosphere effect, but plants also play a role in directly degrading and volatilizing PHCs. Different plant species are particularly effective at stimulating PHC degradation under particular conditions, and members of the Poaceae are often suggested as being most effective because of their extensive adventitious root systems. The ability of plants to stimulate the degradation of aged PHC contamination has frequently been noted. POPs are generally resistant to microbial attack and too hydrophobic for uptake by many plants but, despite this, plants can enhance the attenuation of POPs in soil. For species such as *Morus rubra* (red mulberry), rhizosphere processes result in co-metabolism of PCBs—that is, exudates from the roots (in this case phenolic) stimulate bacterial populations that together with plant enzymes can attack POPs. Plants lack the enzymes necessary to attack the biphenyl bond, and thus cannot completely degrade PCBs, but in conjunction with bacteria such degradation is possible. The Cucurbitaceae, and in particular *Curcurbita pepo* subsp. *pepo*, together with other species such as *Solidago canadensis*, have been shown to take up and translocate PCBs. Thus in the case of organic xenobiotics that are relatively amenable to degradation (such as PHCs), and also recalcitrant POPs, plants can contribute significantly to their attenuation in soil primarily via rhizosphere effects, with volatilization and detoxification sometimes adding to the process.

Plant-enhanced remediation of soils contaminated with xenobiotic toxins is slow, and although it is less expensive than mechanical remediation it still requires resources for monitoring and maintenance. Therefore the finding of value in the biomass produced during phytoremediation, or the manipulation of complex rhizosphere effects to enhance degradation rates, would be advantageous. Rhizosphere effects are significant not only for herbaceous plants but also for trees, which can often also explore greater soil depths. For *Salix*, *Populus*, and several other species, many studies have shown that wood is a valuable by-product during remediation. In many cases, reeds such as *Phragmites australis* can also provide biomass of some value. Mycorrhizal associations and endophytic bacteria frequently contribute significantly to rhizosphere effects, and manipulation of their populations might increase their efficiency. Techniques to enhance contaminant availability in soils include soil amendments and electrokinetics. In many rhizospheres the C:N ratio limits metabolic processes. The addition of either C- or N-rich compounds to optimize the C:N ratio has been used to increase remediation efficiency. This can sometimes be achieved by using legumes to introduce N via rhizobial symbionts, or adding organic wastes. Intercrops of members of the Fabaceae and Poaceae have demonstrated increased degradation rates partly for this reason. Surfactants, of which progressively more environmentally benign examples are becoming available, can be used to increase the bioavailability of many organic toxins. The use of electrokinetics, which involves the application of low direct current between widely spaced electrodes, often in conjunction with electrolyte addition, is well established in soil decontamination of inorganic compounds, and can significantly alter the flow and availability of contaminants, sometimes in conjunction with plant uptake. It has found less use for organic contaminants, but combinations of surfactants and plants that complement electrokinetic effects can be effective for ionic organic compounds in particular. The complex interactions between soils and plants offer great potential for the development of phytoremediation of organic toxins, but also make it difficult to manage (Figure 13.19). In temperate climates, degradation clearly slows down or even stops during winter, and numerous other long-term changes of hydrology or physico-chemical variables can become limiting. There are also significant regulatory and infrastructural challenges involved in implementing phytoremediation of organic xenobiotics, but such is the scale of the problem that it is likely to find at least a significant niche in remediation strategies.

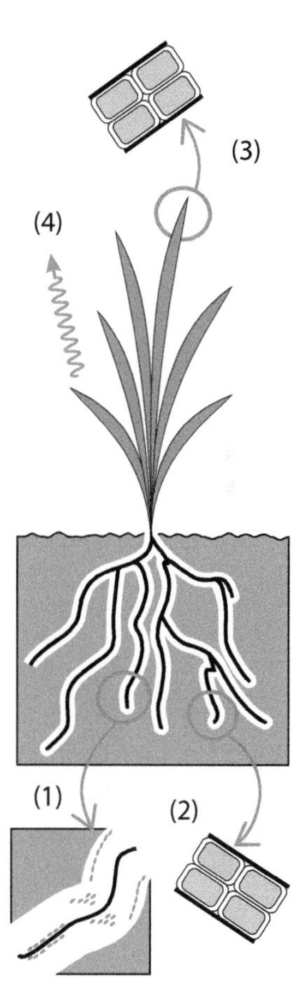

Figure 13.19. Processes of plant-enhanced degradation of organic xenobiotics. Depending on the contaminant and the environmental conditions, a wide range of processes can contribute to plant-enhanced degradation of organic xenobiotics. (1) An oxygenated rhizosphere provides habitats for a wide range of heterotrophic microbes. Many are stimulated by root exudates, and many utilize redox gradients. In some instances, co-metabolism between microbes and plants can degrade contaminants that are resistant to either of these alone. (2) For lipophilic xenobiotics in particular there can be accumulation, degradation, and compartmentalization in roots, whereas for polar and ionic xenobiotics, transport to the shoots can lead to degradation at that site (3). Volatile xenobiotics or metabolites can be volatilized from plant shoots, sometimes via the aerenchyma systems of wetland plants (4).

Manipulation of plant tolerance of organic toxins is of increasing importance

By far the most widely grown GM plants can tolerate herbicides. From their introduction in 1996, the use of GM crops increased to over 150 million hectares within 15 years, a rate of increase that is likely to be sustained well into the twenty-first century. In 2012 about 60% of this area was herbicide-resistant crop, with a further 15% having herbicide resistance as a stacked trait. The most widely grown herbicide-resistant crops are glyphosate resistant. Most glyphosate-resistant crops have been engineered to express EPSP synthase from *Agrobacterium tumefaciens* CP4. The next most important herbicide-resistant crops are resistant to glufosinate, which is a glutamate analog that prevents the condensation of glutamate and ammonia by glutamine synthetase (GS). Glufosinate-resistant crops express a phosphinothricin acetyltransferase (PAT) from *Streptomyces* that acetylates glufosinate, preventing it from inhibiting GS. Clearly plants can be modified and grown to tolerate organic toxins with site-specific effects. The xenobiotic concentrations in many contaminated soils can cause toxicity (for example, TNT at many contaminated sites decreases plant growth), so there is much interest in increasing the non-site-specific resistance of plants to organic toxins. However, this is not the only limitation on toxin phytoremediation. The relatively limited availability and low uptake rates of many organic toxins from soils limit the application of phytoremediation, although not all sites necessitate rapid remediation. Plant degradation of most contaminants is incomplete, and consists of compartmentalization of products of partial degradation, sometimes in forms that can be toxic if other organisms are exposed to them. There are also some organic xenobiotics that cannot be metabolized by plants, but which can be metabolized by microorganisms. Modification of plant tolerance of organic toxins has therefore been the focus of much attention.

Phase I and phase II detoxification has been genetically modified in plants many times. Among the most widely researched CYPs is CYP2E1 from mammals. It has much higher activity than most plant CYPs against a range of xenobiotics. Model plants such as *Arabidopsis* and *Nicotiana* have been modified to express CYP2E1, as have trees such as *Sesbania*. These plants have enhanced capacity to degrade TCE, DDT, lindane, benzene, toluene, and many other xenobiotics. Yeast NADPH reductase expressed in plants has been found to increase the efficacy of CYP2E1. Bacterial CYP encoded by *xplA/B* and many plant CYPs found to have particularly high activity have been expressed in species with phytoremediation potential, increasing their capacity for degradation. Thus the oxygenase activity of phase I can be enhanced in candidate phytoremediation species. As with CYPs, the conjugation activity of several non-plant enzymes is higher than that of plant enzymes. For example, GTs and GSTs from *Trichoderma virens* have been used to increase conjugation activity and help to detoxify anthracene. Increased expression of GSH synthase can also increase the activity of GSTs. Co-expression of CYPs and GT/GST has been achieved to enhance detoxification capacity.

Xenobiotics such as TNT cannot be completely metabolized by plants, but can be by bacteria with nitroreductases. Genes that encode nitroreductases (for example, *pnrA* from *Pseudomonas putida*) have been used to modify plants, including tobacco and aspen, conferring the ability to degrade TNT. The bacterial *atzZ* gene encodes an atrazine chlorohydrolase that has been used to enhance atrazine degradation. PCBs are not degraded in plants, but bacterial dihydroxybiphenyl dioxygenases can be expressed in plants to enable them to do this. Laccases from fungi have also been expressed in plants, conferring the ability to degrade phenols in the rhizosphere. Plants with these novel degradative abilities might find widespread use as the pressure to decontaminate sites increases. In addition, modified plants have been developed that might act as biomonitors of toxin concentrations during phytoremediation. Mammalian cytosolic AhR receptors that when joined to xenobiotics bind to XRE elements, increasing the expression of CYPs, have been engineered into tobacco in conjunction with β-glucuronidase (GUS) reporters to provide an *in-planta* monitor of xenobiotic concentrations. Manipulations to maintain root growth, such as increased expression of ACC deaminase to overcome stress-induced effects of ethylene, and even the production of biosurfactants to increase toxin availability in the rhizosphere, are possible in the laboratory. The modification of plant interactions with xenobiotics was the most successful part of the first phase of use of GM plants, and seems likely to find a variety of roles in the coming decades.

Summary

Organic xenobiotics have a wide range of availability in soil and concentration factors in plants, but there are many instances in which environmental contamination with these organic compounds results in toxicity to plants and/or transfer into human food chains. Plants have the ability to control the activity of many reactive functional groups, especially electrophilic ones, and enzymes that probably evolved to catalyze reactions of primary and secondary metabolism in plants have been adapted to detoxify organic toxins during low-level exposures (Figure 13.20). In contrast to microorganisms, plants do not completely degrade most xenobiotics, there are many xenobiotics that they cannot degrade at all, and in contrast to animals they have only a limited ability to secrete toxins and degradation products, which can therefore accumulate in plants. The widespread application of herbicides to crop plants is perhaps the largest-scale catastrophic stress to which humans have subjected plants. In response, plant populations have evolved site-specific and non-site-specific herbicide resistances remarkably quickly. Although plants tend not to degrade organic toxins completely, they are often quite tolerant of them, and have therefore attracted much attention because of their potential for phytoremediation. Herbicide-tolerant GM plants and the many plants that have been modified to remediate organic toxins have ensured that plant-xenobiotic reactions are now, and will be for some time, central to GM plant development.

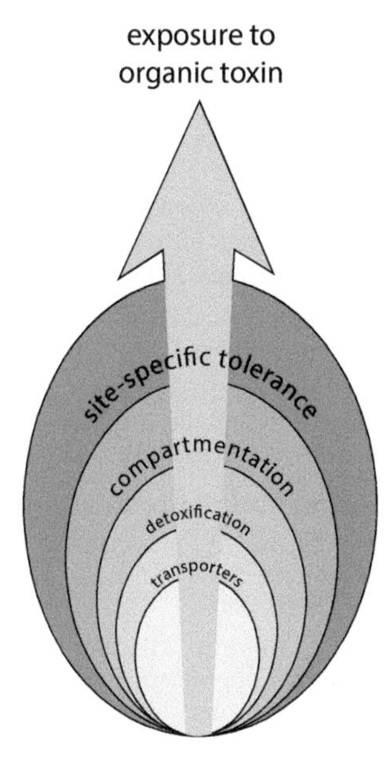

Figure 13.20. Organic toxins and the stress-response hierarchy. Depending on the contaminants and the environmental conditions, organic toxins elicit a hierarchy of responses in plants. Many plants can tolerate low-level exposure to these toxins, but relatively few plants can mount the responses necessary to tolerate high levels of exposure. Widespread use of herbicides is forcing the evolution of plants that are resistant to the effects of extremely toxic organic compounds.

Further reading

Plants and reactive organic compounds

Carroll SP (2011) Conciliation biology: the eco-evolutionary management of permanently invaded biotic systems. *Evol Appl* 4:184–199.

Kostova Z & Wolf DH (2003) Waste disposal in plants: where and how? *Trends Plant Sci* 8:461–462.

Thimmappa R, Geisler K, Louveau T et al. (2014) Triterpene biosynthesis in plants. *Annu Rev Plant Biol* 65:225–257.

Wink M (Ed) *Annual Plant Reviews, Volume 40: Biochemistry of Plant Secondary Metabolism*, 2nd ed. Wiley-Blackwell.

Environmental impact of organic toxins

Richardson SD & Ternes TA (2014) Water analysis: emerging contaminants and current issues. *Anal Chem* 86:2813–2848.

Schaefer RB, von der Ohe PC, Kuehne R et al. (2011) Occurrence and toxicity of 331 organic pollutants in large rivers of north Germany over a decade (1994 to 2004). *Environ Sci Technol* 45:6167–6174.

Tehrani R & Van Aken B (2014) Hydroxylated polychlorinated biphenyls in the environment: sources, fate, and toxicities. *Environ Sci Pollut Res Int* 21:6334–6345.

Teng Y, Xu Z, Luo Y et al. (2012) How do persistent organic pollutants be coupled with biogeochemical cycles of carbon and nutrients in terrestrial ecosystems under global climate change? *J Soils Sediments* 12:411–419.

Uptake of xenobiotics

Bordas B, Belai I & Komives T (2011) Theoretical molecular descriptors relevant to the uptake of persistent organic pollutants from soil by zucchini. A QSAR study. *J Agric Food Chem* 59:2863–2869.

Cropp RA, Hawker DW & Boonsaner M (2010) Predicting the accumulation of organic contaminants from soil by plants. *Bull Environ Contam Toxicol* 85:525–529.

San Miguel A, Ravanel P & Raveton M (2013) A comparative study on the uptake and translocation of organochlorines by *Phragmites australis*. *J Hazard Mater* 244:60–69.

Yin X, Liang X, Xu G et al. (2014) Effect of phenanthrene uptake on membrane potential in roots of soybean, wheat and carrot. *Environ Exp Bot* 99:53–58.

The effects of organic toxins on plants

Belz RG & Duke SO (2014) Herbicides and plant hormesis. *Pest Manag Sci* 70:698–707.

Ramel F, Sulmon C, Serra A et al. (2012) Xenobiotic sensing and signalling in higher plants. *J Exp Bot* 63:3999–4014.

Zezulka S, Kummerova M, Babula P et al. (2013) *Lemna minor* exposed to fluoranthene: growth, biochemical, physiological and histochemical changes. *Aquat Toxicol* 140:37–47.

Zhu B, Peng R, Xiong A et al. (2012) Analysis of gene expression profile of *Arabidopsis* genes under trichloroethylene stresses with the use of a full-length cDNA microarray. *Mol Biol Rep* 39:3799–3806.

Enzymatic transformation of organic toxins

Iwakami S, Endo M, Saika H et al. (2014) Cytochrome P450 CYP81A12 and CYP81A21 are associated with resistance to two acetolactate synthase inhibitors in *Echinochloa phyllopogon*. *Plant Physiol* 165:618–629.

Ling W, Lu X, Gao Y et al. (2012) Polyphenol oxidase activity in subcellular fractions of tall fescue contaminated by polycyclic aromatic hydrocarbons. *J Environ Qual* 41:807–813.

Nelson D & Werck-Reichhart D (2011) A P450-centric view of plant evolution. *Plant J* 66:194–211.

Renault H, Bassard J-E, Hamberger B et al. (2014) Cytochrome P450-mediated metabolic engineering: current progress and future challenges. *Curr Opin Plant Biol* 19:27–34.

Conjugation and detoxification of organic toxins

Cummins I, Dixon DP, Freitag-Pohl S et al. (2011) Multiple roles for plant glutathione transferases in xenobiotic detoxification. *Drug Metab Rev* 43:266–280.

Geu-Flores F, Moldrup ME, Boettcher C et al. (2011) Cytosolic gamma-glutamyl peptidases process glutathione conjugates in the biosynthesis of glucosinolates and camalexin in *Arabidopsis*. *Plant Cell* 23:2456–2469.

Gunning V, Tzafestas K, Sparrow H et al. (2014) *Arabidopsis* glutathione transferases U24 and U25 exhibit a range of detoxification activities with the environmental pollutant and explosive, 2,4,6-trinitrotoluene. *Plant Physiol* 165:854–865.

Jo H-J, Kong J-N, Lim J-K et al. (2014) Site-directed mutagenesis of evolutionarily conserved serine residues in the N-terminal domain of rice Phi-class glutathione *S*-transferase F5. *J Mol Catal B-Enzym* 106:71–75.

Xu Z-S, Xue W, Xiong A-S et al. (2013) Characterization of a bifunctional O- and N-glucosyltransferase from *Vitis vinifera* in glucosylating phenolic compounds and 3,4-dichloroaniline in *Pichia pastoris* and *Arabidopsis thaliana*. *PLoS One* 8:e80449.

Compartmentation and degradation of organic toxins

Ge X, d'Avignon DA, Ackerman JJH et al. (2012) Vacuolar glyphosate-sequestration correlates with glyphosate resistance in ryegrass (*Lolium* spp.) from Australia, South America, and Europe: a P-31 NMR investigation. *J Agric Food Chem* 60:1243–1250.

Landa P, Storchova H, Hodek J et al. (2010) Transferases and transporters mediate the detoxification and capacity to tolerate trinitrotoluene in *Arabidopsis*. *Funct Integr Genomics* 10:547–559.

Ohkama-Ohtsu N, Sasaki-Sekimoto Y, Oikawa A et al. (2011) 12-Oxo-phytodienoic acid–glutathione conjugate is transported into the vacuole in *Arabidopsis*. *Plant Cell Physiol* 52:205–209.

Pang S, Duan L, Liu Z et al. (2012) Co-induction of a glutathione-*S*-transferase, a glutathione transporter and an ABC transporter in maize by xenobiotics. *PLoS One* 7:e40712.

Non-target-site herbicide resistance

Manalil S (2014) Evolution of herbicide resistance in *Lolium rigidum* under low herbicide rates: an Australian experience. *Crop Sci* 54:461–474.

Moustaka J & Moustakas M (2014) Photoprotective mechanism of the non-target organism *Arabidopsis thaliana* to paraquat exposure. *Pestic Biochem Physiol* 111:1–6.

Okada M & Jasieniuk M (2014) Inheritance of glyphosate resistance in hairy fleabane (*Conyza bonariensis*) from California. *Weed Sci* 62:258–266.

Riechers DE, Kreuz K & Zhang Q (2010) Detoxification without intoxication: herbicide safeners activate plant defense gene expression. *Plant Physiol* 153:3–13.

Target-site herbicide resistance

Lee H, Rustgi S, Kumar N et al. (2011) Single nucleotide mutation in the barley *acetohydroxy acid synthase (AHAS)* gene confers resistance to imidazolinone herbicides. *Proc Natl Acad Sci USA* 108:8909–8913.

Li L, Du L, Liu W et al. (2014) Target-site mechanism of ACCase-inhibitors resistance in American sloughgrass (*Beckmannia syzigachne* Steud.) from China. *Pestic Biochem Physiol* 110:57–62.

Shaner DL (2014) Lessons learned from the history of herbicide resistance. *Weed Sci* 62:427–431.

Vila-Aiub MM, Goh SS, Gaines TA et al. (2014) No fitness cost of glyphosate resistance endowed by massive *EPSPS* gene amplification in *Amaranthus palmeri*. *Planta* 239:793–801.

Phytoremediation of organic toxins

Elsaesser D, Blankenberg AB, Geist A et al. (2011) Assessing the influence of vegetation on reduction of pesticide concentration in experimental surface flow constructed wetlands: application of the toxic units approach. *Ecol Eng* 37:955–962.

Fester T, Giebler J, Wick LY et al. (2014) Plant–microbe interactions as drivers of ecosystem functions relevant for the biodegradation of organic contaminants. *Curr Opin Biotechnol* 27:168–175.

Kang JW (2014) Removing environmental organic pollutants with bioremediation and phytoremediation. *Biotechnol Lett* 36:1129–1139.

Wei J, Liu X, Zhang X et al. (2014) Rhizosphere effect of *Scirpus triqueter* on soil microbial structure during phytoremediation of diesel-contaminated wetland. *Environ Technol* 35:514–520.

Modification of interactions between plants and organic toxins

Ervin D & Jussaume R (2014) Integrating social science into managing herbicide-resistant weeds and associated environmental impacts. *Weed Sci* 62:403–414.

Rylott EL, Lorenz A & Bruce NC (2011) Biodegradation and biotransformation of explosives. *Curr Opin Biotechnol* 22:434–440.

Seth CS (2012) A review on mechanisms of plant tolerance and role of transgenic plants in environmental clean-up. *Bot Rev* 78:32–62.

Su Z, Xu Z, Peng R et al. (2012) Phytoremediation of trichlorophenol by phase II metabolism in transgenic *Arabidopsis* overexpressing a *Populus* glucosyltransferase. *Environ Sci Technol* 46:4016–4024.

Chapter 14

Air Pollutants

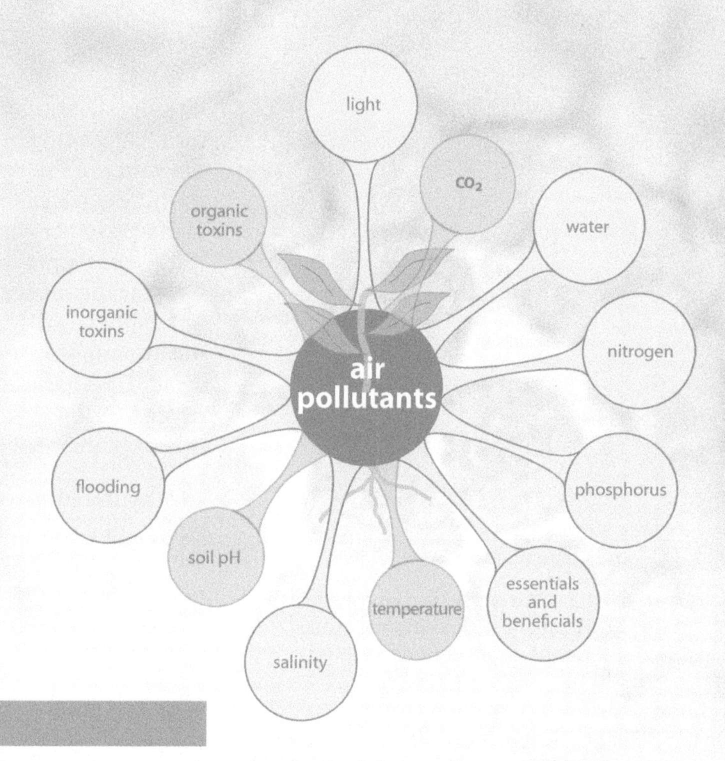

Particular interactions occur between air pollutants and temperature, CO_2 uptake, and soil pH.

Key concepts

- Plants interact extensively with the atmosphere and are affected by its composition.
- Anthropogenic air pollutants have adverse consequences for ecosystems and crops.
- Plants are exposed to significant dry and wet deposition of air pollutants.
- Sulfurous air pollutants can affect plant apoplasts and alter plant S nutrition.
- Nitrogenous air pollutants are widely problematic, with effects on plant pH buffers and N nutrition.
- Volatile compounds affect plants and are produced by them in globally significant quantities.
- Tropospheric ozone has diffuse chronic effects on plant growth, and its levels are increasing.
- Particulates are the most widespread air pollutants, and they interfere with plant growth via both physical and chemical means.

Plants are dependent on an extensive surface area that interacts with the atmosphere

During colonization of the land surface, plants evolved to interact intimately with the atmosphere. This was necessary for the uptake of CO_2. In contrast to many animals, terrestrial plants did not evolve long-distance transport systems for CO_2 or O_2, and there are no gas-binding molecules to carry them in the vascular system—that is, the phloem or xylem. Although in some instances CO_2 dissolved in the xylem fluid can make a significant contribution to CO_2 supply, gas exchange in extant plants is primarily diffuse, as many of the cells in a plant are in contact with the atmosphere. In most plant tissues this gas exchange occurs via air-filled intercellular spaces. In addition to facilitating gas exchange, extensive contact with the atmosphere also utilizes the very low ψ of the atmosphere to drive water flow, to the extent that excessive water loss is a risk. The cuticle of shoots and the bark of stems provide a barrier not only to excessive water loss but also to CO_2 and O_2 exchange. The surfaces of shoots and stems are therefore perforated

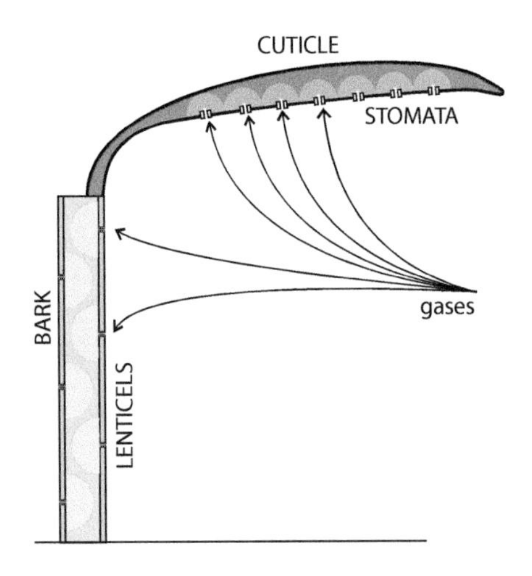

Figure 14.1. The perfusion of gases into plants. The cuticle and, in many plants, the bark present almost impermeable barriers to the movement of many gases into and out of the plant. As photosynthesizing cells require CO_2, to aid its entry the cuticle and bark are perforated by stomata and lenticels, respectively. There is almost no long-distance transport of gases in plants, so in order to obtain the gases that they need plant cells have to be close to pores that perforate a large surface area which is in contact with the atmosphere. The branching shoot morphology of stems and leaves that are perforated by lenticels and stomata meets these requirements.

by stomata and lenticels, respectively, that control gas and water exchange for local groups of cells (Figure 14.1). Plasma membranes in contact with the water of the apoplast provide the gas exchange surface in plants. Shoot morphology in general presents, per unit area of land surface, a large surface area of contact with the atmosphere. Plants fix at least 15% of all the CO_2 in the Earth's atmosphere annually, while O_2 released during photosynthesis is by far the most significant source of the O_2 that now makes up 21% of the atmosphere. In addition, plants release sufficiently large quantities of volatile organic compounds (VOCs) to affect atmospheric chemistry on a global scale. Thus gas exchange with the atmosphere by plants is a planetary-scale process. However, extensive intimate contact with the atmosphere also makes plants susceptible to air pollution.

In a hydrated plant, the apoplast contains a significant amount of water in which all gases must dissolve before they can reach the plasma membrane, where gas exchange occurs. The dissolution of CO_2 is pH dependent, so regulation of apoplastic pH via secretions from the symplast helps to control CO_2 supply. Apoplastic pH is also vital to cell elongation via "acid growth"—during such growth a decrease in apoplastic pH increases cell wall plasticity, allowing plant cells that would normally be constrained by the inflexible cell wall to expand. The pH of the apoplast is also the primary determinant of the proton-motive force (PMF) across the plasma membrane, because the pH of the symplast is highly buffered. This PMF drives the transport of charged solutes across the plasma membrane. Apoplastic pH, together with excretions from the symplast, can also affect the ion-exchange properties of the cell wall, giving apoplastic pH a significant role in controlling the ion supply to plants. Thus although the apoplast is non-living, it has numerous metabolically important roles that depend on its pH. The buffering capacity of the apoplast has been estimated to be about 4 mM pH-unit^{-1}, which is about tenfold less than that of the symplast. At levels of exposure that widely occur in the environment, several air pollutants that are produced as a result of human activity can overwhelm the pH buffer of the apoplast, with adverse consequences for plant growth.

In addition, the apoplast solution is an important site of interaction between plants and pathogens, which helps to explain the effects of some air pollutants. For pathogens that enter plants through stomata and those that spread through the plant body via intercellular spaces, the apoplast provides a key opportunity for the plant to mount a defense before the pathogens reach the symplast. Plant apoplasts have been shown to contain numerous biologically active molecules, often phenolics, which can deter the growth of pathogens. A key property of the apoplast that affects numerous aspects of its functioning is its redox potential, which is actively regulated by secretions from the symplast. In most apoplast solutions, ascorbic acid is the primary non-enzymatic antioxidant, but glutathione and thioredoxin have also been shown to have significant antioxidant roles in some plants. In addition, almost all of the important enzymatic antioxidants have been shown to have significant roles in the apoplast, including superoxide dismutases, peroxidases, and catalase. These molecules usually control apoplastic redox, but when they cannot do so, the resultant changes in ROS provide a key stress signal from the apoplast to the symplast. This can trigger not only the release of defense compounds but also, via enzymes such as NADPH oxidase, an oxidative burst in the apoplast which can attack pathogens and eventually trigger a **hypersensitive response**. Some atmospheric pollutants are powerful oxidants that can both affect the plant's ability to control apoplast redox and trigger the hypersensitive response. Overall, during the colonization of the land and air by plants, the apoplast evolved as a vital site of interaction with the atmosphere and an important line of defense against airborne pathogens (Figure 14.2). Naturally occurring biotic and

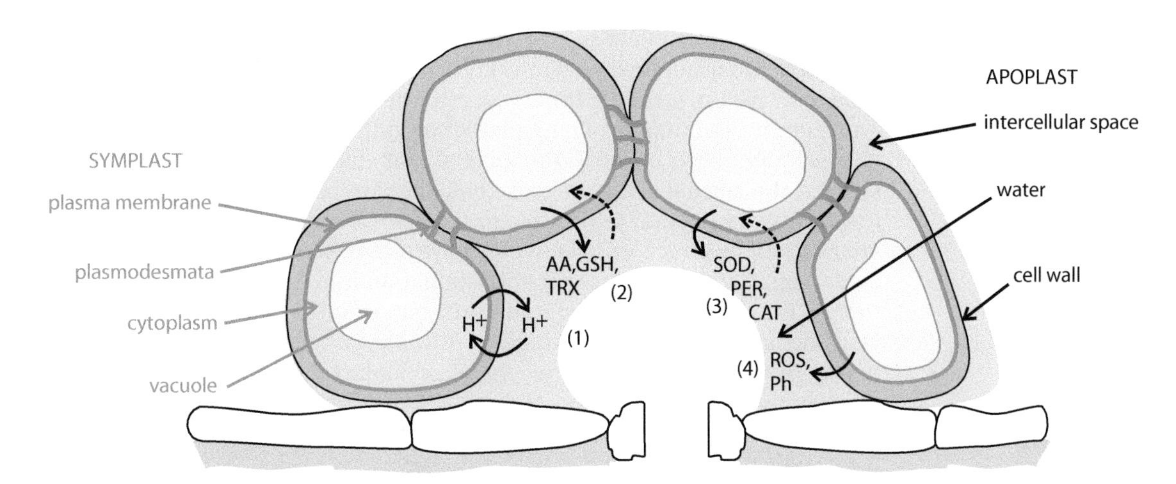

Figure 14.2. Properties of the apoplast solution. The plant body, including the leaf section shown in the figure, can be divided into the symplast (shown in green) and the apoplast (shown in gray), which are living and non-living, respectively. When gases or pathogens penetrate the stomata, their first interface with the plant occurs in the apoplast solution. The properties of the apoplast are actively managed from the symplast. (1) Its pH is buffered by proton exchange, (2) non-enzymatic antioxidants, especially ascorbic acid (AA) but also glutathione (GSH) and thioredoxin (TRX), are excreted into it, as are enzymatic antioxidants (3) such as superoxide disumutases (SOD), peroxidases (PER), and catalase (CAT). Changes in the apoplast solution can signal biotic and abiotic stress to the symplast (dotted lines). Bursts of reactive oxygen species (ROS) and phenolic compounds (Ph) can then be released into the apoplast (4).

abiotic stressors, including toxic gases, probably helped to drive the evolution of apoplast functions, but air pollution produced by human activities can overwhelm them.

One of the largest biological surfaces that interacts with the environment is that between the plant cuticle and the atmosphere. The cuticle, which is excreted from the epidermis and covers the surface of essentially all of the primary shoot organs of land plants, probably first evolved to prevent water loss from the shoot. However, in most extant terrestrial plants it now has numerous additional functions, and damage to the cuticle caused by, for example, atmospheric pollution can have significant consequences. The additional functions of the cuticle include defense against the entry of pathogens, control of organ development, and protection against UV radiation. The thickness of the cuticle varies in different species, and is in the range 0.1–14 μm, with three layers of differing distinctness. Epicuticular waxes form the outermost layer and have a three-dimensional, crystalline structure that can be subdivided into many different types. The hierarchical microstructure of this outer layer produces a wide variety of interactions with water, ranging from **superhydrophobicity** (for example, in lotus leaves) to **superhydrophilicity** (for example, in leaves that absorb mist). These interactions determine how easily leaves are cleaned by the presence of water, and the extent to which droplets, including pollutants, are absorbed from the atmosphere. The next layer consists of a cutin matrix in which cuticular waxes are embedded (Figure 14.3). Organic solvents can solvate the waxes, which are a complex mixture of very-long-chain alkanes, fatty acid derivatives, alcohols, aldehydes, and ketones. In many species, triterpenoids and flavonoids are included in these waxes. Cutin is an insoluble polyester of fatty acids and glycerol, and its long molecules provide the matrix in which the waxes are embedded to form the cuticle. Individual precursors are C16 and C18 fatty

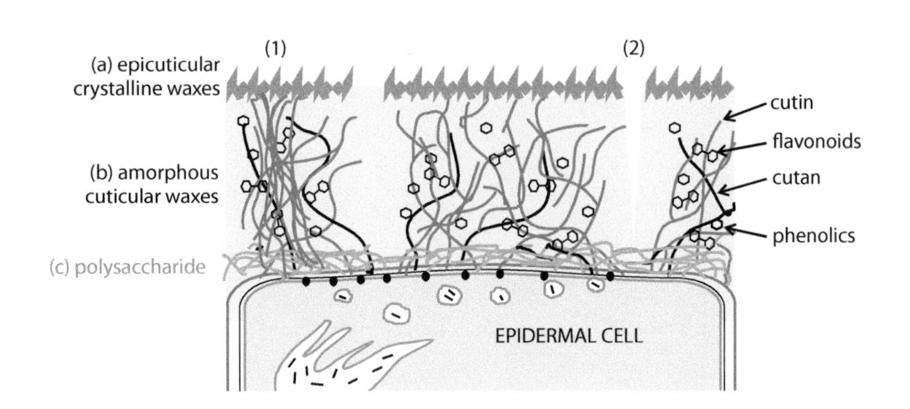

Figure 14.3. The plant cuticle. The cuticle is excreted from epidermal cells and generally has three layers; (a) epicuticular waxes that provide a surface architecture, (b) amorphous waxes, often impregnated with flavonoids and phenolics, that are embedded in a matrix of cutin and cutan, and (c) a layer of polysaccharides that attaches the cuticle to the epidermal cell wall. The passage of water and other molecules through the cuticle can be significantly increased due to channels of cutin (1) and/or cracks caused by physical damage (2). The depth of the cuticular layers varies significantly between species.

acids, which are subject to oxidation in particular, before polymerization and excretion from the epidermis. Polysaccharides attached to the cell wall, especially pectin, form the inner layer to which the cutin matrix is attached. The quantity and quality of cuticular waxes differ significantly between species and often also between the adaxial and abaxial leaf surfaces. In some species, the adaxial leaf surface is considered to be particularly resistant to the effects of wind and rain, whereas the abaxial surface is more resistant to attack by pathogens. Cuticular waxes have often been shown to provide protection against the penetration of UV radiation into the leaf, to be the major barrier to infiltration of exogenous substances into the plant, and to be a vital barrier to infection by pathogens. Thus the cuticle of the plant shoot can be extensively exposed to atmospheric pollutants, helps to determine their infiltration into the plant and, if damaged by air pollutants, can determine their effects on plants.

Adverse effects of air pollution on plants will be important in the twenty-first century

The air pollutants that have the most important effects on plant growth are SO_2, NO_x (NO/NO_2), volatile organic compounds (VOCs), O_3, and particulate matter. All of these occur naturally, so terrestrial plants have been exposed to them throughout their evolution. At low doses, SO_2 and NO_x can even be used as nutrient sources by plants. Although combustion and smelting are known to have produced significant local air pollution from early in human history, it is since the Industrial Revolution, and particularly during the twentieth century, that air pollutants have been released in unprecedented amounts as a result of human activity (Figure 14.4). The dramatic impacts of air pollutants on human and plant health in the second half of the twentieth century led to some of the earliest and most successful environmental legislation (Box 14.1). The sensitivity of plant species, and sometimes varieties, to particular air pollutants differs, and there is evidence that the scale of the environmental challenge has been such that populations of some species have evolved some tolerance to enhanced concentrations of air pollutants. For example, early studies in North America showed that some populations of *Populus tremuloides* (quaking aspen) had evolved some O_3 tolerance, as had some *Plantago major* (greater plantain) populations in the UK. Tolerance of enhanced SO_2 exposure in some plant populations is well established. However, some of the most successful phytomonitors are for air pollutants because, in general, plants are sensitive to concentrations of air pollutants that human activities can easily generate.

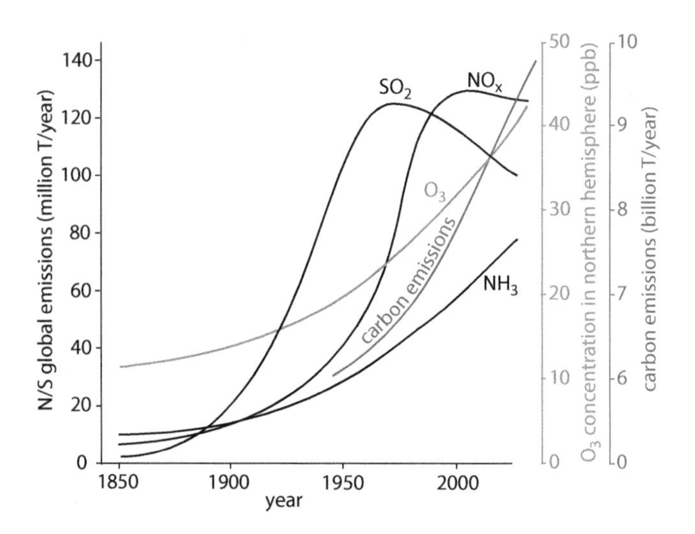

Figure 14.4. Global emissions of air pollutants. There is significant regional variation in emissions of air pollutants, but there was in general a very great increase in emissions of major air pollutants during the twentieth century. For SO_2 in particular, but also for NO_x, by the end of the twentieth century there were overall declines in emissions, but emission levels are increasing in some regions and still significant in many areas, the effects of many twentieth-century emissions are ongoing, and emissions of NO_x will probably increase again before 2050. For NH_3 and C emissions (which include not only CO_2 but also a wide range of other combustion products), global emissions are predicted to continue to increase for at least the first half of the twenty-first century. Particularly in the northern hemisphere, the background concentration of O_3 continues to increase significantly, and will do so in particular in rapidly growing regions of Asia, Africa, and Latin America.

BOX 14.1. ACIDIFICATION OF THE ATMOSPHERE

The atmosphere naturally contains compounds that can make its water slightly acidic. The major constituents of the atmosphere—N_2 and O_2—have little effect on the pH of water when they dissolve in it, but when CO_2 dissolves in atmospheric water, a small proportion of it reacts with H_2O to form carbonic acid (H_2CO_3), a weak acid. In atmospheric water that has not been affected by pollutants from human activity, the dissociation of H_2CO_3 and the presence of a variety of organic acids and other naturally occurring acidic compounds generally produce a pH of about 5.6. Smelting, the burning of fossil fuels, and many other activities associated with industrialization produce atmospheric pollutants that can significantly decrease the pH of atmospheric water. At below pH 5, atmospheric water is generally regarded as producing "acid rain." The two primary pollutants that contribute to acid rain are SO_2 and NO_2, which dissolve in water to produce H_2SO_4 and HNO_3, respectively. Both H_2SO_4 and HNO_3 are strong acids.

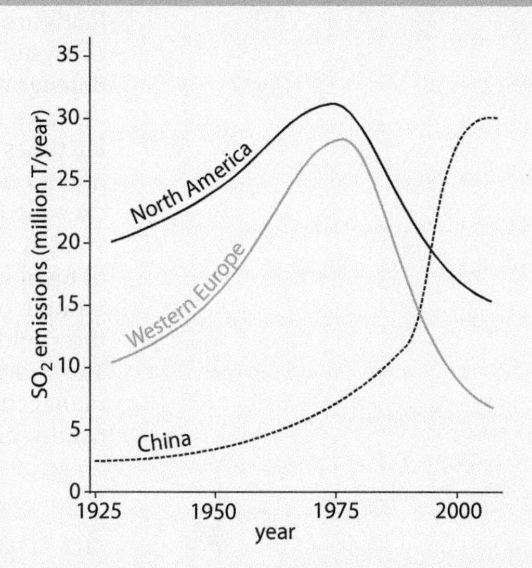

Figure 1. Annual SO_2 emissions. From 1970 onward, legislation to curb emissions of SO_2 in North America and Western Europe successfully brought about a reduction in SO_2 emissions. Since about 2000, similar efforts have been made in China, and these are likely to strengthen, but it will probably be some time before large reductions in emissions are achieved.

It was clear from the operation of early smelters that air pollutants could have adverse effects on plants and ecosystems, but it was not until the second half of the twentieth century that the causes and effects of "acid rain" were fully elucidated. pH values of well below 5, and in some cases of less than 2, were recorded in fog and rain across wide areas of North America and Western Europe. Although it was sometimes difficult to separate out other ecological changes in vegetation from those caused by acid rain, it was clear by the 1960s and 1970s that in many instances there was sufficient acid rain to cause rapid direct effects on vegetation. It was also clear that deposition on vegetation and soils, especially in ecosystems with low base saturation, was changing soil and water pH, with profound ecological consequences across swathes of North America and Western Europe. Acid rain was therefore the primary impetus for environmental legislation to control emissions of air pollutants. In North America and Western Europe this legislation has been successful in decreasing SO_2 emissions in particular (Figure 1). For NO_2, primarily because of the rise in its total emissions from vehicles, reductions have been less dramatic and might stop decreasing. In many other regions, but especially in South Asia, South-East Asia, and China, emissions of SO_2 and other acidifying pollutants have been more challenging to bring under control, and are predicted to keep rising well into the twenty-first century. In addition, in many ecosystems and catchments there is a wave of acidification in soils that has yet to work through the system. Thus the phenomenon of acid rain not only provides a type example of the importance of the effects that air pollutants can have on plants, ecosystems, and catchments, but is also an ongoing challenge in some of the most populous areas on Earth.

Numerous effects of air pollutants on plants, ranging from cellular to ecosystem levels, have been reported. In significant part they arise because at relatively low doses, air pollutants can affect shoot and sometimes root biomass production, whereas at moderate to high doses, observable symptoms such as chlorosis or necrosis can occur. Air pollutants generally initially affect shoot growth, because the pollutants are delivered directly to photosynthetic cells. From the mid-twentieth century onward, many experimental programs, often on a national or regional scale, and using methods that include closed chambers, open-topped chambers, and a variety of field monitoring, have provided data that are used to set empirically derived effects thresholds for pollutants. A **critical level** is an atmospheric concentration of a pollutant which, if exceeded, can have an adverse effect on plants. The extent by which this exposure threshold is exceeded can be important, so the accumulated dose above the threshold is often calculated (for example, for O_3), and sometimes it is peak concentrations that produce adverse effects.

Often the total N or acidity deposited on an ecosystem provides better estimates of when adverse effects on vegetation will occur, so the **critical**

load—that is, the amount deposited per year below which there are no effects on the most sensitive ecosystem component—is calculated. Critical loads are often calculated on the basis of long-term effects on vegetation in ecosystems. Critical levels and loads are often set for an acceptable level of damage rather than for no damage at all. Even though there have been recent declines in emissions of some air pollutants (for example, SO_2), at particular times and places critical levels and loads are still regularly exceeded, and for other pollutants (for example, O_3 and NH_3), critical levels are being exceeded with increasing frequency. The critical levels for semi-volatile and volatile organic contaminants, which are often produced by the combustion of fossil fuels, for example, are less well understood. Thus, globally, there is a range of air pollutants that have established adverse effects on plants at thresholds that are in some cases being exceeded with increasing frequency. The widespread adverse effects of air pollutants on both unmanaged and managed ecosystems will be a significant environmental challenge in the twenty-first century.

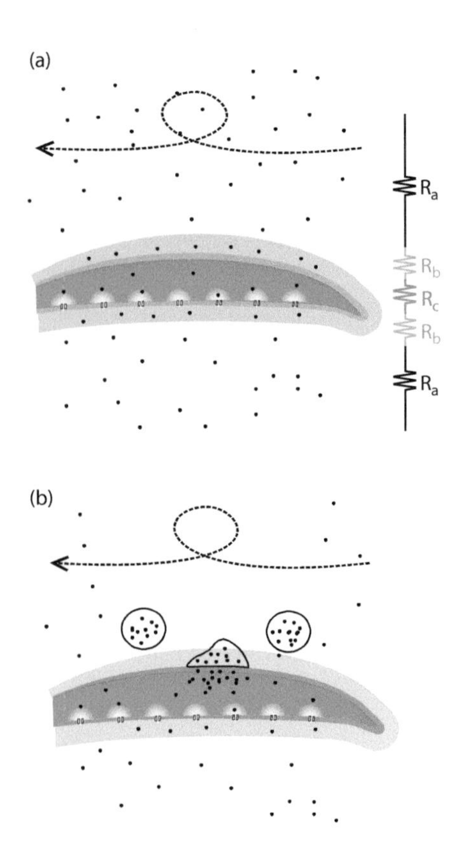

Figure 14.5. **Dry and wet deposition of air pollutants on vegetation.** Pollutant interception by vegetation depends on plant morphology (as measured by, for example, leaf area index) and its effects on airflow. Once the air carrying a pollutant reaches a leaf, it is transferred to plant surfaces and taken up by processes that differ significantly between species. (a) For a given air pollutant, the dry deposition rate depends on resistance to movement in the atmosphere (R_a), resistance to movement in the vegetation's boundary layer (R_b) (shown in light gray) and canopy resistance (R_c) (because of the cuticle, shown in dark gray)—that is, the reciprocal of stomatal conductance. (b) For a given pollutant, wet deposition depends on the scavenging of pollutants by water droplets, the wettability of the cuticle, and the rate of transfer across the cuticle.

The deposition of air pollutants on plants depends on the properties of plants and pollutants

In the absence of rain, the flux (F) of atmospheric pollutants to a surface is a product of their deposition velocity (V_d) and concentration (c)—that is, $F = V_d.c$. The value of V_d is much greater in relation to vegetation than to, for example, a lake surface. This is because the surface roughness of the vegetation generally dramatically decreases the aerodynamic resistance (R_a) to deposition by increasing turbulent transfer. In many models of dry deposition of air pollution, vegetation is therefore categorized according to surface roughness. Specific values of V_d for particular pollutants or plants also depend on the resistance to movement through boundary layers around vegetation (R_b), canopy resistance to entry (R_c), and how the vegetation behaves as a sink (Figure 14.5). The lower the sum of resistances and the more perfect the sink, the higher the V_d will be. In general, for NO_2 and O_3, the most significant determinant of V_d is stomatal conductance. For this reason, physiological models that predict stomatal conductance from C assimilation rates and soil moisture can also be used to estimate V_d for NO_2 and O_3. For SO_2 and VOCs, deposition on the cuticle is also important, and for particulates R_b often determines V_d. Plants generally act as a **perfect sink** for HNO_3, whereas for NH_3 there is a compensation point because, depending on temperature, they can either release or assimilate this gas. Models that link atmospheric air movement, landscape features, and vegetation types to V_d for particular pollutants and plants are now used to estimate the total dry deposition of many air pollutants on vegetation.

The behavior of water in the atmosphere also affects V_d. Wet deposition of air pollutants occurs when water droplets in the atmosphere scavenge air pollutants and deposit them on vegetation. Pollutants can be scavenged when they come into contact with water because of Brownian motion, turbulent airflows, coagulation of droplets, and movement under gravity. The former can be particularly important during in-cloud scavenging and the latter during precipitation scavenging—that is, **washout**. Water scavenging of air pollutants can also lead to significant transfers of pollutants in the atmosphere. In many instances, if washout occurs it can produce the highest deposition rates. For example, during the passage across Europe of the cloud of radioactivity from the Chernobyl accident, dry deposition on vegetation declined significantly with distance from the reactor, but across Europe areas of precipitation incurred significant wet deposition of radioactivity on vegetation. Wet deposition rates generally depend on vegetation characteristics, in particular the interception rates produced by the morphology of leaves and canopies. However, uptake of droplets through stomata is generally minor, so the wettability of the cuticle and the water uptake capacity of the leaves are also significant. Overall, pollutant deposition models that link emission, atmospheric, landscape, and vegetation processes (often including environmentally dependent phenological, developmental, and physiological aspects) predict deposition velocities in relation to vegetation that depend on particular vegetation characteristics for particular pollutants. They also demonstrate that, around the world, the concentrations of air pollutants are frequently high enough to produce deposition rates that can produce adverse effects in plants.

Plants can assimilate some sulfur dioxide, but anthropogenic deposition rates can exceed this capacity

Plants can be exposed to natural sources of SO_2 from the combustion of S-containing compounds during fires and volcanic activity. SO_2 penetrates plant cuticles slowly, but can diffuse rapidly through stomata and into intercellular spaces. From low SO_2 concentrations, S can enter plant metabolism and, in S-deficient environments, contribute significantly to plant S content. In areas where anthropogenic SO_2 emissions have been curbed because of their adverse effects on natural ecosystems and human health, S deficiency in crop plants is now more common than it was during periods of high SO_2 emission. High levels of anthropogenic SO_2 emissions are problematic to plants partly because of the toxic effects of SO_2. Although the reactions of SO_2 with atmospheric water and with NO_3 are significant contributors to acid rain and the formation of aerosols, respectively, direct dry deposition of SO_2 on plants can also cause adverse effects. Some aspects of S metabolism have been interpreted as mechanisms for coping with low-level exposure to SO_2, including the capacity to convert SO_2 to useful S-containing molecules and the ability to produce H_2S to get rid of S, but plant S metabolism can be overwhelmed by SO_2. The smelting of sulfurous metal ores and combustion of fossil fuels, especially coal, have for many centuries been documented as adversely affecting the surrounding vegetation. Some of the earliest studies that investigated the effects of pollutants in the environment showed, in the nineteenth and early twentieth centuries, that SO_2 toxicity was the cause of air pollution effects on trees. In the mid-twentieth century, many studies documented SO_2 toxicity symptoms in a wide range of species in industrialized nations. Often the earliest visible symptoms of SO_2 toxicity are irregular chlorotic leaf spots that spread and turn necrotic, often with a red/brown "bronzed" coloration. In dicots, necrosis often spreads from the leaf margins, whereas in monocots and gymnosperms it spreads from the leaf tips.

Different plant species have different sensitivities to SO_2, but the toxicity of SO_2 has long been associated with acidifying and oxidative effects. SO_2 dissolves in water to produce H^+, HSO_3^- (bisulfite), and SO_3^{2-} (sulfite) (Figure 14.6). The production of H^+ can be sufficient to acidify the apoplast and cytoplasm. Depending on their physiological capacity to control pH, plants that are exposed to elevated SO_2 levels can be harmed by these acidifying effects. More importantly, sulfite in particular is strongly nucleophilic and can attack many important molecules in cells, including lipids, proteins, and DNA, causing **sulfitolysis**. Lipid oxidation has often been reported with excess SO_2, with the classic oxidative stress marker malondialdehyde (produced by the action of reactive oxygen species on polyunsaturated lipids) often being found. The overall result of sulfitolysis is oxidative stress, sometimes indicated by red/brown "bronzing" of the leaves. In humans, the toxic effects of sulfites have been more extensively investigated because of their use in the food industry and their role in some diseases. In kidney cells, the nucleophilic action of SO_3^{2-} on glutathione is associated with increased production of $O_2^{\bullet-}$. The conversion of glutathione between the reduced

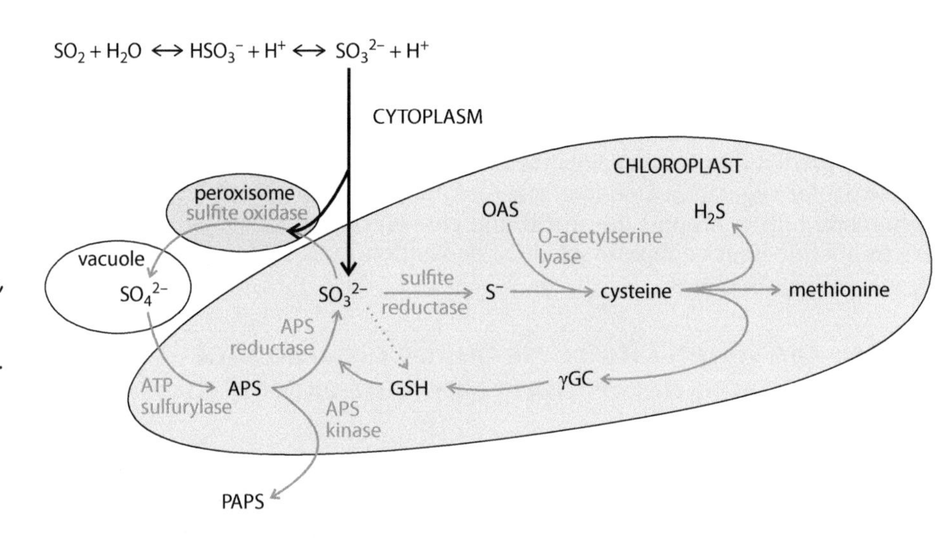

Figure 14.6. The metabolism of SO_2 in cells. SO_2 dissolves in water to produce sulfite (SO_3^{2-}). In peroxisomes this can be oxidized to sulfate (SO_4^{2-}) by sulfite oxidase and transferred to vacuoles. In plastids such as the chloroplast (shown in the figure), sulfite can be reduced to S^- for assimilation into cysteine. Sulfite reductase is similar to nitrate reductase, and is dependent on electrons from ferredoxin supplied by the light-dependent reactions of photosynthesis. Adenosine phosphosulfate (APS) reductase utilizes reducing power from glutathione (GSH) to convert stored SO_4^{2-} into SO_3^{2-} for assimilation. Phosphoadenosine phosphosulfate (PAPS) is the substrate for sulfurous secondary metabolites. OAS = *O*-acetylserine; γGC = γ-glutamylcysteine.

form (GSH) and the oxidized form (GSSG) is vital for cellular redox control, and is dependent on intact thiol groups. In most cells, exposure to sulfites increases the activity of both superoxide dismutases, which produce H_2O_2 from the $O_2^{\bullet-}$ formed during oxidative stress, and of catalase, which breaks down H_2O_2. In plant cells, manipulation of these enzymes alters susceptibility to SO_2 toxicity. In plants, sulfite toxicity has been shown to increase significantly with exposure to light, indicating a role for photosynthetically generated $O_2^{\bullet-}$. In plant cells, disruption of glutathione by the nucleophilic reaction of SO_3^{2-} with its thiol groups probably also decreases the capacity of cells to deal with superoxides such as $O_2^{\bullet-}$, increasing oxidative stress and the activity of enzymes that combat it. Differences in stomatal conductance and leaf anatomy that affect SO_2 penetration into the leaf contribute to differences in SO_2 sensitivity, but genetic manipulations of the antioxidant capacity of plants indicate that it too is important in explaining different sensitivities. SO_2 dissolves to produce SO_3^{2-}, which is a key intermediate of S metabolism, through which it can be incorporated into S-containing compounds in plants (Figure 14.6).

With chronic low-level exposure and in the absence of visible symptoms of toxicity, many long-term effects of SO_2 on plant growth have been demonstrated, particularly in forests. The increased oxidative burden of SO_2 exposure promotes stomatal closure and can affect the photosynthetic machinery directly, in both cases potentially resulting in a decrease in biomass production. This has been shown to alter many aspects of plant growth, with consequent ecosystem effects, particularly in forests that have high rates of dry deposition. SO_2 exposure can also affect leaf stomatal densities and stomatal indices, and this response is sensitive enough to need to be taken into account when using stomatal indices to reconstruct the record of atmospheric CO_2 concentrations. In general, when conditions—including S availability—are optimal for growth, SO_2 toxicity is produced at lower levels of exposure because S assimilation is nearer capacity, and younger leaves are often more susceptible than older leaves because they are more metabolically active. Differences in SO_2 sensitivity can also alter the species composition of ecosystems that are exposed to SO_2. Studies using ^{18}O, in which primary sulfates are enriched relative to ^{16}O, and ^{34}S, which is enriched relative to ^{35}S in emissions, have traced the pathways of emissions-derived S in forests and shown that although there can be some transport into the roots, changes in S metabolites and toxicity effects mainly occur at the sites of exposure. At present, in much of the industrialized world, established legislation—sometimes over several decades—has successfully reduced SO_2 concentrations in the atmosphere to within the assimilatory capacity of plants. However, particularly in India and China, there are continuing efforts to establish and enforce legislation because SO_2 pollution is causing not only adverse effects on human health but also toxicity in natural and agricultural ecosystems. China's Tenth Five-Year Plan (2000–2005) did not bring about the intended decrease in SO_2 emissions, but the Eleventh Five-Year Plan did, so SO_2 emissions in China peaked in about 2006 and seem likely to slowly decrease in the coming years. With an estimated current cost to China's economy of over US$ 10 billion per annum, even as SO_2 emissions decline their effects will be significant for several decades to come. Many other nations will also be trying to limit the toxic effects of SO_2 during this time.

Direct uptake of gaseous reactive nitrogen species can affect plant growth and ecosystem dynamics

Increases in gaseous reactive N ($N_{r(g)}$) emissions (Figure 14.4) are contributing to the altered biogeochemical cycling of N that is resulting from human activity. $N_{r(g)}$ occurs naturally at very low concentrations in the atmosphere, in both oxidized and reduced forms, because of the action of lightning and microbes. Emission of N_2O has been significantly enhanced by human activity, but its effects on plants are much less significant than those of NO and NO_2 ("NO_x"). Global emissions of NO_x and NH_3 resulting from human activity are now significantly greater than those which occur naturally (Figure 14.7). NH_3 is both the primary reduced form of $N_{r(g)}$ and the primary basic compound in the atmosphere, with significant consequences for atmospheric chemistry. The most significant source of NO_x is combustion (the primary fate of fossil fuels and hence the most significant source of NO_x), which initially produces $NO_{(g)}$ from N and O. NO is a free radical, so in the atmosphere it reacts with O_2, and especially with O_3 if this is present, to produce $NO_{2(g)}$. In the atmosphere, NO_2 has a short, temperature-dependent half-life of a few days, in part because of photolysis (which produces NO), but also because of reactions with inorganic ions to produce particulates and with peroxy radicals of gaseous hydrocarbons to produce peroxyacetyl nitrate (PAN). Emissions of NO_2 from automobiles contribute to the wide dispersal of NO_2, and particulates and PAN contribute to the long-range transport of N_r, so the effects of NO and NO_2 emissions can be significant on a regional scale. The majority of NH_3 emissions to the atmosphere are from livestock kept by humans, which are frequently locally concentrated, producing primarily local enhancements in NH_3 emission. $NO_{x(g)}$ and $NH_{3(g)}$ are subject to significant dry deposition directly onto plants, but both also readily dissolve in atmospheric water droplets, in which they can react not only with other soluble atmospheric constituents and be transported significant distances, but also be subject to wet deposition on vegetation. At high concentrations, which in many but not all terrestrial environments are now less frequent for NO_x but increasing in frequency for NH_3, $N_{r(g)}$ can be toxic to plants. However, plants can metabolize the NO_3^-, NO_2^-, and NH_4^+ produced

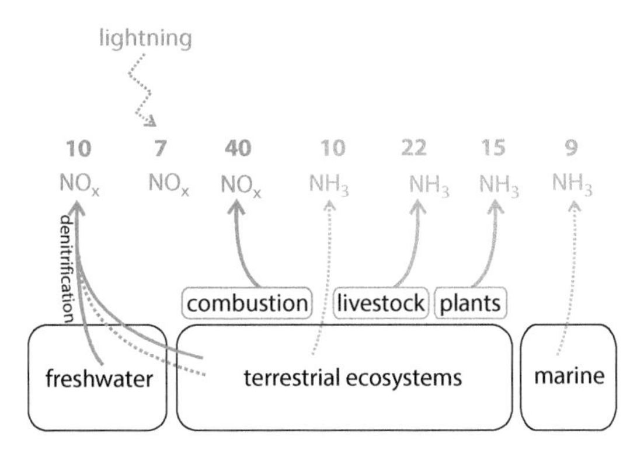

Figure 14.7. Natural and anthropogenic emissions of NO_x and NH_3. There are natural emissions (shown as dotted lines in the figure) of both NO_x and NH_3 from aquatic and terrestrial ecosystems. Denitrification in marine environments results almost entirely in the production of N_2 and N_2O, but in terrestrial and freshwater environments some NO, and hence NO_x, is also produced. Inputs of fixed N to soil from human activity are increasing the production of NO_x from denitrification. The largest emissions of NO_x and NH_3 to the environment are now from human activity (solid lines in the figure), almost all from combustion and domesticated animals and plants. Average values in Tg N $year^{-1}$ for the period 2000–2010 are shown. (Data from Sutton MA, Bleeker A, Howard CM et al [2013] Our Nutrient World, Centre for Ecology & Hydrology, Edinburgh, UK)

by the dissolution of $N_{r(g)}$ in H_2O, and at sub-toxic concentrations incorporate $N_{r(g)}$ into biomass. Given that many terrestrial ecosystems are naturally N limited, there can be major consequences—on scales ranging from the plant organ to whole ecosystems—of N_r input directly from the atmosphere into biomass.

The deposition of $N_{r(g)}$ directly on soil affects soil N regimes, with attendant consequences for terrestrial vegetation, but terrestrial plants also scavenge $N_{r(g)}$ directly from the atmosphere, significantly increasing both dry and wet deposition rates. A proportion of N_r can be deposited externally on vegetation, and can be washed off into the soil, but a proportion can also enter plants. For both NO_2 and NH_3, flux through the stomata into the apoplast solution of mesophyll cells is the principal route of entry. Stomatal conductance and concentration in the sub-stomatal cavity, which are controlled by the rate of dissolution in the apoplastic solution, determine influx rate. NO_2, which is in equilibrium with N_2O_4, dissolves in water to produce HNO_2 (nitrous acid) and HNO_3 (nitric acid). At average apoplastic pH values of 4.5–5.5 and pK_a values of 3.3 and 1.4, respectively, these acids dissociate to produce H^+, NO_2^-, and NO_3^- in the apoplast solution (Figure 14.8). Transporters with NO_2^-/NO_3^- influx capacity are distributed throughout plants, and $\delta^{15}N$ studies often report rapid uptake of NO_2-derived N into leaf mesophyll cells. NO_2^- and NO_3^- that are taken up from the apoplast can be reduced and then assimilated. Dissolution of NH_3 in the apoplast solution, with a pK_a of 11.6, strongly favors the production of NH_4^+, which can also be taken up across the plasma membrane and assimilated. The rates of influx of these molecules, which are driven by rates of assimilation, can affect apoplast concentrations, dissolution rates, and concentrations in the sub-stomatal cavity. NO_x and NH_3 uptake by plants from the atmosphere is therefore generally linearly related to stomatal conductance (as determined in response to environmental variables), with species-specific influx rates explained by a combination of morphological characters that influence, for example, surface area and physiological variables that affect N-sink strength.

In some instances there can be influx of N from NO_2 and NH_3 that does not occur through the stomata. Some plants— for example, tank bromeliads and

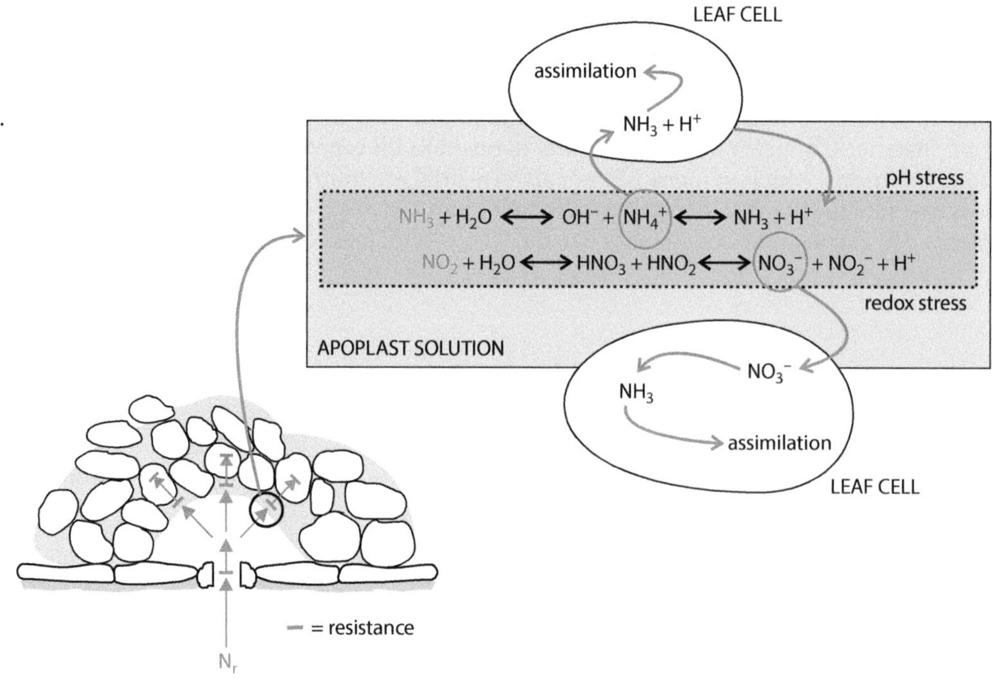

Figure 14.8. Reactive N in the leaf apoplast. The apoplast solution contains dissolution products of both NO_2 and NH_3 that can be assimilated by plants. Assimilation of NO_3^- produces HCO_3^- ions that undergo efflux into the apoplast, neutralizing the acidifying effects of protons produced by the dissolution of NO_2. Excess NO_2 therefore primarily produces oxidative challenges for plant cells. Assimilation of NH_3 produces H^+ that undergoes efflux into the apoplast, exacerbating the pH challenges produced by excess NH_4^+.

intermittently or permanently submerged plants—have adapted to take up NO_3^- and NH_4^+ directly across the leaf surface. These plants have, respectively, specialized trichomes with N-transport capacity, and thin cuticles. Petals have almost no cuticle (so that fragrances can be emitted), and have high rates of N influx from NH_4NO_3. Well-developed cuticles generally present an impermeable barrier to the transport of NO_2, NH_3, and their dissolution products found in water droplets, but there are pores in many cuticles, often water filled, through which NH_4^+ in particular can move, and cuticles are often damaged, not least by the action of acid droplets formed by the dissolution of air pollutants. In temperate forests, NO_2 and NH_3 uptake has been relatively well investigated and is dominated by stomatal uptake, but in tropical forests, in part due to their rich epiphyte floras, there is much canopy recycling of nutrients, and non-stomatal pathways of N_r uptake may be more important. Studies using fluorescent nanoparticles have also revealed flows of external water into the stomata. In general, the hydrophobic cuticle, the surface tension of water, and the diameter of stomata prevent the influx of water through, and blockage of, stomata, but the reduction of hydrophobicity and surface tension by pollutants suggests that water influx can sometimes carry pollutants through the stomata.

The antioxidant and superoxide dismutase capacity of the apoplast solution can generally cope with low atmospheric concentrations of the oxidant NO_2 (for most species, the low ppb range). If this capacity is insufficient because of high NO_2 concentrations (for most species, the ppm range), low species-specific capacity, or the presence of other stressors, NO_2 can increase the concentration of ROS, damaging in the first instance plasma-membrane lipids (as evidenced by increasing lipid saturation and the production of malondialdehyde) and then other cellular components, including DNA. Salicylic acid is an important hormone in the control of systemic acquired resistance (SAR) to pathogens, and its effects are often mediated by redox changes in the apoplast and cytoplasm. Constitutively enhanced salicylic acid production in *Arabidopsis thaliana* increases NO_2 tolerance, and manipulations of various aspects of SAR produce changes in sensitivity to NO_2 exposure that are consistent with the effects of NO_2 being controlled by antioxidant systems. The dissociation of HNO_2 and HNO_3 in the apoplast can also change the apoplastic pH, an effect that is counteracted by the efflux of OH^- ions produced by NO_2^- assimilation, so in general effects on pH are less significant than those of redox. NH_3 and NH_4^+ are toxic to organisms and cannot be stored. The electrophilic NH_4^+ ion can react with biomolecules and disrupt electron flows, but also the production of oxoglutarate (a key tricarboxylic acid cycle intermediate) from glutamate by glutamate dehydrogenase produces NH_4^+, an excess of which inhibits oxoglutarate production and hence aerobic respiration. In addition, NH_3 assimilation releases protons that are pumped out of cells into the apoplast (Figure 14.8). In plants that have a preference for NH_4^+ as their N source for root uptake, NH_3 assimilation occurs primarily in the roots, and protons undergo efflux into the rhizosphere, which is connected to the bulk soil and thus has a significant buffering capacity. The assimilation of NH_4^+-derived N in shoots acidifies the leaf apoplast, which has a buffering capacity that although significant is less than that of the rhizosphere. Visible symptoms of NO_2^- toxicity often begin with a water-soaked appearance and, for both NO_2^- and NH_3, progress to inter-veinal necrosis. In general, legislative control of NO_x emissions has almost eliminated reports of NO_2 toxicity in the field, but in greenhouses that are heated by combustion of fossil fuels it is still reported. Problems caused by NH_3 deposition are likely to be more frequently reported in the coming decades.

In experimental conditions, many species have been grown with a significant proportion (up to 50%) of their N being derived from N_r in the air. In N_r-polluted environments, much lower but often significant contributions to N nutrition have been reported. $\delta^{15}N$ studies have frequently demonstrated that entry of NO_2^- and NH_3-derived N into plant metabolism is rapid. For NH_3 in particular there is a compensation point for NH_3 uptake—that is, when there is no atmospheric NH_3, plants can emit NH_3, and a particular, species-specific atmospheric concentration is necessary for net uptake. In some species, compensation points have been reported for NO_2^-, but at extremely low concentrations. Many agricultural and semi-natural ecosystems that are dominated by annual plants can be significant sources of NH_3, especially in warm dry conditions that favor NH_3 degassing from the apoplast, whereas net uptake occurs in many forest ecosystems. In species that are generally NH_3 feeders and assimilate N in roots, N_r pollutants can provide a higher proportion of the N in shoots than is the case for NO_3^- feeders, in which there is often a higher proportion of N assimilation in shoots. In N-limited ecosystems, supplementary N in shoots affects N and C metabolism, with many reported effects on the concentrations of N and C compounds and their stoichiometry.

Semi-volatile and volatile organic compounds can be absorbed by and released from vegetation

Many anthropogenic contaminants are semi-volatile or volatile organic compounds (SVOCs or VOCs) that, even if they are initially released into the soil, tend to move into the atmosphere. In many ecosystems there can be significant deposition of these SVOCs and VOCs from the atmosphere onto vegetation. For many semi-volatile PAHs, the dynamics of their interaction with vegetation is a significant determinant of their global cycling. For lipophilic SVOCs and VOCs, which are retained in the roots if taken up from the soil, there can be significant transfer from the soil to the shoot, but via volatilization from the soil and uptake from the atmosphere, rather than via translocation from root to shoot. Due to the well-established adverse effects of many SVOCs and VOCs on human health, there are controls on emissions in many regions, but total emissions of anthropogenic VOCs are nevertheless probably in excess of 100 Tg per year. However, this figure is dwarfed by the more than 1000 Tg of biologically generated VOCs (BVOCs) that are emitted by terrestrial vegetation each year. BVOCs have significant biological roles, including regulation of membrane properties, defense against herbivores, and inter-plant communication. The quantities in which they are released help to determine net C fixation by vegetation. They also induce the production of globally significant quantities of secondary organic aerosols (SOAs), react with $OH^\bullet$ (the primary atmospheric oxidant) to reduce the oxidation rates of key atmospheric gases such as methane, and interact with anthropogenic contaminants to produce globally significant amounts of PAN and O_3.

The octanol–air partition coefficient (K_{OA}) of an organic contaminant, an empirical property that is measured with octanol and air at the same temperature and their concentration of contaminant in equilibrium, can be used to make general predictions of SVOC/VOC partitioning between organic compartments of the environment and the air. K_{OA} is dependent on temperature, and many organic compartments have properties that can be taken into account to improve predictions of partitioning based on K_{OA}. Such predictions include SVOC/VOC deposition from the atmosphere onto plants, with three different mechanisms controlling deposition, depending on K_{OA} (**Figure 14.9**). SVOCs can enter plants through the stomata when they are open but, in contrast to deposition of SO_2 and NO_x, deposition rates onto vegetation tend not to correlate with stomatal conductance. In general, SVOCs do not dissolve rapidly in the aqueous apoplastic fluid, so the concentration gradients that drive fluxes through the stomata into the sub-stomatal cavities are minimal. SVOCs are deposited onto the external surfaces, where, depending on their lipophilicity, they can dissolve in the cuticle and

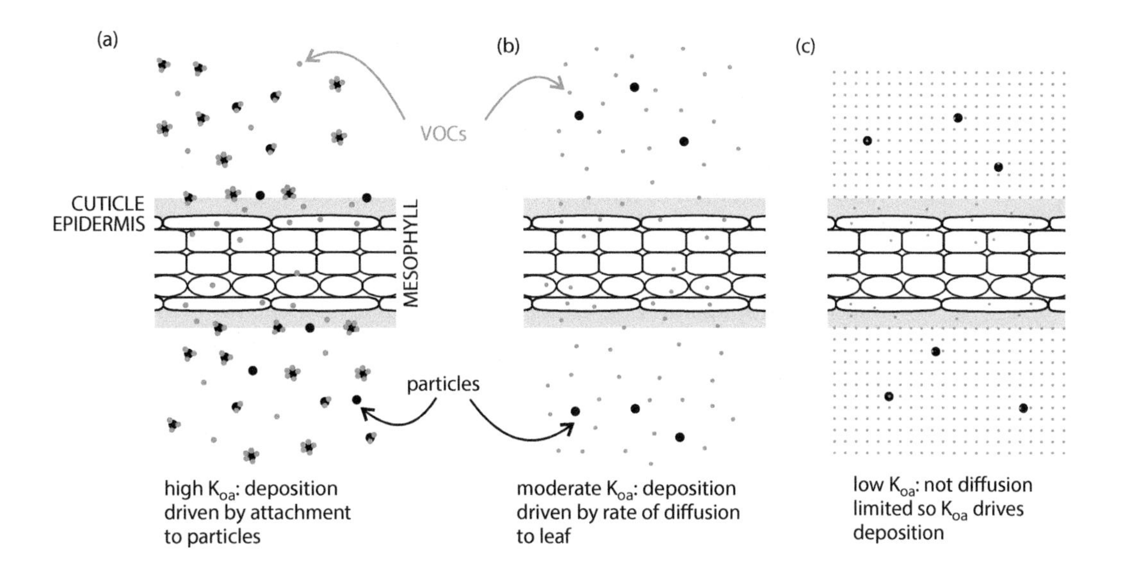

Figure 14.9. The deposition of volatile organic compounds (VOCs) on leaves. (a) Compounds with low volatility—that is, those with high K_{oa}—tend to be scavenged by particles and do not volatilize from them, so deposition of these compounds on leaves is primarily determined by the rate of particle deposition. Compounds that are too volatile to attach to particles must diffuse through the boundary layer to the leaf. (b) For compounds of moderate volatility, movement through the boundary layer is limited by their diffusion coefficients, which therefore determine deposition rates. (c) For highly volatile compounds—that is, those with low K_{oa}—there is no diffusion limitation, so their deposition rate is determined by the equilibrium partitioning between leaf and air—that is, their K_{oa}. (Adapted from Yogui GT, Sericano JL & Montone RC [2011] *Sci Total Environ* 409:3902–3908. With permission from Elsevier)

enter the plant. At a given lipophilicity, the rate of cuticular movement is dependent on molecular weight. Importantly, SVOCs can also volatilize from plant surfaces on which they have been deposited, a process that is accelerated by increases in temperature. Thus vegetation is not only a significant filter of but also—depending on conditions—a significant source of SVOCs. The long-range atmospheric transport of SVOCs is significant, with "cold trapping" of SVOCs at high latitudes decreasing atmospheric concentrations and causing a general drift towards the poles. During this drift to the poles there is significant "hopping" of SVOCs on vegetation—that is, they are deposited and revolatilized several times. The boreal forests of high northerly latitudes provide a significant final filter for SVOCs, and decrease their concentration in air before it reaches the high arctic. The deposition of SVOCs on vegetation can remove a third or more of these compounds from the atmosphere, this effect being particularly pronounced for SVOCs with low K_{OA} values that promote association with particulates.

PCBs are among the least volatile and mostly tightly controlled SVOCs. They have a relatively low K_{OA} (in the range 7–11) and, where they are found in the atmosphere, associate with particulates. Their deposition rates depend on the surface properties of the vegetation, and wind factors. A wide range of PAHs, especially those produced from combustion, contaminate the environment, with those of lower molecular weight (and hence with fewer benzene rings) in particular producing a higher K_{OA} (in the range 6–13) than for PCBs. Many PAHs are lipophilic. The PAHs that are found in plant shoots are therefore primarily derived from atmospheric deposition. Even in experiments with hydroponic solutions contaminated with PAHs, the primary route of these compounds to the shoot is via volatilization from the solution and deposition from the atmosphere. The lipophilicity of many PAHs allows them to penetrate leaf cuticles. Experiments using phenanthrene (a lipophilic PAH with three benzene rings and a K_{OA} of 7.5) have visualized its penetration into the leaf, which in some species can occur within 24–48 h (Figure 14.10). In general, the higher the molecular weight of a PAH, the more slowly it penetrates the cuticle. Photodegradation can be significant while PAHs are in the cuticle in particular, which for some PAHs and in some plant species can be for extended periods, primarily because although much UV radiation does not penetrate the cuticle, UV-A radiation can do so to a significant extent. There are many other SVOC contaminants, such as phenols, carbonyls, and tetrachloroethene, that are also deposited on and can be emitted from plant cuticles at rates that depend on K_{OA}, plant properties, and environmental conditions. For many SVOCs, the plant factors that affect

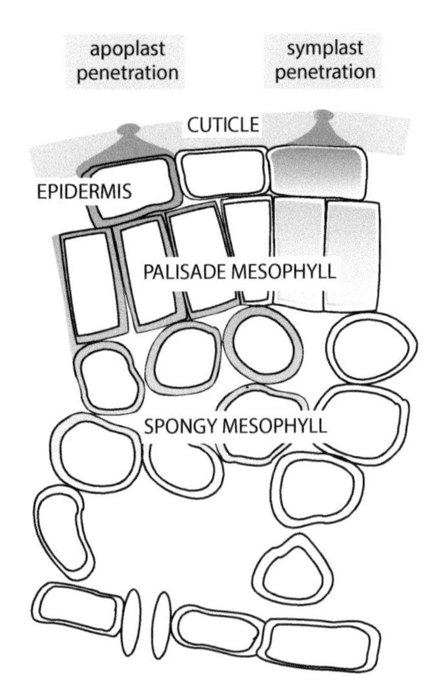

Figure 14.10. The penetration of phenanthrene into leaves. Phenanthrene is a polycyclic aromatic hydrocarbon (PAH) that commonly contaminates the atmosphere. It is primarily deposited on leaves when it is in the gas phase. Techniques for visualizing its penetration into leaves have shown that during gas-phase exposure, over a period of days it can penetrate a significant distance into the apoplast and symplast. Phenanthrene does not therefore remain in the cuticle as long as might be expected from its chemical properties. In maize plants, apoplastic penetration is much more significant, whereas in spinach plants symplastic penetration is more significant. (Adapted from Wild E, Dent J, Thomas GO, & Jones KC [2006] *Environ Sci Technol* 40:907–916. With permission from ACS publications.)

deposition rates are particularly dependent on leaf age, with younger leaves being associated with higher deposition rates. In many instances the correlations between SVOC concentrations in leaves or needles and their concentrations in the environment are sufficiently close to enable plants to be used as biomonitors of SVOC concentrations. This reflects the relatively slow rate of degradation of many SVOCs in plant leaves. SVOCs with the highest K_{OA} (in the range 8–15), such as flame retardants, have deposition rates that correlate with their K_{OA}.

Many anthropogenic VOC contaminants are released into the environment, including, for example, toluene, ethylene, xylene, benzene, and a variety of other alkanes, aldehydes, and chlorinated compounds. Due to their high vapor pressure and low K_{OA}, these compounds are deposited on plants slowly in gaseous form, although particulates can enhance deposition rates. Entry via the stomata does occur, but for many the primary route is via the cuticle, with deposition rates depending on K_{OA}. Species-specific differences in uptake can be related to cuticular and other morphological properties. In general, although there are well-established effects of many of these VOCs on human health, plants are not adversely affected by these compounds unless their concentrations are high, mostly well in excess of those that might be expected in the environment in all but the most polluted cases. For many anthropogenic VOCs there is evidence of metabolism, which although relatively slow helps to provide sinks to drive uptake and explain tolerance. For some VOCs—for example, those that are chlorinated—photodegradation can induce the production of free radicals, with potentially toxic effects. The exception is ethylene, whose role as a plant hormone was first suggested by investigations of the effects of burning coal gas on street trees, and which can have dramatic effects on numerous aspects of plant growth at low concentrations. Overall, however, the importance of anthropogenic VOCs to air pollution pales by comparison with that of BVOCs.

BVOCs that are produced by some plants include, for example, methanol, ethanol, acetone, and acetaldehyde, but by far the most important BVOCs produced by plants are the more than 1700 known isoprenoids from the deoxy-xylulose phosphate/methylerythritol phosphate (DOXP/MEP) pathway (Figure 14.11). These compounds have myriad roles—for example, as attractants, herbivore deterrents, allelopathic chemicals, defense molecules, and alarm signals. They also have roles in numerous stress responses—for example, maintaining membrane integrity when the temperature and oxidative conditions change. Many of their roles evolved because they can be rapidly released from plants to the atmosphere. BVOC emission from plants evolved in atmospheres that were devoid of anthropogenic contaminants, so pollution with non-BVOC compounds can affect their functioning, and hence that of plants and ecosystems, and has been suggested to be an underestimated impact of air pollutants on plants. Many of the environmental fluctuations to which plants are exposed, ranging from attack by herbivores to temperature change, result in the release of large amounts of BVOCs into the atmosphere. The extent of BVOC release (over 1000 Tg $year^{-1}$) is such that they provide the major link between the biosphere and the organic chemistry of the atmosphere. This has global consequences because the BVOCs that are released react with $OH^•$. In particular, the concentration in the atmosphere of one of the most active greenhouse gases, CH_4, is in significant part controlled by the action of $OH^•$. BVOCs reacting with $OH^•$ reduce the degradation of CH_4, prolonging its greenhouse effects on a globally significant scale. The oxidation of the wide range of BVOCs emitted also produces many intermediates that form particulates, increasing the aerosol burden of the atmosphere. In some locations, up to 50% of the organic aerosols in the atmosphere are from isoprenes. The particulate concentration affects air quality, with impacts both on human health and on the physico-chemical

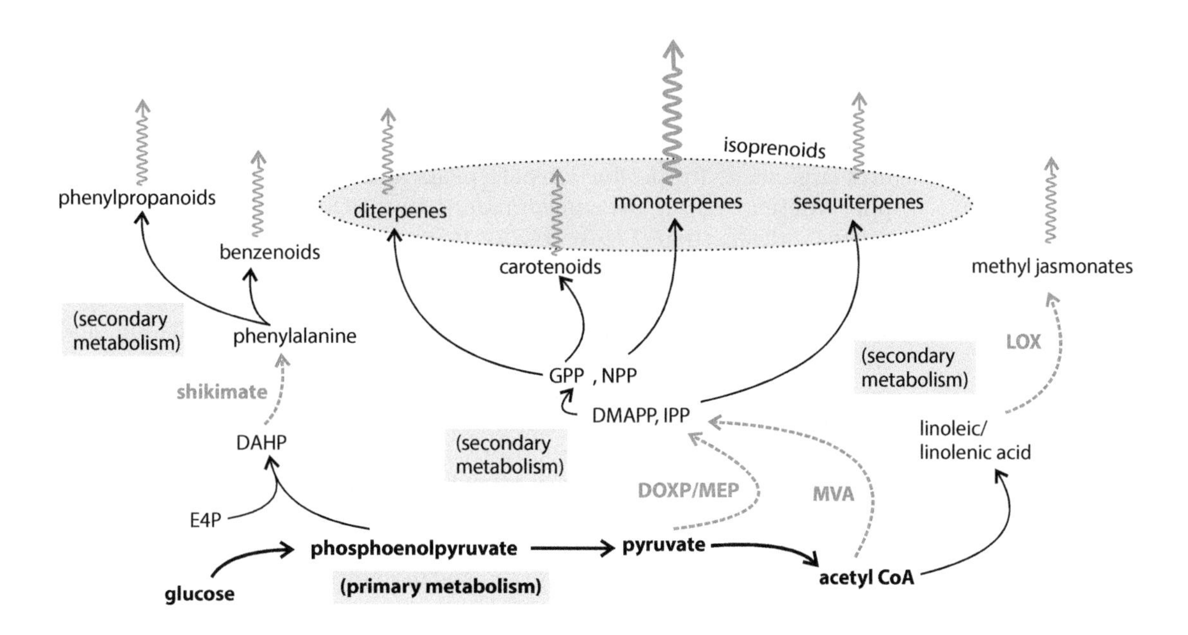

Figure 14.11. **Synthesis of volatile and semi-volatile organic compounds in plants.** There is a plethora of plant secondary metabolites that are volatile organic compounds (VOCs). They are some of the products of four different biosynthetic pathways—the shikimate, DOXP/MEP, MVA, and LOX pathways. Monoterpenes are generally the largest single group emitted, and the isoprenoids in general (shown in the gray ellipse in the figure) dominate emissions. There are numerous steps in the biosynthetic pathways that occur in a variety of cellular compartments. DOXP/MEP = deoxy-xylulose phosphate/methylerythritol phosphate; MVA, mevalonic acid; LOX, lipoxygenase; E4P, erythrose 4-phosphate; DAHP, deoxy-arabinoheptulosonate; IPP, isopentyl pyrophosphate; DMAPP, dimethylallyl-pyrophosphate; GPP, geranyl pyrophosphate; NPP, neryl pyrophosphate.

properties of the atmosphere. In addition, the reaction of NO_2 with BVOCs produces PAN, which makes a significant contribution to urban smog. These reactions also affect the production of O_3 in the atmosphere. PAN has an irritant effect on the eyes in humans, but is unusual in being particularly phytotoxic. It is more stable than NO_x, but breaks down to produce NO_2, and can thus be a significant contributor to long-range transport of N_r, especially from urban to rural environments. Drought, high temperature, elevated CO_2 levels, and aerosols can all cause a significant increase in BVOC emission from plants, which in turn can affect atmospheric properties, so BVOCs are important not only in terms of their effects on pollutant chemistry and air quality, but also for predicting changing global climate (Figure 14.12).

Chronic effects of ozone on terrestrial plants will be significant in the twenty-first century

Ozone is produced in the **stratosphere** via the photolysis of O_2, and a proportion of it can penetrate into the troposphere. However, the majority of tropospheric O_3 is formed by surface or near-ground processes in which NO_x, CO, CH_4, and other VOCs interact to produce O_3 as a pervasive secondary air pollutant. Under certain conditions, O_3 concentrations can peak to produce acute effects on plants, but perhaps one of the most significant air pollution effects in the twenty-first century will be chronic increases in background concentrations of O_3. Although O_3 precursors and the conditions that promote O_3 production can occur on restricted spatial and temporal scales, causing peaks in O_3 production, they are often diffuse, leading to widespread chronic elevation of O_3 levels. There has been at least a doubling

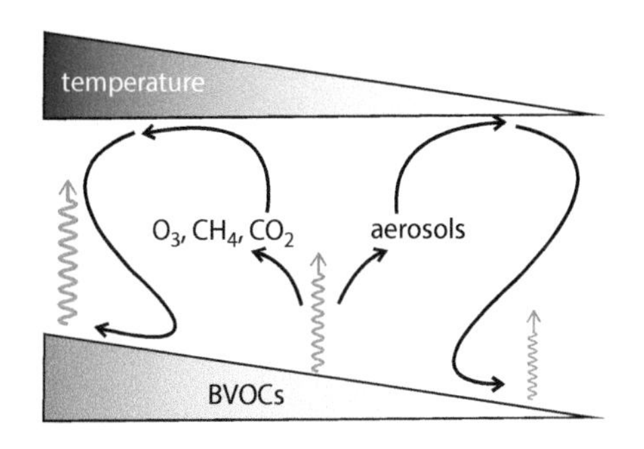

Figure 14.12. **Biologically generated volatile organic compounds (BVOCs) and climate change.** BVOCs derived from plants are emitted in sufficient quantities to affect climate change. They promote the formation of O_3, prolong the atmospheric half-life of CH_4, and produce CO_2. All of these changes can increase the temperature, thereby increasing the amount of BVOC emitted. Aerosols can promote condensation, and hence the release of latent heat into the atmosphere. However, because they intercept and reflect sunlight, overall they decrease the temperature of the atmosphere, hence decreasing the release of BVOCs. A knowledge of the balance of these processes, which are affected by numerous other variables that, for example, affect plant health, is important for understanding climate change.

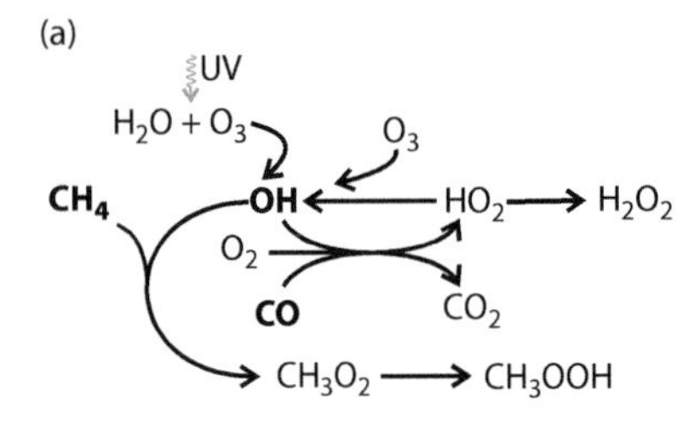

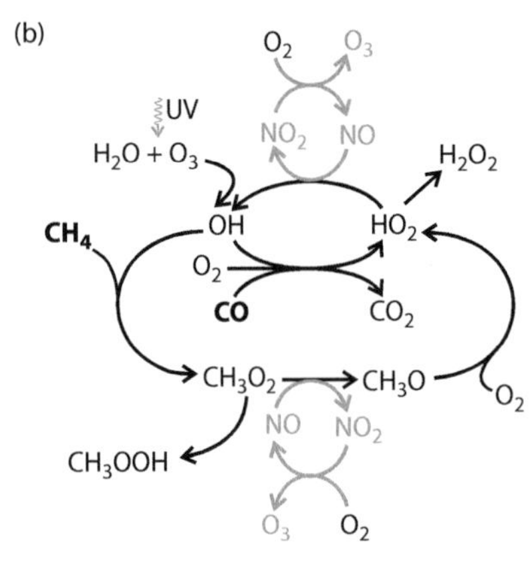

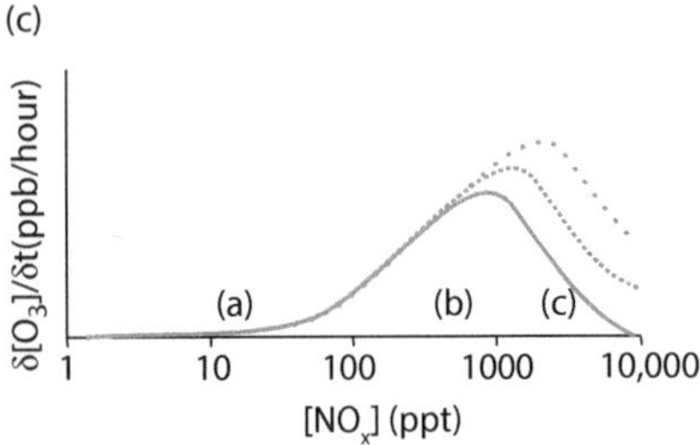

Figure 14.13. The production of ozone in the troposphere. (a) In the absence of NO_2, O_3 in the troposphere is generally consumed by the reaction of oxidants (OH), organic compounds, and CO. This can produce organic aerosols, but decreases O_3 concentration. (b) In the presence of moderately high concentrations of NO_2, O_3 is produced, ultimately, by the reaction of NO_2 and O_2 driven by the reactions of oxidants and organic compounds. (c) At high concentrations of NO_2 the formation of HNO_3 is favored, decreasing O_3 production. The green solid and broken lines show the general effect of increasing VOC concentration. (Data from Royal Society [2008] RS Policy Documant 15/08. Royal Society, London.)

of background O_3 concentration across the land surface of the northern hemisphere since the Industrial Revolution. Due to its production pathways, O_3 often occurs at high concentrations not only in urban areas but also in rural areas, so agricultural crops and natural ecosystems are exposed to O_3 on a large scale. The production of O_3 peaks at an NO_x concentration of 1000 ppb, whereas $OH^\bullet$ at low concentrations, and NO_x at high concentrations, cause O_3 destruction (Figure 14.13). In some locations a decrease in NO_x emissions is contributing to an increase in O_3 production. Conditions that produce changes in O_3 concentration in the atmosphere include dry deposition on vegetation during the day, high levels of incident radiation, and spring warmth coinciding with BVOC production. Predictions of O_3 effects on plants based on peak O_3 concentrations over particular time periods, or cumulative exposure above a threshold (AOT) of 40 ppb, have been developed primarily from chronic exposures in open-top-chamber studies in the USA and Europe, respectively, but flux-based indices that perhaps take better account of effects induced by chronic background O_3 concentrations are also now widely used. In addition to direct effects on vegetation, O_3 is the third most important greenhouse gas, and its contribution to climate change feeds back to effects of O_3 on plant growth.

By the end of the twentieth century it was estimated that, globally, O_3 was probably causing US$ 14–26 billion of lost crop production. This is more than that lost due to climate change, and does not include adverse effects on quality or on disease, pest, and stress susceptibility. In many parts of Asia, despite efforts to increase it, yield per hectare of staple crops has stagnated, and it is often suggested that O_3 is contributing to this effect. Long-term predictions of O_3 concentrations depend on predicting emissions of the precursors of O_3 and on predicting climate. In those locations with controls on primary pollutants, most modeled scenarios suggest that O_3 concentrations are likely to stabilize at their current, harmful concentrations. In significant areas, especially across Asia, substantial increases in O_3 are very likely to occur during the twenty-first century. In some of the most significant food production areas on Earth, high O_3 concentrations could have significant adverse effects (Box 14.2). O_3 concentrations that commonly occur in the environment also have adverse effects on the growth of many wild plants, including trees, and there are numerous reports of effects on interspecies competition, ecosystem production, and biodiversity. For example, lost timber production worth over US$ 50 million per year has been estimated for forests in Sweden. It is clear that species differ in their sensitivity to O_3, and that O_3 sensitivity can vary at different growth stages. For example, the legumes appear to be particularly sensitive to the effects of O_3, and there can be significant reproductive effects on some species.

Acute exposure to O_3 at concentrations higher than 150 ppb can cause visible symptoms on many plants. This is within the range of concentrations that have been measured in many industrialized countries, with symptoms on many plant species having been observed for decades. Such symptoms are primarily caused by necrosis of cells. O_3-induced necrosis has been shown to have similarity with the hypersensitive response induced by many pathogens—that is, the presence of O_3 triggers a type of programmed cell death. However, O_3 can affect plants at concentrations well below those that induce visible symptoms. Exposure to O_3 at sub-necrosis-inducing concentrations can induce effects in the plant cuticle. The long-chain fatty acids of cuticular waxes are oxidized by O_3, increasing the proportion of shorter-chain fatty acids. One function of the plant cuticle is probably to neutralize reactive molecules, such as O_3, that arrive at the plant surface, so that the effects are only adverse if they inhibit cuticle function overall. Some species have been shown to produce larger amounts of epicuticular wax in response to O_3. Ozone that does not react with the cuticle can enter plants via the stomata.

BOX 14.2. OZONE IN THE INDO-GANGETIC PLAIN

The Indo-Gangetic Plain is one of the most significant food production areas on Earth (Figure 1). Over 350 million people are now dependent on food from the rice, wheat, sorghum, and other crops that are widely grown there. The Ganges River, which flows down to the Bay of Bengal from the high Himalayas, has been central to the Hindu belief systems of Northern India for millennia, in significant part because of its role in providing the water that underpins food production. The Indus River and its tributaries provide much of the water for crop production in Pakistan and Northern India. There are significant yield gaps for many of the crops on the Indo-Gangetic Plain—that is, increased yields are theoretically possible. The closing of such yield gaps is likely to make a vital contribution to global food security. In general, between 1960 and 1990 increases in crop production on the Indo-Gangetic Plain more than kept pace with population increases. In many of the high-yield areas in particular, increases in production have slowed significantly since 1990. For the most significant crops on the Indo-Gangetic Plain some current estimates are that 10% or more of yield is lost due to O_3. It is very likely that O_3 concentrations will increase significantly across the Indo-Gangetic Plain during the twenty-first century (due to increases in population, industrial activities, and the burning of fossil fuels), and that this will have further adverse effects on crop production. It is probably necessary to minimize the effects of O_3 on crops in the Indo-Gangetic Plain if food production is to keep pace with population increases. Similar effects on crop yield are detectable in many parts of South Asia, including some of the most significant food production areas in China, so the effects of O_3 on crops will probably have a significant impact on global food security in the twenty-first century.

Further reading: Debaje SB (2014) Estimated crop yield losses due to surface ozone exposure and economic damage in India. *Environ Sci Pollut Res* 21:7329–7338.

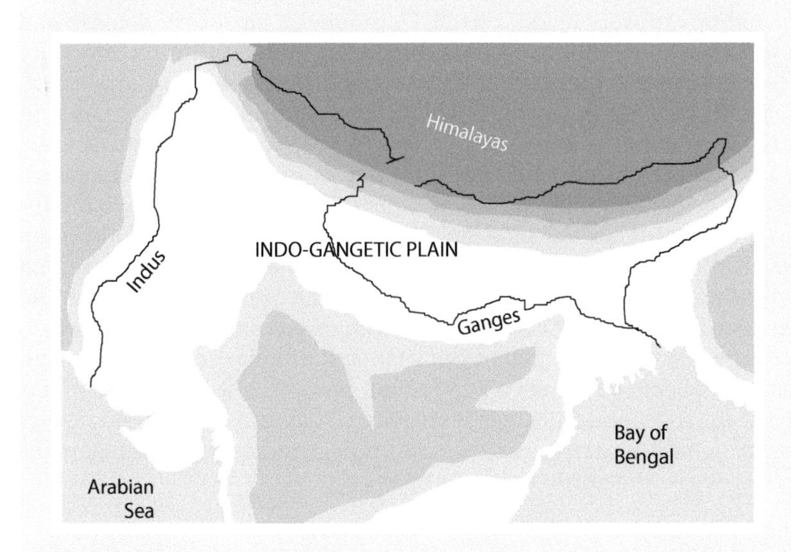

Figure 1. **The Indo-Gangetic Plain.** Most of the important crop-growing areas of Northern Pakistan, Northern India, and Bangladesh are on the Indo-Gangetic Plain.

Stomatal conductance and concentration gradients can be used to obtain accurate predictions of O_3 flux into plants. Acute exposure to O_3 causes rapid decreases in stomatal aperture in many species. This is related to oxidative effects in guard cells, and can often be reversed quite rapidly. Chronic exposure to low concentrations of O_3 can produce a similar decrease in stomatal aperture, but its effects are primarily due to the pernicious effects of chronic O_3 exposure on the balance of primary and secondary metabolism. Effects on overall plant biomass production have frequently been reported at background concentrations of 100 ppb or less. These effects are mainly on photosynthesis, which decreases, and respiration, which increases. The decrease in photosynthesis reduces the demand for CO_2 and hence stomatal aperture. Studies involving photon-flux measurements have described numerous changes in electron transport through the light reactions as a result of O_3 exposure. These are probably driven by changes in many of the enzymes of C fixation, which alter electron sink strength. Numerous enzymes of respiration are changed such that respiration rates increase. These changes are consistent with the frequently observed changes in concentrations of many carbohydrates. They are also consistent with an increase in shoot sink strength for carbohydrates, explaining the significant effects of O_3 on root growth. In plants that produce isoprenes, during O_3 exposure there is a significant increase in isoprene production, and in many plants there is an increase in the synthesis of phenolic compounds, including lignin. These changes reflect altered C allocation and decreased biomass production.

The chronic effects of O_3 arise from its oxidant capacity. O_3 that flows into intercellular spaces can dissolve in the apoplastic fluid surrounding the cells. The apoplastic fluid often contains antioxidants, especially ascorbate, but O_3 is a powerful oxidant and is soluble at the pH values that occur in the apoplastic fluid, so can initiate the production of ROS. At low O_3 fluxes, apoplastic antioxidants can reduce O_3 flux to the plasma membrane, but they probably do not eliminate such fluxes, and they incur considerable energy costs to the plant. **Ozonolysis** of fatty acids can occur in the plasma membrane. O_3 and H_2O_2 can cross the plasma membrane and enter the cytoplasm, increasing its ROS concentration. Increased cellular ROS levels occur with many stressors (although the ROS in these instances are produced within the cell), and are often a signal for plants to initiate the activity of proteins that control antioxidant capacity and the stress response. The defense and stress pathways that are mediated by salicyclic acid, jasmonic acid, and ethylene can all be affected by exposure to O_3. Overall, O_3 induces a mixture of defense and stress responses, depending on the exposure dynamics. Chronic exposure to O_3 is debilitating to plants not only because it increases the rate at which cell death is initiated in leaves, but also because it increases the amount of energy that plants direct towards stress management. Differences in O_3 fluxes into plants, mostly related to differences in stomatal conductance, clearly help to explain the variation in sensitivities to O_3, but so too do differences in stress-response capacity. This has resulted in numerous, often successful efforts to manipulate O_3 sensitivity in crops, and the development of externally applied compounds such as **ethylene diurea** that can increase resistance to the effects of O_3. The effects of O_3 on stomatal aperture have an impact on the water use efficiency of plants and their drought responses, whereas the effects on biomass production have an impact on N dynamics. These impacts, and those on reproductive capacity, have been shown to occur at community and ecosystem levels at O_3 concentrations that regularly occur across the northern hemisphere.

Particulates filtered by plants from the atmosphere can affect their growth

Atmospheric particulates are probably the most widespread air pollutant. Studies from, for example, datable ice in glaciers reveal that the particulate burden of the atmosphere was quite stable for several thousand years, until the marked increase of the last 200 years. There are very significant adverse effects of particulate inhalation on human health, but plant health can also be affected. Coarse particulates over 2.5 μm in diameter include a wide variety of primary compounds, whereas fine particulates less than 2.5 μm in diameter are often secondary compounds, formed by condensation, coagulation, and reactions on their surface. Particles less than 60 μm in diameter are mostly suspended in air, with deposition rates being dependent on diffusion and interception, whereas those over 60 μm in diameter are deposited under gravity (Figure 14.14). Primary particulates can include significant amounts of heavy metals, inorganic compounds produced during industrial activity, and the products of incomplete combustion, including globally significant quantities of black carbon. Secondary particulates can include a wide variety of N, S, and organic compounds. Plants are naturally exposed to a wide variety of particulates, including pollen, spores, volcanic dust, sandstorms, ash from fires, and salt spray. Plants in habitats with naturally high

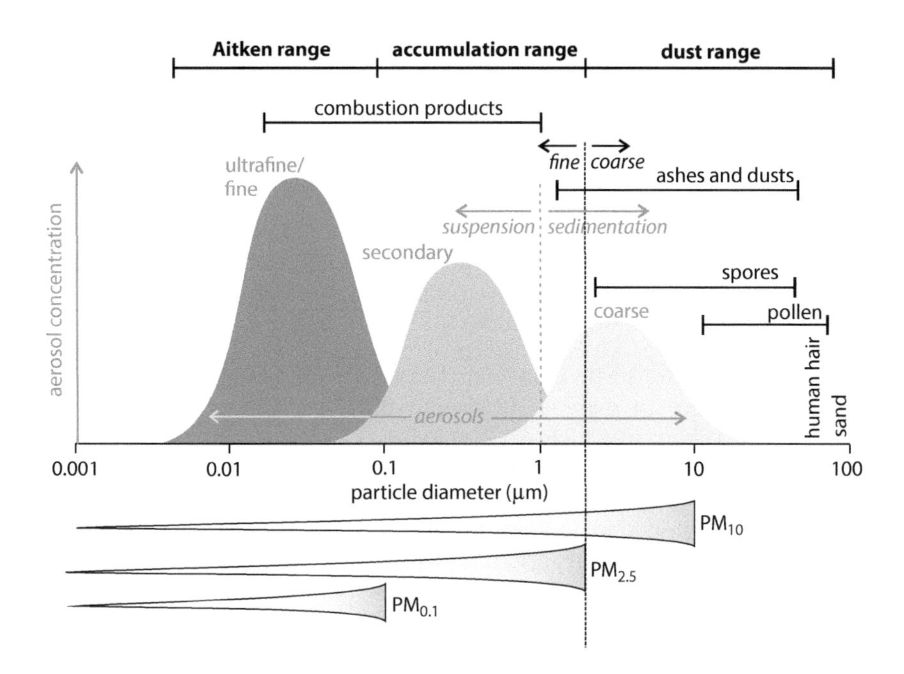

Figure 14.14. The classification of particulates. Particulates can be classified according to their diameter. Many coarse particles are primary particles of ash or dust, which are produced by natural processes as well as by industrial processes. They are finer than sand but mostly in the range of biological particles such as pollen and spores, and they sediment under gravity. Fine particles are often secondary and anthropogenic. Their deposition is primarily dependent on diffusion rates through boundary layers, because they are suspended in air. Particles less than 0.1 μm in diameter, known as the Aitken range, are vital condensation nuclei (and are named after the Scottish meteorologist who first posited their existence). The classification of particulate matter (PM) relates primarily to human health, with PM_{10} able to penetrate the respiratory tract, and $PM_{2.5}$ able to cross the surface of the lungs.

rates of deposition of particulates can have leaf adaptations that minimize their impact. Long-distance transport of particulates can significantly redistribute nutrients—for example, dust from the Sahara is deposited in significant amounts across the northern hemisphere, and contributes significantly to P availability in soils in the eastern USA. Particulates affect the transmission of light through the atmosphere, and thus affect both the photosynthetically active radiation (PAR) available to plants and significant aspects of the greenhouse effect.

Coarse particles have higher deposition velocities than fine particles, and therefore have increased deposition when flow is turbulent. This is particularly the case around trees, which can be effective filters for coarse particles with a diameter of up to 10 μm or more. Conifers are often noted to produce greater deposition velocities than broadleaved trees, but there is wide variation between species. Deposition velocities have been correlated with leaf surface roughness, produced for example by trichomes, but the presence of coarse particles on leaves can also be correlated with the amount of wax available to trap particles (Figure 14.15). In general, the atmosphere has about 1 coarse particle per cm^3, but many human activities increase this figure substantially. With moderate exposure rates, the mass of particles on a leaf can be almost us much as that of the wax, with coarse particles being the most significant contributor. Coarse particles are generally deposited primarily on the adaxial leaf surface. Leaf exposure to an atmosphere containing coarse particles can abrade the leaf, but can also have physiological effects when particulates become deposited on and embedded in the cuticle. Radiation balance is vital to leaf function. The presence of particles on leaves changes photosynthetic photon flux density (PPFD) into the leaf, and also changes the leaf's convective properties through which heat loss can occur. In addition, there have been many observations of particulates lodging in stomata and sub-stomatal cavities, with consequences for gas exchange. Coarse particulates can be associated in particular with heavy metals and some organic pollutants, such as PAHs. Although deposition can be quite localized around sources, and scavenging by vegetation can be significant, long-distance dispersal can be very important. Wash-off of coarse particulates from leaves is often considerable, but depends on the wettability of the leaves. For a water droplet, the difference in bonding angle to the leaf between its downslope and upslope sides will determine whether or not it

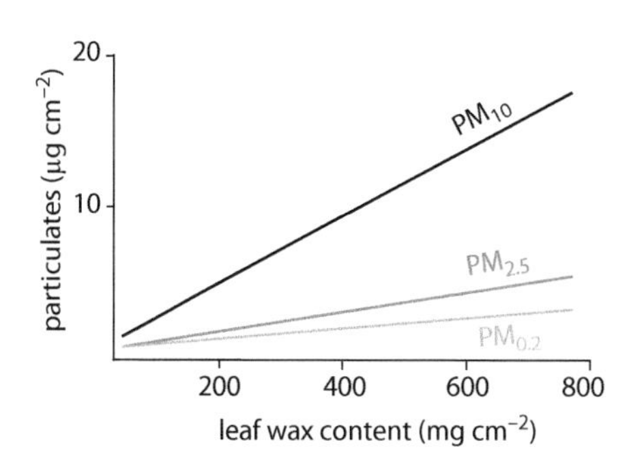

Figure 14.15. The effects of leaf characteristics on accumulation of particulates. In a study of 47 tree species in Poland and Norway, leaf wax content (shown in the figure) and leaf hairiness were found to be better predictors of particulates accumulated than leaf surface roughness. Other studies have found significant effects of leaf surface roughness, leaf height, and leaf area index. (From Sæbø A, Popek R, Nawrot B et al [2012] *Sci Total Environ* 427:347–354. With permission from Elsevier.)

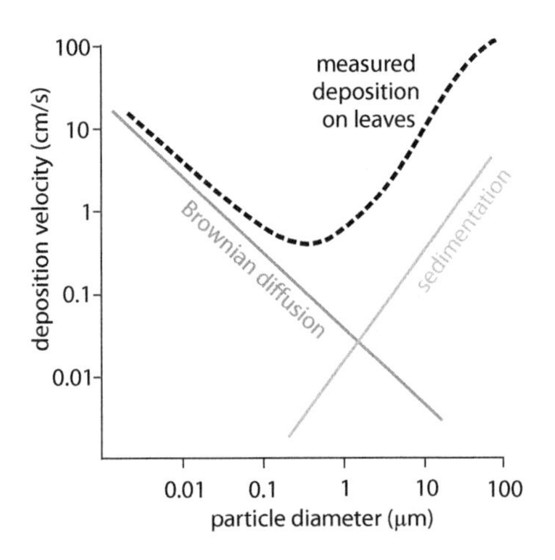

Figure 14.16. The dry deposition of particles. The sedimentation of coarse particles from air can be described by Stokes' law, and decreases almost to zero at particle diameters below 1 μm. The movement of fine particles can be predicted from Brownian diffusion, and increases with decreasing particle size. Measurements of dry deposition on leaves generally fit well with deposition rates predicted from these processes.

runs off the leaf, potentially taking particles with it. For some particles, such as those from cement works, the presence of water on the leaf causes dissolution, from which precipitation can occur upon drying, forming a significant crust that interferes with leaf function. Coarse particulates can also be blown off leaves. Models of dispersion for spores and pollen (which are generally about 1–10 μm and 10–100 μm in diameter, respectively) often show that average wind speeds are insufficient to pick them up, but that speeds at the high end of the usually log-normal distribution of speeds can do so. The gustiness of wind, especially in the canopy, can therefore determine the re-suspension rate of coarse particulates.

Vegetation has long been known to be a sink for particles of diameter less than 2.5 μm. In general, Brownian diffusion across the boundary layer around leaves, in which wind velocity is much reduced, determines dry deposition velocity, which is much less than for coarse particulates (Figure 14.16). The deposition of fine particulates is noticeably higher at, for example, leaf edges where the boundary layer is thinner. Nevertheless, vegetation acts as a filter of fine particles, in significant part because once these particles are adsorbed to the cuticle they are very unlikely to remobilize. Air generally contains a few thousand fine particles per cm^3, but there can be many tens of thousands of fine particulates per cm^2 of leaf. Filtration theory can provide good predictions of the deposition of fine particles on leaves. In particular, as predicted, there can be very significant wet deposition of fine particulates on leaves. Fine particulates are frequently hygroscopic, and at 80% relative humidity can double their weight, and hence change their size, by water absorption. The hygroscopic nature of many fine particles releases some of their constituents into solutions from which, depending on their lipophilicity, they can penetrate the cuticle. The substances released from fine particles can give rise to predictable physiological responses, in particular redox responses. They can also change the surface of leaf cuticles, which has been blamed for several of the adverse effects of air pollution on plants. This change can be due to direct attack on the cuticle, but it has also been suggested that deliquesced particles on cuticles might contribute to observations of "damaged" cuticles. Furthermore, the hygroscopic particles change the surface water characteristics of the leaf, with some observations showing that surface tension can be reduced sufficiently for water films to be formed through stomata to the apoplast, providing a route for direct transfer of molecules into the apoplast.

Plants can be used to monitor and manage air quality

There is a long history of using plants as biomonitors of air pollution in order to understand its distribution and effects. Computer models are now used to make detailed predictions of the distribution of air pollutants, but plant biomonitors still often play a key role in determining the accuracy of such predictions. Mosses, and in particular epiphytic mosses, are used to monitor a variety of air pollutants. As non-vascular plants that lack a cuticle and which have evolved to take up their nutrients from the air, mosses can be effective scavengers of air pollutants. They can provide a record of total accumulated pollutants, which can be particularly important for studies comparing different locations. The chemistry of the pollutants can determine the effectiveness of mosses as biomonitors—for example, they provide good records of accumulated Cd and Pb, but not of the volatile Hg. However, for nitrogenous pollutants and a wide range of organic pollutants, mosses can provide a useful record of deposition. Many trees also accumulate particular pollutants in particular tissues, especially the cuticle. Thus, for some organic air pollutants, measurements of concentrations in plant material can provide a useful record of deposition. In addition, there are established effects of a wide variety of air pollutants on particular plant species, and these effects can in some circumstances provide information on the concentrations to which vegetation has been exposed. Thus leaf morphological characteristics, leaf reflectance properties, and toxicity symptoms have all been used to infer concentrations of particular pollutants. All of these techniques have the same advantages and disadvantages as most biomonitors—that is, they provide a ready-made sampling regime of biological relevance, estimates of cumulative totals, and information about maximum exposures, but they are also affected by a myriad other environmental and biological factors which produce a very noisy signal.

Plants can also be used to actively manage air quality, particularly in urban environments. In some urban environments the overall air quality has improved substantially over the last few decades, but in most urban locations it often still does not meet agreed standards in many respects. However, in many other urban environments, air quality has declined over the last few decades. There is therefore a great incentive to find ways of improving urban air quality, and plants are beginning to be incorporated into strategies to achieve this. NO_x and particulates are both particular problems in urban air quality management, but can both be removed from the air by plants. Depending on the urban environment and the airflows that it generates, the effects of plants in scavenging these pollutants can be significant (Figure 14.17). In many studies, "green walls" have been found to be much more effective than "green roofs" in scavenging air pollutants, probably because of their proximity to the sources of pollution and the airflows that carry them. Plants can also be sources of air pollutants in urban environments, primarily because of the VOCs they can produce, but also because they can affect airflows such that they increase the concentrations of particulates. The sensitivity of a species, ecotype, or cultivar has an important bearing on how it can be used as a monitor or to manage air quality.

Summary

Plants have been naturally exposed to similar, and sometimes the same, compounds as those that human activity produces as air pollutants. However, the high levels of exposure that are now common in many parts of the world can overwhelm plants and cause widespread adverse effects in unmanaged and managed ecosystems, some of which will worsen as the twenty-first century progresses. Dry deposition of pollutants such as SO_2, NO_x, and O_3 occurs through stomata, and can have rapid effects on the apoplast and then

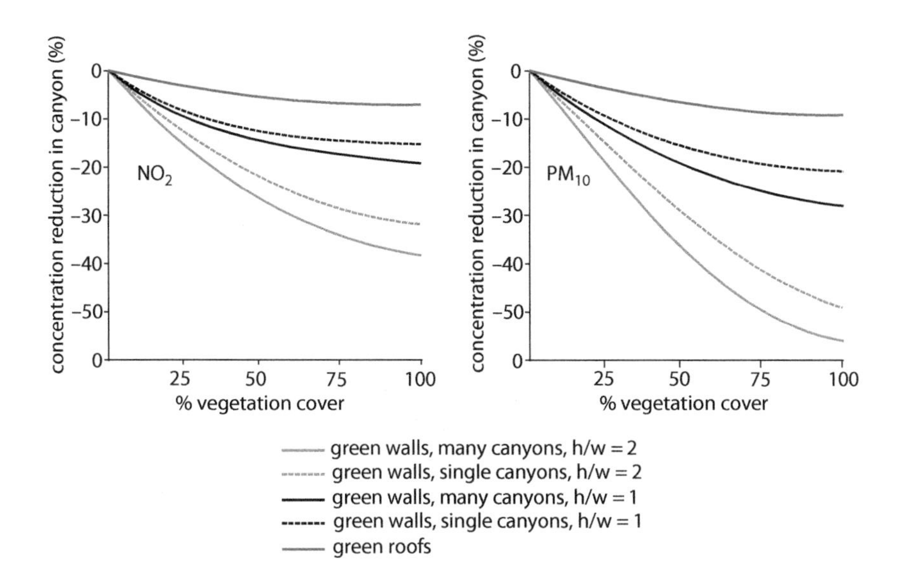

Figure 14.17. The scavenging of air pollutants by plants. In many urban environments, roadways with buildings on either side act as canyons for airflow. The height/width (h/w) ratio affects airflow significantly. Depending on the number of canyons and their h/w ratios, "green walls" can significantly affect the concentration of NO_2 and particulates in urban environments. (From Pugh TAM, MacKenzie AR, Whyatt JD, & Hewitt CN [2012] *Environ Sci Technol* 46:7692–7699. With permission from ACS publications.)

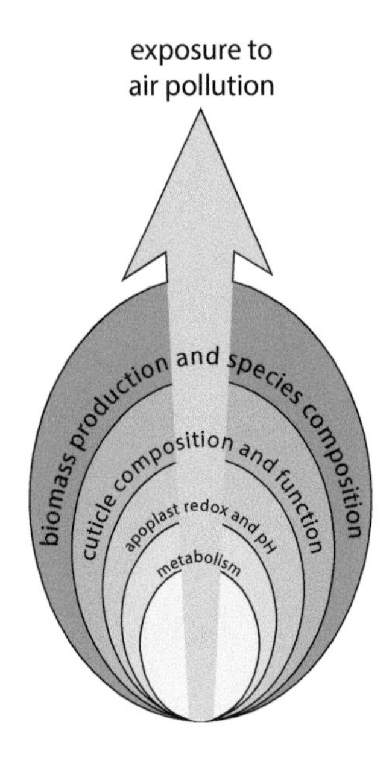

Figure 14.18. A hierarchy of responses to air pollution. Low-intensity air pollutants can exert subtle metabolic effects on plants. For some pollutants these include being used as nutrients. With more intense exposure to air pollutants, adverse effects on the apoplast can occur rapidly, and adverse effects on the cuticle build up over time. Significant exposure to air pollutants—that is, either acute exposure to high concentrations or chronic exposure to low concentrations—affects biomass production and, because of differences in susceptibility between species, the plant species composition of ecosystems. Plants are adapted to withstand some degree of exposure to air pollutants, and there is some evidence of evolution of tolerance of anthropogenic pollutants, but in general no plants are well adapted to significant exposure to air pollutants produced by human activity.

the metabolism of plants. Volatile, lipophilic, and small particulate pollutants can enter the plant via the cuticle, or become embedded in the latter, contributing to damage to the cuticle and consequent loss of its function. The species that have evolved tolerance to air pollutants are adapted in a variety of ways, according to which pollutant is predominantly responsible for the adverse effects (Figure 14.18). During the twenty-first century, management of the effects of air pollutants on plants, production of crop plants that can tolerate air pollutants, and perhaps the use of plants to manage air quality will be vital for some important conservation efforts, for food production in certain regions, and for the health of urban inhabitants.

Further reading

Plant surfaces and the atmosphere

Gjetting KSK, Ytting CK, Schulz A et al. (2012) Live imaging of intra- and extracellular pH in plants using pHusion, a novel genetically encoded biosensor. *J Exp Bot* 63:3207–3218.

Yeats TH, Martin LBB, Viart HM et al. (2012) The identification of cutin synthase: formation of the plant polyester cutin. *Nat Chem Biol* 8:609–611.

Yeats TH & Rose JKC (2013) The formation and function of plant cuticles. *Plant Physiol* 163:5–20.

Zhang C & Guo Y (2012) OsTRXh1 regulates the redox state of the apoplast and influences stress responses in rice. *Plant Signal Behav* 7:440–442.

Impacts of air pollution on plants

Amann M, Klimont Z & Wagner F (2013) Regional and global emissions of air pollutants: recent trends and future scenarios. *Annu Rev Environ Resour* 38:31–55.

Bender J & Weigel H (2011) Changes in atmospheric chemistry and crop health: a review. *Agron Sustain Dev* 31:81–89.

Calvo AI, Alves C, Castro A et al. (2013) Research on aerosol sources and chemical composition: past, current and emerging issues. *Atmos Res* 120:1–28.

Ge BZ, Wang ZF, Xu XB et al. (2014) Wet deposition of acidifying substances in different regions of China and the rest of East Asia: modeling with updated NAQPMS. *Environ Pollut* 187:10–21.

Deposition of air pollutants on plants

Burkhardt J, Basi S, Pariyar S et al. (2012) Stomatal penetration by aqueous solutions—an update involving leaf surface particles. *New Phytol* 196:774–787.

Burkhardt J & Hunsche M (2013) "Breath figures" on leaf surfaces—formation and effects of microscopic leaf wetness. *Front Plant Sci* 4:422.

Pan YP, Wang YS, Tang GQ et al. (2012) Wet and dry deposition of atmospheric nitrogen at ten sites in Northern China. *Atmos Chem Phys* 12:6515–6535.

Pinho P, Theobald MR, Dias T et al. (2012) Critical loads of nitrogen deposition and critical levels of atmospheric ammonia for semi-natural Mediterranean evergreen woodlands. *Biogeosciences* 9:1205–1215.

Plants and SO_2

Giraud E, Ivanova A, Gordon CS et al. (2012) Sulphur dioxide evokes a large scale reprogramming of the grape berry transcriptome associated with oxidative signalling and biotic defence responses. *Plant Cell Environ* 35:405–417.

Randewig D, Hamisch D, Herschbach C et al. (2012) Sulfite oxidase controls sulfur metabolism under SO_2 exposure in *Arabidopsis thaliana*. *Plant Cell Environ* 35:100–115.

Randewig D, Hamisch D, Eiblmeier M et al. (2014) Oxidation and reduction of sulfite contribute to susceptibility and detoxification of SO_2 in *Populus* × *canescens* leaves. *Trees* 28:399–411.

Yarmolinsky D, Brychkova G, Fluhr R et al. (2013) Sulfite reductase protects plants against sulfite toxicity. *Plant Physiol* 161:725–743.

Plants and reactive N

Faust C, Storm C & Schwabe A (2012) Shifts in plant community structure of a threatened sandy grassland over a 9-yr period under experimentally induced nutrient regimes: is there a lag phase? *J Veg Sci* 23:372–386.

Giordani P, Calatayud V, Stofer S et al. (2014) Detecting the nitrogen critical loads on European forests by means of epiphytic lichens. A signal-to-noise evaluation. *Forest Ecol Manag* 311:29–40.

Sutton MA, Reis S, Riddick SN et al. (2013) Towards a climate-dependent paradigm of ammonia emission and deposition. *Philos Trans R Soc Lond B Biol Sci* 368:20130166

Vallano DM, Selmants PC & Zavaleta ES (2012) Simulated nitrogen deposition enhances the performance of an exotic grass relative to native serpentine grassland competitors. *Plant Ecol* 213:1015–1026.

SVOCs, VOCs, and BVOCs

Ahammed GJ, Wang M, Zhou Y et al. (2012) The growth, photosynthesis and antioxidant defence responses of five vegetable crops to phenanthrene stress. *Ecotoxicol Environ Saf* 80:132–139.

Li Q & Chen B (2014) Organic pollutant clustered in the plant cuticular membranes: visualizing the distribution of phenanthrene in leaf cuticle using two-photon confocal scanning laser microscopy. *Environ Sci Technol* 48:4774–4781.

Loreto F, Dicke M, Schnitzler J-P et al. (2014) Plant volatiles and the environment. *Plant Cell Environ* 37:1905–1908.

Sharkey TD & Monson RK (2014) The future of isoprene emission from leaves, canopies and landscapes. *Plant Cell Environ* 37:1727–1740.

Effects of ozone on vegetation

Ainsworth EA, Yendrek CR, Sitch S et al. (2012) The effects of tropospheric ozone on net primary productivity and implications for climate change. *Annu Rev Plant Biol* 63:637–661.

Ainsworth EA, Serbin SP, Skoneczka JA et al. (2014) Using leaf optical properties to detect ozone effects on foliar biochemistry. *Photosynth Res* 119:65–76.

Dizengremel P, Vaultier M, Le Thiec D et al. (2012) Phosphoenolpyruvate is at the crossroads of leaf metabolic responses to ozone stress. *New Phytol* 195:512–517.

Tai APK, Martin MV, Heald CL (2014) Threat to future global food security from climate change and ozone air pollution. *Nature Clim Change* 4:817–821.

Effects of atmospheric particulates on plants

Burkhardt J & Pariyar S (2014) Particulate pollutants are capable to 'degrade' epicuticular waxes and to decrease the drought tolerance of Scots pine (*Pinus sylvestris* L.). *Environ Pollut* 184:659–667.

Lin M, Katul GG & Khlystov A (2012) A branch scale analytical model for predicting the vegetation collection efficiency of ultrafine particles. *Atmos Environ* 51:293–302.

Prospero JM, Collard F-X, Molinie J et al. (2014) Characterizing the annual cycle of African dust transport to the Caribbean Basin and South America and its impact on the environment and air quality. *Global Biogeochem Cycles* 28:757–773.

Saebo A. Popek R, Nawrot B et al. (2012) Plant species differences in particulate matter accumulation on leaf surfaces. *Sci Total Environ* 427:347–354.

Monitoring and management of air pollutants

Harmens H, Foan L, Simon V et al. (2013) Terrestrial mosses as biomonitors of atmospheric POPs pollution: a review. *Environ Pollut* 173:245–254.

Khavanin Zadeh AR, Veroustraete F, Wuyts K et al. (2012) Dorsiventral leaf reflectance properties of *Carpinus betulus* L.: an indicator of urban habitat quality. *Environ Pollut* 162:332–337.

Speak AF, Rothwell JJ, Lindley SJ et al. (2012) Urban particulate pollution reduction by four species of green roof vegetation in a UK city. *Atmos Environ* 61:283–293.

Wania A, Bruse M, Blond N et al. (2012) Analysing the influence of different street vegetation on traffic-induced particle dispersion using microscale simulations. *J Environ Manag* 94:91–101.

Chapter 15
Synopsis and Outlook

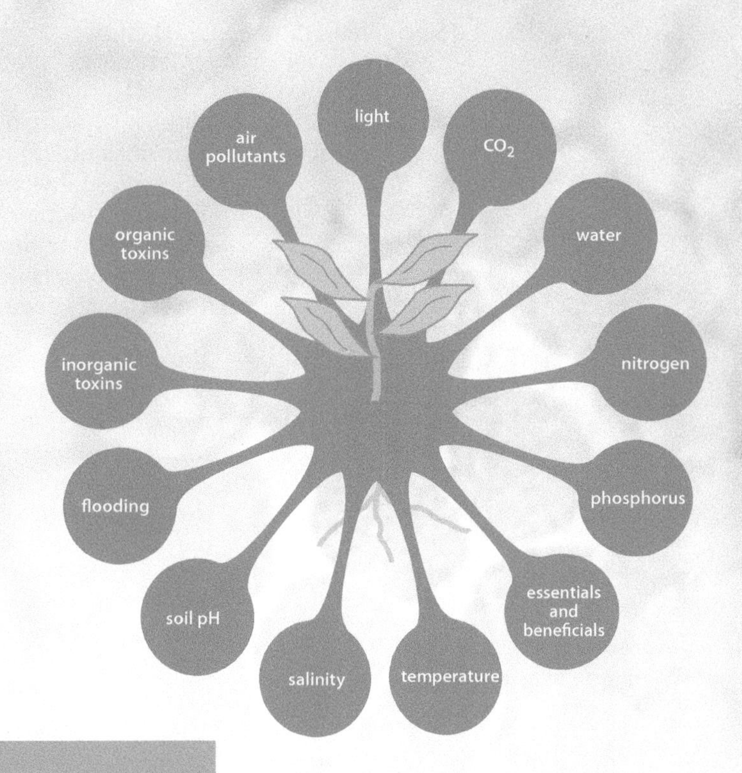

Plant growth and survival are affected by numerous interacting environmental variables.

Key concepts

- Many planetary boundaries for Earth systems are linked to interactions between plants and their environment.
- Advances in research from across the life sciences will extend descriptions of plant response mechanisms.
- Analyses of pattern and scale will improve the modeling of plant adaptations to environmental variation.
- Ecological and evolutionary contexts will provide further insights into plant adaptations.
- For the foreseeable future, an understanding of plant–environment interactions will contribute to enabling the environmental, biological, and agricultural sciences to confront a number of global challenges.

Plant–environment interactions play a significant role in determining the boundaries of non-linear effects in Earth systems

The Quaternary period of the last 2.5 million years has been dominated by the growth and retreat of ice caps. The Quaternary epoch with the most sustained stable conditions for agriculture and human civilization has been the last 11,500 years, namely the Holocene. Human impact on the environment began early in the Holocene and increased after the Industrial Revolution in Europe, but overall assessments of the environment—for example, the Global Environment Outlook-5 (GEO-5) report of the United Nations Environment Program (UNEP)—emphasize the "great acceleration" from 1950 onward. During this period almost all aspects of human activity that have an impact on the environment accelerated at unprecedented rates. The Intergovernmental Panel on Climate Change (IPCC) Fifth Assessment Synthesis Report, published in 2014, showed that human impacts on climate are, for most measures, as predicted in their previous four reports going back 24 years, and that there is no sign that the great acceleration of climatically relevant impacts is slowing down. There is ample evidence that humankind is running the risk of destabilizing, or is in fact already destabilizing, climate and other aspects of the Earth systems on which human life currently depends. The dependence of humankind on the plant component of the environment for food can be illustrated by calculations of sources of human calories, proteins, and nutrients (Figure 15.1). Of all the foodstuffs, minerals, and vitamins that humans need, only vitamin B_{12} and vitamin A are supplied mostly via sources other than terrestrial plants. The current dependence of humankind on terrestrial plant products for food is essentially complete, but is only part of our dependence on the Earth systems that are at risk of becoming unstable.

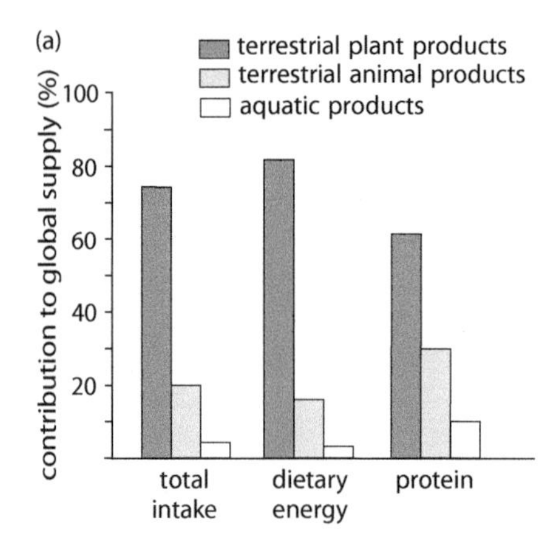

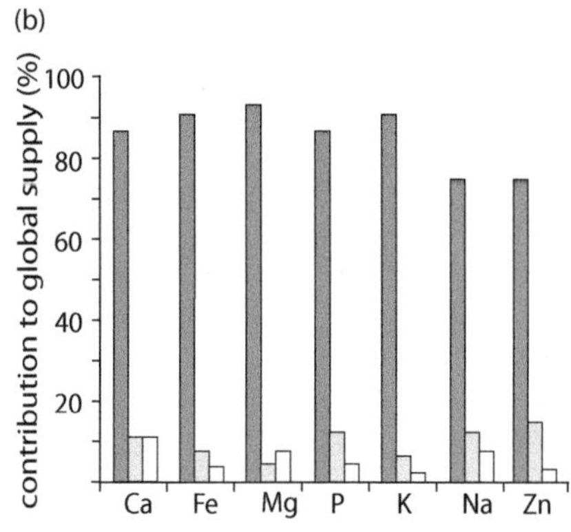

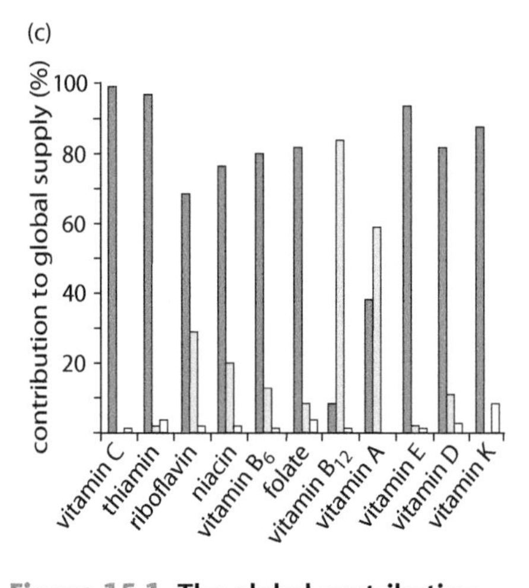

Figure 15.1. The global contribution of plant and other products to human diets. The vast majority of humankind's total (a) dietary, (b) mineral, and (c) vitamin intake is directly from plant products. The supply of many animal products is indirectly dependent on plant production. (From White PJ, George TS, Gregory PJ et al [2013] *Ann Bot* 112:207–222. With permission from Oxford University Press.)

The concept of planetary boundaries helps to emphasize both the importance of Earth systems for human life, and the key roles of plant-environment interactions. Planetary boundaries are impact thresholds beyond which non-linear effects on aspects of Earth systems are likely to occur. The general concept of planetary boundaries is in one sense precautionary—that is, the boundaries are not set where non-linear effects are certain, but where they are likely—but in significant ways it is not precautionary, because they deflect attention from linear impacts that humans have up until the boundary, and they do not emphasize the interrelatedness between boundaries. Non-linear effects beyond a boundary have abrupt, unpredictable, and potentially catastrophic consequences, which might include effects that have an impact on the position in relation to other boundaries. Although the quantification of boundaries and the Earth's condition with respect to them is complex, the scientific consensus suggests that human activity has already caused Earth systems to exceed the planetary boundary for biodiversity loss, the N cycle, and climate change, and is rapidly approaching the planetary boundary for P (Figure 15.2). The chapters of this book have shown that initiatives built on an understanding of plant-environment interactions can contribute significantly to meeting the challenge of biodiversity loss, to moderating human impact on the N and P cycles, and to moderating climate change. The planetary boundaries for land use and freshwater resources are also rapidly being approached, and the topics discussed in this book show that they too are intimately linked to plant-environment interactions. It is less clear where the planetary boundaries lie for chemical pollutants and for atmospheric aerosol loading, but the topics discussed in this book are very relevant to these boundaries as well. Thus if humans are to sustainably inhabit Earth—that is, to live within the planetary boundaries—initiatives aided by an understanding of the topics dealt with in this book are likely to make a significant contribution.

The understanding of plant stress response mechanisms can be extended by comparisons with other organisms

The mechanisms that plants use to respond to variation in environmental conditions provide many significant examples of the common mechanisms that link organisms at the molecular level. These include the signaling pathways that are used to respond to stress, the mechanisms that are used to control gene expression, the genes that are expressed, and the proteins that are synthesized. Thus, for example, transporter-based mechanisms for controlling the concentrations of many nutrients and for controlling the flow of water are clearly related to those used by other organisms. In addition, the proteins that are involved in plant responses to stress are related to several of the most highly conserved stress proteins of life, which are part of the fundamental cellular stress response (Table 15.1). The cellular stress response is thought to consist of proteins that evolved in ancient prokaryotes in response to the challenging conditions that prevailed early in Earth's history. The chaperone, proteasome, free-radical-scavenging, and DNA-repair proteins of the cellular stress response perform functions that play a significant role in responses to several environmental stressors in most organisms, including plants.

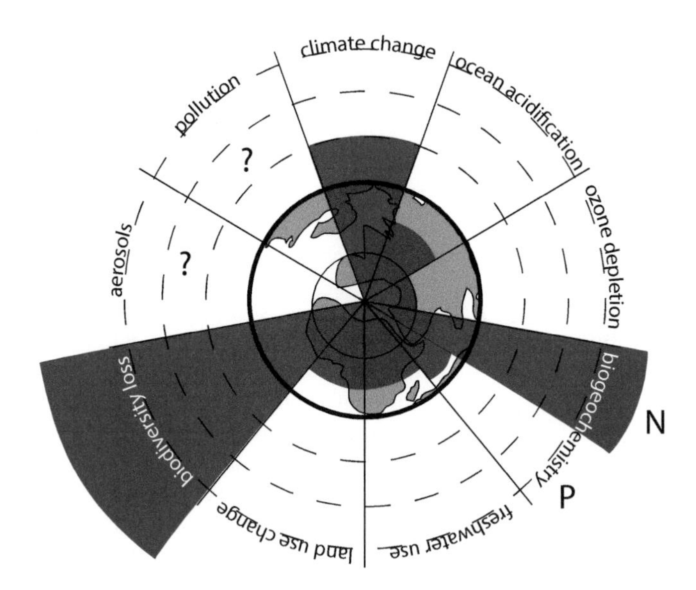

Figure 15.2. Earth's planetary boundaries. In 2009 the Stockholm Resilience Centre identified nine aspects of Earth systems on which human existence depended, and quantified the boundary at which non-linear effects on those aspects were likely. Estimates were then made of human impacts on the environment in relation to these boundaries. In the figure the planetary boundary for each aspect of the Earth system is shown as a bold line around the globe, and human impacts in relation to them as dark shading. The position in relation to atmospheric aerosols and environmental pollution was not quantifiable, but human impacts on biodiversity, the N cycle, and climate are already significantly beyond planetary boundaries, and that for P is rapidly approaching the boundary. Boundaries for other categories are likely to be crossed within a few decades. (From Rockström J, Steffen W, Noone K et al [2009] *Ecol Soc* 14:32. With permission from the authors.)

The fundamental similarities in stress responses across the kingdoms of life mean that advances in research from across the life sciences might provide an insight into plant–environment interactions. For example, advances in our understanding of reactive oxygen species (ROS) signaling are providing insights into how organisms interpret ROS signals, advances in our understanding of the roles of small RNAs are elucidating their roles in many processes, and advances in Ca signaling are revealing it to be among the most fundamental of signaling processes. All of these are involved in

Table 15.1. The most highly conserved stress-response proteins across the superkingdoms of life

Gene product (in humans)	Function
HSP70	Molecular chaperone
HSP40	Molecular chaperone
PRS1	Proteasome pathway, cell-cycle checkpoint
PRS2	Proteasome pathway, cell-cycle checkpoint
SelB	Seleno-cysteine elongation factor
MSH	DNA repair
Lon protease	Stress-response protease
HSP60	Molecular chaperone, cell-cycle regulation
DNA topoisomerase III	Chromosome maintenance, DNA repair
Glutathione reductase	Free-radical scavenging
MLH	DNA repair
Peptide methionine sulfoxide reductase	Free-radical scavenging, protein repair

(Data from Kültz D [2003] *J Exp Bot* 206:3119–3124.)

plant-environment interactions, and advances in understanding of them in any organisms are likely to provide further insights in relation to plants. These advances will perhaps be most important in understanding how cellular conditions are fine-tuned to the environment, but their roles in controlling anatomical and morphological responses to profound environmental variation may also be important. However, many aspects of plant stress responses are unique to plants, and these must also be recognized if our understanding of response mechanisms in plant-environment interactions is to be developed further.

Terrestrial plants provide a unique biological arena in which fundamental processes occur. This is, for example, particularly important in roots. Investigations of many aspects of root biology are more challenging than those of shoots, and an increased understanding of root responses to stress is likely to elucidate plant-environment interactions significantly. Unique and important root adaptations to the soil environment have been quite regularly reported for many decades, so it seems likely that further research might provide further examples. In general, the importance of, for example, rhizosheaths, the exudation of proteases, the uptake of organic N, and root tropisms for water and metals has until recently probably been underestimated. The photosynthesizing leaf also provides a unique context for stress responses. Processes such as photorespiration, for example, are unique to plants and play a role in responses to several stressors. Plants also have a unique suite of hormones, including brassinosteroids and salicylic acid, whose role in stress responses has only relatively recently been discovered, and for which further insights are likely. In addition, secondary metabolites clearly have a role in responses to some stressors at least, and there may be many other roles that have yet to be discovered.

Our understanding of the importance of variation in plant–environment interactions can be extended by modeling that includes the pattern and scale of variation

The environmental variables that affect plant growth, and that are dealt with in this book, vary significantly in time and space. For each variable, significant plant responses have been discussed for which the proximate and ultimate explanations can best be understood by reference to characteristics of the variation. For most of the environmental variables, at low levels of variation homeostatic mechanisms maintain constant cellular conditions. Homeostasis is primarily achieved by regulating the constituents and components of the cell. Transport proteins are used to regulate the ionic environment, metabolic feedback mechanisms are used to regulate the biomolecular environment, and turnover of components of organelles (for example, membranes) is used to regulate the cellular context. Inorganic toxins, especially some toxic metals, provide the one clear example in which, in a few species, massive amplification of homeostatic mechanisms has achieved adaptation to the most extreme environmental exposure. In general, however, more profound variation drives the evolution of some species with anatomical and morphological adaptations that complement intracellular homeostatic mechanisms, and chronic exposure can drive the evolution of plastic and symbiotic responses to complement these. In general, the characteristics of the variation are therefore important in driving the responses, and for a fuller understanding can be positioned within studies that take account of pattern and scale. In addition, models of microorganism adaptation to nutrient variation, for example, reveal interesting interactions between environmental variation and threshold/graded responses that may also be relevant to plant-environment interactions, and that highlight the importance of recent modeling developments to their study.

Fitness is a key biological concept that provides an important example of why understanding the effects of pattern and scale might be useful for extending our understanding of plant-environment interactions. Fitness is a quantitative description, measured using genotypic or phenotypic frequencies, of the ability to survive and reproduce. A fitness landscape describes the fitness—that is, reproductive success—of a range of phenotypes, genotypes, or alleles. Fitness landscapes can be predicted for an invariant environment. This can be useful when studying, for example, the effect of particular genotypes on fitness. However, in terrestrial ecosystems the environment varies. Frequently there are patterns in the variation—that is, it is not random either spatial or temporally—and the scale at which the variation occurs and at which the impacts are investigated can have important effects. Dynamic fitness landscapes can be constructed in which fitness is continually calculated in an environment that varies. Understanding plant-environment interactions in a variable environment is likely to benefit and supplement studies of the importance of pattern and scale in the biosphere. A significant contributor to such understanding is the study of variation in plant phenotypes—that is, phenomics. High-technology, high-throughput screening of massive numbers of individual plants, both in controlled facilities and in the field, provides the data for understanding the plant component of variation in the plant-environment system. Interestingly, studies of this type often reveal that the variable biotic component interacts with the variable environment—that is, "the theatre is being redesigned by the players." Integration of the topics dealt with in this book into such frameworks is likely to significantly extend our understanding of plant-environment interactions. Of particular importance will be the impact of human activity on the characteristics of environmental variation. For example, the variance spectrum for CO_2 variation shows that although the range of CO_2 values within which anthropogenic change is occurring is not unprecedented, the same cannot be said for the scale of change (Figure 15.3). Such understanding for all of the environmental variables discussed in this book is likely to be useful.

Understanding how plant stress responses evolved will provide insights about the plant–environment interface

The most fundamental definition of stress to an organism is evolutionary—that is, a reduction in fitness. This can be difficult to measure and challenging to understand within appropriate time frames, but enables analyses to integrate the phylogenetic constraints on stress responses that clearly exist. It also suggests that responses to different stressors are likely to be linked, which as this book shows is often the case for plant responses to different environmental stressors. These different stressors include not only abiotic

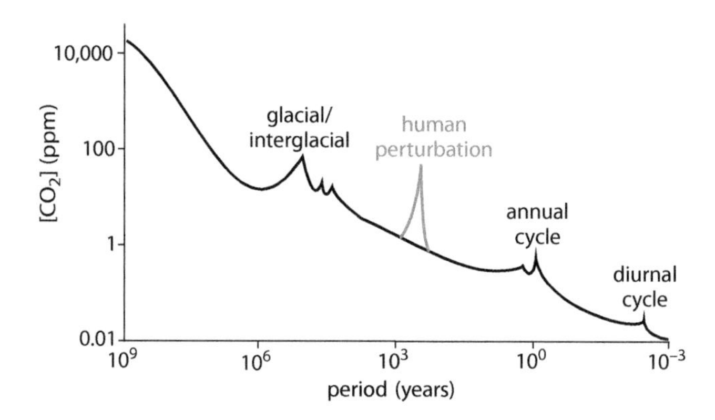

Figure 15.3. The variance spectrum for CO_2 on Earth. CO_2 concentrations on Earth vary significantly, with periodicities ranging from daily to hundreds of thousands of years. The periodicity of human-induced changes in CO_2 over hundreds of years is unique, and probably presents unique challenges to Earth systems. (From Chave J [2013] *Ecol Lett* 16:4–16. With permission from John Wiley & Sons)

but also biotic stressors. A variety of stress matrices have been proposed for plants. These identify which stressors interact in evoking stress responses. It is clear that there are interactions between different stressors, such as drought and heat, but it is less clear how interactions between stressors compare and whether there is one overall matrix that can be identified. For the variables that have been discussed in this book, a stress-response hierarchy, which is as yet untested but perhaps heuristically helpful, can be suggested (Figure 15.4). Stress responses to xenobiotics and to pH, even across all scales and levels of intensity, are primarily biochemical and physiological. It is perhaps the chemical nature of the stressor and, in the case of xenobiotics, the relatively recent widespread exposure that accounts for the fact that there are few other adaptations to these stressors. Those plants that have adapted to stress caused by CO_2, salinity, and temperature variation have, in general, not only biochemical and physiological but also anatomical and morphological adaptations. This is perhaps because these variables tend not to vary on spatial scales but over long time spans. There are anatomical and morphological adaptations to variation in light, water, nutrients, and flooding, but also significant plastic and developmental responses. This is perhaps because of the spatial and temporal scales over which these variables can change. For N and P, all of these types of adaptation are found, but in addition there are significant symbioses that help plants to capture these resources, especially in the case of P. This is probably because the challenge of capturing these resources by other means is so significant. The description of stress-response matrices and their evolutionary origins will be useful for deepening our understanding of plant stress responses. The advances that are being made in understanding stress-response matrices in other organisms are also likely to be useful.

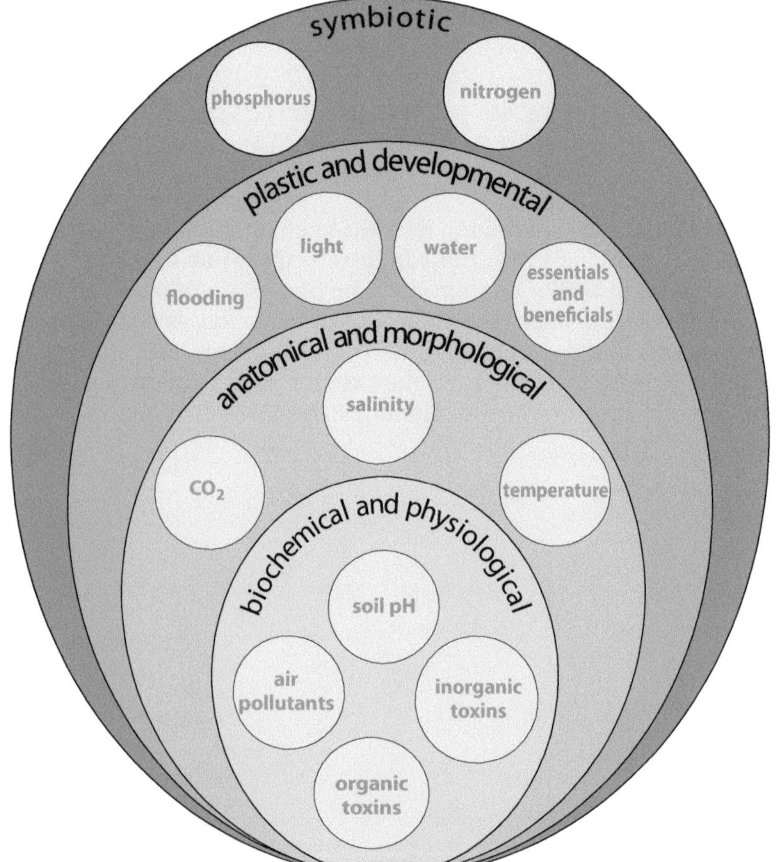

Figure 15.4. **Plant adaptation to significant variation in key environmental variables.** Some environmental variables, including values at the upper and lower limits of their ranges of variation, invoke adaptations in plants that are primarily biochemical and physiological. Other environmental variables invoke further adaptations that are frequently additive. In general, in order to adapt to significant variation in environmental variables, anatomical and morphological adaptations supplement biochemical and physiology adaptations, whereas for other stressors plastic/development and symbiotic adaptations can also be used.

Plant responses to the variables discussed in this book are traits that are a product of evolution and help to determine fitness. These traits clearly help to determine processes at the level of ecosystem and biome, but do not do so independently—they are part of survival strategies. Concepts such as the leaf economics spectrum are useful for analyzing these strategies and their effects at higher levels of organization (Figure 15.5). For example, traits related to water, C, N, and P tend to converge at particular points on a "worldwide fast-slow plant economics spectrum"—that is, some plants take up these resources quickly, utilize them quickly, and provide the basis of ecosystems with fast turnover rates, whereas other plants are located at distinct, slower points on the spectrum. The integration of plant responses to the environmental variables discussed in this book into trait-based ecology can extend our understanding of the traits and their ecological and evolutionary importance.

Understanding plant–environment interactions helps us to confront global challenges

This book has outlined how and why plants respond as they do to the abiotic challenges that they face in the Earth's current environment. In order to provide direct links for those with interests in the environmental, biological, and agricultural sciences, examples are taken from unmanaged and

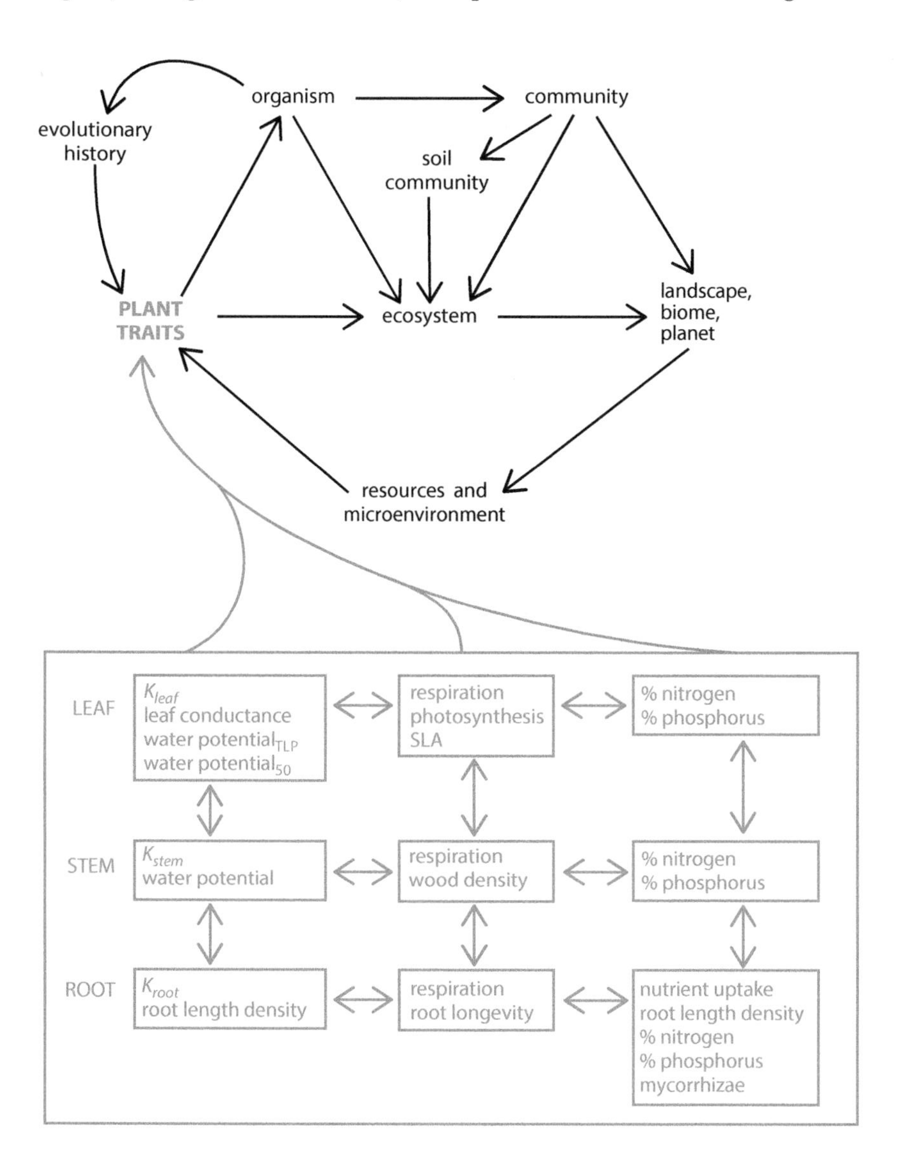

Figure 15.5. The significance of plant traits to ecology. Trait-based ecology uses suites of characteristics to understand processes on a range of scales. Plant traits probably play a key role in determining processes on a range of scales. In the example shown in the figure, plant traits are proposed to be primarily dependent on characteristics related to the capture of C, N, P, and water. Such analyses can determine where plants are located, for example, on a fast–slow economics spectrum. K, hydraulic conductance; SLA, specific leaf area; water potential$_{TLP}$, water potential at turgor loss point; water potential$_{50}$, water potential at 50% conductivity. (Redrawn from Reich PB [2014] *J Ecol* 102:275–301.)

managed ecosystems, and ranging from molecular to evolutionary scales. Some key themes that emerge from the chapters, and that are likely to be important for the foreseeable future, include the need to position the topics within an Earth systems context, the relationship between nature and agriculture, and the need for models across a range of scales. Radical shifts in the relationship between nature and agriculture are possible. For example, a change in dietary preferences of human populations on a large scale would alter the pressure on food production systems significantly, as would a shift towards eating food produced in synthetic biological systems or a shift in markets to prioritize the environmental costs of production. However, for the foreseeable future an improved understanding of plant-environment interactions will be very useful to the environmental, biological, and agricultural sciences in confronting some of the major global challenges of the twenty-first century.

Further reading

Plants, the environment, and planetary boundaries

Gerst MD, Raskin PD & Rockstrom J (2014) Contours of a resilient global future. *Sustainability* 6:123–135.

Laurance WF, Sayer J & Cassman KG (2014) Agricultural expansion and its impacts on tropical nature. *Trends Ecol Evol* 29:107–116.

Odegard IYR & van der Voet E (2014) The future of food—scenarios and the effect on natural resource use in agriculture in 2050. *Ecol Econ* 97:51–59.

White PJ, George TS, Gregory PJ et al. (2013) Matching roots to their environment. *Ann Bot* 112:207–222.

Response mechanisms

Kellermeier F, Armengaud P, Seditas TJ et al. (2014) Analysis of the root system architecture of *Arabidopsis* provides a quantitative readout of crosstalk between nutritional signals. *Plant Cell* 26:1480–1496.

Minocha R, Majumdar R & Minocha SC (2014) Polyamines and abiotic stress in plants: a complex relationship. *Front Plant Sci* 5:175.

Schroeder JI, Delhaize E, Frommer WB et al. (2013) Using membrane transporters to improve crops for sustainable food production. *Nature* 497:60–66.

Vaahtera L, Brosche M, Wrzaczek M et al. (2014) Specificity in ROS signaling and transcript signatures. *Antioxid Redox Signal* 21:1422–1441.

Space, time, scales, and modeling

Chave J (2013) The problem of pattern and scale in ecology: what have we learned in 20 years? *Ecol Lett* 16:4–16.

de Visser JAGM & Krug J (2014) Empirical fitness landscapes and the predictability of evolution. *Nat Rev Genet* 15:480–490.

Luis Araus J & Cairns JE (2014) Field high-throughput phenotyping: the new crop breeding frontier. *Trends Plant Sci* 19:52–61.

Sivak DA & Thomson M (2014) Environmental statistics and optimal regulation. *PLoS Comput Biol* 10:e1003826.

Ecological and evolutionary outlook

Kassahn KS, Crozier RH, Poertner HO et al. (2009) Animal performance and stress: responses and tolerance limits at different levels of biological organisation. *Biol Rev Camb Philos Soc* 84:277–292.

Reich PB (2014) The world-wide 'fast-slow' plant economics spectrum: a traits manifesto. *J Ecol* 102:275–301.

Schulte PM (2014) What is environmental stress? Insights from fish living in a variable environment. *J Exp Biol* 217:23–34.

Suzuki N, Rivero RM, Shulaev V et al. (2014) Abiotic and biotic stress combinations. *New Phytol* 203:32–43.

Confronting current challenges

Baudron F & Giller KE (2014) Agriculture and nature: trouble and strife? *Biol Conserv* 170:232–245.

De La Fuente GN, Frei UK & Luebberstedt T (2013) Accelerating plant breeding. *Trends Plant Sci* 18:667–672.

Freudenberger L, Hobson P, Schluck M et al. (2013) Nature conservation: priority-setting needs a global change. *Biodivers Conserv* 22:1255–1281.

Gunawardena J (2014) Models in biology: 'accurate descriptions of our pathetic thinking.' *BMC Biol* 12:29.

Abbreviations

ABA	abscisic acid
ABRE	ABA-responsive element
ACC	1-aminocyclopropane-1-carboxylic acid
ADH	alcohol dehydrogenase
ADME	adsorption, distribution, metabolism, and excretion
AF	actin filament
AFP	antifreeze protein
ALS	acetolactate synthase
A_{mass}	photosynthetic carbon fixation per unit leaf mass
AM	arbuscular mycorrhiza(s)
AMP	adenosine monophosphate
AOX	alternative oxidase
APS	adenosine phosphosulfate
ATP	adenosine triphosphate
BCF	bioconcentration factor
BER	base excision repair
bHLH	basic helix loop helix
bp	base pair
B:R ratio	ratio of blue to red light
BS	bundle sheath
BVOC	biologically generated volatile organic compound
CAB	chlorophyll *a*/*b* binding
CaM	calmodulin
CAM	Crassulacean acid metabolism
CDF	cation diffusion facilitator
CEF	cyclic electron flow
CFC	chlorofluorocarbon
cGMP	cyclic guanosine monophosphate
Chl	chlorophyll
CLC	chloride channel
COPT	copper transporter
COX	cytochrome *c* oxidase
CYP	cytochrome P450
DAG	diacylglyceride
DCD	dicyandiamide
DCL	Dicer-like protein
DMA	2′-deoxymugineic acid
dsRNA	double-stranded RNA
EC	electrical conductivity
EcM	ectomycorrhiza(s)
ELIP	early light-induced protein
ER	endoplasmic reticulum
ERF	ethylene responsive factor
ERM	ericoid mycorrhiza(s)
FACE	free-air carbon dioxide enrichment
FAD	fatty acid desaturase
Fd	ferredoxin
FFD	first flowering date
FMN	flavin mononucleotide
FR	far red
FRO	ferric reductase/oxidase
G_s	leaf conductivity
GA	gibberellic acid
GABA	γ-aminobutyric acid
GAD	glutamic acid decarboxylase
GDU	growing degree units
GGT	γ-glutamyltranspeptidase
GM	genetically modified
GOGAT	glutamine oxoglutarate aminotransferase
G-POX	glutathione peroxidase
GR	glutathione reductase
GS	glutamate synthetase
GSH	glutathione
GST	glutathione transferase
Gt	gigatonne
GT	glucosyltransferase
GTP	guanosine triphosphate
HAMK	heat-activated MAP kinase
HIPP	heavy-metal-associated isoprenylated plant protein
HLIP	high light-induced protein
HMA	heavy-metal ATPase
HR	homologous recombination
HRP	horseradish peroxidase
HS	heat shock
HSE	heat stress-responsive element
HSF	heat stress transcription factor
HSP	heat stress-responsive protein
IDP	intrinsically disordered protein
IP_3	inositol 1,4,5-trisphosphate
IPCC	Intergovernmental Panel on Climate Change
IPP	isoprenylated plant protein
IR	infrared
KO	knockout
L_r	root hydraulic conductivity
LAI	leaf area index
LEA	late embryogenesis abundant (protein)
LEF	linear electron flow
LHC	light-harvesting complex
LL	leaf longevity
LMA	leaf mass per unit area
LNT	linear no-threshold
LR	lateral root
LSU	large subunit
LT	linear threshold
LUE	light use efficiency
MA	mugineic acid
MAP	mitogen-activated protein
MAPK	mitogen-activated protein kinase
MATE	multidrug and toxic compound extrusion
MBD	metal-binding domain

ME	malic enzyme
MEA	Millennium Ecosystem Assessment
MFS	major facilitator superfamily
miRNA	microRNA
MMR	mismatch repair
Mn-SOD	manganese superoxide dismutase
MT	metallothionein
MTP	metal tolerance protein
N_{mass}	leaf nitrogen concentration per unit leaf mass
NA	nicotianamine
NAC	nascent polypeptide-associated complex
NAD	nicotinamide adenine dinucleotide
NADP	nicotinamide adenine dinucleotide phosphate
NAS	nicotianamine synthase
NCED	9-*cis*-epoxycarotenoid dioxygenase
NDH	NADH oxidase-like complex
NER	nucleotide excision repair
NHEJ	non-homologous end joining
NI	nitrification inhibitor
NiR	nitrite reductase
NLP	NIN-like protein
NORM	naturally occurring radioactive materials
NPP	net primary production
NPQ	non-photochemical quenching
NR	nitrate reductase
NRAMP	natural-resistance-associated macrophage protein
NSCC	non-selective cation channel
OAA	oxaloacetate
OEC	oxygen-evolving complex
PAH	polycyclic aromatic hydrocarbon
PAL	phenylalanine ammonia lyase
PAN	peroxyacetyl nitrate
PAP	purple acid phosphatase
PAPS	phosphoadenosine phosphosulfate
PAR	photosynthetically active radiation
PC	phosphatidylcholine; phytochelatin
PCB	polychlorinated biphenyl
PCS	phytochelatin synthase
PDC	pyruvate decarboxylase
PE	phosphatidylethanolamine
PEP	phosphoenolpyruvate
PEPc	phosphoenolpyruvate carboxylase
PFK	phosphofructokinase
PHC	petroleum hydrocarbon
Pi	inorganic phosphate
PIF	phytochrome interacting factor
PL	phospholipase
PLD	phospholipase D
PMF	proton-motive force
POP	persistent organic pollutant
PPCK	phosphoenolpyruvate carboxylase kinase
PPFD	photosynthetic photon flux density
PPO	protoporphyrinogen oxidase
PQ	photochemical quenching; plastoquinone
PR protein	pathogenesis-related protein
PS	phytosiderophore
PSI	photosystem I
PSII	photosystem II
PTOX	plastid terminal oxidase
PYR	pyrabactin resistance
QSAR	quantitative structure–activity relationship
QTL	quantitative trait locus
RET	resonant energy transfer
R:FR	red to far-red light (ratio)
RGR	relative growth rate
RH	relative humidity
RIL	recombinant inbred line
RNAi	RNA interference
RNS	reactive nitrogen species
ROL	radial oxygen loss
ROS	reactive oxygen species
rRNA	ribosomal RNA
RSG	reference soil group
rubisco	ribulose-1,5-bisphosphate carboxylase/oxygenase
RuBP	ribulose-1,5-bisphosphate
SAM	*S*-adenyl-methionine
SAR	shade avoidance response; sodium absorption ratio; systemic acquired resistance
sHSP	small heat stress-responsive protein
SIK	salt-inducible kinase
siRNA	short-interfering RNA
SOA	secondary organic aerosol
SOD	superoxide dismutase
SOS	salt overly sensitive
SPAR	soil–plant–atmosphere research
SRL	specific root length
ssRNA	single-stranded RNA
SSU	small subunit
SVOC	semi-volatile organic compound
TAG	triacylglyceride
TCA	tricarboxylic acid
TCE	trichloroethylene
TF	transcription factor
TH	thermal hysteresis
TIC	transcription initiation complex
TWh	terawatt hour
UNEP	United Nations Environment Program
USDA	United States Department of Agriculture
UTR	untranslated region
UV	ultraviolet
VOC	volatile organic compound
VPD	vapor pressure deficit
WHO	World Health Organization
WLES	worldwide leaf economics spectrum
WUE	water use efficiency
ZIP	zinc/iron permease

Glossary

14-3-3 protein any of a family of proteins that can bind to a wide range of target proteins involved in cell signaling, and thus play an important role in regulating their activity.

α-crystallin an important protein component of the lens in vertebrate eyes.

AAA proteins "ATPases Associated with diverse Activities", which form oligomers that act as ATP-driven molecular machines that remodel macromolecules.

ABC transporters a ubiquitous ancient superfamily of transporters that all have an ATP-binding cassette (ABC) that ATP binds to and hydrolyzes, releasing energy to drive a range of functions, mostly but not always involving transport.

abaxial the side of a leaf that during development faces away from the main axis or stem and in many species thus becomes the lower surface of the leaf.

acclimation the process whereby some plants can develop the ability to withstand increasingly severe stress, especially cold, through progressive or prior exposure to it.

acetylation the replacement of an H atom in a molecule of an organic compound with an acetyl (C_2H_3O) group.

acid sulfate soils soils that have been flooded by saline sulfur-rich water and subsequently drained by natural or artificial processes to produce an acid substrate with high levels of available sulfur. Raised sea levels caused by glacial melt often cause the initial flooding.

acidosis an increase in acidity, usually in a cell, tissue, or biological substance such as blood plasma.

actin a protein that forms the microfilaments of the cytoskeleton.

activator a protein that increases gene transcription.

active site the part of an enzyme molecule to which a substrate binds, thus enabling a chemical reaction to occur.

adaxial the side of a leaf that during development faces towards the main axis or stem and in many species thus becomes the upper surface of the leaf.

adenosine triphosphate (ATP) the nucleotide molecule that is used to transfer energy in living cells. Dephosphorylation of ATP results in the rapid release of a significant amount of energy, which is used to drive energy-requiring metabolic reactions.

adventitious root any root that develops in an unusual position (for example, from a stem).

aerenchyma plant tissue that contains large air spaces between the cells.

aerobe an organism that cannot live in the absence of oxygen, which it requires as the terminal electron acceptor in respiration.

agroforestry a mixed system of arable and tree crops used to provide a continuous food supply, produce economically useful plant products, and prevent soil erosion.

alanine one of the 20 amino acids that are encoded in DNA and used as the basic building blocks for protein synthesis.

albedo the reflectivity of a surface to solar radiation. It is equivalent to the degree of whiteness, and is the ratio of reflected radiation to incoming radiation—that is, a reflection coefficient that ranges from 1 (pure white) to 0 (pure black).

alkaloid any of the complex alkaline nitrogen-containing compounds that are produced from amino acids in a wide range of plants. Many alkaloids have important pharmacological properties.

alkyl relating to a hydrocarbon group derived by removal of one hydrogen atom from an alkane (that is, a saturated hydrocarbon).

allelopathic relating to chemical effects on plant species other than the species that produce the chemical(s).

allochthonous relating to material that originated from somewhere else, and was introduced to its present position by some process (for example, plant material in a lake sediment).

alternative crops crop species other than the few that generally dominate food production, which can be used to produce common products under different conditions, or to produce new products.

alternative oxidase an oxidase found in the mitochondrial membranes of many plants that provides an alternative route to cytochrome oxidase for electrons.

alternative splicing during gene expression, a controlled process that enables a single gene to code for multiple proteins, as removal of exons in different combinations from the mRNA results in the creation of a wide range of different mRNAs from a single pre-mRNA, and thus a range of different versions of proteins.

amphipathic having both polar and non-polar parts.

amphistomatous having stomata on both sides of the leaf.

anabolism the energy-requiring metabolic reactions that result in the synthesis of complex molecules in living organisms.

anaerobe an organism that can live in the absence of oxygen because it can use other substances as terminal electron acceptors in respiration.

anammox anaerobic ammonium oxidation, an important bacterially mediated reaction of the nitrogen cycle in which ammonium is oxidized using nitrite as an electron acceptor under anaerobic conditions.

anaplerotic relating to the process of replenishing the levels of intermediates in a metabolic pathway, such as the tricarboxylic acid (TCA) cycle.

angiosperms the flowering plants, all of which produce ovaries containing ovules that after fertilization develop into seeds that are entirely enclosed by a fruit.

anhydrobiosis the ability of some living organisms to survive in extremely dry and/or extremely hot or cold conditions in an almost completely desiccated state.

anthesis the flowering time of a plant (that is, the period between the opening of the flower buds and the development of fruit).

anthocyanin any of the water-soluble phenolic flavonoid pigments that often occur in the vacuole of plant cells, and which are responsible for producing red, purple, and blue pigmentation, such as the autumnal colours of the foliage of deciduous plants.

Anthropocene relating to the current epoch, in which human activity has had a major geological impact on the Earth's ecosystems.

anticlinal occurring at right angles to a surface.

antifreeze proteins a diverse set of proteins produced by certain organisms to decrease the freezing point of water and thus reduce the formation of potentially lethal ice crystals, allowing survival at sub-zero temperatures.

apoplast the non-living space in plant tissue, including the plant cell walls and the dead tissue of the xylem. It is one of the main pathways of water movement through the plant.

aquaporin any of the proteins that form pores in cell membranes and mediate the transmembrane transport of water and other uncharged molecules.

aquatic macrophyte a multicellular plant that grows in or near water, and that may be emergent (that is, rooted in shallow water with most of the plant emerging above the surface), floating, or submerged. Aquatic macrophytes include species of flowering plants, ferns, mosses, liverworts, and multicellular algae.

aquifer a body of geological material (for example, rock) that is capable of storing significant amounts of water.

archaea a domain of life that consists of single-celled prokaryotic organisms with a range of unique biochemical features that distinguish them from bacteria and eukaryotes. Members of the archaea include sulfate-reducing prokaryotes, methane-producing prokaryotes, and those that can survive extremes of temperature, pH, or salinity.

Archean relating to the geological eon from about 4 billion to 2.5 billion years ago.

Artiodactyla the order of mammals that characteristically bear their weight evenly on two toes. They include many ruminants, in which microorganisms in the first two stomach compartments have significant levels of phytase activity, and are thus able to break down material with a high cellulose content. In contrast, monogastric animals are unable to digest such material.

aryl in a chemical compound, relating to any functional side group derived from an aromatic ring.

Ascomycota a division of the fungal kingdom, members of which characteristically produce their spores in a sac-like structure known as an ascus.

ATP synthase the enzyme that synthesizes ATP in chloroplasts and mitochondria.

autotroph an organism that can use an external energy source to synthesize all of the complex organic compounds that are essential for life from simple inorganic ions and molecules.

auxilin a protein that cooperates with heat shock proteins to enhance their function.

axenic relating to a culture that contains only one species of organism.

axial of or towards the main axis.

azetidine a saturated organic chemical compound consisting of a ring containing three carbon atoms and one nitrogen atom.

bacteriochlorophyll a pigment that is used in photosynthesis in some bacteria. It is related to chlorophyll, but mostly absorbs longer-wavelength (that is, lower-energy) light.

bacteroid a modified bacterial cell, especially a morphologically distinct form of *Rhizobium* that develops in plant-derived peribacteroid membranes in the root nodules of leguminous plants.

basicole a plant species that tends to be restricted to soils with a high concentration of basic ions.

Basidiomycota a division of the fungal kingdom, members of which characteristically produce spores in a group of four on a specialized cell called a basidium.

basifuge a plant species that does not usually inhabit soils containing a high concentration of basic ions.

biofortification the use of genetic engineering or conventional plant breeding to develop crops with enhanced nutrient concentrations.

biological half-life the time it takes for a molecule in an organism to lose 50% of its chemical activity, usually via a decrease in its concentration.

biomimetic relating to the mimicking of a biological system or structure, especially in the design of structures, processes, or systems.

biosorption the process of binding of metal ions, especially toxic heavy metals, to certain biological materials (for example, binding to plant cell walls due to their high ion-exchange capacity).

biotroph a pathogenic or parasitic organism that depends on a live host organism for its nutrient supply.

Blue Revolution the recent expansion and changes in human use of freshwater resources, especially the emergence of aquaculture as a highly productive form of agriculture. When used in this context the term is analogous with the "Green Revolution" in agriculture.

blue water the water in rivers, lakes, and seas.

bordered pit any of the cavities in plant cell walls that connect adjacent xylem elements, where the cavity of the pit is partially covered by a border which is an extension of the cell wall that is formed by secondary growth.

boundary layer the layer of air adjacent to a surface, such as a leaf, in which air movement is markedly reduced due to surface friction.

Bowen ratio the ratio of sensible heat loss to latent (evaporative) heat loss.

bulliform relating to large bubble-shaped cells, especially those that occur in the epidermis of grasses.

bunded relating to a structure or feature that has had a retaining bank or wall constructed around it.

calamine a historical term for zinc ore, which consists of minerals rich in zinc carbonate and/or zinc silicate.

calcicole a plant species that is confined to or most often inhabits soils with a high concentration of Ca^{2+} ions.

calcifuge a plant species that does not usually inhabit soils with a high concentration of Ca^{2+} ions.

callose an insoluble polysaccharide that occurs in higher plants and is composed of glucose molecules, mostly joined by β(1–3) linkages , but with some β(1–6) links. It is deposited in specialized cell walls. This occurs as part of normal development in certain tissues (for example, pollen grains). Callose is also formed in a range of tissues in response to pathogen attack or abiotic stress. Under these conditions it alters cell wall and plasmodesmatal function, often isolating groups of cells.

calmodulins small calcium-binding proteins, containing 148 amino acids, which occur in all eukaryotic cells and that transduce calcium signals by binding to a wide range of other proteins to affect their function.

canalization the production of a standard phenotype that resists changes induced by the environment.

carbamate a functional group that contains one carbon atom and one nitrogen atom (H_2NCO_2H). It is derived from an amine, with an acidic carboxyl (COOH) group attached to the nitrogen atom.

carbonic anhydrase an enzyme that catalyzes the reversible interconversion of carbon dioxide (CO_2) and water (H_2O) to HCO_3^- and H^+.

Carboniferous relating to the geological period from about 359 million to 299 million years ago, characterized by the finding of fossil seed plants and also the first reptiles in coal seams.

carbonyl in an organic chemical compound, relating to the C=O functional group (that is, a carbon atom double bonded to an oxygen atom).

carboxylation the addition of carbon dioxide (CO_2) or bicarbonate (HCO_3^-) to an organic molecule, resulting in the formation of a carboxyl (COOH) group.

carotenoid any of the water-insoluble non-nitrogenous pigment molecules that have a role as accessory photosynthetic pigments in many higher plants and photosynthetic bacteria. They are tetraterpenoids (that is, based on four terpene molecules, each composed of two isoprene molecules). The two main classes of carotenoids are the xanthophylls (which are oxygenated) and the carotenes (which are non-oxygenated).

carrying capacity the maximum population of a particular organism that a specific environment can sustain.

Casparian strip a band in cell walls waterproofed with lignin, on which suberin can be deposited, that forms on the radial and transverse walls of the endodermal cells in plant roots. Its presence means that water can only enter the pericycle through the cytoplasm of the endodermal cells.

catabolism the metabolic reactions that break down complex molecules and release energy.

catechin a type of flavanol that is a plant secondary metabolite. Catechins are polyphenols that are derived from a skeleton consisting of two benzene rings and a heterocyclic ring.

caulescent having a stem.

cavitation the formation of vapor-containing bubbles or cavities in a liquid, which shows a dramatic increase in frequency if the temperature of the liquid is increased while the pressure remains constant, or if the pressure is reduced at constant temperature.

Cerrado a savannah in eastern and southern Brazil that consists of almost closed woodland, inhabited by many endemic plants and animals. It covers an area of about 2 million km^2, covering about 20% of Brazil.

chalcocite a mineral rich in copper sulfide that is found in copper ores.

chaparral the scrub-like regions, containing many sclerophyllous evergreen plant species and having a Mediterranean-type climate, that are found in South-Western North America, mainly in California.

chaperones proteins that aid the assembly or folding of molecules, especially that of other proteins, but which are not part of the functioning molecule.

charophyte any freshwater green alga belonging to the division Charophyta. Charophytes include the Charales (stoneworts), which contain deposits of calcium carbonate in their cell walls that give them a hard stone-like appearance. Like land plants and the other green algae (the Chlorophyta), charophytes contain chlorophyll *a* and *b*, and use starch as a storage polysaccharide.

chemotaxis a change in the direction of movement of a motile organism or cell in response to a change in the levels of a particular chemical (or chemicals) in its environment.

chitin a polysaccharide composed of linked chains of the linear molecule *N*-acetyl-D-glucosamine, which are arranged in layers to form a lightweight but very strong material. Chitin is a major component of the cell walls of fungi, and the main component of the exoskeletons of arthropods.

chlorophyllous parenchyma unspecialized ground tissue consisting of parenchyma cells containing large numbers of chloroplasts.

choline a quaternary ammonium compound—that is, a compound based on ammonium but in which the hydrogen atoms have been replaced by alkyl or aryl groups. Choline has three ethyl groups and a hydroxyethyl group attached to its nitrogen atom.

chromatin the material, containing protein, DNA, and RNA, of which the chromosomes in the nuclei of eukaryotic cells are composed.

circadian clock the molecular timekeeper of cells, which generally maintains an approximately 24-hourly rhythmic pattern of metabolic activities in living organisms.

clathrin a protein that has a major function in the formation of a lattice-like coating around the small vesicles within which membranes and proteins can be moved around the cell for transport, communication, etc. The clarithrin coating aids the budding of the vesicles from donor membranes.

CNGC channels cyclic nucleotide-gated channels, which are ion channels in the cell membrane that are directly activated (that is, opened or closed) by the binding of cyclic nucleotides.

cofactor a non-protein component that is required by a protein, usually an enzyme, in order for it to function, and that is different to the substrate of the enzyme.

colligative properties properties of solutions that depend on the ratio of the number of molecules of solute to the number of molecules of solvent, in contrast to non-colligative properties, which depend on the type of molecules present.

collimated light light that has parallel rays and therefore shows minimal dispersal as it travels. Light arriving at the Earth's surface from the Sun is essentially collimated.

compartmental model a mathematical model of a process or system (for example, a whole ecosystem) in which the numerical values of variables in different compartments are determined by a measured or estimated constant of proportionality.

compatible solute a small molecule that generates a solute potential while remaining compatible with physiological functioning.

complex II in mitochondria, the part of the electron transport chain that involves succinate dehydrogenase.

complex III in mitochondria, a cytochrome bc_1 complex, which is involved in part of the Q cycle in the electron transport chain.

complex IV in mitochondria, the final part of the electron transport chain, involving cytochrome *c* oxidase, which transfers electrons to oxygen to produce water, a step that is inhibited by cyanide.

conformation the three-dimensional structure of proteins that gives them their catalytic properties and is produced by patterns of binding between their constituent amino acids.

constant of proportionality a measured or estimated constant that reflects the proportional amount of a variable between compartments (for example, in the environment).

coumarin a vanilla-scented phenolic compound that is produced by many plants. The coumarin molecule is based on a benzopyrone skeleton (that is, a structure with one 6-carbon ring joined to a heterocyclic ring containing 5 carbon atoms and 1 oxygen atom).

Cretaceous relating to the geological period from about 144 million to about 66 million years ago. The name (which means "chalky terrain") refers to the extensive chalk deposits of this age, many of which were derived from marine algae.

critical level the atmospheric concentration of a particular pollutant above which direct adverse effects on living organisms are observed.

critical load the level of a particular pollutant below which adverse effects on living organisms are not expected to occur.

cryptochrome any of a group of flavoproteins that are blue-light photoreceptors and have key roles in growth and development, including interaction with phytochrome to regulate photoperiodic and photomorphogenic responses in plants, such as flowering time and hypocotyl elongation.

cultivar any cultivated variety of a plant.

cyanide-resistant respiratory pathway a pathway of respiration which occurs in many plants and that does not use cytochrome oxidase, which is inhibited by cyanide, but an alternative oxidase that is cyanide-insensitive.

cyanobacteria a large group of bacteria that contain chlorophyll *a* and can carry out photosynthesis with the production of oxygen.

cyclic nucleotide a nucleotide containing a single phosphate group, in which there are two single bonds between the phosphate group and the ribose ring.

cytochrome b_6f a two-subunit protein that occurs in the thylakoid membrane of chloroplasts, where it functions as an electron carrier. It is closely related to cytochrome bc_1 of the mitochondrial electron transport chain (complex III).

cytosis a transport mechanism that involves the enveloping of molecules by the formation of membrane vesicles, or by the invagination or fusion of a membrane. It is used to move large polar molecules, such as proteins, which cannot pass through cell membranes, into and out of cells and also from one location to another within the cell.

cytoskeleton a framework which consists of microtubules composed of the protein tubulin and microfilaments composed of the protein actin, and that provides the plant cell with structural form and shape.

dauciform carrot-shaped.

δ^{66}Zn the change in the stable isotope ratio $^{64}Zn/^{66}Zn$, used in physiological studies to determine, for example, the proportion of zinc transport that occurs as the free ion (which favors the lighter ^{64}Zn).

decameric relating to a molecule composed of 10 linked monomer units.

de-etiolating relating to the morphological, physiological, and biochemical changes that a plant shoot undergoes when it is exposed to light after previously being grown in the dark.

deficiency suboptimal availability of one or more essential nutrients, which leads to specific adverse effects in plants and animals. This may be due to inadequate levels of the nutrient(s), or to the fact that they are not in a form that is available to the plant or animal.

Deg proteases proteolytic enzymes that hydrolyze peptide bonds (that is, degrade proteins). They are associated with a wide range of cellular processes, such as protein processing, signaling, and removal of damaged proteins. In plants, many Deg proteases have a high temperature requirement and are therefore also referred to as Htr proteases.

dehydrins the wide range of hydrophilic and thermostable proteins that are produced by plants in response to drought and cold stress.

denitrification the bacterially mediated reduction of nitrate to nitrite and then to gaseous products, mainly nitric oxide (NO), nitrous oxide (N_2O), and nitrogen (N_2).

desaturase an enzyme that removes two hydrogen atoms from a fatty acid to create a C=C double bond.

desertification the process of formation or expansion of desert, which is characterized by increased aridity and decreased available free water. It is usually accompanied by a decrease in plant and animal diversity and biomass.

determinate in plants, relating to any type of growth that ceases once a particular structure has been formed. The term is primarily used to describe the main axis of a shoot, terminating in a flower bud.

Devonian relating to the geological period from about 419 million to 359 million years ago, characterized by a diversity of fossils of fish and terrestrial plants.

diazotroph an organism that can utilize or "fix" atmospheric nitrogen (N_2) as a nitrogen source.

dichotomous branching the repeated division of a branch into two (usually equal) branches.

diffuse light light in which the rays are traveling at many different angles, due to the light being either transmitted from a relatively large light source or reflected off a surface and scattered in different directions.

dimer a structure that is formed from two identical or similar molecules or subunits that are paired together, especially a protein that consists of two polypeptide chains or subunits.

dipole moment a measure of the polarity of a system, which is due to an uneven charge distribution, with one pole having a net positive charge and the other pole having a net negative charge.

dismutation a reaction that produces both oxidized and reduced molecules in equal amounts. For example, the dismutation of superoxide ($O_2^{\bullet-}$) produces hydrogen peroxide (H_2O_2) and oxygen (O_2) in equal amounts.

diterpenes a group of chemicals that are produced in fungi and plants, and which are composed of four isoprene units, so contain 20 carbon atoms.

Dole effect the difference between the ratio of ^{18}O to ^{16}O in the atmosphere and that in seawater.

ecosystem services the benefits gained by humans from ecosystems that underpin and enhance human existence.

ectotherm any organism that cannot regulate its own temperature, and is therefore dependent on the external environment for the gain or loss of heat.

electrical conductivity a measure of the ability of a material to allow the movement of electrically charged particles (that is, to conduct an electric current).

electronegative relating to the tendency of an atom or molecule to attract electrons.

electrophilic attracted to an electron-rich area of a molecule.

embolism the condition that occurs when conduits in a vascular system become blocked by, for example, air bubbles or foreign material.

emergent relating to a property of a system or group that is not exhibited by individual components of the system or members of the group.

enantiomer one of two molecules that are mirror images of each other and so cannot be superimposed on one another.

endemism the restriction of a species to one particular geographical location.

endosome in eukaryotic cells, a membrane-bound vesicle that is formed from the plasma membrane and used to transport or recycle "cargo" in the form of protein and lipid molecules (for example, by conveying them from the biosynthetic or endocytotic pathways to the lysosome, vacuole, or plasma membrane).

enediol an unsaturated organic compound that has a hydroxyl group on both sides of the double bond.

epicotyl the part of a seedling that is above the cotyledons and that will develop into a shoot and produce the true leaves.

epigenetic relating to the heritable changes in genomes that are not due to changes in gene sequence but to changes in the expressibility of genes.

epinasty a difference in the rate of growth of the upper surface of a plant organ. For example, an increase in the rate of growth of the upper surface of the petiole causes the leaf to be directed downward.

epoch an interval of geological time that is a subdivision of a period.

Equisetaceae (also known as the horsetail family). An ancient family of plants that first flourished during the Devonian period. There is a single extant genus, *Equisetum*, and only 15 extant species, although horsetails were once very abundant and diverse, and have a rich fossil record.

erinaceous relating to or resembling a hedgehog.

ester a compound that is formed as a result of the condensation reaction between an acid and an alcohol, with the elimination of water. Most common biological esters are formed from a carboxylic acid and an alcohol.

ethylene diurea an ethylene molecule with two urea side groups, which appears to provide some protection against the harmful effects of ozone on plants.

etiolation the condition, caused by growing plants in the dark, that is characterized by pale green or yellow leaves and stems (due to lack of chlorophyll) and very long internodes.

Euclidean geometry a geometry of absolute three-dimensional space that was developed by the Greek mathematician Euclid as axioms from which many theorems followed. It contrasts with Cartesian geometry, which is based on coordinates, and with geometries in which space is not absolute and more than three dimensions are possible.

euphyllophyte any member of the taxon of vascular plants that includes angiosperms, gymnosperms, and ferns.

evapotranspiration the total amount of water lost from the soil or an open water surface by evaporation and water lost from the plant surface by transpiration.

Everglades the low flat area of plains in the south of Florida, USA, which is a wetland due to periodic flooding in the summer, but is dry in winter.

exaptation a change in the function of a particular trait or characteristic during the course of its evolution.

exodermis the layer of cortical cells, with Casparian bands, immediately beneath the epidermis in the roots of the majority of angiosperms.

exonuclease an enzyme that cleaves nucleotides from a polynucleotide chain.

expansins a family of non-enzymatic proteins that loosen the bonds between the cellulose microfibrils in plant cell walls. They have important roles in cell expansion, fruit ripening, and abscission.

extra-radical outside the root system.

fatty acid desaturase (FAD) an enzyme that removes two hydrogen atoms from a long-chain saturated fatty acid (that is, one which has a chain of carbon atoms linked by single bonds), resulting in the formation of a C=C double bond, and thus an unsaturated fatty acid.

fermentation the conversion of carbohydrate (especially sugar) to alcohol or lactate under anaerobic conditions to produce ATP.

ferredoxin an iron–sulfur protein that functions as an electron carrier in redox reactions (for example, in photosynthesis).

ferritin a large protein with 24 subunits that stores iron. It occurs in almost all living organisms, and is found in the plastids and mitochondria in plants.

Fick's first law the law which states that diffusive flux of particles down a concentration gradient is directly proportional to the steepness of the gradient.

field capacity the amount of water that remains in a soil after all of the excess water has drained freely from the soil. It represents the total amount of water that the soil can hold against the force of gravity.

flavin mononucleotide a molecule produced by phosphorylation of riboflavin. It is a cofactor in phototropins and also in a number of oxidoreductases.

flavonoids a group of plant phenolic compounds that consist of two phenyl rings and a heterocyclic ring. They are important plant pigments, and also have roles in signaling and UV light filtration.

fluorophor a compound that, following exposure to light, will fluoresce (that is, emit light).

fluorescence the emission of radiation, especially light, by an atom or molecule when electrons within it rapidly move from a higher-energy state to their former lower-energy state.

fractal geometry the study of shapes that have self-similarity across a range of scales.

fructans polymers of fructose that are produced by about 15% of plants, especially in more recently evolved families. In some families they are the main storage carbohydrate. Inulins are common in the asterids, whereas levans are mainly found in monocots.

FtsH2/3 proteases ATP-dependent membrane-bound enzymes that degrade proteins (for example, damaged reaction center protein D1 from photosystem II complexes).

fynbos the scrubby heathland, dominated by sclerophyllous shrub species, and with a Mediterranean-type climate, that is found in the Western Cape of Southern Africa.

G-proteins a family of proteins that bind GTP (guanine triphosphate) and hydrolyze it to GDP, which changes their conformation, enabling them to transmit a signal.

galactolipid a glyceride in which one of the carbon atoms of glycerol is attached to one or more galactose molecules. Galactolipids are important components of plant cell membranes.

gametophyte the gamete-producing generation of the plant life cycle.

gated in a biological system, denoting a pathway or channel that can be opened or closed only under specific conditions (for example, a change in membrane potential).

genome all of the genetic material (both coding and non-coding) that is carried by a single set of chromosomes.

geostatistics the branch of statistics that involves the analysis of spatially dependent data.

germplasm the genetic material that is transmitted from parents to offspring via the germ cells. It represents the genetic resources of an organism.

glaucous having a bluish-green or waxy appearance (for example, the highly reflective leaves of certain plant species).

global dimming a decrease in solar radiation at the Earth's surface caused by an increase in the number of particles in the atmosphere resulting from human activities.

glomalin an insoluble glycoprotein with glue-like properties that is produced by arbuscular mycorrhizal fungi. Glomalins bind soil particles together, and can be a significant component of soil organic matter.

Glomeromycota a division of the fungal kingdom, containing about 250 species, all of which form arbuscular mycorrhizae with plants.

gluconeogenesis in living organisms, the synthesis of glucose from molecules other than the usual carbohydrate sources (for example, from amino acids or lactate).

glucosinolate any of over 100 sulfur-containing glucosides produced by members of the Brassicaceae (for example, cabbage, mustard). Glucosinolates and their breakdown products have a pungent aroma and bitter taste, and provide plants with some resistance to herbivores and pathogens.

glutaredoxins redox enzymes that regulate the redox state of the cell and redox-dependent signaling pathways by using the reducing power of glutathione.

glycerolipids a group of lipid molecules that contain glycerol (a 3-carbon molecule) linked by ester bonds to at least one long-chain carboxylic acid ("fatty acid").

glycophyte any plant that will only thrive in a soil or other medium with a low salt content (that is, a salt-sensitive plant).

glycosylation a biochemical reaction in which a glycosyl group (that is, a carbohydrate) is added to another biomolecule (for example, a protein).

Gondwanan relating to Gondwana, the southern supercontinent that existed from about 500 million to 180 million years ago, which broke up to form many of the existing continents of the southern hemisphere, and the Indian subcontinent in the northern hemisphere.

gravitropic relating to a directional response by a plant organ to the force of gravity (for example, the downward growth of a root is a positive gravitropic response).

Green Revolution the dramatic change to intensive high-yielding crop production that occurred between the 1940s and the late 1960s, especially in developing countries. It was driven by advances in technology and plant breeding, and by the development of new chemical fertilizers, pesticides, and herbicides.

green water the proportion of rainfall that infiltrates the soil and can be directly accessed by plants.

guano the excrement of seabirds that accumulates in dry climates under mass seabird colonies on islands or in coastal regions. The term also refers to the excrement of cave-dwelling bats.

guttate the exudate from hydathodes at leaf edges or from cut shoots. It is caused by positive root pressure pushing the xylem contents up through the vascular tissue.

gymnosperms seed plants in which the ovules are not enclosed by ovaries, but are produced on the scales of cones, where they develop into "naked" seeds after fertilization. The living representatives of the gymnosperms include conifers, cycads, and ginkgo.

Hadean relating to the first eon of the Earth's history, from about 4.5 billion to 4 billion years ago, characterized by widespread volcanism (eruption of molten rock onto the Earth's surface), surface instability, and frequent collisions with asteroids.

halophyte a terrestrial plant that is adapted to grow in soils (and atmospheres) with a high salt concentration (that is, a salt-tolerant plant).

Hartig net the network of fungal hyphae that is formed in the outer layers and around the outside of a plant root in ectomycorrhizal associations.

harvest index the proportion of harvested product as a percentage of the total biomass of crop produced.

haustorium (pl. haustoria) in certain parasitic fungi, a structure that penetrates a plant host cell in order to obtain nutrients from it. The term is also used to describe the outgrowths of the roots that are formed by some parasitic flowering plants.

heat stress-responsive element (HSE) any of the characteristic DNA sequences in promoter regions of heat-responsive genes to which heat stress transcription factors bind.

heat stress-responsive protein (HSP) any of the proteins characteristic of a proteome produced by heat stress. Many organisms produce homologous HSPs, and many of these are involved in protein stabilization or repair.

heat stress transcription factor (HSF) any of the proteins that initiate the transcription of heat-responsive genes. Heat stress transcription factors are found in a wide range of organisms.

Henry's law the law which states that, at a constant temperature, the amount of a particular gas that dissolves in a given volume of a particular liquid is directly proportional to the partial pressure of the gas.

heterocyclic ring in a molecule of an organic compound, a ring structure that contains atoms of two or more different elements, usually carbon plus at least one other element.

heterocyst a relatively large specialized cell that is the site of nitrogen fixation in certain cyanobacteria.

heterologous expression the transfer and expression of a gene in an organism other than the one from which it originated.

heterosis (also known as hybrid vigor) the increased vigor that is observed in hybrids, due to a mixing of genetic material.

heterotrimeric relating to a molecule that consists of three different subunits.

hexadecamer relating to a molecule composed of 16 linked monomer units.

histone any of a group of basic proteins that are associated with DNA in the chromatin of eukaryotic cells.

Holocene relating to the current geological epoch, which began around 11,500 years ago, when the glaciers began to retreat during the most recent ice age.

holoenzyme a complete functional enzyme, consisting of both a protein component and a prosthetic group (or cofactor).

homeostasis the process whereby a living organism or a biological system is able to maintain stable internal conditions.

homeotherm an organism that maintains a virtually constant internal body temperature that is largely independent of changes in the temperature of its environment. Homeothermic animals include mammals and birds.

homolog a gene related to another gene by descent from a common ancestral DNA sequence.

hormetic responding positively to exposure to low doses of an environmental toxin that at high doses would have inhibitory or toxic effects.

hydathode a multicellular modified stomatal structure, usually in the epidermis of a leaf, through which water and solutes are exuded during guttation.

hydrogenase an enzyme that catalyzes the reversible oxidation of molecular hydrogen (H_2) to protons (H^+).

hydrophilins a group of proteins that are extremely hydrophilic (that is, strongly attracted to water molecules) and rich in glycine.

hydrothermal vent a fissure in the Earth's surface through which geothermally heated water is forced out.

hydrotropic relating to a plant organ that shows directional growth towards water.

hyperaccumulator a plant that accumulates significantly higher concentrations of a toxin, such as a heavy metal, than those at which the toxin occurs in the soil in which the plant is growing. The amount accumulated is often the highest level of that particular element found in any organism.

hypersensitive response in plants, a response to pathogen attack that includes the rapid death of cells near the site of infection, thus depriving the pathogen of a food source and limiting the spread of infection.

hypocotyl in a seedling, the part of the plumule (embryonic shoot) that is located below the cotyledons and above the radicle (embryonic root).

hypodermis a layer of thick-walled cells immediately below the epidermis that can form in the leaves of certain plant species.

hyponasty a difference in the rate of growth of the lower surface of a plant organ. For example, an increase in the rate of growth of the lower surface of a leaf causes the leaf blade to curve upward.

hypostomatous having stomata on the underside.

hypoxia a condition in which a suboptimal oxygen supply leads to inadequate oxygenation of the tissues of an organism.

infrared thermography the use of an infrared camera to measure the infrared radiation emitted from an object, convert the radiation measurements to temperature measurements, and display images of the temperature distribution.

inhibitor a substance that slows down or inhibits a chemical reaction, especially one catalyzed by an enzyme.

inositol a cyclohexane (6-carbon ring) with a hydroxyl (OH) group attached to each carbon atom.

inositol lipid a molecule that consists of a lipid attached to inositol via a phosphate group.

intercropping the practice of growing two or more crops simultaneously on the same area of land, in order to maximize the yield in terms of both time and space.

internal ribosomal entry segment a nucleotide sequence that allows mRNA to be positioned in a ribosome for initiation of translation.

intra-radical within a root.

intron within a gene, a nucleotide sequence that will not help to encode the final protein and is therefore removed during RNA splicing.

invertase an enzyme that catalyzes the hydrolysis of sucrose to glucose and fructose.

ionome all of the inorganic ions that are present in an organism—that is, the organism's mineral nutrient and trace element composition.

irradiance the amount of electromagnetic radiation, usually visible light, that is incident on a defined area per unit time.

isoflavonoid any of a class of polyphenolic flavonoid compounds that are based on isoflavone molecules.

isoprenylated plant proteins a group of plant metallochaperones with a C-terminal hydrophobic prenyl group which aids protein–protein binding and membrane anchoring.

isothiocyanate a chemical containing the -N=C=S group. Most plant isothiocyanates are found in members of the Brassicaceae, in which they are formed as a result of the hydrolysis of glucosinolates.

Kautsky effect the change in chlorophyll fluorescence that occurs when light is provided to dark-adapted photosynthesizing systems. It is characterized by an initial burst of fluorescence which is then attenuated as the photosystems adapt to the light.

kinase an enzyme that catalyzes the transfer of a phosphate group from a nucleotide triphosphate, especially ATP, to another molecule.

kinesin any of a class of motor proteins that use the energy released by ATP hydrolysis to move along the microtubules of the cytoskeleton.

knock-outs (KOs) organisms in which the expression of a particular gene or genes has been significantly decreased or completely eliminated.

Kwongan an ecoregion of Western Australia that has sclerophyllous vegetation and a Mediterranean-type climate.

lacunae large gaps, often filled with air, between the cells of a tissue.

late embryogenesis abundant (LEA) proteins extremely hydrophilic proteins that are thought to have important roles in desiccation tolerance in plants, bacteria, and invertebrates. When fully hydrated they have no real secondary structure, but upon dehydration they form an α-helix structure around many cellular components, such as other proteins, which are probably thereby stabilized and protected from aggregation.

lateral root any of the roots that extend from the primary root of a dicotyledonous plant.

law of limiting factors the law which proposes that growth is not limited by the total amount of resources available, but by the availability of the resource that is most scarce.

leaching the movement of solutes down through a soil profile, caused by the movement of water.

leaf area index (LAI) the total leaf surface area exposed to light energy per unit area of ground beneath the plant.

lentic relating to a freshwater habitat that has still or standing water, such as a pond or lake.

lenticel any of the pores that perforate the periderm of shoots and roots, allowing gas exchange to occur.

Lewis acid a chemical substance or compound that can accept a pair of electrons from a donor compound. The electron donor is known as a Lewis base.

ligand any atom, molecule, or ion that binds to a central metallic atom in a coordination complex.

light-harvesting complex in the thylakoid membrane of plants, the complex of protein and pigment molecules, mostly chlorophylls, that captures light for photosynthesis and then channels it towards the reaction center.

lignin a polyphenol with which plant cell walls become impregnated during the formation of wood. It is the second most abundant organic polymer on Earth after cellulose.

liverwort a member of a very early phylum of small non-vascular spore-producing terrestrial plants, which form a small procumbent gamete-producing structure in the gametophyte generation and unique spore-producing structures in the sporophyte generation. Liverworts are so called because in many species the shape of the gametophyte thallus resembles that of a lobed liver.

lumogallion a compound that fluoresces when bound to aluminium, and can thus be used to enable visual localization of aluminium in plant tissue.

lycophyte a plant in the division of the plant kingdom that includes the most ancient extant vascular plants. They form spores and free-living gametophytes, and include the clubmosses, quillworts, and spike mosses.

lysigeny in plant tissue, the creation of cavities or lacunae as a result of localized cell death and disintegration (lysis).

M-domain a coiled domain of a heat stress protein that is essential for protein disaggregation.

major intrinsic proteins a large family of proteins that form transmembrane channels. They include aquaporins and channels that mediate the passage of small neutral molecules.

malondialdehyde a compound that is produced by lipid oxidation and can therefore be used as a marker of oxidative stress.

mangrove a tree that produces aerial roots that allow it to inhabit saline coastal waters in the tropics and subtropics.

mannitol a sugar alcohol that is a product of the reduction of mannose, a hexose monosaccharide.

maquis a French term for the scrubby heathland of the Mediterranean biome, which is dominated by sclerophyllous evergreen shrubs and small trees. It is known as macchia in Italy, matorral in Spain, and phrygana in Greece.

mass flow the movement of solutes into a plant due to the movement of water in which they are dissolved from the roots to the leaves in the transpiration stream.

matorral in Spain, a region of heathland scrub that is dominated by sclerophyllous vegetation and has a Mediterranean-type climate. An area of similar vegetation in South America is known as the Chilean Matorral.

maturation zone in a developing root, the region just above the elongation zone, where the cells differentiate and mature.

mechanistic model a model that is based on the assumption that an observed phenomenon is underpinned by particular physico-chemical processes.

megafauna large animal species.

Mehler reaction the photoreduction of oxygen (O_2) in photosystem I (PSI).

mesophytic relating to plants that inhabit environments with moderate amounts of available water.

metallochaperone any of the proteins that are used for moving metal ions in a cell.

metalloenzyme any of the enzymes that contain tightly bound metal ions as cofactors.

metalloprotein any of the proteins that contain a metal ion which acts as a cofactor.

metallothionein any of the small cysteine-rich proteins that bind metals using thiol groups.

metastability the transient state of a compound that can occur between two stable states (for example, between liquid and gas states).

methanogen any microorganism that, under anoxic conditions, releases methane as a product of its metabolism.

methylation the addition of a methyl (CH_3) group to an organic chemical compound.

MFS transporter a transport protein of the major facilitator superfamily. MFS transporters are one of the two largest families of membrane transporters in living organisms. They transport small organic molecules such as sugars, amino acids, and organic anions, and can function as uniporters or antiporters.

microarray a miniaturized two-dimensional array of spots of numerous known molecules (for example, oligonucleotides or antibodies) attached to a solid surface, used to assay a biological sample.

microphyll a small leaf with a single unbranched vein. It is characteristic of the early plant groups, and their extant relatives, before the evolution of leaves with complex patterns of venation.

micro-RNA a small non-coding RNA molecule that contains about 22 nucleotides.

Millennium Ecosystem Assessment a study commissioned by the UN in 2001 to assess the state of the Earth's ecosystems and also the consequences of changes in those ecosystems for human well-being.

mineralization the decomposition of tissues composed of organic compounds to form simple inorganic compounds. This process is mediated by microorganisms.

Miocene relating to the geological epoch from about 23 million to 5.5 million years ago, which was notable for the expansion of the grasslands.

mitogen a substance that induces cell division.

mixotrophic relating to an organism that uses more than one source of energy, especially one that is both autotrophic and heterotrophic.

moiety a part or a functional group of a molecule.

monogastric relating to an organism that has a single-chambered stomach.

monolignol a derivative of phenylalanine that is the starting material for the synthesis of lignins.

Monte Carlo methods statistical or computational methods that use repeated random sampling to describe probabilistic phenomena.

mycorrhiza a symbiotic relationship between a fungus and the roots of a plant.

myosin an ATP-dependent motor protein that acts on the actin filaments of the cytoskeleton.

myrosinase an enzyme that catalyzes the hydrolysis of glucosinolates by cleaving the thio-linked glucose.

NAD nicotinamide adenine dinucleotide.

nastic relating to the response of a plant organ to a non-directional stimulus (for example, the opening of a flower in response to an increase in light intensity).

nectary a nectar-secreting gland in a plant.

net primary productivity (NPP) the total amount of biomass produced minus that used for respiration by plants.

network model a model that treats objects or values and their relationships as interrelated in complex but unbounded ways (for example, not bound by hierarchies).

neutral theory in molecular evolution, the theory that the majority of the genetic variation in populations is caused by genetic drift and random mutations, rather than by natural selection.

neutrophil the most abundant type of white blood cell.

niche segregators the phenomena whereby plant or animal species are able to coexist and avoid competing with each other for the same resources by inhabiting different ecological niches.

nicotinamide adenine dinucleotide (NAD) a coenzyme made from two nucleotides (adenine and nicotinamide) that has an important role in redox reactions in all living organisms via its interconversion between oxidized (NAD^+) and reduced (NADH) forms.

nicotianamine a chemical compound that is present in all higher plants and which can chelate a variety of metals, thus facilitating the intra- and intercellular transport of cations such as Fe^{2+}, Fe^{3+}, and Zn^{2+}.

nitrate reductase an enzyme that catalyzes the reduction of nitrate to nitrite.

nitrification the microbially mediated oxidation of ammonium to nitrite, and of nitrite to nitrate.

nitrogenase complex in prokaryotes, the complex of metalloenzymes that are responsible for the conversion of dinitrogen (N_2) into ammonia (NH_3).

nucleophilic having a tendency to donate an electron pair to an electrophilic compound.

nucleosome the basic subunit of chromatin. It consists of just under two turns of DNA wrapped around eight histone proteins.

nucleotide the monomer of nucleic acids. It contains a nitrogenous base, a sugar (either ribose or deoxyribose), and at least one phosphate group.

nugget the semivariance that occurs at a spatial scale smaller than that of the sampling regime (that is, the semivariance that occurs at any one place).

nyctinastic relating to diurnal movements of plant parts (that is, movements that are regulated by the circadian clock).

octulose an 8-carbon sugar containing a ketone group.

Okavango Delta a large inland delta formed where the Okavango River runs into a large low-lying area in the Kalahari Desert in Botswana.

oligomer a polymer that contains a relatively small number of structural units or monomers.

oligonucleotide a polymer that contains only up to about 20 nucleotides. These short sequences of DNA or RNA have many applications in molecular biology.

oligotrophic relating to water that has a low nutrient content.

optimal foraging strategy a strategy of foraging for food that maximizes either food energy intake per unit time, or food energy intake relative to the resources expended in obtaining it.

Ordovician relating to the geological period that lasted from about 485 million to 443 million years ago, characterized by mollusc and arthropod fossils, including many trilobites, as well as algae.

organic anion transferase an organic anion-transporting protein that transfers organic anions across the cell membrane.

ortholog a gene with the same function and ancestry that is found in two or more different species.

outgroup a monophyletic group that is less closely related to a group of two or more clades than they are to each other.

oxyacid an acid that contains oxygen, to which at least one hydrogen atom is bound and can dissociate to release a proton (H^+).

oxygenase an enzyme that catalyzes the addition of oxygen to a substrate molecule.

ozonolysis the oxidative cleavage by ozone of an alkene, in which the C=C double bond is replaced by two carbonyl (C=O) groups.

^{32}P a β-emitting radioisotope of phosphorus with a half-life of 14 days.

^{33}P a β-emitting radioisotope of phosphorus with a half-life of 25 days.

PACMAD clade one of the two major lineages of the true grasses (Poaceae), which contains the Panicoideae, Arundinoideae, Chloridoideae, Micrairoideae, Aristidoideae, and Danthonioideae, and includes many C_3 and all of the C_4 grass species. The other major lineage, namely the BEP clade, consisting of the Bambusoideae, Ehrhartoideae, and Pooideae, contains only C_3 grasses.

palindromic sequence a nucleic acid (DNA or RNA) sequence in which the 5' to 3' sequence on one strand is the same as the 3' to 5' sequence on the complementary strand, enabling them to bind to form a hairpin if the sequence is repeated.

palisade mesophyll tightly packed columnar mesophyll cells, usually on the upper side of the leaf, and so called because in longitudinal section they resemble a palisade or fence. Mesophyll cells contain many chloroplasts and are the site of photosynthesis.

paramo an Arctic-alpine region that occurs between the treeline and the snowline in the Andes. It consists of meadow and scrub, and has a high humidity level due to cloud and mist.

patch clamp a technique that is used to study the ion-channel properties of an area of cell membrane, in which the open end of a micropipette is positioned on a small patch of membrane, thus enclosing only a few membrane transporters, or even a single one.

pectic polysaccharides the diverse group of polysaccharides of which pectin is composed. They are rich in galacturonic acid.

pedosphere the outermost layer of the Earth, consisting of the soil and all its chemical constituents and biological inhabitants.

pentamer a molecule composed of five subunits or five linked monomer units.

pentose a sugar that has five carbon atoms.

peptide a short polymer consisting of two or more amino acids. It is much shorter than a protein, and does not act as an enzyme or structural molecule.

perfect sink a sink in which the tendency of a substance to absorb does not change with the amount absorbed.

periclinal parallel to the surface.

pericycle the layer of plant tissue, consisting mainly of parenchyma cells, that lies between the endodermis and the phloem, and from which the lateral roots originate.

periderm in plants that undergo secondary thickening, the outer layer of the bark that consists of an outer cork layer (phellem) and an inner phelloderm that are both produced by cell division in the phellogen that is sandwiched between them.

peroxidase an enzyme that catalyzes the oxidation of certain organic molecules, and is so called because it uses hydrogen peroxide as an electron acceptor. Most peroxidases contain a heme cofactor at their active site.

petiolule the stalk of a leaflet in a compound leaf.

phaeophytin a chlorophyll molecule that lacks the central Mg^{2+} ion.

phellem in plants that undergo secondary thickening, the outer corky protective layer of periderm, which forms the outermost layer of the bark.

phellogen in plants that undergo secondary thickening, the meristematic tissue that gives rise to the phelloderm (bark) on its inner side and the phellem (cork layer) on its outer side. It is also known as cork cambium.

phenolic relating to compounds based on phenol, an aromatic organic compound that has one or more hydroxyl (OH) groups.

phenological relating to the study of the timing or seasonal occurrence of life events in plants and animals (for example, flowering period, migration).

phenylalanine ammonia lyase the enzyme that catalyzes the conversion of phenylalanine to ammonia and cinnamic acid, which is the starting point for the synthesis of phenylpropanoid compounds in plants.

phenylpropanoid pathway the metabolic pathway that generates a wide range of secondary metabolites in plants.

phi cell a cell with thickened walls that can be impregnated with lignin and suberin. Phi cells are often found adjacent to the endodermis.

phosphatase an enzyme that catalyzes the removal of one or more phosphate groups from a biomolecule.

phosphoenolpyruvate carboxylase an enzyme that carboxylates phosphoenolpyruvate by adding bicarbonate ions to it. It helps to regulate the TCA cycle in respiration, and is used to fix carbon in plants that have C_4 or CAM pathways of photosynthesis.

phosphoester an ester that is formed by the reaction of phosphoric acid with hydroxyl groups on another molecule.

phosphoglycerate the phosphorylated form of glycerate, a small 3-carbon organic acid that has important roles in glycolysis and also in the Calvin–Benson cycle, where it is the initial product of CO_2 fixation to ribulose 1,5-bisphosphate (RuBP).

phosphoglycolate the phosphorylated form of glycolate, a small 2-carbon hydroxy acid that is a product of O_2 fixation to ribulose 1,5-bisphosphate (RuBP) during photorespiration.

phosphogypsum the mixture of H_3PO_4 and $CaSO_4$ (gypsum) that is a by-product of the processing of calcium phosphates from rock phosphate with sulfuric acid during the manufacture of superphosphate fertilizer.

phospholipase any enzyme that breaks down a phospholipid by hydrolysis.

phospholipid a polar lipid, with a hydrophilic "head" and a hydrophobic "tail", that contains one or more phosphate groups (for example, phosphatidylcholine).

phosphoribulokinase an enzyme that catalyzes the phosphorylation of ribulose-5-phosphate to ribulose 1,5-bisphosphate (RuBP).

phosphorylation the addition of a phosphate group to an organic molecule (for example, a protein), often to make it physiologically active.

photoautotroph an organism that obtains its energy from sunlight, and uses carbon dioxide as its main source of carbon.

photoinhibition the inhibition of a plant process (e.g. germination, photosynthesis) by excess light.

photomorphogenesis the effect of light on the control of plant growth and development.

photosynthetically active radiation the wavelengths of light that can be used by plants for photosynthesis.

photosystem either of the two complexes of proteins and chlorophyll that mediate the absorption of light and the excitation of electrons in the light-dependent reactions of photosynthesis.

phototropin a blue-light receptor in plants that controls a range of responses (for example, phototropism, and chloroplast movements in response to changes in light intensity) so as to optimize the plant's photosynthetic efficiency.

phycobilisome the light-harvesting complex in red algae and cyanobacteria that traps and then transfers light energy to photosystem II complexes. The pigments present include phycocyanin and phycobilin.

phylogenetics the study of evolutionary relationships based on genetic information.

phylogeny the sequence of events involved in the development or evolutionary history of a particular species or taxonomic group of organisms.

physiotype the set of physiological features that are characteristic of a particular species or taxonomic group (for example, the unique nutrient requirements of certain plants).

phytase an enzyme that catalyzes the hydrolysis of phytate (a major phosphorus storage compound in plants) to form inositol and phosphoric acid. Many phosphatases have phytase activity, but those with highest activity and specificity are found in fungi and microorganisms.

phytate any of the salts of phytic acid (myo-inositol hexakisphosphate) that are major phosphorus storage compounds in plants. Phytate is primarily digested by the enzyme phytase, which is most abundant in microorganisms. When phytate binds to cations such as iron and zinc it makes them much less readily available as nutrients for humans and animals (that is, it has anti-nutrient activity).

phytoalexin any of a wide range of antimicrobial compounds that are produced by plants in response to damage or infection, in order to inhibit the growth of microorganisms and thus provide protection against disease.

phytochelatin a plant-derived metal-chelating compound that is a polymer of glutathione. Phytochelatins are thought to play a role in heavy metal detoxification in plants.

phytochemical any of a large group of chemical compounds that are synthesized only in plants.

phytochrome a photoreversible receptor of red and far red light that is found in all of the major plant groups. It is a protein with a chromophore that consists of an open chain of four pyrrole rings.

pinitol a cyclic polyol—that is, a 6-carbon ring with a hydroxyl (OH) group attached to each carbon atom, except for one H atom which is substituted by a methyl (CH_3) group. It was first identified in pine trees.

pK_a the acid dissociation constant, which is a measure of the tendency of an acid to produce H^+ ions . The lower the pK_a value, the stronger the acid.

plasmid a circular particle of double-stranded DNA that occurs in the cytoplasm of the cells of many bacteria, and normally remains separate from the chromosome. It can replicate separately from the chromosome.

plasticity the ability of an organism to change its phenotype in response to environmental change.

plastid terminal oxidase an enzyme that is bound to the thylakoid membrane in plant chloroplasts, and that catalyzes the oxidation of the plastoquinine pool.

plastocyanin a small copper-containing protein that is involved in photosynthetic electron transport, and which is generally confined to the plastids and their progenitors.

plastoquinone an electron transport molecule derived from quinone that is involved in photosynthetic electron transport, and which is generally confined to the plastids and their progenitors.

pneumatophore a specialized "air-breathing root" that develops in some plant species (for example, black mangroves). It is negatively gravitropic and grows up above the substrate surface, penetrating the air or water above it.

poikilotherm an organism whose internal temperature tends to vary significantly, mainly in response to fluctuations in the temperature of its environment.

polarized epithelium a surface layer of cells (the epithelium) in which the inner and outer surfaces of the cells show a distinct apical–basal polarity.

polder a low-lying area of land that is separated hydrologically from the surrounding low-lying land by banks or dikes. The term is often applied to land that has been reclaimed from the sea or a lake.

polyamine an organic chemical compound that contains two or more amine groups.

polyol an alcohol that contains multiple hydroxyl (OH) groups.

polyphenol an organic chemical compound that contains two or more phenolic groups.

polyphyletic relating to a taxonomic group of organisms that have multiple evolutionary origins, so do not represent a natural grouping.

potash any of various salts or mixtures of salts that contain potassium in a readily available form that can be used as a plant fertilizer. The name derives from the plant ashes ("pot ash") that were traditionally collected as a source of this fertilizer.

pre-mRNA incompletely processed RNA that has been synthesized from a DNA template, prior to the removal of introns and the production of the mature RNA.

primordial radionuclide radioactive isotopes that were produced during the Big Bang and subsequent cosmic processes, and which have such long half-lives that they have existed on Earth since its formation.

programmed cell death (PCD) a predetermined closely regulated sequence of cellular events that is triggered under specific conditions and which leads to the death of a cell.

proline a proteinogenic amino acid. Proline is not an essential amino acid for the human diet.

promoter a specific DNA sequence to which RNA polymerase binds before it then initiates the transcription of mRNA in the first step of gene expression.

propidium iodide a dye containing a fluorescent molecule that is used to stain DNA.

protease an enzyme that can cleave proteins by catalyzing the hydrolysis of the peptide bonds within the protein molecule.

proteinogenic relating to an amino acid that is used to synthesize proteins. A total of 20 amino acids are encoded in the genomes of eukaryotes and are used to manufacture almost all proteins. Plants contain over 200 amino acids that are non-proteinogenic and which are rarely found in other organisms.

proteome the entire set of proteins expressed by a particular organism or cell, usually at a particular time.

protoplast every part of a plant cell except the cell wall.

protoxylem the first xylem tissue to develop, consisting of narrow thin-walled cells, in contrast to the mature metaxylem that is formed later in development.

PTS trisodium 3-hydroxy-5,8,10-pyrenetrisulfonate, a fluorescent compound that is used to measure water flux through the apoplast.

pubescent covered with fine hairs or down.

pulvinus the thickened area at the base of a petiole, which can swell or contract as its cells rapidly move water in or out of their vacuoles, resulting in movement of the petiole and thus of the leaf.

purine a basic nitrogenous compound formed from a pyrimidine ring and an imidazole ring, and which therefore contains both carbon and nitrogen atoms. The two main purines, namely guanine and adenine, are major components of nucleic acids.

pyrabactin a synthetic compound that binds to abscisic acid (ABA) receptors and thus mimics ABA, a naturally occurring stress hormone that can increase the drought tolerance of plants by inhibiting growth.

pyrimidine a basic heterocyclic compound containing a single benzene ring with nitrogen atoms attached to carbon atoms 1 and 3. The three main pyrimidines, namely uracil, cytosine, and thymine, are major components of nucleic acids.

Q_{10} value the ratio of the rate of a reaction (for example, an enzyme-catalyzed reaction) at a particular temperature to the rate of the same reaction at a temperature 10°C lower.

Q/I plot for a soil, a plot of the quantity of potentially available but adsorbed ion (Q) in a soil against its soil solution concentration, or intensity (I). It provides one means of estimating the soil's supply capacity for the ion.

quantitative trait locus (QTL) any of the sequences of DNA that help to determine the value of a quantitative trait—that is, a phenotype that is determined by two or more genes and that can have a range of values, in contrast to a trait that occurs in two or more states with no intermediate forms.

quinone an aromatic compound synthesized in plants from the prototypical benzoquinone, which is an aromatic ring containing two hydroxyl (OH) groups.

radial relating to a radius or ray. Organisms or structures that have radial symmetry can be divided equally about a central point.

radicle the part of the plant embryo that will develop into the root, and is generally the first to emerge from the seed.

raffia a fibrous tissue from the exceptionally long leaves of the palm *Raphia vinifera*, which can be used to produce fibers and other products.

raffinose a trisaccharide composed of glucose, galactose, and fructose.

ramet any of the individuals in a group of clones.

random coils in a polymer, especially a protein, irregular coils or folds due to monomers being oriented randomly.

Rayleigh scattering the scattering of light by particles up to about a tenth of the wavelength of light. The light causes charges in the particle to move with the same frequency as the light, resulting in radiation of some light energy. The scattering of light in the sky and hence its blue color is caused by Rayleigh scattering off the molecules of the air.

RBOHF respiratory burst oxidase homolog protein F, which is an NADPH oxidase.

reactive nitrogen species a family of reactive molecules derived from nitric oxide that can damage cellular components.

reactive oxygen species chemically reactive molecules that contain oxygen and are therefore strongly oxidizing.

recombinant inbred lines lines of an organism, especially a plant, that have recombined chromosomes which are maintained through inbreeding. Different recombinations of true breeding lines can be used to map quantitative trait loci (that is, to work out the location of sequences linked to traits).

Redfield ratio in phytoplankton, the atomic ratio of carbon to nitrogen to phosphorus (C:N:P). It is named after Andrew Redfield, who first discovered that this ratio is almost constant throughout the oceans.

redox potential reduction/oxidation potential, which is the tendency of a chemical species or chemical system to gain or lose electrons. Positive values indicate a tendency to gain electrons, whereas negative values indicate a tendency to lose electrons.

reductant a substance that is capable of reducing another substance by donating electrons to it.

redundancy the situation that occurs when two or more genes perform the same function, so they are not all essential, and the inactivation of one of them (or more, if there are more than two such genes) will have little or no effect on the phenotype.

regolith the layer of unconsolidated, weathered, and/or decomposed material that lies on the surface of solid bedrock. It includes soil (which is defined as regolith that can support the growth of plants) as well as rock fragments and mineral grains.

resilience the ability of an ecosystem to withstand change caused by disturbance or damage.

resonant transfer the fluorescence resonance energy transfer that occurs between two light-sensitive molecules when they are closer together than the wavelength of the light and there is dipole–dipole coupling but not emission of radiation.

restios a group of rush-like plants belonging to the genus *Restio*, which are endemic to southern Africa.

rhamnogalacturonan II an important polysaccharide that occurs in the primary cell walls of all flowering plants.

rhizoid a simple root-like organ which is found in primitive plants that have no vascular system. It serves to anchor the plant and absorb water and minerals, and is multicellular in mosses and unicellular in liverworts.

rhizomorph in certain fungi, a root-like structure consisting of many hyphae packed together in an almost parallel orientation, often with dark pigmentation.

rhizosheath the cylindrical sheathing structure around a plant root, which represents the interface between the plant and the soil, as soil particles are bound to each other and to the root in mucilage secreted by the root.

rhizosphere the zone of soil in the immediate vicinity of plant roots. Its properties are influenced by the metabolic and physiological activity of the roots.

Rhynie chert an early Devonian sedimentary rock in Scotland, consisting of chert which is so fine that it contains exceptionally well preserved early terrestrial plants, in which the anatomical detail of the cells can be seen.

ribosome any of the subcellular structures composed of RNA and protein which are found in the cytoplasm and that translate the sequence of mRNA into a polypeptide chain of amino acids by joining together free amino acids using tRNA. Ribosomes are thus the site of protein synthesis.

rice blast a serious disease of rice caused by the ascomycete fungus *Magnaporthe grisea*.

riparian relating to the shore of a lake or the bank of a river.

RNA-induced silencing complex (RISC) a multiprotein complex that contains one strand of small interfering RNA (siRNA) or microRNA (miRNA). When the siRNA or miRNA binds to a complementary sequence on mRNA, it recruits one of the RISC proteins, known as Argonaute, to cleave the RNA.

RNA polymerase II in eukaryotes, the enzyme that catalyzes the transcription of DNA to form precursors of mRNA.

RNase ribonuclease, an enzyme that catalyzes the breakdown of RNA.

rock phosphate a sedimentary deposit that contains large amounts of calcium phosphates, especially apatites. The rock phosphates that produce the most effective fertilizers also contain significant levels of P_2O_5.

root hair a thin-walled outgrowth (up to 1 mm or more in length in some plants) from an individual root epidermal cell.

rubisco ribulose 1,5-bisphosphate carboxylase/oxygenase, the enzyme involved in the first main step of CO_2 fixation in the Calvin cycle.

salt marsh the ecosystems, consisting of vegetation that is frequently flooded by seawater, which develop in low-lying coastal fringes in the temperate zone.

saltpetre a mineral composed mainly of potassium nitrate and sometimes sodium nitrate (also known as Chile saltpetre) and other nitrogen-containing minerals. The term is often used to refer specifically to potassium nitrate.

saprotroph a heterotrophic organism that absorbs nutrients from a non-living source (for example, dung, or dead plant or animal material).

schizogeny the process whereby lacunae are produced in a tissue (for example, in the aerenchyma of a wetland plant) as a result of the separation of cells at the middle lamella of cell walls.

scion the plant that provides the shoot-derived material for grafting onto a rootstock.

sclerified relating to cell walls in which lignin has been deposited, thus making them rigid and sealing them. If the main purpose of the tissue is to provide mechanical support, the protoplasts usually die, and the tissue contains predominantly sclereid cells and fibers.

sclerophyllous relating to vegetation with evergreen leaves that are small, hard, and leathery. These plants consist of trees and shrubs that typically inhabit scrub in regions with a Mediterranean-type climate.

secondary growth (also called secondary thickening) in woody plants, the addition of successive woody layers to a stem or root that occurs as a result of repeated cell division in the cambium.

secondary metabolism the metabolic pathways that produce an extremely diverse range of useful although technically non-essential chemicals in plants (for example, alkaloid, phenolic, and terpenoid compounds).

self-similarity the property of appearing superficially similar on any scale.

semivariogram a diagram that describes the spatial variance of a variable.

serine/threonine kinase an enzyme that phosphorylates the hydroxyl (OH) group of serine or threonine. It plays a key role in the regulation of cellular processes by changing the conformation of receptor molecules.

serpentine a type of rock that contains significant amounts of a group of minerals rich in magnesium and iron silicates.

shade leaves the characteristic leaves that some plants develop in the shade, which show anatomical and morphological differences (for example, they are larger, thinner, and have fewer stomata) compared with leaves that develop in full sun.

short-interfering RNAs short double-stranded RNA molecules that are involved in RNA interference with protein translation, and hence the suppression of gene expression.

sill the maximum semivariance (that is, the plateau representing the distance beyond which the semivariance no longer increases).

Silurian relating to the geological period from about 444 million to 416 million years ago, which is notable for the appearance of terrestrial vascular plants, the first plants to colonize significantly beyond the land–sea interface.

singlet oxygen oxygen in an excited or higher-energy state due to the spin of a pair of electrons in opposite directions.

sink leaves leaves that are net consumers of carbohydrates.

siroheme a heme-like prosthetic group that is found in the enzymes that catalyze the reduction of nitrite to ammonia and of sulfite to sulfide.

skotomorphogenesis the control of plant growth and development in response to darkness (for example, the elongation of the hypocotyl of a germinating seedling that enables it to grow up above the soil surface and reach the light).

small RNAs small non-coding RNA molecules that have a role in the regulation of translation of mRNA.

soda ash anhydrous sodium carbonate (Na_2CO_3), so called because it was originally obtained from the ashes of harvested halophytes. It is now mined or synthetically manufactured from other sources.

solar constant the total amount of energy received from the Sun in the form of electromagnetic radiation per unit time per unit area of the Earth's surface perpendicular to the Sun's rays.

source leaves leaves that are net producers of carbohydrates.

sphingolipids a class of organic chemical compounds in which a sphingosine molecule is linked to a polar head group, often choline or serine, and a long-chain fatty acid.

spin-forbidden relating to a transition of electron states that is not possible because there is a change in the number of unpaired electron spins.

spongy mesophyll in a leaf, the mesophyll tissue that contains loosely packed, mostly chlorophyllous cells, with large air spaces between them. It is the site of gaseous exchange in photosynthesis.

sporophyte the spore-producing generation in the life cycle of a plant.

spring varieties varieties of crop plants that are planted in spring and harvested the same year.

stachyose a tetrasaccharide formed from glucose, fructose, and two galactose molecules.

stele the vascular tissue located at the center of a root or shoot. It contains the xylem and phloem, and consists of one or more strands that extend down the longitudinal axis of the plant.

stochastic relating to a system or process that involves a random variable and whose outcome is therefore determined on the basis of chance or probability.

stoichiometry the study of the ratios between different elements and compounds that take part in chemical or biochemical reactions.

stratification the exposure of a seed to a period of low temperature to increase the likelihood of subsequent germination.

stratosphere the layer of the Earth's atmosphere above the troposphere. The higher layers are hotter than the lower ones (in contrast to the troposphere, where the higher layers are cooler than the layers near to the Earth's surface).

strigolactone any of a class of terpenoid lactones that are plant hormones and signaling molecules. They stimulate germination in parasitic plants (and are so named because they were first identified as the signal for *Striga* seeds to germinate), control shoot branching in plants, and stimulate the growth of mycorrhizal fungi.

suberin a waxy waterproof substance that is the main component of the cell walls of cork and of the Casparian strip.

Sudd in South Sudan, a large area of lowland swamp formed by the White Nile.

Suess effect the decrease in the atmospheric concentrations of ^{13}C and ^{14}C caused by the introduction of large amounts of CO_2 derived from fossil fuel, which is depleted in ^{13}C and contains no ^{14}C. It is named after the Austrian chemist Hans Suess, who discovered its effect on the accuracy of radiocarbon dating.

sulfitolysis the cleavage of a disulfide (S–S) bond, especially in proteins.

sulfolipid any of a class of lipids that have a sulfur-containing functional group.

sulfoquinovose a monosaccharide sugar with a sulfonic acid functional group, which has a major role in sulfur metabolism in plant cells.

sumoylation post-translational modification that involves the attachment of a small ubiquitin-related modifier (SUMO) to a substrate protein in order to alter its structure or tag it for a specific fate, especially with regard to its subcellular localization.

Sunderbans the extensive area of mangrove swamp in the Ganges–Brahmaputra Delta.

sun leaves the characteristic leaves that some plants develop in full sunlight, which show anatomical and morphological differences (for example, they are smaller, thicker, and have more stomata) compared with leaves that develop in the shade.

supercoiling the over-twisting (positive supercoiling) or under-twisting (negative supercoiling) of a section of DNA compared with its natural helical coil (the normal number of base pairs per turn of the helix is 10.5).

superhydrophilicity an exceptionally strong affinity for water. Water on superhydrophilic surfaces, which include the cuticle of some plants (for example, the pitcher plant), spreads very rapidly.

superhydrophobicity the property of repelling water so strongly as to be almost unwettable. Water on superhydrophobic surfaces (for example, the leaf of the lotus) forms spherical droplets that readily roll off the surface.

SWR1 complex an ATP-dependent protein complex that is required for the remodeling of chromatin by insertion of histones. It is involved in the regulation of transcription and in DNA damage repair.

symplast the continuum of interconnected cytoplasm of adjacent plant cells, which are linked via "cytoplasmic bridges", known as plasmodesmata, across cell walls.

sympodial branching a type of branching in which what appears to be a main axis in fact consists of many lateral branches, each arising from the one before, as growth from the original tip is suppressed.

tannins a generic term for a wide range of phenolic compounds produced by plants, some of which were formerly used for tanning leather.

TATA box a DNA sequence that indicates to other molecules where a genetic sequence can be read and decoded (that is, the precise point at which transcription begins), and in what direction this should occur. It is so called because its conserved DNA sequence is most commonly TATAAA.

teratogenic capable of causing developmental abnormalities.

tetrad a group of four or a four-part structure.

tetrapyrrole any of a class of pigment molecules that have four joined pyrrole rings, which can surround and chelate a metal ion.

thermal hysteresis the phenomenon whereby a temperature-dependent property of a body depends on whether the temperature of that body is increasing or decreasing.

thermogenesis the production of heat by a living organism.

thermoregulation in a living organism, the maintenance of an internal temperature within a narrow range despite variation in the temperature of its environment.

thioredoxin a redox protein (that is, one that participates in oxidation–reduction reactions) which has a disulfide bridge at its active site.

tomography an internal imaging technique that views cross-sections of solid objects using penetrating waves.

topology the study of shapes.

toxicity adverse effects on living organisms caused by the availability of a chemical element or compound at above optimal levels.

tracheids the elongated cylindrical dead cells of the xylem that are used for water transport and also provide structural support. They are found in all vascular plants, and are less efficient for water transport than vessels because they are of smaller diameter and have plates at each end.

transceptor a molecule that is both a receptor and a transporter.

transcription the process by which a DNA sequence is used to produce RNA with a sequence complementary to that of a single strand of the DNA.

transcription factor a protein that binds to a specific DNA sequence and helps to initiate transcription into mRNA.

transcription initiation complex a complex consisting of RNA polymerase II and transcription factors that helps to position the RNA polymerase II for initiation of transcription.

transpiration ratio the mass of water transpired by a plant per unit mass of dry matter produced.

treeline the maximum altitude at which trees will grow.

trehalose a disaccharide consisting of two linked glucose units.

trimer a structure that is formed from three identical or similar molecules or subunits that are linked together, especially a protein that consists of three polypeptide chains or subunits.

triplet state the existence of three quantum states in a molecule, usually providing an excited reactive molecule (the exception being oxygen, in which the triplet state is stable).

triterpene any of a group of terpenes that consist of six isoprene units and are found in plant gums and resins.

tropic relating to any directional response by a plant organ to a stimulus (for example, a gravitropic response to gravity, or a phototropic response to light).

troposphere the layer of the atmosphere that is closest to the Earth's surface.

turgid relating to a cell that is rigid because it has a maximal water content and so positive hydrostatic pressure is exerted on all of the cell walls.

tyrosine an amino acid that is synthesized from phenylalanine and that is a constituent of most proteins.

ubiquitin a small protein that binds to a target protein in order to tag it for destruction.

ubiquitin–proteasome system the complex that targets proteins for destruction by tagging them with ubiquitin, and then degrades them via the proteasomal pathway.

uronic acid any of a class of organic acids that contain both carboxyl and carbonyl functional groups, and are derived from the oxidation of sugars.

valence the number of electrons that an atom loses, shares, or gains when it reacts with another chemical element.

Venturi effect the reduction in pressure that occurs when a liquid or gas passes through a constricted section of a pipe or vessel. The phenomenon is named after the Italian physicist Giovanni Battista Venturi.

vernalization the prolonged exposure of seeds or plants to low temperatures in order to induce or hasten flowering.

vessels the xylem transport elements that occur in angiosperms and which provide more efficient water transport than tracheids because of their greater width and length.

virulence factors substances produced by pathogens that enable them to become established in a host and that increase their pathogenicity.

Walker-type nucleotide-binding domain a motif in proteins that is associated with phosphate binding. It is present in many ATP-binding proteins.

washout the removal of air pollutants from the atmosphere by rainfall.

water storage parenchyma ground tissue consisting of predominantly non-chlorophyllous cells with very large vacuoles for storing water.

watershed an area of land that forms a drainage basin in which all the water courses flow into one river. The term is also used to refer to the high narrow ridge of land between two such drainage basins.

water use efficiency the amount of biomass produced by a plant or a crop per unit of water used.

winter varieties varieties of crop plants that are planted in autumn, then overwinter, and finally mature in late spring or early summer of the following year.

xanthophyll a yellow plant pigment that is an oxygenated carotenoid and does not require light for its synthesis.

xenobiotic a chemical compound that does not occur naturally in an organism, or which does not occur naturally at the concentrations in which it is found in that organism.

xerophytic relating to a plant that is adapted to grow in very dry conditions and can withstand periods of drought.

zinc-binding domain in a protein, a sequence of amino acids responsible for binding zinc in particular.

zwitterionic relating to a molecule that has no net charge because it has areas of positive charge and negative charge.

Zygomycota a division of the fungal kingdom, members of which characteristically produce a unique type of spore known as a zygospore during sexual reproduction.

Index

Note. The index covers the main text but not the Further Reading sections, Abbreviations list or Glossary. The letter B, F and T after a page number indicate that a subject is mentioned there only in a Box, Figure or Table. However, on pages where a text treatment has already been indexed, non-text material is not separately listed. Chemical entities are listed at their full names, thus Na^+ at sodium and CH_4 at methane.

B

C

D

E